Advances in Intelligent Systems and Computing

Volume 1322

AF892815

The series "Advances in Intelligent Systems and Computing" contains publications on theory, applications, and design methods of Intelligent Systems and Intelligent Computing. Virtually all disciplines such as engineering, natural sciences, computer and information science, ICT, economics, business, e-commerce, environment, healthcare, life science are covered. The list of topics spans all the areas of modern intelligent systems and computing such as: computational intelligence, soft computing including neural networks, fuzzy systems, evolutionary computing and the fusion of these paradigms, social intelligence, ambient intelligence, computational neuroscience, artificial life, virtual worlds and society, cognitive science and systems, Perception and Vision, DNA and immune based systems, self-organizing and adaptive systems, e-Learning and teaching, human-centered and human-centric computing, recommender systems, intelligent control, robotics and mechatronics including human-machine teaming, knowledge-based paradigms, learning paradigms, machine ethics, intelligent data analysis, knowledge management, intelligent agents, intelligent decision making and support, intelligent network security, trust management, interactive entertainment, Web intelligence and multimedia.

The publications within "Advances in Intelligent Systems and Computing" are primarily proceedings of important conferences, symposia and congresses. They cover significant recent developments in the field, both of a foundational and applicable character. An important characteristic feature of the series is the short publication time and world-wide distribution. This permits a rapid and broad dissemination of research results.

Indexed by SCOPUS, DBLP, EI Compendex, INSPEC, WTI Frankfurt eG, zbMATH, Japanese Science and Technology Agency (JST), SCImago.

All books published in the series are submitted for consideration in Web of Science.

More information about this series at http://www.springer.com/series/11156

Dario Russo · Tareq Ahram ·
Waldemar Karwowski ·
Giuseppe Di Bucchianico ·
Redha Taiar
Editors

Intelligent Human Systems Integration 2021

Proceedings of the 4th International Conference on Intelligent Human Systems Integration (IHSI 2021): Integrating People and Intelligent Systems, February 22–24, 2021, Palermo, Italy

IHSI
International
Intelligent Human
Systems Integration
Conference
2021
www.ihsint.org

Springer

Editors
Dario Russo
Università degli Studi di Palermo
Palermo, Palermo, Italy

Tareq Ahram
Institute for Advanced Systems Engineering
Florida, FL, USA

Waldemar Karwowski
University of Central Florida
Orlando, FL, USA

Giuseppe Di Bucchianico
Università degli Studi di Chieti-Pescara
Pescara, Italy

Redha Taiar
Université de Reims Champagne-Ardenne
Reims, France

ISSN 2194-5357 ISSN 2194-5365 (electronic)
Advances in Intelligent Systems and Computing
ISBN 978-3-030-68016-9 ISBN 978-3-030-68017-6 (eBook)
https://doi.org/10.1007/978-3-030-68017-6

This Springer imprint is published by the registered company Springer Nature Switzerland AG
The registered company address is: Gewerbestrasse 11, 6330 Cham, Switzerland

Preface

This volume, entitled *Intelligent Human Systems Integration* 2021, provides a global forum for introducing and discussing novel approaches, design tools, methodologies, techniques, and solutions for integrating people with intelligent technologies, automation, and artificial cognitive systems in all areas of human endeavor in industry, economy, government, and education. Some of the notable areas of application include, but are not limited to, energy, transportation, urbanization and infrastructure development, digital manufacturing, social development, human health, sustainability, a new generation of service systems, as well as developments in safety, risk assurance, and cybersecurity in both civilian and military contexts. Indeed, rapid progress in developments in ambient intelligence, including cognitive computing, modeling, and simulation, as well as smart sensor technology, weaves together the human and artificial intelligence and will have a profound effect on the nature of their collaboration at both the individual and societal levels in the near future.

As applications of artificial intelligence and cognitive computing become more prevalent in our daily lives, they also bring new social and economic challenges and opportunities that must be addressed at all levels of contemporary society. Many of the traditional human jobs that require high levels of physical or cognitive abilities, including human motor skills, reasoning, and decision-making abilities, as well as training capacity, are now being automated. While such trends might boost economic efficiency, they can also negatively impact the user experience and bring about many unintended social consequences and ethical concerns.

The intelligent human systems integration is, to a large extent, affected by the forces shaping the nature of future computing and artificial system development. This book discusses the needs and requirements for the symbiotic collaboration between humans and artificially intelligent systems, with due consideration of the software and hardware characteristics allowing for such cooperation from the societal and human-centered design perspectives, with the focus on the design of intelligent products, systems, and services that will revolutionize future human-technology interactions. This book also presents many innovative studies of ambient artificial technology and its applications, including the human–machine interfaces with a particular emphasis on infusing intelligence into the development of technology throughout the lifecycle development process, with due consideration of user experience and the design of interfaces for virtual, augmented, and mixed reality applications of artificial intelligence.

Reflecting on the above-outlined perspective, the papers contained in this volume are organized into eight main sections, including:

Section 1 Human-Autonomy Teaming
Section 2 Automotive Design and Transportation Engineering
Section 3 Humans and Artificial Cognitive Systems
Section 4 Intelligence, Technology and Analytics
Section 5 Computational Modeling and Simulation
Section 6 Humans and Artificial Systems Complexity
Section 7 Technology, Materials and Inclusive Human Systems
Section 8 Applications and Future Trends

We would like to extend our sincere thanks to Axel Schulte and Stefania Camplone, for leading a part of the technical program that focuses on Human-Autonomy Teaming and Inclusive Human Systems. Our appreciation also goes to the members of Scientific Program Advisory Board who have reviewed the accepted papers that are presented in this volume, including the following individuals:

D. Andujar, USA
D. Băilă, Romania
H. Blaschke, Germany
S. Camplone, Italy
J. Chen, USA
G. Coppin, France
M. Draper, USA
A. Ebert, Germany
M. Ferrara, Italy
M. Hou, Canada
M. Jipp, Germany
E. Karana, The Netherlands
A. Kluge, Germany
D. Lange, USA
F. Leali, Italy
S. Lucibello, Italy
E. Macioszek, Poland
S. Nazir, Norway
M. Neerincx, The Netherlands
R. Philipsen, Germany
J. Platts, UK
D. Popov, USA

A. Ratti, Italy
V. Rognoli, Italy
U. Schmid, Germany
A. Schulte, Germany
G. Sierpiński, Poland
N. Stanton, UK
A. Vergnano, Italy

We hope that this book, which presents the current state of the art in Intelligent Human Systems Integration, will be a valuable source of both theoretical and applied knowledge enabling the design and applications of a variety of intelligent products, services, and systems for their safe, effective, and pleasurable collaboration with people.

Dario Russo
Tareq Ahram
Waldemar Karwowski
Giuseppe Di Bucchianico
Redha Taiar

Contents

Computational Modeling and Simulation

Human-Autonomy Teaming

Anticipating Human Decision for an Optimal Teaming Between Manned and Unmanned Systems

Jane Jean Kiam, Marius Dudek, and Axel Schulte(✉)

Institute of Flight Systems, Bundeswehr University Munich,
Werner-Heisenberg-Weg 39, 85579 Neubiberg, Germany
{jane.kiam, marius.dudek, axel.schulte}@unibw.de

Abstract. We propose to include "anticipatory thinking" ability to predict human decision-making in a manned-unmanned teaming (MUM-T) system. In particular, the prediction will be leveraged to augment the "preparedness" of the unmanned vehicles in reacting optimally (with minimal delay) to the subsequent tasks to be assigned by the human operator. With this, we intend to reduce the persistent use of the automated suggestions of complex plans for task assignment, often exploited to compensate for human incapacity in "looking ahead". By doing so, we optimize human autonomy in decision making for complex MUM-T operations.

Keywords: Human decision-making · Manned-unmanned teaming · Prospect theory · Cognitive agent · Anticipatory thinking

1 Background and Motivation

Manned-unmanned teaming (MUM-T) can be exploited for future combat systems [1, 2] and first-responder operations [3]. To state the obvious, the inclusion of unmanned vehicles (UV) as enablers in these applications is to exploit them for performing routines and dangerous tasks. Leveraging the increasing autonomy of UV, simpler mission segments such as point-to-point navigation can be executed autonomously.

Most UV in MUM-T rely on tasking by either the human driving the manned system or a remote human operator. In this article, we refer to this human actor simply as an "operator". Tasking for the UV can either be performed either with a longer-term plan for routine operations, or sequentially by reacting to observations, which is more suitable in an agile environment. The latter, which we refer to as "reactive tasking", can be challenging, especially if the environment is hazardous and uncertain. Furthermore, in such an environment, anticipatory thinking [4] can be beneficial so that the UV can be better prepared for coping with the operator's sequential decisions. However, reactive tasking of the UV with anticipatory thinking can be overwhelming for the operator, since human has a bounded level of iterative (rational) reasoning [5], causing therefore delays or effectiveness in looking ahead. Having recognized this limitation, the operator of a MUM-T is often assisted by an automated cognitive agent [3],

D. Russo et al. (Eds.): IHSI 2021, AISC 1322, pp. 3–9, 2021.
https://doi.org/10.1007/978-3-030-68017-6_1

specifically for looking ahead to anticipate unfolding events by suggesting a plan of a longer time horizon, but will, in the long run, cause an overreliance and complacency of the operator, i.e. the suggested plans can be conveniently assumed the perfect "solutions".

In this article, we propose to include as an ability of the cognitive agent, the anticipation of human decision-making, in order to "prepare" the UV for optimal teaming (i.e. minimal delay in response to hazard and opportunities) with the manned system. By doing so, reactive tasking of the UV by the operator (without suggestion of longer-term plans) is still possible, without depriving the MUM-T system of efficiency and safety. Furthermore, by avoiding persisting and automated suggestions of longer-term plans, we believe this new concept of a cognitive agent gears the MUM-T system towards more respect of human autonomy. As a straightforward example, if a team of UV is tasked by the operator to accompany the manned system to carry out a search and rescue (SAR) mission in an unexplored area, the cognitive agent will place the UV with camera to the leading positions of the team, as it predicts that the operator will want to perform reconnaissance once the MUM-T system comes closer to the unexplored area.

2 Scope and Scenario

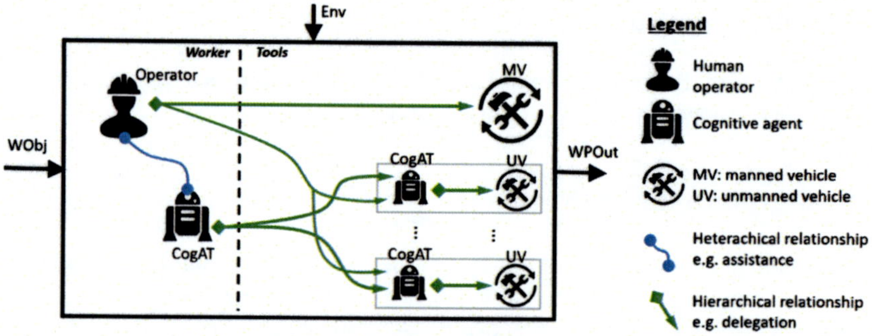

Fig. 1. A typical MUM-T system.

MUM-T represents the interoperability of manned and unmanned vehicles destined to pursue the same mission goal(s) [3]. In future MUM-T systems, the unmanned vehicles are no longer simply remote-controlled, but are instead equipped with onboard intelligence, capable of carrying out the assigned tasks in a highly automated manner [3]. Fig. 1 depicts a typical example using the semantical graphical representation for human-autonomy teaming (HAT) defined in [6], meant for a top-down systematic description of a "human-in-the-loop" system design. An onboard cognitive agent collocates with and supervises each UV. Each agent-UV ensemble in the Tools is in a hierarchical relationship with or supervised by either the onboard operator of the manned system (e.g. pilot for an airplane, driver for a car, etc.); the cognitive agent in Worker is in a heterarchical relationship with the operator, i.e. it works as an associate of the onboard operator. This associate, as explained in [6], helps to correct human errors, moderate mental workload and increase situation awareness for the human. *In this work, we propose to equip this associate cognitive agent with anticipatory thinking in order to augment the "preparedness" of the UV to react efficiently to the (human) operator's future decisions.* We briefly present two scenarios with agile and uncertain environment in which this concept can be beneficial for MUM-T while avoiding high workload on the operator without compromising human autonomy in decision making.

2.1 Scenario A: First-Responder Rescue Mission

MUM-T system can be leveraged to increase operational efficiency in a rescue mission. As shown in Fig. 2, the piloted helicopter enters an unknown area (in terms of the gravity of the disaster). On the left, distressed human voice is detected by the sensitive microphone that UAV-Mic carries, while no information is available from the area of the right. The pilot onboard of the helicopter task the UAV to divide into two teams and search for survivors in both areas. The associate cognitive agent onboard of the helicopter assists the pilot in coordinating the UAV for the task. Based on its prediction, the pilot is likely to decide for, as the next steps, counselling and medical relief for distressed people found in the area on the left, as well as for information communication with the base control station. Therefore, the associate cognitive agent assigns at least one UAV-C (a UAV with counseling ability), one UAV-M (a UAV that carries first-aid medical supplies) and the UAV-5G to be deployed in that area.

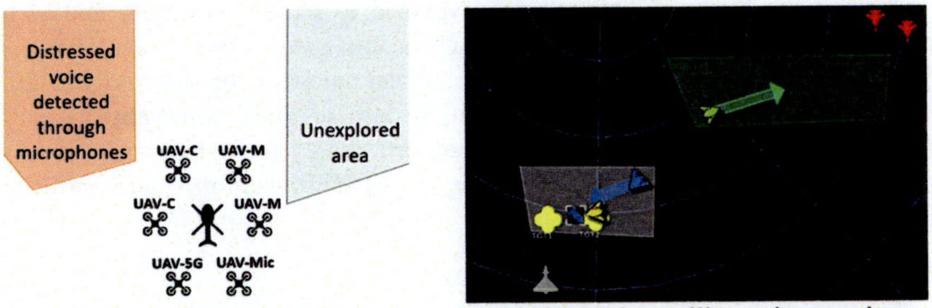

a. MUM-T in a rescue mission. b. MUM-T in a military air operation.

Fig. 2. Applicable scenarios for MUM-T.

2.2 Scenario B: Military Air Operation

A fighter pilot (see grey aircraft in Fig. 3) deployed in a MUM-T system has to guide several unmanned aircraft in addition to its own in a highly dynamic environment. The blue UAV is assigned with the suppression of enemy air defenses, while the green UAV is tasked to keep the airspace clear of enemy aircraft (in red) and the yellow UAV for the investigation of two yellow buildings. The cognitive agent that assists the pilot in coordinating the UAV has a certain degree of freedom in the overall formation (i.e. waypoint planning) of the UAV. The ideal positions of the UAV depend on the subsequent tasks to be assigned by the pilot. Potential tasks could be to engage one of the investigated buildings, to support the air-to-air UAV or to return to base. If the pilot's decision for subsequent tasks can be predicted, the prediction can be exploited for planning the waypoints to coordinate the UAV, so that they are strategically well positioned for optimal "preparedness" for the subsequent tasks.

3 Consideration of Human Decision-Making in MUM-T

In this section, a model of human decision-making is provided. Some insights on the use of the model for anticipatory thinking for the cognitive agent in Worker in Fig. 1 is provided.

3.1 Model of Human Decision-Making

Prospect Theory that has earned Kahneman and Tversky their Nobel Prize [7] proposes a human decision-making model that is considered to be more relevant than the previously most popular model based on the expected utility. The underlying observations that have motivated the development of Prospect Theory are:

1. Human measures losses and gains subjectively, by relating to a certain reference point;
2. Human sees losses more severely than gains;
3. Human tends to overweigh low probabilities and underweigh high probabilities.

Although the Prospect Theory was developed to reason on wealth and welfare, these three aspects above are highly relevant for predicting in real-time human decision-making in a MUM-T system, since 1) total rationality or objectivity is difficult to achieve during a time-critical operation, 2) loss of asset, especially loss of life represents an increasingly severe societal impact, and 3) the optimism level of each human is different. The prospect of a strategy $V(\sigma)$ of decision σ is used and is given by [7]:

$$V(\sigma) = \sum_{\{x \in X\}} \pi(x) \cdot v(\Delta x), \tag{1}$$

where v is a value defined on the differential outcome Δx with respect to a reference point (e.g. current world state or outcome of another decision choice), and π is the decision weighting function, that is a transformation of the probabilities of the outcomes.

3.2 Applicability in MUM-T

In the MUM-T scenarios described in Sect. 2, Δx can represent 1) the distance to the risk areas, 2) the negative distance to a goal position (so that an increase of it is considered a gain), 3) the level of situation awareness (e.g. information coverage), etc. Tab. 1 suggests the gains Δx can represent in each of the scenarios in Sect. 2.

Table. 1. Possible modelling of Δx for each scenario

	Scenario A	Scenario B
Δx	Δx_1: Lives to rescue Δx_2: Security of infrastructure Δx_3: Safety of the rescuer Δx_4: Security of rescuer's assets Δx_5: Situation awareness on the disaster area Δx_6: Remaining operator's (workload) effort of the pilot	Δx_1: Distance of the UV to enemy Δx_2: Distance of the fighter jet to enemy Δx_3: Negative threat level of ground-to-air attack from the enemy Δx_4: Situation awareness on target(s) Δx_5: Remaining (workload) effort of the pilot

According to [8], the value function can be expressed by a two-part power function, where a and b are the gain and loss satiation coefficient respectively, and λ is the coefficient of loss aversion:

$$v(\Delta x) = \begin{cases} \Delta x^a, & \textit{if } \Delta x \geq 0 \\ -\lambda(-\Delta x)^b, & \textit{otherwise.} \end{cases} \tag{2}$$

[9] states that, if the value function in Eq. (1) is "riskless", i.e. $v(\Delta x)$ is increasing and monotonic, and v exhibits loss aversion, i.e. $\frac{|v(|\Delta x|)|}{|v(-|\Delta x|)|} < 1$, then $\lambda > 1$, and $a = b > 0$. Figure 3 shows how the value function behaves with different coefficients. The value over gain (concave) and loss (convex) are flatter as the value of $a = b$ decreases (see Fig. 3a), while for the same gain and loss, if $a = b$, the factor of $\frac{|v(-|\Delta x|)|}{|v(|\Delta x|)|}$ equals exactly λ. The knowledge on the behavior of v makes it easier to devise an expert-driven model of the value function. For instance, the value function on Δx_1 for Scenario 2 as described in Table 1 can bear a much smaller λ than Δx_2, since the loss of a piloted fighter jet is considered more severe than the loss of a UV.

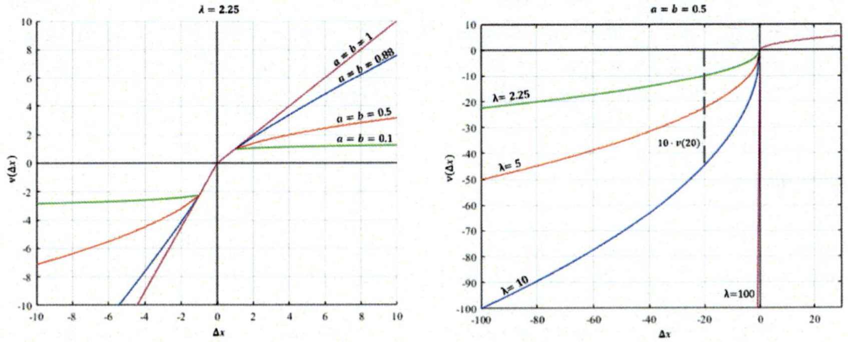

a. $\lambda = 2.25$, with varying a and b. b. $a = b = 0.5$, with varying λ.

Fig. 3. Behavior of the value function with different coefficients

The decision weight in Eq. (1) transforms the probability of an outcome $p(x)$:

$$\pi(\mathrm{x}) = \frac{p(x)^c}{\left(p(x)^c + (1 - p(x))^c\right)^{\frac{1}{c}}}, \tag{3}$$

where c is a coefficient that characterizes the transformation and can take a different value for the "gain" and the "loss" outcomes. π can be used to quantify the optimism (or pessimism) level of the human, which can be "learned" by observing behavior of the human in different scenarios. Otherwise, c can also represent the reliability of the predicted probability distribution of the outcome.

However, the Prospect Theory is usually used for predicting a single-criterion human decision-making. For a realistic prediction of the operator's decision, the listed criteria in Table 1 must be considered collectively. This is possible by applying the generalization of the Prospect Theory to a multiple-criteria decision-making as described in [10].

4 Discussion and Future Work

In this work, a concept is proposed to augment the "preparedness" of the UV in a MUM-T system by anticipating human decision-making. With this concept, sequential tasking of the UV by the operator can be performed without drop in overall efficiency or safety, since with the ability of the cognitive agent to predict human decision-making, the UV can be better "prepared" to respond to human decisions in a reactive-tasking setting with minimal delay. Overreliance on lookahead plans suggested by an automated planner can be avoided, reducing therefore human complacency and automation bias [11]. The concept will be implemented and tested using the scenarios described in Sect. 2. In particular, human-in-the-loop tests will be performed to evaluate the degree of respect for human autonomy in real-time decision-making [12] compared to the previous approach that relies on suggestion of lookahead plans.

Nevertheless, we must be cautious against "sneaky" automation. More concretely, while the cognitive agent coordinates the UV to execute the current task assigned by the operator in view of also optimizing the UV's preparedness for the predicted subsequent task, the resulting world state can also indirectly influence the operator's subsequent decision. This must be taken into account in future work.

References

1. Lashley, H., Thorpe, A., Tylor, R., Grabham, A: Measuring effectiveness of human autonomy teaming. In: NATO STO HFM-300 (2019)
2. Perelma, B.S., Lakhmani, S., Wright, J.L.: Human-autonomy teaming essential research program project 2: transparent multimodal crew interface designs. In. Technical Note 1: General Overview (ARL-TN-1003) DEVCOM Army Research Laboratory (2020)

3. Schmitt, F., Brand, Y., Rudnick, G., Schulte, A: Experimental evaluation of a cooperative automation approach for manned-unmanned teaming in future military helicopter missions. In: NATO STO HFM-300 Symposium on Human Autonomy Teaming (2018)
4. Klein, G., Snowden, D., Chew L.P.: Anticipatory thinking. In: Proceedings of the Eighth International NDM Conference, June 2007
5. Gill, D., Prowse, V.: Cognitive Ability, Character Skills, and Learning to Play Equilibrium: A Level-k Analysis. J. Polit. Econ. **124**(6), 1619–1676 (2016)
6. Schulte, A., Donath, D., Douglas, S.L.: Design patterns for human-cognitive agent teaming. In: International Conference on Engineering Psychology and Cognitive Ergonomics (2016)
7. Kahneman, D., Tversky, A.: Prospect theory: an analysis of decision under risk. In: Econometrica, vol. 47, no. 2 (1979)
8. Tversky, A., Kahneman, D.: Advances in prospect theory: cumulative representation of uncertainty. J Risk Uncertainty **5**, 297–323 (1992)
9. Al-Nowaihi, A., Bradley, I., Dhami, S.: A Note on the Utility Function Under Prospect Theory. Econ. Lett. **99**(2), 337–339 (2008)
10. Korhonen, P., Moskowitz, H.: Wallenius, J: Choice behavior in interactive multiple-criteria decision-making. Ann. Oper. Res. **23**, 161–179 (1990)
11. Cummings, M.L.: Automation bias in intelligent time critical decision support systems. In: AIAA Intelligent Systems Technical Conference (2013)
12. AI high-level expert group on artificial intelligence (HLEG): ethics guidelines for trustworthy AI. In: Publication by the European Commission (2018)

Delegation in Human-Machine Teaming: Progress, Challenges and Prospects

Jurriaan van Diggelen(✉), Jonathan Barnhoorn, Ruben Post(✉), Joris Sijs, Nanda van der Stap, and Jasper van der Waa

TNO, The Hague, The Netherlands
{jurriaan.vandiggelen, jonathan.barnhoorn, ruben.post, joris.sijs, nanda.vanderstap, jasper.vanderwaa}@tno.nl

Abstract. This paper describes DASH: Delegation to Autonomous Systems within Human-machine teams. The purpose of DASH is to combine ideas from recent progress in human machine interaction to confront current and prospected challenges of distributed delegation and meaningful human control. We describe the design considerations of DASH and illustrate how delegation using plays, a web of interaction modules, and interdependency graphs are used in the context of a military human-robot team.

Keywords: Human factors · Human-systems integration · Systems engineering

1 Introduction

For over two decades, delegation has been recognized as a crucial notion within the context of autonomy, cooperation, and collaboration [1]. It allows a delegator to set an agenda either broadly or specifically, but leaves some authority to the subordinate to decide exactly how to achieve the commands supplied by the delegator [2]. Research has made significant progress on a wide range of issues that are essential for enabling humans to delegate tasks to machines, such as developing appropriate levels of trust [3], allowing the human to have situation awareness of a diverse range of aspects of the work domain [4], and characterizing different levels of automation between human delegator and the machine [5].

Recent developments in artificial intelligence and robotics have significantly increased the application possibilities of delegation [6], yet raised new challenges as well:

Meaningful Human Control. As AI becomes ever more capable, it is applied to an increasingly wide range of problems, including ethically sensitive ones such as medical diagnosis, balancing driving speed and safety, and warfighting. While the topic is heavily discussed in many respects [7], there is general consensus that the delegation of task execution, responsibility, and accountability from humans to machines is not without limits. The human must always remain in control of ethically sensitive decisions. Designing delegation that safeguards this is one of present time's profound challenges.

D. Russo et al. (Eds.): IHSI 2021, AISC 1322, pp. 10–16, 2021.
https://doi.org/10.1007/978-3-030-68017-6_2

Distributed Delegation. AI is hardly used by one user [8, 9], nor does it take place at one particular moment. For example, an organization might set behavioral constraints to a system during deployment, while an operator commands the AI system to pursue a certain goal during operation. Both of these instructions determine the AI system's behavior. Furthermore, the machine may also act as a delegator, e.g. for delegating subtasks to other machines or even back to humans (as is the case when humans and machines work together as equal teammates). Most current delegation systems assume a one to one relation between delegator and subordinate. To be future proof, these mechanisms should be extended to enable various forms of delegation that are distributed in time and source of the command (see Fig. 1). As shown in Fig. 1d, the most advanced form of delegation in a HAT also allows machines to delegate tasks to humans, leading to complex delegation chains.

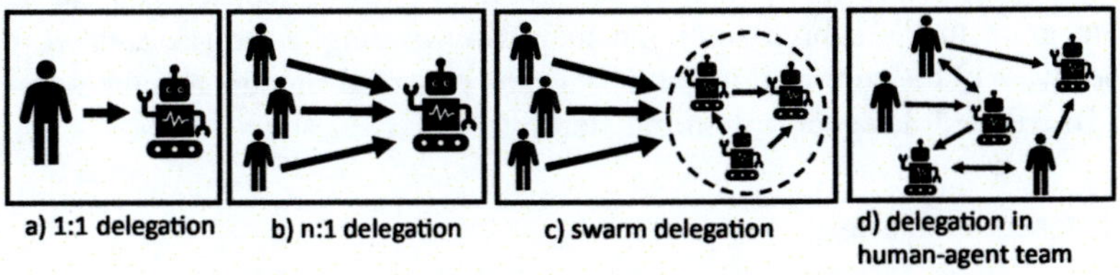

Fig. 1. Various forms of distributed delegation.

The challenges of distributed delegation and meaningful human control are closely related. A useful analysis of this dynamic was written by Chen et al. [3], describing various complexities such as development of trust over time, and mutual understanding of each other's capabilities.

This paper presents our experiences confronting these challenges while designing a uniform delegation system (called DASH) for controlling multiple intelligent assets in an ISR (Intelligence, Surveillance and Reconnaissance) task. DASH is based upon insights from ecological interface design [4], and aims at increasing the human's awareness of the sociotechnical system and environment by providing customizable views at different levels of abstraction, tailored to the desired level of involvement. Furthermore, DASH supports various ways of distributed delegation by connecting multiple humans and machines to the interface. This paper lays out the design principles of DASH, our initial experience with them, and describes their relation to the current challenges of delegation systems.

2 DASH System Design

The challenges described above require us to rethink delegation on a functional, knowledge, system, and interaction level. The main design considerations at each of these levels is specified below.

2.1 Functional Design

The table below describes the most important requirements for Observability, Predictability and Directability (resp. OR, PR, and DR).

OR1	DASH provides insight in the work system from the perspectives of tasks, goals, constraints, and resources
OR2	DASH provides insight in all directives that were issued, by whom, and how
PR3	DASH provides insights in which behaviors are expected in the future
DR4	DASH allows delegation by multiple actors at multiple moments
DR5	DASH allows delegation at various levels of abstraction

DASH should allow the user to issue a directive at an abstract level (DR5), but to drill down if needed to set the details (depending on how much freedom the delegator wants to grant to the subordinate). To maintain *meaningful human control*, humans must be aware of the current situation (OR1) and expected situations in the future (PR3, OR2). Distributed delegation is formulated by DR4, OR2.

2.2 Interaction Design

To translate the functional requirements described above into design solutions, we propose three components:

- **Web of DIMs** Understanding the workspace (OR1) completely and from all points of view would be too complicated and would lead to overly complex interfaces that require fulltime attention of a supervisory control operator (e.g. [4]). Dependent on specific user-needs, the web of DIMs (Dash Interaction Modules) provides insight in one particular aspect from any particular point of view by adjusting the workspace's perspective (task, goal, resources, constraints), visualization (e.g., map or temporal) and level of detail (OR1, OR2).
- **Plays:** Whereas DIMs serve to satisfy the observability requirements, plays are used to sets things in motion (directed at DR4, DR5). A play (as originally proposed in [2]), allows for rapidly communicating a directive. This can be done with a *low level of human control*, e.g. by specifying which goal must be achieved to a highly capable machine, or with a *high level of human control,* e.g. by specifying exactly which resources must perform which tasks (OR1). Because all types of plays are known by the different actors in advance, this allows them to anticipate on each other's actions (PR3).
- **Interdependency graphs:** Large sociotechnical systems comprise a complicated network of interdependencies between its members (human or machine), which can be made transparent by so-called *interdependency graphs.* For example, they facilitate coordinating the use of a shared asset for multiple purposes (OR2, DR4).

2.3 Knowledge Design

DASH follows a knowledge-based systems engineering approach [10], meaning that the interactions between humans and machines are for the most part defined by a shared knowledge model or ontology. The ontology must be aligned with both the human mental model of the work domain, as well as with the machine internal workings. This led us to adopt the following main concepts (adapted from the well-established theory CWA [11], and ecological interface design [4]):

- **resources** are agents, tools and consumables, that are required to perform tasks in an environment. They exist at various levels of aggregation (e.g. *Human Machine Teams, humans, robots, cameras, locomotion systems*, etc.).
- **goals** are the objectives that agents aim to achieve. An example of a high-level goals is *provide safety*. Examples of low-level goals include *obtain-optical-image*, or *located-at-waypoint-X*.
- **tasks** describe activities that agents perform to achieve their goals. Again, these exist at various levels of detail, e.g. *carry-out-air-support*, or *move-forward*.
- **constraints** are limitations on how goals should be achieved and tasks may be performed. These may be legal constraints (e.g. *rules of engagement*) that apply to the whole human machine team, or more specific user-defined constraints (e.g. on the usage of certain sensors), and can be aggregated on different levels.

The concepts above allows a *passive* understanding of the work system but are not yet sufficient to initiate actual activity. Action occurs when goals are required to perform above a certain ***threshold*** (according to a well-defined ***performance measure***); or when **tasks** are scheduled to be executed within a ***plan***. This is done by initiating **Plays** which can be understood as parameterizable templates for commonly occurring forms of delegation. For example, a play might state a sequence of **tasks** that a UAV must visit a specified location and record camera-imagery. A play may also be more abstract in terms of a high level **goal** and **constraints**, e.g. by stating that threat-level must remain minimal and use of force must be avoided at all times. Specifying multiple plays simultaneously (by multiple actors), or specifying static constraints in the work system and calling plays afterwards, can be regarded as ways to achieve distributed delegation.

3 Scenario

The following scenario illustrates some of the requirements and proposed functionality that we implemented in an early DASH prototype (Fig. 2).

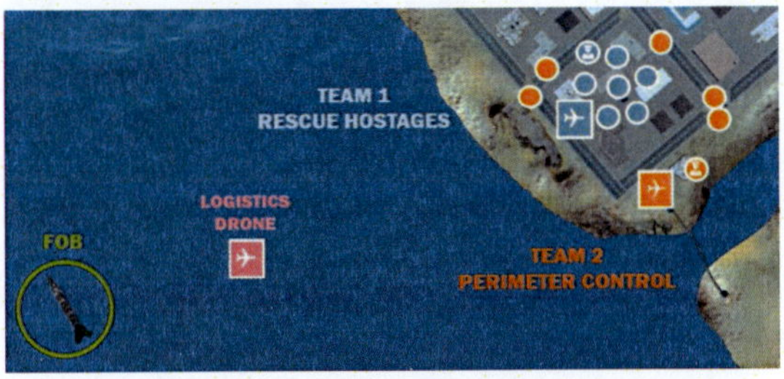

Fig. 2. An overview of the scenario.

An embassy has been ambushed. A VIP is held hostage and a rescue operation is initiated and coordinated from a Forward Operating Base (FOB) on sea. One of two teams on the shore performs perimeter control. The other team rescues the hostages from the embassy terrain. One of the latter team's members is "John", whom we will focus on. Three UAVs support the operation: a "Logistics" UAV supporting both teams when requested, a "Spotter1" UAV providing SA to the perimeter team, and a "Spotter2" UAV providing, via John, SA to the team raiding the embassy. Personnel on the FOB supports the UAVs for activities that the machines cannot perform independently, among which task selection (in case of conflicting requests) and verifying identification of contacts. Since the teams are working in a contested electromagnetic environment, dropping radio connections are a considerable risk.

Event 1: Connections between the FOB and the shore become unstable.
FOB personnel uses interdependency graphs to establish which UAV activities currently depend on interaction with the FOB. Personnel from the perimeter team is requested to temporarily take-over supporting the UAVs in these tasks. The task of verifying identification of contacts cannot be performed from land and thus John is warned that this functionality will be disabled in case of lost connections.

- In this event, the interdependency graphs provide insight into the current composition of the network based on an operational need (Fig. 3). Gaining understanding of the problems that would result of a connection-drop allows humans to respond pro-actively (see OR1).
- Losing Spotter1's ability to identify contacts is ethically sensitive and by making this known to the relevant actors, ***meaningful human control*** is supported.

Event 2: John finds a hostage.
Using the plays-interface, he quickly re-tasks Spotter1 to first scan the route to the drop zone at the embassy entrance, return and guide John to the entrance, and then return to the next building to be searched (Fig. 4).

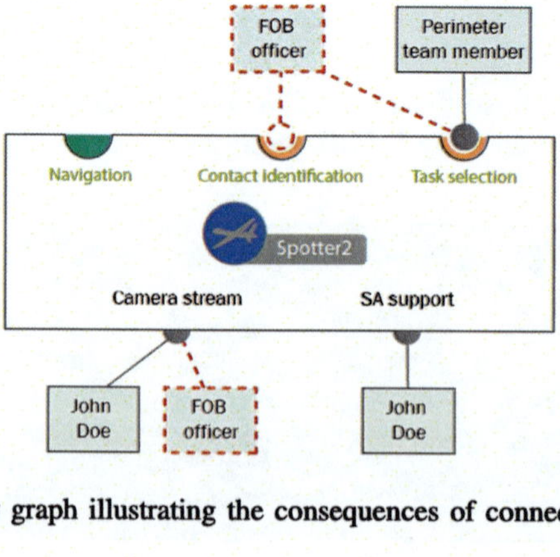

Fig. 3. Interdependency graph illustrating the consequences of connection loss between shore and FOB.

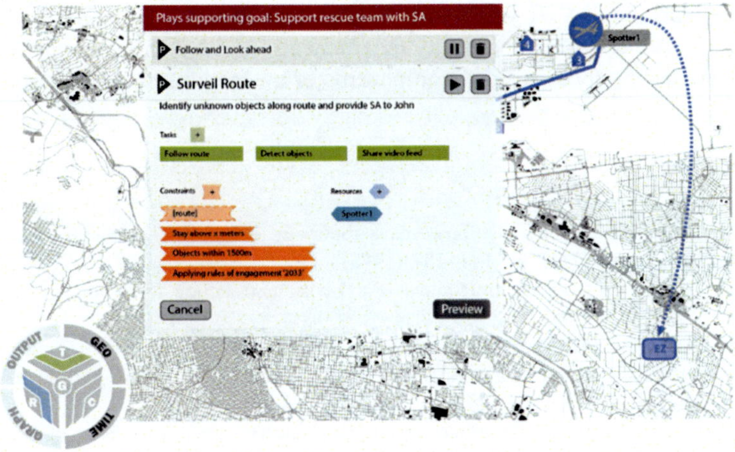

Fig. 4. An overview of the play interface, which is overlayed on a DIM the user is currently viewing.

- Using the DASH interface (see Fig. 4) allows John to delegate a set of tasks immediately using plays, increasing efficiency and reducing attentional demands (DR5).
- The initial play 'follow and look ahead' comprises a large set of tasks, called for using a single abstract play with many details filled in by the UAV. The 'surveil route' task is an example of a much more specific play (DR5).

Event 3: Connections between FOB and the UAVs are re-established.
John is informed about this. He wants to know whether Spotter1 is receiving support in identifying contacts again and checks the interdependency graph for Spotter1.

- Each actor in the network has different needs in terms of understanding the work system, here, John finds information relevant for him easily by selecting the right combination of Resource (Spotter1), Constraints (specifically: tasks constrained by a requirement for human-support), and Level of detail (see OR1).

4 Conclusion

This paper discusses progress, challenges and prospects in delegation architectures. In the light of rapidly advancing AI technology, distributed delegation and meaningful human control can be identified as two main challenges. We believe that the following areas of progress in intelligent human systems integration will play a vital role: playbased technology that allows delegation at multiple levels of abstraction (thereby varying the level of human control) by different users, and observability technology which allows humans to observe a work system from different perspectives at different levels of detail. We described DASH as an example that combines the two technologies and illustrated its merits in a multi-robot military mission.

Future work on DASH will focus on evaluating the system against the observability, predictability, and directability requirements, and designing an appropriate set of plays and testing them against requirements of meaningful human control.

References

1. Castelfranchi, C., Falcone, R.: Towards a theory of delegation for agent-based systems. Robot. Auton. Syst. **24**(3–4), 141–157 (1998)
2. Miller, C.A.: Delegation Architectures: Playbooks and Policy for Keeping Operators in Charge. WS3, 28 (2005)
3. Chen, J.Y.C., Barnes, M.J., Harper-Sciarini, M.: Supervisory control of multiple robots: human-performance issues and user-interface design. IEEE Trans. Syst. Man Cybern. Part C (Appl. Rev.) **41**(4), 435–454 (2010)
4. Bennett, K.B.: Ecological interface design and system safety: one facet of Rasmussen's legacy. Appl. Ergon. **59**, 625–636 (2017)
5. Parasuraman, R., Sheridan, T.B., Wickens, C.D.: A model for types and levels of human interaction with automation. IEEE Trans. Syst. Man Cybern. Part A Syst. Hum. **30**(3), 286–297 (2000)
6. Peeters, M.M., van Diggelen, J., Van Den Bosch, K., Bronkhorst, A., Neerincx, M.A., Schraagen, J.M., Raaijmakers, S.: Hybrid collective intelligence in a human–AI society. AI Soc. 1–22 (2020)
7. Horowitz, M., Scharre, P.: Meaningful human control in weapon systems: a primer. Center for a New American Security, Washington (2015)
8. Roth, G., Schulte, A.: A concept on the shared use of unmanned assets by multiple users in a manned-unmanned-teaming application. In: International Conference on Human-Computer Interaction, pp. 189–202. Springer, Cham, July 2020
9. Voshell, M., Tittle, J., Roth, E.: Multi-level human-autonomy teams for distributed mission management. In: AAAI Spring Symposia (2016)
10. Beckers, G., et al.: Intelligent autonomous vehicles with an extendable knowledge base and meaningful human control. In: Counterterrorism, Crime Fighting, Forensics, and Surveillance Technologies III, vol. 11166, p. 111660C. International Society for Optics and Photonics (2019)
11. Naikar, N., Hopcroft, R., Moylan, A.: Work domain analysis: Theoretical concepts and methodology (No. DSTO-TR-1665). Defence Science and Technology Organisation Victoria (Australia) Air Operations Div (2005)

Human-Centered Design in an Automated World

Claire Blackett(✉)

Institute for Energy Technology, Os Alle 5, 1777 Halden, Norway
claire.blackett@ife.no

Abstract. Human-centered design is widely recognized as a basic principle for the successful implementation of a system or technology in society. The concept of human-centered design is assumed to be utilized by developers of advanced technologies, but there are worrying indications that developers of such systems do not adequately consider the role or reaction of the human using their products. Recent high-profile automotive and aviation accidents illustrate what can go wrong when automation is introduced to a system without suitable consideration of the humans that interact with that system. In this paper, I will take a closer look at three of these accidents and discuss what we can learn regarding how humans should be considered during the development of automated systems. I link these accidents to crucial topics such as trust, communication, and transparency, and describe how these are fundamental to ensuring that future developers of automated technologies are appropriately human-centered.

Keywords: Human factors · Human-centered design · Autonomous systems

1 Introduction

We are living in an unprecedented age of technological development, and this may be most evident in the field of transportation where sophisticated automation is already commonplace. Considering the rapid pace of this development, it is not unreasonable to think that full, Level 5 [1] autonomy of transport vehicles could be achievable within our lifetime, although some experts predict that it may be several decades yet before this becomes a reality [2]. Regardless, it is indisputable that we are already enjoying the benefits of advanced automation. For example, modern airplanes incorporate automation in flight management, control, navigation, and landing systems [3]. Road vehicles routinely feature Advanced Driver-Assistance Systems (ADAS) such as adaptive cruise control (ACC), lane departure warning (LDW), and autonomous emergency braking (AEB) [4, 5]. Most new cars sold today incorporate several ADAS as standard [6], whether the car owners are aware of it or not [7].

Also on the rise, unfortunately, are the numbers of accidents linked to new technologies that appear to indicate a mismatch between user expectations and how these technologies are designed to operate. Of particular concern are those accidents where users were inadequately informed about a technological feature that played a significant role in the event. There are also accidents where users have demonstrated a much higher expectation of the capabilities of the technology than designed. The number of

D. Russo et al. (Eds.): IHSI 2021, AISC 1322, pp. 17–23, 2021.
https://doi.org/10.1007/978-3-030-68017-6_3

accidents with probable causes such as these indicates that something has gone wrong in the human-centered design process that we assume to be utilized by the developers of the advanced technological systems that we now use daily.

In this paper, I present examples of three high-profile transportation accidents that illustrate these worrying trends. I discuss how these accidents point to issues related to trust, communication, and transparency, and what this may mean for the future of human-centered design in the world of automation technology development.

2 Examples of Accidents Involving Advanced Technology

2.1 Example 1: The Volvo Auto Brake Accident

In 2015, a video[1] surfaced on the internet showing a Volvo XC60 driving into a group of people standing nearby. It appears from the video that this was meant to be a demonstration of Volvo's Pedestrian Detection feature which automatically stops the car if a person walks out in front of it. The video shows a group of people standing in front of an XC60 which is stopped. The car starts reversing slowly away from the group of bystanders and then pauses for a few moments. The car then accelerates and drives straight into the group of people at speed, colliding with two and throwing them up onto the hood of the car. The fact that nobody in the group attempted to get out of the way as the car was speeding towards them is a strong indication that they fully expected the car to brake before impacting the people. According to the blog that first published the video, the two persons hit were "bruised but are ok" [8].

Although no official reports of this event could be located, in the aftermath of the accident a Volvo spokesperson pointed to a fundamental flaw in the demonstration; this car did not have the Pedestrian Detection feature installed. It seems that it was not a standard feature for that model at that time and was sold as an optional extra [9]. Furthermore, the Volvo spokesperson noted that even if this feature had been installed, it would not have prevented the car from driving into the group of people because the behavior of the driver would have overridden the safety feature. The feature is designed to avoid collisions at speeds of up to 35 km/h [10], but (according to the spokesperson), the fact that the driver was "intentionally and actively accelerating" [11] deactivated Pedestrian Detection and overrode the auto-braking function.

2.2 Example 2: The Tesla Autopilot Crashes

In 2017, a Tesla Model X Sport Utility Vehicle (SUV) was being driven on US Highway 101 in Mountain View, California, when it hit a concrete barrier at a speed of about 71 mph. System performance data downloaded from the Tesla after the crash indicated that the driver had engaged the Traffic-Aware Cruise Control (an adaptive cruise control system) and the Autosteer lane-keeping system, both of which are ADAS in Tesla's "Autopilot" suite. Although the driver survived the initial crash, he later died at a local hospital as a result of blunt-force trauma injuries sustained during the crash.

[1] Retrieved on Sep 29 2020 from: https://www.youtube.com/watch?v=_47utWAoupo.

The United States National Transportation Safety Board (NTSB) released a report on their investigation which noted the probable cause of the crash as being "the Tesla Autopilot system steering the sport utility vehicle into a highway gore area due to system limitations, and the driver's lack of response due to distraction likely from a cell phone game application and overreliance on the Autopilot partial driving automation system" [12]. Additionally, the report identified contributing factors such as ineffective monitoring of driver behavior by the vehicle, which created the conditions for driver complacency and inattention.

According to the NTSB, because the driver was distracted by his cell phone, he did not notice that the car had driven into an area that is unsuitable for automated driving (the highway gore area). The vehicle had no means to monitor the driver's level of engagement, so any alerts or warnings that may have sounded would not have been within a sufficient time window for the driver to refocus on the driving task and prevent the collision. It also appears that the Tesla Model X software had not been programmed to detect the type of concrete barrier that the car drove into in this accident, and thus the collision avoidance system did not provide any alert to the driver, nor was the emergency braking system activated [13].

The probable cause and findings recorded by NTSB are reminiscent of another fatal crash involving a Tesla vehicle in Autopilot mode, which occurred in 2016 in Williston, Florida. In this crash, a Tesla Model S 70D collided with a Freightliner Cascadia truck-tractor; the probable cause was listed as "the truck driver's failure to yield the right of way to the car, combined with the car driver's inattention due to overreliance on vehicle automation, which resulted in the car driver's lack of reaction to the presence of the truck". A contributing factor to the driver's overreliance on automation was "its operational design, which permitted his prolonged disengagement from the driving task and his use of the automation in ways inconsistent with guidance and warnings from the manufacturer" [14]. The driver of the Tesla died in the crash; the driver of the truck-tractor was unharmed.

2.3 Example 3: The Boeing 737 MAX 8 Crashes

On October 29th, 2018, Lion Air flight 610 took off from Jakarta, India bound for Pangkal Pinang City in Indonesia. Shortly after takeoff, the flight crew communicated to the air traffic control that they were experiencing problems with flight control and altitude. Twelve minutes after takeoff, the airplane crashed into the Java Sea, killing the 189 passengers and crew on board. On March 10th, 2019, Ethiopian Airlines flight 302 departed from Addis Ababa, Ethiopia bound for Nairobi, Kenya. Just six minutes after takeoff, flight 302 crashed near the town of Bishoftu, Ethiopia, killing all 157 passengers and crew on board.

Both accidents involved the new Boeing 737 Max 8 series of aircraft, which first entered service in 2017 as a more fuel-efficient craft with longer range and lower operating costs. A significant promotional point for the new Max aircraft is that pilots would require "minimal additional training" as the similarities with previous models of aircraft were so great [15]. There were some major design changes, however; the engines on the Max were larger and positioned higher up on the wing than on previous generations of the 737 series. This change caused the craft to fly differently, and there was a higher likelihood that the nose of the craft could be pushed upwards during low-speed or manual flight, causing the airplane to stall [16].

To counteract potential stall, and to help make the handling of the Max 8 feel more like previous generations of 737, Boeing developed an automated software tool called the Maneuvering Characteristics Augmentation System (MCAS). If MCAS detected that the nose of the airplane had pitched too far upwards, it would adjust the horizontal tail trim to effectively push the nose back down and stabilize the aircraft, thus preventing stall. MCAS was designed to operate on the inputs from just one of two Angle of Attack (AOA) sensors located on either side of the airplane nose, meaning that if one sensor was faulty, MCAS could still engage if the other sensor detected a pitch problem. Furthermore, MCAS was designed to re-engage repeatedly if the sensor(s) detected that the pitch problem was not resolved [17].

MCAS was designed to be invisible to pilots, working "discreetly in the background" to correct the AOA "if the pilots were messing up" [15]. Thus, MCAS was not mentioned in the pilot manual for the Max 8 series, and the pilots did not receive any simulator training for this new system, meaning that they were completely unaware of its' existence or functionality [17]. Investigations of both accidents revealed that (in addition to countless latent errors around regulation, oversight, economic pressure, and safety culture) faulty data from the AOA sensors incorrectly triggered MCAS to push down the nose of the aircraft. Since the pilots had no knowledge or understanding of MCAS, and since MCAS was designed to continuously engage until the AOA sensors detected that the pitch was stabilized, the pilots were effectively fighting an invisible force that kept pushing the nose of the airplane downwards until they crashed.

3 Discussion: Trust, Communication, and Transparency

In my introduction, I noted three issues – trust, communication, and transparency – which I believe are at the heart of an appropriately human-centered design process. These three issues are linked [18] and, in combination, can have a powerful impact on the successful and safe use of advanced automated technologies, as illustrated by the accident examples described in the previous section.

Probably the most immediately striking aspect of the Volvo and Tesla crashes is the high level of trust placed by the humans in the automated systems. In the Volvo accident, it is evident that both the driver and the bystanders fully trusted the automatic braking system that they assumed to be present in the car. The Tesla accidents demonstrated such high levels of trust in the Autopilot system that they felt comfortable not only taking their eyes off the road but completely disengaging from the driving task and paying attention to something else.

I point this out not to place blame on the drivers, but rather to illustrate that there appears to be a false sense of security for drivers regarding automation. How have we become so trusting of automation that we are willing to place our lives in the hands of such systems? The answer could be related to the next issue, which is communication. In recent years, some car companies have been brought to task over how they communicate the abilities and limitations of ADAS in their vehicles. Both Mercedes [19] and Tesla [20] have been accused of running misleading advertising campaigns that imply their cars' self-driving abilities are far more advanced than they are. The self-driving abilities of these cars remains at Level 2 [1] for now, but the highly competitive car industry advertises these as if they are already at Level 5 automation, despite this being many decades away in reality.

In addition to communication *about* the automation, communication *with* the automation is also problematic. In the Volvo case, there was no dashboard indication that could inform the driver of either the presence or status of the Pedestrian Detection system. The reports of the Tesla accidents highlight insufficient warning systems to alert the drivers that the cars were being driven in ways that were not intended by the designers (i.e. drivers completely disengaged) or of the upcoming dangers. In the Boeing 737 Max 8 crashes, there were no interfaces in the cockpit that could have informed the pilots that MCAS was malfunctioning and that it was MCAS that kept pushing the nose of the airplanes down (besides the pilots not being told that MCAS existed).

This leads us to the next issue which is automation transparency. The Boeing 737 Max 8 accidents are the most obvious example of how the implementation of "invisible" automation can have disastrous results, especially when that automation can significantly affect the handling of the airplane. The deliberate decision to not tell pilots about MCAS to avoid additional training costs, left the pilots completely helpless when the craft started acting in an unexpected way. As stated in [21] "Progress is always good to strive for, but flying is one activity in which two pilots that have years of experience using that skill and judgment need to be fully in command of the flight." Insufficient automation transparency also contributed to the car crashes because the drivers did not have an understanding of the limitations (or even the presence) of the automated systems that they were relying on to keep them safe, which in turn meant that their level of trust in the systems was misplaced.

4 Conclusions

In this paper, I have attempted to illustrate that three groups of recent high-profile accidents have underlying causes that point to a worrying trend regarding an insufficiently human-centered design process for automated systems. Could this be because the rapid development of automated systems is changing how humans interact with and operate such systems? If so, it may be that traditional human-centered design methods are no longer adequate and may no longer be applicable as we move closer to Level 5 autonomy. I believe that a fundamental rethink of human-centered design principles is required to achieve the levels of safety, efficiency, and autonomy that are aimed at. We also need to consider more carefully the effects that trust, communication, and transparency can have on the use of advanced technologies to ensure that human-centered design is upheld as a vital best practice in this new age of technological transformation.

References

1. Society of Automotive Engineers International: SAE J3016: Taxonomy and Definitions for Terms Related to Driving Automation Systems for On-Road Motor Vehicles. SAE International (2018)
2. Leonard, J.J., Mindell, D.A., Stayton, E.L.: Autonomous Vehicles, Mobility, and Employment Policy: The Roads Ahead. MIT Work of the Future (2020)
3. Tsang, P., Vidulich, M.: Principles and Practices of Aviation Psychology. Lawrence Erlbaum Associates, New Jersey (2002)
4. Bengler, K., Dietmayer, K., Farber, B., Maurer, M., Stiller, C., Winner, H.: Three decades of driver assistance systems review and future perspectives. IEEE Intell. Transp. Syst. Mag. **6** (4), 6–22 (2014). IEEE
5. Høye, A., Hesjevoll, I.S., Vaa, T.: Advanced driver assistance systems – status and future potential [in Norwegian]. TØI report 1450. Transportøkonomisk institutt (2015)
6. American Automobile Association: Advanced Driver Assistance Technology Names; AAA's recommendation for common naming of advanced safety systems. AAA (2019)
7. Boelhouwer, A., van den Beukel, A.P., van der Voort, M.C., Hottentot, C., de Wit, R.Q., Martens, M.H.: How are car buyers and car sellers currently informed about ADAS? An investigation among drivers and car sellers in the Netherlands. Transp. Res. Interdiscip. Perspect. **4**, 100103 (2020). Elsevier Ltd.
8. New York Daily News: Self-braking Volvo fails test drive, slams into group of onlookers. https://www.nydailynews.com/news/world/self-braking-volvo-fails-test-drive-slams-onlookers-article-1.2238171 (2015). Accessed 28 Sept 2020
9. Bansal, R.: Sharing the Road. IEEE Antennas Propag. Mag. **55**(2), 190 (2013). IEEE
10. Volvo Cars Global Newsroom: Now the Volvo XC60 also gets Pedestrian Detection and the new infotainment system. https://www.media.vol-vocars.com/global/en-gb/media/pressreleases/35580. Accessed 28 Sept 2020
11. Splinter: Volvo says horrible 'self-parking car accident' happened because driver didn't have 'pedestrian detection'. https://splinternews.com/volvo-says-horrible-self-parking-car-accident-happened-1793847943. Accessed 28 Sept 2020
12. National Transportation Safety Board: Highway Accident Report; Collision Between a Sport Utility Vehicle Operating With Partial Driving Automation and a Crash Attenuator Mountain View, California, 23 March 2018. Report no. NTSB/HAR-20/01, p. 58. NTSB, Washington DC (2020)
13. Vox: Tesla needs to fix its deadly Autopilot problem. https://www.vox.com/recode/2020/2/26/21154502/tesla-autopilot-fatal-crashes. Accessed 28 Sept 2020
14. National Transportation Safety Board: Collision between a Car Operating with Automated Vehicle Control Systems and a Tractor-Semitrailer Truck Williston, FL. Report Abstract. https://www.ntsb.gov/news/events/Documents/2017-HWY16FH018-BMG-abstract.pdf. Accessed 28 Sept 2020
15. Palmer, C.: The Boeing 737 MAX saga: automating failure. Engineering **6**(1), 2–3 (2020). Elsevier Ltd.
16. Sgobba, T.: B-737 MAX and the crash of the regulatory system. J. Space Saf. Eng. **6**(4), 299–303 (2019). Elsevier Ltd.
17. Endsley, M.R.: Human Factors & Aviation Safety, Testimony to the United States House of Representatives Hearing on Boeing 737-Max8 Crashes. https://transportation.house.gov/imo/media/doc/Endsley%20Testimony.pdf. Accessed 29 Sept 2020

18. Kunze, A., Summerskill, S.J., Marshall, R., Filtness, A.J.: Automation Transparency: Implications of Uncertainty Communication for Human-Automation Interaction and Interfaces. Ergonomics **62**(3), 345–360 (2019). Taylor & Francis Online
19. Reuters: Mercedes rejects claims about 'misleading' self-driving car ads. https://de.reuters.com/article/us-mercedes-marketing-idUKKCN1081VV. Accessed 28 Sept 2020
20. CNBC: German court rules that Tesla misled consumers on Autopilot and Full Self Driving. https://www.cnbc.com/2020/07/14/tesla-autopilot-self-driving-false-advertising-germany.html. Accessed 29 Sept 2020
21. SAE Mobilus: How has automation transformed aviation? https://saemobilus.sae.org/automated-connected/feature/2019/07/how-has-automation-transformed-aviation. Accessed 29 Sept 2020

Sensor Fusion-Based Supervised Learning Approach to Developing Collaborative Manipulation System with Variable Autonomy

Stefan Wheeless and S. M. Mizanoor Rahman(✉)

Hal Marcus College of Science and Engineering, University of West Florida, Pensacola, FL, USA
smrahman@uwf.edu

Abstract. The objective is to create and program a demonstration of a manipulator arm that is capable of detecting objects, distinguishing between them, and relocating target objects to corresponding locations. The experiments are carried out to evaluate and improve the arm's design, and to gather preliminary data from the sensors to better program the system's behavior. The proposed manipulation system design makes the robot arm capable of performing a predetermined series of object relocation tasks with or without outside (human) commands from an unknown initial state. During the experiments, the performance of the robotic manipulation system is compared between two separate sensing conditions: (i) use of an ultrasonic sensor alone, and (ii) use of an ultrasonic sensor plus a light sensor. The human operator has a varying role in the manipulation. The operator may reset system inputs, put command, observe the operation, and collaborate with the system as a complementary performer or co-worker when the system needs human's support. A survey is conducted to determine the potential human involvement and human factors associated with the manipulation system, which helps determine and vary the autonomy levels of the system. Then, an experiment is conducted to determine the autonomy levels of the system based on assessing the varying contribution of the human operator. A supervised learning approach is then proposed that may learn from previous events, predict required autonomy levels for future events, and thus may help maintain appropriate autonomy levels with task requirements. The results may help develop intelligent robotic manipulation systems in industries that may work independently or in collaboration with human workers with varying levels of autonomy.

Keywords: Human-robot collaboration · Object manipulation · Variable autonomy · Sensor fusion · Supervised learning

1 Introduction

Picking up objects and moving them to where they are expected is a fundamental task in a great many industrial processes – including assembly and manufacture of anything with physical components such as appliances, vehicles, circuit boards, item storage and

D. Russo et al. (Eds.): IHSI 2021, AISC 1322, pp. 24–30, 2021.
https://doi.org/10.1007/978-3-030-68017-6_4

recall, package sorting, and a great deal more [1–3]. There are many common methods for automating these manipulation tasks such as conveyor beltways and cargo vehicles, but the best options for precise manipulation are robotic systems [4–7]. The literature shows many contributions in robotic manipulation [1–7]. In many cases, human users are also included in the manipulation system design to create a human-robot collaborative manipulation, which can enhance flexibility and performance [1, 4]. Detecting objects during the tasks, distinguishing among them, and relocating them to corresponding locations are the keys to be successful in robotic manipulation [2]. However, integrated and proven methods for successfully detecting objects distinguishing between them, and relocating them to corresponding locations are rare. It is assumed that sensor-fusion techniques could facilitate this effort [9], but such techniques are not used so often. In addition, the opportunity to change the autonomy level of human-robot collaboration system could help adjust the autonomy of the user and of the system depending on task requirements [1]. It is assumed that a supervised learning model could be used to predict such adjustment [10], which could help enhance team fluency, dynamic engagement, mixed-initiatives, and overall co-manipulation performance [1–7]. However, the literature does not show suitable learning models for this purpose.

The objective is to develop a manipulator arm that is capable of detecting objects, distinguishing among them, and relocating them to corresponding locations following a sensor fusion method. We also develop a variable autonomy model in a human-robot collaborative scenario that can demonstrate fully autonomous manipulation and human-robot collaborative manipulation with varying autonomy. A supervised learning method is proposed to predict the autonomy levels at different task situations to satisfy the task requirements. Finally, the methods are partly evaluated through experiments.

2 Development of the Robotic Manipulation System

A simple LEGO Mindstorms robotic system was developed taking inspiration from [8] as a proof-of-concept system for object manipulation. We developed a robotic arm capable of vertical and angular articulation with a manipulator end-effector capable of performing a grabbing action using servomotors. The configuration included light emitting diodes (LEDs), ultrasonic sensors and light sensors [8]. We developed the system capable of detecting the presence of objects at a distance using an ultrasonic sensor. The initial ultrasonic sensor might be able to determine the presence of objects, but an overly wide field of "vision" combined with some inconsistent sensitivity due to age and lacking sensitivity could make the sonar system difficult. Thus, an ambient light sensor was used in the development for allowing the system to more precisely detect and grasp objects. In the design, the ultrasonic sensor was affixed to a bracket mounted to the first arm segment on a mild hinge [8]. The light sensor was later mounted on a long three-piece extension attached to the bracket after testing. The robotic arm terminated in a grabbing claw-type end-effector, which was operated by a servomotor paired with a high-reduction gearing system to grasp objects. The basis of the design was derived taking inspiration from NXT STEP's CraneBot [8]. The robot's motion and behavior were programmed using a custom language created by Carnegie Mellon called Robot C. We used modular functions to perform a variety of tasks.

3 Experiments and Results

Experiments were conducted during early development of the manipulation system to evaluate and modify the arm's design, and to collect preliminary data from the sensors to be used to set thresholds in the software. Once the robot manipulator was fully developed, individual functions were then programmed and evaluated (e.g., moving objects to designated locations, grabbing and releasing objects, searching for objects, and so on), and then integrated to a working demonstration program. The initial programming effort for the search function resulted in complications probably due to the ultrasonic sensor's lack of angular precision, which caused the arm to miss objects in the yaw axis. To address this problem, a close-in ambient light sensor was added to the sensor module at the end of the robotic arm, which was used to more accurately detect an object's location by seeking a state where the test area's illumination was occluded. The robot manipulator was programmed to find out some colored test objects and relocate them to a pedestal located at the starting point with and without outside (human) interference or command. In a successful test run, the robot was able to detect Object 1, reject it, detect Object 2, accept it, and then deposit it at the target location back at the origin. The test was run 20 times: 10 with using the ultrasonic sensor only, and 10 with using the ultrasonic sensor plus an additional light sensor. The tests were conducted entirely autonomously, however human interaction was only involved in resetting the experiment between runs. In a separate experiment, the human was integrated as a part of the manipulation system, but the human had varying roles.

3.1 Results for the Trials Completed Without Human Integration

The results are shown in Table 1 for the ultrasonic sensor only and in Table 2 for the ultrasonic plus the light sensor. The 'Y' indicates success, and 'N' indicates failure.

Table 1. Results for the ultrasonic sensor only

Trials	Detect 1	Grasp 1	Detect 2	Grasp 2	Delivery	Time (s)
1	N	N	N	N	N	1:12
2	N	N	Y	Y	Y	0:55
3	N	N	N	N	N	1:02
4	N	N	N	N	N	1:01
5	N	N	N	N	N	1:02
6	N	N	Y	N	N	1:15
7	Y	Y	N	N	N	1:18
8	N	N	Y	Y	Y	0:56
9	Y	N	N	N	N	0:53
10	Y	Y	Y	Y	Y	2:48

The results showed that the ultrasonic sensor alone was found highly unreliable for detecting the target objects. At times when it would recognize the presence of one it

Table 2. Results for the ultrasonic sensor integrated with a light sensor (sensor fusion)

Trials	Detect 1	Grasp 1	Detect 2	Grasp 2	Delivery	Time (s)
1	Y	Y	Y	Y	Y	1:28
2	Y	Y	Y	Y	Y	1:31
3	Y	Y	Y	Y	Y	1:29
4	Y	Y	Y	Y	Y	1:30
5	Y	Y	Y	Y	Y	1:15
6	Y	Y	Y	Y	Y	1:13
7	Y	Y	Y	Y	Y	1:15
8	Y	Y	Y	Y	Y	1:31
9	Y	Y	Y	Y	Y	1:12
10	Y	Y	Y	Y	Y	1:28

would do so too early or late, which caused the end-effector to miss the target. Runs 1-9 were aborted and then sent back to start again due to failing to detect the target object. On run 10, it was permitted to explore repeatedly until both target objects (blue and red balls) were detected. The probable reason for the two sets of times $\sim$16 s apart in the light sensor assisted experimental scenarios were probably due to the sensor detecting and investigating the empty pedestal that the blue target object was to be deposited on in some cases. The integration of the close proximity light sensor was able to increase the reliability and repeatability significantly, which proved the advantages of the sensor fusion technique implemented in object co-manipulation as presented herein [9].

3.2 Results for the Trials Completed With Human Integration

In order to evaluate the impact of moderate and heavy human involvement in the manipulation system, two additional tests were conducted: (i) the human operator was controlling the timing of the grasp in place of the automated object detection, and (ii) the human operator was controlling every servo manually via a Bluetooth that was remote controlled. The "Crane Game" [8] was used to replace the searching function with a wait loop that was able to terminate when a button was activated by the operator. The operator waited until the robotic arm was positioned above the target, whereupon they pressed the button to cue the robotic arm to stop rotating, descend, check the target object's validity, and if found valid, grasp and relocate the object as in the autonomous test. The results showed that the timing was consistent in tests due to the consistent location of the target object, however the success or failure entirely depended on the operator's sense of timing, e.g., had they slipped up and triggered the robot arm's descent too late, or in a place where it would have knocked the target object off its stand, the test would have been rendered impossible to complete. The results showed that times were shorter than that in the fully autonomous trials as the operator could see that the first target object was invalid, and thus declined to grasp it for validation, which was similar to the early pre-light-sensor attempts where the robotic system failed to notice the presence of the first target object. The "Backhoe" [8] trials involved forgoing the previous coded routine in favor of direct control of the servomotors by the human operator. The "NXT Remote Control" (By Jacek Fedorynski on the Google Play Store) was installed onto the human operator's smartphone, and was connected to the manipulation system via Bluetooth. The "Three Motor Controls" screen was selected by the operator. The operator was tasked with opening the hand, raising it, rotating the robot arm above the target object, lowering and grasping the object, lifting it back up, and dropping it onto the designated location. The results showed that the operator successfully completed the trials on all attempts.

The results were as follows as shown in Tables 3 and 4. The results showed that the human-operated test was completed in dramatically less time than the automated ones probably due to the operator's greater understanding of the target object's location eliminating the need for the robotic arm to rotate in a slow, cautious sweep (note: the programming assigned motion to 20% of the maximum for most of the processes to ensure good detections). However, the manipulation completion time varied due to minor human errors and behaviors: fine alignment of the end effector with the target object for pickup and with the drop-off podium took varying numbers of stoppages and readjustments to get right and occasional slip-ups such as moving a motor in the wrong direction created a great deal of inconsistency, as the observations showed.

Table 3. Results for minor human involvement trials "Crane Game"

Trials	Grasp	Color detect	Delivery	Time (s)
1	Y	Y	Y	0:53
2	Y	Y	Y	0:54
3	Y	Y	Y	0:54
4	Y	Y	Y	0:55
5	Y	Y	Y	0:53

Table 4. Results for major human involvement trials "Backhoe"

Trials	Grasp	Delivery	Time (s)
1	Y	Y	0:17.1
2	Y	Y	0:19.7
3	Y	Y	0:16.6
4	Y	Y	0:14.2
5	Y	Y	0:21.3

4 The Proposed Supervised Learning Approach

For the ultrasonic plus light sensing (sensor fusion) case, the autonomy levels may be expressed in a 10-point scale (10 indicating the highest autonomy or fully autonomous, and 1 indicating the lowest autonomy or manual manipulation). Similarly, the task complexity levels may be expressed in a 10-point scale (10 indicating the most complex, and 1 indicating the least complex task). The expected manipulation performance levels can be expressed as highly satisfactory, satisfactory and unsatisfactory. A training dataset Eq. (1) may be produced for n number of trials that can use the autonomy and complexity levels as the inputs vector (x), and the experimenter's assessment/labeling of the performance as the outputs vector (y) [10]. Table 5 partly illustrates the training dataset. A target function or general formula (f) as in Eq. (2) can be learned based on the input-output relationships in Eq. (3) that can be used to predict the performance level for different levels of task complexity through varying the desired autonomy levels, and then an appropriate autonomy level can be determined before starting the collaborative tasks that can render appropriate levels of performance for a given level of task complexity [10]. The learning can also be used to detect anomaly in the sensor readings.

$$D = \{(x_1, y_1), (x_2, y_2), \ldots\ldots\ldots\ldots\ldots (x_n, y_n)\} \tag{1}$$

$$f : X \rightarrow Y \tag{2}$$

$$Y = f(X) \tag{3}$$

Table 5. Supervised learning training dataset (partial, for illustration only)

Trials (n)	Inputs (x)	Labeling/Outputs (y)
1	Autonomy level 3, complexity level 4	Unsatisfactory
2	Autonomy level 7, complexity level 3	Satisfactory

5 Conclusions and Future Work

A proof-of-concept robotic manipulation system was developed for object co-manipulation. The results show that the sensor fusion-based manipulation (ultrasonic plus light sensor) produced better manipulation performance over the single sensor-based (ultrasonic only) manipulation. The results also show varying human involvement providing variable autonomy causes variation in manipulation performance, which can be adjusted with task requirements. A simple machine learning approach was proposed that can train the system learn appropriate autonomy levels for different tasks, and then autonomously decide required autonomy levels for future tasks verifying the task requirements. In the future, a collaborative industrial robot arm will be used to verify the results. The effectiveness of the machine learning approach will be evaluated.

References

1. Rahman, S.M.M., Wang, Y.: Mutual trust-based subtask allocation for human-robot collaboration in flexible lightweight assembly in manufacturing. Mechatronics **54**, 94–109 (2018)

2. Rahman, S.M.M., Liao, Z., Jiang, L., Wang, Y.: A regret-based autonomy allocation scheme for human-robot shared vision systems in collaborative assembly in manufacturing. In: Proceedings of the 12th IEEE International Conference on Automation Science and Engineering (IEEE CASE 2016), pp. 897–902 (2016)
3. Rahman, S.M.M., Wang, Y.: Dynamic affection-based motion control of a humanoid robot to collaborate with human in flexible assembly in manufacturing. In: Proceedings of ASME Dynamic Systems and Controls Conference, pp. V003T40A005 (2015)
4. Rahman, S. M.M., Ikeura, R.: Improving interactions between a power assist robot system and its human user in horizontal transfer of objects using a novel adaptive control method. Adv. Hum. Comput. Interact. **2012**, 1–12 (2012). ID 745216
5. Rahman, S.M.M., Ikeura, R.: Cognition-based control and optimization algorithms for optimizing human-robot interactions in power assisted object manipulation. J. Inf. Sci. Eng. **32**(5), 1325–1344 (2016)
6. Rahman, S.M.M., Ikeura, R.: Investigating the factors affecting human's weight perception in lifting objects with a power assist robot. In: Proceedings of 2012 21st IEEE International Symposium on Robot and Human Interactive Communication, pp. 227–233 (2012)
7. Rahman, S.M.M., Ikeura, R.: Cognition-based variable admittance control for active compliance in flexible manipulation of heavy objects with a power assist robotic system. Robot. Biomimetics **5**(7), 1–25 (2018)
8. Boogaarts, M., et al.: The LEGO MINDSTORMS NXT Idea Book: Design, Invent, and Build (2007)
9. Fung, M.L., Chen, M.ZQ., Chen, Y.H.: Sensor fusion: a review of methods and applications. In: Proceedings of Chinese Control and Decision Conference, pp. 3853–3860 (2017)
10. Bzdok, D., Krzywinski, M., Altman, N.: Machine learning: supervised methods. Nat. Methods **15**, 5–6 (2018)

Towards a Balanced Analysis for a More Intelligent Human Systems Integration

Frank Flemisch[1,2], Michael Preutenborbeck[1], Marcel Baltzer[2], Joscha Wasser[2], Ronald Meyer[1(✉)], Nicolas Herzberger[1], Marten Bloch[2], Marcel Usai[1], and Daniel Lopez[2]

[1] Institute of Industrial Engineering and Ergonomics (IAW), RWTH Aachen University, Eilfschornsteinstr. 18, 52062 Aachen, Germany
{f.flemisch,m.preutenborbeck,r.meyer,n.herzberger, m.usai}@iaw.rwth-aachen.de

[2] Fraunhofer Institute for Communication, Information, Processing and Ergonomics (FKIE), Fraunhoferstr. 20, 53343 Wachtberg, Germany
{marcel.baltzer,joscha.wasser,marten.bloch, daniel.lopez}@fkie.fraunhofer.de

Abstract. With the increasing intelligence and power of machines, an intelligent integration of these machines with people, organizations and the environment becomes ever more crucial. This results in a number of tension fields, for both scientists and practitioner: Being too critical or analytical without sufficient synthesis and constructiveness will slow down technical and societal progress, being not critical enough might be faster in the beginning, but might have an even larger bad effect in the long run. Only a proper, dynamic balance of these approaches can bring real progress. Within the critical and analytical part itself are tension fields as well; e.g. different system qualities or different stakeholders which need to be balanced intelligently by us human systems integrators. This paper therefore discusses the tension fields within the critical and analytical aspect and discusses them with examples from vehicle automation and human autonomy teaming.

Keywords: Human systems integration · Human-machine systems · Human-machine cooperation · Balanced analysis · Vehicle automation · Human autonomy teaming

1 Introduction

From the many challenges of the 21st century, the balanced integration of increasingly powerful technology like automation and AI with increasingly individual humans, increasingly agile organizations, complex societies and finally an overheating planet earth is one of the most crucial challenges we face today. Human Systems Integration addresses this challenge by bridging humans, technology, organization and environment in an intelligent way and with the goal of intelligent systems not only on an individual human-machine scale, but increasingly on an organizational up to a global scale.

D. Russo et al. (Eds.): IHSI 2021, AISC 1322, pp. 31–37, 2021.
https://doi.org/10.1007/978-3-030-68017-6_5

One way to structure this challenge is to decompose our complex systems into subsystems and relationships, which is already at the very core of HSI as a fusion of Systems Engineering and Human Factors. An additional, orthogonal way for us to understand the challenges of HMS and HSI can be tension fields. Philosophically, tension fields can be traced back to Hegel's Dialectic, a "method of philosophical argument that involves some sort of contradictory process between opposing sides [1]. The back-and-forth debate between opposing sides produces a kind of linear progression or evolution in philosophical views or positions." Hegel's dialectic strives to get from a first moment of understanding, together with a second "dialectical" moment where the first understanding is questioned with increasing tension, via a self-sublation of conflicts to a speculative moment, which "grasps the unity of the first two understandings, and leads to a self-sublation of the conflict and a more sophisticated understanding of the world [1]. Figure 1 (right) describes an extended version of Hegel's dialectic process applied to a dialectic of HSI, where from a starting point, which could also be diffuse, tension fields of two or more theses should be polarized, balanced and synthesized again into an integrated system model. Figure 1 (right) shows a more detailed version with examples of potential tension fields in Human Systems Integration.

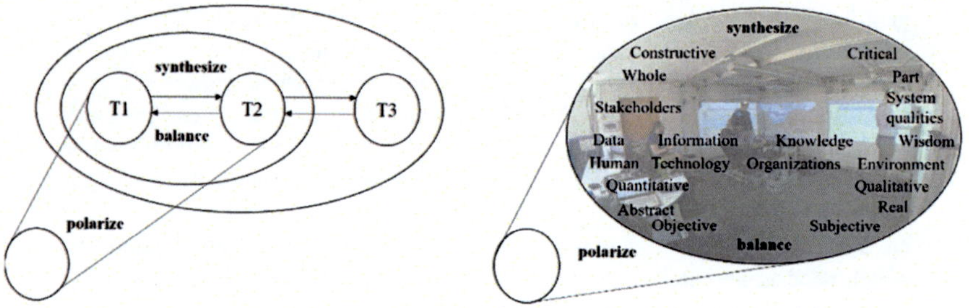

Fig. 1. Left: Potential dialectic process for Human Systems Integration (extended from [1]); Right: Basic tension fields for a dialectic of Human Systems Integration, applied to a real HSI situation (extended from [2])

2 Basic Tension Fields of HSI and Balanced Analysis

2.1 Tension Field "Constructive and Critical" "Decomposition and Analysis and Synthesis: Towards a Balanced Analysis

In general, system analysis is only one part or phase of any scientific or engineering activity. System analysis starts with observing and de-composing reality into discrete elements and relationships to gain data, information, knowledge and hopefully wisdom about the system. Based on this, a recomposing, which is called synthesis can bring this knowledge back into reality. **Applied to HSI**, decomposing human machine systems into their subsystems like people, hardware and software, and relationships like cooperation and interaction patterns allow to analyze the human machine system regarding its present or future state in relation to pre-defined system qualities. Based on this analysis, subsystems and relationships can be changed and recomposed, e.g. a new assistant system: A new invention that may change reality to the better, e.g. with decreasing the number of accidents. On the one hand, only a proper decomposition and analysis leads to insight. On the other hand, only with proper synthesis and re-composition, the result of the analysis will have an effect on reality.

What creates the tension is that in one part of the scientific community, only the de-composing and analyzing activities are considered proper science, while in other parts of the community, only the recomposing and synthesizing in form of building real systems is respected. **What could balance the tension** is the insight and agreement, that with the proper combination of both directions of thinking and acting, decomposing & analyzing and then acting by synthesizing and recomposing, both knowledge and its quality of impact on the world can be increased. A proper orchestration of the phases and a fair distribution of time and resources increases the chances to use the energy of this tension field to the mutual benefit.

Applied to vehicle automation and human autonomy teaming, it is already common sense since many decades that decomposing complex human machine systems into a digital model, sometimes also called digital shadow, using this shadow to recompose new versions in form of simulator or real-world prototypes and once again test those prototypes, can be an efficient way to both increase knowledge and know-how and improve the quality of impact of this technology on the world. Figure 2 (right) shows an example of the usage of decomposing and analyzing to recompose a simulated reality, in which a user can immerse so much that the original reality "shines" through the data. Figure 2 (right) shows an example from the project EnUSI, where an identification and attach process of a combat tank is decomposed, analysed and recomposed in a tangible XR approach so that new designs and processes can be tested [3].

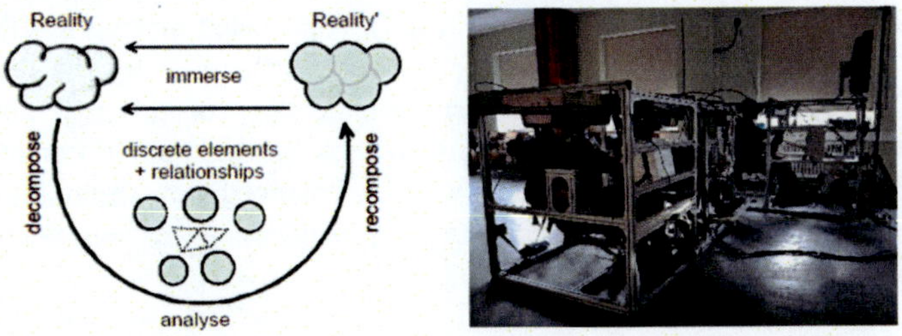

Fig. 2. Left: Iteration of Decomposing, Analysis and Synthesis (adapted from [2]); Right: Balanced Analysis and re-composition as part of a balanced HSI: Example Tank System EnUSi.

2.2 Tension Field "The Whole and the Part": Human Machine System and Meta-System

In general, a part of this tension field is already described with Aristotle's "The Whole is Greater than the Sum of its Parts" (Aristotle 348 and 322 BC), a phrase which still has a massive impact on our modern understanding of systems. **Applied to human machine systems (HMS)**, it indicates that especially the relations between the sub-systems and the meta-system characterize the HMS itself. The way in which these relations are designed depends mainly on the chosen system borders. **What creates the tension** is that in designing HMS, different stakeholders draw different system borders and choose different levels of detail (LoD). The appropriate choice of the system borders is closely linked to an appropriate LoD in the development activities. Due to the ambiguity of the required LoD, a tension field arises, which then has to be balanced.

A low LoD allows to create a 'Big Picture' of the HMS. Creating a 'Big Picture' allows to understand the relationship of the system with external components and other meta-systems. It can help to place the system within the context of its environment and help to recognize the impact it may have on aspects such as society or the environment. In general, it prevents getting lost in the details too early in the development process and forgetting about the bigger picture. On the other hand, a high LoD enables a more accurate planning and designing of the overall system, which ensures that intricate details and functions are given sufficient consideration. This, however, often leads to a narrow window in which the system is designed as the creators seek to limit the scope in order to avoid overwhelming themselves with options. Another source of tension is limited resources, especially in terms of time and money, which create a challenge to identify the required LoD in the development activities. Capacities must be divided or shared between the understanding of the overall system incl. the metasystem, and the understanding of individual subsystems. **What could balance the tension** is, to choose the appropriate approach at the right time, initially scoping a large system at a low LoD can ensure an understanding of the system in the wider context. After this, the LoD needs to be gradually increased whilst narrowing the system components that are addressed in the development.

Applied to vehicle automation, the overall picture includes aspects in the environment in which the vehicle operates, such as road safety as well as aspects that are more general at a city, national, a continental or even at a global level, such as legislation or climate protection (see Fig. 3, left). Automated driving has the potential to increase road safety, reduce the number of accidents, relieving the legislator and protecting the climate through a more sustainable way of driving. This example demonstrates how systems with increasing system borders and different LoDs are interrelated and influence each other. As a result, a balanced view of the individual part as part of the whole becomes necessary, in contrast to a completely ego- or reproduction-centric perspective (see Fig. 3, right).

2.3 Tension Field "Stakeholders and System Qualities"

In general, project management as an important part of HSI is about balancing the resources time, cost, function and quality. In 1985, Harry Sneed described the relationship between these competing resources using the Devil's Square, which he used to represent the tradeoffs between these four opposing variables [4, 5]. For example, increasing the quality and functions of a system leads to an increase in costs and development time. However, competing stakeholder interests can easily turn this balance into an imbalance. Due to limited resources a tension field arises, which has to be balanced.

Fig. 3. Left: Balancing LoDs with increasing world awareness; Right: Example from traffic systems ("Who is your turtle.. And who gets which bunnies?" Commercial Mercedes Benz/WSJ, 2015)

Diverging stakeholder interests might lead to the fact that the four resources time, cost, function and quality are difficult to be kept in perfect balance. As a result, the balance of the Devil´s Square resources can be heavily distorted in a particular direction. This unbalance can seriously compromise the system's integrity by prioritizing some system qualities over the others (e.g. low cost over higher safety). This problem can be solved by developing all system qualities to the maximum possible. This is obviously an unrealistic expectation since such system would require an unlimited amount of time and money; it would also be highly complex. Therefore, it is necessary to identify the most important qualities of the system and the extent to which they should be developed for the system to function as intended in the environment it is designed for.

It is therefore crucial to choose the appropriate approach for identifying the required system qualities. In order to assess this, Flemisch et al. extended the devil's square by adding second order system qualities such as usability, safety, trust and joy of use, etc. [6]. These criteria can be used to evaluate the competing requirements and wishes of individual stakeholders and bring them back into balance. The front of Fig. 4 depicts the extended devil's square, the corresponding sublevels as well as the different stakeholders and their competing interests.

Applied to human autonomy teaming, e.g. in aviation, the criticality of this tension field between the stakeholder's interests and the system qualities becomes even clearer. The tragic accident of a Boeing 737 Max in March 2019 shows how time and cost pressure during the development of the Maneuvering Characteristics Augmentation System (MCAS) caused people to lose their lives. The MCAS system was initially developed to assist pilots in adjusting the pitch angle and thereby increasing the safety of the aircraft and its passengers [7]. Balanced analysis helps to reduce tensions in advance and thus contributes to a balanced and safe system design.

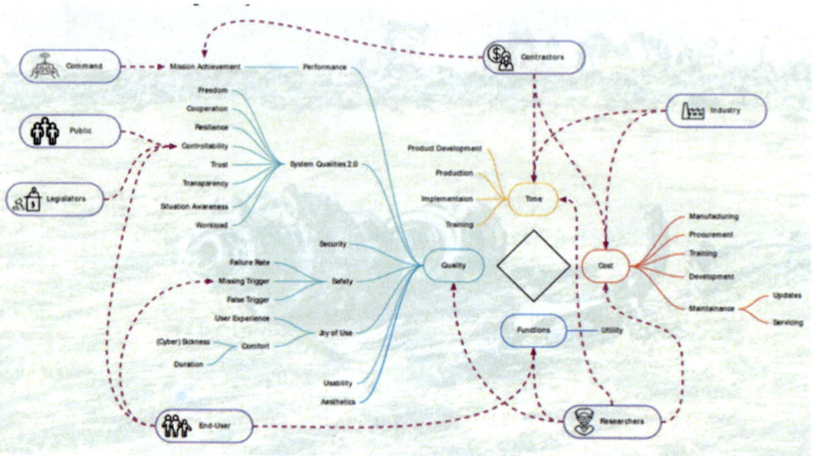

Fig. 4. Front: The Extended Devil's Square 2.0 (based on [4] and [6]) with the resources time, cost, function and quality as well as their associated sublevels; Background: Example of inappropriate Balance: Crash site of the Ethiopian Airline Crash in March 2019 [7]

3 Outlook: Towards Principles and Methods of Balancing Analysis and Balanced Human Systems Integration

This article can only sketch briefly, how a balanced analysis, as a part of a balanced Human Systems Integration, can be structured and applied. Many more tension fields should be structured and systematically described to reach well-balanced systems. It looks quite promising to structure fundamental challenges of HSI in tension fields, and condense from the vast design and application space the fundamental principles of balancing.

If we take Hegel serious, "all genuine, non-external elevation above the finite is to be found in this principle [of dialectics]" [1], there is no real progress without enough tension of thesis and anti-thesis and a proper use of this tensions towards balance and synthesis. To create, moderate and orchestrate these tensions, so that they do not lead to imbalance and destruction but to well-balanced systems should be at the very core of any intelligent system engineering, project management and Human Systems Integration.

References

1. Maybee, J.E.: Hegel's Dialectics. Von Stanford Encyclopedia of Philosophy (2016). https://plato.stanford.edu/entries/hegel-dialectics/. Accsessed 30 Sept 2020
2. Flemisch, F., Onken, R.: Open a window to the cognitive work process! pointillist analysis of man-machine interaction. Cogn. Technol. Work **4**, 160–170 (2002). https://doi.org/10.1007/s101110200015

3. Flemisch, F., Bielecki, K., López Hernández, D., Meyer, R., Baier, R., Herzberger, N., Wasser, J.: Let's get in touch again: tangible AI andtangible XR for a more tangible, balanced human systems integration. In: Ahram, T., Karwowski, W., Vergnano, A., Leali, F., Taiar, R. (eds) Intelligent Human Systems Integration 2020. IHSI 2020. Advances in Intelligent Systems and Computing, vol 1131. Springer, Cham (2020). https://doi.org/10.1007/978-3-030-39512-4_153
4. Sneed, H.M., Mérey, A.: Automated software quality assurance. IEEE Trans. Softw. Eng. **9**, 909–916 (1985)
5. Freeman, R.E.: Strategic Management. A Stakeholder Approach. Pitman **1984**, 31 (1984)
6. Flemisch, F., Meyer, R., Baltzer, M., Sadeghian, S.: Arbeitssysteme interdisziplinär analysieren, bewerten und gestalten am Beispiel der automatisierten Fahrzeugführung. In. GfA Dortmund (Hrsg.). Arbeit interdisziplinär analysieren - bewerten- gestalten. Dresden (2019)
7. Leung, H.: A U.S. Lawsuit Targets Boeing Over the Deadly Ethiopian Airlines Crash. Von TIME (2019). https://time.com/5561035/boeing-jackson-musoni-lawsuit-ethiopia-airlines-crash/. Accessed 30 Sept 2020

Multi-agent Collaboration in an Adversarial Turret Reconnaissance Task

Rolando Fernandez[1(✉)], Anjon Basak[1], Bryson Howell[2], Christopher Hsu[1], Erin Zaroukian[1], Jake Perret[3], James Humann[1], Michael Dorothy[1], Piyush K. Sharma[1], Scott Nivison[4], Zachary Bell[4], and Derrik Asher[1]

[1] US CCDC Army Research Laboratory, Adelphi, MD, USA
{rolando.fernandezl.civ, christopher.d.hsu.civ, erin.g.zaroukian.civ, james.d.humann.civ, michael.r.dorothy.civ, piyush.k.sharma.civ, derrik.e.asher.civ}@mail.mil, anjon.sunny@gmail.com

[2] Computer Science Department, University of Tennessee, Knoxville, TN, USA
bhowell3@vols.utk.edu

[3] Computer Science Department, University of Maryland, College Park, MD, USA
djakeperret@gmail.com

[4] Air Force Research Laboratory, Wright-Patterson Air Force Base, OH, USA
{scott.nivison, zachary.bell.10}@us.af.mil

Abstract. Although we have previously observed the emergence of collaboration, or more specifically, coordination between cooperative agents in a Predator-Prey Pursuit task, we have yet to demonstrate this interdependence of agents in a different task domain, where the coordination between cooperative agents is expected to depend on specific environmental parameters that may not support collaboration. In this work, we focus on a multi-agent Turret Reconnaissance task (T-RECON) to explore agent interdependence identified through a state-space perturbation technique. This work aims to 1) identify collaboration, and 2) quantify the magnitude of coordination between cooperating agents.

Keywords: Multi-agent reinforcement learning · Simulation experiments · Adversarial games · Collaboration

1 Introduction

In multi-agent adversarial games, cooperative agents typically must work together against an opposing team (who may cooperate among themselves) to achieve an objective. If we understand how environmental parameters and agent observations lead to collaboration across a team of artificial intelligence (AI) agents in these multi-agent adversarial games, we may have a basis for integrating novel partners, including humans, "on the fly," into a general scenario without total degradation of team performance.

D. Russo et al. (Eds.): IHSI 2021, AISC 1322, pp. 38–43, 2021.
https://doi.org/10.1007/978-3-030-68017-6_6

Collaboration in multi-agent learning paradigms has primarily been measured through the performance of agents in a cooperative task [1]. However, this superficial quantification of collaboration does not capture the interdependencies between agents' behaviors, nor does it help us understand how different agent behaviors may "fit together" to form a team strategy. Therefore, we argue here that a fundamental component of collaboration is the measurable interdependencies between agents performing a cooperative task; we call this coordination. Prior work in a Predator-Prey Pursuit task has demonstrated coordination between cooperating agents using Convergent Cross Mapping (CCM) [2–4] and collaboration with the aforementioned state-space perturbation technique [5]. Here, we show that the quantification of collaboration can generalize to other adversarial game environments where the outcomes may not require coordination between teammates. To our knowledge, the state-space perturbation technique is the only reliable, outcome-agnostic method for quantifying collaboration in cooperative multi-agent tasks.

In this work, we focus on a multi-agent adversarial Turret Reconnaissance (T-RECON) task to explore collaboration via quantifiable agent interdependence identified through a state-space perturbation technique. Importantly, this method for quantifying collaboration is independent of the success or failure of the objective, which is ideal in tasks where it is not clear how good performance is measured, as is the case in most novel tasks.

2 Methods

A continuous 2D multi-agent T-RECON simulation environment (Fig. 1, left) was utilized to train the agents (2 Recon and 1 Turret per model). The environment consisted of two Recons that had a shared objective of entering a goal region around the Turret during an episode (fixed number of timesteps). Conversely, the Turret's goal was to orient towards a Recon once they entered the Turret's sensing region (Fig. 1, left). Once a Recon has entered the goal region or been hit by the Turret it no longer has any effect on the simulation. When the Recons have been removed from the simulation the episode terminated early (without executing all episode timesteps).

Each Recon had the same velocity and acceleration movement limits, which were 7.5% of the Turret's angular movement limits (see Fig. 1, right, x-axis, red dot). The Turret was limited to rotating around a fixed position. The angle at which the Turret was oriented simulated a projectile or laser traveling in a ray from the Turret. If a Recon had at any point entered the Turret's sensing region, it could then be neutralized (even if at the moment of attack it was again outside the sensing region).

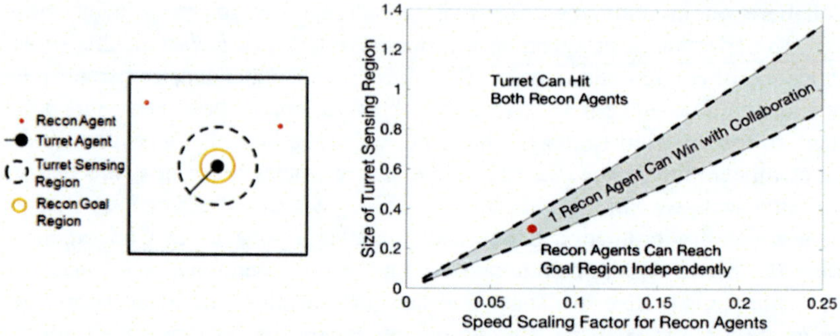

Fig. 1. LEFT: Turret reconnaissance task shown with legend. RIGHT: Segmentation of parameter space based on expected outcomes from differential game theory optimal behavior from [7]. The red dot at [0.075, 0.30] represents our choice of parameters.

The multi-agent deep deterministic policy gradient (MADDPG) algorithm was used to train all agents simultaneously [6]. Prior to perturbations, agents were trained for 150k episodes at 50 time steps per episode for the selected set of environmental parameters (Fig. 1) that were selected from ongoing work in T-RECON analytical solutions (shaded area in Fig. 1 phase space plot) [7]. These parameters were selected to favor collaboration between Recons. Test data was collected from ten independently trained models (150k episodes at 50 timesteps).

Utilizing the perturbation technique [5] we manipulate the state of a single Recon (i.e., relative distance to partner Recon) and calculate how those manipulations altered their partner's velocity. Perturbations were applied to Recon by adding the perturbation value, ±0.3, ±0.15, to the x or y component of its state. 500k state-action pairs (test data) were collected for all coordinating agents (two Recons per model). We only consider data points where both Recon agents are outside the sense region and remove perturbed states where an agent was moved out of the environment or into the sense region of the Turret, resulting in approximately 100k valid data points per x and y (total 200k data points). The resulting coordination profiles for the x and y components were combined to reduce random individual differences between the single dimension perturbation results as was done previously [5]. Importantly, the measured difference in behavior provides a metric for comparing how one Recon's velocity depended on another Recon's state. Indeed, if a Recon is primarily accounting for the orientation of the Turret, then perturbations to another Recon's state should have little to no impact on its velocity.

3 Results

The purpose of these experiments is to measure the interdependence (coordination) between MARL trained cooperative agents (Recons) in the T-RECON task using the state space perturbation technique [5]. However, before measuring Recon coordination we first evaluate the outcomes from training a set of 10 MARL models with the same environmental parameters (see Fig. 1 right, red dot in gray shaded region) with the collection of 10k test episodes (after 150k training episodes) at 100 timesteps per episode (Note, episodes could end early if both Recons were either hit or reached the goal region).

The outcomes of episodes discretely fall into 1 of 9 possible categories (Fig. 2), which are shown for two representative models (Fig. 2A and 2B) and the population of all 10 models together (Fig. 2C).

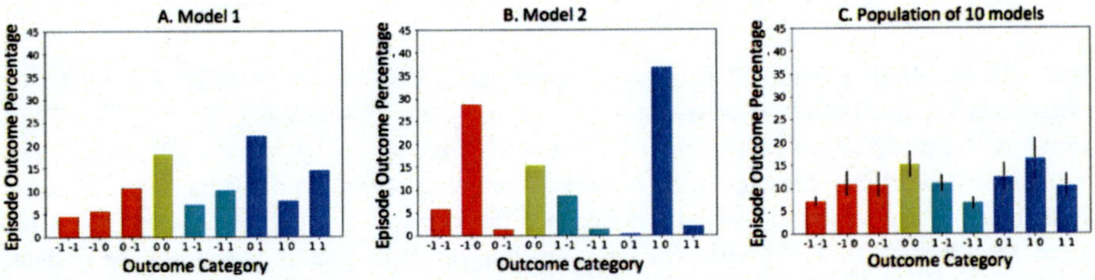

Fig. 2. Outcome plots showing the percentage of episodes (y-axes) from 10k episodes per model across the 9 outcome categories (top) for 2 independently trained models and the population of all 10 trained models. Episode outcome categories are labeled with respect to the Recons individual outcomes [Recon 1, Recon 2], signified by a -1 (Recon got hit by the Turret), 0 (Recon did not reach the goal region or get hit), or 1 (Recon reached the goal region). Thus there are 9 possible outcome pairs in our T-RECON task (x-axes). A. Model 1. B. Model 2. C. Population of all 10 Models with error bars showing standard error of the mean (SEM).

State-space perturbation plots are shown for pairs of Recons from 2 independently trained models (Fig. 3) which correspond to the outcome plots shown in Fig. 2A–B. Approximately 200k data samples were collected per agent for each perturbation value (see Fig. 3, x-axes). The data points in Fig. 3 are the means of the 200k data samples with the error bars showing standard error of the mean (SEM). The red and black colors correspond to the result of perturbing the state-space of the indicated Recon's partner (e.g., "Recon 1" label represents the impact of perturbing Recon 2's x and y positions). It is important to note that the labeling of the agents (i.e., Recon 1 or Recon 2) is arbitrary per model and should not be taken as meaningful. The Recons are identical before training and develop their policies in accordance to the MARL algorithm utilized (MADDPG). Therefore, the fact that Recon 2 in Model 1 (see Fig. 3A, red) and Recon 1 in Model 2 (see Fig. 3B, black) both show nearly zero impact from perturbing the state-space of their respective partners, is not an important observation.

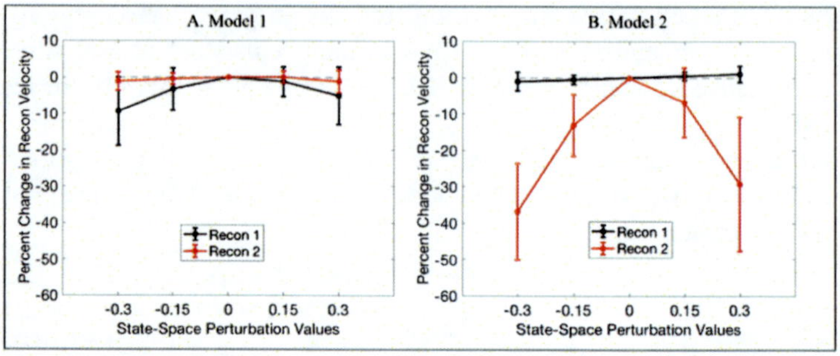

Fig. 3. Impact of state-space (x or y position) perturbations between cooperative Recons for two independently trained models (corresponding outcome plots shown in Fig. 2A and 2B). The x-axes show the perturbation values applied to the state-space of a cooperative agent (either Recon 1 or 2) with the colors representing the impact of the perturbations on the velocity of the partner agent (black is the impact of perturbing Recon 2 on Recon 1, red is the impact of perturbing Recon 1 on Recon 2). The y-axes show the resulting average percent change in the impacted Recon's velocity with negative values representing reduced velocity and positive values showing increased velocity. Zero (no coordination) is shown as a dashed line.

It was shown that the collaboration profiles between 2 independently trained models were both asymmetric within a model (see Recon 1 versus Recon 2, Fig. 3A or B), but both models showed one Recon did not have a strong dependence on its partner's actions (see Fig. 3A Recon 2 and Fig. 3B Recon 1).

4 Discussion

Utilizing the adversarial T-RECON task and the MADDPG MARL algorithm to train pairs of Recons to perform this task, then implementing the state-space perturbation technique to elucidate coordination between cooperative agents (i.e., collaboration). We have quantified the magnitude of coordination (and given that the Recons were cooperative, it is conclusive that this coordination magnitude is also the magnitude of collaboration) with percent change of Recon velocity associated with approximately 200k state-space perturbations (exclusively outside of the Turret sense region) of the Recon's partner.

In this work it is shown that the perturbation technique in conjunction with the categorical representations of episode outcomes in the T-RECON task revealed that agents did not need to coordinate *outside the Turret sensing region* to achieve their cooperative goal (both reach goal region). In Model 1, Recon 2 was able to reach the goal region across the greatest proportion of test episodes without coordination (connect Fig. 2A, x-label: [0, 1] to Fig. 3A, red line near zero). Similarly, in Model 2, Recon 1 was able to reach the goal region the greatest proportion of episodes without coordination (connect Fig. 2B, x-label: [1, 0] to Fig. 3B, black line near zero). Therefore, in the T-RECON task, it appears that coordination and in turn collaboration outside the sensing region may not be critical to the Recon agents to achieve their objective.

In contrast, Recon 2 appeared to be a mostly inactive agent (connect Fig. 2B, x-label: [1, 0], to Fig. 3B). This indicates that the Turret got distracted by one of the Recons. We also observed that between model 1 and model 2, the Recon agents learned a closer behavior, where the magnitude of coordination is less (Fig. 2A and 3A), resulting in better goal achievement because that forced the Turret to be indifferent between the Recons, which impedes the exploitation of the Recon by the Turret. On the other hand, we observed that the magnitude of coordination is higher and the Recons take in much-fixed roles which were learned to be exploited by the Turret (Fig. 2B and 3B) resulting in a better performance.

This environment gives us many avenues that we can explore for future work such as analyzing the potential spatial dependence between agents and analyzing a larger population to overcome any underlying issues with observed variance across the ten models we trained. Further, we apply the perturbation technique to fixed-policy agents defined explicitly to coordinate or not coordinate to illustrate the clear difference of interdependence between the two. Lastly, we can modify different environmental parameters (Fig. 1) and agent parameters (number of agents, heterogeneous abilities, and sensing parameters) to analyze how they affect performance and coordination. One particular future direction we might be interested in is whether the Recons learn to distract the Turret with uni-direction coordination of their motions *inside and outside the sensing region* in a manner similar to the role selection described in [7] so that one of the Recons can reach the goal region.

References

1. Baker, B., Kanitscheider, I., Markov, T., Wu, Y., Powell, G., McGrew, B., Mordatch, I.: Emergent tool use from multi-agent autocurricula. arXiv preprint arXiv:1909.07528 (2019)
2. Barton, S.L., Waytowich, N.R., Zaroukian, E., Asher, D.E.: Measuring collaborative emergent behavior in multi-agent reinforcement learning. In: International Conference on Human Systems Engineering and Design: Future Trends and Applications, pp. 422–427. Springer (2018)
3. Barton, S.L., Zaroukian, E., Asher, D.E., Waytowich, N.R.: Evaluating the coordi-nation of agents in multi-agent reinforcement learning. In: International Conference on Intelligent Human Systems Integration, pp. 765–770. Springer (2019)
4. Barton, S.L., Waytowich, N.R., Asher, D.E.: Coordination-driven learning in multi-agent problem spaces. arXiv preprint arXiv:1809.04918 (2018)
5. Asher, D., Garber-Barron, M., Rodriguez, S., Zaroukian, E., Waytowich, N.: Multi-agent coordination profiles through state space perturbations. In: 2019 International Conference on Computational Science and Computational Intelligence (CSCI), pp. 249–252. IEEE (2019)
6. Lowe, R., Wu, Y.I., Tamar, A., Harb, J., Abbeel, O.P., Mordatch, I.: Multi-agent actor-critic for mixed cooperative-competitive environments. In: Advances in Neural Information Processing Systems, pp. 6379–6390 (2017)
7. Von Moll, A., Shishika, D., Fuchs, Z.E., Dorothy, M.: The turret-runner-penetrator differential game. In: 2021 American Control Conference. IEEE (2021, in Process)

Comparison of a Logistic and SVM Model to Detect Discomfort in Automated Driving

Paul Dommel[1(✉)], Alois Pichler[1], and Matthias Beggiato[2]

[1] Faculty of Mathematics, Chemnitz University of Technology, Reichenhainer Straße 41, 09111 Chemnitz, Germany
paul.dommel@s2013.tu-chemnitz.de
[2] Cognitive and Engineering Psychology, Chemnitz University of Technology, Wilhelm-Raabe-Str. 43, 09120 Chemnitz, Germany

Abstract. Continuous monitoring of users' comfort level in automated driving could allow for optimizing human-automation teaming in this domain. Physiological parameters such as heart rate, eye blink frequency and pupil diameter are promising potential indicators for discomfort. In a driving simulator study, 20 participants experienced three automated close approach situations to a truck driving ahead and could report discomfort continuously by a handset control. Heart rate was measured by a smartband, and eye related parameters by eye tracking glasses. Two mathematical models, a logistic regression model and a Support Vector Machine (SVM) model, were compared for estimating discomfort by combing these physiological parameters. Both models showed similar prediction performance with slightly better prediction accuracy for the logistic model, even if the number of parameters (model complexity) contained in the logistic model was far less than in the SVM model.

Keywords: Discomfort · Automated driving · Logistic regression · Support vector machine (SVM) · Heart rate · Pupil diameter

1 Introduction

Ensuring and enhancing driving comfort is considered one of the key drivers for broad public acceptance of automated vehicles [1]. Due to the role change in automated driving entailing less control for the driver/user, new comfort issues become relevant. Familiarity of driving maneuvers, apparent safety, trust in the system, motion sickness and information about system states and actions are some of new factors in this changed setting [2]. As these novel comfort aspects are mainly related to specific and dynamic situations, constant evaluation is required. The automated vehicle is therefore increasingly becoming a team member in the driver-vehicle interaction. Each team player should know about each other's strengths, limitations and current states to act in a coordinated, safe and comfortable manner. The coordinated research centre "Hybrid Societies" (https://www.tu-chemnitz.de/hybrid-societies/) investigates how this teaming between humans and automated agents should be shaped to be effective, accepted and smooth. Automated driving is one of the rather dynamic and safety-critical domains, in which constant coordination at relatively short time periods is required to

D. Russo et al. (Eds.): IHSI 2021, AISC 1322, pp. 44–49, 2021.
https://doi.org/10.1007/978-3-030-68017-6_7

maintain safety and comfort. Enhanced driver monitoring including comfort aspects could detect and prevent emerging discomfort (by, e.g., adjusting driving style parameters and/or information presentation) to avoid disuse or dangerous and not necessary takeover situations [3].

Physiological parameters are one potential source for estimating drivers' comfort level in automated driving. Eye related measures such as pupil diameter and eye blink frequency as well as heart rate have been identified as promising potential indicators for discomfort [2]. However, even though each single parameter shows discomfort-related changes, the integration of these parameters by adequate mathematical models should offer best discomfort detection. Thus, the present study aimed at testing and comparing two mathematical models, a logistic regression model and a support vector machine (SVM) model for combing the physiological parameters to estimate discomfort. These models are key representatives of parametric and non-parametric, respectively, machine learning approaches. The logistic model shows low computational effort and good outlier stability. However, its intrinsic linear structure limits its prediction ability in more complex settings. Conversely, the SVM model commonly shows high prediction power in trade off for higher computational effort and vulnerability for outliers.

2 Methods

Experiment. A total of 40 participants between 25 to 84 years (25 male, 15 female) took part in a driving simulator study, consisting of two distinct driving sessions with about two months delay in between. In each session, all participants experienced an identical three minutes highly automated drive with three potentially critical and discomfort-inducing approach situations to a truck driving ahead. The truck drove at a constant speed of 80 km/h, whereas the own car approached in automated mode with 100 km/h. Automated braking started roughly late at a distance of 9 m, reaching a minimum distance of 4.2 m and minimum time to contact of 1.1 s. A handset control was integrated into the driving simulator, allowing the participants to report perceived discomfort continuously during the whole drive. Participants were instructed to press the handset control lever in accordance with the extent of perceived discomfort. There were no possibilities to intervene by pedals or the steering wheel. Further details with figures on the method, sensors and measures are reported in [4].

Sensors and Physiological Measures. Heart rate was measured continuously by the smartband Microsoft Band 2. The SMI Eye Tracking Glasses 2 were used to assess pupil diameter and eye blinks. The eye tracking glasses were not applied in the second driving session because of testing camera-based facial expressions recognition [5]. In addition, participants already wearing eyeglasses could not wear the eye tracking glasses, which resulted in a final sample of 20 trips from 20 participants. To correct for signal fluctuations, especially close to eye blinks, a moving average over $\pm$300 ms was calculated for pupil diameter. Eye blinks were transformed into a running "interblink interval time" to obtain a continuous blink rate at each moment of time. A timer was set to zero every moment a new blink was detected, and timer values increased until the subsequent eye blink started. The gradual handset control signal was transformed into the two distinct categories discomfort (values > 0) and no discomfort (0).

Training Set. The analyzed dataset was extracted from the measurements of 20 drives from of 20 different participants. The measured quantities were heart rate, interblink interval time, pupil diameter and handset control. The first three quantities were used as features and the last for labeling the data point. As the features contain individual components, as heartrate at rest, e.g., they have to be standardized. They were therefore transformed for each person's drive by

$$z = \frac{x - \bar{x}}{s}$$

where $\bar{x}$ is the mean and $s = \sqrt{s_n^2}$ the estimated standard deviation of the corresponding drive. The transformed point z is commonly referred to as the z-score of x. The interpretation of the z-score z is the (signed) distance from the individual mean, measured in units of individual standard deviation.

3 Models

Two important models, the most recognized parametric and nonparametric models have been compared.

The Logistic Model
Model. The parametric logistic model is based on the logistic regression function

$$f(z_{HR}, z_{IB}, z_{PD}) = \sigma(b + a_{HR}z_{HR} + a_{IB}z_{IB} + a_{PD}z_{PD}),$$

where $\sigma(x) = \frac{1}{1+\exp(-x)}$ is the sigmoid function. z_{HR}, z_{IB}, z_{PD} are the input features, i.e., the z-scores of heart rate, interblink interval time and pupil diameter, respectively. The coefficients b, a_{HR}, a_{IR}, a_{PD} are the regression parameters which are determined during the training of the model.

Advantages of the parametric logistic model include its small model complexity and its four parameters have a clear interpretation. They can be trained comparably fast. Further, its nonlinear structure allows controlling the outliers efficiently.

Training and Test. The logistic model was applied for each participant individually. The model parameters were determined by the maximum likelihood method ([6] p. 205–208). The parameters were validated by leave-one-out-cross-validation, employing the data of the rest of the participants as a training set. The corresponding training set is extremely unbalanced as it contains much more comfort points than discomfort points. Hence, it is vulnerable for overfitting. This means the model only fits the data points that constitute the larger part of the data set, i.e., the comfort points. In order to avoid this overfitting, the discomfort records were employed multiple times, such that the amounts of both labels were similar.

The SVM Model

Model. The kernel support vector machine (SVM) is a nonparametric machine learning approach based on a kernel function k; see (cf. [6] p. 293–294) for a detailed exposition. The kernel employed in our study was $k(x, y) = \exp(-x - y)$. It is worth notable, that the SVM model contains parameters equal to the amount of stored training points. Thus, the model gets more powerful with an increasing training set but has also a higher computational effort during the evaluation.

Kernel SVM is a learning approach which maps the training set into the feature space (a high dimensional vector space) and then separates the differently labeled data with a hyperplane; here, the labels are comfort and discomfort. Every new data point is then classified based on which side of the hyperplane it lies.

Training. As in the logistic model, the training was performed by leave-one-out-cross validation. Instead of the entire data set, 500 training points of each label (i.e., comfort and discomfort) were randomly chosen from all participants, except the one to be forecasted. Note again that the amount of training data equals the amount of features generated by the model. Thus, the employed SVM model contains 1000 features for each participant.

4 Results

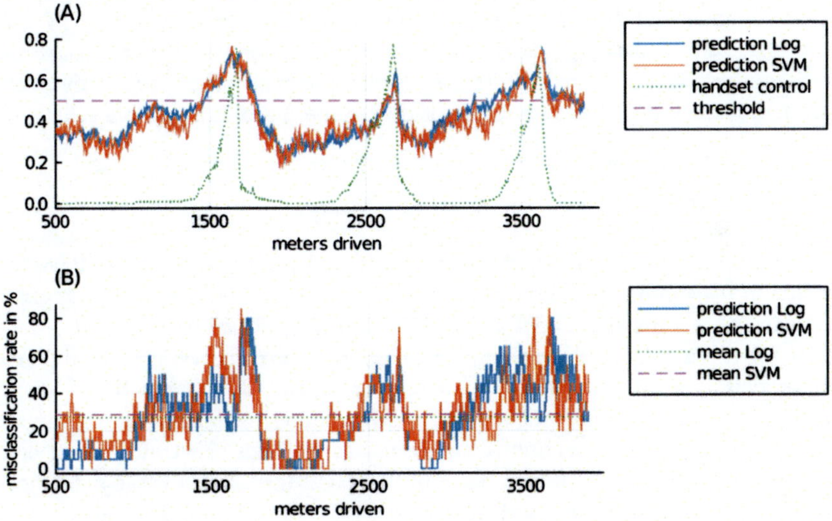

Fig. 1. (A) Predictions of the logistic model (Log) as well as the SVM model, averaged over all 20 participants. (B) Percentage of misclassified participants

Figure 1(A) shows predictions of the logistic model, the SVM model and the handset control averaged over all 20 participants. The threshold triggering discomfort is set to 0.5: values below (exceeding, resp.) the threshold correspond to a prediction of comfort (discomfort, resp.). Note that the plot starts at 500 rather than 0 driven meters. This is caused by the fact that the participants tested their handset control at the beginning of the drive. Thus, these data points were excluded from this study and hence not considered.

It is visible that the curves of both models behave similarly in terms of when they predict comfort or discomfort. Furthermore, both curves show a similar behavior compared with the "real" discomfort indicated by the handset control. During the intervals, where no participant felt discomfort, the curves mostly lie below the threshold. This implies that the prediction, at least for the majority of participants, tends to label these points correctly as comfort. Then, close to the intervals where most of the participants felt discomfort, the prediction curves exceed the threshold and they decrease afterwards. The methods correctly predict discomfort during these intervals. This indicates that both models are able to distinguish comfort and discomfort based on z-scores of heart rate, interblink interval time and pupil diameter. The first and the last discomfort interval are classified as discomfort prematurely and then show a delayed decease afterwards. This leads to an overly risk-aware assessment before and after these discomfort phases. Figure 1(B) shows the percentage of misclassified participants. As in Fig. 1(A), the logistic model and the SVM model display a very similar behavior. However, most of the times, the SVM curve lies above the logistic curve. Additionally, the logistic model has a lower mean misclassification rate (27,2%) than the SVM model (28,7%). This implies that the logistic model has slightly better prediction accuracy (72,8%) than the SVM model (71,3%). This is a very surprising result as the number of parameters contained in the logistic model is far less than in the SVM model. Figure 1(B) further shows inequity of the misclassification rate between comfortable and discomfortable situations. While the misclassification rate is low in situations of comfort, it increases significantly close or during the discomfort intervals.

5 Conclusion

The present study investigated different machine learning approaches towards the detection of discomfort, based on data from a driving simulator study. The approaches considered were a logistic model and a SVM with a radial kernel function.

The input features are z-scores of heart rate, interblink interval and pupil diameter. Both models generally correctly predict comfort and discomfort based on these parameters so that both methods are suitable prediction methods. The logistic model shows a slightly better prediction accuracy and as well has substantial lower computational efforts. This is a surprising result of this investigation and the logistic model is therefore preferable over the SVM model.

The mean accuracy of the model is about 72,8%. Both models predict the discomfort situations correctly, so that the accuracy regarding discomfort is 100%, at least for our data on average. The model parameters are therefore risk-averse.

To obtain predictions for comfort at a higher level, more discomfort related input features should be added to the present features heart rate, interblink interval and pupil diameter. Promising candidates could be facial expressions [5] or body movements [4]. We have shifted this search for adequate additional features for future investigations.

Acknowledgement. All authors acknowledge funding by Deutsche Forschungsgemeinschaft (DFG, German Research Foundation). Project-ID 416228727 – SFB 1410.

References

1. ERTRAC: Connected Automated Driving Roadmap. European Road Transport Research Advisory Council, European Road Transport Research Advisory Council (2019). https://www.ertrac.org/uploads/documentsearch/id57/ERTRAC-CAD-Roadmap-2019.pdf
2. Beggiato, M., Hartwich, F., Krems, J.: Physiological correlates of discomfort in automated driving. Transp. Res. Part F Traffic Psychol. Behav. **66**, 445–458 (2019). https://doi.org/10.1016/j.trf.2019.09.018
3. Techer, F., Ojeda, L., Barat, D., Marteau, J.-Y., Rampillon, F., Feron, S., Dogan, E.: Anger and highly automated driving in urban areas: the role of time pressure. Transp. Res. Part F Traffic Psychol. Behav. **64**, 353–360 (2019). https://doi.org/10.1016/j.trf.2019.05.016
4. Beggiato, M., Hartwich, F., Krems, J.: Using smartbands, pupillometry and body motion to detect discomfort in automated driving. Front. Hum. Neurosci. **12**, 3138 (2018). https://doi.org/10.3389/fnhum.2018.00338
5. Beggiato, M., Rauh, N., Krems, J.: Facial expressions as indicator for discomfort in automated driving. In: Ahram, T., et al. (eds.) Intelligent Human Systems Integration, IHSI 2020. AISC 1131. Springer Nature Switzerland (2020). https://doi.org/10.1007/978-3-030-39512-4_142
6. Bishop, C.: Pattern Recognition and Machine Learning. Springer New York Inc. (2006). ISBN 0387310738

Evaluation of Swarm Supervision Complexity

Sebastian Lindner(✉) and Axel Schulte

Institute of Flight Systems (IFS), Universität der Bundeswehr Munich (UBM), Werner-Heisenberg-Weg 39, 85577 Neubiberg, Germany
{sebastian.lindner,axel.schulte}@unibw.de

Abstract. Swarming is an operational concept to enable multi-robot guidance. With a human-swarm interface, called the avatar-agent, the complexity of guiding multiple unmanned vehicles (UVs) becomes independent of size. We experimentally verified in a human-in-the-loop experiment that this operational concept keeps the demand for attention of the human supervisor low. With a questionnaire we evaluated, that the avatar, as single-point-of-contact, maintained trust in automation and transparency of task execution.

Keywords: Swarming · Multi-robot guidance · Human-in-the-loop · Manned-unmanned teaming

1 Introduction

To guide and monitor unmanned vehicles (UVs), a human user must devote his attention to supervision. Human mental resources are limited, restricting the number of UVs in his responsibility. Kolling [1] summarizes the complexity of human-multi-robot systems guidance using organizational schemes. Supervising individual entities yields a complexity $O(\text{n})$ linearly increasing with the participants n. The swarming organizational scheme clusters unmanned vehicles to treat them as a single asset resulting in a constant complexity $O(1)$. To achieve this size independent complexity, both commands and feedback of the swarm must be generated by a centralized human-swarm interface – what we call **avatar**.

2 Background

A robot swarm is a group of multiple homogeneous platforms, which follow a dedicated task by means of decentralized, temporally, and spatially limited coordination. The task execution only emerges due to interaction whereas their number is irrelevant. The notion of swarming stems from biology [2], describing the movement of a large or dense group of small, rather simple beings. The observed complexity of swarming behavior emerges from frequent and parallel, but usually simple and local interactions of the swarm members, based upon exchange and adaptation of behavioral parameters (e.g. direction and speed) [3]. Usually, all swarm members follow the same goal. The primary challenge of human-swarm-interaction is *"to reconcile the distributed nature*

D. Russo et al. (Eds.): IHSI 2021, AISC 1322, pp. 50–55, 2021.
https://doi.org/10.1007/978-3-030-68017-6_8

of swarms with a central human element of control and the ability to collect information about the swarm and environment" [1].

In military applications, swarming is a historical battlefield tactic to accelerate task execution or to enable an objective by saturation. The further development and miniaturization of robot hardware and its falling costs have brought real applications for swarms within reach. One expects that unmanned swarms operate autonomously to reach a defined goal and promise to be useful in a wide range of applications. Operational benefits can be in simultaneous attacks [4] or scanning large fields within search and rescue missions [5]. A major element in military warfare is to gain air superiority in the electromagnetic spectrum [6]. For this, swarming promises to be useful as a stand-in jammer [7, 8]. Due the inherent resilience a swarm is capable of penetrating denied environment through saturation. Regardless the application, the human user must remain the decision-making authority. Therefore, the operator must be able to monitor and evaluate the resulting behavior. Due to nature of swarming, no human being can be efficiently integrated into a swarm structure. *"However, even though the human may be taken out of the physical control loop in these systems, it will still be critical to include the human [...]"* [9]. On the other hand, Coppin [10] identified that swarm performance breaks down or significantly decreases when direct human interventions in swarming algorithms are allowed. So how to create a prolific human-machine interaction between a human user and a swarm? If we want to use the characteristics of a swarm in a human-machine system, an automation interface is required that abstracts the unpredictable characteristics to a plannable and transparent, and therefore human-understandable instance.

3 Human-Swarm-Interface

The avatar is a human-swarm-interface with the function to translate the pilot's intention into parameters suitable for utilized swarm algorithms. Our implementation of the avatar consists of two modules, the delegation system and the feedback system (Fig. 1). Both system contribute to maintain the $O(1)$ supervision complexity.

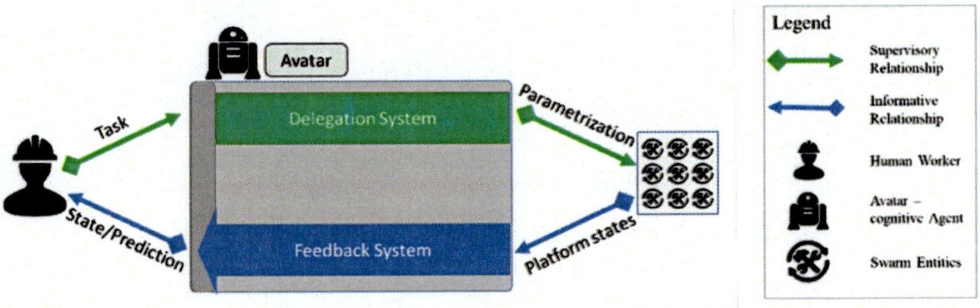

Fig. 1. Schematic of a human user using the avatar to control multiple robots

For each swarm, tasks can only be assigned to the avatar, while direct access to swarm members by the human user is not provided. The **delegation systems** chooses a suitable algorithm with adequate parameters to steer the UVs. For formation flight, as an example, we implemented a modified version of the original Boids model [11] in which each UV acts individually according to three simple rules (Separation, Alignment Cohesion).

After delegation, the **feedback system** keeps the human user informed about the current overall state of the swarm, for formation flight the center of mass and the standard deviation will be shown as proposed in [12].

4 Application for Fighter Jet Mission

In future air operations, a composite of manned and unmanned aerial vehicles shall operate as a team. In Manned-unmanned teaming (MUM-T), *"the unmanned platform(s), as well as its/their mission payloads will be commanded by the manned asset(s). From this, the major challenge for MUM-T technologies arises, i.e. to master the high work demands posed on the human user(s) arising from the multi-platform mission management and task execution*" [13]. Thus, the challenge is to enable the pilot to command unmanned systems in addition to the own aircraft. Workload is due to the high task load the essential aspect for a work system design. Thus, swarming as an operational concept supports to decrease workload by simplifying the supervision process. To visualize the relationship between human and automation, we use Schulte's modelling language for human-autonomy teaming [14]. An exemplary configuration of a MUM-T system is shown in Fig. 2. The pilot has supervisory control over his own aircraft and commands a swarm as a single team asset.

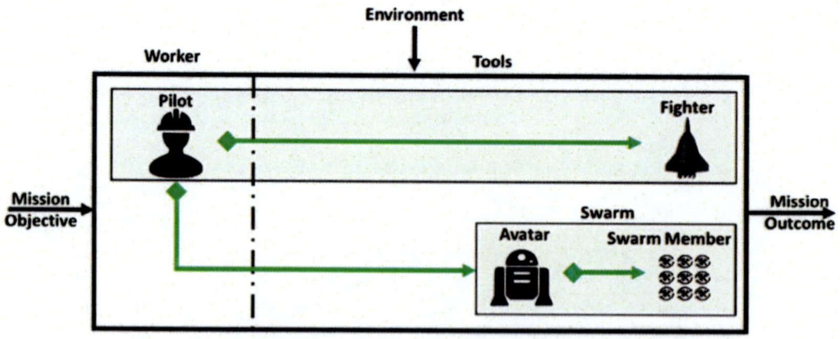

Fig. 2. Work system view of a pilot supervising a swarm.

4.1 Human-in-the-Loop Experiments

We developed a laboratory cockpit prototype for mission simulation to perform human-in-the-loop (HITL) experiments (Fig. 3). Meaningful results can only be obtained by validating concepts in a naturalistic task environment and with suitable human subjects.

Thus, we conducted the experiment with eight active German Air Force pilots. For a more detailed experimental description refer to Lindner et. al. [15]. The field of application represented a military air combat scenario where pilots had to use multiple UVs to gain tactical advantages. The participants could either guide UVs as team member, so called unmanned aerial combat vehicles (UCAV) or use the avatar to guide UVs as a swarm.

We investigated the effects on human attention when guiding UVs within a MUM-T system for two configurations.

1. Manned Fighter, UCAV, Swarm
2. Manned Fighter, UCAV, UCAV

We use this attention measure as an indicator for guidance complexity. The hypothesis is that both force approaches occupy the pilots' attention in a similar way. With a fully integrated eye-tracking system, the visual activity of the pilot is recorded [16]. In short words, we use a model that evaluates fixations on display elements to generate observations. These observations are connected to pilot activity by a task model. We assume the pilot focuses his attention on supervision, if the observations can be attributed to UVs.

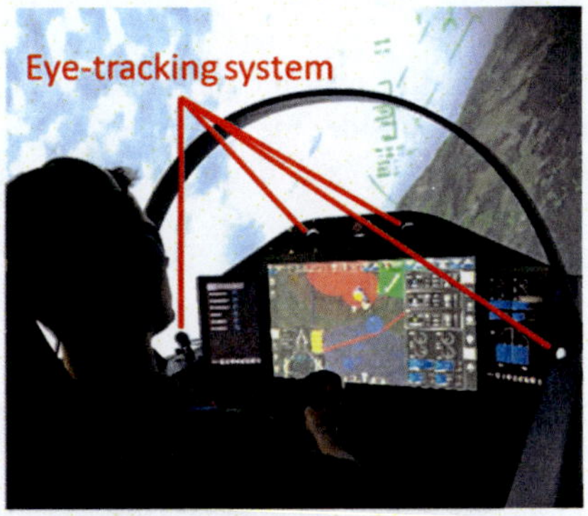

Fig. 3. Mission simulator at the Institute of Flight System.

4.2 Results

In the box-plot of Fig. 4a) the results of this activity measurement are depicted. The percentage indicates the time the pilot focused on UV monitoring in respect to the total mission time. For the first configuration the median is 36.2%. For the second configuration without a swarm it is 39.1%. Although the pilots operate more than ten UVs (Conf. 1) instead of two (Conf. 2), they applied comparable amount of resources for UV supervision.

Additionally, we surveyed the pilots about acceptance, transparency, and trust regarding the avatar. The results are depicted in the Likert-Scale of Fig. 4b). Despite the small number of test persons, their statements were consistent.

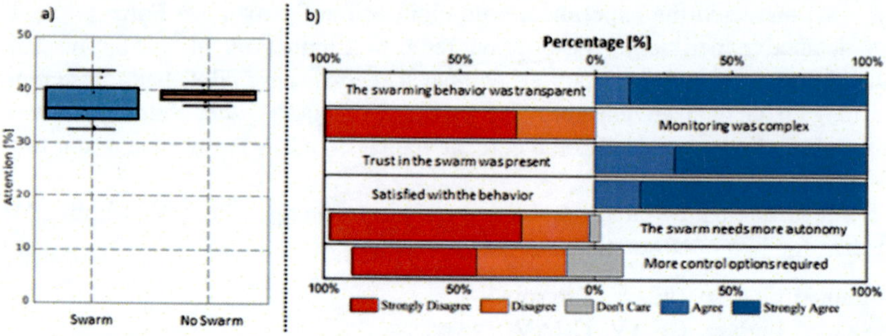

Fig. 4. a) Percentage of mission time spent by the human to monitor a swarm (left) or a single team asset (right). b) Results of the questionnaire concerning the swarm asset.

4.3 Discussion

As expected, the avatar concept induces a reduction in guidance complexity. The measured attention, as well as the questionnaire result, generate evidence that the presented avatar concept is suitable to meaningfully guide a multitude of UVs within a tactical mission. The avatar translates the unpredictable nature of swarming to a transparent trustworthy asset.

The decrease in attention for the swarm configuration may be contributed to two points. First, the pilots might not consider the swarm as a high-value asset since it was disposable in the missions. Second, it is difficult for the enemy to eliminate a swarm due to its number.

5 Conclusion

This work shows how multi-robot system guidance can be realized using swarming as an operational concept. The conducted experiments confirmed that the swarm guidance has a complexity of $O(1)$ and occupies the human being in a similar way as guiding a single UV. Using a single-point-of-contact, the avatar, upgrades a swarm to an intuitive, plannable asset. This is a necessity especially for MUM-T application, in which the integration of a human being demands a reduction of complexity to cope with the workload.

Future work focus on providing a human user with a more tactical human-swarm interface. I.e. it might be favorable to split a swarm, if the swarm operator is facing multiple temporal-overlapping objectives. We assume that guiding multiple swarms n_s leads to a complexity of $O(n_s)$.

We also want to improve the feedback system to make it a workload-adaptive interface. Depending on the information needs of the pilot, different detail levels are selectable. This will further improve the transparency of the swarm for a human user.

References

1. Kolling, A., Walker, P., Chakraborty, N., Sycara, K., Lewis, M.: Human interaction with robot swarms a survey. IEEE Trans. Hum.-Mach. Syst. **46**, 9–26 (2015)
2. Bonabeau, E., Dorigo, M., Theraulaz, G.: Swarm Intelligence: From Natural to Artificial Systems. Oxford University Press, Oxford (1999)
3. Clough, B.T.: UAV swarming? So what are those swarms, what are the implications, and how do we handle them? Air Force Research Lab Wright-Patterson AFB OH Air Vehicles Directorate (2002)
4. Pham, L.V., et al.: UAV swarm attack: protection system alternatives for destroyers. Naval Postgraduate School, Monterey, California (2012)
5. Arnold, R.D., Yamaguchi, H., Tanaka, T.: Search and rescue with autonomous flying robots through behavior-based cooperative intelligence. J. Int. Humanitarian Action **3**(1), 18 (2018)
6. Stillion, J.: Trends in air-to-air combat: implications for future air superiority. Center for Strategic and Budgetary Assessments (2015)
7. Mears, M.J.: Cooperative electronic attack using unmanned air vehicles. In: IEEE American Control Conference (2005)
8. Cevik, P., Kocaman, I., Akgul, A.S., Akca, B.: The small and silent force multiplier: a swarm UAV—electronic attack. J. Intell. Rob. Syst. **70**(1–4), 595–608 (2013)
9. Cummings, M.L.: Human supervisory control of swarming networks. In: 2nd Annual Swarming: Autonomous Intelligent Networked Systems Conference, pp. 1–9 (2004)
10. Coppin, G., Legras, F.: Autonomy spectrum and performance perception issues in swarm supervisory control. Proc. IEEE **100**(3), 590–603 (2012)
11. Reynolds, C.W.: Flocks, herds and schools: a distributed behavioral model. ACM SIGGRAPH Comput. Graph. **21**(4), 25–34 (1987)
12. Nunnally, S., et al.: Human influence of robotic swarms with bandwidth and localization issues. In: 2012 IEEE International Conference on Systems, Man, and Cybernetics (SMC), pp. 333–338 (2012)
13. Schulte, A., Heilemann, F., Lindner, S., Donath, D.: Tasking, teaming, swarming: design patterns for human delegation of unmanned vehicles. In: International Conference on Applied Human 2020, pp. 3–9 (2020)
14. Schulte, A., Donath, D.: A design and description method for human-autonomy teaming systems. In: International Conference on Intelligent Human Systems Integration, pp. 3–9 (2018)
15. Lindner, S., Schulte, A.: Human-in-the-loop evaluation of a manned-unmanned system approach to derive operational requirements for military air missions. In: International Conference on Human-Computer Interaction, pp. 341–356 (2020)
16. Mund, D., Schulte, A.: Model- and observation- based workload assessment and activity determination in manned-unmanned teaming missions. In: EAAP Conference (European Association for Aviation Psychology) (2018)

Psychophysics-Based Cognitive Reinforcement Learning to Optimize Human-Robot Interaction in Power-Assisted Object Manipulation

S. M. Mizanoor Rahman(✉)

University of West Florida, Pensacola, FL, USA
smrahman@uwf.edu

Abstract. This paper introduces a novel method of inclusion of human cognition in the form of weight perception in the dynamics and control of a power assist robotic system (PARS) for object manipulation. A 1-DOF test-bed PARS is developed. The dynamics for human-robot co-manipulation of objects is derived that includes weight perception. Then, an admittance control scheme with position feedback and velocity controller is derived from the weight-perception-based dynamics. In the control model, the mass parameter of the inertial force is considered different from that of the gravitational force. The system is simulated in MATLAB/Simulink for 36 different pairs of inertial and gravitational mass parameters. Human subjects lift an object with the system for each pair of parameters separately. The levels of human-robot interaction (HRI) is psychophysically evaluated by subjects separately using a Likert scale. In each trial, the subject evaluates the system for appropriate level of HRI. Then, a training database is generated following the reinforcement learning approach that includes inputs (pairs of mass parameter values) and corresponding outputs (levels of HRI). Then, the labeled database is used to predict a condition where subjects feel the highest level of HRI. Then, the mass parameters for the best HRI pattern are selected as the mass parameters to be used in the control system. In the testing phase, the best mass parameters are used in the control system, and the subjects evaluate the HRI for the system. The results show that the best mass parameters predicted through the reinforcement learning method produce satisfactory HRI, and in the contrary, the mass parameters deviated from the best mass parameters do not produce satisfactory HRI. The results show that inclusion of weight perception in the dynamics and control and optimization of HRI through reinforcement learning are effective.

Keywords: Psychophysics · Cognition · Reinforcement learning · Power assist robot · Object manipulation · Human-robot interaction

1 Introduction

Workers in various industries manipulate heavy materials and objects for manufacturing and other operations [1–3]. It is thought that instead of fully manual or fully autonomous manipulation, human-robot co-manipulation such as manipulation with

D. Russo et al. (Eds.): IHSI 2021, AISC 1322, pp. 56–62, 2021.
https://doi.org/10.1007/978-3-030-68017-6_9

Power Assist Robotic Systems (PARSs) may be more convenient and effective [4]. In PARSs, proper combination of physical capabilities of robots and intelligence of humans can make human-robot collaborative systems superior to individual robots or individual humans [5]. Currently, a good number of PARSs have been proposed for manipulating objects [4–7]. However, a fundamental limitation with the PARSs is that there is a mismatch between visual and haptic weight perceptions, which causes poor human-robot interaction (HRI) and manipulation performance.

A human operator/user perceives reduced weight when he/she lifts an object with a PARS [7]. The user usually feed-forwardly estimates the vertical lifting force (load force) to lift an object with the PARS, which depends on the visually perceived weight of the object. The haptically perceived weight is lower than the visually perceived weight, and thus the load force estimated by the user bases on the visual weight perception, and it becomes larger than the actually required load force. This cognitive mismatch causes undesired motions, lack of stability and safety, and may cause injuries/accidents to users. All these can adversely affect HRI and co-manipulation performance. It is thought that the estimation of load force in PARSs is a perceptual phenomenon, and the load force should be adjusted psychologically through reflecting human cognition such as weight illusion in power-assist system dynamics and controls. Hence, a novel cognition-based psychophysical approach is required to solve this cognitive problem [12]. It is assumed that a cognitive approach reflecting human's weight perception in the power-assist dynamics and control can solve this problem [8]. In addition, a machine learning approach such as the reinforcement learning may be a tool that can be used here to predict the satisfactory HRI and manipulation performance for variations in human's cognitive states reflected through variations in weight perceptions [9, 10]. However, the state-of-the-art PARSs do not consider weight illusion in their dynamics and controls, and machine learning approach is almost absent in PARS design and control.

The objective of this paper is to apply reinforcement learning through psychophysical approach to human cognition-based power-assist system dynamics and control, and try to predict favorable cognitive conditions with respect to power-assisted object manipulation that can render favorable HRI and co-manipulation performance.

2 The Proof-of-Concept PARS: Configuration, Dynamics and Control System

A 1-DOF proof-of-concept PARS for lifting (here, lifting is considered as a case of manipulation) an object was developed, as Fig. 1 shows. As in Fig. 1(a), an AC servomotor and a ball screw were coaxially fixed on a rectangular metal plate, and then the plate was vertically attached to a wall of the laboratory where the research was conducted. A foil strain gauge type force sensor was attached to the ball nut of the ball screw system through a wooden holder. An object (a rectangular box) was built bending aluminum sheet (the thickness was 0.0005 m, the self-weight was 0.016 kg, and the dimensions were 0.06 × 0.05 × 0.12 m). Two rectangular aluminum blocks with a hole in the center of each were made and then attached inside the box to help tie the box with the wooden holder (see Fig. 1(b)). The device was such that a human could grip the box (object) using a power grip and lift it vertically (see Fig. 1(c)). The force sensor was used to measure the load force applied to the power-assisted object by the human operator when lifting the object with the PARS. The (optical) encoder attached to the servomotor was used to measure the lifting motion (e.g., displacement, velocity and acceleration) for the object lifted with the assist system.

(a) (b) (c)

Fig. 1. (a) Configuration of the PARS, (b) the object to be lifted with the PARS, and (c) the scenario of lifting an object with the PARS by a human operator/user.

Dynamics for lifting an object with the PARS in collaboration with a human user can be expressed as Eq. (1), where m is mass, x_d is desired displacement, K is viscosity, g is acceleration of gravity, F is friction force, f_a is actuation force, and f_h is human force (load force) [4–7]. In order to imagine free motion dynamics, the viscosity, friction, actuating force and noises/disturbances can be ignored [11, 13]. Thus, the dynamics model can be simplified to Eq. (2).

$$m\ddot{x}_d + K\dot{x}_d + mg + F = f_h + f_a \tag{1}$$

$$m\ddot{x}_d + mg = f_h \tag{2}$$

As explained earlier, the perception of weight due to inertia may be different from the perception of weight due to gravity for the case of lifting an object with the PARS. This may happen because the visually perceived weight may be different from the haptically perceived weight [4–7]. To address this problem and to reflect such weight illusion in the system dynamics, it can be hypothesized that the mass parameter of the inertial force in Eq. (2) may be considered different from that of the gravitational force, and if so, the model in Eq. (2) can be modified to as in Eq. (3). In Eq. (3), m_1 is considered as the mass parameter of the inertial force, and m_2 is considered as the mass parameter of the gravitational force, and it is considered that $m_1 \neq m_2$. This can be treated as a novel approach to modeling system dynamics for a PARS reflecting human cognition such as weight perception. An admittance-based feedback control scheme considering positional feedback was designed for the PARS for lifting objects with it based on the cognition-based dynamics modeled in Eq. (3). The control system is shown in Fig. 2 as a block diagram.

$$m_1\ddot{x}_d + m_2 g = f_h \tag{3}$$

The proposed cognitive control in Fig. 2 falls in the admittance control category because the force is the input and the displacement is the output [11]. For lifting tasks, positional accuracy seems to be extremely desired that can come from the admittance control. The position-based control may compensate friction, viscosity, inertia, etc. that may not be compensated by force control methods. In addition, dynamic effects and nonlinear forces may influence force controls especially in multi-degree of freedom manipulation systems. These effects may not be so significant for the position control. All these factors motivated us to select the position-based control for the proposed PARS.

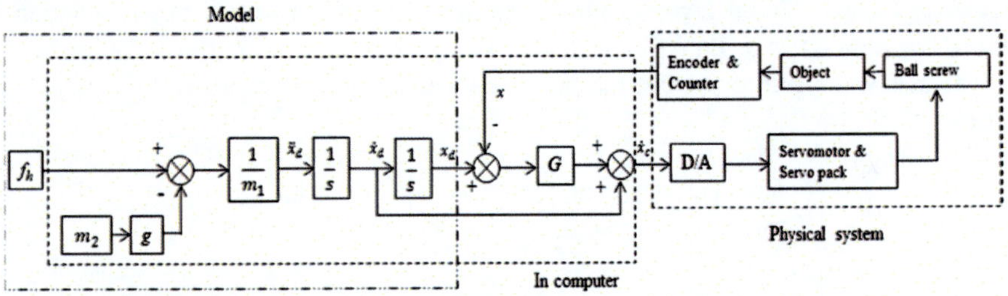

Fig. 2. The weight-perception-based feedback position control scheme for the PARS.

3 Experiments

The position control system for the PARS exhibited in Fig. 2 was implemented using MATLAB/Simulink environment. Table 1 shows 36 pairs of m_1 and m_2 values that we selected based on our experience. In each experimental trial, a set (pair) of values of m_1 and m_2 from Table 1 was randomly selected and put in the control system. However, the exact values of m_1 and m_2 were confidential to the subject. Then, the assist system was run, a human subject lifted an object with the system up to a certain height (about 0.1 m), maintained the lift for a few seconds, and then released the object (see Fig. 1(c) for detailed procedures). For that pair of values of m_1 and m_2, the trial was repeated for three times by the same subject. Then, the subject subjectively evaluated (scored) the HRI following a rating scale between 1 and 10 where 1 indicated the least favorable HRI and 10 indicated the most favorable HRI. The HRI was expressed in terms of maneuverability, safety and stability, and such expression was communicated to the subjects before they lifted the object. The subject perceived the maneuverability, safety and stability of the collaborative system during the lifting task. The combined perception created an integrated feeling in the subject, and the subject rated the system depending on that feeling about the HRI. In total, 30 subjects participated in the experiment separately, and each subject lifted the object with the system for 36×3 times for 36 pairs of m_1 and m_2 values. The grand total trials were $36 \times 3 \times 30 = 3240$, and thus 3240 rating values were obtained.

Table 1. Values of m_1 and m_2 used in the experiment (6 × 6 pairs)

m_1 (kg)	0.25	0.50	0.75	1.0	1.25	1.5
m_2 (kg)	0.25	0.50	0.75	1.0	1.25	1.5

4 Experimental Results and Analyses

The grand total 3240 trials (3240 pairs of m_1 and m_2 values) and 3240 rating values were used as the training dataset of a machine learning problem. The training dataset may be elaborated as in Eq. (4) (where n is the number of trials) that used the pairs of m_1 and m_2 values as the inputs vector (x), and the subjective rating values as the outputs vector (y) [9, 10]. Table 2 partly illustrates the training dataset. A target function or general formula (f) as in Eq. (5) was learned based on the input-output relationships in Eq. (6) that was used to predict the performance rating for given pairs of m_1 and m_2 values.

$$D = \{(x_1, y_1), (x_2, y_2), \ldots\ldots\ldots\ldots\ldots(x_n, y_n)\} \tag{4}$$

$$f : X \to Y \tag{5}$$

$$Y = f(X) \tag{6}$$

Table 2. The sample machine learning training dataset (partial list for illustration only)

Trials	Pairs of m_1 and m_2 values (inputs, x)	Performance rating (outputs, y)
1	m_1 = 1.0 kg, m_2 = 1.0 kg	5
2	m_1 = 1.0 kg, m_2 = 1.5 kg	2

The 3240 pairs of m_1 and m_2 values were available from the beginning of the experiments. However, how the users felt and evaluated the HRI for 3240 pairs of m_1 and m_2 values were not available in the beginning. Then, we obtained the 3240 rating values for 3240 pairs of m_1 and m_2 values, which completed the training dataset. As a result, this can be considered as a reinforcement learning model [9, 10]. We then developed a simple frequency-based optimization algorithm that was used to pick a pair of m_1 and m_2 values or a few pairs of m_1 and m_2 values that produced the highest levels of ratings (for example, 9 and 10 ratings), i.e., the target function f in Eq. (5). The results showed that m_1 = 0.5 kg and m_2 = 0.5 kg produced most of the 10 ratings, and m_1 = 1.0 kg and m_2 = 0.5 kg produced most of the 9 and 10 ratings. Hence, m_1 = 0.5 kg and m_2 = 0.5 kg were considered as the learned control system parameters that could predict the highest levels of HRI.

As the testing phase, we randomly selected 10 subjects. We put $m_1 = 0.5$ kg and $m_2 = 0.5$ kg in the control system shown in Fig. 2. However, the information about the $m_1 = 0.5$ kg and $m_2 = 0.5$ kg was confidential to the subjects. Each subject separately lifted the object with the PARS once, and then rated his/her perceived HRI in the 10-point rating scale. The results showed that the rating values were 10 out of 10 for 9 times (for 9 subjects), and the rating value was 9 out of 10 for only one time (for only one subject). Thus, it is posited that the control system parameters learned in the training phase was able to successfully predict the highest levels of HRI, as proved in the testing phase.

5 Conclusions and Future Work

We developed a 1-DOF proof-of-concept PARS for lifting objects. We considered human cognition in the form of weight perception, and modeled the system dynamics and control being influenced by the weight perception. We developed a reinforcement learning approach to learn control system parameters of the PARS that could predict the highest levels of HRI. The experimental results showed that the weight-perception-based system dynamics and control was able to produce satisfactory HRI, and the reinforcement learning approach was the tool to successfully learn the control system parameters of the PARS to predict the highest levels of HRI. In the future, we will use a 6-DOF robotic manipulator to verify and validate the results, and use more subjects and trials to create bigger training dataset to achieve more accurate reinforcement learning outcomes.

References

1. Rahman, S.M.M., Wang, Y.: Mutual trust-based subtask allocation for human-robot collaboration in flexible lightweight assembly in manufacturing. Mechatronics **54**, 94–109 (2018)
2. Rahman, S.M.M., Liao, Z., Jiang, L., Wang, Y.: A regret-based autonomy allocation scheme for human-robot shared vision systems in collaborative assembly in manufacturing. In: Proceedings of the 12th IEEE International Conference on Automation Science and Engineering (IEEE CASE 2016), pp. 897–902 (2016)
3. Rahman, S.M.M., Wang, Y.: Dynamic affection-based motion control of a humanoid robot to collaborate with human in flexible assembly in manufacturing. In: Proceedings of ASME Dynamic Systems and Controls Conference, pp. V003T40A005 (2015)
4. Rahman, S.M.M., Ikeura, R.: Improving interactions between a power assist robot system and its human user in horizontal transfer of objects using a novel adaptive control method. Adv. Hum.-Comput. Interact. **2012**, 1–12 (2012). Article no. 745216
5. Rahman, S.M.M., Ikeura, R.: Cognition-based control and optimization algorithms for optimizing human-robot interactions in power assisted object manipulation. J. Inf. Sci. Eng. **32**(5), 1325–1344 (2016)
6. Rahman, S.M.M., Ikeura, R.: Investigating the factors affecting human's weight perception in lifting objects with a power assist robot. In: Proceedings of 2012 21st IEEE International Symposium on Robot and Human Interactive Communication, pp. 227–233 (2012)

7. Rahman, S.M.M., Ikeura, R.: Cognition-based variable admittance control for active compliance in flexible manipulation of heavy objects with a power assist robotic system. Robot. Biomim. **5**(7), 1–25 (2018)
8. Rahman, S.M.M.: Admittance-based bio-inspired cognitive PID control to optimize human-robot interaction in power-assisted object manipulation. In: Proceedings of IEEE/ASME International Conference on Advanced Intelligent Mechatronics, pp. 679–684 (2020)
9. Neftci, E.O., Averbeck, B.B.: Reinforcement learning in artificial and biological systems. Nat. Mach. Intell. **1**, 133–143 (2019)
10. Botvinick, M., Ritter, S., Wang, J., Kurth-Nelson, Z., Blundell, C., Hassabis, D.: Reinforcement learning, fast and slow. Trends Cogn. Sci. **23**(5), 408–422 (2019)
11. Rahman, S.M.M., Ikeura, R.: Weight-perception-based novel control of a power-assist robot for the cooperative lifting of lightweight objects. Int. J. Adv. Robot. Syst. **9**(118), 1–3 (2012)
12. Rahman, S.M.M., Ikeura, R., Nobe, M., Sawai, H.: A psychophysical model of the power assist system for lifting objects. In: Proceedings of 2009 IEEE International Conference on Systems, Man, and Cybernetics, USA, pp. 4125–4130 (2009)
13. Rahman, S.M.M., Ikeura, R.: Weight-prediction-based predictive optimal position and force controls of a power assist robotic system for object manipulation. IEEE Trans. Ind. Electron. **63**(9), 5964–5975 (2016)

Automotive Design and Transportation Engineering

System Architecture for Gesture Control of Maneuvers in Automated Driving

Marcel Usai[1(✉)], Ronald Meyer[1], Ralph Baier[1], Nicolas Herzberger[1], Kristian Lebold[1], and Frank Flemisch[1,2]

[1] Institute of Industrial Engineering and Ergonomics (IAW), RWTH Aachen University, Eilfschornsteinstr. 18, 52062 Aachen, Germany
{m.usai, r.meyer, r.baier, n.herzberger, k.lebold, f.flemisch}@iaw.rwth-aachen.de

[2] Fraunhofer Institute for Communication, Information Processing and Ergonomics (FKIE), Fraunhoferstr. 20, 53343 Wachtberg, Germany

Abstract. This contribution presents a concept that allows to steer partially and highly automated vehicles with maneuver-based gestures on a haptic device like a steering wheel, sidestick etc. Maneuver-based driving aims to shift driver's tasks from the direct vehicle control (control layer) onto maneuvers (guidance layer) and allows the driver to interact with a cooperatively automated system even when he or she is not required to control the vehicle manually. Two sets of gestures were developed that are based on an abstract gesture set applicable to any input device. As a crucial part of cooperative vehicle guidance, a human-machine mediator arbitrates conflicts that possibly arise from driver's gesture input. Both gesture sets were implemented and tested in a fixed based driving simulator for cooperative vehicle guidance and control.

Keywords: Steer-by-wire · Alternative controls · Maneuver-based driving · Steering gestures · Automated driving · Human-machine cooperation

1 Introduction

Automated driving, sometimes misleadingly called autonomous driving, is at the edge of being implemented on public roads. While autonomous driving focuses on the autonomous capability of the machine to drive by itself while the driver is a passenger, automated driving uses autonomous capabilities, but binds it to a human operator in a cooperative way (e.g. 0). In partially automated driving, the driver still has control and responsibility for the driving task, but is strongly supported by the automation. In highly automated driving, the automation is doing most or all of the driving task, but the driver can still be involved and influence the driving, while the manufacturer is mainly responsible for the driving task 0. Between SAE Level 2 and Level 4 an "unsafe valley of automation" exists, where drivers expect more than the automation can deliver, might be lured into an unsafe behavior and crash 0, 0. An intuitive interaction concept is necessary to avoid the unsafe valley, and allow a good way to use a powerful driving automation, but is largely missing. This gap is addressed with our concept of steering gestures for maneuver-based driving. The basic idea is that both in SAE Level

D. Russo et al. (Eds.): IHSI 2021, AISC 1322, pp. 65–71, 2021.
https://doi.org/10.1007/978-3-030-68017-6_10

2 or 4, the automation offers possible maneuvers, which the driver can activate, influence and interrupt with a light twist or push on the steering device, or by swiping the steering device, e.g. a steering wheel or a side stick (see Fig. 1).

The presented work extends previously reported findings. Details on the use and

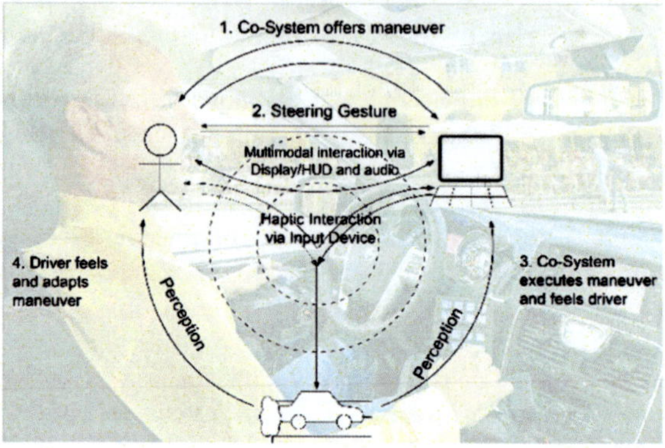

Fig. 1. Gestures for maneuver based driving. (0; modified)

design space for steering gestures, a legal discussion as well as other aspects of the user study have already been published in 0. This contribution focuses on the system architecture and recognition of steering gestures for a maneuver-based driving system.

1.1 Maneuver-Based Driving

Control of movement of e.g. a vehicle can be separated into three different layers of control, i.e. a strategic, tactical and operational layer. In addition to traditional three layer models, 0 defines a cooperational layer, which is not about driving, but cooperation itself, e.g. the level of automation and transitions. In manual driving, the human operator makes all strategic and tactical decisions, sets a target and plans the next maneuvers to reach the destination. In a highly automated vehicle, the vehicle is able to make tactical (maneuver, trajectory) decisions and give operational (stabilization) input to achieve a strategic (navigation) goal set by the human. It is important to notice that highly automated does not necessarily mean that the driver cannot influence the operational and tactical part of driving anymore. It can be reasonable to give the driver still influence and allow her or him to initiate and adapt driving behavior. With the introduction of gestures, the driver is given a tool allowing her or him to consciously intervene on the tactical layer without having to change navigational decisions and without having to take back operational control in order to execute a maneuver. This setup enables the machine to safely execute a maneuver selected by the driver. Therefore, the driver does not need to be constantly aware of relevant information (ARI) 0,0.

2 System Design

A gesture gives the co-system a hint on the actual intention of the driver. Based on the maneuver space as described in 0 (see Fig. 2, left), an abstracted gesture set was defined (Fig. 2, center). It is based on a pars pro toto movement of the vehicle and applicable to any input device. Applied to pedals and steering wheel, a twist/push and applied to steering wheel, a swipe gesture set was developed.

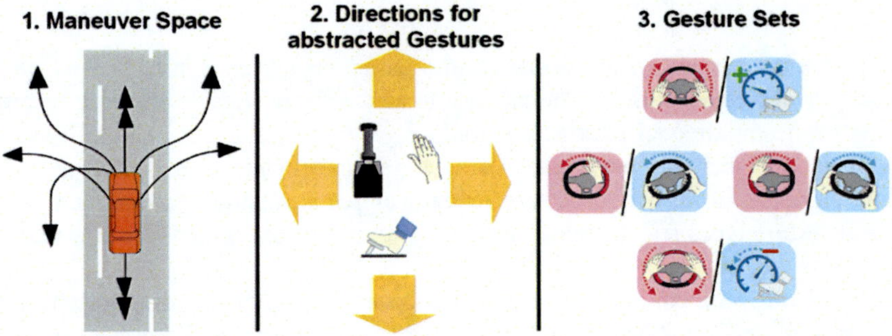

Fig. 2. Driver Use Cases related to the relative vehicle position (left); directions for abstracted gesture set (center); twist/push gesture set in blue, swipe gesture set in magenta (right). 0

The twist/push gestures require the driver to apply a small deflection in the corresponding direction on respecting input device. When reaching a threshold, the gesture is then recognized. All gestures are assigned a gesture progress and intermediate levels are recognized to generate live feedback on the gesture execution status, which is displayed in a Head-Up Display (HUD) 0. For swipe gestures, a touch-sensitive steering wheel was divided into ten evenly distributed zones. For the input of lateral gestures, the driver had to swipe in respecting direction with one hand, and with two hands for longitudinal gestures. As depicted in Fig. 3 (left), the progress of a gesture was captured by a state machine. Whenever the driver touches a zone, a new state machine is spawned and all state transitions are switched. The highest progress of any state machine in either direction represents the overall lateral gesture progress. Results of all active state machines of the first type, which reflect the progress of single-handed gestures, are fed into another state machine (see Fig. 3, right) to determine the progress of two-handed gestures.

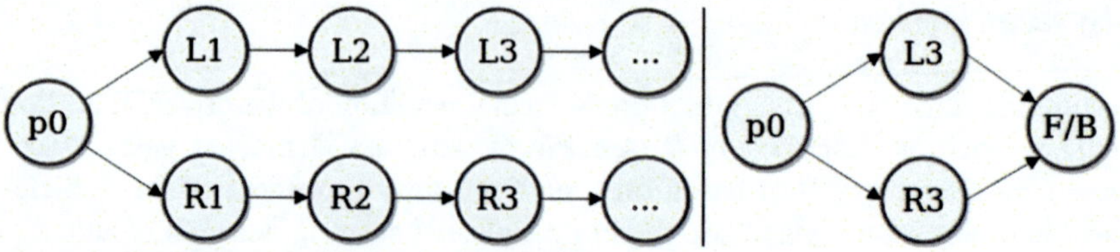

Fig. 3. State machine for recognition of single-handed swipe gestures (left) and consecutive state machine to detect two-handed gestures (right).

The overall system setup of the gesture-based vehicle in a fixed based driving simulator is depicted in Fig. 4. The driver interacts with the vehicle automation system using a haptic-multimodal interface, including gesture input devices, as well as vehicle's visual and auditory feedbacks. Using a gesture module to create an abstracted gesture signal from all active gesture recognizer parts provides the ability to merge multiple gesture types and design all further system components independent from any gesture sets.

The interaction mediator is implemented based on 0. It recognizes gesture input, paires gesture and situation information to a given maneuver and presents this to the driver while adapting the automation to execute the recognized maneuver. Detailed behavior of the interaction mediator for gesture-based control was published in 0. The automation is based on works of 0. Based on its information on the environment, the

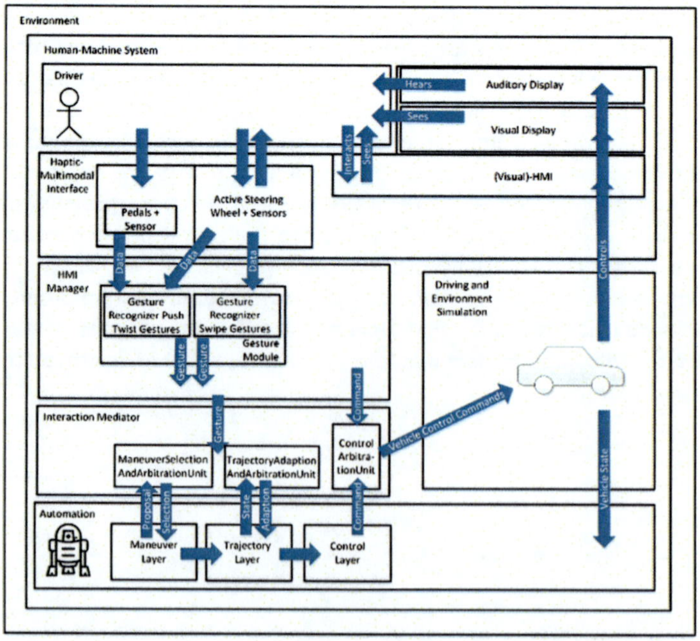

Fig. 4. Overview of full system setup that the gesture-based driving was implemented on.

automation provides the interaction mediator with a set of possible maneuvers and receives a maneuver to execute from the mediator, which decides based on gesture input and driving situation. To enable partially and highly automated driving, the interaction mediator is able to arbitrate control between driver and automation.

3 Experimental Validation and Discussion

In a driving simulation study, the gesture sets were tested on drivers representing the mass market as well as on novice and disabled drivers 0. The goal of the user study was to check whether the developed concepts are suitable for all drivers, can be operated safely and are well understood and accepted. $N = 26$ test persons participated in the study, testing both gesture sets in SAE Level 2 and 4 and a baseline with conventional vehicle controls. The order of the overall five variants was randomized. All trials started with a naive run in which the participants received no explanation or instruction. Afterwards, the participants received an introduction to the respective gesture set followed by a second run. Both runs were followed by an initial subjective assessment. Questionnaires used to collect the subjective evaluations were based on the dimensions of the extended devil's square (e.g. joy of use, controllability, usability) 0.

Regardless of the level of automation, the twist/push gestures were preferred to the swipe gestures by the participants - the twist/push gestures in level 4 were even more preferred than the baseline (manual driving) 0. These subjective evaluations are also backed up by the objective measurement data; For example, the percentage of “false activations” and “false non-activations” for twist/push gestures in level 4 were lower than for the other control concepts in the trained run (see Fig. 5). Results in Fig. 5 (left) shows data for the naive (N) and trained (T) runs. Figure 5 left shows single gesture activations on the pedals (for twist/push gestures “to front” and “to back” on both levels of automation) or any other gestures on the steering wheel. All inputs of each participant were analyzed on the participant’s intention and how the co-system reacted. Figure 5 (right) shows the analogous evaluation of the driving situations. A driving situation describes a certain time window, in which the driver has the intention and possibility to activate a maneuver through a gesture. This time window closes, once the driver gives up trying or the opportunity to activate the maneuver has passed.

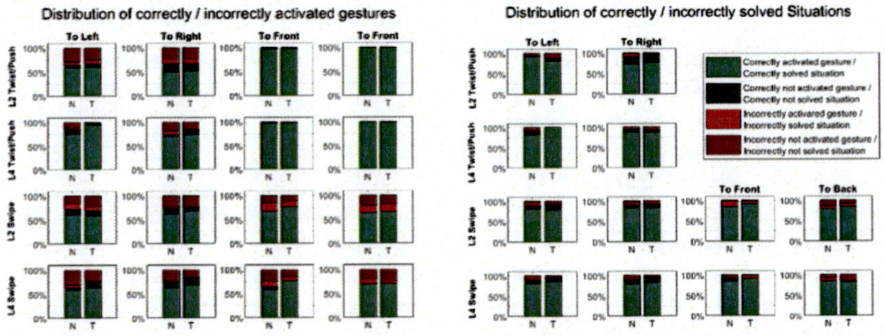

Fig. 5. Results show system performance on single gesture activations (left) and driving situations, which may contain multiple gestures.

Results show that even if some gesture input of the driver was not correctly recognized on the first attempt, the driver could still correctly solve most situations with the input of another gesture. For some gestures, however, especially swipe gestures, the execution is not intuitive for all participants without a training. A high rate of incorrectly not activated gestures indicates that the algorithm is not flexible enough to recognize individual adaptions of thresholds for gesture execution. Most "incorrectly not activated" gestures are results of "unfinished" input from a driver, suggesting that drivers prefer lower execution thresholds. But this would affect the robustness of gesture recognition algorithms, thus increase incorrectly activated gestures and therefore has to be tested. In the case of swipe gestures, a few cases were observed where the correct gesture activation was followed by an incorrect one due to a misinterpretation of a second hand on the steering wheel when touching the second hand while performing the gesture. This behavior can be reduced by increasing the number of segments available for swipe gesture detection. Overall, however, steering gestures worked well for most participants, even in naive runs, i.e. only with gesture suggestion in the HUD but without an explanation or training of the steering gesture mechanism.

4 Conclusion

As mentioned before, with increasing vehicle automation there are opportunities to expand the target groups significantly, for example to include older or physically limited people. The user study showed that there was still a need for adaptation, since, for example, restrictions of the legs meant that not all gestures could easily be used by all participants. For this reason, the twist/push gestures were enhanced as part of an iterative development process regarding the longitudinal controls, so that all gestures can be entered via pushing/twisting the steering wheel. Data suggests that an adaption of the gesture thresholds based on driver input preference might reduce the number of missing triggers. In order to be able to test the gestures in real vehicles, a Wizard-of-Oz vehicle was specified and built 0. This vehicle will allow for the first time to test cooperative gesture interaction and other use cases in real traffic situations.

References

1. Flemisch, F., Winner, H., Bruder, R., Bengler, K.: Cooperative vehicle guidance & control. In: Winner, H., Hakuli, S., Lotz, F., Singer, C. (eds.) Handbook of Driver Assistance Systems, Wiesbaden. Springer Nature (2016)
2. SAE: Taxonomy and Definitions for Terms Related to Driving Automation Systems for On-Road Motor Vehicles (2016)
3. Flemisch, F., Altendorf, E., Canpolat,Y., Weßel, G., Baltzer, M., Lopez, D., Herzberger, N., Voß, G., Schwalm, M., Schutte, P.: Uncanny and Unsafe Valley of Assistance and Automation: First Sketch and Application to Vehicle Automation; C. Schlick et al. (eds.), Advances in Ergonomic Design of Systems, Products and Processes, DOI https://doi.org/10.1007/978-3-662-53305-5_23; Springer 2016
4. Claybrook, J., Kildare, S.: Autonomous vehicles: No driver… no regulation? Science **361** (6397), 36–37 (2018)

5. Flemisch, F., Abbink, D.A., Itoh, M., Pacaux-Lemoine, M.P., Weßel, G.: Joining the blunt and the pointy end of the spear: towards a common framework of joint action, human–machine cooperation, cooperative guidance and control, shared, traded and supervisory control. Cogn. Technol. Work **21**(4), 555–568 (2019)
6. Herzberger, N.D., Voß, G.M.I., Becker, F.K., Grazioli, F., Altendorf, E., Canpolat, Y., Flemisch, F., Schwalm, M.: Derivation of a model of safety critical transitions between driver and vehicle in automated driving. In International Conference on Applied Human Factors and Ergonomics (pp. 421–433). Springer, Cham (2018)
7. Kim, N., Yang, M.Y., Lee, J., Kim, J.: A study on the effect of information types on Drivers in Takeover period of automated vehicles. J. Digital Contents Soc. **19**(1), 113–122 (2018)
8. Flemisch, F., Diederichs, F., Meyer, R., Herzberger, N., Baier, R., Altendorf, E., Spies, J., Kaim, V., Doentgen, B., Bopp-Bertenbreiter, A.V., Widlroither, H., Ruth-Schumacher, S., Arzt, C., Bozbayir, E., Bischoff, S., Kaiser, F.: Vorreiter: manoeuvre-based steering gestures for partially and highly automated driving. In: Meixner, G. (ed.) Smart Automotive Mobility, pp. 231–304. Springer, Cham (2020)
9. Baltzer, M.C.A.: Interaktionsmuster der Kooperativen Bewegungsführung von Fahrzeugen – Interaction Patterns for Cooperative Guidance and Control of Vehicles (2020)
10. Altendorf, E., Baltzer, M., Kienle, M., Meier, S., Weißgerber, T., Heesen, M., Flemisch, F.: H-Mode 2D. In: Winner, H., Hakuli, S., Lotz, F., Singer, C. (eds.) Handbuch Fahrerassistenzsysteme. ATZ/MTZ-Fachbuch, 3rd ed. Springer Vieweg, Wiesbaden (2015)

Check for updates

Service Oriented Software Architecture for Vehicle Diagnostics

Lorenz Görne[1(✉)] and Hans-Christian Reuss[2]

[1] FKFS, Research Institute for Automotive and Engine Design, Pfaffenwaldring 12, 70569 Stuttgart, Germany
Lorenz.Goerne@fkfs.de

[2] IFS Stuttgart, Institute for Automotive Engineering, Pfaffenwaldring 12, 70569 Stuttgart, Germany
Hans-Christian.Reuss@ifs.uni-stuttgart.de

Abstract. The growing number of software functions in the vehicle means an exponential increase in software complexity. All components must be testable, which makes the use of appropriate vehicle testers indispensable. These comprise hardware and software for finding and solving issues that arise during the development of a new vehicle. The FKFS in Stuttgart has developed a new approach to software architecture for vehicle testers that combines various data sources to provide more insights for automotive developers. Incorporating agile design patterns, this new software architecture is very well suited for the rapidly changing requirements of the modern automotive industry. This paves the way for new more versatile and competitive vehicle testing products for the car manufacturing companies.

Keywords: Vehicle diagnostics · Vehicle testers · Software architecture · Service oriented architecture

1 Introduction

The rapid development of new car features has greatly increased the amount of software and microcontrollers in vehicles. However, as the number of functions and complexity increases, so does the cost to achieve reliability [1]. To diagnose problems within the complex and time-critical system that makes up the modern vehicle, original equipment manufacturers (OEMs) may spend as much as 17% of the overall project's costs [2]. Although there are norms for regular testing of each subsystem of a vehicle, for example how to validate the integrity of the motor control unit [3], many problems will only appear in field-testing by driving thousands of miles and going through all kind of situations.

This is where supplier companies sell vehicle testers that comprise hardware and software to aid in finding issues. Due to the rapidly changing automotive industry, the vehicle testing technology has to keep pace. Especially newer functions like automated driving demand, that more versatile and interconnected test equipment reduce the workload on test engineers.

D. Russo et al. (Eds.): IHSI 2021, AISC 1322, pp. 72–77, 2021.
https://doi.org/10.1007/978-3-030-68017-6_11

Common vehicle testers usually consist of powerful test-racks that can monitor and analyse great amount of data. They are however space consuming and costly to install, which is why it is challenging to equip a larger fleet of vehicles with such system. In the FKFS, a small electronic device has been developed, which is specialized upon vehicle diagnostics, making use of the fact that small electronic boards incorporate all major components (like CPU, RAM, HDD, and display ports) of a computer and that they are very affordable nowadays [4]. In this paper, a service-oriented software architecture for said vehicle testers, is presented that fills the gap between the high-performance test-racks and the flexible desktop software.

2 Requirements to the Software Architecture

As former projects in corporation with major OEMs have shown, there are sophisticated vehicle testing products on the market provided from different companies that each, on its own right, fulfill a certain diagnostic tasks. However, there are distinctive boundaries between the capabilities of each product, being it software or hardware. As companies develop independently and do not share common interface, interconnected features, like exchanging data from different testing equipment, prove to be impossible or complicated at least.

Therefore, there is the need for a new vehicle tester that incorporates different testing protocols and interfaces towards its surroundings in order for a more complete test of the vehicle. The minimal set of hardware features that need to be present is shown in Fig. 1. However, the computer of the tester can be expanded in the future via USB-modules. In this study, the focus lays on testers that understand the two main communication protocols, the Controller Area Network (CAN) and BroadR-Reach. With the CAN protocol, the tester can listen to each communication message in the vehicle's network. BroadR-Reach is based on Ethernet and has found its usage in broadband communication like camera data for automated driving.

Additionally, the human machine interface (HMI) needs to be taken into account. It provides visible indication if the tester detects misbehavior of the vehicle software or other issues and gives the possibility to react to that event to the testing person. A touchscreen provides interactivity and high information density, but LEDs or push buttons can be considered to lower the overall cost of the tester, too.

Finally, to introduce big-data analysis, the system can incorporate broadband internet connections like WIFI or LTE via expansion modules. This also eases off-board access to the data, for example the download of the data to a laptop or server for intensive data analysis.

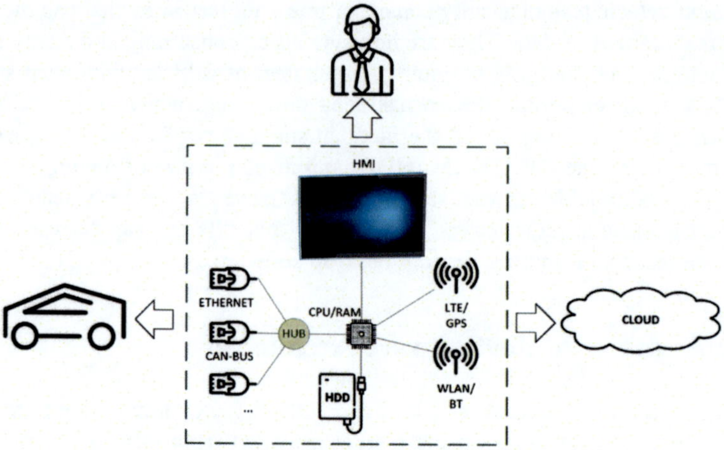

Fig. 1. The vehicle tester (dotted box) interfaces through different protocols with its surrounding world. There are three distinctive paths: Vehicle, Human and Cloud

This is why one key requirement for the software architecture is the possibility to incorporate different tasks and to connect those using defined interfaces. One exemplarily use-case may be to record a set of measurement values while driving, and when the user detects a malfunction in the vehicle, to store a record of the last few seconds before that event. Obviously, this would already demand an interface between the HMI and recorder module. Additionally, if at the same time a service on the tester downloads a new definition for measurement values of the recorder, inconsistency and conflicts are inevitable between the recorder module and the download module that needs to be resolved [5].

3 Service-Oriented Software Architecture

In the early phase of software development projects, the software architecture needs to be defined, although the requirements are still vague. To avoid problems downstream, it is desirable to choose an architecture that provides enough flexibility for further expansion, since decisions being made on the architecture govern all future development and the implementation details later on [6] (see Fig. 2). Service-oriented architectures (SOA) are well suited to organize different entities (like tasks or devices) that may not be under the control of a single organization or team [7] in a way that intercommunication and dependencies are minimized. The SOA approach leads to a modular design when in after-sales, new features arrive via software updates [8].

The technology used for service-based architectures normally consists of programming-language independent and cross-platform communication channels. Each function of the system maps to one service. As an example, it is one function to monitor the vehicle communication and indicate that an error is present. Each service can run without the presence of other services, and during runtime, the different services will discover each other and start to communicate. At the beginning of a project, this gives a lot of flexibility for project planning as each team can develop and test their service before other services are finished, since they are not necessarily depending on the input of other services, or the input can be simulated [9].

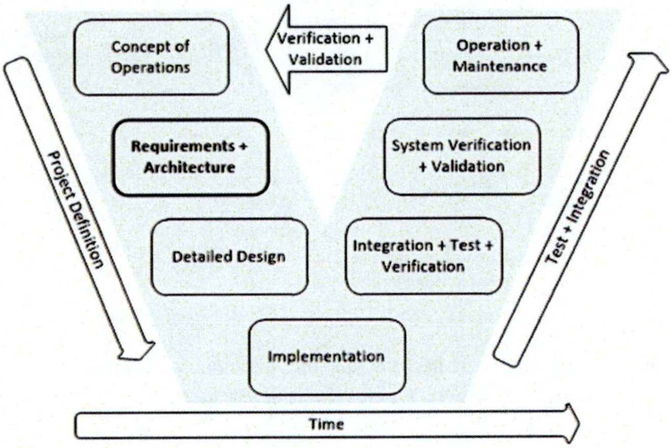

Fig. 2. Given by the V-model for the software development process, the definition of the architecture is the cornerstone for detailed design choices that follow [10]

In order to deliver their functionality independently, each service has to incorporate defensive mechanisms to protect itself against certain scenarios like the crash of another service. This generates a certain overhead for each module in terms of performance and complexity, so the overall design should take hardware limitations into consideration [11].

3.1 Implementation

To address the concerns and requirements mentioned, the proposed implementation (see Fig. 3) of the architecture in software uses on-board tools of a Linux operating system to achieve the desired effects of a SOA. Each service corresponds to a system task. This concept has multiple advantages: The operating system will manage memory and CPU time automatically, and such allows multiple instances of the same service to run at the same time, if necessary. As each service maps to one system task, they can each run concurrently on the hardware and crashes will not shut down the whole operating system.

Different services like "Monitor" or "Recorder" will access the vehicle communication lines via a shared memory entity, which may be necessary due to the high bandwidth of the sum of all vehicle communication. Data that flows within the vehicle tester architecture does so in controlled ways through an application interface (API). For this task, the hypertext-transfer-protocol (HTTP) is used as it is simple to implement. In addition, it lends itself well for web-based services that may be located outside the vehicle tester, like in the cloud or on another vehicle tester.

A common problem in operating systems is the exhaustion of system-resources like memory or CPU time. That is why a so-called "Watchdog" service makes sure that all other services are within their respective limits or otherwise, terminates them.

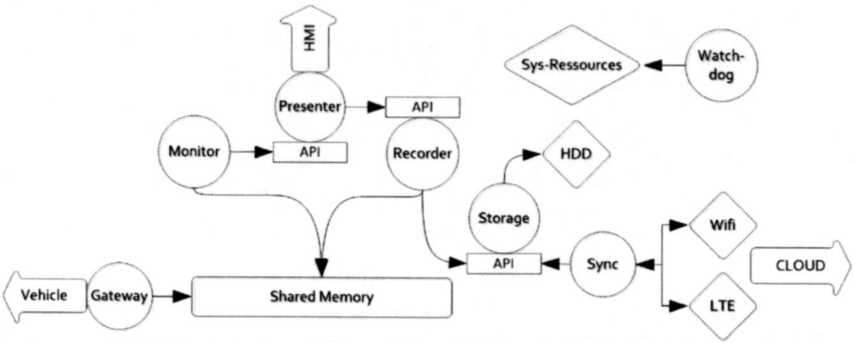

Fig. 3. A proposed software architecture for the exemplary use-case of monitoring and recording the vehicle communication. Circles correspond to system tasks, and diamonds to physical resources.

The service-oriented architecture is what gives this concept the necessary flexibility. When booting up the tester, the operating system will load each service one by one. The services will then discover which other services are running, and expand their functionality accordingly. For example, at boot-up, the "Monitor" service is not yet ready to indicate if errors are present in the vehicle, however the "Presenter" module can already display a loading screen for the test engineer.

Ultimately, the independence of all services from another makes it possible to develop and test a service, and the versatile API gives each development unit the possibility to choose the programming language and paradigms that they deem to be suited the best.

4 Conclusions

FKFS Stuttgart achieved to generate a versatile platform for various vehicle diagnostic tasks by applying the principles of a service-based architecture. It has been shown that the novel vehicle tester has great capabilities in terms of flexibility, incorporating more and more tasks. There is ongoing development work for the proposed software architecture so it will increase in size and complexity. In future work, it should be investigated how more complex diagnostic tasks can be carried out. The integration of additional data sources like audio notes or sensor data might be of great interest.

As the architecture will include more diagnostic tasks, safety attacks will become a concern. In future work, safety mechanisms to restrict what individual services are allowed to do, will be investigated.

References

1. Sha, L.: Using simplicity to control complexity. In: IEEE Software, pp. 20–28 (2001)
2. Bennet, L., Burton, S., Tompkins, G.E.: Characteristics of application software maintenance. Commun. ACM **21**, 466–471 (1978)
3. VDA QMC Working Group 13/Automotive, "Automotive SPICE® Process Reference Model, Process Assessment Model" (2017). https://www.automotivespice.com/download/, Accessed 20 Sept 2020
4. Johnston, S.J., Basford, P., Perkins, C.S., Herry, H.: Commodity single board computer clusters and their applications. Fut. Gener. Comput. Syst. **8**, 201–212 (2018)
5. Brewer, E.: CAP twelve years later: how the "'rules"' have changed. IEEE Comput. **45**, 23–29 (2012)
6. Balaji, S., Murugaiyan, S.: Waterfallvs vs. Model vs. agile: a comparative study ON SDLC. Int. J. Inf. Tech. Bus. Manag. **2**, 26–29 (2012)
7. Valipour, H.M., Bavar, A., Khashayar, M.N., Negin, D.: A brief survey of software architecture concepts and service oriented architecture. In: 2009 2nd IEEE International Conference on Computer Science and Information Technology, Beijing (2009)
8. Nadareishvili, I., Mitra, R., McLarty, M., Amundsen, M.: Microservice Architecture. O'Reilly, Sebastopol (2016)
9. Bozkurt, M., Harman, M., Hassoun, Y.: Testing and verification in service-oriented architecture: a survey. Softw. Test. Verification Reliab. **23**, 12 (2012)
10. Osborne, L., Brummond, J., Hart, R., Mohsen, Z., Conger, S.: Clarus: concept of operations. In: Federal Highway Administration (FHWA), Washington, DC (2005)
11. Östberg, P.O., Elmroth, E.: Mediation of service overhead in service-oriented grid architectures. In: 2011 IEEE/ACM 12th International Conference on Grid Computing, pp. 9–18 (2011)

Investigation of Personality Traits and Driving Styles for Individualization of Autonomous Vehicles

Yvonne Brück(✉), Dario Niermann, Alexander Trende, and Andreas Lüdtke

OFFIS e.V., Escherweg 2, 26121 Oldenburg, Germany
{yvonne.brueck, dario.niermann, alexander.trende, andreas.luedtke}@offis.de

Abstract. The individualization of autonomous vehicles (AV) promises great improvements in AV acceptance and reduction of uncertainty about the safe functioning of the vehicle. To individualize the AV as quickly as possible, a priori knowledge about the passengers could be used to set up the vehicle behavior. We performed a driving simulator study to derive maneuver-based driving styles. Theses driving styles were presented to users as video sequences during an online study. The online study focuses on items regarding driving style, trust in automation and personality traits. We analyzed correlations between these factors and preferences in different traffic scenarios. Our results suggest that the usage of a priori knowledge about the passenger including a subset of demographic data, personality traits and driving styles can produce a good preset for a driving behavior to quickly increase the acceptance of automation.

Keywords: Individualization of autonomous driving · Driving style · User profiles · Human centered design

1 Introduction

Individualizing autonomous vehicles (AV) should lead to a faster acceptance of the technology and reduction of uncertainty about the safe functioning of the vehicle [1]. The manual as well as automatic individualization is a difficult task, since a) most users themselves do not know their preferred parameters, b) setting up various individualizations (e.g. driving styles) by hand would take expert knowledge and a lot of time and c) automatic individualization would need to be done via a feedback loop, which would lead to many uncomfortable situations until the optimal parameters are found. To individualize the AV as quickly as possible, a priori knowledge about the passengers could be used to set up the vehicle parameters.

As proposed previously [2], this knowledge could be gathered via a short survey and saved into a user profile. Studies show that the own driving style has an effect on

Y. Brück, D. Niermann and A. Trende—Equally contributing.

D. Russo et al. (Eds.): IHSI 2021, AISC 1322, pp. 78–83, 2021.
https://doi.org/10.1007/978-3-030-68017-6_12

the preferred automated driving style [3]. Further studies showed that different personality traits (e.g. type A personality) correlate with the manual driving style [4–6]. Also, sociodemographic characteristics such as gender, age and driving years influence the manual driving style [5, 6]. Furthermore, variables like trust in AVs and technical affinity might have an impact on the preferred automated driving style. To infer the driving style from such a user profile, we investigate correlations between user information and the preferred driving style. We first investigate and parametrize differences in driving styles in a simulator study. These driving styles were then presented to participants of a larger online survey featuring questions about above mentioned factors to investigate correlations between these factors and the preferred driving style.

2 Driving Style Study

We performed a driving style study with N = 25 (33.6 + −12.03y, 10 female) in a fixed-base driving simulator, which offered a 150° field of view and a realistic vehicle mock-up. During a 15-min drive the participants performed different manoeuvres such as overtaking a slower vehicle or following a vehicle on the same lane. Although more manoeuvres have been performed during the study we decided to focus on car-following and overtaking manoeuvres for the rest of this project. We defined driving styles based on a set of specific parameters for each manoeuvre. As an example, a driving style for an overtaking manoeuvre is defined by the time-headway, when a participant crossed the middle line of the road, his speed, and the lateral acceleration during the lane change. We calculated the 10th, 50th and 90th percentile of these parameters based on the experimental data (s. Fig. 1). From here on these three driving styles, will be called careful, normal, and risky. After the data analysis we implemented these parameters into our driving simulation software's vehicle automation. We subsequently recorded videos of the overtaking and car-following manoeuvre for all three driving styles to present them during the online survey.

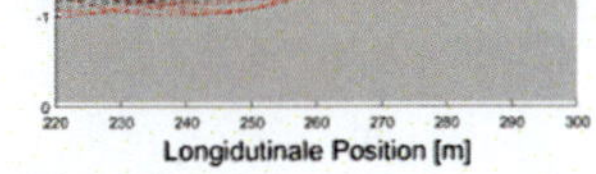

Maneuver	Parameter	Careful	Normal	Risky
Car-following	Time Headway [s]	2.86	2.40	1.62
Overtaking	Time Headway [s]	4.21	3.25	2.04
	Maximum lateral acceleration [m/s*s]	0.41	0.89	1.71

Fig. 1. Left: Example of the extraction of the driving styles for the overtaking maneuver. 10th and 90th percentile and below/above (blue and red) of the time-headways for initiation of the maneuver are highlighted. Right: Overview of the extracted parameters for car-following and overtaking maneuver.

3 Online Survey

3.1 Participants

308 subjects (35.51 ± 13.29y, 187 female) participated in our online survey; three responses are not included in the evaluation due to inconsistent answers. Most participants (58.8%) drive regularly (daily or almost daily). 94.2% have had their driver's license for at least 3 years. On a 5-point Likert scale (where 1 is not familiar/never and 5 is very familiar/very frequently) the familiarity with driver assistance systems is (MD = 2, IQR = 1–4) and the frequency of use (MD = 2, IQR = 1–3).

3.2 Methodology

Participants answered questions about demographic data like age, sex and driving experience. Additionally, the following factors were investigated using the cited questionnaires: personality traits (the Big Five [7], type A/B [8]), technical affinity [9], trust in AVs (adapted items from [10]) and driving styles [11]. We measured all these items on a 5-point Likert scale.

Furthermore, we presented different traffic scenarios, visually and textually, which the participant had to evaluate in terms of risk assessment and own preferences on a 5-point Likert scale. These scenarios reflect certain driving style choices that an AV could make including the overtaking and car following maneuver. For the different maneuvers we took the parameters time headway und lateral acceleration from our previous study. Further, we asked whether the participants would like to overtake vehicles with various speeds (40, 80, 90 km/h) and if they want to give way to a group of bicycles despite the right of way.

For the data analysis, we first evaluated the different constructs according to the used questionnaires. Then, we calculated the Pearson correlations between the constructs. To analyse the videos, we only took the data from participants who reportedly saw a difference in the video sequences, of which there where 209 people. Reasons why the other test persons reported to have not seen any differences could have been their display quality or size. We calculated the average risk perception (Av. Risk perception) over all video sequences. In addition, we calculated the difference in risk perception (Δ risk perception) for the shown videos, such that the video with the most careful driving style served as a baseline and the differences in risk perception should always be positive. For the textually presented overtake intention scenarios we also calculated the mean overtake intention (Av. Overtake intention) and the difference in overtake intention (Δ overtake intention).

3.3 Results

Table 1 shows the correlations between the presented scenarios and other factors. Correlations that are not significant are left out. Overall, our results suggest that the own driving style and other personality traits correlate with the preferred driving behavior of an AV. However, the influence with respect to different driving styles of an AV is limited. People with different personality traits, age and gender have other risk perception in maneuvers, but only the own driving style seems to have an influence for the preferences which driving style the AV should drive. This is consistent with previous research from Bellem et al. [12]. Besides the driving style of an AV, our findings suggest that various factors have a small or medium influence on the preferred driving behavior of an AV (e.g. whether an AV should overtake). The most promising factors to adjusting an initial driving behavior of an AV to increase the acceptance of its use also seem to be the manual driving style. Especially the factors anxious, risky, angry and careful driving have an influence on the preferred driving behavior. In addition, technical affinity, gender or an extraverted and neurotic personality seems to be promising factors to individualize the AV.

Table 1. Correlations between personality traits / demographic data / driving styles and different scenarios (only existent (> 0.1) and significant (p > 0.05) correlations).

		Av. Risk perception	Δ risk perception	Av. Overtake intention	Δ overtake intention	Give way to bicycles
Personalities	Extraversion	–	–	0.117*	0.120*	0.137*
	Neuroticism	0.233**	–	−0.140*	–	–
	Openness	–	–	0.118*	–	−0.140*
	Conscientiousness	–	–	–	–	0.131*
	Agreeableness	–	–	–	–	–
	Type A	–	–	0.143*	0.158**	–
Trust in AVs		–	–	0.144*	–	–
Technical affinity		–	–	0.185**	0.124*	−0.122*
Age		−0.258***	–	–	–	–
Gender		0.147*	–	−0.244***	–	0.137*
Driving styles	Dissociative	0.150*	–	−0.153**	–	–
	Anxious	0.151*	–	−0.331***	–	–
	Risky	−0.148*	−0.154*	0.391***	0.319***	–
	Angry	–	–	0.227***	0.248***	−0.165**
	Distress-reduction	–	–	–	–	–
	Careful	–	–	−0.333***	−0.262***	–

(* $p < 0.05$; ** $p < 0.01$; ***$p < 0.001$)

These results have several implications for a predefined individualization of an AV based on user information. For example, how the car handles overtake situations. For users who drive riskier, an AV can handle situations similar regardless of the speed of the vehicle being overtaken. If the user tends to drive carefully, the AV's behavior should be more differentiated in how it handles situations, e.g. not automatically overtake another car with a speed of 90 km/h. However, the user should still be able to tell the car when to overtake, since the precision of a model build on such subjective factors will not be exact. Moreover, the AV could differentiate when interacting with the environment. The AV could give way to bicycles despite right of way if the driver has a more extraverted personality and the traffic situation allows this.

Overall, we suggest that the following factors should be collected and processed before the first drive to increase the acceptance of AVs: gender, technical affinity, extraverted and neurotic personality as well as anxious, risky, angry and careful driving style.

4 Discussion and Outlook

Overall, meaningful and significant correlations between the investigated factors and risk assessment as well as preferences in different traffic scenarios were found. These findings suggest that the usage of a user profile including a subset of demographic data, personality traits and driving styles can produce a good preset for a driving behavior to quickly increase the acceptance of automation. Therefore, we believe that the use of a user profile could help to increase the acceptance of an AV, as the vehicle behaves in a preferred way. Nevertheless, these results should be treated with caution, since the correlations are significant but low or moderate. Consequently, they can help to set an initial preset of driving behavior, but the driver should be able to change these presets according to his individual preferences. A feedback loop could also help to improve these presets while using the AV.

The present online survey has some methodological limitations that should be taken into account. First, we measured the driving style with a modified version of the Multidimensional Driving Style Inventory (self-reported driving behavior), which may differ from the actual behavior in vehicles and driving simulators, but can serve as a first indicator of people's typical driving behavior in a driving simulator [13]. Second, the number of maneuvers shown in video sequences were limited and there was no control how the participants watched the videos (e.g. display size). Therefore, the online survey can only serve as an initial indicator of which user characteristics correlate with which preferred driving style. A study in a driving simulator or a test track will be required to evaluate whether the correlations found are suitable for different maneuvers.

In future, we plan to further test these findings in a field study with a real autonomous vehicle and to build a model that uses as few questions as possible to predict meaningful presets for the driving behavior.

Acknowledgments. This work was supported by the German Federal Ministry of Transport and Digital Infrastructure in the funding program Automated and Connected Driving, AutoAkzept and by the DFG-grant "Learning from Humans – Building for Humans" (project number: 433 524 510).

References

1. Drewitz, U., Ihme, K., Bahnmüller, C., Fleischer, T., La, H., Pape, A.A., Trende, A.: Towards user-focused vehicle automation: the architectural approach of the AutoAkzept project. In: International Conference on Human-Computer Interaction, pp. 15–30. Springer, Cham (2020)

2. Trende, A., Gräfing, D., Weber, L.: Personalized user profiles for autonomous vehicles. In: Proceedings of the 11th International Conference on Automotive User Interfaces and Interactive Vehicular Applications: Adjunct Proceedings, pp. 287–291 (2019)
3. Griesche, S., et al.: Should my car drive as I do? what kind of driving style do drivers prefer for the design of automated driving functions. In: Braunschweiger Symposium, vol. 10, no. 11 (2016)
4. Poó, F.M., Ledesma, R.D.: A study on the relationship between personality and driving styles. Traffic Inj. Prev. **14**(4), 346–352 (2013)
5. Wang, Y., et al.: Effect of personality traits on driving style: psychometric adaption of the multidimensional driving style inventory in a Chinese sample. PLoS one **13**(9), e0202126 (2018)
6. Boyce, T.E., Geller, E.S.: An instrumented vehicle assessment of problem behavior and driving style: do younger males really take more risks? Accid. Anal. Prev. **34**(1), 51–64 (2002)
7. Rammstedt, B., et al.: Eine kurze Skala zur Messung der fünf Dimensionen der Persönlichkeit [A short scale for assessing the big five dimensions of personality]. Methoden Daten Analysen **7**(2), 233–249 (2013)
8. Glazer, H.: Stress prone type A personality questionnaire. In: Philip Goldberg's Executive Health (1978)
9. Franke, T., Attig, C., Wessel, D.: A personal resource for technology interaction: development and validation of the affinity for technology interaction (ATI) scale. Int. J. Hum.-Comput. Interact. **35**(6), 456–467 (2019)
10. Choi, J.K., Ji, Y.G.: Investigating the importance of trust on adopting an autonomous vehicle. Int. J. Hum.-Comput. Interact. **31**(10), 692–702 (2015)
11. van Huysduynen, H.H., et al.: Measuring driving styles: a validation of the multidimensional driving style inventory. In: Proceedings of the 7th International Conference on Automotive User Interfaces and Interactive Vehicular Applications (2015)
12. Bellem, H., et al.: Comfort in automated driving: an analysis of preferences for different automated driving styles and their dependence on personality traits. Transp. Res. Part F Traffic Psychol. Behav. **55**, 90–100 (2018)
13. van Huysduynen, H.H., Terken, J., Eggen, B.: The relation between self-reported driving style and driving behaviour: a simulator study. Transp. Res. Part F Traffic Psychol. Behav. **56**, 245–255 (2018)

Predicting Takeover Quality in Conditionally Automated Vehicles Using Machine Learning and Genetic Algorithms

Emmanuel de Salis[1(✉)], Quentin Meteier[2], Marine Capallera[2], Leonardo Angelini[2], Andreas Sonderegger[3], Omar Abou Khaled[2], Elena Mugellini[2], Marino Widmer[4], and Stefano Carrino[1]

[1] Haute Ecole Arc Ingénierie, HES-SO, Saint-Imier, Switzerland
{emmanuel.desalis, stefano.carrino}@he-arc.ch
[2] HumanTech Institute, HES-SO//University of Applied Sciences and Arts Western Switzerland, Fribourg, Switzerland
{quentin.meteier, marine.capallera, leonardo.angelini, omar.aboukhaled, elena.mugellini}@hes-so.ch
[3] Department of Psychology, University of Fribourg, Fribourg, Switzerland
andreas.sonderegger@unifr.ch
[4] DIUF, University of Fribourg, Fribourg, Switzerland
marino.widmer@unifr.ch

Abstract. Takeover requests in conditionally automated vehicles are a critical point in time that can lead to accidents, and as such should be transmitted with care. Currently, several studies have shown the impact of using different modalities for different psychophysiological states, but no model exists to predict the takeover quality depending on the psychophysiological state of the driver and takeover request modalities. In this paper, we propose a machine learning model able to predict the maximum steering wheel angle and the reaction time of the driver, two takeover quality metrics. Our model is able to achieve a gain of 42.26% on the reaction time and 8.92% on the maximum steering wheel angle compared to our baseline. This was achieved using up to 150 s of psychophysiological data prior to the takeover. Impacts of using such a model to choose takeover modalities instead of using standard takeover requests should be investigated.

Keywords: Human machine interaction · Machine learning · Conditionally automated vehicles · Genetic algorithms · Artificial intelligence · Takeover request

1 Introduction

Nowadays, much progress is being made to create fully automated vehicles, but we are not there yet. If we can maybe see fully autonomous cars working in research labs as early as this year (2020) [1], we will probably see in the meantime conditionally automated vehicles populating our roads, corresponding to level 3 of the SAE taxonomy [2]. At this level of automation, it is not mandatory for the driver to monitor the

D. Russo et al. (Eds.): IHSI 2021, AISC 1322, pp. 84–89, 2021.
https://doi.org/10.1007/978-3-030-68017-6_13

environment while the car is driving, but they must be able to regain control of the car if the vehicle issues a takeover request (TOR). This transfer of control from the car to the driver is a critical moment that can lead to accidents if not executed correctly. Therefore, the Human-Machine Interface (HMI) used to convey the TOR (called the TOR modalities from now on) must be chosen carefully.

Currently, TOR are standardized in modern cars, meaning they depend on the model of the car and do not change depending on the situation, despite research having highlighted that some TOR modalities are better suited for different situations (cf. Related Work). However, there are currently no models able to foresee takeover quality based on inputs known to influence takeover quality, such as the driver psychophysiological state. In this paper, we aim at creating a Machine Learning solution able to predict takeover quality based on driver psychophysiological state and TOR modalities, thus allowing us to refine the TOR modalities if needed to ensure the best takeover possible. This will also provide a tool able to estimate takeover quality to researchers and car manufacturers, hopefully useful in later research.

2 Objective

This study aims to investigate if Machine Learning (ML) can be used to predict the takeover quality before a TOR is issued, using the driver's physiological data and the TOR modalities. This would allow a smart HMI to select the TOR modalities providing the best takeover quality, thus reducing the risk of accident. In particular, we aim at predicting two widely used metrics: Reaction Time (RT) [3] and Maximum Steering Wheel Angle (MaxSWA) [4]. RT is the time between the TOR and the effective takeover of the driver, either by braking or steering the wheel upon a certain threshold. MaxSWA is the maximum angle reached by the steering wheel during the takeover procedure, until control is given back to the car. Both RT and MaxSWA should be minimized to ensure the best takeover quality.

3 Related Work

The driver state has an impact on takeover quality, as shown in various study. For example, drivers are slower to disengage the autopilot when performing a visual-manual texting task with high cognitive load [5]. In addition, drowsiness leads to slower first braking reaction [6].

TOR modalities also impact takeover quality, with multimodal TOR perceived as more urgent and leading to shorter RT but more stress on the driver [7]. Visual modalities were shown to be less effective than auditory or haptic modality [7] when used as unique modality.

At the extent of our knowledge, no research studied the possibility of using ML to predict takeover quality for MaxSWA. RT was studied by Gold [8] (called Take-Over Time in the cited research) as well as lateral acceleration, longitudinal acceleration, and Time-to-Collision. Gold showed that computing the regression of RT was possible, opening the way to more complex models.

4 Methodology

4.1 Experiment

80 participants (54 females) were recruited for this study. Participants' age ranged from 19 to 66 years old (M = 23.94 years old; SD = 8.25 years old). Having a valid driving license was mandatory to participate in this study. They reported on average driving 6312 km per year (SD = 14 415 km) and have held a driving license for about 5 years and a half (SD = 8.10 years). The written informed consent of all participants was obtained.

The experiment consisted in a driving session on a fixed-base driving simulator. The driving scenario consisted in an hour-long driving session in a rural environment without traffic, with five takeover situations. Each takeover was caused by a different limitation of automation, depicting five of the six categories of the taxonomy proposed by Capallera et al. [9]. The causes of each takeover and their associated category are: a steep slope limiting visibility (Road category), fading lane markings (Lanes category), massive rock obstructing the right lane (Obstacle category), heavy rain (Environment category) and a deer crossing the road from the right side (External Human Factor category).

The participants were asked to participate in cognitive non-driving secondary tasks (NDST), in the form of N-Back tasks, to ensure a wide variety of driver psychophysiological states. NDST and causes of takeovers were randomized throughout the study in order to minimize bias.

The TOR was conveyed using audiovisual modalities through a handheld device for half the participants and through a dashboard located behind the steering wheel for the others. The participants' physiological data (respiration, electrodermal activity and electrocardiogram) were recorded during the experiment.

4.2 Machine Learning

Driver physiological data were processed using Neurokit [10] to be used as features, based on five periods of different lengths, from 0 to 30 s (Dataset30), 0 to 60 s (Dataset60), 0 to 90 s (Dataset90), 0 to 120 s (Dataset120) and 0 to 150 s (Dataset150) before the TOR. This allowed us to refine what is the relevant time window needed prior to the TOR to predict its quality. The TOR modality, either through a handheld device or through the dashboard, was also used as input. Missing values were dropped and not interpolated, since every instance of takeover was independent. Datasets size were 307 takeovers for Dataset30, 332 takeovers for Dataset60, 338 takeovers for Dataset90, 339 takeovers for Dataset120 and 341 takeovers for Dataset150.

Various regression models were then trained using a grid search with a genetic algorithm approach for the selection of the models and their parameters. This was achieved using TPOT [11], which was shown to work well with biomedical data [12]. Population of models were trained and tested on every dataset independently, in a 5–fold cross validation manner on 75% of the data, and validated after the genetic process on the remaining data. Population size was 50 and trained during 5 generations. The validation score was then used to compare models across datasets. The metrics used to compare scores is the Mean Square Error (MSE).

5 Results

The regression models were compared to a constant prediction of the mean of the metrics (baseline). Improvements is calculated as described in the following equation (Eq. 1):

$$\text{Improvement} = (\text{New score} - \text{baseline})/\text{baseline}. \quad (1)$$

The best model for every dataset and their score can be seen in Table 1 for RT, and Table 2 for MaxSWA.

Table 1. Datasets and their corresponding best model for RT, along with their score. Improvements over baseline is indicated in parenthesis if there is any

Dataset	Model	Score (improvement)
Dataset30	Random Forest Regressor (RFR) applied on a selection of features by Variance Threshold and Principal Component Analysis	2.4120 (17.61%)
Dataset60	Combination of RFR and ElasticNet	2.4200 (17.35%)
Dataset90	Combination of RFR, ElasticNet and Gradient Boosting	3.2113 (–)
Dataset120	Combination of RFR, Linear SVR and Lasso	3.4150 (–)
Dataset150	**Extra Trees Regressor**	**1.6906 (42.26%)**

Table 2. Datasets and their corresponding best model for MaxSWA, along with their score. Improvements over baseline is indicated in parenthesis if there is any

Dataset	Model	Score (improvement)
Dataset30	**K-Nearest Neighbors (k = 86)**	**161.93 (8.92%)**
Dataset60	Combination of ElasticNet, RBFSampler [13] and Independent Component Analysis	189.32 (–)
Dataset90	K-Nearest Neighbors (k = 67)	175.20 (1.45%)
Dataset120	Combination of RFR and ElasticNet	182.83 (–)
Dataset150	Gradient Boosting	186.65 (–)

We can see that overall, an Extra Trees Regressor achieved an MSE of 1.6906 for the RT, (baseline = 2.9279), an improvement of 42.26%. For the MaxSWA, a K-Nearest Neighbors regressor achieved an MSE of 161.93 (baseline = 177.79), an improvement of 8.92%. RT seems easier to predict than MaxSWA, with multiple models achieving more than 17% improvement over the baseline, while MaxSWA did not get higher than an 8.92% improvement. We can also see that Dataset150 was more suited for RT prediction, in contrast with MaxSWA which got better results for Dataset30. This leads to think that using multiple time windows for prediction of TOR quality is advisable, instead of choosing only one.

6 Conclusion

This study showed promising results of using ML to predict TOR quality, with models able to predict the RT and MaxSWA. Further research is necessary to evaluate the total gain of takeover quality achievable using this model instead of standard HMI of car models available on the market. Comparing the impact on the user experience of using a dynamic TOR, which modalities will change depending on the situation, instead of a fixed one is also an interesting possible research.

Acknowledgments. This work is part of the AdVitam project funded by the Hasler Foundation. We would also like to thank our colleagues who helped us during this project.

References

1. Goh, B., Sun, Y.: Tesla 'very close' to level 5 autonomous driving technology, Musk says. https://www.reuters.com/article/us-tesla-autonomous-idUSKBN24A0HE, Accessed 09 July 2020
2. SAE International: Taxonomy and Definitions for Terms Related to Driving Automation Systems for On-Road Motor Vehicles (2018)
3. Gold, C., Körber, M., Lechner, D., Bengler, K.: Taking over control from highly automated vehicles in complex traffic situations: the role of traffic density. Hum. Fact. J. Hum. Fact. Ergon. Soc. **58**(4), 642–652 (2016). https://doi.org/10.1177/0018720816634226
4. Bueno, M., Dogan, E., Selem, F. H., Monacelli, E., Boverie, S., Guillaume, A.: How different mental workload levels affect the take-over control after automated driving. In: IEEE 19th International Conference on Intelligent Transportation Systems (ITSC) (2016). https://doi.org/10.1109/itsc.2016.7795886
5. Wandtner, B., Schömig, N., Schmidt, G.: Secondary task engagement and disengagement in the context of highly automated driving. Transp. Res. Part F Traffic Psychol. Behav. **58**, 253–263 (2018). ISSN 1369–8478, https://doi.org/10.1016/j.trf.2018.06.001
6. Jarosch, O., Bellem, H., Bengler, K.: Effects of task-induced fatigue in prolonged conditional automated driving. Hum. Fact. J. Hum. Fact. Ergon. Soc. **61**(7), 1186–1199 (2019). https://doi.org/10.1177/0018720818816226
7. Politis, I., Brewster, S., Pollick, F.: Language-based multimodal displays for the handover of control in autonomous cars. In: Proceedings of the 7th International Conference on Automotive User Interfaces and Interactive Vehicular Applications - AutomotiveUI 2015 (2015). https://doi.org/10.1145/2799250.2799262

8. Gold, C.: Modeling of take-over performance in highly automated vehicle guidance. In: SAE International, 2018 Taxonomy and Definitions for Terms Related to Driving Automation Systems for On-Road Motor Vehicles (2017)
9. Capallera, M., Meteier, Q., de Salis, E., Angelini, L., Carrino, S., Abou Khaled, O., Mugellini, E.: Owner manuals review and taxonomy of ADAS limitations in partially automated vehicles. In: Proceedings of the 11th International Conference on Automotive User Interfaces and Interactive Vehicular Applications (AutomotiveUI 2019), pp. 156–164. Association for Computing Machinery (2019). https://doi.org/10.1145/3342197.3344530
10. Makowski, D., Pham, T., Lau, Z.J., Brammer, J.C., Lespinasse, F., Pham, H., Schölzel, C., Chen, A.S.H.: NeuroKit2: A Python Toolbox for Neurophysiological Signal Processing. Accessed 28 Mar 2020, https://github.com/neuropsychology/NeuroKit
11. Fu, W., Olson, R., Nathan, Jena, G., Gijsbers, P., Tom, A., Carnevale, R.: EpistasisLab/tpot: v0.11.5 (Version v0.11.5), Zenodo (2020). https://doi.org/10.5281/zenodo.3872281
12. Olson, R.S., Urbanowicz, R.J., Andrews, P.C., Lavender, N.A., Kidd, L.C., Moore, J.H.: Automating biomedical data science through tree-based pipeline optimization. In: Squillero, G., Burelli, P. (eds.) Applications of Evolutionary Computation. EvoApplications 2016. Lecture Notes in Computer Science, vol. 9597. Springer, Cham (2016). https://doi.org/10.1007/978-3-319-31204-0_9
13. Rahimi, A., Recht, B.: Weighted sums of random kitchen sinks: replacing minimization with randomization in learning. In: Advances in Neural Information Processing Systems 21 - Proceedings of the 2008 Conference, pp. 1313–1320 (2008)

"Automated but Not Alone": How the Possible Forms of Future Human Activity Are Analyzed in the Advent of Automated Vehicles?

Daniel Silva[1(✉)] and Liliana Cunha[1,2]

[1] Centre for Psychology at University of Porto (CPUP), Porto, Portugal
danielsilva@fpce.up.pt
[2] Faculty of Psychology and Educational Sciences of the University of Porto (FPCEUP), Porto, Portugal
lcunha@fpce.up.pt

Abstract. Automation is transforming the human's role in driving situations. Yet, the utopian imaginary of a "vehicle without the human driver" is far away from being a reality. The driver's activity remains poorly understood in a context which tends to be polarized by its technological content. The aims of this study are to identify how the human activity is being analyzed on automated driving research and determine the analysis criteria used. Two distinctive paradigms were identified in the literature: the first one is focused on the behavioral measurable changes induced by automation functioning; the second paradigm puts at the center of analysis the construct of "activity" when interacting with technology. Based on the work psychology knowledge regarding the work activity within automated systems, the need to bring to the fore the drivers' point of view is argued as a real means to design automated vehicles taking human aspects into account.

Keywords: Automated driving · Human activity · Automation · Work automated systems · Work psychology

1 Introduction

The road transport and mobility system as a whole are crossing a revolution in relation to traditional systems with the advent of automated driving (AD) [1]. In this context, automated vehicles (AVs) are expected to influence and shape our future mobility, the road transport safety and efficiency, and our quality of life.

AVs hold the promise to transform the driving task into a "hands and feet free" activity. Notwithstanding all remarkable technological advancements, in the existing automation architectures, the human driver continues to play an active role in system safety by resuming manual control of the vehicle when technology detects that its operational limits are about to be exceeded [2]; or when it is needed to deal with the unpredictability of the environment and its dynamic elements [3]. Despite acknowledging this centrality of the role of human drivers in the safe implementation of AD, the driver's intervention remains as a contentious issue and poorly understood, or at least underestimated [4, 5]. On the current AD research, how the possible forms of

D. Russo et al. (Eds.): IHSI 2021, AISC 1322, pp. 90–96, 2021.
https://doi.org/10.1007/978-3-030-68017-6_14

human activity are investigated? In the analysis of how human drivers interact with AD technologies, what is taken as "priority" dimensions of inquiry? And which dimensions tend to be kept as "discrete" areas? Our paper addresses these questions through a literature survey following an integrative strategy [6].

In the fields of work psychology and activity ergonomics, this kind of theoretical review and integration has been conducted mainly in work contexts where human operators perform their activities in advanced work automated systems (e.g., continuous process industries; air traffic or railway control) [7–9]. Nonetheless, this analysis is lacking in the literature on the automatization of driving tasks. Thereby, the purpose of this integrative review is to fill this absence, mostly when increasingly complex levels of automation are being introduced in vehicles.

2 Automated Vehicles: The Human as a "Back-Up Driver"?

Automation in driving apparently relies on a simple concept: a vehicle operating without human intervention. Nevertheless, automated driving reveals a much more complex reality for which definitions remain uncertain [1]. While there is no generally agreed-upon definition on what an AV is, the classification framework provided by the Society of Automotive Engineers (SAE) is commonly taken as a reference. This classification is a taxonomy that describes five levels of automation, from driver-only control (manual driving - L0) to fully automated control (L5). A core issue regarding safety in AD is how well machine-to-human handovers occur in SAE L2 and SAE L3 when automation cannot handle with its environment [10]. At the SAE L2, automation assumes vehicle control, but the human is still expected to monitor the driving environment and the system functioning in order to regain manual control when needed. At the SAE L3, within certain conditions, the driver may be involved in non-driving tasks, since automation is in charge of driving tasks. Nonetheless, it is expected that the human resumes the manual control in a safe and timely way in the case of critical situations, after a request to intervene from the system. Therefore, at least up to level 3 of vehicle automation, the human acts as a "back-up driver" [10], or a "fallback-ready user" [2, 11].

3 Integrative Literature Review Methodology

An integrative review was conducted to explore our research questions. The choice for this kind of review is justified by the fact that an integrative review uses a broad approach and diverse sampling that includes empirical and theoretical literature, reviewing qualitative and/or quantitative studies on a specific topic [6].

The literature search was conducted in three multidisciplinary bibliographic databases: Scopus; Elsevier ScienceDirect; DOAJ. These databases were selected based on their relevance to human – automation research, integrating articles from different scientific domains, ranging from engineering to ergonomics/human factors and psychology. The search was performed with an open-ended search with the following overarching terms: "automated driving", "automated vehicles", and "human activity". To combine the searched terms, the Boolean operators (AND, OR) were used.

4 Results and Discussion

To explore how the human activity is currently analyzed when interacting with vehicle automation, a thematic dispersion matrix was created. All the reviewed empirical studies were categorized according to two dimensions: (i) the nature of research questions explored; and (ii) the analysis criteria used to assess the human-automation relationship in the context of AD. Figure 1 presents the results' matrix.

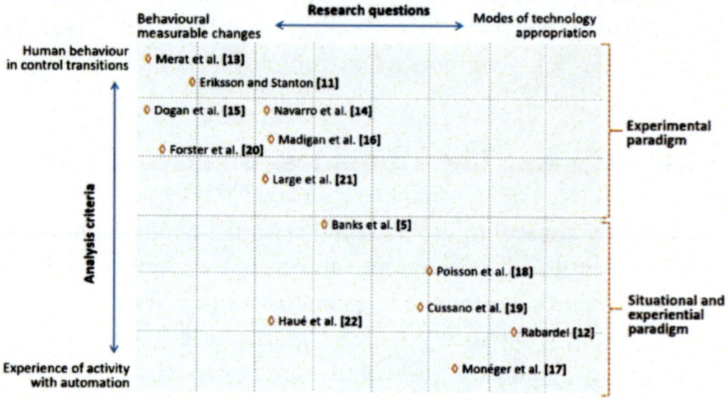

Fig. 1. Dispersion matrix of studies selected for the analysis.

4.1 The Nature of the Research Questions

The first dimension that emerged from the analysis is related to the nature of the research questions about how human drivers relate to AD systems. On the one hand, the studies that explore the changes in the human behavior given the interaction with the automation stand out, focusing the cognitive processes (e.g., perception, attention, situation awareness, mental workload) involved in the performance of driving tasks. These studies compared the (visible) human behavior and performance in AD situations with conventional manual driving situations, assessing concrete measures of the human-machine relationship [11, 13, 16, 20].

On the other hand, the studies placed in the opposite pole explored research questions related to the transformations in the human activity *in* the use and *by* the use of technology. These studies sought to understand the subjects' development modalities when they carry out processes to appropriate the AD technologies [12, 18, 19]. These research questions are inscribed in a developmental perspective [9], whose purpose is to understand the processes through which the subjects develop internal and external resources (representations, utilization schemes, other modes of know-how, systems of rules and values). It is through these resources that the subjects transform the technological artefacts to deal with the specificities of the driving situations.

4.2 Analysis Criteria Adopted

The second dimension considered in the matrix refers to the analysis criteria used to investigate the human-automated vehicle relationship. One of the poles concentrates the studies where the central analysis criterium is the adaptation of the human behavior after machine-to-human handovers (mostly control transitions in SAE L2 and SAE L3). These control transitions between automation and the human driver have been addressed by several experimental studies using driving simulators [13–16, 21], exploring some issues that could compromise the safe operation of the vehicle, such as mental overload, loss of situation awareness, and manual skill degradation. These studies focused mainly on objective behavior measures, such as:

- The time needed to gain vehicle control, considering measures associated with reaction times in machine-to-human handovers [11, 13, 15, 20].
- Decision times and the characteristics of situations at which human drivers activate and deactivate the automated driving system [5].
- Deviations on vehicle lateral position, speed, distances to a vehicle ahead or to an unexpected obstacle in the road, and the maneuver quality [14, 16].
- Drivers' attention and distraction measures while performing the driving task through eye movements tracking [13–15].

In the opposite pole of the matrix, there is another set of studies that adopted as analysis criterium the point of view of the subjects performing the activity, i.e., the experience the human develops in the relationship with automation. These studies are mainly conducted under the scope of activity ergonomics and work psychology [12, 17–19], scientific fields that deal with epistemological models of human activity that go beyond the cognitive dimensions. Unlike the approaches that model the human activity as a set of responses to external stimuli received more or less passively; the capacity of humans to actively explore, interpret, use and transform their technical and social environment is a property of the processes of humans' appropriation of technological artefacts [9, 12]. Based on this epistemological view, the analysis of the driver's activity becomes a reference criterium, seeking to address the lived experience [17, 22]. One of the main interest points regards the analysis of the ways AD technologies are appropriated and used, looking thoroughly to how the variability in the real driving situations interferes with the development of such usage. Thus, the "possible forms of human activity" [23] in the context AD are not analyzed exclusively through the lens of objective behavior measures; instead, they are analyzed from the drivers' point of view, from their experience and working strategies [e.g., 12] that may have not been anticipated in experimental settings.

5 Conclusions

These results show the selected studies are clustered in two main thematic quadrants (see Fig. 1). The matrix upper left quadrant concentrates the greater number of studies, dedicated to analyzing the driver's behavior in simulator experimental situations. Here, the research paradigm may be classified as imminently "experimental", focused on the observable dimensions of behavior, and that seems to have today a greater expression on AD research. However, our review also shows that a different set of studies follow a situational and experiential paradigm. The studies placed in the matrix lower right quadrant addressed the modes of appropriation and usage of AD technologies developed by the drivers. The analysis criteria adopted by these studies are related to the drivers' experience. The human activity encompasses a dimension of lived experience, which tends to be considered as a "residual", or a "discrete", dimension by the experimental research paradigm [7, 8].

These two research paradigms shall not be considered antagonist; on the contrary, they may benefit from reciprocal enrichment. In fact, they ground different research strategies that have in common the potential to contribute to assure that all ergonomic/human factors aspects are taken into consideration in the design processes of AVs. The experimental approach is useful considering it provides descriptive representations of human behavior with automation. However, at the same time, it tends to produce a set of "abstractions" about how humans appropriate and use technological artefacts, i.e., the drivers' point of view. We underline it is not about dropping the experimental logic but producing a "continuum balance" in AD research, through the introduction of the drivers' experience built when interacting with automation.

This "balance" we hereby suggest becomes even more needed in the case of AD in professional contexts. Driving vehicles in a professional context is subject to determinants, demands, and conflict of values quite different from those that take place in non-professional driving: follow tight schedules; service quality assurance; guarantee the passengers' safety and/or the clients satisfaction [24]. The management and the situated arbitrations professional drivers make upon such demands shall not continue to be considered "discrete" areas of inquiry, and, for that reason, be excluded from the analysis on how professional drivers interact with AD technologies.

Acknowledgments. This work is supported by the FCT - Portuguese Foundation for Science and Technology under Grant SFRH/BD/139135/2018; by the Project PTDC/ECI-TRA/28526/2017 "AUTODRIVING - Modelling the Driver Behaviour in Autonomous Driving Using a Driving Simulator" (POCI-01-0145-FEDER-028526) funded by FEDER funds through COMPETE2020 - Programa Operacional Competitividade e Internacionalização (POCI) and by national funds (PIDDAC) through FCT/MCTES; and by the Centre for Psychology at University of Porto under Grant FCT UIDB/00050/2020. The authors thank the funders for their support.

References

1. Orfeuil, J.-P., Leriche, Y.: Piloter le véhicule autonome au service de la ville. Descartes & Cie, Paris (2019)
2. Hancock, P., Kajaks, T., Caird, J., Chignell, M., Mizobuchi, S., Burns, P., Feng, J., Fernie, G.R., Lavallière, M., Noy, I.Y., Redelmeier, D.A., Vrkljan, B.: Challenges to human drivers in increasingly automated vehicles. Hum. Factors **62**(2), 310–328 (2020)
3. Simões, A., Cunha, L., Ferreira, S., Carvalhais, J., Tavares, J.P., Lobo, A., Couto, A., Silva, D.: The user and the automated driving: a state-of-the-art. In: Stanton, N. (ed.) AHFE 2019. AISC, vol. 964, pp. 190–201. Springer, Cham (2020)

4. Kyriakidis, M., de Winter, J., Stanton, N., Bellet, T., van Arem, B., Brookhuis, K., Martens, M.H., Bengler, K., Andersson, J., Merat, N., Reed, N., Flament, M., Hagenzieker, M.P., Happee, R.: A human factors perspective on automated driving. Theor. Issues Ergon. Sci. **20** (3), 223–249 (2017)
5. Banks, V., Eriksson, A., O'Donoghue, J., Stanton, N.: Is partially automated driving a bad idea? Observations from an on-road study. Appl. Ergon. **68**, 138–145 (2018)
6. Toronto, R., Remington, R.: A Step-by-Step Guide to Conducting An Integrative Review. Springer, Cham (2020)
7. Clot, Y.: Le travail sans l'homme? Pour une psychologie des milieux de travail et de vie. La Découverte/Poche, Paris (1995)
8. Cavestro, W.: Beyond the deskilling controversy. Comput. Integr. Manuf. Syst. **3**, 38–46 (1990)
9. Rabardel, P.: Les hommes et les technologies: approche cognitive des instruments contemporains. Armand Colin, Paris (1995)
10. International Transport Forum: Safer roads with automated vehicles? OECD, Paris (2018)
11. Eriksson, A., Stanton, N.: Takeover time in highly automated vehicles: noncritical transitions to and from manual control. Hum. Factors **59**(4), 1–17 (2017)
12. Rabardel, P.: Instrument mediated activity in situations. In: Blandford, A., Vanderdonckt, J., Gray, P. (eds.) People and Computers XV - Interaction Without Frontiers, pp. 17–30. Springer, London (2001)
13. Merat, N., Jamson, A., Lai, F., Dalay, M., Carsten, O.: Transition to manual: driver behaviour when resuming control from a highly automated vehicle. Transp. Res. Part F Traffic Psychol. Behav. **27**, 274–282 (2014)
14. Navarro, J., François, M., Mars, F.: Obstacle avoidance under automated steering: impact on driving and gaze behaviours. Transp. Res. Part F Traffic Psychol. Behav. **43**, 315–324 (2016)
15. Dogan, E., Rahal, M-C., Deborne, R., Delhomme, P., Kemeny, A., Perrin, J.: Transition of control in a partially automated vehicle: effects of anticipation and non-driving-related task involvement. Transp. Res. Part F Traffic Psychol. Behav. **46**, 205–215 (2017)
16. Madigan, R., Louw, T., Merat, N.: The effect of varying levels of vehicle automation on drivers' lane changing behaviour. PLoS ONE **13**(2), 1–17 (2018)
17. Móneger, F., Coutarel, F., Moták, L., Chambers, P., Izaute, M., Dhome, M.: L'expérience vécue et les valeurs en acte des accompagnants pour la conception d'un service de transport par navettes destinées à être autonome. Activités **15**, 1–35 (2018)
18. Poisson, C., Bourmaud, G., Decortis, F.: Vers une compréhension des genèses instrumentales dans la conduite autonome. In: Actes du 53ème Congrès de la SELF, L'Ergonomie à quelles échelles? Bourdeaux, pp. 903–910 (2018)
19. Cussano, H., Vidal-Gomel, C., Le Bellu, S.: Comment guider les genèses instrumentales pour la prise en main du véhicule autonome? In: Actes du 5ème Colloque International de Didactique Professionnelle, Montréal, pp. 145–158 (2019)
20. Forster, T., Hergeth, S., Naujoks, F., Beggiato, M., Krems, J., Keinath, A.: Learning to use automation: behavioural changes in interaction with automated driving systems. Transp. Res. Part F Traffic Psychol. Behav. **62**, 599–614 (2019)
21. Large, D., Burnett, G., Salanitri, D., Lawson, A., Box, E.: A longitudinal simulator study to explore drivers' behaviour in Level 3 automated vehicles. In: Proceedings of 11th International ACM Conference, pp. 222–232 (2019)

22. Haué, J.-B., Le Bellu, S., Barbier, C.: Le véhicle autonome: se désengager et se réenagger dans la conduite. Activités **17**, 1–26 (2020)
23. Daniellou, F.: Simulating future work activity is not only a way of improving workstation design. Activités **4**, 1–8 (2007)
24. Silva, D., Cunha, L., Barros, C., Baylina, P.: Preparing the future scenario of automated vehicles: recommendations drawn from the analysis of the work activity of road transport workers. In: Cotrim, T.P., Serranheira, F., Sousa, P., Hignett, S., Albolino, S., Tartaglia, R. (eds.) HEPS 2019. AISC, vol. 1012, pp. 301–310. Springer, Cham (2019)

The Impact of Connected and Autonomous Trucks on Freeway Traffic Flow

Yue Qiao and Yongju Hu(✉)

Key Laboratory of Urban Rail Transit Intelligent Operation and Maintenance Technology and Equipment of Zhejiang Province, Zhejiang Normal University, Zhejiang 321004, China
stargazers71@outlook.com, huyongju@163.com

Abstract. Using computer simulation to study the impact of connected and autonomous trucks on freeway mixed traffic flow. The future freeway vehicle models are divided into 4 types, conventional passenger cars, connected and autonomous passenger cars, conventional trucks, and connected and autonomous trucks. Based on the characteristics of the four models, the Nasch model was improved, including slowing down probability, driving style, maximum speed, lane changing behavior, and safety distance determination. Created a new mixed traffic flow model.

Keywords: Connected and autonomous truck · Cellular automata · Mixed traffic flow

1 Introduction

Cargo freight plays an important role in transportation and promoting economic development, especially in China where freeway construction is fast. However, truck accidents on the freeway also followed. Compared with passenger cars, trucks have the characteristics of large size and mass, difficult steering operations, lower safety, and truck drivers are more prone to symptoms such as fatigue and anxiety. As a result: The proportion of trucks responsible for road traffic accidents is much higher than the proportion of trucks in motor vehicles in China.

Fortunately, the development of Connected and Autonomous Vehicle (CAV) will change the driving environment on freeways. Compared with conventional vehicles, CAVs can obtain more accurate driving conditions faster and never get tired. Moreover, CAVs not only strictly abide by traffic rules, but also shorten the distance between two consecutive CAVs. For a long time in the future, CAVs and conventional vehicles will coexist on freeways [1].

In response to this phenomenon in the future, many studies have focused on the impact of CAV penetration rates on mixed traffic flow, and put some insights on the mixed traffic flow of CAVs and conventional vehicles [2, 3]. Taking into account the characteristics of trucks and the proportion of truck accidents mentioned earlier, the innovation of this article is to use the cellular automaton car-following model to distinguish trucks from passenger cars, so as to specifically explore the impact of connected and autonomous trucks (CAT) on freeways mixed traffic flow. This article

D. Russo et al. (Eds.): IHSI 2021, AISC 1322, pp. 97–103, 2021.
https://doi.org/10.1007/978-3-030-68017-6_15

classifies the vehicles in the mixed traffic flow, and specifically divides them into four types: conventional passenger cars,conventional trucks, connected and autonomous passenger cars (CAP), connected and autonomous trucks (CAT). This division is more complete than the conventional model, and distinguishes the driving characteristics of the four types of vehicles.

2 Research Methods

Cellular automata are an important tool for traffic micro-simulation, which is widely used in many fields, including society, politics, military, etc. The cellular automata model has simple rules and fast calculation speed and has achieved rapid development in the field of transportation.

The model in this paper is based on the Nasch model [4] and the two-lane Nasch model [5]. In 1992, Nagel and Schreckenberg introduced the slowing down probability to the 184 rule and proposed the Nasch model, which focused on the study of freeways. The Nasch model has four steps:

Step 1: Acceleration.

$$v_n = \min(v_n + 1, v_{max}). \quad (1)$$

It corresponds to the characteristics that the driver expects to travel at maximum speed.

Step 2: Deceleration.

$$v_n = \min(v_n, d_n). \quad (2)$$

It means that the driver slows down to prevent a collision with the front vehicle. $d_n = x_{n+1} - x_n$, it denotes the number of empty cells between the current car n and the preceding car n + 1.

Step 3: Random slowdown.

$$v_n = \max(v_n - 1, 0). \quad (3)$$

Corresponds to the random deceleration caused by various uncertain factors in reality.

Step 4: Position update.

$$x_n = x_n + v_n. \quad (4)$$

The vehicle moves forward at the updated speed in the above steps. Here, x_n and v_n represent the position and speed of the nth vehicle, respectively.

2.1 Lane-Changing Model

Since this article considers a two-lane freeway, lane changing behavior needs to be added. For both conventional vehicles and CAV, three conditions need to be met when changing lanes.

1. Whether it is necessary to change lanes and whether the current speed of the target vehicle is greater than the distance to the preceding vehicle.
2. Check whether there is enough distance in front of the corresponding position of the target lane.
3. Check whether there is enough distance behind the corresponding position of the target lane to ensure that you are not rear-ended by the vehicle behind the target lane.

In particular, in the conventional driving lane change model, the driver's driving style is usually considered, such as conservative, aggressive and other factors. The lane change probability P will be introduced into the lane change model, and the vehicle will use the probability P to change lanes. In this article, based on the research experience, the conventional passenger cars change lanes with a probability of 0.8. The conventional truck changes lanes with a probability of 0.5. For CAV, it will change lanes as long as the conditions are met, and the lane change probability is 1. When the CAT changes lanes, it uses a larger safety distance judgment to ensure its safety. Too many lane changes may cause safety hazards. Therefore, we define that its lane-changing probability under the conditions is less than that of conventional trucks.

2.2 Feature Modeling

Based on the Nasch model, this article distinguishes four types of vehicles.

Conventional Passenger Car. The maximum speed is 5 cells. Taking into account the uncertainty of the driver behavior, the slowdown probability is usually set to 0.2 in the Nasch model.

Conventional Truck. The structure of trucks is large, and the length can be 3 to 5 times that of ordinary cars. In this study, the length of trucks is set to be 3 times that of passenger cars, occupying 3 cells. Due to the difference in vehicle models and their powerful performance, trucks and passenger cars show obvious differences during driving. This article will adjust the maximum speed of the truck based on the research experience. The maximum speed of the truck is 4 cells.

Connected and Autonomous Passenger Car (CAP). The response time of the CAP is extremely short, but considering the road conditions and uncertain factors, the slowdown probability P of the CAP is set to 0.05. In the process of CAP following cars, it can keenly capture the acceleration and deceleration information of the preceding vehicle, and make real-time adjustments, making it more sensitive to road traffic conditions, reducing many unnecessary decelerations, and shortening the distance. Therefore, this article modifies the step2 deceleration model of the CAV based on the Nasch model.

$$\text{if } v_n > d_n,\ v_n = \min(v_n, d_n + v_{n+1} - 1). \tag{5}$$

During the deceleration phase, the CAV decelerates according to the speed of the preceding vehicle and the distance from the preceding vehicle, and reduces unnecessary deceleration behavior while ensuring the distance. Taking the following picture as an example (the second car is a CAV and the others are conventional vehicles) (Fig. 1):

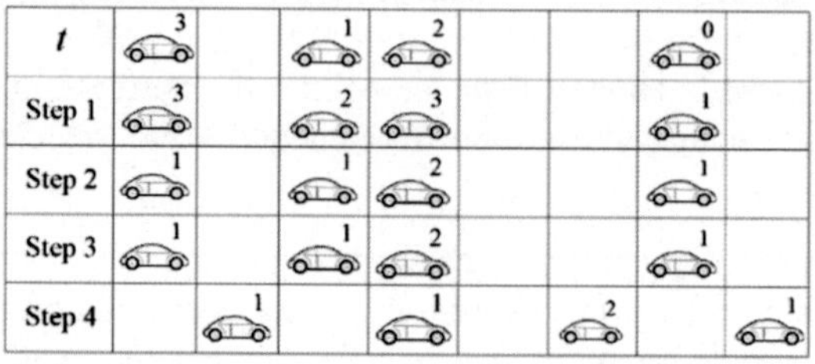

Fig. 1. Improved Nasch model step diagram.

Connected and Autonomous Truck (CAT). It has the same characteristics as the above-mentioned CAP. In particular, for CAT, safety should first be ensured. Because of its large size, the acceleration and deceleration performance are less than passenger cars. At the step of deceleration process, the required safety distance is greater, which is more conservative than CAP.

Therefore, this paper takes the slowdown probability, driving style, maximum speed, lane-changing behavior, and other factors into account. Improving the classic Nasch model, and establishing the mixed traffic of passenger cars and trucks in the connected and autonomous environment. Using MATLAB to simulate the new model to further study and analyzing the impact of connected and autonomous trucks on freeway traffic flow.

3 Experimental Procedure

In this CA model, a two-lane model with a length of 1.5 km is established. The conventional passenger car and CAP are one cell, and the conventional truck and CAT are three cells in length. The boundary is set as a periodic cycle. In the initial state, various models are randomly distributed on the freeway according to the proportion, and the initial speed is random. The simulation evolves 1000-time steps for each run, records the speed of all vehicles in the last 100-time steps, and obtains the average speed of the vehicles in each time step. Finally, the obtained speed value is time-averaged to get the average speed v of the vehicles.

This research sets the CAV penetration rates: 30%, 60%, and 90%. The ratio of passenger cars to trucks on most freeways in China is about 3:1, therefore, each type of situation is subdivided into three types:

1. Give priority to increasing the proportion of CAP.
2. The number of CAP and CAT is weighted and distributed according to the proportion of the original models, that is, the number of CAP and CAT is distributed at a ratio of 3:1;
3. Give priority to increasing the ratio of CAT.

The purpose is to explore the impact of different proportions of connected and autonomous trucks (CAT) on freeway mixed traffic flow under the condition of a certain CAV penetration rates.

4 Main Result

Figure 2 shows the relationship between mixed traffic flow density and speed when the CAV penetration rate is 30%. It can be found that when the density is from 0 to 20, the speed change is relatively small, the mixed traffic flow is in a free flow stage, and various types of vehicles can run at the maximum speed. Due to the difference in maximum speed and the existence of random slowing down, the curve fluctuates slightly. When the density is greater than 20, the average speed of the system begins to drop significantly, and the system begins to congest. Give priority to increasing the number of the CAT, the average speed of the system has been reduced. This is because to ensure the safety of the CAT, a conservative condition has been set, and the probability of lane changing is lower than the other three kinds of vehicles.

Figure 4 shows the simulation situation where the CAV penetration rate is 90%. When 90% of the vehicles on the road are CAVs, priority is given to increasing the proportion of the CAT, resulting in a significant decrease in the overall speed of the system. This is because the maximum speed of the truck is less than that of the passenger car, so the black curve is significantly lower than the other two curves.

Comparing Fig. 2, Fig. 3, and Fig. 4, it can be found that the higher the CAV penetration rate, the slower the decline of the curve, and the average speed of mixed traffic flow will also increase. This is because CAVs have more advantages than conventional vehicles. They can interact with surrounding vehicles, perceive the road environment, and move forward at a faster speed.

Figure 5 is the raw data of the simulation when the CAV penetration rate is 30%, without data fitting. Pink represents the conventional ratio of the CAP and the CAT, black represents priority in increasing the ratio of the CAT, and green represents priority in increasing the ratio of the CAP. It can be found that the black line has small fluctuations and the green line has large fluctuations. After calculating the variance, it is found that the variance of the black line is the smallest, which is 1.10. This means that priority is given to increasing the number of the CAT, the system speed fluctuates less, and the stability is stronger. Because the driving style of the CAT is conservative, there is no excessive lane changing behavior, and the safety distance judgment is relatively large, making the average speed of the system more stable than the other two experiments.

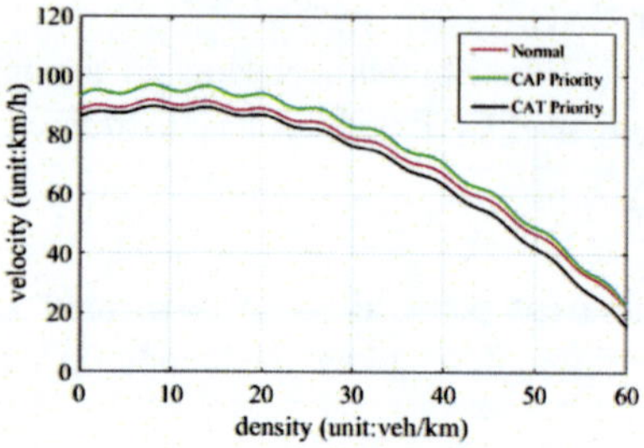

Fig. 2. 30% penetration rate.

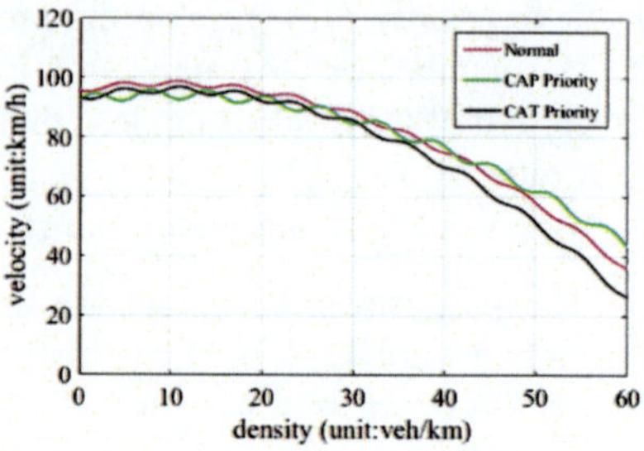

Fig. 3. 60% penetration rate.

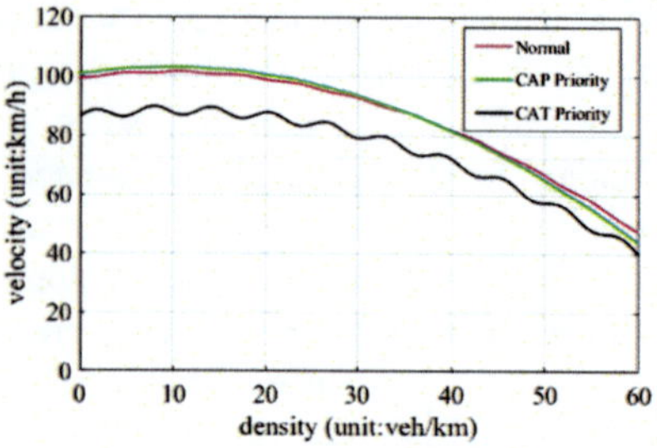

Fig. 4. 90% penetration rate.

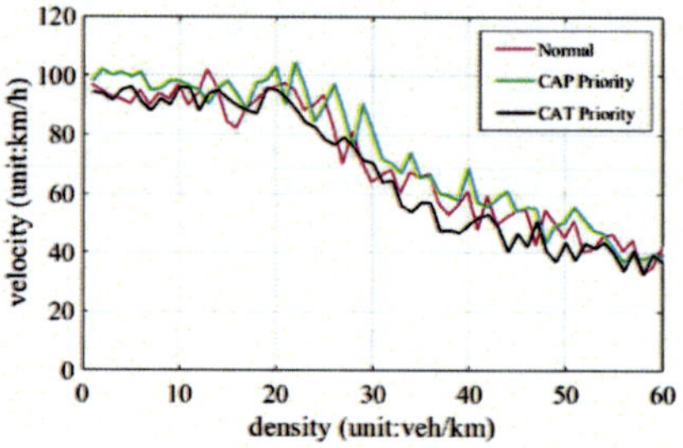

Fig. 5. Raw data of 30% penetration rate.

5 Conclusion

The results of the experiment found that: 1. With a certain CAV penetration rate, priority is given to increasing the proportion of CAT. The average speed of the mixed traffic flow system is reduced, but the speed volatility is smaller. 2. As the CAV penetration rate increases, the average speed of mixed traffic flow will also increase.

Our country has always put road safety and driving efficiency in an important position. With the development of CAVs, the study of freeway mixed traffic flow is necessary and beneficial. This article aims to provide some insights for the stability study of mixed traffic flow in a connected and autonomous environment. Through comparative experiments, the research explores the relationship between the density and speed of mixed traffic in the connected and autonomous environment; it reveals the changes brought about by giving priority to improving the priority of different connected and autonomous vehicles under different CAV penetration rate. This result can provide some insights for the research of CATs, and explore more reasonable and effective driving characteristics of CATs. The results of this study have important reference value for the study of improving the stability of road traffic flow in the future.

References

1. Fagnant, D.J., Kockelman, K.: Preparing a nation for autonomous vehicles: opportunities, barriers and policy recommendations. Transp. Res. Part A Policy Pract. **77**, 167–181 (2015)
2. Tian, J., Li, G., Treiber, M., Jiang, R., Jia, N., Ma, S.: Cellular automaton model simulating spatiotemporal patterns, phase transitions and concave growth pattern of oscillations in traffic flow. Transp. Res. B **93**, 560–575 (2016)
3. Yuan, Y., Jiang, R., Hu, M.: Traffic flow characteristics in a mixed traffic system consisting of ACC vehicles and manual vehicles: a hybrid modeling approach. In: World Congress on Intelligent Transport Systems & Its Americas Meeting (2008)
4. Nagel, K., Schreckenberg, M.: A cellular automaton model for freeway traffic. J. Phys. I **2** (12), 2221–2229 (1992)
5. Rickert, M., Nagel, K., Schreckenberg, M., Latour, M.: Two lane traffic simulations using cellular automata. Physica A **231**(4), 534–550 (1996)

Application of the Algorithm for the Recognition of Pedestrian Disturbance Patterns by Lucas-Kanade Method in Real Time

Wendy Quispe[1], Josue Tinoco[1], Grimaldo Quispe[2], and Carlos Raymundo[3](✉)

[1] Facultad de Ingenieria, Universidad Continental, Huancayo, Peru
{72168152,45096745}@continental.edu.pe
[2] Facultad de Ingenieria, Universidad Nacional Autonoma Altoandina de Tarma, Tarma, Peru
gquispe@unaat.edu.pe
[3] Direccion de Investigacion, Universidad Peruana de Ciencias Aplicadas, Lima 15023, Peru
carlos.raymundo@upc.edu.pe

Abstract. The use of security cameras allows obtaining relevant information that through video processing can provide a solution to the problem that every country faces, crime. In the following work an algorithm is developed for the detection of criminal acts, taking into account the disturbance pattern, with the use of the Python 3 programming language. For the identification of people, the area occupied by the body was taken into account, which allowed to discard any other object seen in the video. On the other hand, the monitoring of people and the analysis of changes in speed through distance was the product of tracking a corner of the body identified in a set of superimposed images, this was done with the help of the OpenCV library. The results show that the proposed algorithm can be improved with the additional use of patterns to increase its precision.

Keywords: OpenCV · Pedestrian disturbance · Lucas-Kanade · Image processing

1 Introduction

Crime is often related to economic problems and South America is not alien to this phenomenon [1]. Peru shows high levels of crime victimization, which is why it was considered the second most insecure country in Latin America according to the 2017 Americas Barometer [2], also, between 2018 and June 2019 crime remained second place as one of the main problems in Peru followed by corruption [3], which is reflected in a high rate of perception of insecurity among citizens (85.3%) [4].

In previous works they highlighted the importance of tracking objects in video surveillance cameras because the implementation of this system would help to identify

D. Russo et al. (Eds.): IHSI 2021, AISC 1322, pp. 104–110, 2021.
https://doi.org/10.1007/978-3-030-68017-6_16

actions contrary to what is established by law [5]. Likewise, many methods have been developed for the detection and monitoring of people, of which the most used are: the optical flow that offers the movement of the surfaces and edges of the objects, the particle filter that uses characteristics of color, texture and contours [6]. The tracking method based on silhouettes for the precise description of objects subsequently searched in following tables [7]. Finally, the Lucas-Kanade method that relates the speed and translation of bodies to the flow or displacement of colors through pixels from objects [8, 9].

2 State of the Art

The research work presents similarities with respect to previous works regarding the procedure that was used, for this a table will be used to compare with the most similar ones (Table 1).

Table 1. Percentage of precision of previous methods

Author	Detection method	Tracking method	Accuracy percentage
Ma, Lu y Zhang (2008) [10]	Bidirectional projection histogram	Directional Neighbor Matching	57%
Yeh et al. 2010 [11]	Background extraction	Similarity of silhouette and particle filter	95%
Benfold y Reid (2011) [8]	Markov-Chain Monte-Carlo	Kanade, Lucas and Tomasi	73.6%
Brendel, Amer y Todorovic (2011) [12]	Frame analysis	Independent set of maximum weight	70%
W. Hu, X. Li, W. Luo [13]	Riemannian log-Euclidean subspace learning	Particle filter	95%
Chu y Yang (2014) [14]	Background extraction	Fourier transform	94%
Tsai y Chang (2015) [15]	Adaptive search method	Grouping method	94%
Yoon et al. (2015) [16]	Frame analysis	Bayesian filter	96.9%
Cao et al. (2016) [17]	Merge-split count		95%
Gómez-Silva, Armingol y De Escalera (2017) [7]	Optical flow	Silhouette method	94.4%
Funde et al. (2019) [18]	Background extraction	Silhouette method	95%

3 Methodology

3.1 Optical Flow by Lucas-Kanade

The optical flow is the features of apparent movement of objects between two consecutive frames of a video, which is caused by their movement. Optical flow can be represented by a vector that shows the variation in position that exists between consecutive frames.

The Lucas-Kanade method assumes that the displacement of the colors of the bodies contained in the image between two consecutive frames is small and approximately constant.

Also, the intensity of the pixels between two consecutive frames does not change. To study these pixels, it is necessary to add a third parameter: time. Thus, the intensity of a pixel is given by:

$$I(x, y, t) \tag{1}$$

Consider a following frame in which the pixel under study moved a certain distance (dx, dy) in a certain time interval dt.

$$I(x + dx, y + dy, t + dt) \tag{2}$$

As mentioned, the intensity between two consecutive instants does not change. So, we have:

$$I(x, y, t) = I(x + dx, y + dy, t + dt) \tag{3}$$

Approximating, by Taylor series with respect to time and removing the common terms, we have the following:

$$I_x u + I_y v + I_t = 0 \tag{4}$$

This is the equation for optical flow. Where *Ix* and *Iy* are gradients of the image, and It is the gradient with respect to time. They are unknown u and v, and it should be mentioned that they are components of the vector of the pixel's optical flow, but they do not have a strong relationship with the speed intensity of the objects that are represented in the frames.

$$I_x u + I_y v + \; = -I_t \tag{5}$$

4 Design

4.1 Optical Flow Displacement Assignment

In the previous step, the velocity vector corresponding to the optical flow of the pixel being tracked is mentioned. This pixel velocity vector of a tracked object lacks a strong relationship with the velocity magnitude of the same object in our 3d environment for three reasons: small displacements are lost in the analysis, large displacements are converted to other undefined scales, and the Lucas-Kanade method is designed to determine the direction and direction of the pixel velocity vector, but not its magnitude.

However, to determine the pedestrian disturbance, it is necessary to obtain a relationship of what happens in our 3D environment with what happens in the representation of the frames. This answer is found in the tracking points of the pixels.

The change in position of the pixels represents the point of this object is related to the change in position with respect to time of the same object in the 3D environment that we live, even more so if the reception of the camera is tangential to the movement of the body. It can be demonstrated as follows:

$$m = \left| \overleftarrow{P_2} - \overleftarrow{P_1} \right| \tag{6}$$

Where m is the magnitude of the change in position.

$$\textit{Pixel position variation with respect to Time} = \frac{\sqrt{(x_2 - x_1)^2 + (y_2 - y_1)^2}}{dt} \tag{7}$$

$$\textit{Variation of body position with respect to time} = \frac{m}{dt} \tag{8}$$

$$\frac{m}{dt} = k \cdot \frac{\sqrt{(x_2 - x_1)^2 + (y_2 - y_1)^2}}{dt} \tag{9}$$

Where k is a constant of relationship between pixel and object displacement.

$$\frac{\frac{m}{k}}{dt} = \frac{\sqrt{(x_2 - x_1)^2 + (y_2 - y_1)^2}}{dt} \tag{10}$$

If we want to analyze a disturbance, in this case pedestrian, it is not necessary to scale the displacement of the pixels. Rather, it is necessary to limit the maximum allowed for which there is no disturbance.

$$G = \frac{\sqrt{(x_2 - x_1)^2 + (y_2 - y_1)^2}}{dt} \tag{11}$$

Where G will our disturbance metric be. Y is the variation that occurs from one frame to the next; therefore, it can be considered zero.

$$G = \sqrt{(x_2 - x_1)^2 + (y_2 - y_1)^2} \tag{12}$$

What is analyzed is the shift in displacement of a pixel that occurs between two consecutive frames. The faster the object goes, the faster the pixel scrolling will be (Figs. 1 and 2).

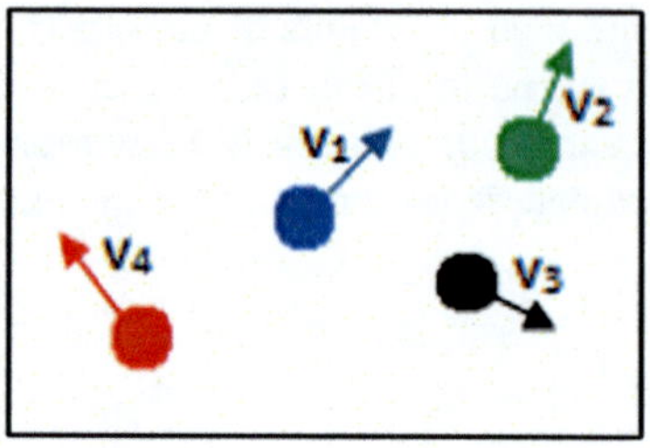

Fig. 1. Velocity analysis.

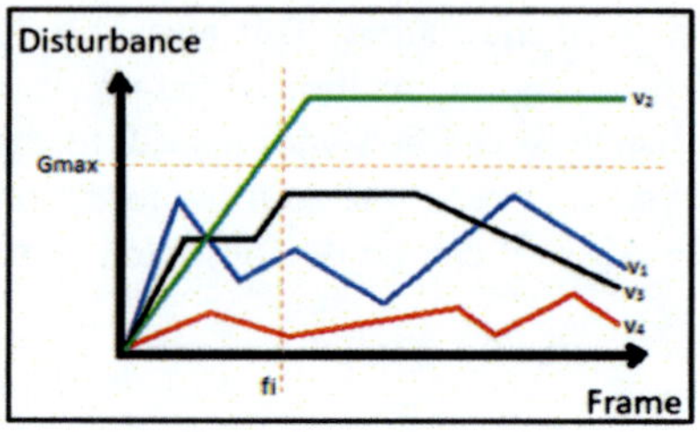

Fig. 2. Disturbance analysis.

At one point, when the frame is analyzed, the magnitude pixel exceeds the maximum value that is allowed and that is when the disturbance is identified.

5 Results

To measure the efficiency of the algorithm, we have determined two types of errors characteristic of the Lucas-Kanade technique and of other types in computer vision. There are cases in which tracking is lost as in Fig. 08; in this situation, if the object's displacement is above the disturbance metric G, the algorithm output shows the negative disturbance signal. It is considered a false positive or type I error.

The other case, is when the tracking of the objects is shown in other areas outside the area of pixels that they should be. Consequently, the intensity of increasing movement is altered and the displacement of the object is above the disturbance metric G, and the output of the algorithm shows the positive signal of the disturbance. It is considered a false negative or type II error.

The "continuous detection" is an indicator in which it is affirmed that before all the reproduction of the video the objects have been detected. False positives, false negatives, true positives, and true negatives are understood as FP, FN, VP and VN respectively.

The videos are arranged in such a way that the riots start after 1.5 s into the video. Videos are recorded at 20 fps; that is, for every second 20 frames are stored. Therefore, the average VN is 30 frames for all tests.

The ROC is, Receptor Operational Characteristic, indicator of graphic representation of sensitivity versus specificity for an algorithm that, in our case, is found as follows:

$$AUC_{ROC} = \sum_{i=1}^{n}\left[\frac{(TPR_i + TPR_{i-1}) * (FPR_i + FPR_{i-1})}{2}\right] \tag{13}$$

If the Area under the Curve () is closer to "1" it means that it is efficient; Otherwise, if it is close to 0.5 it means that the algorithm is not optimal to discriminate when there is disturbance or not (Fig. 3, Table 2).

Fig. 3. Disturbance detection

Table 2. Comparison of resolutions

Resolution	ROC
360P	0.88
480P	0.91
720P	0.93
1080P	0.94

The results are optimal for all cases. The higher resolution the videos contain, the more effective object detection will be.

6 Conclusions

Today, new applications are being sought for machine vision techniques, due to their popularity, versatility and simplicity. In this work, a new algorithm was presented to offer a solution to the increase in crime, for which three Harris methods were taken into consideration for the detection and location of pedestrians based on points, Shi-Tomasi for edge detection and Lucas- Kanade for pedestrian tracking. The videos considered as the database were extracted from static security cameras. The objectives are seen from different points of view and interact in different settings. In addition, a data set of 21 videos with an average duration of 1 min was used for the quantitative evaluation. The data set presents a wide range of challenges: the receptions of the images captured by the camera must be approximately perpendicular for the movement of the pedestrian to be detected.

References

1. Lissardy, G.: Por qué América Latina es la región más violenta del mundo (y qué lecciones puede tomar de la historia de Europa) - BBC News Mundo, BBC News Mundo (2019). https://www.bbc.com/mundo/noticias-america-latina-48960255. Accessed 09 Oct 2019
2. VI Ronda del Barómetro de las Américas en el Perú: informe, presentación y vídeosInstituto de Estudios Peruanos. IEP (2018). https://iep.org.pe/noticias/iep-presenta-barometro-de-las-americas-2016-2017/. Accessed 09 Oct 2019
3. Soto, C.M., et al.: PLAN NACIONAL DE SEGURIDAD CIUDADANA 2019–2023. El Perú Primero (2018). www.mininter.gob.pe. Accessed 09 Oct 2019

4. Sánchez Aguilar, A., Hidalgo Calle, N., Gutiérrez Espino, C., Olazabal, M.V., Erazo, J.G., Guerra, K.O., Huapaya, M.O.: PERÚ: PERCEPCIÓN CIUDADANA SOBRE GOBERNABILIDAD, DEMOCRACIA Y CONFIANZA EN LAS INSTITUCIONES, ENERO-JUNIO 2019, INEI (2019). https://www.inei.gob.pe/media/MenuRecursivo/boletines/boletin_percepcion_gobernabilidad_2.pdf. Accessed 09 Oct 2019
5. Anuradha, K., Jebin Matthew, J., Balaji, J.: Motion detection and object tracking based on frame comparison. In: 2019 International Conference on Computer Communication and Informatics, pp. 1–4 (2019)
6. Tokta, A., Hocaoglu, A.K.: A fast people counting method based on optical flow. In: 2018 International Conference on Artificial Intelligence and Data Processing, IDAP 2018, pp. 1–4 (2019)
7. Ǵomez-Silva, M.J., Armingol, J.M., De La Escalera, A.: Multi-object tracking errors minimisation by visual similarity and human joints detection. IET Semin. Dig. **2017**(5), 25–30 (2017)
8. Benfold, B., Reid, I.: Stable multi-target tracking in real-time surveillance video. In: Proceedings of IEEE Computer Society Conference on Computer Vision and Pattern Recognition, pp. 3457–3464 (2011)
9. Condori, L., Quispe, G., Carhuachín, G., Huamaní, A., Carhuapom, J.: PRINCIPALES INDICADORES DE SEGURIDAD CIUDADANA A NIVEL REGIONAL, INEI (2019). https://www.inei.gob.pe/media/MenuRecursivo/boletines/boletin_seguridad_ciudadana_departamental2013_2019.pdf. Accessed 09 Oct 2019
10. Ma, H., Lu, H., Zhang, M.: A real-time effective system for tracking passing people using a single camera. In: Proceedings of World Congress on Intelligent Control and Automation, pp. 6173–6177 (2008)
11. Yeh, H.H., Chen, J.Y., Huang, C.R., Chen, C.S.: An adaptive approach for overlapping people tracking based on foreground silhouettes. In: Proceedings - International Conference on Image Processing,, ICIP, no. 1, pp. 3489–3492 (2010)
12. Brendel, W., Amer, M., Todorovic, S.: Multiobject tracking as maximum weight independent set. In: Proceedings of the IEEE Computer Society Conference on Computer Vision and Pattern Recognition, pp. 1273–1280 (2011)
13. Hu, W., Li, X., Luo, W., Zhang, X., Maybank, S., Zhang, Z.: Single and multiple object tracking using log-euclidean riemannian subspace and block-division appearance model. IEEE Trans. Pattern Anal. Mach. Intell. **34**(12), 2420–2440 (2012)
14. Chu, H.C., Yang, H.: A simple image-based object velocity estimation approach. In: Proceedings of 11th IEEE International Conference on Networking, Sensing and Control, ICNSC 2014, pp. 102–107 (2014)
15. Tsai, T.H., Chang, C.H.: A high performance object tracking technique with an adaptive search method in surveillance system. In: Proceedings - 2014 IEEE International Symposium Multimedia, ISM 2014, pp. 353–356 (2015)
16. Yoon, J.H., Yang, M.H., Lim, J., Yoon, K.J.: Bayesian multi-object tracking using motion context from multiple objects. In: Proceedings - 2015 IEEE Winter Conference on Applications of Computer Vision, WACV 2015, pp. 33–40 (2015)
17. Cao, J., Sun, L., Odoom, M.G., Luan, F., Song, X.: Counting people by using a single camera without calibration. In: Proceedings of 28th Chinese Control and Decision Conference, CCDC 2016, no. 1, pp. 2048–2051 (2016)
18. Funde, N., Paranjape, P., Ram, K., Magde, P., Dhabu, M.: Object detection and tracking approaches for video surveillance over camera network. In: 2019 5th International Conference on Advanced Computing and Communication Systems, ICACCS 2019, pp. 1171–1176 (2019)

Comparison the Ultrasonic Distance Sensor with the Lidar in Different Conditions

Łukasz Karbowiak, Mariusz Kubanek(✉), and Janusz Bobulski

Department of Computer Science, Czestochowa University of Technology, Dabrowskiego Street 69, 42-201 Czestochowa, Poland
{lukasz.karbowiak, mariusz.kubanek, janusz.bobulski}@icis.pcz.pl

Abstract. Distance measurement can be performed basically by two types of sensors. The first type is ultrasonic sensor, which uses sound waves with a frequency of 40 kHz to measure distance. Second type is lidar, laser-based optical ranging sensor. The main idea in this study is to compare two types of distance sensor for the different distances between 50 cm–300 cm. Additional parameters are environment: laboratory and natural, surfaces: wall and glass in front of the wall. All tests contain 1000 measurements from which minimum value, maximum value and median are picked. Sensors are connected into Raspberry Pi 4B 4 GB RAM. The ultrasonic distance sensor which is used in this study is US-015. The sensor's operating range is between 2 and 400 cm. Connected to the hardware as recommended by the manufacturer. The test code was obtained from an official manufacturer site. The laser sensor used in this paper is Lidar-lite v3 with a declared distance range between 5 and 4000 cm. The test code with defaults parameters and connection setup are downloaded from the official site. Maximum distance update rates can be set to 500 Hz. Tests cases prepared for both sensors consist of stable place in which sensors are placed. In front of the sensors are set walls at different distances such as 50, 150 and 300 cm. In the last test case an glass obstacle has been added at distance 50 cm from sensor. In that test sensors are placed 300 cm from the wall. Preliminary results of these tests show interesting aspects of sensor works.

Keywords: Lidar · Radar · US-015 · Lidar-lite v3

1 Introduction

In autonomous vehicles redundancy and overlapping of sensors is normal and important. It helps to provide correct real data without disruptions. Primary distance sensors are lidar and radar, which works together. In some specific environmental conditions one type is better than another and this is the main idea of this article.

D. Russo et al. (Eds.): IHSI 2021, AISC 1322, pp. 111–115, 2021.
https://doi.org/10.1007/978-3-030-68017-6_17

2 Building a Test System

Distance measurement can be performed basically by two types of sensor. The first type is ultrasonic sensor [1], which uses sound waves with a frequency of 40 kHz to measure distance. Second type is lidar, laser-based optical ranging sensor.

The main idea in this study is to compare two types of distance sensor for the different distances between 50 cm–300 cm. Additional parameters are environment: laboratory and natural, surfaces: wall and glass in front of the wall. All tests contain 1000 measurements from which minimum value, maximum value and median are picked. Sensors are connected into Raspberry Pi 4B 4 GB RAM.

The ultrasonic distance sensor which is used in this study is US-015. The sensor's operating range is between 2 and 400 cm. Connected to the hardware as recommended by the manufacturer. The test code was obtained from an official manufacturer site [2].

The laser sensor used in this paper is Lidar-lite v3 with a declared distance range between 5 and 4000 cm. The test code with defaults parameters and connection setup are downloaded from the official site [3]. The connections are presented on Fig. 1.

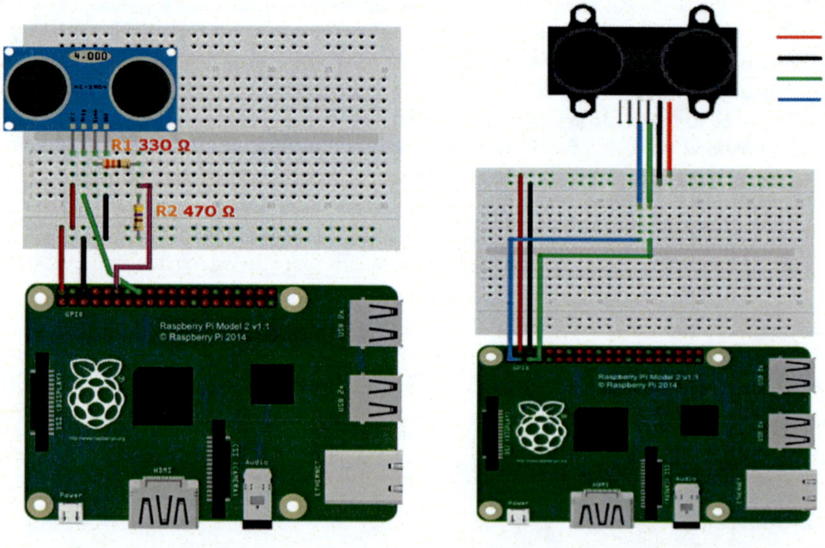

Fig. 1. An example of connecting radar and lidar to the Raspberry device [4]

3 Research

Tests cases prepared for both sensors consist of stable place in which sensors are placed. In front of the sensors are set walls at di_erent distances such as 50, 150 and 300 cm. In the last test case an glass obstacle has been added at distance 50 cm from sensor. In that test sensors are placed 300 cm from the wall.

The first distance is equal 50 cm. In that test case all radar measurements in the laboratory give a correct distance. In the outdoor test the sensor received few incorrect data but median is correct (50 cm). Figure 2 show the radar results.

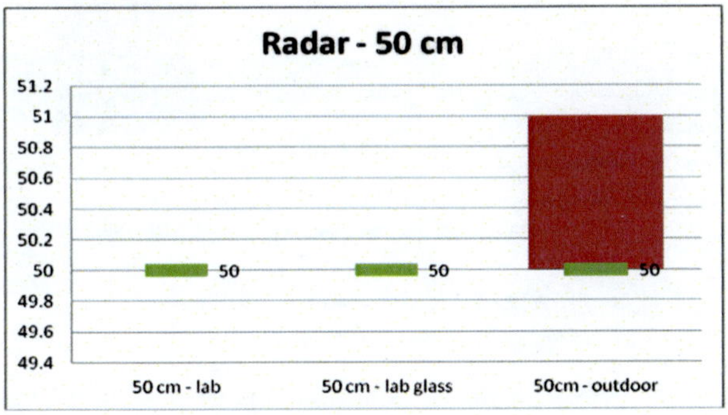

Fig. 2. Radar result – 50 cm

The lidar returned much worse results. The measurement error in some samples was over 30%. The results are shown on Fig. 3.

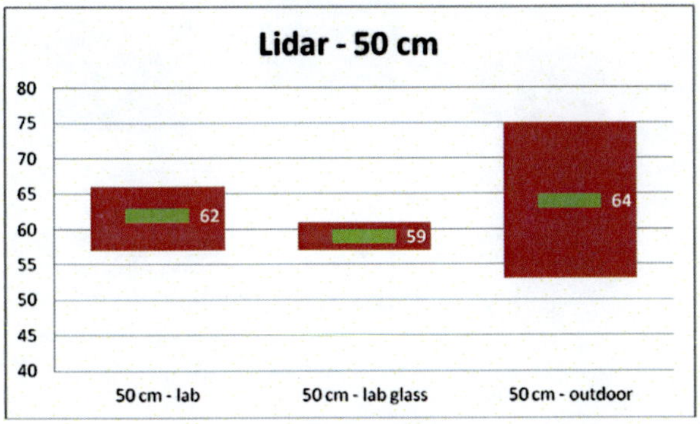

Fig. 3. Lidar result – 50 cm

In the second scenario the distance is set to 150 cm. In that distance the radar also has small discrepancy in the results (Fig. 4). The outdoor test provide information about 30% measurement errors for few samples.

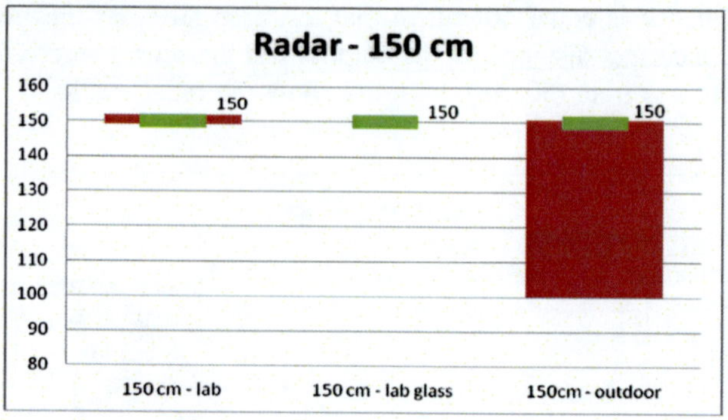

Fig. 4. Radar result – 150 cm

The lidar has still worse distance value, but the results have smaller differences between minimum and maximum (Fig. 5).

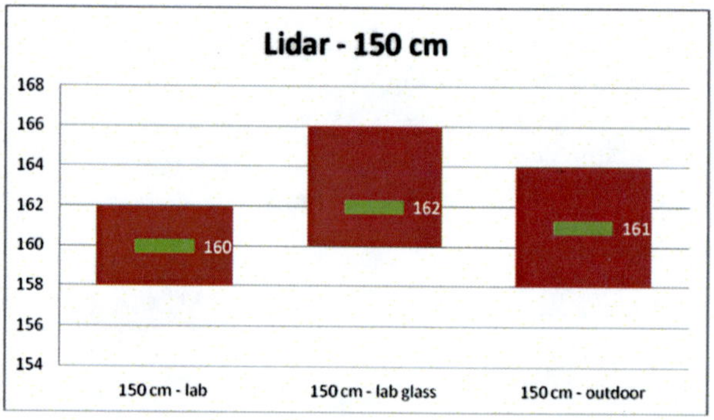

Fig. 5. Lidar result – 150 cm

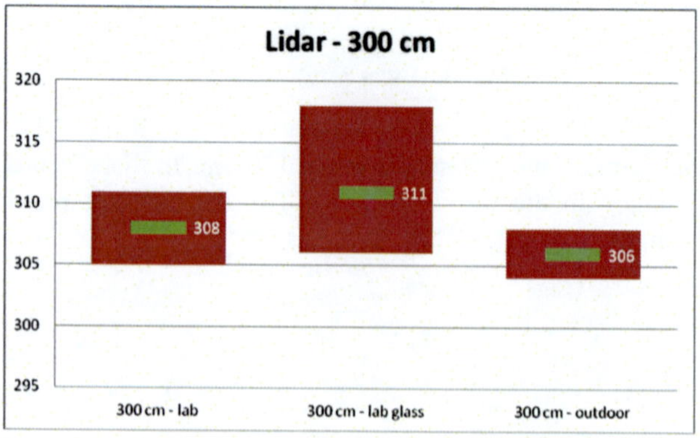

Fig. 6. Lidar result – 300 cm

In the last (third) scenario the distance is set to 300 cm. Despite of the declared maximum distance the value by manufacturer radar didn't get correct data. The lidar in that case gets the best accuracy from all tested distances. Figure 6 shown lidar results.

Different type of test case is an obstacle positioned 50 cm from the sensor. The sensor is located 300 cm from the wall. The radar returns exactly 50 cm in all measurements. The lidar didn't see the obstacle and returned values in the range 302–313 cm with median 308 cm.

4 Conclusion

The radar is a good choice for low distance. It is not possible to get many results in 1s, so fast moving obstacle will not be detected by the radar. That type of sensor can detect glass obstacle correctly. The lidar gets better results with higher distance. The maximum update rates can be set to 500 Hz so fast an obstacle is not a problem for that sensor. With a higher distance accuracy is improved. The problem for this sensor is detection of a glass obstacle. Both types of sensors are good but in the different usage.

Acknowledgments. The project financed under the program of the Polish Minister of Science and Higher Education under the name "Regional Initiative of Excellence" in the years 2019–2022 project number 020/RID/2018/19, the amount of financing 12,000,000 PLN.

References

1. Zhmud, V.A., Kondratiev, N.O., Kuznetsov, K.A., Trubin, V.G., Dimitrov, L.V.: Application of ultrasonic sensor for measuring distances in robotics. J. Phys: Conf. Ser. **1015**(3), 032189 (2018)
2. US-015 connection. https://botland.com.pl/en/content/168-Distance-measurement-using-Raspberry-Pi-and-HC-SR04-or-US-015. Accessed 24 Sept 2020
3. Lidar-lite v3 connection. https://github.com/garmin/LIDARLiteRaspberryPiLibrary/. Accessed 24 Sept 2020
4. https://botland.com.pl/img/art/inne/cms168_sch.jpg. Accessed 24 Sept 2020

Humans and Artificial Cognitive Systems

Taxonomy for Individualized and Adaptive Human-Centered Workplace Design in Industrial Site Assembly

Patrick Rupprecht(✉) and Sebastian Schlund

TU WIEN, Institute of Management Science,
Theresianumgasse 27, 1040 Vienna, Austria
{patrick.rupprecht, sebastian.schlund}@tuwien.ac.at

Abstract. The industrial assembly of large equipment is usually organized as site assembly, requires a high degree of manual work. Humans, with their cognitive skills and abilities, still carry out a significant share of manual and cognitive tasks in this context. For process-efficient as well as human-centered design of workplaces, anthropometric characteristics and needs of the employees have to be considered. These specific attributes differ from human to human and require new adaptation concepts for workstations. In this paper, the individualization dimensions of site assembly are specified and extended with the focus on user-adaptive design. In addition, adaptive concepts for the implementation of individual configuration options, such as capturing of human characteristics and adaptive execution of the adjustment, are identified and presented in an extended taxonomy.

Keywords: Taxonomy of user-adaptive design · Site assembly · Human-systems interaction · Individualizable workstations · Adaptive workplace

1 Introduction

The industrial assembly of large equipment is mostly organized as site assembly and involves a high degree of manual work and a low degree of automation compared to other assembly modes [1, 2]. Figure 1 gives a schematic overview of industrial site assembly.

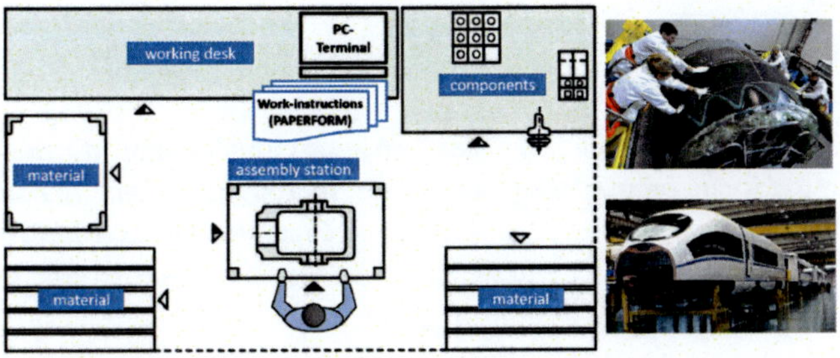

Fig. 1. Schematic overview of industrial site assembly [1, 2]

D. Russo et al. (Eds.): IHSI 2021, AISC 1322, pp. 119–126, 2021.
https://doi.org/10.1007/978-3-030-68017-6_18

In site assembly the components stay on a stationary position during the whole assembly process and all parts, as well as the employees, move to that location [1]. Hence, humans with their cognitive abilities and skills play an important role in this form of assembly [1, 3–5]. For human-centered, as well as process-efficient design of the assembly, the anthropometric characteristics and individual needs of the person have to be considered with new design option in the best possible way [6, 7]. These specific attributes differ from human to human and require concepts for the easy adaptive adjustment of the assembly station [8–10]. The "self-configured workstation", which adapts dependently to the specific anthropometric characteristics data is one of the basic features of the assembly station 4.0. It supports assembly operations, reduces process time and improves ergonomics [11]. The consideration of individual dimensions also offers the possibility of solving the issue of designing workplaces according to the principle of "one-size-fits-all" and indeed of establishing ideal conditions for individual users of a work system [12].

In this context, a taxonomy was developed to show which possible dimensions can be individualized in site assembly and which advantages ensue for employees and industrial companies [9, 12]. Figure 2 shows a first rendering of potential individualization concepts in site assembly [9, 12].

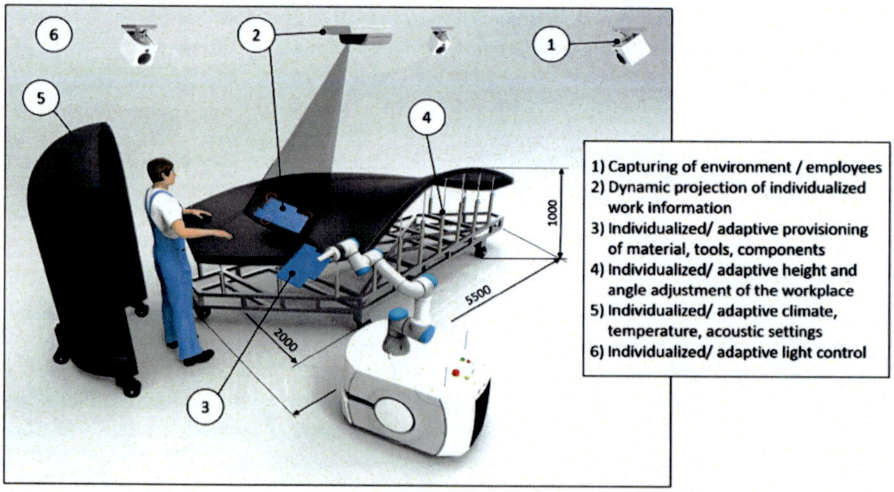

Fig. 2. Individualization concept for a site assembly workplace [9, 12]

This paper presents results of specified individualization dimensions in site assembly with the focus on user-adaptive design. In addition, adaptive concepts for the capturing of human characteristics, as well as for execution of the adaptive adjustment are identified and presented to extend the taxonomy with innovative concepts of human centered workplace design.

2 Taxonomy of User Adaptive Design

The current taxonomy of individualization possibilities is divided into seven dimensions that consider the assembly workstation and its environment [9, 12]:

1) working height
2) range of vision, gripping area, handedness
3) lighting
4) ventilation and air conditioning
5) acoustic situation
6) use of information and assistance systems
7) further aspects (organization of work, sense of security, visual privacy, etc.)

These dimensions and the related configuration options are described in our previous work [9, 12]. Based on these initial considerations, this paper concretizes, expands and restructures the dimensions and configuration options and presents them in an updated structure of the taxonomy. Further, it is analyzed which concepts can be used to adaptively capture and identify the user and their individual characteristics, how this data can be saved digitally and how the specific adaptive configuration options on the workplace can be executed. The adaptive concepts and considerations are prepared in order to integrate them into the taxonomy as an extension.

2.1 Updated Structure of Dimensions and Configuration Options

The question *"WHAT can be designed user-adaptively at site assembly?"* is used to concretize the generally formulated dimensions from the current taxonomy.

The dimensions are subdivided for more details or extended with additional aspects. The result is a new structure of the taxonomy with ten dimensions and related configuration options (DIM1 to DIM 10), as shown in Table 1.

Table 1. Updated structure of dimensions and configuration options

Dimension	Configuration options
DIM 1 – Working desktop	- Working height - Working angle/rotation
DIM 2 – Arrangement of physical objects and provision of materials	- Physical objects at the workplace (tools, containers, aids,...) - Arrangement of physical material supply - Arrangement of physical control elements
DIM 3 – Arrangement and design of digital elements/user interface	- Structure/arrangement/design of User interface/control elements - Software selection
DIM 4 – Lighting parameters	- Illuminance - Type of lighting (passive/active/natural/artificial) - Light color/Light temperature - Arrangement of lighting sources

(*continued*)

Table 1. (*continued*)

Dimension	Configuration options
DIM 5 – Climatic parameters	- Temperature - Air humidity - Aeration parameters (direction, strength, flow type) - Purity and foreign particle concentration
DIM 6 – Acoustic parameters	- Noise level (dB), Frequencies - Music/acoustic information
DIM 7 – (Work-) Information Systems/ Digital Assistance Systems	- Information content - Type of information (visual, auditory, kinesthetic, olfactory) - Information provision location (screen, wearable, data-glasses, ...) - Scope/granularity of information (frequency, depth of detail, ...)
DIM 8 – Physical assistance systems	- Duration of the assistance (frequency, interval) - Use of the assistance (Yes/No) - Strength of the assistance - Type of assistance systems
DIM 9 – Human-system interaction	- Type of interaction (input/output) - Design of the interaction - Height of the system default - Level of division of labor between man & machine
DIM 10 – Work organization	- Cycle length/cycle time - Working hours - Work activity/work content - Division of labor in general

2.2 Extension: User-Adaptive Recognition and Identification Options

For user-adaptive design, it is necessary to define relevant user characteristics and subsequent modes to capture and identify the specific users. It also must be defined how this data can be saved and requested for further processing. For this purpose, a few aspects of differentiation and related variants are considered to get potential options for the user-adaptive capturing and identification. The following presented options were found to be the most appropriate based upon a morphological analysis.

Option 1: Adaptive Identification of Human and Request of the Pre-saved Individual Characteristics From Digital Person Profiles: With this option, the individual characteristics, preferences and needs of the humans are initially recorded and saved in digital personal profiles. After identification of the employee at the workstation, the corresponding saved data is retrieved from the personal profiles and the assembly adapts itself to each digital profile. The adaptive identification of the employee can take place with a RFID tag in personalized work clothing and recognition via RFID-gateway or with recognition/identification of the person by facial-, iris-, fingerprint scanner.

Option 2: Adaptive Recognition of the Human and their Individual Characteristics: Camera systems and image recognition algorithms (e.g. YOLO, OpenPose) are used to identify humans. In addition, image recognition analyzes movements, gestures and anthropometric characteristics and configurates the workstation accordingly to these parameters. In comparison to the other option, only live detection and no pre-defined data is used in this case. Consequently, personal preferences cannot be considered so well with this option. In summary, both recognition and identification options are suitable for the user-adaptive design of the individual adaptive adjustment of the workplace in site assembly.

These two options of user-adaptive recognition and identification are an important extension for the taxonomy, to bring the human in the center of the design and to ensure adaptive concepts for the adjustment of the workstation.

2.3 Extension: User-Adaptive Execution Options for Adjusting the Workstation

After presenting the possibilities for the recognition of humans, it is necessary to determine concepts for the user-adaptive execution for adjusting the workstation.

For the analysis, the question "HOW/WHEREBY can the adjustment options be executed accordingly to individual characteristics and preferences?" was used to obtain concrete results. Table 2 represents the potential adaptive adjustment possibilities for the assembly station.

Table 2. User-adaptive execution options for adjusting the workstation

Dimension	User-adaptive execution options for adjusting the workstation
DIM 1 – Adaptive working desktop	Automatic adjustment of the working surface in height, angle, rotation with controlled -) electric engines/drives -) electro-pneumatic or electro-hydraulic cylinders and drives
DIM 2 - Adaptive Arrangement of physical objects and provision of materials	Physical arrangement/material supply -) with self-drive material carriers -) by flexible and collaborative robots -) by automated guided vehicle systems -) by automatic cable pulls/overhead conveyors

(*continued*)

Table 2. (*continued*)

Dimension	User-adaptive execution options for adjusting the workstation
DIM 3 - Adaptive Arrangement and design of digital elements/user interface	Adaptive arrangement of digital elements with -) dynamic projection systems/Spatial Augmented Reality systems -) individual user interfaces, designs on tablets/screens, etc. -) individual elements when using data glasses/AR systems, wearables -) individual APPs for Smartphones, etc.
DIM 4 - Adaptive lighting parameters	Adaptive lighting parameters through -) individual, digital lighting control systems and APPs -) intelligent, smart LED panels, LED inserts -) modern daylight supply systems -) sensors for weather, day and night -) adaptive blue light filters, gray light filters
DIM 5 - Adaptive Climatic parameters	Adaptive climatic parameters through -) digital ventilation and temperature control -) coupled sensors at the workplace, digital personal profiles -) self-propelled mobile ventilation systems on AGV, robot arms -) customizable, digitally networked air humidifiers and dehumidifiers, fans -) customizable digitally networked heating and cooling devices
DIM 6 - Adaptive Acoustic parameters	Adaptive acoustic parameters through -) digital control of noise protection measures -) self-driving noise protection systems on e.g. AGV -) moving noise protection devices with robots -) sound sensors at the workplace -) automatic Noise Cancelling Headphones -) digital control of acoustic information systems
DIM 7 - Adaptive (Work) Information Systems/Digital Assistance Systems	Adaptive work-instructions -) through individual information systems (Spatial AR, AR, wearables, screens) -) on individual information provision locations (directly at the workplace, central terminals, mobile with wearables, etc.) -) including individual information content, granularity, depending on qualifications and experience)

(*continued*)

Table 2. (*continued*)

Dimension	User-adaptive execution options for adjusting the workstation
DIM 8 - Adaptive Physical assistance systems	Usage of adaptive physical assistance systems -) with individual useful lifetime and frequency -) as decision assistance for use YES/NO/Optional -) with individual strength of the assistance through automatic adjustment actuators -) as automated just-in-time assistance system
DIM 9 – Adaptive Human-system interaction	Adaptive Human-system interaction with -) individual type of interaction (input/output system) -) individual setting of the system default -) individual setting of the division of labor
DIM 10 - Adaptive Work organization	Adaptive work organization through -) individual cycle length through flexible AGV, roller carts -) individual working time with flexible shift models, flextime, control by mobile app -) randomized division of labor, job rotation, enlargement, use of assistance systems

2.4 Conclusion and Future Works

The identified options and concepts to extend the taxonomy bring new possibilities for human-centered and user-adaptive designs to industrial site assembly. Through the integration and combination of these different perspectives, the human characteristics and needs can be captured and used adaptively to adapt the work station accordingly. In further research, elected adaptive concepts will be implemented and evaluated using various demonstrators in the TU Wien pilot factory Industry 4.0.

Acknowledgments. This research was supported by the Austrian Research Promotion Agency (FFG), the Austrian Federal Ministry for Climate Action, Environment, Energy, Mobility, Innovation and Technology (BMK) and the TU WIEN through the endowed professorship "HCCPPAS" (FFG-852789).

References

1. Lotter, B., Wiendahl, H.-P.: Montage in der industriellen Produktion: Ein Handbuch für die Praxis (2. Aufl.). VDI-Buch. Springer (2012)

2. Rupprecht, P., Kueffner-McCauley, H., Schlund, S.: Information provision utilizing a dynamic projection system in industrial site assembly. In: Proceedings of the 53rd CIRP Conference on Manufacturing Systems (2020)
3. Gorecky, D., Schmitt, M., Loskyll, M., Zuhlke, D.: Human-machine-interaction in the industry 4.0 era. In Pereira, C.E. (Hg.) 2014 12th IEEE International Conference on Industrial Informatics, Porto Alegre - RS, Brazil, 27–30 July 2014, S. 289–294. IEEE (2014)
4. Jensen, P.L., Alting, L.: Human factors in the management of production. CIRP Ann. **55**(1), 457–460 (2006)
5. Bauer, W., Ganschar, O., Gerlach, S., Hämmerle, M., Krause, T., Schlund, S.: Industrie 4.0–flexiblere und reaktionsfähigere Produktionsarbeit–Ergebnisse der Industrie-4.0-Leitstudie des Fraunhofer IAO. Werkstatttech Online **104**(3), 134–138 (2014)
6. Romero, D., Bernus, P., Noran, O., Stahre, J., Fast-Berglund, Å.: The operator 4.0: human cyber-physical systems & adaptive automation towards human-automation symbiosis work systems. In: Advances in Information and Communication Technology. Advances in Production Management Systems, Bd. 488, S. 677–686. Springer (2016)
7. Bauer, W., Schlund, S., Vocke, C.: Working life within a hybrid world–how digital transformation and agile structures affect human functions and increase quality of work and business performance. In: International Conference on Applied Human Factors and Ergonomics, pp. 3–10. Springer, Cham (2017)
8. Kaasinen, E., Aromaa, S., Heikkilä, P., Liinasuo, M.: Empowering and engaging solutions for operator 4.0 - acceptance and foreseen impacts by factory workers. In: Advances in Production Management Systems. Production Management for the Factory of the Future. Springer (2019)
9. Schlund, S., Mayrhofer, W., Rupprecht, P.: Möglichkeiten der Gestaltung individualisierbarer Montagearbeitsplätze vor dem Hintergrund aktueller technologischer Entwicklungen. Zeitschrift für Arbeitswissenschaft **72**(4), 276–286 (2018)
10. Reinhart, G.: Handbuch Industrie 4.0: Geschäftsmodelle, Prozesse, Technik. Hanser (2017)
11. Bortolini, M., Ferrari, E., Gamberi, M., Pilati, F., Faccio, M.: Assembly system design in the Industry 4.0 era: a general framework. IFAC-PapersOnLine, **50**(1), 5700–5705 (2017)
12. Mayrhofer, W., Rupprecht, P., Schlund, S.: One-Fits-all vs. Tailor-made: user-centered workstations for field assembly with an application in aircraft parts manufacturing. Procedia Manuf. **39**, 149–157 (2019)

Cyber-Therapy: The Use of Artificial Intelligence in Psychological Practice

Chiara Lucifora[1(✉)], Leonardo Angelini[2], Quentin Meteier[2], Carmelo M. Vicario[1], Omar Abou Khaled[2], Elena Mugellini[2], and Giorgio M. Grasso[1]

[1] Department of Cognitive Science, University of Messina, 98122 Messina, Italy
{clucifora, cvicario, gmgrasso}@unime.it

[2] Technology for Human Well-Being Institute, University of Applied Sciences of Western Switzerland, 1700 Fribourg, Switzerland
{leonardo.angelini, quentin.meteier, omar.aboukhaled, elena.mugellini}@hefr.ch

Abstract. Cyber Therapy is a research project based on the relationship between Computer Science and Psychology. We are working on post-traumatic stress disorder (PTSD) due to severe traffic accidents using a virtual reality driving simulator and ECG, EDA and breathing sensors. With the help of virtual reality (VR) our goal is to build one software that can process the user's biofeedback signals - heart rate, body temperature, state of tension, etc. - in real time, to make the phobic stimulus autonomous. To this purpose, we developed a platform capable of adapting the phobic stimulus based on the user's biofeedback signals. We believe that this human-computer integrated system could be useful to patients as it would allow them to face fear autonomously, and to the psychotherapist, as it would allow a real time - physiologically based - knowledge of fear symptoms severity able to promote a timely and more appropriate program of intervention.

Keywords: Virtual reality · Psychology · PTSD · Cyber therapy

1 Introduction

Cyber therapy can be defined as the use of innovative technologies that help traditional therapy in the clinical field of psychology [1]. Its effectiveness is mainly due to its imaginative power, which allows patients to experience real situations; as well as the possibility of actively perceiving one's own body within a simulated environment (i.e., Virtual reality - VR - based); and its connectivity [1]. VR allows the subject to feel immersed in a temporally and spatially different place, thanks to the technological setup capable of processing the visual-sound information and returning it to the subject in real time. Within immersive VR, mediated by the use of a 3D helmet, it is possible to experience feelings of telepresence [2], which is due to the possibility of making real movements in a virtual environment. For these reasons, bringing a person into a stressful virtual situation causes increased heart rate, increased skin conductance, and other body responses like the real-world situation [3]. Recent studies show that cyber

D. Russo et al. (Eds.): IHSI 2021, AISC 1322, pp. 127–132, 2021.
https://doi.org/10.1007/978-3-030-68017-6_19

therapy allows the patient to obtain better results than traditional therapy [4, 5]. It is used to treat anxiety or specific phobias [6] such as fear of public speaking [7] arachnophobia [8, 9] and PTSD [10, 11]. For example, in arachnophobia Hofmann et al. [8] show that VR treatment, associated with increased touch (the ability to touch a silicon spider in the real world, while the patient sees a real spider in the virtual world) reduced the fear after only three sessions (N = 36. 8 subjects with clinical phobia). In a recent study, Rothbaum et al. [12] showed a reduction in phobia in 9 Vietnam veterans with PTSD in response to Cyber therapy; Moreover, Wald [13] showed a 60% reduction in phobia severity on 5 subjects with driving phobias; similarly, Beck et al. [11] reported that 10 treatment sessions with VR reduced symptoms severity c in 6 patients affected by PTSD due to a car accident.

Overall, the use of VR technology in the clinical setting is a valid support for memory or imagination helping the user to face the feared situation. In PTSD mental images i.e. flashbacks, intrusive images, memories [14] play an important influence in the amplification of emotions [15]. VR can be used as a useful imaginal system to induce positive responses [16] that allow the user to face the fear.

Based on this idea, we aim to build up a virtual environment where the exposure to fear related scenario is dynamically calibrated to the level of tolerance of the patients, established from the analysis of their psychophysiological response. This tool can be used by psychologists to create an Avoidance Hierarchy of stressful situations [17] based on the analysis of psychophysiological signals, which can be automatically used to establish the exposure to progressively stressful (i.e., anxiety generator) scenarios, as the level of tolerance to the previous one increases. In alternative, the psychologist can determine the Avoidance Hierarchy in a traditional way and define in the system the anxiety level for the patient, who can then be exposed progressively to the intervention, based on the severity level suggested by the psychologist and on the user's reaction to the phobic stimuli.

2 Our Research

2.1 Instrumental

In our experimental setup we use a virtual reality driving simulator (based on the VR headset "Oculus Rift"), equipped with rotation and position sensors and integrated headphones, that provide a 3D audio effect. The driving simulator is composed of a steering wheel with force feedback and pedals (Logitech G27), combined around a real car seat, used to provide a full driving experience during the simulation.

Moreover, we use the BioSignalsPlux Kit, and specifically the ECG, Respiration and EDA sensors – Fig. 1-, whose data are recorded using Python 3.7 program and post-processed using the Neurokit library. Unity 3D 18.4.22 was used to build highly immersive and realistic virtual scenarios (based on Windridge City for AirSim assets).

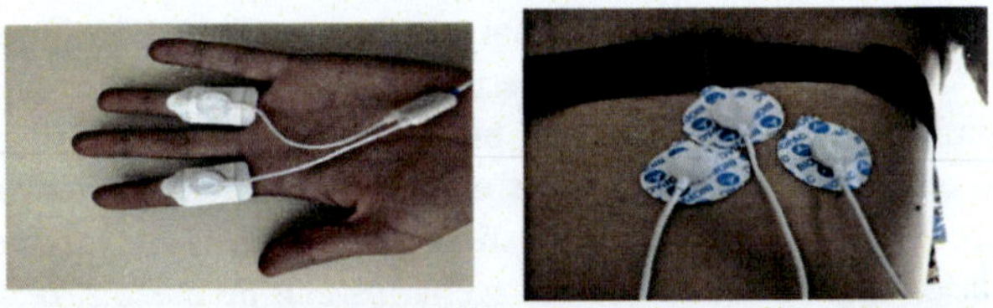

Fig. 1. BioSignalsPlux Kit. On the left EDA sensor; on the right ECG and respiration sensors

We use ECG, EDA and respiration data because the electrocardiogram allows us to measure the electrical activity of the heart using electrodes placed on the skin, which enables the detection of small electrical changes during each cardiac cycle; while electrodermal activity (EDA) records attentional, affective and motivational processes through bodily responses. In particular, the skin conductance level (SCL) and the respiration rate are reliable measures of the degree of stress [18].

2.2 Procedure

In our study we want to test a sample of 200 subjects, divided into two subgroups, each consisting of 100 subjects. The first group has a PTSD condition, due to a road accident, whereas the second group is made of 100 healthy subjects (control group).

Subjects will be asked to complete a 30-min VR driving session, in which they will be involved in situations of unavoidable road accidents. The subject's biological signals are continuously recorded during the driving session.

Currently, for our prototype, we have outlined 15 scenarios of stressful driving situations, divided into 3 groups:

1. Soft group: driving in adverse weather and/or heavy traffic conditions - Fig. 2A;
2. Mid-Hard group: driving while an accident occurs that does not involve the driver – Fig. 2B;
3. Difficult group: driving into an accident that involves the driver – Fig. 2C.

Fig. 2. A: Level easy. Driving with the fog; B: Level medium. Driving with accidents that do not involve the driver and C: Level Hard. Driving with an accident involving the driver

Generally, VR sessions proceed through gradual exposure, in order to respect the patient's personal rhythm [19]. Our scenarios will be combined creating an order of increasing difficulty (easy-medium-hard), but the order of execution of the scenarios can be changed according to the patient's anxiety state. This allows us to re-propose less stressful scenarios as soon as the patient's vital signs are above average. Furthermore, the order of presentation of the scenarios can be customized on the basis of the prior traumatic experience of the subject. For this aim, we ask the psychologist to adjust, according to the patient experience, each scenario as easy, medium or difficult before entering the virtual world, following a similar approach to the Avoidance Hierarchy [17].

2.3 Implementation

The system works through continuous communication between its constituent parts, namely a Python application, collecting and analyzing the ECG, EDA and respiration rate from the BioSignalsPlux kit, and the Unity 3D application running the simulated environment. At the beginning of the experimentation, the system calculates the baseline of the 3 biosignals over 60 s. After baseline calibration, the vital sign acquisition application continuously reads the subjects' values and reports them to Unity every 10 s, using windows of 30 s and compares them to the baseline values. When the difference between the current value and the baseline is higher than a predetermined threshold, the status of the user is set to stressed, otherwise it is set to relaxed. These values were established empirically, based on the analysis of an existing dataset of biosignals acquired in VR. If the user is relaxed, the simulation environment proposes phobic stimuli with an increasing level of difficulty, otherwise it keeps proposing stimuli that are on the same level.

2.4 Single-Case

So far, we have performed tests on only one person from the non-clinical population.

The user was subjected to 3 scenarios belonging to 2 different levels (easy and medium). Specifically, the simulation began with driving in adverse weather conditions (dense fog) and continued with road accident conditions without the involvement of the user (side impact between two vehicles, falling of a tree on a car in front of the user). The results showed a user reaction to the proposed scenarios, recorded through changes in physiological signals. In particular, there is a slight increase in ECG signals and a noticeable change in SCL – Fig. 3.

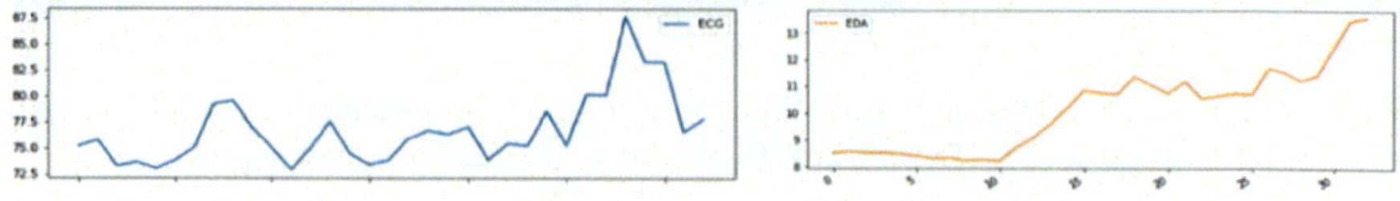

Fig. 3. ECG and EDA of the user, during the simulation

During the test the user experienced some nausea, probably due to motion sickness, which suggests that such experience should be run in a dynamic driving simulator, rather than on a VR headset.

3 Discussion

The use of VR as a therapeutic tool has provided good clinical results for the treatment of anxiety disorders [20]. Our aim is to improve the effectiveness of this treatment by creating a phobic scenario that is dynamically calibrated to the level of tolerance of the patients, established via psychophysiological response (i.e., biofeedback signals).

This could be useful both for the user, to face his own fear independently, and for the psychologist, to know in real time the user's degree of fear and to intervene easily. We believe that building a phobic stimulus that changes based on the user's stress can allow the patient to feel safe, within the environment, which is essential for the success of the therapy [1]. Currently, the system for assessing the stress level of the user is very basic and might not be robust enough to detect subjective variability. We are currently implementing a machine learning algorithm that will allow detecting the stress level with a higher degree of accuracy. In order to train this algorithm, we will build a dataset of annotated stress data in driving conditions, with both healthy subjects and subjects affected by PTSD. Importantly, while the CyberTherapy platform has been built to treat PTSD due to road accidents, it can be easily adapted to other anxiety related scenario, thanks to the modular architecture of the system. Depending on the scenario (fixed or moving camera), the subject can experience the phobic stimuli either in VR or on a screen, respectively.

Acknowledgments. We thank the students Daniel D'Inca and Giovanni D'Italia for having participated in carrying out this work, as a support to the IT implementation.

References

1. Ventura, S., Baños, R.M., Botella, C., Mohamudally, N.: Virtual and augmented reality: new frontiers for clinical psychology. In: State of the Art Virtual Reality and Augmented Reality Knowhow, pp. 99–118. InTech (2018)
2. Steuer, J.: Defining virtual reality: dimensions determining telepresence. J. Commun. **42**(4), 73–93 (1992)
3. Meehan, M., Insko, B., Whitton, M., Brooks, F.P.: Physiological measures of presence in stressful virtual environments. In: Proceedings of SIGGRAPH 2002, San Antonio, TX, USA (2002)
4. Opriş, D., Pintea, S., García-Palacios, A., Botella, C., Szamosközi, Ş, David, D.: Virtual reality exposure therapy in anxiety disorders: a quantitative meta-analysis. Depression Anxiety **29**(2), 85–93 (2012)
5. Botella, C., Fernández-Álvarez, J., Guillén, V., García-Palacios, A., Baños, R.: Recent progress in virtual reality exposure therapy for phobias: a systematic review. Curr. Psychiatry Rep. **19**(7), 42 (2017)

6. Parsons, T.D., Rizzo, A.A.: Affective outcomes of virtual reality exposure therapy for anxiety and specific phobias: a meta-analysis. J. Behav. Ther. Exp. Psychiatry **39**(3), 250–261 (2008)
7. Anderson, P.L., Zimand, E., Hodges, L.F., Rothbaum, B.O.: Cognitive behavioral therapy for public-speaking anxiety using virtual reality for exposure. Depression Anxiety **22**(3), 156–158 (2005)
8. Hoffman, H.G., Garcia-Palacios, A., Carlin, A., Furness Iii, T.A., Botella-Arbona, C.: Interfaces that heal: coupling real and virtual objects to treat spider phobia. Int. J. Hum.-Comput. Interact. **16**(2), 283–300 (2003)
9. Minns, S., Levihn-Coon, A., Carl, E., Smits, J.A., Miller, W., Howard, D., Papini, S., Quiroz, S., Lee-Furman, E., Telch, M., Carlbring, P.: Immersive 3D exposure-based treatment for spider fear: a randomized controlled trial. J. Anxiety Disord. **58**, 1–7 (2018)
10. Rizzo, A., Pair, J., Graap, K., Manson, B., McNerney, P.J., Wiederhold, B., Wiederhold, M., Spira, J.: A virtual reality exposure therapy application for Iraq War military personnel with post traumatic stress disorder: from training to toy to treatment. NATO Secur. Through Sci. Ser. E Hum. Soc. Dyn. **6**, 235 (2006)
11. Beck, J.G., Palyo, S.A., Winer, E.H., Schwagler, B.E., Ang, E.J.: Virtual reality exposure therapy for PTSD symptoms after a road accident: an uncontrolled case series. Behav. Ther. **38**(1), 39–48 (2007)
12. Rothbaum, B., Hodges, L., Ready, D., Graap, K., Alarcon, R.: Virtual reality exposure therapy for Vietnam veterans with posttraumatic stress disorder. J. Clin. Psychiatry **62**, 617–622 (2001)
13. Wald, J.: Efficacy of virtual reality exposure therapy for driving phobia: a multiple baseline across-subjects design. Behav. Ther. **35**, 621–635 (2004)
14. Clark, I.A., James, E.L., Iyadurai, L., Holmes, E.A.: Mental imagery in psychopathology: From the lab to the clinic. In: Watson, L.A., Berntsen, D. (eds.) Clinical Perspectives on Autobiographical Memory [Internet]. Cambridge University Press, Cambridge (2015)
15. Holmes, E.A., Mathews, A.: Mental imagery and emotion: a special relationship? Emotion **5** (4), 489–497 (2005)
16. Day, S., Holmes, E., Hackmann, A.: Occurrence of imagery and its link with early memories in agoraphobia. Memory **12**(4), 416–427 (2004)
17. Kircanski, K., Mortazavi, A., Castriotta, N., et al.: Challenges to the traditional fear hierarchy in exposure therapy. J. Behav. Ther. Exp. Psychiatry **43**, 745–751 (2012)
18. Braithwaite, J.J., Watson, D.G., Jones, R., Rowe, M.: A guide for analysing electrodermal activity (EDA) & skin conductance responses (SCRs) for psychological experiments. Psychophysiology **49**(1), 1017–1034 (2013)
19. Maples-Keller, J.L., Bunnell, B.E., Kim, S.J., Rothbaum, B.O.: The use of virtual reality technology in the treatment of anxiety and other psychiatric disorders. Harvard Rev. Psychiatry **25**(3), 103 (2017)
20. Powers, M.B., Emmelkamp, P.M.: Virtual reality exposure therapy for anxiety disorders: a meta-analysis. J. Anxiety Disord. **22**(3), 561–569 (2008)

Integrating Voice Based Interaction with Massive Data Process Description and Execution

Constantin Nandra(✉), Sonia Grigor, and Dorian Gorgan

Computer Science Department, Technical University of Cluj-Napoca,
G. Baritiu street 26-28, 400027 Cluj-Napoca, Romania
{constantin.nandra,dorian.gorgan}@cs.utcluj.ro,
sonia.grigor@gmail.com

Abstract. Processing Earth Observation data is a compute-intensive task, due to their large size and increasing acquisition rates, often requiring the kind of computing power offered by Cloud solutions. To bridge the gap between the Earth data scientist and the programming expertise required to effectively use such solutions, we have developed a workflow-based process description model. It uses a description language to define algorithms in an intuitive manner, while also increasing the potential for exploiting parallelism. Within this paper, we present the results of an attempt at integrating the capabilities of the Amazon Alexa personal assistant with the process description solution that we have developed. It is meant as a first step towards a multi-modal interface that would provide support for new users to describe data processing algorithms using natural language, and potentially improve accessibility for users in the field who rely on mobile devices to visualize data.

Keywords: Voice-controlled GIS · Workflow description · Distributed processing

1 Introduction

Earth Observation (EO) data pose a significant challenge to individuals and organizations, in terms of storage and processing capabilities. The most problematic aspects are the large size of the data and their increasing acquisition rates. EO data often consist of large images of hundreds of millions of pixels with a given area of the planet represented by multiple layers, each depicting different spectral domains. On top of this, the newest satellite constellations, like the Sentinel 2 program [1], are able to map the entire surface of the planet once every five days. Processing such amounts of data is often beyond the capabilities of standalone computers, either because of limited storage or inadequate computational power. To meet the challenge, various distributed processing architectures have been proposed and implemented over the last decades, with the Cloud being the latest and most mature iteration [2].

The main benefit of employing Cloud solutions in the processing of EO data stems from the possibility of parallel execution, by taking advantage of the multiple processing units available within a Cloud. As part of the BigEarth research project [3], we

D. Russo et al. (Eds.): IHSI 2021, AISC 1322, pp. 133–139, 2021.
https://doi.org/10.1007/978-3-030-68017-6_20

have implemented such a solution, with the dual goal of shortening the execution time of the algorithms through parallelism exploitation and minimizing user training time. To simplify the definition of algorithms, we have developed a workflow-based description language (WorDeL) [4], in which complex processes are built using pre-defined operators. The language is integrated within an editor application, offering features such as code navigation and autocomplete, graphical visualization of the workflow layout and detailed feedback on errors [5]. While usability tests on both the language and its supporting software indicate superior results in terms of model simplicity and ease of understanding when compared to programmatic descriptions, there were still issues with minor syntax mistakes in otherwise valid model descriptions [6]. In this context, we have implemented a new, voice-based interaction model for the existing algorithm description methodology. Based on the voice-recognition services offered by *Amazon Alexa*, it is meant as an add-on, paving the way for a multi-modal interface to our GIS solution. In this paper, we will present our work on the development of the voice-based interaction model and command set required to allow *Alexa* to control the workflow definition environment. We will mention some of the basic functioning principles and features, as well as the main challenges faced and potential future developments.

2 Related Work

GPS car navigation systems have been some of the first mass-produced devices to effectively employ voice feedback, in a bid to allow drivers to focus their attention on the road. Advancements in voice recognition technology have allowed for some basic input from the user such as setting waypoints and simple address searching [7]. Recent developments were looking into the possibility of combining speech recognition with natural language processing to accommodate more complex user queries. Such a system is described within [8], which employs the Stanford NLP Framework to extract relevant navigation data from the user's query and then feeds them to the Google Maps API to receive navigational instructions.

The processing and visualization of geospatial data is another area to potentially benefit from the addition of a voice-based input model. Supplementing more conventional input methods and lending itself to the development of multimodal interfaces, the speech interaction model has the potential of bringing advanced GIS functionality into the hands of the end user. The authors of [9] describe such a multimodal interface, making a case for the benefits of integrating multi-user access with voice and gesture recognition into GIS solutions. A similar approach is described within [10], where the authors propose voice interaction as an enhancement to GIS accessibility and usability. While pointing out the disadvantages of conventional input devices for users with disabilities, the authors also challenge the assumption that the interaction with GIS solutions should take place in office environments. In light of these arguments, they present the potential advantages of utilizing a multi-modal interface within a web-GIS application. The authors of [11] explore the potential of voice-based interaction in the context of mobile GIS applications. Their proposed system offers multiple interaction modes, allowing its users to choose between them according to specific tasks and scenarios. It also records the individual interactions, building user profiles and employing the gathered information to provide personalized spatial data to its users.

3 Voice Interaction Model

GIS solutions typically offer graphical interfaces that are complex and full of features, as they were originally intended for expert users [11]. This was the main problem that we tried to address when developing WorDeL. While the workflow model has proven intuitive and simple to understand [6], new users still need to practice with the syntax. To meet this need, we envisaged the addition of an interaction model based on natural language. This would offer new users the chance to practice and learn the syntax by example, from the code generated as a result of the voice commands. Furthermore, this interaction mode could also benefit users in the field relying on mobile devices to visualize and do light processing on image data, without the need to write code.

To implement the voice-based interaction model, we employed *Alexa*'s configurable command functionality, creating a custom command set for defining and editing workflows. With *Alexa* handling speech recognition though *Amazon*'s server side AI, our voice interaction module registers its responses within a *DynamoDB* database and associates them to the appropriate actions to be performed – Fig. 1. These consist of various code generation scripts, controlled by a simple state machine that is keeping track of the user's current position within the code.

To demonstrate the functionality provided by the voice-interaction module, we have used its command set to generate an algorithm description for the computation of a vegetation index (NDVI). This consists of three pixel-wise operations on the red and near-infrared spectral bands of a satellite image: NDVI = (NIR − ED)/(NIR + RED). We have previously demonstrated our system's ability to process such workflows in [12]. To better exemplify the use-case, we have provided a transcript (Fig. 2) of the commands that were employed. They consist of keywords and user-chosen values. To differentiate them, the latter are encased in quotation marks.

The first four commands are used to generate the definition of the workflow (algorithm description), together with its input and output ports. They are fairly simple and similar, requiring the user to choose unique identifiers for the description elements. When specifying a new input/output, the user must also choose from a predefined set of data types. If no type is specified, a "File" type is assumed by default. Figure 3 shows a screen capture of the generated description code, along with its visual representation. This is taken from the description tool that we have implemented for our platform. The code from the figure was generated and validated through voice commands. As such, the first four commands (shown in Fig. 2) resulted in the code lines 1 to 3 (from Fig. 3) being generated.

Sets of three commands (5 to 7 and 8 to 10) are used to define the first two pixel-wise arithmetic operations and set their inputs. For the second operation, we used a variation of "set input", which inserts the input arguments in order. These six commands, shown in the lines 5–10 from Fig. 2, result in the code lines 5 and 6 (Fig. 3) being generated.

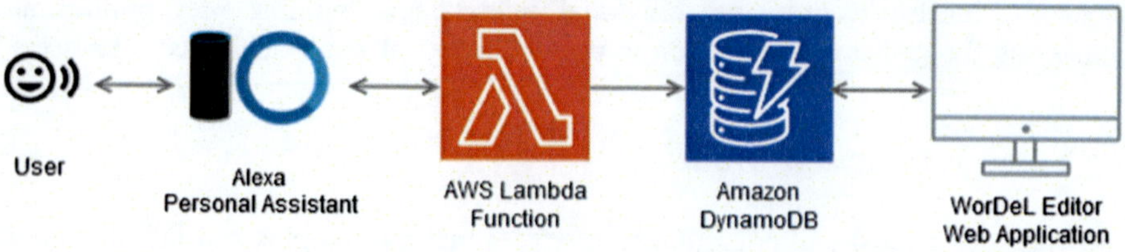

Fig. 1. Architecture of the voice-interaction module

```
1  create workflow "ndvi"
2  create input "red" "File"
3  create input "infrared" "File"
4  create output "result" "File"
5  insert operator "sum"
6  set input 1 "red"
7  set input 2 "infrared"
8  insert operator "difference"
9  set input "red"
10 set input "infrared"
11 insert operator "divide"
12 link input 1 to "operator0"
13 link input 2 to "operator1"
14 insert operator "color"
15 link input 1 to "operator3"
16 set output to "result"
17 create process for "ndvi"
18 create workflow parameter "file1
19 create workflow parameter "file2
```

Fig. 2. Transcript of the use-case voice commands

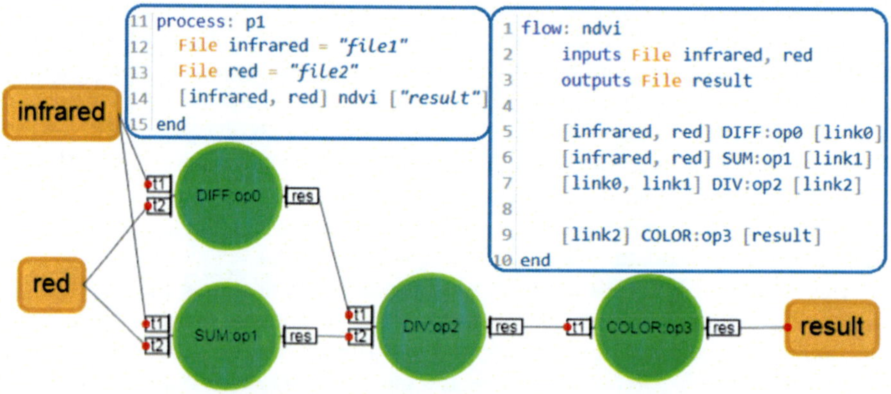

Fig. 3. NDVI workflow and process – collated from the Editor Application

For the third operation – "divide" – we employed the "link" command to create a data connection between its inputs and the outputs of the previous two operations. A connection is created by using the same argument for two different operations – once as an output and then as input. Thus, one "link" command is equivalent to two "set" commands. The "divide" operation was set up with commands 11 through 13, generating code line number 7 and setting the output arguments on the previous 2 code lines.

The last operation – "pseudo-coloring" – is linked to the third and, with a "set output" command, its output is routed to the workflow's output. Commands 14 to 16 generate code line 9.

Finally, to pair the algorithm description with the data it should process, we've created a "process" element for the defined workflow. For this, we have used commands 17 to 19, generating the code lines 11 through 15. These create an instance of the workflow and bind to it two "File" values, representing the two image bands.

When specifying the two input files, we had to use aliases – "file1" and "file2" – for their filepaths. Specifying these using voice commands was simply not practical, both because of their potential length and because of Alexa's limited capability of recognizing words that are not part of the English language. While using simple file names could work if the user would select an input directory beforehand and name the input files using English words, we are of the opinion that input data should be interactively selected by the user, while being guided by voice feedback from *Alexa* – a feature that is yet to be implemented within the system.

With the two input files and their path preset, the description code generated (shown in Fig. 3) is enough to compute a vegetation map of a given area when sent to the BigEarth platform. Figure 4 shows the result of the "NDVI" processing algorithm applied on a satellite image depicting a region of the Danube Delta.

Overall, a total of 19 commands were necessary to create the description and link it to a set of input data. Out of these, 9 were unique, with the remaining 10 being repeated commands, with different parameters. An average of three commands per algorithm operation was achieved by using the "link" command for the three inter-operator connections. This removed the need to individually set the output parameters for the first three operators.

The average time, computed after three attempts, required for the complete vocal interaction amounts to 3 min and 35 s. This includes the time required for feedback vocalization by *Alexa*. The equivalent code to that generated by the voice interaction module took an average of 2 min and 30 s to write manually, having as reference the graphical workflow representation from Fig. 3 – bottom. The results of both tests were obtained by having experienced users interact with the system.

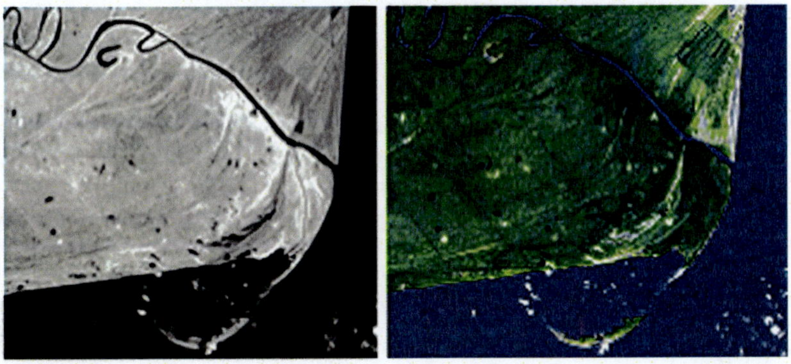

Fig. 4. Result of the "ndvi" workflow – left: NIR image band; right: colored NDVI image

4 Conclusions

The integration of a voice-based interaction model into the existing EO data processing platform has the potential to improve accessibility for new users, who lack sufficient knowledge on WorDeL's syntax. In this context, the voice interaction model can help them become familiar with the structure and syntax of the algorithm description, by learning from the generated code examples.

Voice interaction could also help in providing a multi-modal interface, with potential uses in the field, where writing code on mobile devices, such as smartphones or tablets, could prove impractical. In this type of scenario, a user could visualize the data on a mobile device, while having Alexa (available as a service on the device) create processing workflows and send them for execution within the Cloud-based solution.

Going forward, there are a number of incremental improvements that we could bring to the voice interaction module. The first one would be to address the input file specification problem through an interactive file browsing and selection mechanism. Further developments might also see the expansion of the "help" command to include basic tutorials on how to use the basic commands, with spoken examples. After adding proper input data selection mechanisms, an extensive usability study should be conducted, to assess the efficacy of the voice interaction module when employed by users who are unfamiliar with WorDeL syntax.

References

1. Sentinel 2 - Earth Online (European Space Agency). https://sentinel.esa.int/web/sentinel/missions/sentinel-2
2. Armbrust, M., Fox, A., Griffith, R., Joseph, A.D., Katz, R., Konwinski, A., Lee, G., Patterson, D.A., Rabkin, A., Stoica, I., Zaharia, M.: A view of cloud computing. Commun. ACM **53**(4), 50–58 (2010)
3. BIGEARTH Project. http://cgis.utcluj.ro/bigearth
4. Nandra, C., Gorgan, D.: Workflow description language for defining big earth data processing tasks. In: Proceedings of the 11th International Conference on Intelligent Computer Communication and Processing, Cluj-Napoca, pp. 461–468 (2015)
5. Nandra, C., Gorgan, D.: Workflow editor for definition and execution of parallelized earth data processing algorithms. In Proceedings of the National Conference of Human-Computer Interaction, Cluj-Napoca, pp. 76–83 (2018)
6. Nandra, C., Gorgan, D.: Usability evaluation of a domain-specific language for defining aggregated processing tasks. In: Proceedings of the 15th International Conference on Intelligent Computer Communication and Processing, Cluj-Napoca, pp. 87–94 (2019)
7. Dobesova, Z.: Voice control of maps. In: Proceedings of the 35th International Conference on Telecommunications and Signal Processing, Prague, pp. 460–464 (2012)
8. Withanage, P., Liyanage, T., Deeyakaduwe, N., Dias, E., Thelijjagoda, S.: Road navigation system using automatic speech recognition (ASR) and natural language processing (NLP). In: Proceedings of the IEEE Region 10 Humanitarian Technology Conference, Malambe, pp. 1–6 (2018)

9. Rauschert, I., Agarwal, P., Sharma, R., Fuhrmann, S., MacEachren, A., Wang, H., Cai, G.: Designing a human-centered, multimodal GIS interface to support emergency management. In: Proceedings of the 10th ACM International Symposium on Advances in Geographic Information Systems, pp. 119–124 (2002)
10. Jacobson, D., Sam, K.: Multimodal WebGIS: augmenting map navigation and spatial data visualization with voice control. In: Electronic Proceedings of AutoCarto (2006)
11. Doyle, J., Weakliam, J., Bertolotto, M., Wilson, D.C.: A multimodal interface for personalising spatial data in mobile GIS. In: Proceedings of the Eighth International Conference on Enterprise Information Systems: Databases and Information Systems Integration, Paphos, pp. 23–27 (2006)
12. Nandra, C., Bacu, V., Gorgan, D.: Parallel earth data tasks processing on a distributed cloud based computing architecture. In: Proceedings of the 21st International Conference on Control Systems and Computer Science, Bucharest, pp. 677–684 (2017)

Analysis of Employee Unsafe Behavior Based on Cellular Automata Model

Yan-mei Wang[1], Yunqi Dong[2], and Xue-bo Chen[3](✉)

[1] School of Control Science and Engineering, University of Science and Technology Liaoning, Anshan 114051, Liaoning, People's Republic of China
ml7865353148@163.com

[2] Graduate School, University of Science and Technology Liaoning, Anshan 114051, Liaoning, People's Republic of China
602173858@qq.com

[3] Graduate School, University of Science and Technology Liaoning, Anshan 114051, Liaoning, People's Republic of China
xuebochen@126.com

Abstract. At present, many enterprises at home and abroad have more and more attention to the safety awareness of employees, the fundamental reason is that the level of employees' safety awareness in the enterprise directly affects the production safety of the enterprise. However, some unsafe behaviors still occur from time to time. Furthermore, the hexagon cellular automata model is established to analyze the influence of high, medium or low safety awareness and number of employees on unsafe behaviors. Therefore, this paper proposes to adopt safety management system to strengthen effective internal communication.

Keywords: Cellular automate model · Safety consciousness · Effective communication

1 Introduction

The unsafe behaviors of employees in enterprises often bring great losses, and the unsafe behaviors of employees have a great relationship with their safety awareness. Therefore, improving employees' safety awareness is the key to reduce unsafe behaviors.

At the same time, the aggregation of employees can be regarded as a complex system, so the cellular automata model is used for analysis. Cellular automata model is a powerful tool for the study of complex systems. It can be applied in many fields, such as society, economy and science [1]. Apply the security management system to study how to increase the effective communication within the enterprise. At present, the research on security management is mainly about network security and enterprise production security.

D. Russo et al. (Eds.): IHSI 2021, AISC 1322, pp. 140–145, 2021.
https://doi.org/10.1007/978-3-030-68017-6_21

2 Model Constructing

Cellular automata is a kind of mesh dynamic model in which time, space and state are discrete and spatial interaction and temporal causality are local [2–5]. The CA model is composed of cellular individuals, and the specific model is as follows:

$$C = \{\mathrm{L, S, N, f}\} \tag{1}$$

where, L represents all employees, which can be expressed as:

$$L = \left\{\mathrm{C}_{(\mathrm{i,j})} | \mathrm{i, j} \in \mathrm{Z}, 0 \leq \mathrm{i} \leq \mathrm{L}, 0 \leq \mathrm{j} \leq \mathrm{L}\right\} \tag{2}$$

Where (i, j) represents the coordinates of unary cells in cellular space.In this paper, S is defined as the safety awareness of employees. $S\{L1, L2, L3\}$ can be set, where $L1$ stands for high safety awareness of employees, $L2$ stands for low safety awareness of employees, and $L3$ stands for medium safety awareness of employees. N represents the number of neighbors of each individual. In this paper, hexagon-type adjacency relation, namely hexagon cellular automata, is adopted to represent that the core cellular individual has 6 adjacent individuals, and the distance between them and each adjacent individual is equal. The following is the schematic diagram of the hexagonal cell model (Fig. 1):

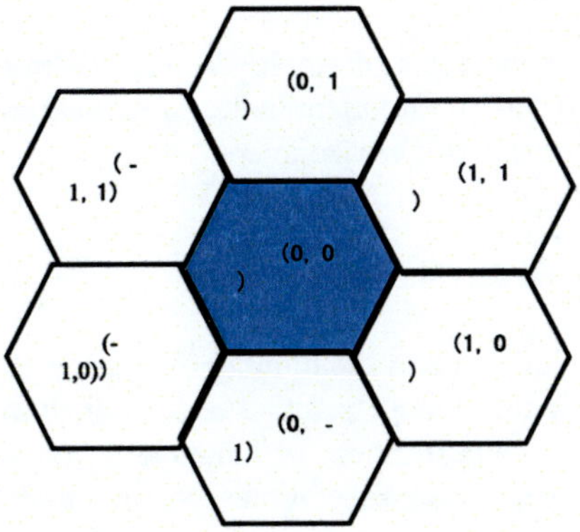

Fig. 1. The hexagonal cell model

Among them, the middle blue represents the core cellular individuals, and the rest are the adjacent individuals of the core cellular. For the core cell, its adjacent cells can be defined as:

$$N_{(i,j)} = \{C_{(i,j+1)}, C_{(i,j-1)}, C_{(i+1,j+1)}, C_{(i-1,j+1)}, C_{(i-1,j)}, C_{(i+1,j)}\} \tag{3}$$

f stands for rule for security state change. In the CA model, the security state of the core cell at time $t + 1$ is determined by the security state of the core cell itself and the adjacent cell individuals at time *t*. Therefore, it can be expressed as:

$$S_{(i,j)}^{t+1} = f(N_{(i,j)}^{t}, S_{(i,j)}^{t}) \tag{4}$$

Here, specific quantization values can also be given for the safe state *S* [6], namely:

$$W = ran\,\text{int}[0, 50] \tag{5}$$

Within this value range, an integer can be randomly selected. For example, 0–10 indicates low safety awareness and employees have low safety awareness. 11–30 indicates that the safety awareness of employees is between high and low, and the safety status is medium. 31–50 are high safety status, that is, employees have high safety awareness. The higher the value, the higher the safety status, the higher the safety awareness of employees.

The total safety awareness of employees is the total safety state *V*, which is the sum of the quantized safety state values of all employees in the enterprise working environment at time *t* [7, 8]. It can be expressed as:

$$V^{t} = \sum_{i=0}^{l} \sum_{j=0}^{l} W_{(i,j)}^{t} \tag{6}$$

By quantifying the safety status of employees and the improvement of the overall safety status, it can be used as an important indicator to measure the safety management and control in the enterprise work environment.

3 The Simulation Analysis

Through simulation, the influence of the number of employees with high, medium and low safety awareness on the overall safety status is analyzed. Among them, green represents the high safety consciousness of employees, red represents the low safety consciousness of employees, and blue represents the middle state of safety consciousness of employees.

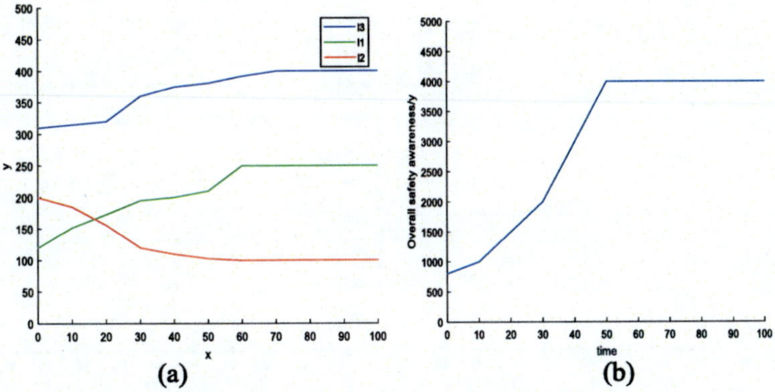

Fig. 2. The number of employees *(a)* and Overall safety awareness *(b)*

As shown in the figure above, Fig. 2 shows the trend of employees with high and low safety awareness and the number of employees. The number of employees with high and medium safety awareness gradually increases, and the number of employees with medium safety awareness is higher than that of employees with high safety awareness. The overall security situation increased slowly, but the total amount was low.

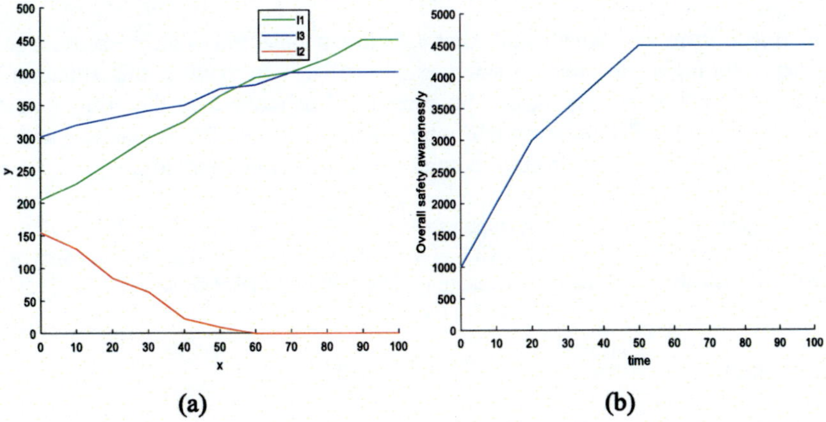

Fig. 3. The number of employees (a) and Overall safety awareness (b)

It can be seen from Fig. 3(a) that the number of people with high safety awareness and medium safety awareness increases rapidly, while the number of people with low safety awareness decreases gradually. Therefore, the overall safety state increases rapidly and the total amount is high in Fig. 3(b).

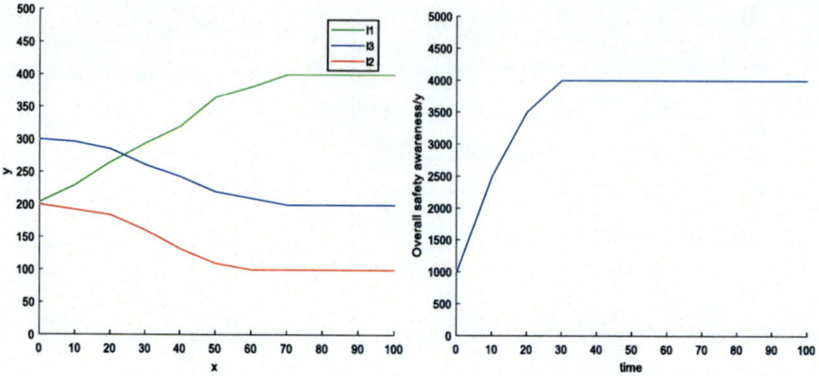

(a) (b)

Fig. 4. The number of employees (a) and Overall safety awareness (b)

Figure 4 shows that the number of employees with high safety awareness gradually increases, while the number of employees with medium safety awareness and the number of employees with low safety awareness slowly decreases. Correspondingly, the overall security posture increases rapidly, but in small quantities.

4 Conclusion

It can be seen from the above results that the more people with high or medium safety awareness, the higher the overall safety status will be. In view of this situation, the enterprise security management can be considered to strengthen the communication between employees. Take appropriate punishment measures for employees with low safety awareness to ensure effective communication between employees.

Acknowledgments. This work was supported in part by the Funds of the National Natural Science Foundation of China under Grants 71571091 and 71771112, and in part by the University of Science and Technology Liaoning Talent Project under Grant No. 601011507-03.

References

1. Jin, Z., Liu, Q.X., Mainul, H.: Chin. Phys. **16**, 1267 (2007)
2. Robben, L.: Cage reactions in sodalites – a phenomenological approach using cellular automata. Microporous Mesoporous Mater. **294**, 109874 (2019)
3. Chen, Y., Wang, C., Li, H., Yap, J.B.H., Tang, R., Xu, B.: Cellular automaton model for social forces interaction in building evacuation for sustainable society. Sustain. Cities Soc. **53**, 101913 (2020)
4. Shan, X.M., Liu, F., Ren, Y.: Acta Phys. Sin. **51**, 1175 (2002)
5. Xuebo, C., Qiubai, S.: Research on emergence suppression of unsafe behaviors based on safety consciousness. J. Univ. Sci. Technol. Liaoning **38**(06), 430–439 (2015)

6. Zhang, X.: Research on modeling and simulation of agent-based complex system. University of Science and Technology Liaoning (2018)
7. Sirakoulis, G.C., Karafyllidis, I., Thanailakis, A.: Ecol. Model. 133209 (2000)
8. Gagliardi, H., Alves, D.: Math. Popul. Stud. **17**, 79 (2010)

Technological Innovations for Executive Functions Stimulation

María Judith López[1] and Carlos Ramos-Galarza[1,2](✉)

[1] Facultad de Psicología, Pontificia Universidad Católica del Ecuador, Av. 12 de Octubre y Roca, Quito, Ecuador
maria.judith.lopez@outlook.com, caramos@puce.edu.ec
[2] Centro de Investigación en Mecatrónica y Sistemas Interactivos MIST/Carrera de Psicología, Universidad Tecnológica Indoamérica, Av. Machala y Sabanilla, Quito, Ecuador

Abstract. Executive functions are complex, high-level mental abilities associated with the frontal lobe brain function. These skills have been described as the inhibitory control ability, working memory, cognitive flexibility, planning, monitoring, among others. One of the most important skills of executive functioning is the ability to control behavior and cognition. Its adequate stimulation level is key to this process, highlighting the importance of generating motivational elements for the work and development of these mental skills, as proposed in the context of technological development. In this context, this article proposes a review of the main technological innovations that are currently available for stimulation.

Keywords: Executive functions · Technological innovations · Cognitive stimulation · Gamification · Frontal lobe

1 Introduction

Executive functions (EF) were originally described in 1868 by Harlow, who discovered cognitive changes, variations in behavior and personality when there is an injury to the frontal lobes. He supported his claim in his exceptional case: Phineas Gage [1]. On the contrary, it is mentioned that the merit of the conceptualization of EF is due to Alexander Luria, as he found the correlation between the frontal lobes and the inhibition of immediate responses, motor programming, the integration and orientation of behaviors aimed at a goal, personality and consciousness [2].

Furthermore, functional neuroimaging studies have verified the fundamental role of EF as integrating planning, behavioral and cognitive regulation and organization processes that enable voluntary and complex activities [3]. In general, it could be said that EF is a process in which ideas, actions and movements that allow conscious attentional control, the organization of elements to solve tasks, the regulation of emotions and behavior to obtain appropriate behaviors, flexibility in cognitive work and the ability to think in the future [4] are associated.

Due to the broad theorization around EF, it can be concluded that there are key elements that allow human beings to respond to their environment in an adaptive,

D. Russo et al. (Eds.): IHSI 2021, AISC 1322, pp. 146–151, 2021.
https://doi.org/10.1007/978-3-030-68017-6_22

effective, and strategic way. These elements result in the following classification: Emotional Control, Cognitive Flexibility, Initiative, Inhibition, Working Memory, Monitoring, Materials Organization and Planning.

EF take great importance in the development of intelligence and mental activity, as they act as an instrument that incorporates the pieces that regulate higher-order mental activities, which allow the subject to connect and relate information from the past and present with a view to the future [5]. Similarly, EF develops from childhood to early adulthood and continues to strengthen and decrease as older adulthood ages, which is why critical the stimuli the brain receives to its optimal development. For this reason, the maturation of executive functioning will influence and will be affected by people's experience, such as their childhood, environmental factors, among others, where cognitive stimulation becomes a fundamental tool for the maintenance and preservation of cognitive functions that have been affected by illness or over the years [6].

Within the context proposed, this paper proposes a review of the main technological innovations that are used for the stimulation of executive functions and that can be a contribution to the neuropsychological work with patients with brain disorders.

2 Technological Innovations for the Stimulation of Executive Functions

2.1 Lumosity

Lumosity [7] is an app that can be used on a desktop or a mobile, it is aimed at improving memory and concentration through personalized cognitive training for users. The Lumosity experience provides the customer with an initial assessment that allows to set a starting point and to adapt a specialized program. There are a cognitive training of speed, memory, attention, flexibility, problem-solving and math skills (Fig. 1).

Fig. 1. Luminosity screenshot

2.2 Elevate Brain Training

Elevate [8] is a brain training program designed to improve verbal fluency, attention, processing speed, memory, math skills, etc. Each user receives a personalized training program that adapts over time to maximize results. There are more than 15 games for cognitive skills such as concentration, memory, processing, and comprehension. It offers the opportunity to train the brain in both English and Spanish (Fig. 2).

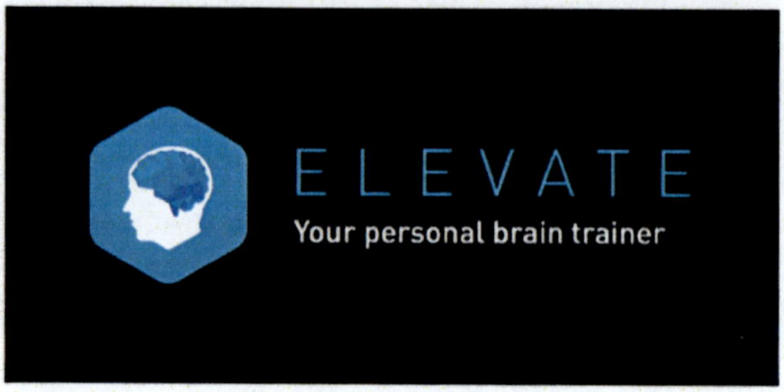

Fig. 2. Elevate screenshot

2.3 NeuroNation

NeuroNation [9] is an application aimed at brain training, it has received the Leonardo AOK Health Award for digital prevention sponsored by the Federal Ministry of Health in Germany. The app has 27 exercises in 250 levels for brain cognitive training and is aimed at each user through a customized plan according to the needs of each person (Fig. 3).

Fig. 3. NeuroNation screenshot

2.4 Kwido Mementia

Kwido [10] is a platform that is available for mobiles and digital platforms, which has been designed for the cognitive stimulation of seniors or older adults and allows it to be used both at home and in consultation with specialists. Kwido is a solution for elderly care composed of flexible modules that adapt to the needs of care centers or organizations. There are 3 complementary modules: Health, Home, and Cognitive Training. The first provides the sensor-based home senior monitoring solution, the second works as telemedicine for monitoring health variables and the third functions as a digital cognitive stimulation program that has been designed to combat and detect cognitive decline from home (Fig. 4).

Fig. 4. Kwido Mementia screenshot

2.5 Stimulus Home

Stimulus Home [11] is a mobile app that provides exercises in cognitive areas such as executive functions, calculus, language, long-term memory, reasoning, and vision-motor skills. Within the executive function training, the app has 10 activities. It is a preventive application for people who have some form of cognitive impairment (Fig. 5).

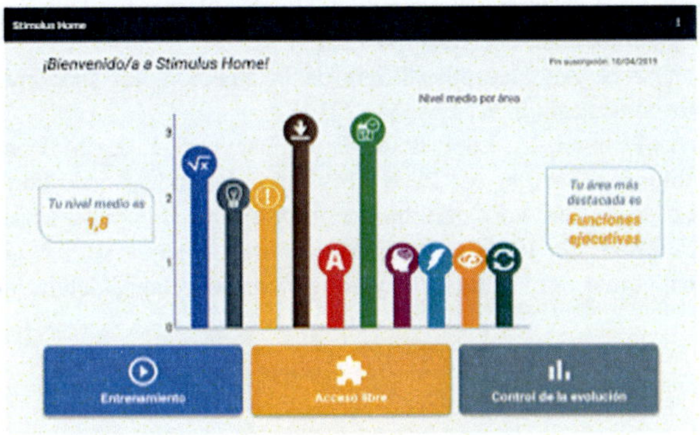

Fig. 5. Stimulus Home screenshot

3 Conclusions

The neuropsychological stimulation process as well as work in executive functions is a challenge in the mental health area. For this reason, there are currently several technological proposals that facilitate the clinician's work in their purpose of providing a solution to reduce the impact of cognitive impairment to a patient with acquired brain alterations [12].

The review began by describing the Lumosity app, which is aimed for memory training, short-term memory, processing speed and in general cognitive functions. Next, was described the Elevate Brain Training app, which is designed to improve verbal fluency, attention, processing speed, memory, math skills, etc. Followed was the description of the NeuroNation app, which is aimed at brain cognitive training. Afterwards, was described the Kiwido Mementia app, which has been designed for the elderly cognitive stimulation. Finally, was described the Stimulus Home app, which provides exercises in the cognitive areas such as executive functions, calculus, language, long-term memory, reasoning and vision-motor skills.

A future research line, we set ourselves two objectives: first, the development of technological applications to stimulate executive functions with ecological validity within the reality that the neuropsychological patient lives in the Latin American context. Second, verify the contribution of each of the apps described in this work through experimental studies with pre- and post-test measurements that allow measuring the contribution of each of these technological innovations.

References

1. Mujica, A.: El llamado síndrome del lóbulo frontal, actualmente llamado síndrome disejecutivo. Revista Argentina de Clínica Neuropsiquiátrica **17**(1), 42–47 (2011)
2. Ardila, A., Rosselli, M.: Neuropsicología Clínica. Editorial Manual Moderno, México (2007)
3. González, M.: Desarrollo neuropsicológico de las funciones ejecutivas en la edad preescolar. Editorial Manual Moderno, México (2015)
4. Hernández-Lalinde, J., Carrillo, S.: Funciones ejecutivas en estudiantes de undécimo grado de colegios oficiales de Cúcuta y Envigado, Colombia Resumen. Archivos Venezolanos de Farmacología y Terapéutica **38**, 124–131 (2019)
5. Portellano, J., García, A.: Neuropsicología de la atención, las funciones ejecutivas y la memoria. Editorial Síntesis S. A, España (2014)
6. Matijasevich, A., Pearson, R., Loret de Mola, C., Mayumi, J., La Maison, C., Munhoz, T., Barros, F., Santos, I., Barros, A.: Early child stimulation and attention-related executive functions at 11 years: 2004 Pelotas birth cohort study. European Child and Adolescent Psychiatry, vol. 29(9), pp. 1265–1276 (2019)
7. Lumosity. Disponible en: https://play.google.com/store/apps/details?id=com.lumoslabs.lumosity

8. Elevate. Disponible en: https://play.google.com/store/apps/details?id=com.wonder
9. NeuroNation. Disponible en: https://play.google.com/store/apps/details?id=air.nn.mobile.app.main
10. Kwido. Disponible en: https://play.google.com/store/apps/details?id=com.kwido.mementia
11. Stimulus Home. Disponible en: https://play.google.com/store/apps/details?id=mobi.stimulus.stimulushome
12. Ramos, C.: Secuelas neuropsicológicas en el daño cerebral adquirido. Universidad Tecnológica Indoamérica, Quito, Ecuador (2018)

Influence of Size and Depth Perception on Ray-Casting Interaction in Virtual Reality

Xiaolei Lv, Chengqi Xue(✉), and Weiye Xiao

School of Mechanical Engineering, Southeast University,
Nanjing 211189, China
770881060@qq.com, {101000270,230189776}@seu.edu.cn

Abstract. This research examined the factors that influence the performance of target-pointing task in virtual reality, and proposed an extended model to incorporate the target depth into the original Fitts' law, which improved the predictive ability of the model compared with previous methods. The research was carried out through two parts of experiments. The pre-experiment was a priori on the relationship between depth and size perception. The formal experiment mainly discussed the effect of size and depth perception of targets at different depths on the performance of ray-casting interaction when the target retinal image size were same. The experimental results supported the size-distance invariance hypothesis. It has been found that it takes less time to point to a target with a larger depth, that is, a larger size. Our findings confirmed the effect of size and depth perception on ray-casting interactions in virtual reality.

Keywords: Fitts' law · 3D pointing · Size perception · Depth perception

1 Introduction

Virtual reality (VR) is widely employed in military research, healthcare, game design, architectural design and other fields because of its various advantages such as telepresence and expressiveness. As virtual reality differs from reality, myriad researches point to perception characteristics in VR. Most of them focus on the depth perception and little empirical research in the relationship of size and depth perception. Machuca et al. [1] found that the misunderstanding of the visual size and the misjudgment of the depth may cause users to aim at the wrong target, which required additional corrective motions, and pointing performance was adversely affected. It remains unclear how perceived differences in target size and depth affect performance.

Visual perception is the result of a series of calculations and analyses in brain. Cognitive psychology has always paid attention to visual perception. The classic research on the theory of the moon illusion is about the discussion between the size-distance invariance hypothesis and the size-distance paradox. J.T. Enright [2] pointed out that the distance paradox needed to be understood by separating the perception of distance from the conscious and subconscious. Lloyd Kaufman et al. [3] found through experimental research that the perception size was proportional to the perception distance, which supported the size and distance invariance hypothesis. However, most researches on cognitive psychology employed qualitative estimation.

D. Russo et al. (Eds.): IHSI 2021, AISC 1322, pp. 152–158, 2021.
https://doi.org/10.1007/978-3-030-68017-6_23

The common research methods for size and depth perception in VR include perceptual matching method [4] and blind walking method [5], generally subjective. The primary objective of this thesis is to explore the impact of the perceived size or depth on performance. Fitts' law is a good choice. It is a significant model in the field of human-computer interaction research, widely used to quantitatively evaluate computer technology and human performance. An important point of Fitts' law is that the difficulty (ID) and throughput (TP) are only related to the ratio of the distance and size. However, when using ray-casting technology to point three-dimensional targets, the difference in depth sometimes results in the difference between the actual and perceived spacing. The effect on the Fitts' law is still unknown.

Therefore, this research was to explore the impact of size and depth perception on the performance of ray-casting pointing task in VR. Three hypotheses were formulated:

H1 The performance of pointing tasks using ray-casting in VR is influenced by the proportion of actual size and target spacing.

H2 The performance of pointing tasks using ray-casting in VR is influenced by the proportion of size and target spacing in retinal imaging.

H3 The performance of pointing tasks using ray-casting in VR is influenced not only by retinal imaging size and target spacing ratio, but also by depth and size perception.

2 Methodology

2.1 Participants

The experiment invited 21 subjects, including 12 males and 9 females, all graduate students, age ranging 23–28 (Mean = 24.76, SD = 1.06). All subjects had normal vision acuity or normal after correction with normal stereo vision.

2.2 Apparatus

In terms of hardware, the experiment was conducted on an Intel Core i7-7700K processor equipped with NVIDIA GeForce GTX 1080 Ti graphics card. The device used in the experiment was HTC Vive Pro virtual reality head-mounted display with a resolution of 1400 * 1600 per eye, and a 110° diagonal field of view.

In terms of software, the program was developed in unity 3D, using c# language. The environment was a virtual space with no depth cues except binocular parallax. The target sphere was made of unlit material. The initial color was green (0.6f, 0.8f, 0.4f, 1f), and the target color was red (1f, 0.4f, 0.2f, 1f). Color changes were shown legible.

2.3 Procedure

2.3.1 The Perceptual Matching Pre-experiment

The pre-experiment was a priori to the relationship between depth and size perception and served as the basis for choosing appropriate factor level. It was set up as a factorial experiment of 2 * 15 * 2. The center target depth had 2 levels, 2 and 30 m, and the surrounding target depth had 15 levels, respectively from 2 to 30 m. The retinal imaging size of all target spheres is the same, with a radius of 0.050 (Fig. 1).

Subjects were required to compare the size of the central and surrounding targets. If it was judged almost the same, pressed the space key; if the surrounding object was smaller, pressed the F key; if the surrounding object was larger, pressed the J key.

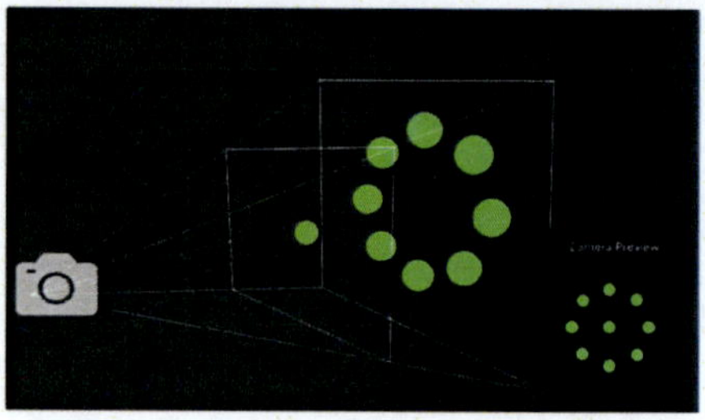

Fig. 1. Experimental schematic diagram

2.3.2 The Ray-Casting Pointing Formal Experiment

Step1: Filled out the basic statistics Form.
Step2: Practiced to familiarize with VR equipment and the requirements.
Step3: Pointing experiment.

Modified ISO9241–9 target-pointing design. Started from the central goal and pointed to the randomly appearing feedback, then returned to the central point. In this way, a subtask was completed, and a group of tasks included 8 subtasks. Participants were required to click on the center of the target as quickly as possible and could decide the rest time independently after one group was completed.

The experiment adopted a 3 * 3 * 3 * 3 * 2 factorial experiment design, 54 groups were divided into two stages. The target retinal size factor W has 3 levels, 0.05rad, 0.1rad and 0.2rad, the center target depth and the surrounding target depth factors each have 3 levels, 2, 16 and 30 m, and the test was repeated twice. The retina size of the distance between the center target and the surrounding targets A was 0.3rad.

Phase I: Coronal section pointing task (18 groups). The center and surrounding target depths of each group of subjects were maintained consistent.

Phase II: Targets at different depths pointing task (36 groups). When the target spacing and the retinal size remained unchanged, the center and surrounding targets were set at different depths.

Step4: Filled out the subjective evaluation form.

3 Results

3.1 The Pre-experiment

According to Fig. 2, the correct rate of the subjects' judgment on the size of different depth targets was shown as follows. From the overall trend, when the center depth was 2, the surrounding depth increased from 4 to 30, and when the center depth was 30, the surrounding depth decreased from 28 to 2, that was, when the difference of depths gradually got larger, the correct judgment rate increased.

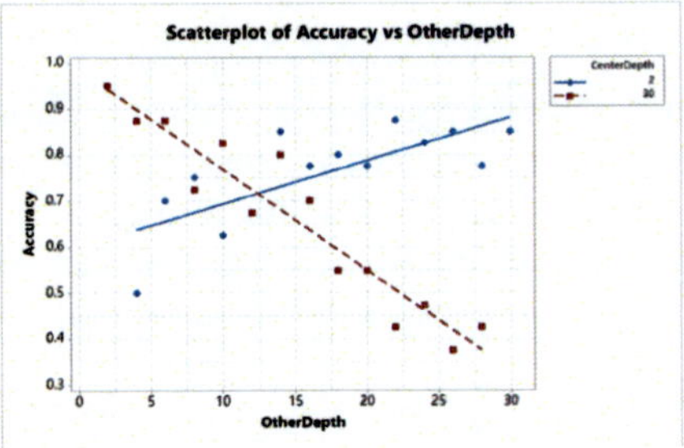

Fig. 2. Correct rate of size judgment

3.2 The Formal Experiment

Consulted the Pauta criterion, and the data larger than $\mu + 3\sigma$ or smaller than $\mu - 3\sigma$ were regarded as outliers and eliminated.

3.3 Phase I: Coronal Section Pointing Task

A one-way analysis of variance was performed on the depth, the adjusted P-value was 0.933 > 0.05, which meant target depth did not show a significant interaction. According to formula (1), substituted the length and radian values to obtain ID(len) and ID(rad). The fitting results of regression equations were both excellent as shown in (2) (3), indicating H1 and H2 true in the coronal section.

$$\mathrm{ID} = \log 2(\mathrm{A}/\mathrm{W} + 1)\,. \tag{1}$$

$$\mathrm{MT} = 0.4244 + 0.2745 * \mathrm{ID}\ (\mathrm{len}),\ \mathrm{R{-}Sq} = 93.77\%\,. \tag{2}$$

$$\mathrm{MT} = 0.4298 + 0.2763 * \mathrm{ID}(\mathrm{rad}),\ \mathrm{R{-}Sq} = 93.79\%. \tag{3}$$

The data were grouped according to the target depth regression as follows (Fig. 3). It was presented that when the target depth was 2 m, the movement time is the longest, followed by the target at 30 m depth and 16 m depth.

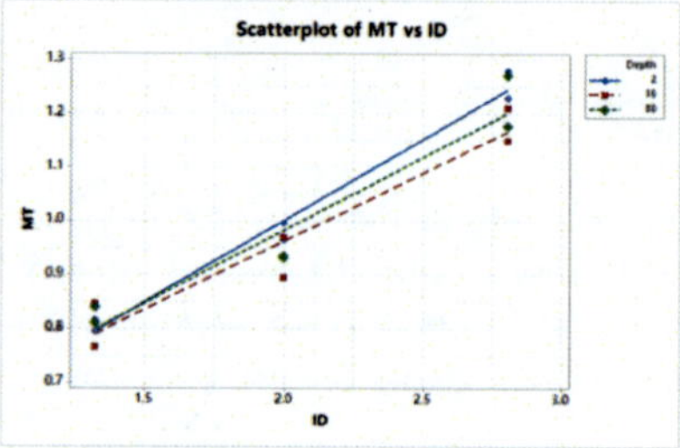

Fig. 3. Scatterplot with regression and groups

3.3.1 Phase II: Targets at Different Depths Pointing Task

There was a big difference when expressed in length and radians respectively. The ID (len) calculated by the length value was substituted regression into formula (4), indicating that the model need to be corrected, which proved that H1 was not valid. The ID (rad) calculated by the radian value was substituted into (5).

$$\mathrm{MT} = 0.8347 + 0.0567 * \mathrm{ID(len)},\ \mathrm{R{-}Sq} = 33.05\% \tag{4}$$

$$\mathrm{MT} = 0.4669 + 0.2802 * \mathrm{ID(rad)},\ \mathrm{R{-}Sq} = 85.39\% \tag{5}$$

Machuca and Stuerzlinger [1] incorporated the target depth change (CTD) as a predictor of movement time, and proposed an extended Fitts' law model:

$$\mathrm{MT} = a + b(\mathrm{ID}) + c(\mathrm{CTD}) \tag{6}$$

Compared with the traditional Fitts' law, this model significantly predicted the movement time. It showed that including the change of target depth as a predictor can predict the time more accurately. Therefore, according to the formula proposed by Machuca et al. to perform linear regression on MT, got (7), and the fitting effect was not ideal. The p value of CTD is $0.390 > 0.05$, indicating that the target depth difference CTD had no significant effect, which was different from the conclusion drawn by Machuca et al.

$$\mathrm{MT} = 0.4669 + 0.2802 * \mathrm{ID} - 0.000525 * \mathrm{CTD},\ \mathrm{R{-}Sq(adj)} = 85.03\% \tag{7}$$

Different depth target pointing mainly changed the depth difference and target depth. From the main effect diagram, more obviously it had a strong correlation with the depth of the central and surrounding targets. Therefore, based on the research of Machuca et al., another extended Fitts' law model was proposed:

$$\mathrm{MT} = a + b(\mathrm{ID}) + c(\mathrm{CD}) + d(\mathrm{OD}) \tag{8}$$

CD refers to the depth of the central target, and *OD* refers to the depth of the surrounding targets. Got the regression Eq. (9). The p value of *ID(rad)*, *CD* and *OD* were all $0.000 < 0.05$, Which meant that the three factors all had significant effects on

MT, and R-Sq(adj) can reach 93.83%, which was much higher than the other several fitting methods. Thus falsified H2, the performance of the pointing task was not only affected by the ratio of target retinal size and spacing, but also by the depth of the targets.

$$MT = 0.6370 + 0.2821 * ID(rad) - 0.004209 * CD - 0.005259 * OD, \; R - Sq(adj) = 93.83\% \tag{9}$$

The interaction diagram (Fig. 4) showed that when the center depth was constant and the surrounding target depth increased, the movement time tended to decrease, and when the surrounding target depth was the same, the center target had a larger depth and movement time was shorter.

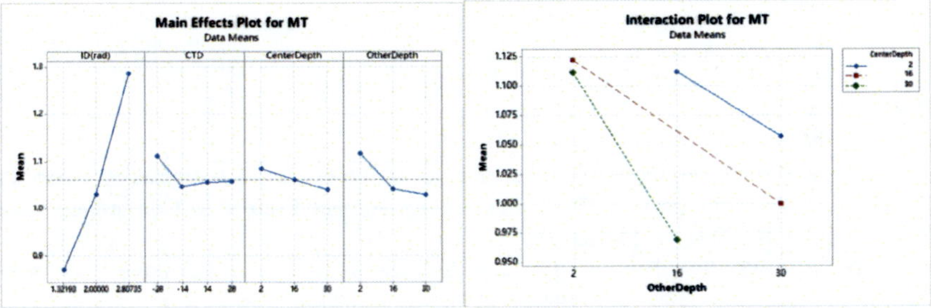

Fig. 4. Main effects plot and interaction plot for MT

4 Discussion

The focus of this article was to study the pointing performance between targets at different depths in Phase II. Under the current experimental factor level, as the depth of the central or surrounding targets increases, the motion time shrank. It was speculated that the perception of size affected interactive performance. This result was consistent with the size-distance invariance hypothesis: the perceived size was directly proportional to the perceived depth, and the pointing time was inversely proportional to the perceived depth and size, confirming H3. In Phase I, movement time should be the shortest when the target depth was 2 m and the largest when that was 30. However, the group regression analysis showed that the MT is the shortest when the target depth is 16 m. The possible reason was that the light beam indicated by the controller was blurred due to the distance, and the visual feedback was weakened.

In Phase II the ID (len) fitting effect of different depth pointing was poor. It is necessary to understand the deep meaning of Fitts' law about *A*. In some cases, *A* can refer to the target distance, but more essentially, *A* is the amplitude of motion. Therefore, in a more general case, the radian value should be used for calculation.

5 Conclusion

This research mainly discussed the impact of the size and depth perception of subjects at different depths when the target retinal size were the same. An extended model was proposed to incorporate the depth into the original Fitts' law, which improved the predictive ability of the model. It was confirmed that the performance of pointing tasks was not only affected by the ratio of retinal size and spacing, but also by the distance and size perception judgment. The research on size and depth perception also had certain guiding significance for how to balance size and depth in menu design under VR. However, the division of the two phases in the formal experiment may make the results lacking convincing and scientificity. The following research can be further optimized.

Acknowledgments. This research was supported by National Natural Science Foundation of China (No. 71901061).

References

1. Machuca, M.D.B., Stuerzlinger, W.: The effect of stereo display deficiencies on virtual hand pointing. In: Proceedings of CHI Conference on Human Factors in Computing Systems Proceedings (CHI2019), p. 14 (2019)
2. Mon-Williams, M., Tresilian, J.: The size-distance paradox is a cognitive phenomenon. Exp. Brain Res. **126**, 578–582 (1999)
3. Kaufman, L., Vassiliades, V., Noble, R., Alexander, R., Kaufman, J., Edlund, S.: Perceptual distance and the moon illusion. Spat. Vis. **20**(1–2), 155–175 (2007)
4. Ping, J, Weng, D, Liu, Y, Wang, Y.: Depth perception in shuffleboard: Depth cues effect on depth perception in virtual and augmented reality system. J. Soc. Inf. Display **28,** 164–176 (2020)
5. Kelly, J.W., Cherep, L.A., Siegel, Z.D.: Perceived Space in the HTC Vive. ACM Trans. Appl. Percept. 15, 1, Article 2 (November 2017), 16 pages (2017)

Technological Resources to Stimulate Multiple Intelligences: Verbal-Linguistic and Logical-Mathematical

Mónica Bolaños-Pasquel[1], Micaela Silva-Barragán[2], Pamela Acosta-Rodas[2], Omar Cóndor-Herrera[1], Jorge Cruz-Cárdenas[1], and Carlos Ramos-Galarza[1,2](✉)

[1] Centro de Investigación en Mecatrónica y Sistemas Interactivos MIST, Carrera de Psicología y Centro de Investigación ESTec, Universidad Tecnológica Indoamérica, Av. Machala y Sabanilla, Quito, Ecuador
{monicabolanos, omarcondor, jorgecruz}@uti.edu.ec, caramos@puce.edu.ec

[2] Facultad de Psicología, Pontificia Universidad Católica del Ecuador, Av. 12 de Octubre y Roca, Quito, Ecuador
{pmsilva, mpacosta, caramos}@puce.edu.ec

Abstract. Multiple intelligences encompass the different types of intelligences that a human being possesses; which are the verbal-linguistic and logical-mathematical intelligences. These two types of intelligence represent, in a significant part, the axis of the human cognitive procedure and are activated when there is a problem that needs to be solved. This paper aims to provide a summary of technological resources which allow stimulation in various contexts.

Keywords: Multiple intelligences · Gamification · Serious games · Cognitive stimulation · Educative resources

1 Introduction

The idea of multiple intelligences was proposed by Howard Gardner in the 80's decade, with the objective of stepping away of the prevailing perspective in the educative context and understanding that every student has unique abilities that will set a trace on the students' interests and performance. This conceptualization derives from leaving aside the unique concept of intelligence to understand it as an abilities integration or independent capacities between themselves that present potentially in every uman being. Gardner identified eight intelligences, verbal-linguistic, logical-mathematical, spatial-visual, bodily-kinesthetic, musical, interpersonal, intrapersonal, and naturalist [1, 2].

The impact that generates this new perspective is essential since it has demonstrated that when different intelligences are stimulated, being those dominant or potential, it is possible to impulse learning, leadership abilities and cooperation, interest, and engagement. Also, it allows diminishing behavioral problems [3]. All of these are framed in an adaptation and respect logic towards students' learning and knowledge

D. Russo et al. (Eds.): IHSI 2021, AISC 1322, pp. 159–164, 2021.
https://doi.org/10.1007/978-3-030-68017-6_24

acquisition processes [1]. It is important to highlight that intelligence stimulation is not limited to privileged scholar systems, but can be realized even from home. The primary instruction is to make activities that encompass multiple stimuli and bring those to daily-life and institutional contexts [4]. In this sense, the present article shows technological innovations that stimulate two of the eight Gardner's multiple intelligences: verbal-linguistic and logical-mathematical.

2 Technological Innovations to Stimulate Verbal-Linguistic Intelligence

Verbal-linguistic intelligence implies the significant employment of words, speaking, and writing, including handling syntax phonology, semantic, and language pragmatic dimensions. Technological resources for its stimulation are described as following.

2.1 Neobook

It is a software that designs multimedia interactive resources easily, there are not programming previous knowledge requirements. This tool allows the creation of applications easily accessible, through the combination of text, graphs, sounds and animations. This application could motivate the creation of spoken presentations with interactive text, being the ideal space for the stimulation of verbal-linguistic intelligence [5] (Fig 1).

Fig. 1. Neobook program interface

2.2 StoryJumper

It is a tool that allows the development of writing skills through the origination of creative stories. It is a site for tales' publishing that could be on-line or in a hardcover version. In this platform, it is possible to insert various multimedia resources where verbal-linguistic intelligence is highly stimulated with this technological resource (Fig. 2) [6].

Fig. 2. Storyjumper application screenshoot

2.3 iA Writer

It is an application for creative writing that diminishes distractions when focusing on writing a text (Fig. 3). It represents a paper sheet with robust and simple characteristics. Thus, the author lets his/her imagination fly and develop their writing in a technological and ordered environment with every feature where linguistic characteristics flow [7].

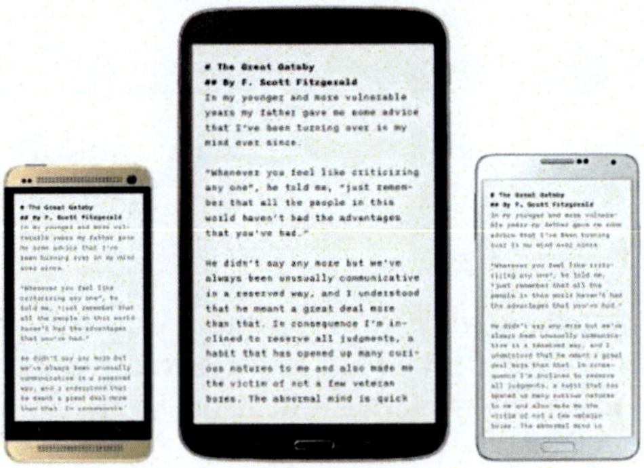

Fig. 3. iA Writer Application

3 Technological Innovations to Stimulate Logical-Mathematical Intelligence

Logical-mathematical intelligence refers to the capacity for using numbers and mathematical reasoning effectively. This intelligence is sensitized with patterns and logic relations, affirmations and cause-effect proposals, inductive and deductive reasoning, problem-resolutions, and identification of patterns and sequences.

3.1 Descartes Network

It is a technological tool that helps develop educative interactive resources for mathematical study (Fig. 4). These are amicable to those who use it and allow the interactions between the user and numerical calculation activities [8].

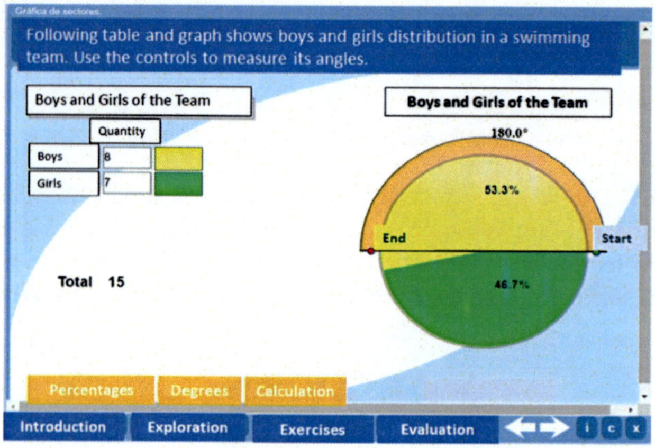

Fig. 4. Descartes Network Platform screenshoot

3.2 Khan Academy

This platform counts with a variety of practical exercises to stimulate logical-mathematical intelligence (Fig. 5), through active learning and generating motivation when participants are executing numerical calculations [9].

3.3 Retomates

This platform counts with many different interactive games for learning mathematical skills (Fig. 6). One of the elements that generate students' interest and motivation for learning math is the objective that creates the proposed challenges in this platform, making the ideal environment for logical-mathematical intelligence [10].

Fig. 5. Plataforma Khan Academy

Fig. 6. Retomates Screenshoot

4 Conclusions

Multiple intelligences is a wide cognitive conceptualization that encompasses human intelligences, such as verbal-linguistic, logical-mathematical, spatial-visual, bodily-kinesthetic, musical, interpersonal, intrapersonal, and naturalist [11].

The development of these intelligence types depends on the stimulation that the environment generates in the human being. In this article, it is analyzed some technological devices that could contribute in favor of this objective. Benefits obtained from the utilization of technologies like the ones already described in this article imply children's and adolescents' higher motivation since technological resources are based on the gamification, generating a higher attractiveness for its usage, achieving higher intelligence stimulation.This work has been focused on two types of intelligence: verbal-linguistic and logical-mathematical; with this purpose, it has been described mobile applications and internet platforms that counts with many different activities to improve the level of these two cognitive abilities.

Future work approaches to identify technological innovations for other intelligence types and its analysis in longitudinal studies with measurements pre and post-test determining its benefits in many contexts where a person develops daily.

References

1. Garmen, P., Rodríguez, C., García-Redondo, P., San-Pedro-Veledo, J.: Multiple intelligences and video games: assessment and intervention with TOI software. [Inteligencias múltiples y videojuegos: Evaluación e intervención con software TOI]. Revista Comunicar **58**, 95–104 (2019)
2. Castaño, F., Tocoche, Y.: Multiple Intelligences and emotional competences in university students. [Inteligencias múltiples y competencias emocionales en estudiantes universitarios]. Campo Abierto **37**(1), 33–55 (2018)
3. Suarez, J., Maiz, F., Meza, M.: Inteligencias múltiples: la teoría en la práctica. Una pedagógica innovación para potenciar el proceso enseñanza-aprendizaje. Revista Investigación y Postgrado **25**(1), 81–94 (2010)
4. Antunes, C.: Las inteligencias múltiples. Cómo estimularlas y desarrollarlas. Madrid: Narcea, S.A. de Ediciones (2002)
5. SinLios Soluciones Digitales S.L. NeoBook Rapid Application Builder. http://www.neosoftware.com/
6. Story Jumper. http://www.neosoftware.com/
7. iA Writter. https://ia.net/es/writer
8. Red educativa digital Descartes. Red Descartes. https://proyectodescartes.org/descartescms/
9. Khan Academy: matemáticas. https://es.khanacademy.org/math
10. Retomates. http://www.retomates.es/
11. Gardner, H.: Inteligencias Múltiples. Paidos, La Teoría en la Práctica, Barcelona (1995)

NeuroDesignScience: An fNIRS-Based System Designed to Help Pilots Sustain Attention During Transmeridian Flights

Amanda Liu[1], Binbin Li[2], Xiaohan Wang[2], Songyang Zhang[3], Yancong Zhu[2], and Wei Liu[2](✉)

[1] Duke University, Durham, NC 27708-0187, USA
amanda.liu@duke.edu

[2] Beijing Normal University, Xinjiekouwai Street 19, Haidian district, Beijing 100875, China
{binbin.li,xiaohan.wang}@mail.bnu.edu.cn, {yancong.zhu,wei.liu}@bnu.edu.cn

[3] New York University, New York, NY 10003, USA
songyang.zhang@nyu.edu

Abstract. The transmeridian flight required pilots to perform tasks that are high in mental workload for long periods. Consequently, they are likely to experience fatigue and have difficulty in concentrating, which could lead to flight safety risks since pilots' emergency response ability would be weakened. This study proposes a system to help pilots maintain attention on tasks over time to reduce air accidents. The neuroimaging method functional near-infrared spectroscopy (fNIRS) would be used to assess the attention level and mental workload in this system. Based on the results, appropriate feedback, such as a reminder to re-engage the pilot if disengagement detected, would be provided. For testing the effectiveness of this system on sustained attention, an experiment was designed to examine whether using this system could reduce pilots' missed instructions when communicating with ground control.

Keywords: NeuroDesignScience · Human-computer interaction · Passive brain-computer interface · fNIRS

1 Introduction

Pilots' inattention can significantly endanger commercial flight safety. Among the 431 plane crashes investigated by Baker et al. [1], 188 crashes were caused by inattention. Pilots errors caused by this disengagement problem include inputting wrong parameters into the on-board computer or mishearing Air Traffic Control (ATC) instructions. Besides, evidence shows that inattention relates to reduced reaction speed and impaired decision-making ability. These may cause pilots to fail to respond in time in an emergency [2, 13]. Because of the threat to aviation safety, pilots' inattention urgently needs a solution.

Physical and mental fatigue are the main causes of inattention, and they are inevitable for pilots, especially for whom fly transmeridian routes. This kind of flight

D. Russo et al. (Eds.): IHSI 2021, AISC 1322, pp. 165–170, 2021.
https://doi.org/10.1007/978-3-030-68017-6_25

tasks disrupts their circadian timing, resulting in poor sleep quality and physical fatigue. Meanwhile, merely operating a flight demands a high Mental WorkLoad (MWL) since it requires tremendous mental effort. For long flights, pilots would face sustained high MWL, which can induce severe mental fatigue [2]. Also, people who continue to work under high MWL will have worse performance and higher error rates, so this mental state also threatens flight safety [3].

In order to solve the inattention problem, many regulatory agencies have issued relevant regulations. An example is the pilot fatigue rule issued by the US Federal Aviation Administration. It requires pilots to rest for ten hours before flying and bans them from working more than nine consecutive hours [4]. However, because flying demands high MWL, a fifty-minute flight is enough to induce severe mental fatigue [2]. Thus, the rule cannot protect pilots from being mentally exhausted, thereby the inattention problem still exists.

As far as we know, there has been no previous research to help pilots maintain concentration or adjust MWL while flying. However, in non-flight scenarios, a solution to inattention has been found. Researchers Yang et al. [5] developed a feedback system, that would immediately remind the subjects when their attention begins to drift. This system is NeuroDesignScience-based since it combines neuroimaging methods and Human-Computer Interaction (HCI) design. It uses Electroencephalography (EEG) to timely monitor attention state changes, which humans can hardly detect in time. As soon as an attention drift episode is detected, the system would re-engage the subjects by playing voice feedback.

Due to its limitation, EEG could not be used on pilots, making this system unsuitable for real flight conditions. (Detailed explanation is in Sect. 2, Assessing Mental States.) Fortunately, an alternative, fNIRS, was found fitting for this scenario. It has been used in the flight simulator and real aircraft to monitor mental states such as attention and MWL [6].

Therefore, to tackle pilots' inattention problem in real flight situations, we will develop a NeuroDesignScience-based feedback system that incorporates fNIRS and HCI design. This would be the first study that applies the NeuroDesignScience approach to aviation safety. Our system would re-engage pilots as soon as attention drift is detected; meanwhile, it would also help them control their MWL. Avoiding long hours of high MWL could stave off cognitive fatigue, making pilots less prone to mind-wandering. This function would also alleviate other adverse effects of this MWL state, such as impairing decision-making ability [3].

2 System Design

To help pilots maintain focus while flying, we developed a system to (1) re-engage pilots immediately when their attention begins to drift; (2) prevent pilots from working on big MWL tasks for too long (in order to stave off cognitive fatigue which induces disengagement) [7].

To realize these functions, the system should be capable of tracking the changes in pilots' mental states in real-time. In this way, if an undesired state is detected, the system would immediately interact with the subject to revert it.

Assessing the Mental States. The neuroimaging method fNIRS would be used since it can monitor mental states in real-time while allowing pilots to function normally. Methods other than neuroimaging, such as the survey method, cannot provide real-time mental states measurement. Among neuroimaging techniques, both EEG and fNIRS are portable and lightweight. However, only fNIRS can be used in real flight conditions. EEG is extremely sensitive to motion artifacts. While using it, body movements such as turning head must be restricted, which would hinder pilots from checking different screens [6, 12]. Single-channel EEG is less susceptible to the movements, but it cannot accurately measure MWL [5]. Therefore, fNIRS, which is robust to movements, is chosen to study pilots in real flight conditions.

Using Feedback to Revert Undesired Mental States. If disengagement is detected, our system will send a reminder to shift the subject's attention back to work.

To prevent pilots from working under high MWL for too long, when it detects a 10-min period of high MWL, the system will call other aircrew members to 'lend a hand' to reduce the pilot's workload.

System Procedures. Since there are two mental states of interest, the system has two modes. The attention mode investigates weather a pilot (the subject) is concentrating. The MWL mode monitors the subject's current MWL level.

The system consists of three parts: (1) an fNIRS device for measuring neural behavior; (2) a computer for processing fNIRS data, deriving attention state or MWL from it, and giving feedback; and (3) a small display screen for showing the feedbacks (Fig. 1).

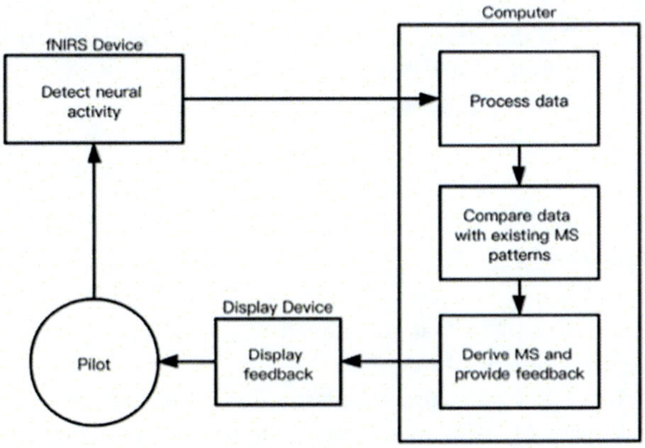

Fig. 1. Diagram of the NeuroDesignScience-based system.

For data acquisition, fNIR100 (Biopac Systems Inc.) is used. The device is a headband so that it can be attached to a person's forehead, and it includes four light sources, ten detectors, and sixteen information channels. The light sources are adjusted to emit two wavelengths (730 nm and 850 nm) [9].

The computer in the system assists in fNIRS data acquisition using COBI control unit software (Biopac Systems Inc.) [8]. Then, to remove physiological and measurement noises, the computer first preprocesses the data using the bandpass filter and PCA filter. Then, software NIRS-SPM is used to derive data regarding brain hemodynamic responses [10]. Details of the calculation procedures are in [6, 11].

To accurately derive the pilot's attention state or MWL from the processed data, we implement a pretest to collect the data patterns that correspond to the pilot's mental state changes. This pretest is conducted before the regular use of our system. During the test, the subject is asked to solve different levels of SART tasks or n-back tasks, which induce changes in attention state or MWL, respectively. Details of the two tasks are in [11, 12]. These mental state changes are measured using the fNIRS device in our system, and mathematical patterns are extracted from the output data.

When the system is formally used in real flight conditions, the system continuously monitors the change in fNIRS data. If the data dynamics in a 30-s epoch match a mental state change pattern discovered during the pretest, then the subject's current mental state can be identified.

According to the result, the computer chooses a piece of feedback to display. For the rules, see Table 1. The pilot can see the feedback on our system's display device.

Table 1. Feedback displayed on screen based on the mental state detected.

Mental state	Feedback displayed on screen
Attention Mode: Disengaged	"Please focus" would be displayed
Attention Mode: Focused	No message would be displayed
MWL Mode: High MWL	If this mental state lasts for 10 min, "Other aircrew members please lend a hand" would be displayed
MWL Mode: Low MWL	No message would be displayed

3 Experiment Design

An experiment is designed to test the effectiveness of our system on tackling the fatigue-induced disengagement.

In order to achieve this goal, the recruited pilots must first be in a state of fatigue. Thus, we will ask each of them to operate a flight simulator three hours before the formal test. Meanwhile, the content of the formal test will be repeating ATC instructions while flying. It demands high MWL since it requires multi-tasking (auditory processing instructions from ATC, memorizing flight parameters, operating flight, etc.) Thus, it can quickly induce cognitive fatigue, causing disengagement [7].

We predict that, while communicating with ATC, our system's use can help the fatigued pilots concentrate and perform better, thereby enhancing the safety of commercial flight.

For this study, the independent variable is whether our system is used; the dependent variables include subjects' attention state, MWL, and the number of missed instruction (which indicates the quality of each pilot's performance.) The two mental states will be measured using the fNIRS device in our system and experimenters will count missed instructions during the test. The working time and the workload will be set identical for each participant for eliminating confounding variables.

Participants and Material. The study will involve fifteen pilots, and all should have three to five years of flying experience. Other requirements include normal vision, no hearing problems, and no history of neurological or psychiatric disorders.

Experiment equipment includes a computer on which the subjects will take the pretest, a flight simulator, and our system. The components of the system are a fNIR100 device (Biopac Systems Inc.), a computer running COBI, fNIRSoft and NIRS-SPM and a screen for displaying feedbacks.

Experimental Procedures. After giving informed consent, all participants will be randomly assigned to three groups. The first two groups will use the system's attention and MWL, respectively; the third will be the control group, which will not use the system.

Before entering the experiment, all pilots (including the control group) will take a computerized pretest (described in Sect. 2.1). Subjects in the attention group will do a series of SART tasks; the MWL group will perform n-back tasks; the control group can choose to do either SART or n-back tasks.

Then, each pilot will enter a flight simulator, and the experiment will formally begin. Members in the attention group and MWL group will start using our system; people in the control group will only wear fNIR100 for acquiring mental state data.

Each subject will first operate the flight simulator for three hours, solving identical flying tasks. This preparation stage is for inducing fatigue. In this stage, a co-pilot will accompany each subject in the MWL group and provide support according to the system's feedback. The co-pilot will leave as soon as this stage is completed.

The formal test stage will begin immediately after the preparation stage. Each pilot will hear 500 pre-recorded ATC instructions (selected from real-world ATC communications.) Among all the instructions, 150 instructions are valid, and the rest are nontargets. If and only if a valid instruction is heard, the pilot must repeat it. An experimenter will record the number of failures of each pilot.

After the experiment, the acquired fNIRS data will be pre-processed following established procedures (described in Sect. 2.1). Then, ANOVA, t-test, and multiple linear regression will be applied to examine the data acquired.

4 Conclusion

In transmeridian flights, commercial pilots inevitably face the problem of inattention. This kind of task demands high MWL for long periods, resulting in fatigue, a significant cause of inattention. This concentration problem can cause pilots to make errors or fail to respond in time in an emergency, thereby endangering flight safety. This research developed a developed a system based on NeuroDesignScience for helping pilots improving concentration. It can immediately re-engage a pilot if his or her attention diverts. Moreover, it can also monitor MWL. If the pilot continues to work under high MWL, it will call for support to help the pilot reduce workload and avoid of fatigue.

The advantage of this system is its high usability. The primary design principle of this HCI system is to minimize distractions when interacting with pilots. It adopts fNIRS to investigate pilots' mental states, which allowing pilots to move and operate normally while collecting data. There are also limitations. Although fNIRS was used as a real-time measuring device by Afergan et al. [12], it has a delay of two to six seconds [9]. Also, the current solution to reduce high MWL is to ask other aircrew members to help. However, this would not work when other people are not available.

Future research can improve our system by solving this problem. Instead of asking humans to 'lend a hand,' the system may handle some aviation tasks on its own with the help of technologies such as artificial intelligence.

References

1. Baker, S.P., Lamb, M.W., Grabowski, J.G., Rebok, G., Li, G.: Characteristics of general aviation crashes involving mature male and female pilots. Aviat. Space Environ. Med. **72**, 447–452 (2001)
2. Dehais, F., Dupres, A., Di Flumeri, G., Verdiere, K., Borghini, G., Babiloni, F., Roy, R.: Monitoring pilot's cognitive fatigue with engagement features in simulated and actual flight conditions using a hybrid fNIRS-EEG Passive BCI. In: International Conference on Systems, Man, and Cybernetics (SMC), pp. 544–549 (2018)
3. Matthews, G., Desmond, P.A.: Task-induced fatigue states and simulated driving performance. Q J Exp Psychol A. **55**, 659–686 (2002)
4. Fact Sheet – Pilot Fatigue Rule Comparison. https://www.faa.gov/news/fact_sheets/news_story.cfm?newsKey=12445
5. Yang, X., Lin, L., Cheng, P.Y., Yang, X., Ren, Y.: Which EEG feedback works better for creativity performance in immersive virtual reality: the reminder or encouraging feedback? Comput. Hum. Behav. **99**, 345–351 (2019)
6. Gateau, T., Ayaz, H., Dehais, F.: In silico vs. over the clouds: on-the-fly mental state estimation of aircraft pilots, using a functional near infrared spectroscopy based passive-BCI. Front. Hum. Neurosci. **12**, 187 (2018)
7. Gateau, T., Durantin, G., Lancelot, F., Scannella, S., Dehais, F.: Real-time state estimation in a flight simulator using fNIRS. PLoS ONE **10**, e0121279 (2015)
8. Ayaz, H.: Analytical Software and Stimulus-Presentation Platform to Utilize, Visualize and Analyze Near-Infrared Spectroscopy Measures. MS thesis (2005)
9. Biopac Systems, Inc. https://www.biopac.com
10. NIRS-SPM. https://www.nitrc.org/projects/nirs_spm/
11. Durantin, G., Dehais, F., Delorme, A.: Characterization of mind wandering using fNIRS. Front. Syst. Neurosci. **9**, 45 (2015)
12. Afergan, D., Peck, E.M., Solovey, E.T., Jenkins, A., Hincks, S.W., Brown, E.T., Jacob, R.J.: Dynamic Difficulty Using Brain Metrics of Workload. SIGCHI Conference on Human Factors in Computing Systems. 3797–3806 (2014)
13. Lim, J., Wu, W.C., Wang, J., Detre, J.A., Dinges, D.F., Rao, H.: Imaging brain fatigue from sustained mental workload: an ASL perfusion study of the time-on-task effect. Neuroimage. **49**, 3426–3435 (2010)

Reading Multiple EEG Frequency-Band Networks in Developmental Dyslexia

Tihomir Taskov and Juliana Dushanova(✉)

Institute of Neurobiology, Bulgarian Academy of Sciences,
23 Acad G. Bonchev Street, Sofia, Bulgaria
tihokibertron@gmail.com, juliana@bio.bas.bg

Abstract. EEG studies using graph-theoretic analysis have found aberrations in functional connectivity (FC) in dyslexics. How the visual training (VT) can change FC of the reading neural network (RNN) in developmental dyslexia is still unclear. We studied differences in local and global topological properties of FC-RNN between controls and dyslexic children before and after VT (pre-D, post-D). The minimum spanning tree method was used to construct RNN in multiple EEG frequency-bands. Pre-D had higher leaf fraction, tree hierarchy, kappa, and smaller diameter (θ-γ), therefore more loaded RNNs than at controls. The RNN-metrics of post-D became similar to controls. In β1 and γ-frequency bands, pre-D exhibited reduced degree and betweenness centrality of hubs in superior, middle, inferior frontal areas in both hemispheres compared to controls. Dyslexics relied on left anterior temporal (β1,γ1), dorsolateral prefrontal cortex (γ1), while in right hemisphere - occipitotemporal, parietal (β1), motor (β2,γ1), somatosensory cortices (γ1). Post-D, hubs appeared in both hemispheric middle occipital (β), parietal (β1), somatosensory (γ1), dorsolateral prefrontal cortices (γ2), while in left hemisphere - at middle temporal, motor (β1), intermediate (γ2), inferior frontal cortices (γ1,β2). Language-related brain regions, more active after VT, contribute to lexical and sublexical understanding, as well areas, important for articulatory reading processes.

Keywords: EEG · Functional connectivity · Developmental dyslexia · Frequency oscillations · Reading of single words · Visual training tasks · Post-training network

1 Introduction

Reading is a multifaceted process that is supported by sublexical and lexical routes. The brain applies these two reading strategies along two neuronal pathways: dorsal (occipitoparietal sublexical) and ventral (occipitotemporal lexical) routes [1]. When reading aloud, in contrast to silent comprehension tasks, children can rely on orthographic and phonological information and less on semantic information, thus reducing the strength of the semantic facilitation effect. Extracting words from memory is a major component in the production of language and articulation. The initial stages of learning to read are related to children with problems with orthographic code violations [2]. Phonological decoding is the basis of reading comprehension [3]. After decoding skill acquirement, explicit teaching is being replaced by self-learning and begins to

D. Russo et al. (Eds.): IHSI 2021, AISC 1322, pp. 171–180, 2021.
https://doi.org/10.1007/978-3-030-68017-6_26

decode the words automatically [3]. The phonological deficit, which implies a deficit in phonological awareness, followed by a visual deficit related to poor spelling due to poor coding of the position of letters (e.g. inversion of letters), and global noise, which suggests the overall inefficiency of processing due to noisy calculations [4], are underlying developmental dyslexia. The learning of reading has great interpersonal differences in vocabulary, phonology, and orthographic skills. Knowing the effectiveness of the RNN is useful for understanding reading difficulties. Considering the brain as a complex network of neuronal populations and the connections between them can lead to the emergence of complex patterns of connectivity between functionally diverse components. Brain networks (BN) are often studied in terms of functional segregation and integration. Functional segregation reflects the ability of the brain to specify locally information about the process, i.e. within a brain region or interconnected group of adjacent areas, while functional integration is the ability to combine information from different brain areas [5]. Functional analysis describes the BN as a set of nodes and their connections. Brain network integration and segregation can be characterized by graphical measures. The brain has been shown to exhibit properties of the small world [6]. Small world networks combine high local connectivity with high global integration. The small-world model was used in the study of the topological reorganization of functional brain networks during normal brain development [7] when the BN shift from a random topology to a more segregated small-world topology. Recent studies show that BN contain areas with tightly interconnected hubs, which process information in segregated modules, while the most important nodes, hubs, play a role in integrating information into the network [8]. The topology of functional BN plays an important role in understanding the human brain, normal functioning, pathology, and development. The BN of developmental dyslexia (DD) at rest have been well studied using an EEG [7, 9]. EEG studies of the functional neural network of DD during the performance of tasks are missing. Although DD has been thoroughly studied at the behavioral level, there is no consensus on its causes. Different behavioral studies of DD have found various deficits in the sensitivity to a coherence motion perception [10], velocity discrimination [11], motion direction encoding [12, 13], contrast sensitivity to stimuli with low-/high-spatial frequency in external noise [14, 15]. These deficits are selectively associated with low accuracy or with slow performance on reading sub-skills [16], problems with clearly seeing letters and their order, orienting and focusing of visual-spatial attention [13]. The effectiveness of intervention efforts has also been studied [17–19]. However, more research is needed to determine the neurophysiological causes of dyslexia. Some studies have focused on examining changes in the activity of specific brain regions [20], but recent research has shown that the causes of deficits may lie in a disrupted relationship between specific brain regions [9]. The question is whether the theory of functional analysis of DD can shed light on the neurophysiological reasons for the observed effectiveness of training with visual nonverbal tasks [16, 17, 21]. The hypothesis is that training with corrective visual tasks in children with DD may lead to changes in their BNs so that they are more similar to those in controls.

2 Participants and Methods

A longitudinal study was conducted in the schools that involved repeated observations of the same children with DD over a long period. In this observational study with the exposure of visual intervention, the dyslexics are followed out over time to observe the outcome from the VT and to evaluate the extent to which the visual tasks contribute to the alteration of this childhood disorder. Reliable EEG data were obtained from 43 children (8–9 years old): 22 children with DD (12 boys) and 21 normal children (11 boys). All children and their parents gave informed consent for an EEG following the Helsinki Declaration. The participants in the study spoke Bulgarian as their first language. All children were right-handed and had a normal or corrected-to-normal vision. Visual stimuli (words) were presented on a laptop at a viewing distance of 57 cm for 800 ms in a pseudo-random sequence with interstimulus interval (ISI: 1.5–2.5 s). The font was Microsoft Sans Serif (black letters on a white background). Each letter has an angular size of about 1 degree. The stimuli were presented in at least two blocks in daily EEG sessions, each block containing 40 words. Participants were asked not to blink, only during the ISI to prevent artifacts in the EEG recordings during word reading. The voice reaction time (vRT: the time from the appearance of the word and the beginning of its reading) was assessed for each child. The vRT of the pre-D and post-D were compared as well as those of dyslexics and normal readers using the Kruskal-Wallis test. To examine whether VT could affect the RNN of children with DD, we recorded an EEG session during reading of words one month later after the three-month VT with five visual programs. Hence, irrespective of the word-reading task, the experimental group received an intensive procedure with VT tasks, presented in an arbitrary order and divided twice a week in individual sessions lasting 45 min over a period of three months. The program designs and parameter thresholds have been described in previous work [15] and in literature [10, 13, 16]. The EEG is recorded with an internally developed 40-channel Wi-Fi EEG system dry EEG sensors (Brain Rhythm Inc., Taiwan) with head positions to the system 10–20: F3-4, C3-4, T7-8, P3-4, O1-2, Fz, Cz, Pz, Oz, and additional positions according to system 10–10: AF3-4, F7-8, FT9-10, FC3-4, FC5-6, C1-2, C5-6, CP1-2, CP3-4, TP7-8, P7-8, PO3-04, PO7-08. The reference sensors were placed on both the processi mastoidei and the ground sensor on the forehead. The skin impedance was controlled to be less than 5 kΩ. The EEG sampling rate was 250 Hz. Continuous EEG data is filtered by passing the band in the frequency bands: $\delta = 1.5$–4, $\theta = 4$–8, $\alpha = 8$–13, $\beta 1 = 13$–20, $\beta 2 = 20$–30, $\gamma 1 = 30$–48, $\gamma 2 = 52$–70 Hz. The data were segmented into trials locked at the time of onset of the stimulus, each with duration of 800 ms. Artifact-free EEG trials ($\leq \pm 200$ μV) with correct answers are included.

2.1 Minimum Spanning Tree

The functional connectivity for all possible pairs of electrodes was determined using the Phase Lag Index (PLI) separately for each frequency band and trial [22]. The PLI gives information about the phase synchronization of two signals by measuring the asymmetry of the distribution of their instantaneous phase differences, based on the Hilbert transform. PLI = 0 indicates that two signals are not phase-locked (or that their phase difference is centered on 0 mod π). They are perfectly phase-locked with a phase difference different from 0 mod π and PLI = 1. PLI does not depend on the amplitude of the signal and is less sensitive to volume conduction in the brain, as well as to spurious correlations because of common sources. The calculated connectivities, using the PLI, between each pair of channels can be used to construct an adjacency matrix (a graph). Due to methodological limitations, the comparison of different BN can be problematic [5]. To avoid this problem, some authors have proposed the use of the minimum spanning tree (MST) to define an unbiased sub-graph of the original network. A separate MST sub-graph was constructed from each of the PLI matrices, i.e. one for each non-rejected trial. The MST is a unique sub-graph that connects all the nodes of the graph without forming loops, such that the wiring cost (the weights) is minimized. The MST was constructed using Kruskal's algorithm. Since the interest of the strongest connections, before using the algorithm, all the original weights (PLI) were converted to distances (1/PLI). The procedure is to order all the links in ascending order. After that, the link with the shortest distance (highest PLI) is added to the sub-network. The link with the second shortest distance, which does not form any loops, is added and this is repeated until all nodes are connected in an acyclic graph. In the end, all the links present in the MST are set to 1, while all the other connections are set to 0. The MST has a fixed density – M = N-1, where N is the number of nodes. There are two extreme MST topologies: (1) line-like topology, where each node is connected to only two nodes, with the exception of the two leaf nodes at either end of the line (2) star-like topology, in which, there is a single central node to which the other nodes are directly connected. The global MST measures, like conventional graph measures, can provide information about network integration and segregation [5]. Four global MST measures were used in this study: diameter (*D*), leaf fraction (*LF*), tree hierarchy (*TH*) and kappa (*K*). The *D* in the MST is the shortest path along the MST. The shortest path between two nodes in the network is the path that involves the fewest number of links between them. *LF* is the number of leaves (nodes with degree = 1) in the MST divided by the total number of nodes. The *TH* characterizes the balance of having high network integration without an overloading the most important nodes in the network. The *TH* (0, 1) is defined as $TH = L/2mBC_{max}$, where m = N-1 links in the MST, L is the number of leaves and BC_{max} is the maximal BC in the MST. On one extreme if the tree has a line-like topology, i.e., L = 2, then if m approaches infinity, *TH* will approach 0. If the tree is star-like, L = m and *TH* approaches 0.5. For topologies between these two extreme cases, *TH* will have higher values. Kappa measures the broadness of the degree distribution in the network and has higher values for scale-free graphs and lower values for more random graphs. *K* reflects the resilience of the network against attacks related to targeted hub removal, specifically a change in the degrees of the connected nodes. High *K* means the network is less vulnerable to random attacks. The nodal measures give information about the importance of individual nodes in the network. The degree of a node is equal to the number of nodes it's connected to. Betweenness centrality (BC) of a node is the fraction of all shortest paths in the network that pass through that node. The networks are more integrated when the nodes have a higher maximum degree or maximum BC [23, 24]. Hubs were defined as nodes with a degree or BC of at least 1 standard deviation above the mean. The links are the most important edges with at least 1 standard deviation above the mean. For each band, the global MST measures and the hubs determined by the degree of local measures were compared between the groups using nonparametric permutation tests [25]. Sensor indices were chosen so that statistical tests for hubs were most sensitive to hemisphere differences. To compensate for the effects of multiple comparisons, Bonferroni adjustment to the significance level was applied separately for the global tests ($p = \alpha/4 = 0.0125$) and the local (hub) tests ($p = \alpha/2 = 0.025$).

3 Global Measures of MST

Both dyslexic groups (pre-D: 1497.96 ± 30.5; post-D: 1266.16 ± 27.9 ms) showed slower vRT compared to the controls (Con: 873.5 ± 12.7, $p < 0.0001$, $\chi^2 > 238.7$). Post-D showed an improvement in vRT ($\chi^2 = 30.4$, $p < 0.0001$). The MST-graph examines the topology of functional BN, derived from EEG data for both Con and children with DD when reading aloud single words (Table 1). Before VT, the global topological organization of the pre-D is different from that of the Con. After VT, their global topology becomes more similar to the Con. The smaller *D* and higher *LF* in pre-D for most frequency bands is indicative of a more integrated star-like topology than controls, presenting a more decentralized network. Compared to controls, dyslexics exhibited a different global brain topology when reading words. The more integrated network of pre-D could reflect a less optimal global organization with an overloading of central connectivity hubs (brain areas). The change in network topology after VT could be the result of a compensatory mechanism. The higher *D* and the lower *LF*, *TH* and *K* found in the Con, compared to the pre-D, could reflect a more mature brain. After VT, the changes in the MST measures of dyslexics were similar to the changes observed in the process of brain maturation [7]. The between-group differences in MST measures were mostly frequency-independent. They had a similar profile in most of the bands, suggesting that similar network constraints occur in different neural circuits. The dyslexics may exhibit differences in brain connectivity suggesting deficits in general sensory functions and attention that are associated with higher frequency EEG activity (β, γ). The increase of topological segregation after VT decreases the load on the important, in terms of connectivity, brain regions, leading to a more efficient BN, analogous to the processes observed in the brain development of children.

Table 1. Nonparametric comparisons of global metrics (D, LF, TH, K) of BN of Con, pre-D and post-D during word reading for frequency bands.

		Controls	Pre-D	Post-D	Con/Pre-D p	Con/Post- D p	Pre-D/ Post-D p
δ	*D*	0.263 ± 0.006	0.271 ± 0.003	0.276 ± 0.007	0.174	0.105	0.53
	LF	0.618 ± 0.009	0.623 ± 0.005	0.608 ± 0.011	0.945	0.451	0.38
	TH	0.413 ± 0.006	0.427 ± 0.003	0.419 ± 0.008	0.034	0.204	0.50
	K	3.934 ± 0.133	4.004 ± 0.099	3.798 ± 0.145	0.968	0. 379	0.23
θ	*D*	0.337 ± 0.004	0.311 ± 0.003	0.317 ± 0.004	**<0.0001**	**0.0025**	0.17
	LF	0.531 ± 0.004	0.569 ± 0.004	0.555 ± 0.005	**0.0001**	**0.0014**	0.067
	TH	0.394 ± 0.004	0.409 ± 0.003	0.404 ± 0.004	**0.004**	0. 036	0.60
	K	3.019 ± 0.042	3.321 ± 0.041	3.195 ± 0.044	**0.0001**	**0.002**	0.058
α	*D*	0.332 ± 0.003	0.316 ± 0.003	0.324 ± 0.004	**0.001**	0.088	0.27
	LF	0.523 ± 0.004	0.552 ± 0.003	0.538 ± 0.004	**0.0001**	0.035	**0.011**
	TH	0.391 ± 0.003	0.402 ± 0.003	0.398 ± 0.003	0.026	0.275	0.333
	K	2.943 ± 0.025	3.155 ± 0.031	3.041 ± 0.034	**0.0001**	0.077	0.017
β1	*D*	0.333 ± 0.003	0.318 ± 0.002	0.333 ± 0.004	**0.0002**	0.762	**0.0012**
	LF	0.507 ± 0.003	0.539 ± 0.003	0.519 ± 0.004	**0.0001**	0.044	**<0.0001**
	TH	0.374 ± 0.003	0.393 ± 0.003	0.387 ± 0.003	**0.0001**	**0.011**	0.11
	K	2.854 ± 0.021	3.056 ± 0.021	2.925 ± 0.023	**0.0001**	0.024	**<0.0001**
β2	*D*	0.331 ± 0.003	0.315 ± 0.002	0.328 ± 0.003	**0.0006**	0.591	**0.006**
	LF	0.506 ± 0.003	0.542 ± 0.003	0.517 ± 0.004	**0.0001**	0.057	**<0.0001**
	TH	0.377 ± 0.003	0.396 ± 0.003	0.385 ± 0.003	**0.0001**	0.061	0.0238
	K	2.829 ± 0.019	3.068 ± 0.024	2.898 ± 0.022	**0.0001**	0.0238	**<0.0001**
γ1	*D*	0.329 ± 0.003	0.301 ± 0.003	0.315 ± 0.003	**0.0001**	**0.003**	**0.0025**
	LF	0.517 ± 0.004	0.566 ± 0.004	0.529 ± 0.004	**0.0001**	0.085	**<0.0001**
	TH	0.384 ± 0.003	0.409 ± 0.003	0.387 ± 0.003	**0.0001**	0.635	**<0.0001**
	K	2.866 ± 0.019	3.263 ± 0.033	2.958 ± 0.027	**0.0001**	0.020	**<0.0001**
γ2	*D*	0.280 ± 0.002	0.262 ± 0.003	0.267 ± 0.003	**0.0001**	**0.001**	0.188
	LF	0.596 ± 0.003	0.636 ± 0.003	0.623 ± 0.004	**0.0001**	**<0.0001**	0.055
	TH	0.422 ± 0.003	0.441 ± 0.002	0.434 ± 0.003	**0.0001**	**0.009**	0.207
	K	3.427 ± 0.037	3.903 ± 0.048	3.708 ± 0.053	**0.0001**	**<0.0001**	0.037

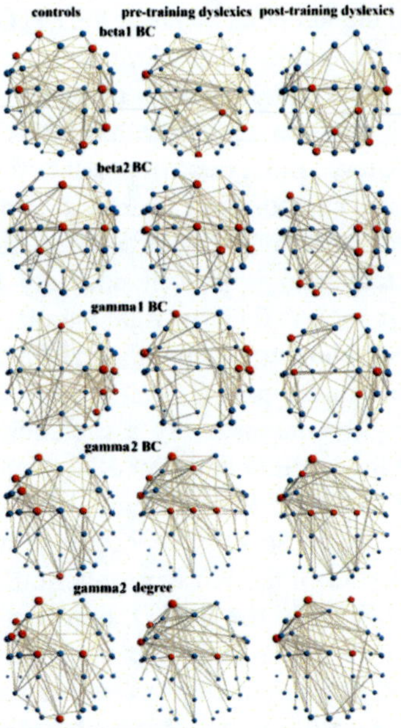

Fig. 1. Group-level hubs for selected frequency bands. Each node corresponds to an EEG sensor: (A, 1st graph) hubs (BC) in β1 for controls: AF3, F4, F7, C3-4, P8, PO4; (A, 2nd graph) pre-training dyslexics: FT9, CP2, P8, Oz; (A, 3rd graph) post-training dyslexics: C1, C6, CP2, Pz, T8, PO8, O1; (B, 1st) Hubs (BC) in β2 for controls: Fz, FC3, Cz, C4, CP1; (B, 2nd) pre-training dyslexics: Fz, C5, Cz, C4, CP2, TP8; (B, 3rd) post-training dyslexics: F7, C2, C4, CP2, P4, PO7, O1. (C, 1st) hubs (BC) in γ1 for controls: Fz, C4, C6, CP4, TP8, P4; (C, 2nd) pre-training dyslexics: AF3, FC6, FT9-10, C4, C6, T8; (C, 3rd) post-training dyslexics: AF4, F7, C3-4, CP2. (D, 1st) Hubs (BC) in γ2 for controls: AF3, FC3, F7, C1-2, Oz; (D, 2nd) pre- training dyslexics: Fz, AF3, F7, FT9, C1-2, Cz; (D, 3rd) post-training dyslexics: AF3, F3, FT9, C1-2, Cz. (E, 1st) hubs (degree) in γ2 for controls: AF3, F7, FC3, FC5, C1-2, Oz; (E, 2nd) pre-training dyslexics: AF3, F7, FT9, C1-2; (E, 3rd) post-training dyslexics: AF3-4, F3, FT9, C2.

3.1 Distribution of Connectivity Hubs

The basic RNN in normally reading children at β1 includes frontotempororietal chains primarily in the left hemisphere (LH; Fig. 1). During word reading, early visual processing occurs in the right inferior occipitotemporal cortex and the superior occipital gyrus, the starting point for the two main hubs transmitting information forward in the RNN. After the right occipitotemporal area, reading includes the bilateral PstCG (motor cortex MC, primary somatosensory cortex S1) and left IFG (pars triangularis), dorsolateral prefrontal cortex DLPFC. These brain regions allow for the understanding of lexical and sublexical phonological representation and play a role in articulatory processes important for reading. Functional abnormalities in the β1 observed in pre-D demonstrate the absence of hubs in the SFG, MFG, and IFG in both hemispheres. Hubs predominate in the anterior temporal (ATL) in the both hemispheres at the superior parietal (SPG) and inferior occipitotemporal areas (included fusiform area). After VT, hubs in the post-D appear in the left hemisphere at the MTG, bilateral SPL and bilateral MOG, right S1 and supramarginal cortices (SMG), left MC, premotor (PMC), supplementary motor cortices (SMA). The β2-RNN of normal children consists of the left SPG, right PstCG, bilateral MC and PMC, the left SMA, and right S1. These brain regions involve in articulatory processes important for reading. The functional abnormalities observed in pre-D demonstrate deactivation of the left frontal areas in the IFG, accompanied by overactivation in the right hemisphere (RH) at the MC and MTG. For the post-D, maximal activation is present not only in the right somatosensory association cortex, but also in the left IFG, the right angular (AG), and the left MOG. In normal children, the γ1-RNN consists of the bilateral PMC and at the RH in the S1, SMG, AG, and MTG. While in the post-D, the γ1-RNN includes the left IFG, right DLPFC, bilateral S1, and right somatosensory association cortex. The pre-D is characterized by deactivation of the left IFG in the γ1-RNN, and by the appearance of hubs in the right IFG, S1, SMG, MTG, and bilateral ATL, left DLPFC. During word reading, early visual processing in the γ2-RNN occurs in the bilateral occipital cortex, transmitting information forward in the left hemispheric RNN to the PMC and SMA, bihemispheric MC, IFG, DLPFC. In the post-D, hubs appeared also in the left intermediate frontal cortex as well as in both hemispheres in the DLPFC. Functional abnormalities in γ2 observed in pre-D demonstrated the absence of hubs in the LH of the MFG, PMC and SMA, as well as in both hemispheres of the visual brain areas, but overactivation was observed in the left IFG. Reading as a complex task involves a number of different areas and functions of the brain in which visual, phonological, orthographic, and semantic systems. The occipital cortex, SMG, fusiform gyrus, and ATL represent these systems. Areas related to reading, such as the frontal opercular and temporal cortices are key regions for the reading process. Beyond the RNN, the auditory cortex is also associated with reading achievement as well as phonological awareness in the acquisition of reading. The presentation of aloud reading is closely related to the activity of the left MC and somatosensory cortices. Visual areas also contribute to reading, as reading written words require visual word processing. There are also language-related areas in RNNs, especially semantic areas such as inferior frontal, inferior temporal and superior temporal cortices. The two hemispheres work together to carry out the reading process in different frequency ranges of the acoustic speech envelope. The two hemispheres contribute to reading but in very specific ways. Most functional connections associated with reading occur in the parietal and frontal cortices in both hemispheres of the brain, and are responsible for higher-order cognitive functions. The least reading-related functional connections dominate the right hemisphere. The density of connections inside reading networks is higher than outside reading networks. Interconnection between the language areas in the left hemisphere and those for attention in the right hemisphere is underlying of both the lexical and the sublexical reading. The reverse lateralization of the frontal areas in pre-D compared to those after VT suggests functional impairment of a major language center such as the left frontal cortex. The presence of hubs in post-D in the dorsal visual network, including the inferior parietal, middle temporal, visual associative cortex, and the dorsal superior frontal cortex, suggests that nonverbal VT could help the child's development. In general, key roles of individual hubs may be disrupted by changes in brain system segregation. Some methodological limitations of the MST study are related to some measures sensitive to the network size, which may affect the relative importance of the nodes in the operation of the network. The main limitation of the segregation analysis is related to the choice of a threshold for the distribution of hubs, which emphasizes the regions with medium or BC at least one standard deviation above the mean and avoids the choice of these nodes with lower intensity links.

Acknowledgments. This work was financially supported by the Bulgarian Science Fund, grant No DN05-14-2016.

References

1. Dehaene, S., Cohen, L., Morais, J., Kolinsky, R.: Illiterate to literate: behavioural and cerebral changes induced by reading acquisition. Nat. Rev. Neurosci. **16**(4), 234–244 (2015)
2. Castles, A., Rastle, K., Nation, K.: Ending the reading wars: reading acquisition from novice to expert. Psychol. Sci. Publ. Interest **19**(1), 5–51 (2018)
3. Share: Phonological recoding and self-teaching: sine qua non of reading acquisition. Cognition **55**(2), 151–218 (1995)
4. Hancock, R., Pugh, K.R., Hoeft, F.: Neural noise hypothesis of developmental dyslexia. Trends Cogn. Sci. **21**(11), 909 (2017)
5. Tewarie, P., van Dellen, E., Hillebrand, A., Stam, C.J.: The minimum spanning tree: an unbiased method for brain network analysis. Neuroimage **104**, 177–188 (2015)
6. Bassett, D.S., Bullmore, E.: Small-world brain networks. Neuroscientist **12**, 512–523 (2006)
7. He, W., Sowman, P.F., Brock, J., Etchell, A.C., Stam, C.J., Hillebrand, A.: Increased segregation of functional networks in developing brains. Neuroimage **200**, 607–620 (2019)
8. Sporns, O.: Structure and function of complex brain networks. Dialogues Clin. Neurosci. **15**, 247–262 (2013)
9. Fraga González, G., Van der Molen, M.J.W., Žarić, G., Bonte, M., Tijms, J., Blomert, L., Stam, C.J., Van der Molen, M.W.: Graph analysis of EEG resting state functional networks in dyslexic readers. Clin. Neurophysiol. **127**, 3165–3175 (2016)
10. Boets, B., Vandermosten, M., Cornelissen, P., Wouters, J., Ghesquière, P.: Coherent motion sensitivity and reading development in the transition from prereading to reading stage. Child Dev. **82**, 854–869 (2011)
11. Demb, J.B., Boynton, G.M., Best, M., Heeger, D.J.: Psychophysical evidence for a magnocellular pathway deficit in dyslexia. Vis. Res. **38**(11), 1555–1559 (1998)
12. Cornelissen, P.L., Hansen, P.C., Hutton, J.L., Evangelinou, V., Stein, J.F.: Magnocellular visual function and children's single word reading. Vis. Res. **38**(3), 471–482 (1998)
13. Stein, J.: Dyslexia: the role of vision and visual attention. Curr. Dev. Disord. Rep. **1**, 267–280 (2014)
14. Sperling, A.J., Lu, Z.-L., Manis, F.R., Seidenberg, M.S.: Deficits in perceptual noise exclusion in developmental dyslexia. Nat. Neurosci. **8**(7), 862–863 (2005)
15. Lalova, J., Dushanova, J., Kalonkina, A., Tsokov, S.: Application of specialised psychometric tests and training practices in children with developmental dyslexia. Psych. Res. **22**, 271–283 (2019)
16. Wilmer, J.B., Richardson, A.J., Chen, Y., Stein, J.F.: Two visual motion processing deficits in developmental dyslexia associated with different reading skills deficits. J. Cogn. Neurosci. **16**, 528–540 (2004)
17. Lawton, T.: Improving magnocellular function in the dorsal stream remediates reading deficits. Optom. Vis. Dev. **42**, 142–154 (2011)

18. Chouake, T., Levy, T., Javitt, D.C., Lavidor, M.: Magnocellular training improves visual word recognition. Front. Hum. Neurosci. **10**, 6–14 (2012)
19. Lawton, T., Shelley-Tremblay, J.: Training on movement figure-ground discrimination remediates low-level visual timing deficits in the dorsal stream, improving high-level cognitive functioning, including attention, reading fluency, and working memory. Front. Hum. Neurosci. **11**, 236 (2017)
20. Habib, M.: The neurological basis of developmental dyslexia: an overview and working hypothesis. Brain **123**(12), 2373–2399 (2000)
21. Ebrahimi, L., Pouretemad, H., Khatibi, A., Stein, J.: Magnocellular based visual motion training improves reading in persian. Sci. Rep. **9**(1), 1142 (2019)
22. Stam, C.J., Nolte, G., Daffertshofer, A.: Phase lag index: assessment of functional connectivity from multi channel EEG and MEG with diminished bias from common sources. Hum. Brain Mapp. **28**, 1178–1193 (2007)
23. Bullmore, E., Sporns, O.: Complex brain networks: graph theoretical analysis of structural and functional systems. Nat. Rev. Neurosci. **10**, 186–198 (2009)
24. Stam, C.J., Tewarie, P., Van Dellen, E., van Straaten, E.C., Hillebrand, A., Van Mieghem, P.: The trees and the forest: characterization of complex brain networks with minimum spanning trees. Int. J. Psychophysiol. **92**, 129–138 (2014)
25. Maris, E., Oostenveld, R.: Nonparametric statistical testing of EEG-and MEG-data. J. Neurosci. Methods **164**, 177–190 (2007)

Technological Resources for Neuropsychological Rehabilitation

Carlos Ramos-Galarza[1,2(✉)], Micaela Silva-Barragán[1], Omar Cóndor-Herrera[2], Mónica Bolaños-Pasquel[2], Valentina Ramos[3], and Jorge Cruz-Cárdenas[2]

[1] Facultad de Psicología, Pontificia Universidad Católica del Ecuador, Av. 12 de Octubre y Roca, Quito, Ecuador
caramos@puce.edu.ec

[2] Centro de Investigación en Mecatrónica y Sistemas Interactivos MIST/Carrera de Psicología/Centro de Investigación ESTec, Universidad Tecnológica Indoamérica, Av. Machala y Sabanilla, Quito, Ecuador
italo_omar@hotmail.com, monika_martina@hotmail.com, jorgecruz@uti.edu.ec

[3] Sistemas de Información, Gestión de la Tecnología e Innovación (SIGTI-Research Group), Escuela Politécnica Nacional, Av. Ladrón de Guevara 253, Quito, Ecuador
valentina.ramos@epn.edu.ec

Abstract. Neuropsychological rehabilitation is a high complexity process that involves returning the functionality lost to a patient with acquired brain damage. In this context, technological development provides several possibilities for the stimulation of various brain functions involved in cognitive processing and behavioral regulation of a patient. For this reason, these types of treatments, including technological input, allow the patient to improve their levels of motivation and performance of neuropsychological rehabilitation activities. This article reviews some technological devices that provide improvements in the context of neuropsychology for the care of patients with acquired brain damage.

Keywords: Rehabilitation · Neuropsychology · Technology · Innovation · Brain function · Gamification · Digital game-based learning

1 Introduction

Neuropsychological rehabilitation is a process that helps people who have suffered acquired brain damage [1]. It is based on three strategies: restoration, compensation and substitution of brain functions affected in the damage. Restoration consists of recovering a specific brain function through iterations of an exercise based on a determined mental ability. Compensation is a strategy that improves a damaged function by working on a conserved brain function. Substitution is a strategy which uses an external resource that assumes the function of brain damage ability [2].

One of the limitations of neuropsychological rehabilitation is that most of its resources have been developed to use pen and paper because the main contribution of

D. Russo et al. (Eds.): IHSI 2021, AISC 1322, pp. 181–186, 2021.
https://doi.org/10.1007/978-3-030-68017-6_27

neuropsychological rehabilitation was in the 80's. Thus, technology was not in service of neuropsychology, however, today some technological proposals try to help people with brain problems [3].

In this context, the present paper presents a revision study of the principal technological resources developed for the neuropsychological rehabilitation process. This work significance is describing the contribution of the eHealth (electronic health) in neuropsychology.

2 Technological Applications for Neuropsychological Rehabilitation

2.1 Dr Lobus: Smartphone App for Cognitive Stimulation

Smartphone apps are a great resource for cognitive stimulation involved in neuropsychological rehabilitation [4, 5]. In this context, Ecuadorian scientists interested in the adults' neuropsychological stimulation and memory developed a mobile application, called Dr. Lobus (Fig. 1), to be freely used by clinicians, patients and anyone interested in improving their cognitive functions, as well as to be implemented in neuropsychology research projects [10].

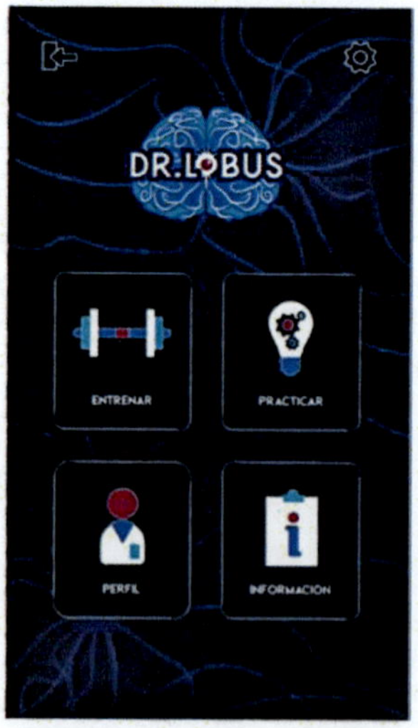

Fig. 1. Dr. Lobus screenshot

2.2 Neuropsychological Rehabilitation and Cognitive Stimulation Program: Dr. Kawashima

The program called Dr. Kawashima (Fig. 2) allows cognitive stimulation for a patient with acquired brain damage. One of the benefits of this app from a clinical standpoint is the content for the neuropsychological patient and the automatic performance level adaptation (as needed in adult patients suffering cognitive impairment) in stimulating brain functions such as memory, attention, understanding, problem-solving and others [6, 7].

Fig. 2. Dr. Kawashima app screenshot

2.3 Virtual Reality Boardgame

This technological device (Fig. 3) aims to complement neuropsychological rehabilitation for brain damage from nervous system conditions, such as strokes and Alzheimer. Also, it is recommended for children with attention and learning disorders. It is a virtual board game, in which the patients can interact with worksheets stimulating their ability to plan, sequencing, categorization, memory and constructive vision strategies [8].

Fig. 3. Virtual reality environment

2.4 Stimulus

Stimulus is a professional cognitive stimulation and rehabilitation app designed for tablets (Fig. 4). Its goal is to train cognitive processes within 10 functional areas: attention, perception, working memory, long-term memory, calculation, reasoning, executive functions, vision-motor functions, language, and speed. It can be used in people with cognitive deficits that resulted from injuries, disabilities, or neurodegenerative processes [9].

Fig. 4. Stimulus screenshot

2.5 Sound and Speech Recognition System for Health Smart Homes: AUDIT HIS

One of the biggest challenges in neuropsychological rehabilitation of patients with acquired brain damage is the loss of autonomy. This problem is addressed by 'Audit His' system (Fig. 5), which has a patient's speech recognition system that allows the patient to perform various actions in his home without an assistant [11].

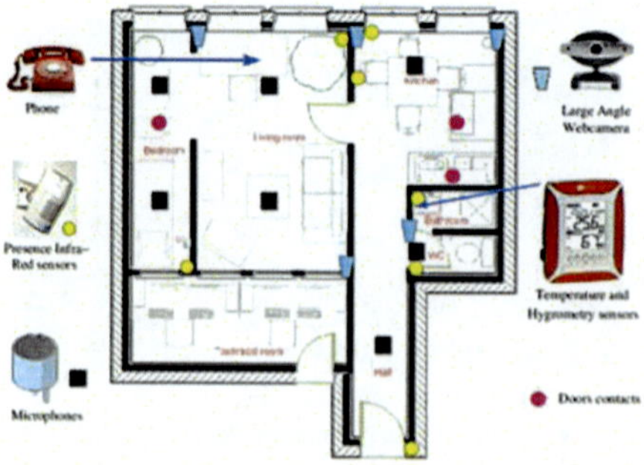

Fig. 5. Sensors arrangement in the environment of a person with brain disturbances for autonomy

3 Conclusions

This article shows a novel review of technological proposals for the neuropsychological rehabilitation of patients with some type of brain impairment. As has been evident throughout this work, the use of technology generates a significant contribution to the neuropsychological treatment process of patients with some type of cognitive deficit from acquired brain damage [12].

The review started describing the elements needed within the neuropsychological rehabilitation process that generates motivation and interest in patients receiving this type of intervention. Within the technological resources available to the clinician highlight apps, computer programs, virtual reality platforms and technological devices that compensate for daily life activities faced by the neuropsychological patient.

One of the challenges projected from this review is the implementation of these devices in experimental studies with pre- and post-test measurements to verify the effectiveness of their implementation on the neuropsychological rehabilitation process in patients with impaired cognitive functions.

The next step in this line of research is the development of technology for patients who need to improve the performance of their brain's function, which are free and patient-friendly and thus improve the neuropsychological rehabilitation process.

Finally, in future research, we propose the application of this type of technologies and the analysis of their impact on the brain's function in their various stages of development, as well as in different conditions of normality or cognitive impairment resulting from brain impairment.

References

1. Ramos-Galarza, C.: Neuropsychological Consequences in the Acquired Brain Damage: case analyses (Secuelas neuropsicológicas en el daño cerebral adquirido: análisis de casos). Ed. Indoamérica, Quito-Ecuador (2018)
2. Ramos-Galarza, C., Ramos, V., Jadán-Guerrero, J., Lepe-Martínez, N., Paredes-Núñez, L., Gómez-García, A., Bolaños-Pasquel, A.: Fundamental concepts in the neuropsychological theory (conceptos fundamentales en la teoría neuropsicológica). Rev. Ecuat. Neurol. **26**(1), 53–60 (2017)
3. Ramos-Galarza, C., Acosta-Rodas, M., Sánchez-Gordón, S., Calle-Jiménez, T.: Mobile technological apps to improve frontal lobe functioning. In: Ayaz, H., Asgher, U. (eds). Advances in Neuroergonomics and Cognitive Engineering. AHFE 2020. Advances in Intelligent Systems and Computing, vol. 1201, Springer, Cham (2021)
4. Sánchez, M., Collado, S., Martín, P., Cano, R.: Neurorehabilitation and apps: a systematic review of mobile applications (Apps en neurorrehabilitación. Una revisión sistemática de aplicaciones móviles). Neurol **33**(5), 313–326 (2018)
5. Vilanou, R.: mHealth: definition, risks and uses. http://www.pmfarma.es/articulos/2390-mhealth-definicion-riesgos-y-usos.html
6. Kawahima, R.: Nuevo Brain Trainer. Ediciones Planeta, Madrid (2012)

7. Jeremic, J., Zhang, F., Kaufman, D.: Older adults' perceptions about commercially available xbox kinect exergames. In: Zhou, J., Salvendy, G. (eds.) Human Aspects of IT for the Aged Population. Design for the Elderly and Technology Acceptance. HCII 2019. Lecture Notes in Computer Science, vol. 11592. Springer, Cham. https://doi.org/10.1007/978-3-030-22012-9_14
8. Gómez-Mora, R.: Application of virtual reality cognitive rehabilitation (Aplicación de realidad virtual en la rehabilitación cognitiva). Rev. Vínculos. **10**(1), 130–135 (2013)
9. Stimulus: App profesional de estimulación cognitiva. [Internet]. Stimulus Pro. https://stimuluspro.com/inicio/
10. Ramos-Galarza, C., Jadán-Guerrero, J., Arias-Flores, H.: Dr. Lobus: aplicación móvil para la estimulación neuropsicológica. Universidad Tecnológica Indoamérica: Quito, Ecuador (2020)
11. Vacher, M., Fleury, A., Portet, J., Noury, N.: Complete sound and speech recognition of activities of daily living. New Developments in Biomedical Engineering. Intech: London, UK (2010)
12. Templeton, J., Poellabauer, C., Schneider, S.: Enhancement of neurocognitive assessments using smartphone capabilities: systematic review. JMIR Mhealth Unhealth **8**(6), e15517 (2020)

A Modeling Method to Evaluate the Finger-Click Interactive Task in the Virtual Environment

Jiarui Li, Xiaozhou Zhou(✉), Lesong Jia, Weiye Xiao, Yu Jin, and Chengqi Xue

School of Mechanical Engineering, Southeast University, Nanjing 211189, China
{220184300, zxz, lesong, esuka_studio, 220184324, ipd_xcq}@seu.edu.cn

Abstract. A modeling method is introduced in this paper to evaluate the finger-click interactive task in the virtual environment. The problem that the finger-click interactive task in the virtual environment currently cannot be quantitatively evaluated has been solved. The method mainly includes five processes: analysis, decomposition, configuration, acquisition, and evaluation. The proposal of this method can help relevant practitioners design the interactive process of user's finger-click interactive task in the virtual environment more scientifically, optimize the interactive efficiency.

Keywords: GOMS · Virtual environment · Evaluation

1 Introduction

In recent years, the emergence and development of technologies such as virtual reality, augmented reality, and mixed reality have broadened people's research horizons on the direction of human-computer interaction (HCI) [1]. In the research of evaluation methods related to HCI, a quantitative evaluation method based on GOMS model with keyboard and mouse input has been proposed for 2D HCI interfaces [2]. However, in the virtual environment, the user interface is transferred from a 2D plane to a 3D space, although the HCI in the virtual environment has a sense of reality and immersion [3], it is usually difficult to quantitatively evaluated.

Marsh studied whether existing methods could be used to evaluate the usability of virtual reality and formulated future research plans [4]. Smith et al. tried to propose a method combining virtual environment simulation technologies with unique human performance modeling software to design, test and evaluate procedures for future biology experiments aboard the International Space Station (ISS) [5]. Chang et al. showed an example of using GOMS to model industrial robot programming tasks, and proposed the research method in the flowchart [6]. After our investigation, although there have been some researches on the evaluation methods in the 3D HCI environment, no one has proposed a quantifiable evaluation method for the finger-click interactive task in the virtual environment.

D. Russo et al. (Eds.): IHSI 2021, AISC 1322, pp. 187–193, 2021.
https://doi.org/10.1007/978-3-030-68017-6_28

2 A Modeling Method to Evaluate the Finger-Click Interactive Task in the Virtual Environment

2.1 Overview

The modeling method proposed in this paper to evaluate the finger-click interactive task in the virtual environment can predict the performance of the corresponding interactive scenes. The method creates a quantitative evaluation model through the five processes of analysis, decomposition, configuration, acquisition and evaluation, which provides a usable guidance for human interaction based on the finger-click interactive task in the virtual environment.

2.2 Implementations

Analysis

User interaction system analysis: First, the user interaction system based on finger-click interactive tasks in the virtual environment is needed to be analyzed, which includes task, user, hardware, software, and database.

User interaction task analysis: Task analysis is a method used to understand how a user performs a specific task and describe user behavior. Therefore, analyze the user interaction tasks based on finger clicks in the virtual environment is necessary.

Decomposition

We can decompose the user's entire interaction process through the Model Human Processor (MHP) to obtain the basic user interaction unit, combining MHP with the user interaction system and the user interaction task analysis, a series of basic user interaction units based on the finger-click interactive task in the virtual environment can be summarized as follows:

Cognition phase

Task Thinking Operator (A): The user thinks about how to operate

Perception phase

Perceptual Reaction Operator (E): The user's reaction and preparation to start operation

Motor phase

Moving In Operator (I): The user moves his hand from outside the trigger area into the trigger area in a natural state

Homing Operator (H): The user moves his hand out of the trigger area and restores to its natural state

Pointing Operator (P): The user points to a 3D virtual target component in the trigger area

Clicking Operator (C): Finger click on a 3D virtual target component.

We categorize these four operators as hand operators.

Others

System Operator (R):

System operation R1: The system receives operations such as clicking, command sending

System feedback R2: System feedback response

Configuration

After determining the basic user interaction unit, we configure interaction rules

Rule 1: Insert task thinking operator A and perceptual reaction operator E before moving in operator I and pointing operator P, and insert perceptual reaction operator E after clicking operator C;

Rule 2: If the hand operators before the task thinking operator A can fully anticipate the hand operators after the task thinking operator A, delete the task thinking operator A and the perceptual reaction operator E after the task thinking operator A;

Rule 3: After clicking operator C is executed, wait for system R, which includes system operation R1 and system feedback R2.

Rule 4:

(1) There is overlap between system feedback R2 and perceptual reaction operator E. If the system feedback R2 time is less than the perceptual reaction operator E time, then the system feedback R2 time is ignored, otherwise the perceptual reaction operator E time is ignored;
(2) There is overlap between system feedback R2 and homing operator H. If the system feedback R2 time is less than the homing operator H, then the system feedback R2 time is ignored, otherwise the homing operator H time is ignored;
(3) The system feedback R2 overlaps with the sequential perceptual reaction operator E and homing operator H. If the system feedback R2 time is less than the sum of the perceptual reaction operator E time and the homing operator H time, then the system feedback R2 time is ignored, otherwise the perceptual reaction operator E time and the homing operator H time are ignored.

Acquisition

The time required for a series of basic user interaction units based on finger-click interactive tasks in the virtual environment can be measured based on the verification experience and cognitive psychology experiments. The specific time is shown in Table 1:

Table 1. The time required for the basic user interaction unit based on the finger-click interactive tasks in the virtual environment

Name	Typical value	Definition
Task Thinking Operator Time	1.2 s	The time to think about how to operate
Perceptual Reaction Operator Time	0.24 s	The time required for the user to react and to prepare to initiate the operation
Moving In Operator Time	0.5 s	The time required for the user to move his hand from outside the trigger area into the trigger area in a natural state

(*continued*)

Table 1. (*continued*)

Name	Typical value	Definition
Homing Operator Time	0.5 s	The time required for the user to move the hand out of the trigger area and restore to its natural state
Pointing Operator Time	$0.381 \times \log_2\left(\frac{D}{W}+1\right)+0.2323$ s	The time required for the user to point to a 3D virtual target component in the trigger area
Clicking Operator Time	0.3 s	The time required for a finger to click on a 3D virtual target component
System feedback Time	0.1 s	The duration of the system feedback response presentation, which varies with the different feedback modes of the system and the presentation form of the visual feedback animation
System operation Time	0.1 s	The time required for the system to receive operations such as clicking, command sending, which varies with the system performance

Evaluation

According to the interaction time, interaction rules and system time, the operation sequence set for the user to complete the current task could be determined and listed to estimate the total execution time of the task. According to the total execution time of the task, the user's interaction process based on the finger-click interactive task could be evaluated.

3 Case Application

3.1 Apparatus

This case is performed using HTC vive. The prototype used to perform the task was developed using LeapMotion's interactive engine module, the software used was Unity, and the programming language used was C#.

3.2 Procedure

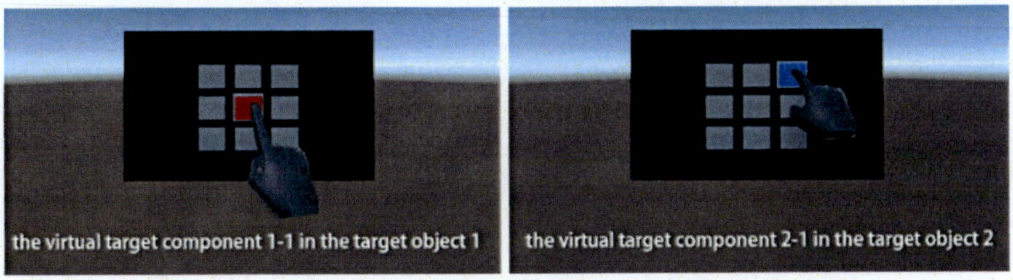

Fig. 1. Case based on the finger-click interactive task

As shown in Fig. 1, the user interaction task is that the user clicks the virtual target component 1–1 in the target object 1 with one hand to enter the next target object 2, and then clicks the virtual target component 2–1 in the target object 2. The quantitative decomposition based on the user interactive task in Fig. 1 is shown in Table 2:

Table 2. The quantitative decomposition based on the user interaction task in Fig. 1

Decomposition of the user interaction task	Model operator decomposition
1. Move the hand from the initial position into the gesture trigger area in a natural state	I
2. Point to the virtual target component red button 1–1 in the target object 1	IP
3. Click the red button	IPC
4. Move the hand out of the gesture trigger area and restore it to its natural state	IPCH
5. Move the hand from outside the gesture trigger area into the gesture trigger area in a natural state again	IPCHI
6. Point to the virtual target component blue button 2–1 in the target object 2	IPCHIP
7. Click the blue button	IPCHIPC
8. Move the hand out of the gesture trigger area and restore it to its natural state	IPCHIPCH

Applying the configured interaction rules to this case, we can determine and list the operation sequence set for the user to complete the task of this case.

According to rule 1, perform the initial insertion of task thinking operator A and perceptual reaction operator E, and get AEIAEPCEHAEIAEPCEH.

According to rule 2, perform the deletion of the expected task thinking operator A and perceptual reaction operator E, and get: AEIPCEHAEIPCEH.

According to rule 3, after the clicking operator C is executed, wait for the system operator R which includes system operation R1 and system feedback R2, and get: AEIPC (R1 + R2) EHAEIPC (R1 + R2) EH.

According to rule 4: the system feedback R2 overlaps with the perceptual reaction operator E and the homing operator H. In this case, the system feedback time is less than the sum of the perceptual reaction operator time and the homing operator time. Therefore, accumulate all the operators, calculate according to the operator time value in Table 1, and get the total time: $A + E + I + P + C + R1 + E + H + A + E + I + P + C + R1 + E + H = 1.2s + 0.24s + 0.5s + \left[b \times \log_2\left(\frac{D}{W} + 1\right) + a\right]s + 0.3s + 0.1s + 0.24s + 0.5s + 1.2s + 0.24s + 0.5s + \left[b \times \log_2\left(\frac{D}{W} + 1\right) + a\right]s + 0.3s + 0.1s + 0.24s + 0.5s = 6.16 + 2P = 7.717s$, where D = 0.085 m, W = 0.05 m, b = 0.381, a = 0.2323. It can be seen that the user interaction task described in Table 2 is evaluated using the method proposed in this paper, and the predicted total time for executing the user interaction task is 7.717 s.

4 Discussion and Future Work

The behavior described using GOMS task analysis is the behavior of experienced users, which means that the task time of these people will always remain stable, and the error rate is also the smallest [2]. Therefore, the main target group of the evaluation method proposed in this paper are also experienced users, and the execution time of any task predicted by this method will be the best execution time under the task. This paper has studied the modeling method to evaluate the finger-click interactive task in the virtual environment. However, the interaction in the virtual environment is not limited to this, future research can verify and analyze the effectiveness of this evaluation method, and may extend it to the virtual environment based on other interaction modes.

5 Conclusion

Based on the GOMS model, this paper proposes an evaluation model that can be used for the finger-click interactive task in the virtual environment. The evaluation model is a quantitative evaluation model that can quantify the user's interactive operation process. The behavior model constructed by the quantitative evaluation model represents the behavior model of each skilled user, the model can be verified, and the verification of the model can use actual data.

Acknowledgment. This work is supported in part by the National Natural Science Foundation of China (No. 71901061, No. 71871056), Science and Technology on Avionics Integration Laboratory and Aeronautical Science Fund (No. 20185569008), Equipment Pre-Research Foundation of China (No. 41412040304), Fundamental Research Funds for the Central Universities (No. 2242019k1G016).

References

1. Milgram, P., Kishino, F.: A taxonomy of mixed reality visual displays. IEICE Trans. Inf. Syst. **77**(12), 1–15 (1994)
2. Card, S.K., Moran, T.P., Newell, A.: The Psychology of Human-Computer Interaction (1983)
3. Gilbert, S.B.: Perceived realism of virtual environments depends on authenticity. Presence **25** (4), 322–324 (2016)
4. Marsh, T.: Evaluation of virtual reality systems for usability. In: Conference on Human Factors in Computing Systems, pp. 61–62. Association for Computing Machinery, New York (1999)
5. Smith, J.D., Gore, B.F., Dalal, K.M., Boyle, R.: Optimizing biology research tasks in space using human performance modeling and virtual reality simulation systems here on earth. In: 32nd International Conference on Environmental Systems, SAE Technical Paper Series, San Antonio (2002)
6. Chang, Y., Devine, K.L.: Modeling cognitive activities in a virtual reality- assisted industrial robot programming environment. In: 73rd EDGD Midyear Conference, pp. 99–104. ASEE, Berkeley (2019)

User Cognitive Abilities-Human Computer Interaction Tasks Model

Jinshou Shi[1(✉)], Wenzhe Tang[2], Ning Li[1], Yingwei Zhou[1], Tuoyang Zhou[1], Ziang Chen[1], and Kaili Yin[1]

[1] China Institute of Marine Technology & Economy, Beijing 100000, China
beihangsjs@163.com, 15210504416@163.com, zhouyingwei_bh@126.com, tuotuo0207@163.com, 18612787694@163.com, yinkaili1012@163.com
[2] School of Mechanical Engineering, Southeast University, Nanjing 211189, China
wenzhe.tang@seu.edu.cn

Abstract. At present, human-computer interaction tasks are complex, and the user abilities required for tasks are not clear. The interface design has problems such as high design cost and inconvenient evaluation. This research decomposes the ship's human-computer interaction tasks into therbligs, establishes the mapping relationship between therbligs and user cognitive abilities, and further establishes a model for human-computer interaction tasks and user cognitive abilities through analytic hierarchy process. This research has clarified the correspondence between human-computer interaction tasks and users' cognitive abilities and measured the performance value of each ability under certain conditions through experiments. Users can obtain the performance of a certain cognitive ability in a specific interface and environment by looking up the table, and thus determine whether the interface design is reasonable. This model can quickly and easily find out the abilities required for each task, and quickly evaluate the interface design.

Keywords: Cognitive ability · Human-computer interaction · Model

1 Introduction

User abilities are affected by three factors, namely environment, human-computer interface and user's physiology and psychology. Therefore, this study mainly considers these three types of factors.

The environment of the crew is a special environment different from that of the land, which will affect the crew's physiology and psychology, and then affect the crew's various cognitive abilities. The working conditions of the ship are special, the cabin working space is closed and narrow, the light is limited, the environment is noisy, and it is accompanied by the sway of the ship. These environmental factors affect the safety, comfort and work efficiency of the crew, of which the sway of the ship is the most critical. When a ship is performing its mission at sea, it is swayed by the

D. Russo et al. (Eds.): IHSI 2021, AISC 1322, pp. 194–199, 2021.
https://doi.org/10.1007/978-3-030-68017-6_29

disturbance of the waves. If the ship is treated as a rigid body, its motion on the sea has 6 forms, 3 of which are linear motions: surge, sway and heave; 3 types are rotary motions: pitch, roll and yaw [1]. Theses motions of the ship are important reasons that affects the crew's work quality and physical discomfort.

The information flow of the ship's human-computer interaction interface is complex, with many constituent elements, which is a typical human-computer interaction interface of a complex information system. Navigation management and document management in the ship control system are important basis for the crew to obtain information and make decisions. The elements of the ship's human-computer interaction interface mainly include layout, controls, icons, colors, text and interaction.

Human abilities can be divided into cognitive ability, operational ability and social ability. Among them, cognitive ability has the greatest impact on human-computer interaction tasks. Cognitive ability is the ability of the brain to process, store and retrieve information, including observation, perception, attention, memory, and thinking. The typical representatives of cognitive abilities are visual search ability, managing cognitive load ability, selective attention ability, visual working memory ability and situation awareness ability [2–4].

Human-computer interaction tasks are diverse, and each task has fixed compositions, namely therbligs. The therbligs are matched with abilities through expert questionnaires, and the required abilities for each task are determined by the type and number of therbligs contained in the task [5].

For example, a human-computer interaction task is shown in Table 1.

Table 1. Task flow of a human-computer interaction task

Task	Operations
Task1	D1
	D2
	D2
	D3
	D5
Task2	D1
	D2
	D3
	D2
	D2
	D2
	D4
	D2

There are five therbligs in this task: D1, D2, D3, D4 and D5. The abilities to match each therblig are shown in Table 2.

Through the corresponding relationship between therbligs and abilities, the importance ranking of five abilities in the human-computer interaction task can be obtained. Use analytic hierarchy process to obtain the weight of each ability in the human-computer interaction task [6]. The result can be obtained through calculation. The weight of each ability in the human-computer interaction task is as follows.

Table 2. Therbligs and corresponding abilities

Task	Therblig	A	V	S	M	L
Task1	D1	○	○	●	○	○
	D2	○	○	●	○	○
	D3	○	●	●	●	●
	D5	●	○	●	○	●
Task2	D1	○	○	●	○	○
	D2	○	○	●	○	○
	D3	○	●	●	●	●
	D4	○	●	●	●	○

Note:1. ● indicates that the ability is necessary for this task, ○ indicates that the ability is not necessary for this task
2. A: situation awareness ability; V: visual search ability; S: selective attention ability; M: visual working memory ability; L: Managing cognitive load ability.

Selective attention ability (0.3940) > Visual search ability (0.2341) > Visual working memory ability (0.1240) = Managing cognitive load ability (0.1240) = Situation awareness ability (0.1240). Therefore, selective attention ability is the most important for a human-computer interaction task. That is to say, a new task only needs to be analyzed according to the model as the above steps, and the user ability that the task requires the most can be obtained according to the proportion of different therbligs. Next, we need to determine under what circumstances the five abilities have the highest performance value.

2 Methods

In order to determine under what circumstances one ability has the highest performance value, it is necessary to conduct experimental research on the ability. Take the selective attention ability as an example. The experiment is divided into two groups: shape complexity and color hue. The shape complexity is characterized by graph entropy. According to the graph entropy difference 0.045bit as an interval, five factor levels are designed [7–9]. The color hue takes 72° as an interval, and five factor levels are designed. According to the guidance of "Environmental Test and Engineering Guidelines for Ship Equipment", pitch and roll are the main sway forms, and three factor levels are designed accordingly. The number of factor stimuli at each level in the experiment is 5.

The experimental participants consisted of 30 graduate students, including 15 males and 15 females, with an average age of 25. They were physically and mentally healthy and had no history of mental illness. They were all right-handed, with normal or corrected vision, and no color blindness.

In the formal experiment, participants press the Space key to start the experiment after reading the instruction. The interface will first present the detection items for 2000 ms, and then present the stimulus group after a delay of 1000 ms. At the same time, the interface simulates the environment of pitch, roll or still. Participants need to determine whether the detection item has appeared in the stimulus group. Press the "F" key to indicate that it has occurred, and the "J" key to indicate that it has not occurred. Each level of the experiment material was repeated 6 times, and 60 invalid trials are included, for a total of 240 trials. The experiment was divided into 4 groups, with 1-minute rest time between each group. Each participant needs 25 min to complete the experiment. The experimental flowchart is shown in Fig. 1.

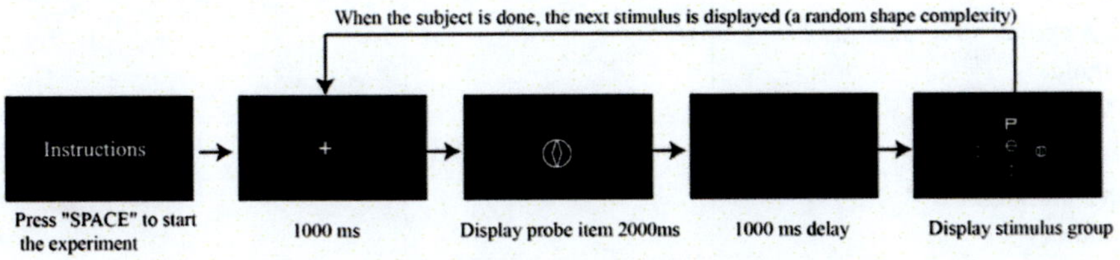

(a) Experimental flowchart (shape complexity)

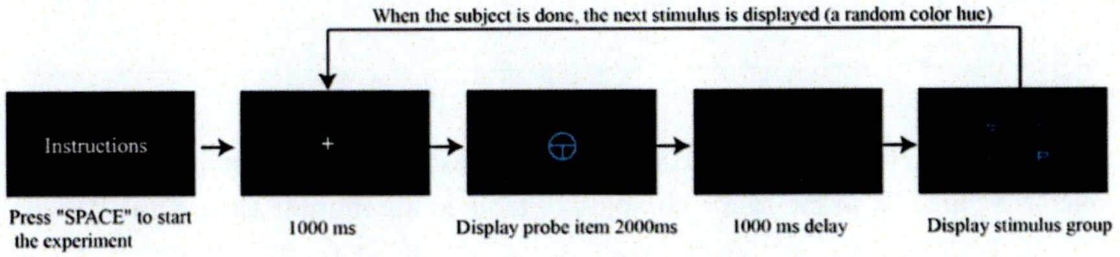

(b) Experimental flowchart (Hue)

Fig. 1. Experimental flowchart

3 Result

There is a total of 30 valid data of the experimental results. The response time and accuracy rate of the participants are calculated, and the performance table of selective attention ability is drawn as shown in Table 3.

Table 3. Selective attention ability performance

Shape complexity			
Performance (ACC/RT)	X_2^1(Pitch $\leq \pm 6.19°$)	X_2^2(Roll $\leq \pm 22.7°$)	X_2^3(Still)
X_1^1(5 stimuli 0.096–0.141 bit per stimulus)	1.25	1.31	1.30
X_1^2(5 stimuli 0.142–0.187 bit per stimulus)	1.09	0.92	1.12
X_1^3(5 stimuli 0.188–0.233 bit per stimulus)	1.09	1.08	1.19
X_1^4(5 stimuli 0.234–0.279 bit per stimulus)	0.97	0.93	1.04
X_1^5(5 stimuli 0.280–0.325 bit per stimulus)	1.01	0.87	1.02
Hue			
Performance (ACC/RT)	X_4^1(Pitch $\leq \pm 6.19°$)	X_4^2(Roll $\leq \pm 22.7°$)	X_4^3(Still)
X_3^1(5 stimuli H = 67.2 per stimulus)	1.25	1.07	1.19
X_3^2(5 stimuli H = 139.2 per stimulus)	0.94	0.85	0.99
X_3^3(5 stimuli H = 211.2 per stimulus)	1.38	1.31	1.20
X_3^4(5 stimuli H = 283.2 per stimulus)	1.07	1.18	0.98
X_3^5(5 stimuli H = 355.2 per stimulus)	1.18	1.03	1.19

It can be seen from the performance results that the performance under the conditions of X_1^1 and X_2^2 in the shape complexity table is the highest, that is, the condition for the maximum selective attention ability is that the icon shape complexity is 5 stimuli, 0.096–0.141 bit per stimulus, roll angle is less than or equal to $\pm$ 22.7°. In the hue table, the performance under the conditions of X_3^3 and X_4^1 is the highest, that is, the condition for the maximum selective attention ability is that the hue is 5 stimuli, H is 211.2 per stimulus, and the environmental condition is that the pitch angle is less than or equal to $\pm$6.19°. The table user can also evaluate existing interfaces according to the table to determine whether the interface design is conducive to the selective attention ability of the user.

4 Discussion

Our research results indicate under what conditions the ability will achieve the best performance, and the conditions for different abilities to achieve the best performance are different, mainly because the cognitive resources used by each ability are different, and the interface elements investigated in the experiment are different. The performance table obtained in this research matches the interface design elements with the user's ability in a standardized form, which is convenient for users to look up the table. The method can also be extended to other fields and has important academic and economic value. The shortcoming of this research is that there are fewer interface elements. Subsequent research can expand the interface elements and make related performance tables to make the interface design better.

5 Conclusion

This paper uses the decomposition of interactive tasks to associate cognitive abilities with interactive tasks and uses analytic hierarchy process to calculate the weight of each cognitive ability in interactive tasks. And through experiments, the conditions for cognitive ability to achieve optimal performance are clarified, which provides guidance for the design of human-computer interaction interface and the configuration of interactive task environment. In addition, quantifying the correlation factors between cognitive ability and design elements through experiments can also assist users in evaluating the applicability of existing interfaces to interactive tasks.

Acknowledgments. This work was supported by Equipment Development Pre-Research Joint Fund (No. 6141B03020602).

References

1. Ghadimi, P., Tavakoli, S., Dashtimanesh, A.: Coupled heave and pitch motions of planing hulls at non-zero heel angle. Appl. Ocean Res. **59**, 286–303 (2016)
2. Burggraaf, R., van der Geest, J.N., Hooge, I.T.C., et al.: Developmental changes in visual search are determined by changing visuospatial abilities and task repetition: a longitudinal study in adolescents. Applied Neuropsychology: Child, 1–11 (2019)
3. Yuan, X., Zhang, X., Wang, Y., et al.: The visual cocktail party problem: enhancement of visual selective attention through phase entrainment to auditory streams (2019)
4. Stasielowicz, L.: How important is cognitive ability when adapting to changes? A meta-analysis of the performance adaptation literature. Pers. Individ. Differ. **166**, 110178 (2020)
5. Hirano, Y., Kushida, D., Matsumoto, H.: Contactless motion analysis system using a kinect and musculoskeletal model. In: 2017 IEEE Life Sciences Conference (LSC), pp. 308–311. IEEE (2017)
6. Liu, Y., Eckert, C.M., Earl, C.: A review of fuzzy AHP methods for decision-making with subjective judgements. Expert Syst. Appl. **161**, 113738 (2020)
7. Ning, J., Lu, T., Liu, L., et al.: The optimization and design of the auto-exposure algorithm based on image entropy. In: 2015 8th International Congress on Image and Signal Processing (CISP), pp. 1020–1025. IEEE (2015)
8. Chen, J., He, J., Su, G.: Combining image entropy with the pulse coupled neural network in edge detection. In: 2010 IEEE International Conference on Image Processing, pp. 1637–1640. IEEE (2010)
9. Wang, L., Wang, Y., Zhang, H., et al.: Bundle of paper money recognition based on image entropy. In: International Conference on Image Processing and Pattern Recognition in Industrial Engineering. International Society for Optics and Photonics, vol. 7820, p. 782026 (2010)

The Fear of Learning Statistics in Latin America: Computational Solutions

Carlos Ramos-Galarza[1,2](✉), Omar Cóndor-Herrera[2], Valentina Ramos[3], Mónica Bolaños-Pasquel[2], Pamela Acosta-Rodas[1], Janio Jadán-Guerrero[2], and Hugo Arias-Flores[2]

[1] Facultad de Psicología, Pontificia Universidad Católica del Ecuador, Av. 12 de Octubre y Roca, Quito, Ecuador
caramos@puce.edu.ec

[2] Centro de Investigación en Mecatrónica y Sistemas Interactivos MIST/Carrera de Ingeniería en Ciencias de la Computación/Maestría en Educación Liderazgo e Innovación/Carrera de Psicología/Carrera de Administración de Empresas, Universidad Tecnológica Indoamérica. Av. Machala y Sabanilla, Quito, Ecuador
{omarcondor, janiojadan, hugoarias}@uti.edu.ec

[3] Sistemas de Información Gestión de la Tecnología e Innovación (SIGTI-Research Group), Escuela Politécnica Nacional, Quito, Ecuador
valentina.ramos@epn.edu.ec

Abstract. In Latin American countries, it has been noticed that a high percentage of university students experience negative feelings such as fear, rejection, and inadequate beliefs about the contents taught in statistical courses, resulting in avoidant behaviors towards their learning and student's low performance. The majority of students are able to obtain the minimum grade required to pass the subject, but, in the end, they do not obtain significant knowledge. This fact has caused many professions in Latin American countries, especially those related to social and human sciences, to show a scientific delay. It is identifiable by the low rankings achieved by these countries in scientific production. In this context, this work focuses on the solutions that computer science offers and must be considered to improve the experience of learning statistics. In this sense, misconceptions, avoidant behaviors, and negative beliefs towards learning statistics, would be reverted for students and future professionals, as well as the improvement of inappropriate emotional reactions about learning statistical skills. Thanks to technological advances, for example, applications and social networks, which nowadays determine Latin American university student's daily-life behavior.

Keywords: Statistics · University education · Computational solutions · Fear of statistics

D. Russo et al. (Eds.): IHSI 2021, AISC 1322, pp. 200–205, 2021.
https://doi.org/10.1007/978-3-030-68017-6_30

1 The Fear of Learning Statistics: The Latin American Context

In Latin America, it is common for university students to present a negative behavior regarding learning Statistics. A familiar feeling related to this course is fear, as well as the presence of ideas against the benefits that learning statistics will bring to their professional future, and, in this sense, university students of human and social fields think they do not fit for this type of learning [1].

One of the main problems that these thoughts and feelings generate in university students is the avoidant behavior that is usually observed towards statistical learning; thus, even when the student can pass the course with the required qualification, the knowledge acquired is not significant [2]. Figure 1 shows a university student's avoidant behavior and fear of statistics.

Fig. 1. This figure represents the **emotional** reaction of fear and anxiety that a university student usually manifests when learning statistics.

These students' cognitive beliefs, feelings, and behaviors when learning statistics have caused that several professions that are formed at the university level in Latin American, especially those related to the social and human sciences, to experience a scientific delay, evidenced by being located at the last places of the world's scientific production rankings [3, 4].

Based on the description made, and as a proposal to develop in this article, computer sciences that seek to offer a solution to this problematic affecting the Latin American educational context, there are considered some challenges and problems to be solved in favor of statistics' significant learning: changing negative beliefs about learning statistics and highlighting the benefits that statistical knowledge will bring to the student's professional future, managing inappropriate emotional reactions when learning statistical skills, promoting real applications of statistical knowledge and skills, making this possible thanks to the usage of technological advances such as applications and social networks that determines Latin American university students' daily life [5].

2 Computational Solutions in Favor of Learning Statistics

Various technological proposals have been developed in favor of learning statistics. As it may be found below, several games, mobile applications, and web platforms are used with a statistical learning purpose.

2.1 Statistics Learning Platform Based on Harry Potter's Movie

An interesting proposal for learning statistics is based on challenges from Harry Potter's film, which presents situations to be solved based on this story, creating a favorable psychological condition for learning this subject [6] (Fig. 2).

Fig. 2. Graphical illustration applied to the platform for learning statistics based on the Harry Potter movie.

The main benefit of this type of application is the physical and emotional state that this simulated situation provides in favor of the students learning since it generates a condition in which the student will have to play to learn and will have to solve challenges as one of the film's main characters allowing the students to enjoy learning statistics, leaving behind the feelings of fear, the avoidant behavior and obtaining meaningful learning.

2.2 Probability Distributions APP for Learning Statistical Skills

This free application allows students from different levels to access the learning of statistical probability analysis. On this application, the student will be able to perform statistical analysis and to solve statistical problems. By conducting these procedures, statistical skills will develop, and learning statistics will be more exciting and more manageable when the student uses his/her mobile device. Figure 3 presents the content of this application [7].

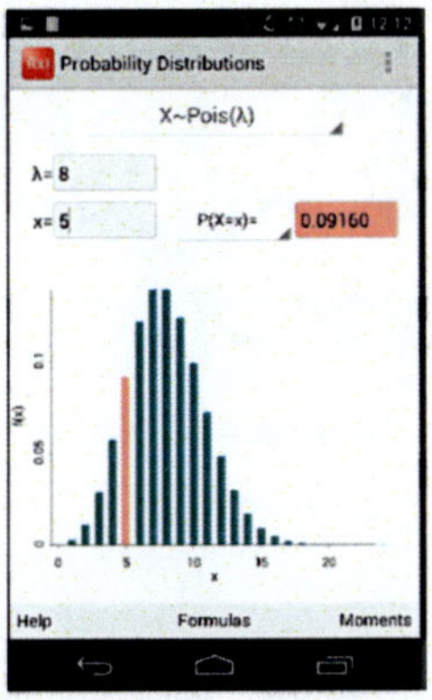

Fig. 3. Screenshot of this application

2.3 Mobile App for Psycho-Statistics Learning

The application for learning psycho-statistics addresses psychology students, who will perform better in this subject. Figure 4 presents a screenshot of this smartphone application [5].

One of the benefits of using this type of mobile application lies in the technological naturalness that exists nowadays for students, especially when learning statistics. The reason that explains this phenomenon relies on the technical usage that is currently provided to carry out most of the educational activities. Therefore, learning statistics using the own smartphone generates an ideal learning environment for the student [8].

2.4 Website: Educatina

On this website, students will find a series of exercises to improve their statistical and probability skills. This platform provides excellent content to the teaching-learning process in favor of students learning statistics [9] (Fig. 5).

Fig. 4. Screenshot of this application

Fig. 5. Screenshot of this application

3 Conclusions

In n this article, a review of computational solutions in favor of the statistical learning process has been realized. For this analysis, there has been taken into account and identified virtual platforms, as well as the contribution of applications developed for smartphones and websites that contribute to the process of learning and developing statistical skills.

A very important aspect that justifies the use of technology when learning statistics relies on the regular usage of technological devices that university students handle nowadays, for example, the easy accessibility and usage of smartphone applications and websites, making learning statistics interesting, challenging, and even fun, where there is no place for fears nor anxiety or avoidant behavior towards this subject. It explains why these devices and software must be involved in their learning process, overcoming the traditionalist method, and engaging students through their active participation when learning statistics, raising feelings of satisfaction. In the end, their learning becomes significant.

The future great challenge for the research team is to develop technological innovations that benefit the process of learning statistics in the Latin American university context. For example, we plan to create mobile applications that contribute to the learning process on developing numerical skills and, in this sense, change negative and inappropriate beliefs and behaviors towards learning statistics.

References

1. Aparicio, A., Bazán, J.: Affective aspects involved in learning statistics, attitudes and their forms of evaluation. In: Lestón, P. (ed.) Acta Latinoamericana de Matemática Educativa. Comité Latinoamericano de Matemática Educativa AC, México DF (2008)
2. Comas, C., Martins, J., Nascimento, M., Estrada, A.: Study of attitudes towards statistics in psychology students. Bolema-Math. Educ. Bull. **31**(57), 479–496 (2017). https://doi.org/doi.org/10.1590/1980-4415v31n57a23
3. Ramos-Galarza, C., Bolaños-Pasquel, M., Ramos, V., Moscoso-Salazar, J., Jadán-Guerrero, J.: Critical status of research in ecuadorian psychology: the abandonment of statistics as a basis of scientific production. Psychol. Soc. Educ. **11**(3), 281–298 (2019)
4. Ramos-Galarza, C.: The abandonment of statistics in psychology Ecuador. Revista Chilena de Neuro-Psiquiatría **55**(2), 135–137 (2017). https://doi.org/doi.org/10.4067/S0717-92272017000200008
5. Acosta-Rodas, P., Bolaños-Pasquel, M., Jadán-Guerrero, J.: Mobile App for Psycho-Statistics Learning. Applied Human Factor and Ergonomics (2020). (in press)
6. Jadán-Guerrero, J.: Technological Platform for Learning Statistics. Universidad Indoamérica, Quito-Ecuador (2019)
7. Probability Distributions. https://play.google.com/store/apps/details?id=com.mbognar.probdist&hl=es_EC
8. Ruiz, E.: Use of technological applications in the treatment of probability and statistics subjects. Difficulties presented by students in formulating correct approaches. Ride **8**(16), 1–30 (2018)

PEST Approach to Managing the Bulgarian Media Ecosystem

Lilia Raycheva(✉)

The St. Kliment Ohridsky University, 15 Tsar Osvoboditel Blvd.,
1000 Sofia, Bulgaria
lraycheva@yahoo.com

Abstract. Contemporary societies are undergoing significant transformations which correlate with the dynamic developments of the information and communication technologies (ICTs). Today these transformations are being catalyzed by the intensity of the media ecosystem, combining the traditional media with the potential of the blogosphere, social networks and mobile communications. Situated in the context of globalization processes, the media themselves are undergoing multi-layered transformations. Using PEST analysis, the article examines the political, economic, social and technological management practices in the Bulgarian media ecosystem.

Keywords: Bulgaria · Management · Media ecosystem · PEST analysis

1 Introduction

The technology breakthrough in fields such as artificial intelligence, robotics, the Internet of things, 3D printing, etc. pose significant challenges to the contemporary media (traditional and social) ecosystem. In the age of the Fourth Industrial Revolution the fusion of physical, digital and biological worlds impacts strongly communication developments.

Rapid progress in ICTs stimulates the diversity of platforms through which audio-visual content is disseminated. Of particular importance is the consideration of whether the European democratic model pillars will retain their resilience within the highly competitive communication markets. No matter how positive the impact of ICT applications and media developments on progress in all areas of life might be, it is no less true that they pose challenges to the social stratification of society.

Based on PEST analysis of the media ecosystem in Bulgaria, this article concentrates on the political and regulatory commitment to fundamental social values when managing the new trajectories in the domains of technology, business and governance.

D. Russo et al. (Eds.): IHSI 2021, AISC 1322, pp. 206–211, 2021.
https://doi.org/10.1007/978-3-030-68017-6_31

2 Pest Analysis

Political Aspects

For the proper functioning of the contemporary media ecosystem, a number of political issues are becoming increasingly important, such as: freedom of expression and access to information; pluralism of opinion and variety of content; professional standards and journalistic ethics; transparency of ownership and accountability to the audience; protection of underage and vulnerable social groups; cooperation between regulation, self-regulation and co-regulation; the expansion of social media, etc. The sustainability of these principles is a decisive factor for the democratic functioning of these human-centred societal developments.

Despite the rapid development of ICT and online services, television continues to be the most preferred source of information and entertainment for most European households [1]. Therefore, the main internationally acknowledged instruments in the field of TV broadcasting in the European Union (EU) continue to be the European Convention on Transfrontier Television (ECTT) of the Council of Europe [2] and the Audiovisual Media Service Directive (AVMSD) of the European Commission [3]. The national broadcasting frameworks of the EU member states are harmonized with the instruments of the ECTT and AVMSD.

Current media legislation in Bulgaria has been closely aligned with EU regulations. The two national institutions that regulate the electronic media are the Council for Electronic Media (CEM) and the Communications Regulation Commission (CRC). CEM is the regulatory body that monitors compliance with Radio and Television Act [4]. CRC enforces the Law of the telecommunications and manages the radio frequency spectrum [5]. Self-regulation is managed by the Ethical Code of the Bulgarian Media [6].

Currently 223 newspapers (32 dailies) with annual circulation of 169 261 000 and 595 magazines and bulletins with annual circulation of 9 016 000 are on the market place. In 2019, 82 radio stations and 128 television stations have been listed, operating on national, regional and local level terrestrially, via cable or via satellite. The privately owned radio- and TV stations undoubtedly challenged the monopoly of the state-owned public television [7].

In addition to traditional media and online-only news sites, using of other social media platforms, as well as networking and microblogging services such as Facebook, Google Plus, Instagram, Twitter and hashtags, is becoming more and more popular.

The dynamics of the developments in communication is so intense that it is difficult to define whether because of technological and economic convergence it will evolve in the context of ever-increasing deregulation in favour of the market or in the process of a co-regulation, which serves the public interest.

Economic Aspects

In contemporary times, internationalization of economy and convergence of modern communications favour the prevalence of multi-sector and multinational corporations. At the same time, the globality and uniformity of the Information Society are dissolving into the separatism of glocality. The media and telecommunications sectors remain among the industries that have felt the strongest impact because of digital transformation [8].

Media pluralism in the contemporary e-communication environment can be assessed through the number and type of media outlets, the number and structure of their owners, the editorial content and the access of different societal groups to media channels. The World Economic Forum, in its White Paper on *Digital Transformation of Industries: Media, Entertainment and Information* (MEI) (2016), points out the widespread recognition among different stakeholders that the role of digital technology is rapidly shifting from being a driver of marginal efficiency to an enabler of fundamental innovation and disruption. In the process of digital transformation, some basic challenges occur, among them customer expectations; demographic and cultural attitudes; and media and communication literacy skills [9].

The trend towards media concentration in Bulgaria resulted from the process of media market build-up and structuring during the three decades following the political and economic changes of 1989. This trend is in pace with world market developments. In Bulgaria it was characterized by the inflow of foreign capital and players, especially after the country's accession to the European Union in 2007. It helped the establishment of successful business models across the entire range of media production, but it depersonalized media diversity in the smaller inhabited places.

In an open and liberalized market, consumers are the ones who ultimately determine the speed of penetration and dissemination of digital services. That is why pan-European and national measures stimulating further efficient use of the radio frequency spectrum are of great economic, social and cultural importance.

Social Aspects

We live in information age [10] when telecommunication networks are being defined as the digital nervous system of modern economic and social life [11].

In times of intensive social, economic and technological transformation, the very paradigm of the media changes as well. The media are becoming convergent phenomena with respect to content and production and dissemination of content. The rapid development of ICTs has transformed the media – from linear to non-linear services – and the audience – from passive consumers to active prosumers.

The inclusion of audiences in the process of content creation, perception and distribution conditions the remodelling of the media ecosystem. Converged communication reality and usage of hybrid products prompt energetic moves towards personalization, contextualization and fragmentation of content. These moves are significant to understanding such phenomena as information overload and digital fatigue that consumers are facing. In addition, artificial intelligence of the future will significantly affect the creative process within the media and challenge its human-centred design of communication.

The creative potentials of the new information and communication environment appear to be a key factor in the development of the Bulgarian media reality. More than 21% of the Bulgarians use YouTube, while 14.3% use Google+, 7.9% use Twitter and 5% use LinkedIn [12]. According to the forecasts in 2022 the number of monthly active Facebook users is projected to reach 2.40 million individuals. This would be an increase of approximately 250,000 new users from 2.15 million users in 2016 [13].

After the massive civil protests of 2013 that were fomented by the widespread involvement of social networks, the influencing effects of various communication channels were acknowledged by the Bulgarian citizens. It is indicative that Facebook (as an effective tool for the organization of protest demonstrations), Julian Assange and Edward Snowden, as well as the two bloggers from North Africa (Egypt and Tunisia) who provoked the Arab Spring, were among those nominated for the Noble Peace Prize in 2013. Traditional media, even radio and television, despite their simultaneousness, are beginning to dramatically lag behind in the race for consumer attention.

Technological Aspects

Of all the factors affecting the construction rate of the new Global information society, technology is undoubtedly the most dynamic. The ICTs have outpaced regulation and have set up an economic basis for the convergence of entire industries: electronics, entertainment, media. Santiago Lorente identified two stages in technological development: convergence between telecommunications and informatics (telematics) and convergence between telematics and audio-vision (mediamatics) [14]. The notions of 'total technocracy' [15] and the 'technetronic' [16] internationalization of communication processes imply the maximum involvement of all inhabitants of the Earth in the construction of the Global Information Society. Therefore, media and information literacy skills acquire additional importance in today's intercultural dialogue in the communication environment.

Bulgaria is one of the countries with the fastest Internet connections in the world [17]. According to the 2017 ranking among 149 countries worldwide Bulgaria ranked 23rd (and 11th among EU member states) with an average Internet connection speed of 15.5 Mbps. As per indicator 4 Mbps Broadband Adoption (IPv4), Bulgaria ranked 2nd in the EU after Malta and 7th in the world with 97%. According to the same report, Bulgaria ranked 22nd with 9.5 Mbps among EU countries in regard to the usage of smartphones, tablets, computers and other devices that connect to the Internet through mobile network providers.

Regarding open data access Bulgaria ranks among the 10 countries defining the new trends in the EU [18].

Today, transformations in the communication environment are catalysed by the emerging capabilities of the blogosphere and social networks, and of mobile communications, which have also led to a paradigm shift in the media – from mass media means to individual media services.

3 Conclusion

The rapid development of ICTs has outstripped the regulatory framework, regulatory practices and theoretical rationalization of the media transformation process. This new 'mosaic culture' is characterized by demassification of media and of society itself [20]. A virtual online culture has been created which, due to its interactive nature, acts as integrating while having an alienating and restrictive impact on people, destroying 'live' communication.

This article attempts to summarize and analyse the Bulgarian media and communication environment in several aspects:

Politically: The development of free and unhindered transmission and distribution of audio-visual services needs improvement in the legal framework and in the regulatory-selfregulatoty-coregulatoty mechanisms for sustaining it.

Economically: The expanding tendency towards deregulation and privatization in online and offline production and distribution of media content requires specific measures to promote media pluralism. It is important to consider how the businesses and the audiences of the media and entertainment industries (broadcasting, publishing, advertising and gaming) will gain more value than disruption in the contemporary media and communication ecosystem.

Socially: Technological innovations will bring forth the need for more serious protection of the fundamental rights and freedoms of citizens (access to information, freedom of expression, diversity of sources, pluralism of viewpoints, protection of minors, etc.) and professional and ethical journalistic standards (provision of informed choice in electoral campaigns, avoidance of hate speech and infringement of personal integrity, etc.).

Technologically: The contemporary ICT developments are redefining many of the traditional parameters of the communication process, which need special policies.

In the 1930s, Aldous Huxley warned in his *Brave New World* that the time would come when mankind would perish uninformed, wallowing in a sea of information [20]. That time may not have come yet, but it becomes ever more obvious how the compression of historical time dictates the new pace of the management in the communication process, along with the good, the bad and the unexpected challenges of ICT.

Acknowledgments. The article has been developed within the framework of the EC COST Action CA16226 of the European Commission: *Indoor Living Space Improvement: Smart Habitat for the Elderly (SHELD-ON),* supported by the research project KP-06-COST/5-18/06/2019 of the National Research Fund of Bulgaria.

References

1. European Commission.: Standard Eurobarometer: Report 86: Media Use in the European Union. https://ec.europa.eu/commfrontoffice/publicopinion/index.cfm/ResultDoc/download/Document/79405
2. Council of Europe.: European convention on transfrontier television. http://conventions.coe.int/Treaty/EN/Treaties/Html/132.htm
3. European Commission Directive 2007/65/EC of the European Parliament and of the Council Amending Council Directive 89/552/EEC on the Coordination of Certain Provisions Laid Down by Law, Regulation or Administrative Action in Member States Concerning the Pursuit of Television Broadcasting Activities. http://eur-lex.europa.eu/LexUriServ/LexUriServ.do?uri=CONSLEG:1989L0552:20071219:EN:PDF
4. Radio and Television Act. http://www.cem.bg

5. Telecommunications Act: http://www.crc.bg/files/_en/LAW_OF_THE_TELECOMMUNICATIONS.htm
6. Union of Bulgarian Journalists.: Code of Ethics of Bulgarian Media. http://www.sbj-bg.eu/index.php?t=58
7. National Statistical Institute.: Culture. http://www.nsi.bg/en/content/4526/culture
8. Raycheva L.: The digital notion of the citizen-centered media ecosystem. Int. J. Digital Telev. **9**(3) (2018)
9. World Economic Forum.: World economic forum white paper: digital transformation of industries: media, entertainment and information. http://reports.weforum.org/digital-transformation/wp-content/blogs.dir/94/mp/files/pages/files/wef-dti-mediawhitepaper-final-january-2016.pdf
10. Castells, M.: The Rise of the Network Society: Economy, Society, and Culture. LIK, Sofia (2004)
11. Gates, B.: Business with the Speed of Light: Digital Nervous Systems. Ciela, Sofia (1999)
12. Passport to Trade 2.0.: Social Media Guide for Bulgaria. https://businessculture.org/eastern-europe/bulgaria/social-media-guide/
13. Statista.: Forecast of Facebook User Numbers in Bulgaria from 2015 to 2022 (in million users)", The Statistics Portal. https://www.statista.com/statistics/568753/forecast-of-facebook-user-numbers-in-bulgaria
14. Lorente, S.: The Global House: New User Opportunities in Automation and Information. Mimeo, New York, p. 119 (1996)
15. Postman, N.: Technopoly. Vintage Books, New York (1992)
16. Brzezinski, Z.: Between Two Ages: America's Role in the Technetronic Era. Praeger, New York (1982)
17. Akamai Technologies.: State of the Internet – Connectivity Report, Cambridge, MA, USA. https://www.akamai.com/uk/en/multimedia/documents/state-of-the-internet/q1-2017-state-of-the-internet-connectivity-report.pdf
18. Open Data Bulgaria.: Open data Portal of the Republic of Bulgaria. https://opendata.government.bg/
19. Toffler, A.: The Third Wave. Peyo K. Yavorov Publ. House, Sofia, vol. 302, pp. 211 (1991)
20. Huxley, A.: Brave New World, Biblioteka Galaktika, Sofia, No 104 (1990)

Implementation of Augmented Reality into Student Practical Skills Training

Dana Dobrovská(✉) and David Vaněček

Czech Technical University in Prague, Masaryk Institute of Advanced Studies, Kolejní 2a, 160 00 Prague 6, Czech Republic
{dana.dobrovska, david.vanecek}@cvut.cz

Abstract. Augmented reality is eligible to be used in educational environments, and many applications successfully apply it to improve learning. It provides new possibilities for different spheres of education, but in these environments it is not yet investigated in its entirety. This paper's objective is to present experience with utilization of augmented reality in professional skills training at a secondary technical school and to discuss its advantages and limits.

Keywords: Augmented reality · Technical subjects · Teaching and training

1 Introduction

AR is one of the most emerging technologies in education these days. AR applications can be used in many different platforms such as desktops, notebooks and mobile devices. In the recent years, augmented reality has been given different meanings. Reference [3] defines AR based on the reality-virtuality continuum as augmenting natural feedback to the operator with simulated cues. The reality-virtuality continuum allows distinguishing between the concept of AR and concepts such as Virtual Environments (also known as Virtual Reality (VR)) and Augmented Virtuality (AV)). While VR deals with settings where the participant observer is totally immersed in a completely synthetic world, AV is concerned with environments in which "the primary world being experienced is in fact [2] predominantly 'virtual'" [4] and augmented with information from the real world. Additionally, reference [1] mentions a more restricted definition where AR is seen as form of virtual reality where the participant's head-mounted display is transparent, allowing a clear view of the real world.

AR is eligible to be used in educational environments, and many applications successfully apply it to improve learning. It provides new possibilities for different spheres of education, but in these environments it is not yet investigated in its entirety. Researchers and educators face different directions of AR applications, which may differ regarding their potential benefits.

D. Russo et al. (Eds.): IHSI 2021, AISC 1322, pp. 212–217, 2021.
https://doi.org/10.1007/978-3-030-68017-6_32

Some references suggest a classification of AR applications within five groups [6, 7]:

- **Discovery-Based Learning**
  A user is provided with information about a real-world place while simultaneously considering the object of interest. This type of application is often used in museums, in astronomical education, and at historical places.
  (In our view, teachers should include principle of discovery-based learning into everyday class practice as psychology of learning has brought evidence of improved cognition quality).
- **Objects Modeling**
  Augmented reality can also be used in objects modeling applications. Such applications allow students to receive immediate visual feedback on how a given item would look in a different setting. Some applications also allow students to design virtual objects in order to investigate their physical properties or interactions between objects. (This type of application can be also used in most technical educational programs).
- **AR Books**
  An immersive, interactive and active learning platform can be formed by applying AR technology into printed books, where despite their many superior qualities, information is provided in a static and non-interactive manner.
  AR books are books which offer students 3D presentations and interactive learning experiences through AR technology. The applications are often realized with head-mounted displays and are augmented with the help of technological devices such as special glasses. The first implementations of AR Books show that this kind of medium is likely to appeal to digital native learners, which makes it an appropriate educational medium even at the primary level.
- **Skills Training**
  Support of training individuals in specific tasks, especially in mechanical skills training, is described in this category. Such applications are, for instance, used in airplane maintenance, where each step of a repair is displayed, necessary tools are identified, and textual instructions are included. The applications are often realized with head-mounted displays.
- **AR Gaming**
  Video Games offer powerful new opportunities for educators - they have recognized and often use the power of games in educational environments. AR technology enables the development of games which take place in the real world and are augmented with virtual information. AR Games can give educators powerful new ways to show relationships and connections. Additionally, they provide educators at technically oriented schools with highly interactive and visual forms of learning.

In recent years, there has been an increasing interest in applying Augmented Reality (AR) to create unique educational settings. We may find some excellent examples of augmented reality in education worldwide. Ability to connect reality and digital content has been steadily improving, opening more options for teachers and students.

However, there is still a lack of review studies with focus on investigating factors such as: the uses, advantages, limitations, effectiveness, challenges and features of augmented reality in educational settings. Anyway, 14 different benefits of AR in our source literature of which two (Improved Learning Curve and Increased Motivation) account for more than 20% of all benefits mentioned [3].

2 Use of AR in Teaching Technical Subjects at a Secondary Technical School

We would like to discuss experience with utilization of AR in teaching a course *"Technical documentation"* at a secondary technical school in a study program *"Electronics mechanics"*. The course is implemented into the 3rd year of study program, the course plan covers 4 teaching units, one of them called *"3D modeling of a complex assembly*", with 7 teaching topics, with time allowance of 90 min each. There are 4 subtopics in the teaching unit: *Component set creation, Manipulation with glasses, Complex sets creation and Construction of suction cup frame*. Students are trained in groups of 10–12 participants. After the training experience, the learner is expected to be able to input various components into the assembly after the components had been assorted from component catalog.

In the introductory phase of training, students are prepared how to create components sets: they are informed how to work with the catalog of components, how to assort them, how to choose the proper components and how to put them into the set. Then they learn how to manipulate with glasses (deployment, function control), installation of data, model completion according to the displayed data and real model control (Fig. 1).

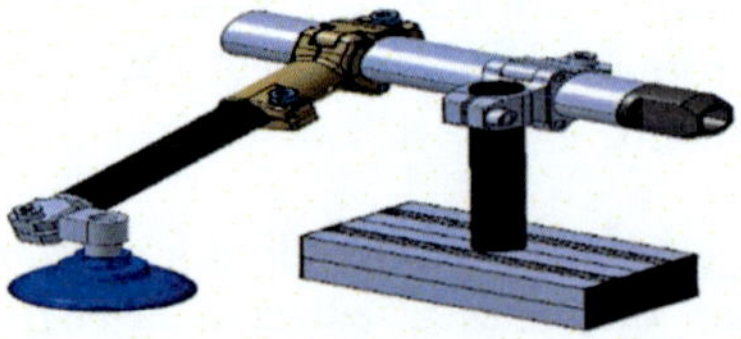

Fig. 1. Student training task [5]

In the second phase of training, students gradually learn how to display basic geometrical entities, to understand principles of quotation of geometrical shapes, creation of 3D components by protrusion or rotation, input standardized components from catalog, and prepare future model of complex machine parts (Fig. 2).

After students had been taught how to model components of the assembly they are subsequently asked to complete the assembly in their individual work in class. They are expected to work with database of components, choose the right components, define

Fig. 2. Example of a student work I [5]

positions of components, define mutual distance and angles of components, reflect completion of assembly and test the quality of assembly by deconstruction and reconstruction. Role of the class teacher is dominantly encouraging in this phase of training: he should stimulate and/or gently correct students in their individual work and back their suggestions and ideas.

In the last training phase students are supposed to construct *the suction cup frame*: select data for the model, set them into running, specify their reference point, complete model with the 3d glasses support and control final model (Fig. 3).

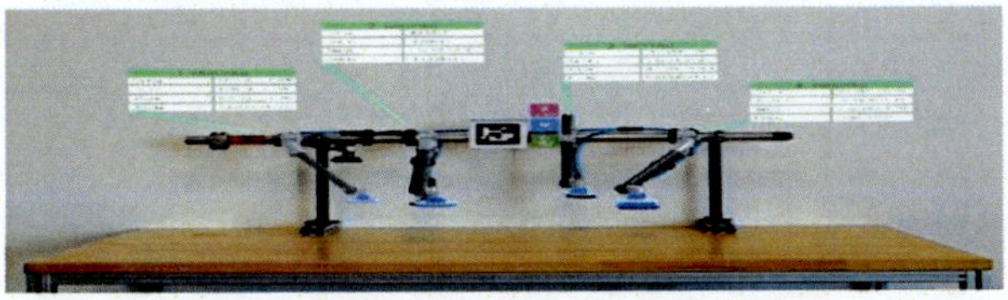

Fig. 3. Final model – a suction cup frame

3 Discussion

We held interviews with teachers who had prepared training course supported by AR at a secondary school, we discussed their experiences and tried to systemize their opinions how to improve teaching with AR.

Our first topic in discussion was the cost of AR equipment – buying all components still seems to be too expensive and few schools are able to invest into complete AR equipment (the cost of quality 3d glasses suitable for training was 6000 dollars in 2017, plus specialized software, computer equipment and components for student training cost additional several thousand dollars). For most schools, the cost is too high unless they apply for funding from government, industry or international resources.

Other point of discussion with teachers was their opinion on benefits and limits using AR in training. These were their remarks: humor moments occur during training…. new technology brings a strong motivational effect…. students are pulled in work, they get excited about it and motivated as many had enjoyed experience with AR and VR gaming. They are prepared to collaborate… they give advice to each other.

However, there are some limits, too:…students must be divided into small groups, 10–12 students are maximum, and it would be more convenient for an educator if the group were even smaller…. a good preparatory work of teacher is a condition for successful management of training in class, especially in the first phase when students are introduced into the AR use.

Once students lose their uncertainty about work with new technology more spontaneity and activity is observed. Teacher should encourage those students who remain passive. As to cognitive qualities, students seem to remember facts better and their readiness to complete the task seems higher than in "normal" classes.

As a result of the interview, we suggest a classification of the impact of AR on learning for further discussion, with special attention to possible educational and psychological effects on students:

Positive aspects of training backed by AR – the pedagogical potential of AR apps:

- **Cognition**
  - supporting deeper understanding *("aha" moment)*
  - improving conceptualizing complex (abstract) information
  - deeper understanding level
- **Emotions and motivation**
  - highly motivating for students
  - reducing learning stress (*imposter syndrome*) if backed by teacher's help
  - enjoying moments of humor when funny errors occur
- **Personality development**
  - encouraging student collaboration
  - building healthy assertiveness (asking for help)
  - improving individual persistence level when completing the task
- **Other**
  - AR usable for different courses and different school programs

Negative aspects of training backed by AR

- cost of equipment
- suitable for small groups of students only
- limits in mutual communication of apps
- existing data in the system cannot be easily actualized
- longer work with 3d glasses can be unpleasant for users
- computer performance should be faster and smoother if to be closer to reality

4 Conclusions

AR is eligible to be used in educational environments, and many applications successfully apply it to improve learning. But it should not be considered a magic bullet in educational environments. Each AR application is in its own way unique and therefore the identified benefits may not apply in each context. Each application has to be implemented thoroughly to prevent drawbacks in user interaction or system failures in order to profit from benefits. Special user groups can benefit in different as well as additional ways due to their requirements to learning methods and the characteristics of AR.

Acknowledgments. We take this opportunity to express our sincere thanks to Václav Šáfr for sharing his experiences with using AR in training students with technical background.

References

1. Bacca-Accosta, J., et al.: Augmented reality trends in education: a systematic review of research and applications. Educ. Technol. Soc. **17**(4), 133–149 (2014)
2. Barak, M., Ziv, S.: Wandering: a web-based platform for the creation of location-based interactive learning objects. Comput. Educ. **62**, 159–170 (2013). https://doi.org/10.1016/j.compedu.2012.10.015
3. Diegman, P., Schmidt-Kraepelin, M., Van den Eynden, S., Basten, D.: Benefits of augmented reality in educational environments – a systematic literature review. In: Conference: 12. Internationale Tagung Wirtschaftsinformatik, Osnabrück, Germany, pp. 1542–1556 (2015)
4. Milgram, P., Kishino, F.: A Taxonomy of mixed reality visual displays. IEICE Trans. Inf. Syst. E77-D (1994)
5. Šáfr, V.: Use of modern methodology in teaching technical subjects, Unpublished (2020)
6. Wrzesien, M., Raya, M.A.: Learning in serious virtual worlds: evaluation of learning effectiveness and appeal to students in the E-Junior project. Comput. Educ. **55**(1), 178–187 (2010). https://doi.org/10.1016/j.compedu.2010.01.003
7. Wu, H.-K., Lee, S.W.-Y., Chang, H.-Y., Liang, J.-C.: Current status, opportunities and challenges of augmented reality in education. Comput. Educ. **62**, 41–49 (2013a). https://doi.org/10.1016/j.compedu.2012.10.024

Multimodal Affective Pedagogical Agents for Different Types of Learners

Nicoletta Adamo[1(✉)], Bedrich Benes[1], Richard E. Mayer[2], Xingyu Lei[1], Zhiquan Wang[1], Zachary Meyer[1], and Alyssa Lawson[2]

[1] Department of Computer Graphics Technology, Purdue University, West Lafayette, IN 47907, USA
{nadamovi,bbenes,lei64,wang4490,meyer204}@purdue.edu
[2] Department of Psychological and Brain Sciences, University of California Santa Barbara, Santa Barbara, CA 93106-9660, USA
{rich.mayer,alyssa.lawson}@psych.ucsb.edu

Abstract. The paper reports progress on an NSF-funded project whose goal is to research and develop multimodal affective animated pedagogical agents (APA) for different types of learners. Although the preponderance of research on APA tends to focus on the cognitive aspects of online learning, this project explores the less-studied role of affective features. More specifically, the objectives of the work are to: (1) research and develop novel algorithms for emotion recognition and for life-like emotion representation in embodied agents, which will be integrated in a new system for creating APA to be embedded in digital lessons; and (2) develop an empirically grounded research base that will guide the design of affective APA that are effective for different types of learners. This involves conducting a series of experiments to determine the effects of the agent's emotional style and emotional intelligence on a diverse population of students. The paper outlines the work conducted so far, e.g., development of a new system (and underlying algorithms) for producing affective APA. It also reports the findings from two preliminary studies.

Keywords: Animated pedagogical agents · Computer animation · Emotions · Affective computing · Intelligent tutoring systems

1 Introduction

Research has shown that animated pedagogical agents (APA) can be effective in promoting learning, but not equally for all learner populations, learning subjects, and contexts. Despite growing evidence in support of the positive value of APA, many questions still remain unanswered, particularly concerning the emotional design of APA. For instance, it is unclear which types of emotional expression, degree of embodiment and personalization benefit a particular leaner population and why. To advance knowledge in this field and maximize the agent's positive impact on learning, there is a need to further investigate the effects of the agent's affective features, and whether they are moderated by learner characteristics, learning topics, and contexts.

D. Russo et al. (Eds.): IHSI 2021, AISC 1322, pp. 218–224, 2021.
https://doi.org/10.1007/978-3-030-68017-6_33

Specifically, with the growing understanding of the complex interplay between emotions and cognition, there is a need to develop life-like, convincing agents that not only provide effective expert guidance, but also believable emotional interactions with the learner. Although a few researchers have tried to equip intelligent agent systems with affective capabilities [2, 8, 9], much work still needs to be done in this area. In short, there is a need to explore the benefits of adding emotional intelligence to online agents, that is the ability to express emotions and to recognize and adapt to the emotional state of others [5]. The work reported in the paper is a first step to filling this need.

Further, the animation quality of existing APA is low. Existing agents either fall in the uncanny valley [10], or are too stylized to display clear and believable personalities and emotions. Learners may fail to connect with such agents and fail to engage in the learning activity. On the other hand, animated characters that captivate and engage by displaying convincing personality and emotions are now possible, as demonstrated by the enormous popularity of computer graphics applications in entertainment, such as movies and games. Nonetheless, these highly engaging characters have not yet had a substantial impact on the educational domain. The work reported in the paper aims to improve on the state-of-the-art in agent animation.

In particular, the paper describes the development of a new animation system (and underlying algorithms) for generating life-like, believable pedagogical agents that display clearly recognizable emotions and that will respond and adapt to the learner's detected emotional state (the emotion detection module of the system is still under development, results will be reported in a future publication). The paper also reports findings from two preliminary studies which examined students' perception of the affective agents generated with the system.

Once completed, the research work described in the paper will have important implications. First, it will contribute to our understanding of the effects of animated affective agents on people's learning and the interaction between agent's affective features, learner's characteristics and learning effects. Second, it will contribute to the development of new, effective algorithmic approaches to emotion representation in animated agents and to emotion recognition.

2 Affective Agents: State-of-the-Art

Pedagogical agents are animated characters embedded within a computer-based learning environment to facilitate student learning. Many studies confirm the positive learning effects of systems using these agents. A meta-analysis by Schroeder showed that lessons with APA led to statistically significant learning improvements compared to lessons without them [12].

Affective pedagogical agents are agents with emotional capabilities. They can support simulated social interactions between learner and computer. According to Kim, Baylor and Shen [6], it is "*the provision of such simulated social interactions that may distinguish pedagogical [affective] agents from traditional computer-based tutoring, seemingly offering a unique instructional impact*". A few affective agent systems have been developed to date. The system by Lisetti and Nasoz [8] combines facial expressions and physiological signals to recognize a limited set of user's emotions, like fear and anger. A multimodal anthropomorphic agent then adapts its interface by responding to the user's emotional states, and provides multi-modal feedback to the user. The IA3 system by Huang et al. [4] was an early attempt at developing intelligent affective agents that recognize human emotion, and based on their understanding of speech and emotional state, decide on how to respond.

D'Mello and Graesser [2] have shown that students display a variety of emotions during learning with the intelligent tutoring system, AutoTutor (such as boredom, flow/engagement, confusion, and frustration) and they have shown how AutoTutor can be designed to detect and respond to them. SimSensei [9] is a virtual agent that engages in interviews with the user in order to elicit behaviors that can be automatically measured and analyzed. SimSensei uses a multimodal sensing system that captures a variety of signals that are used to assess the user's affective state, as well as to inform the agent so she/he can provide appropriate feedback. Research indicates that the manipulation of the APAs' affective states can significantly influence learner beliefs and learning efficacy [13]. A study by Kim et al. [6] showed that an agent's empathetic responses to the student's emotional states had a positive influence on learner self-efficacy for the task. A meta-analytic review that examined findings from studies on the efficacy of affective APAs showed that the use of affect in APAs has a significant and moderate impact on students' motivation, knowledge retention and knowledge transfer [3].

The problem with existing systems is that the pedagogical agents display a limited set of emotions, which often appear robotic and unnatural. Equipping agents with a wide range of life-like emotions provides more believability in the agent's behavior, and might lead to increased student learning. However, such a result cannot be attained without a set of rules that determine which affective behavior the agent should exhibit. The goal of the project described in the paper is to improve on the state of the art by developing new algorithms that produce life-like agents that display a large variety of natural emotions, and integrating them with rules for emotional design and emotion regulation in the context of online learning.

3 The Affective APA System

The system uses the following as the building blocks of emotion animation: (1) whole body poses, (2) specific arm/hand gestures, (3) facial expressions, (4) speech, and (5) motion dynamics.

(1) Whole body poses include the 4 types of body postures commonly used in animation of acting: open, closed, forward, and back poses. We use a continuum of body poses ranging from open to closed to express Russell's valence dimension (from positive to negative), and a continuum of body poses ranging from back to forward to express Russell's arousal dimension (from deactivation to activation) [11].
(2) For specific arm/hand gestures we use McNeill's classification of communicative gestures (e.g. Iconics, Metaphorics, Deictícts, Beats, Emblems).

(3) Facial expressions: the algorithms make use of a face model comprised of 30 joints with 55 DOF which allows for representing 36 Action Units+tongue, teeth, chin, ears movements.
(4) Speech animation is generated using a standard set of 20 visemes (mouth shapes), 12 representing the consonants, 7 representing the monophthongs, and 1 viseme representing the neutral pose, i.e. silence.
(5) Motion parameters such as velocity, acceleration, amplitude, frequency are manipulated to better express the agent's emotional state and to enforce adherence to fundamental principles of animation.

The system takes as input (1) a rigged character, (2) a text script and recorded audio, which determine what the agent says and the specific gestures the agent makes in support of the speech, and (3) the agent emotional style and the detected learner emotion, which determine the types of body poses, facial expressions and movement dynamics.

During the interpretation phase the animation algorithms analyze the input elements and identify the animation components necessary to compute the agent's animation. During animation instantiation, the identified animation elements are transformed into animation data segments. In the compositing phase the algorithms compute realistic transitions between consecutive character poses, enforce adherence to 6 animation principles and perform movement modulation. The animation data is applied to the agent rigged model using animation retargeting and is rendered in real time.

A flowchart illustrating the system is included in Fig. 1.

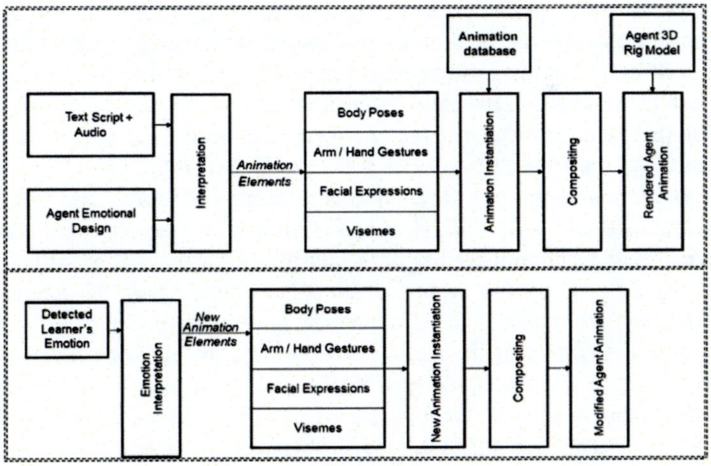

Fig. 1. Overview of APA system.

Videos of APA produced by the system can be accessed using the links below:
Happy: https://youtu.be/JjOjuYSJ8pk; **Frustrated:** https://youtu.be/oLYlFq6IRi8
Content: https://youtu.be/lEZy-LR0XGU; **Bored:** https://youtu.be/_8-Ooazd9bI

When a leaner's emotional state different from neutral is detected, the emotion is interpreted by the algorithms and new animation elements are selected, instantiated, composited and rendered in order to express the agent's response to the learner. As previously mentioned, the emotion recognition module of the system is still under development, a first version of the emotion detection module is described in [14].

4 Findings from Preliminary Studies

The objectives of the two studies conducted so far were to examine people perceptions of the affective agents generated with the system.

The first study [1] explored the effects of certain body gestures on viewer perception of the agent's emotion in terms of valence and arousal. More specifically, the study investigated the extent to which modifications to the range of movement of three beat gestures, e.g., both arms synchronous outward gesture, both arms synchronous forward gesture, and upper body lean, and the agent gender had significant effects on viewer perception of the agent's emotion along Russel's valence and arousal dimensions [11]. The study included two experiments and in each experiment the range of movement of the agent gestures was varied at two discrete levels. The stimuli for both experiments were sets of 12-s animation clips generated with the APA system. Each clip showed a different combination of body gestures and ranges of motion. In the first experiment subjects were asked to watch the stimuli clips and rank them from highest to lowest arousal and from positive to negative valence. In the second experiment subjects were divided into two groups: one group was assigned the clips featuring the female agent and one group the clips featuring the male agent. All subjects were asked to watch the stimuli clips and rate the valence and arousal using a 7-point Likert scale.

Findings showed that the more open and forward the gestures the agent makes, the more positive the perceived emotional valence and the higher the perceived emotional arousal. Findings also showed that female agents' emotions are perceived as having higher arousal and more positive valence that male agents' emotions. These results are important, as one main goal of our research is to create APAs that can convey clearly perceivable emotions not only through speech and facial expressions, but also through body gestures. Furthermore, there has been considerable debate as to whether posture and movement reliably convey emotions, or rather convey only the intensity of the emotion. Findings from the study demonstrate how body gestures are effective at expressing both the quality of the agent's emotion and its level of activation.

The second study [7] examined how well participants were able to perceive different emotions portrayed by a human instructor and by an animated pedagogical agent generated by our system in a video lecture on statistics. Participants were shown short video clips of either a human instructor or a pedagogical agent displaying four different emotions: happy, content, bored, and frustrated. The participants were asked to rate how well each video clip displayed each of those four emotions. Findings showed that subjects were able to recognize each of the emotions displayed by the instructor for both the human instructor and the animated agent. Subjects were much better at distinguishing between positive (happy and content) and negative (bored and frustrated) emotions than between active (happy and frustrated) and passive (content and bored) emotions. Emotions that involved higher activity (happy and frustrated) were more easily recognized in a human instructor than an animated agent. These results demonstrate that the emotions displayed by the agents generated with our system are recognizable, however they are not as clear as the ones conveyed by a real teacher. In future work, we will examine whether implementing exaggeration, one of the 12 principles of animation, will improve the clarity and hence the recognition of the emotions portrayed by our animated agents.

Acknowledgments. The work reported in the paper is supported in part by NSF – Cyberlearning Collaborative Research: Multimodal Affective Pedagogical Agents for Different Types of Learners, Award #1821894.

References

1. Cheng, J., Zhou, W., Lei, X., Adamo, N., Benes, B.: The effects of body gestures and gender on viewer's perception of animated pedagogical agent's emotions. In: Kurosu, M. (ed.) Human-Computer Interaction. Multimodal and Natural Interaction. HCII 2020. Lecture Notes in Computer Science, vol. 12182. Springer, Cham (2020)
2. D'Mello, S., Graesser, A.: AutoTutor and affective AutoTutor: learning by talking with cognitively and emotionally intelligent computers that talk back. ACM Trans. Interact. Intell. Syst. **2**(4), Article 23 (2012)
3. Guo, Y.R., Goh, D.H.L.: Affect in embodied pedagogical agents: meta-analytic review. J. Educ. Comput. Res. **53**(1), 124–149 (2015)
4. Huang, T.S., Cohen, I., Hong, P., Li, Y.: IA3: intelligent affective animated agents. In: ICIG 2000 (2000)
5. Johnson, W.L., Lester, J.C.: Face-to-face interaction with pedagogical agents, twenty years later. Int. J. Artif. Intell. Educ. **26**(1), 25–36 (2016)
6. Kim, Y., Baylor, A.L., Shen, E: Pedagogical agents as learning companions: the impact of agent emotion and gender. J. Comput. Assist. Learn. **23**, 220–234 (2007)
7. Lawson, A., Mayer, R.E., Adamo-Villani, N., Benes, B., Cheng, J., Lei, X.: Recognizing the emotional state of human and virtual instructors. Comput. Hum. Behav. **114**, 106554 (2021)
8. Lisetti, C.L., Nasoz, F.: MAUI: a multimodal affective user interface. In: 10th ACM International Conference on Multimedia (Multimedia 2002), pp. 161–170 (2002)
9. Morency, L.P., Stratou, G., DeVault, D., Hartholt, A., Lhommet, M., Lucas, G., Morbini, F., Georgila, K., Scherer, S., Gratch, J., Marsella, S., Traum, D., Rizzo, A.: Demonstration: a perceptive virtual human interviewer for healthcare applications. In: 29th AAAI Conference on Artificial Intelligence (2015)
10. Mori, M.: The uncanny valley. IEEE Rob. Autom. **19**(2), 98–100 (2012). https://doi.org/10.1109/mra.2012.2192811
11. Russell, J.A.: Core affect and the psychological construct of emotion. Psychol. Rev. **110**, 145–172 (2003)
12. Schroeder, N., Adesope, O.O., Gilbert, R.B.: How effective are pedagogical agents for learning? A meta-analytic review. J. Educ. Comput. Res. **49**(1) (2013)

13. Zhou, L., Mohammed, A.S., Zhang, D.: Mobile personal information management agent: supporting natural language interface and application integration. Inf. Process. Manage. **48** (1), 23–31 (2012)
14. Zhou, W., Cheng, J., Lei, X., Benes, B., Adamo, N.: Deep learning-based emotion recognition from real-time videos. In: Kurosu, M. (ed.) Human-Computer Interaction. Multimodal and Natural Interaction. HCII 2020. Lecture Notes in Computer Science, vol. 12182. Springer, Cham (2020)

Teaching E-learning for Students with Visual Impairments

Theresa Lobo[1,2](✉)

[1] CIAUD/Universidade de Lisboa, Rua Sá Nogueira, Pólo Universitário, Lisbon, Portugal
theresa.lobo@universidadeeuropeia.pt
[2] UNIDCOM/IADE, Av. D. Carlos I 4, Lisbon, Portugal

Abstract. The paper presents an IT network program developed specifically for visually impaired students. Innovative in its conception, this program contains all the learning required equipment, such as classroom (virtual), accessible learning materials, laboratory, and teachers trained to teach visually impaired students [1]. Additional teaching aids, like iNetSim, a universally accessible network simulator, created to allow vision-impaired and sighted students to complete Cisco Certified Network, and Microsoft HoloLens 2 (a new vision for computing, is used to transform abstract concepts into 3D experiences in the virtual classroom), which enables students to research and explore more deeply, see more clearly, and learn by doing [2]. This article aims to allow visually impaired students, as well as sighted students, to perform with similar results.

Keywords: Teaching · Visual impairments · Classroom

1 Introduction

For 30 years, a great number of educational institutions, which offer e-learning programs that have been developed considerably. Although, forthcoming the digital era, often has not become the reality for most students with visual impairments [3]. "Visually impaired", also known as vision impairment or vision loss, is a decreased ability to see to a degree that causes problems not fixable by usual means, such as glasses. This scientific work was established in 2018 for students visually impaired. Scientific and technological contents are rapidly changing in the education system, and the Internet and web became essential media to access information and learning materials. The new learning methods, based on Internet and Web extent use, provide extraordinary opportunities in design. However, visually impaired students are a group deeply affected by this change and find serious problems to deal with it, which goes from simple work, as typing a letter or draw with the use of computers. The objective of this scientific work is to explore an e-learning course challenged by students with visually impaired and sight students. This is reached equally by tracking the best practice eLearning program. On the year of 2017, we started using Windows Mixed Reality. The new project (2018) contains available materials, a laboratory, and a network for the design and the testing and Microsoft HoloLens 2 for all students). Two teachers trained for the visually impaired students and teachers that instruct sighted

D. Russo et al. (Eds.): IHSI 2021, AISC 1322, pp. 225–234, 2021.
https://doi.org/10.1007/978-3-030-68017-6_34

students. This new project (2018) with the Cisco Certified Network Associate (CCNA) course with the iNetSim application [4], offers a visually impaired certification in IT networks and could also be offered a design course. In this last year of 2018, we also started using the iNetSim, which is perceptible by vision-impaired users. It is based on Mac OS X Tiger, a system software with an incorporated screen reader called Voice Over and Microsoft HoloLens, that transforms abstract concepts into 3D experiences in the classroom, permitting students to investigate more deeply, see more clearly, and learn by doing. The project contains available learning tools, an online learning environment, and a remote virtual network for testing of a computer network. It was selected specific colors for text and backgrounds of web and e-learning tools because it could reinforce the readability of on-screen text. The program was also supported by the Web Accessibility Initiative (WAI) and Section 508 for creating web pages usable by students with visual impairments and recommends several online tools to select appropriate page background and text colors. The Cisco Network Academy Program (CNAP) computer networking, a certification course, giving the users with vision impairment relevant qualifications.

2 Students with Vision Impairment

Vision impairment presents a significant variety of forms, from congenitally blind, without vision from birth **or** very early age, to those who lost their vision in variable levels for various reasons, clinical or other. Temporary blind students and partially sighted students are two very different groups that present specific study patterns and difficulties. Each group has expectations of their own and require special staff support. Scientific and technologic fields such as IT, engineering, or design are strongly rooted in visual supports making visual abilities crucial and affecting negatively impaired students who face serious problems in comprehending essential parts of the program.

2.1 Learning Barriers

Low vision students challenge different kinds of obstacles: to fully access the matters, in fact, the first category needs to depend on screen readers, while the second category, lean on optical supports and/or to precise customization selections, may approach a larger variety of software functions, involving, frequently, those with graphic interface. Among the problems faced by students with vision impairments, one is related to the fear of damage to the equipment, usually expensive and sensitive. They feel they cannot fulfill their need to 'play' with the equipment and computer design programs even if this is essential to the educational course [4]. A useful application is offered, as an essential element of the educational course method. Limited vision depends on sequential learning using tactile, proprioceptive, and acoustic perceptions to create spatial associations. Those students require alternative means of developing and understanding the third dimension. The Cisco Network Academy Program (CNAP) has been altered to integrate innovative visualization approaches to assistance the conceptualization, as specific Barriers, with deficiency of accessible e-learning education and nature of vision incapacities, require advanced instruction, and preparing establishments should convert a significant multimodal (using several modes or methods) operator interfaces to a design e-learning tools more accessible to all users.

3 Accessible Learning Management System

It's important to consider in creating an accessible e-learning environments': echnological and methodological subjects. The accessible learning management system presented, is to offer the students with visual impairment the accessibility of iNetSim and the use of Microsoft HoloLens. [5] Cisco Certified Network Associate level two (CCNA 2) is a teaching e-learning program for students with visual impairments. The Cisco curriculum is well known as an international program. This project investigates the efficacy of the available systems and adds the most active methods into an understandable education course. The students with vision impairment, that benefited from this course, they could get results within 5% of the sighted students. Education Institutes offer computer networking career education through certificate, associate degree, or bachelor's degree programs. The purpose of this course was to contribute to students with vision impairments, that could get access to computer networking career education, and they will have a chance to start a curriculum and be able to get a work. A computer networking skill could be important for people with visual impairment for some purposes. Principally, for collaboration with computer networking and computer workstations at the actions stage through text interface [6]. Text interface uses text input and output, and text is simply transformed in audio output by screen reading software. A screen reader is an essential piece of software for the visually impaired person. Simply put, a screen reader transmits whatever text is displayed on the computer screen into a form that a visually impaired user can process (usually tactile, auditory, or a combination of both). Most basic screen readers with additional features can give people with visual impairment much more independence. While the physical network, (When installing and optimizing networks, most people overlook the physical components, which can have a great impact on performance) may include wiring (electronic sketch book and hardware electronics for developing ideas. It is a context for learning fundamentals of computer programming). IT Network Solutions, (offers a comprehensive set of voice and data solutions developed by a team of engineers) is reasonably shaped and is implement rationally through a workspace. Finally, the students with vision impairments and sighted students, that use computers, internet, and HoloLens, as a support, this practice is indispensable not only to support a computer network, but also to allow them to repair the computer systems. To accomplish the purpose of the course, it was required to establish a course, accessible to users with vision problems.

3.1 The E-Learning Project

The E-Learning Project was accomplished over a semester in 2017 at IADE-Creative University, Lisbon and the course was established, and the program year was improved (2018) Universidade Europeia, Lisbon. This plan included about 5 students with vision impairment in India and 5 visually impaired students in Portugal (all with about $< 10\%$

visual insight), recommended by APA, (Association for Blind). The Cisco Systems Inc. Establish the essential network tools for the teachers to improve and instruct the students conveniently. In the live teaching operation phase, where the matters were separated into two sets, sighted students and visually impaired students exposed for purposes of comparison. Cisco Learning Network Space (a digital learning platform built to offer instant user access to training information, course materials, and exam preparation resources. It's designated, as a practical tool operated by visually impaired permitted them to access no more than 20% of the regular Cisco e-learning tools. The result supported a requirement showing the supplies for alternate procedures. The project conclusions were positive. In all Cisco examinations, PassMark Software is 75%. Students with Visual impairment verified 90% in the first test series. The result was identical to sighted students, who also reached 90%. The normal differences in results for the two groups were also very alike.

3.2 Creating a Program for Students with Vision Impairment

It was created a curriculum, especially to contain the requirements of students with vision impairments, the following topics were established: 1. A teaching environment accessible for resident learners; 2. A test center for computer network and research; 3. An offered program inclosing a group of electronic tools for design programs; 4. The use of HoloLens's; 5. Distribution by trained instructors to teach students with vision impairment and regular instructors. These learners were aided with the necessary supplies and offered with the effective instructional strategies.

3.3 Basic Classroom

The basic classroom comprises a physical environment (teachers should design the environment by organizing its spaces, furnishings, and materials to maximize the learning opportunities and the engagement of every student) for students and instructors and a virtual classroom (used by far away (Indian students). The virtual environment (an online learning environment that allows for live interaction between the tutor and the learners as they are participating in learning activities. The most common tools in a virtual classroom are Videoconferencing, TV on-line, and call management application (is the process of designing and implementing inbound telephone call parameters, which govern the routing of these calls through a network), that certifies instructors to broadcast network (is a group of radio stations, television stations, or other electronic media outlets, that form an agreement to air, or broadcast, content from a centralized source) to classes local and distant education. VoIP (software that integrates most frequently used communications applications and services from a single, multimedia interface on PC or Mac). VoIP call management systems were Cisco Call Manager and Ventrilo; offer adequate assistance. It was important, that sighted teaching assistant, achieve the broadcast network, by radio buttons on a visual display by screen reading software, Students were listed into the online learning environment to attend classes, to assist classroom exercises. The online space (where the learners and the tutor work together simultaneously), provides advanced tools for students to communicate with other users and interact with instructors, comparable to a regular environment classroom. All classes were reported and published in audio file format along with other instruction tools for simple learning by all students. Long-Distant communications appeared through VoIP, where mainly students were using freeware (software) functions, like Skype for communication.

The physical environment comprised workrooms equipped with operational technology (OT) (a category of hardware and software that monitors and controls how physical devices perform). Operational technology (OT) (in e shape of screen readers and screen magnifier, specific colors, and contrasts for the computers' screens for the visually impaired), router bundles, and network equipment for the remote laboratory (is the use of telecommunications to remotely conduct real experiments, at the physical location of the operating technology from a separate geographical location), network simulation tool (a system of computers that are connected so that they can share resources and communicate internally and even globally), and learning management system (LMS) to support teaching and learning processes. Additionally, haptic technology, (is any technology that can create an experience of touch by applying forces, vibrations, or motions to the user). Besides this tool, it adds a touch helpful technology that was validated. All students could manage the studying context through the Internet and could approach similar tools and equipment, as those students in the physical environment (the overall design and layout of a given classroom and its learning centers).

3.4 Supplementary Test Center

The Supplementary Test Center was designed to allow both local and faraway students to check their system developments. This center consisted of some system groups and three servers (are a computer program or a device that provides functionality for other programs or devices). Operating VMWare to provide virtual networks (is a software product that provides operations management across physical, virtual, and cloud environments), where students could set up the system to meet requirements of the proposals they established using the system recreation device and check them in the test room. This expertise provided the learners a chance to test different systems proposals and understand how the modifications they introduced in their works, that would be important for the success of their plans in a physical environment. The distant center also allows students positioned far-away to have similar aids, as the students joining the physical environment.

3.5 Cisco Course

During 2017, the course was in a long-lasting development, with the upgrading of the supplies and instruments with changes to each new edition of education systems emitted using the Cisco Learning Network. The Cisco course was created to be operating two years, where the users should acquire expertise: Organize, set up; manage systems; Resolution of network subjects; Settle different types of operating systems. The program gives requirements for all students, such as these tools: substitute keyboard navigation is important for users with visual impairments, that couldn't depend on pointing gadgets, such as a mouse; the basic accessibility features provided by operating techniques, plus preservation of highlights which are important for low vision users, as extending fonts, convert color design, adjusting color brightness and contrast, improving the contrast between text and background; complete assistance for screen reading tools, that is essential for graphic components and written texts, that benefits not only to permit a major understanding of screen subjects, but also to make viable the written interaction between the software application and a user who, may insist on a speech synthesizer; images, cartoons: they should not be the only way to submit material, as they characterize a difficult for the visually impaired; features which may confuse vision and/or reading, such as high – frequency flickering elements (is a frequency at which an intermittent light stimulus appears to be completely steady to the average human observer), should be avoided; the position of the focus point (the point where waves of light or sound that are moving towards each other meet), should be clearly visible by human eye and by assistive technology (assistive, adaptive, and rehabilitative devices for people with disabilities); documentation and other tools sustaining the use of software functions should be accessible in digital format, so that they can be entirely read by any type of users, including those with visual impairments. In this program, the screen reading applications of the computer can convert the text to audio, as the holoLents and the students with vision impairments could have the capabilities to study for further university education and get a job in the networking area.

3.6 Technology Support

One of the difficulties established during the experimental project of 2017, it was about hiring regular teachers, instead of trained professors. Because of the regular instructors, they were not entirely sensible of the requirements of visually impaired user's association with evidence of concepts and supplies. In this 2018 e-learning program were hired two teachers trained for the visually impaired students. They were capable to provide the 2018 e-learning program sources and they offer more methods of showing knowledge sources. The audio Notetaker, (the note-taking app and software for student) was saved into the administration device to following leaner contact. During the distribution of the class, far away students could do interrogations. Tests were accomplished frequently to see the evaluation of the work enclosed and regulate the efficacy of the education methods. Students with vision impairments operated together with the teaching team to shape a successful communication system. All the students adopted equivalent tools and test centers; the lessons planned at change hours in the day, because of the noise produced by the sighted students, this could be hard for users with vision impairment catch audio production from screen reader applications. Screen readers are software applications that try to deliver what users with normal vision see on exposure to their users through non-visual ways, like text-to-speech or sound icons. The students completing each program number of the 2017 program: 5 visually impaired and 5 sighted. Regular teachers organized the classes for sighted students and trained teachers for visually impaired students. iNetSim, available system simulation, designed to permit visually impaired and sighted learners to finished Cisco Certified Network Associate level two (CCNA 2) test center phases. On the 2018 project, the iNetSim, was developed for this new program for student's complete network performs and operated particularly in a teaching router outline, network management, routing protocol configuration, and programming access records to manage approach routers. It approved the arrangement of virtual network tools involving routers, switches, hubs, Macs, and PCs. Control and feedback above the virtual reality were supplied by a command-line interface to tools. The interface operated in a related system to the functioning procedure for that gadget sort, for example, a basic DOS similar method for Macs PCs and Cisco IOS for routers. [7] iNetSim was created to functioned uniquely with the keyboard, easing the eye and hand matters confronted by users with visual impairment. A GUI (graphical user interface) (is interactive components such as icons and other graphical objects that help a user interact with computer software, such as an operating system), was also available. It was provided also to the screen readers, which are software applications that are installed on the computer to provide the translation of the information on the computer screen to an audio output format. The translation is passed to the speech synthesizer and the words are spoken out loud. Currently, fully functional screen readers are only available for Linux in console mode. Linux supports develops an important position of the programs, and a voice synthesizer tool for Linux was advanced precisely intent (comprising both hardware and software).

4 Results

The students with vision impairment reached comparable results regarding sighted students, then it required them extra time to embrace tools in provision for these evaluation procedures. [8] To prepare for the final examination in each course, the visually impaired worked 4–5 times more hours than the other learners. The system tests have a pass mark of 75%, and learners need re-sit exams until they accomplish at least 75% correct answers. The sighted estimated to increase results than visually impaired users in tests; still, the two groups performed with no significant difference concerning average scores and in the number of examination attempts. At the first evaluation, the visually impaired and sighted students achieved a normal 92%. At the second assessment, the visually impaired students gained 87/% compared to 84% for the sighted group in identical evaluations. The grades were closer in the last valuation, the students with visual impairment achieving 88% equaled 87% for the sighted students. Advanced results could be clarified by the devotion of users with visual disabilities as a result of a course precisely planned to increase their abilities. These users used expressively extra time interpretation and concluding performs out of class time than the sighted students. Finally, the program for visually impaired users provides a complete questionnaire concerning matters, with the accessibility of remodeled plan, the competence of tools suggested in the test center, the efficacy of operative equipment utilized (screen reading and text enlargement applications), and the supply of materials by instructors. The users adopted to attend the voice synthesizer tool. One more essential reflection was the length of time the users with visual disabilities were required to embrace the program tools. As a result, a few students with visual disabilities argued that they wish to continue to postgraduate studies. Finishing this program allowed the visually impaired students to feel confident in their capacity to successfully complete their courses.

5 The Use of HoloLents

This work can improve the quality for learning and benefit with environmental orientation and interaction. Visually impaired students confront various challenges when relating with different materials. Principally in unfamiliar environments, this could be hard with unpredictable and indefinite experiences, as the misplacement of an object. Within this program, an example to discover and identify matters in digital images and it was develop adopting intense learning technologies. The work includes a HoloLens system for all students and can be managed to identify items and described them through voice output. Microsoft HoloLens offers not only a mobile mixed reality solution (mixed reality solution, the augmented reality enhanced wearable hybrid simulation IV trainer (AR-RLS), replicates real-world clinical skills training), that covers image catching, microphone and 3D audio playback functionalities. It also offers a spatial model, which can be used for distance calculation between an item in the spatial model and the HoloLens. The latest technology improvements in profound studying and windows mixed reality PC, hardware permit quicker progress of accessible tools. The recommended program proposes various options to increase the learning of the visually impaired and can be utilized without any preceding instruction on the tool. The last version of the HoloLens with help a lot the inclusive education. Seeing that the computing power of the new version is potent enough, an item recognition could be achieved precisely on the HoloLens., The three-dimensional understanding of the HoloLens would make it viable to inform the user if is standing in front of a barrier at a certain distance. The solution of profundity data can be development utilizing an external depth sensor (With optical elements separated the same distance as human eyes, both Structure Sensor and Structure Core bring human-scale depth perception to a multitude of applications). [9] Other applications already identify signs or bank notes [10], applying diverse Optical Character Recognition (OCR) structures.

6 Future Training Supports

Therefore, designing a model specifically for people with special needs requires a precise definition of the target group, which helps to determine the criteria for selecting the appropriate content for them. Teachers in inclusive education may face challenges in evaluating the effectiveness of e-learning program content design for several reasons: The best solution for this problem may be to design flexible programs that can be shaped to each situation. Especially, when the project could permit a precise method, concerning to different assembly of different categories of leaners, one visually impaired and the other sighted. [11] The course could gain from wider study in the forthcoming on the best applicable processes of providing the knowledge supplies to various sorts of people disregarding of their sight capacities.

7 Conclusion

Students with disabilities can be included equally in classroom learning activities, and they could cooperate with and compete with other regular students without any barriers. It is clear, that technology is a powerful collaborative learning tool because it enhances communication skills, which can help the students to maximize the benefits they gain from the learning setting. Furthermore, it is an efficient tool to help students who have special needs to join their peers in the same classroom and focus on the same content. This project provides a safe, therapeutic setting that is unlimited in scope; the size of an e-learning course can far exceed the physical confines of the learning setting and, through employing the power of the internet, can involve participation from all corners of the globe. This e-learning project included extremely good tools to require the stage of availability needed and offer the students with vision impairments support in their capacity to acquire and accomplish the equal education results as the regular students [12]. The excellence of the education supplies appears to have improved during the stage of the investigation [13]. Concluding, it seems that was no differentiation in the results between the sighted students and the students with vision impairment using this course. It signifies such contribution with a proper background and comprehensible program resources, people with visual disability could be able to reach equal results as the sighted students.

References

1. Arditi, A., Knoblauch, K.: Effective color contrast and low vision. In: Rosenthal, B., Cole, R. (eds.) Functional Assessment of Low Vision, Mosby. St. Louis, pp 129–135 (1996)
2. Huang, J.: Dartmouth Computer Science Technical Report TR2017–821 (2017)
3. Harper, S., Goble, C., Stevens, R.: Web mobility guidelines for visually impaired surfers. J. Res. Pract. Infor. Tech. **33**(1), 30–41 (2001)
4. Cisco Systems: CCNA 2: Router and Routing Basics (2004)
5. Bigelow, A.E.: Blind and sighted children's spatial knowledge of the home environments. Int. J. Behav. Develop. **19**, 797–808 (1996)
6. Eckert, M., Blex, M., Friedrich, C.M.: Object detection featuring 3D audio localization for Microsoft HoloLens. In: HEALTHINF 2018 (2018). https://doi.org/10.5220/0006655605550561
7. Meddaugh, J.J.: Tackling. Paper: Tips and Tools for Success for People with Vision Loss". World, July 2011
8. Gregory, R.: Eye and Brain, pp. 185–187. World University Library, New York (1998)
9. Garon, M., Boulet, P.-O., Doironz, J.-P., Beaulieu, L., Lalonde, J.-F.: Real-time high resolution 3D Data on the HoloLens. In: Proceedings of the IEEE International Symposium on Mixed and Augmented Reality (ISMAR-Adjunct), 2016, pp. 189–191 (2016)
10. Sudol, J., Dialameh, O., Blanchard, C., Dorcey, T.: Looktel—A comprehensive platform for computer-aided visual assistance. In: Proceedings of the IEEE Conference on Computer Vision and Pattern Recognition Workshops (CVPRW), pp. 73–80 (2010)
11. Hope, J., von Konsky, B., Murray, I.: A cisco education tool accessible to the vision impaired. In: ASSETS06, Portland, Oregon, USA, 23–25 October 2006

12. Knoblauch, K., et al.: Effects of chromatic and luminance contrast on reading. J. Opt. Soc. Am. A. **8**(2), 428 (1991)
13. Lazar, J., D-Sponaugle, A., Greenidge, D.: Improving web accessibility: a study of webmaster perceptions. Comput. Human Behav. **20**(2), 269–288 (2004)

Gamification for Teaching - Learning Mathematics in Students of Basic Education

Jean Paul Molina-Villarroel(✉), Cesar Guevara, and Roilys Suarez-Abrahante

Maestría en Educación Mención en Pedagogía en Entornos Digitales, Universidad Tecnológica Indoamérica, Bolivar 2035 y Guayaquil, Ambato, Ecuador
jepamovi@gmail.com, {cesarguevara, roilyssuarez}@uti.edu.ec

Abstract. In education, there are many factors that affect a student's academic performance, one of them is poverty in rural areas. Illiteracy and school dropout are a consequence of poverty, this problem increases due to the fact that young people and children of school age see the need to work, in order to contribute economically to their homes. Another factor is the outdated curriculum and the lack of use of technological tools by teachers in the teaching-learning process, class hours become monotonous and boring, where the student limits himself to listen to his teacher becoming the teacher in the main axis of learning, the student does not get involved in his school training process, does not have attractive learning spaces, does not have a true motivation to want to learn, and improve his quality of life. This study proposes the design of a scratch prototype, based on gamification methodology for Mathematics teaching-learning process in "Marco Aurelio Subía" basic level students' school, in "Tanicuchí" parish, Cotopaxi province. With the implementation of this research, it is expected to awaken the interest of students in learning Mathematics from basic levels of education and strengthen their knowledge for future years of schooling.

Keywords: Gamification · Use of technology · Interesting activities · Education improvement

1 Introduction

In education field there are many factors that affect the academic performance of a student. One of these factors is poverty, being the rural sector where this problem is most aggravated. According to the National Statistics and Censuses Institute of Ecuador (INEC), in the month of March 2015, poverty in rural areas was 43.35% and extreme rural poverty was 19.74%. Illiteracy is a direct consequence of extreme poverty due to the inability to access education. According to data from the Ministry of Education, the illiteracy rate in 2004 at the country level was 6.7%, 3.9% at the urban level, and 12.9% at the rural level. Another problem presented for this reason is school dropout, mainly at ages where there is the transition from basic education to high

D. Russo et al. (Eds.): IHSI 2021, AISC 1322, pp. 235–240, 2021.
https://doi.org/10.1007/978-3-030-68017-6_35

school, where this problem deepens in rural areas, as young people are forced to join the labor market to contribute to family income [1].

One of the challenges at present is to eliminate some educational barriers that occur mainly in the rural sector, where it is necessary to expand teaching coverage and regulate the current educational offer, on the other hand it is necessary to invest in their training, as well as provide more attention to initial education. There are also the challenges of aligning the curriculum, which generally come from urban paradigms with the rural context, reducing the access and quality gap, and implementing appropriate infrastructure and technologies [1]. The world is nourished and changes thanks to various innovations introduced in everyday life and in the workplace. This forces different professionals to adapt to new epochs, therefore professional updating is more than a necessity, it is an obligation, to take advantage of the technological advances in the market.

In the work carried out by Bosch [2], it shows that the teaching of exact sciences or Mathematics has not been any change following the traditional parameters where it is limited to an exposure and transmission of knowledge by the teacher. In recent times, there have been ideas about the need for a substantial paradigm shift. By adopting this new paradigm, new technologies and methods are introduced for an experimental learning of Mathematics through the assistance of ICTs (information and communication technologies). Experiments will be carried out whose data allow inducing a model of the phenomenon under study, which is validated by adjusting the experimental values with the prediction of the model. Through the application of these didactic units, it is intended that students acquire the basic scientific competences, set forth in some publications. For each active learning session, a sequence of experiences is presented, in such a way that each of them gives reason to the next.

Currently the use of technology, the internet, social networks, etc. are present in most of the activities we carry out on a daily basis, becoming a social phenomenon. From the point of view of education, the innovations brought by technology are also very remarkable, and as educational professionals we must take advantage of these to apply them in the classroom. Today, our students belong to an era where they are born with technology. For this reason, students have a great facility to acquire and adapt to the digital challenges that may be offered to them from the educational field. This challenge requires re-considering new teaching and learning strategies, therefore introducing methodologies that bring us closer to the requirements of the 21st Century School. It is through these new teaching-learning models and strategies that we can help improve the learning of our students so that they are able to face the new challenges and socio-educational demands of the new millennium, which are set by the use of information and communication technologies. These ICTs constitute instruments and tools that will allow students to generate epistemological knowledge, in any scientific discipline [3].

2 Methods and Materials

Learning methodologies are processes by which students acquire new ways of learning. They open their thinking through new concepts and information. Their ages are a very important aspect to take into account. Teachers are in charge of facilitating their meaningful knowledge. The student is in charge of receiving that knowledge [4].

Mathematics is a fundamental subject in the education curriculum. Children learn this subject from the first school years. This subject provides fundamental tools to be used in other subjects. Mathematics is still considered a purely mental subject. The student often acquires a purely theoretical content. He must simply fulfill a plan. Mathematics is not just numbers and calculations. We must relate its concepts to operations. Answers will be established based on analysis. The methodology is the way to teach Mathematics. An example is the traditional way of teaching Mathematics. The explanation is done on a blackboard, the exercises done in a workbook [5].

The Singapore method is a methodology for teaching Mathematics. This methodology, based on the experiences of various researchers, was developed by the National Institute of Education of Singapore. It arose from the need to improve the teaching of Mathematics in public schools. Singapore government committed to educational renewal, becoming a fundamental pillar in the development of the country; that's why it remains in the first places of educational evaluations [6]. This methodology for learning Mathematics works by focusing mainly on the method, visualizing the problem through diagrams or also called blocks that encourage the student to solve by looking and feeling these blocks.

The term gamification was used by the British Nick Pelling, this software programmer focused on designing and implementing office applications using games. These applications were used in the educational environment, this allowed to enhance creativity in students. Gamification applies dynamics used in games in a playful educational context, its main objective is to attract the interest of the student, and in this way solves specific problems, in a specific situation, in solving problems of a certain situation or specific topic, this will be achieved by implementing an interactive, creative and participatory game. Gamification is used as a methodology in the teaching-learning process, it improves aspects of participation, learning and its subsequent evaluation [7].

The syllabus is an instrument that guides both, the teacher and the student in the teaching-learning process, allowing to plan the contents of a subject. It fulfills the function of guide and orientation, in aspects that are developed in a subject. It is a document that guides the academic process. A syllabus must ensure compliance with the teaching-learning process, contributes to forming a competent professional, integrates their different knowledge, skills and abilities, and must respond to the purpose of the subject. The syllabus proposes contents and teaching-learning strategies, and an orderly evaluation [8].

Computer science and audiovisual media have a very strong influence in our daily life, and so influence our behaviors. Computers have attractive interfaces, with playful elements that can be used in education. The use of games on the computer favors the learning process. Educational robotics can be a very effective tool in this process, where technology plays a very important role where interaction will allow to learn while playing [9]. The technological tool called Makey Makey will be used, an element that will be used to implement digital skills in the student at an early age and facilitate their learning. Makey Mickey is an electronic board invented by two PhD students from MIT Media Lab, belonging to the Lifelong Kindergarten group, Jay Silver and Eric Rosenbaum and whose first prototype appears in 2010 although the final version would not arrive until 2012.

3 Mat Game Prototype Proposal

There are different ways to integrate technology in the classroom, and create a technological learning environment, innovate in learning processes, restructuring the basic functions of the classroom. According to the study carried out, it was determined that it is necessary to strengthen the students' competences in Mathematics as a subject, through the design and implementation of a teaching-learning prototype based on gamification. In this section we will explain how this prototype is developed, each of its steps for its design and its respective implementation.

The designed modules present an environment based on gamification, where participants must face questions or mathematical problems using their ingenuity, awakening their creativity and imagination. In this way, players will acquire new knowledge or reinforce what they already have. Participants will develop new Math skills and even changes in attitudes in the classroom. Through this playful learning methodology, the student acquires a very fun learning experience, where the entrusted activity offers spaces where it is possible to fail but learn at the same time. These modules are designed to hold students′ attention, attracting their interest in participating and learning while playing, obtaining an attractive learning experience. This learning must be accompanied by their respective feedback from the teacher, to transform the school environment into a much more exciting environment. A school educational environment enriched with playful elements allows the student to advance at their own pace and develop collaboration and communication skills with their peers.

This prototype is designed not only as an element of distraction or entertainment for the student, but it must be considered a persuasion so that the student sees his learning in a different way, describing thought processes through a game or playful activity to be solved. These processes are applied to the educational environment in which the student operates.

Module 1. "Movement"
This module's functionality is to handle attractive and interesting elements for a better conception of addition and subtraction. It is planned to be carried out by children of 2nd and 3rd year of elementary education using the method handling elements. These elements are manipulated in a friendly and attractive environment for the student; with this, it will be possible to graphically demonstrate concepts of addition and subtraction. For the use of the elements in this module, principles of addition and subtraction were taken into account through attractive and interactive elements such as basketball players, animals, objects, etc., which are more convenient to achieve the proposed

objective, using pedagogical resources used daily in the classroom such as pencils, erasers, foamy (Ethylene Vinyl Acetate), etc., and technological resources (Makey Makey board, scratch, etc.) that support its design. This module has a window in which the student has an addition or subtraction question, which is read by the teacher and then the student determines by manipulating the graphic elements available on the screen whether to add or remove elements to obtain an answer. If the student succeeds, he will get a reward for the assignment, otherwise the interface will show an image indicating that must try again. The student will be able to manipulate the elements according to his criteria to search for correct alternative solutions. This playful activity will strengthen concepts of where and when to use addition or subtraction. This activity can be developed in work groups, promoting collaborative teamwork.

Module 2. "Block 10"

This module is based on the Singapore method. Its functionality is to develop comprehension skills, retention, and pleasure for Mathematics; without seeking memorization but generating a deep and lasting understanding. The main objective of this work module is to achieve concepts in the student. The abstract environment will lead the student to discover a particular knowledge. This module will be called "BLOCK 10" and is developed for children of elementary education level. This module is designed through a user-friendly interface for the student according to the level of education. It presents an initial screen with a mathematical operation of addition or subtraction with two digits. With the help of the teacher, the student will begin to analyze the operation and will have available elements that will allow him to form the corresponding quantities. The student will find a solution through an open manipulation of the elements. The student will be able to compose or decompose numbers into units, tens, etc. The student must form the two quantities and analyze the answer. Once the student determines the answer, must type it by pressing the quantity using a response module that will be controlled by the Makey Makey board. This module is designed with numbers elaborated in foamy, which when pressed, will determine the answer. If the player or student entered the correct answer, a sound of joy and happiness will be emitted, otherwise a message as "TRY AGAIN" will be presented.

Module 3. "Multiplication Tables"

This interface is very user-friendly. In this case, the elementary education student has the role to develop learning activities in a different and fun way without generating any kind of concern in the student learning the multiplication tables. There are multimedia elements where the student can easily interact with these elements. The module III design, is based on game-based learning that works as a learning strategy through the use of a playful module, achieving satisfactory cognitive learning, through procedural and attitudinal knowledge where the environment allows discovering new knowledge through playing. The game is designed with elements that can be chosen at random. These elements integrate themes related to multiplication tables review where the student interacts with multiplication elements through an attractive and innovative environment. The student develops cognitive skills through the interest in developing the activity correctly.

Module 4. "Get to the Goal"
This module is designed to have a collaborative and participatory work, it means, as a team. It is constituted by an open and flexible method of learning, in which two groups of students must develop mathematical exercises through an athletic-mathematical competition in a virtual way. In this module, a challenge will be presented to each group of participants, to face a virtual race solving mathematical operations. They will look for the different strategies for their group to be the winner. This environment is also developed in scratch, and its objective is to create a playful interface combining physical activity with the knowledge of Mathematics. It consists on, in the application screen will appear mathematical exercises of multiplication of two digits which should be answering correctly and progressing to the goal, through a Makey control connected to a mat that will have each of the natural numbers to solve the mathematical operation posed. Students will seek the order of participation to reach the goal and seek the most appropriate way to solve the problem. This will help the student maintain a learning pace appropriate to their age and interest in learning through recreational activities. Applying these, it is intended the teaching-learning process stops being monotonous and the knowledge transmitted not to be dogmas. The students will check for themselves multiple solutions to a formulated problem, understanding, reasoning and visually checking the mathematical operations they use or that should be used. The student will put aside the rote method of learning, encouraging the student to detach each mathematical operation adapting it to his level of knowledge.

References

1. Contreras, A.C.: Informe de asistencia técnica grupos diálogo rural | impactos a gran escala Situación de la Educación Rural en Ecuador (2015)
2. Innovaciones didácticas para ciencias y matemática asistida por TIC. https://sedici.unlp.edu.ar/handle/10915/61261. Accessed 29 Sep 2020
3. El programa Scratch como estrategia de aprendizaje cooperativo en el tercer ciclo de Educación Primaria - Dialnet. https://dialnet.unirioja.es/servlet/articulo?codigo=6245750. Accessed 29 Sep 2020
4. Abella Molés Tutor de TFG, P., Alcalde Esteban, M.: Metodología en la enseñanza de las matemáticas en Primaria, Universitat Jaume I, Jun 2015
5. Estrategias de enseñanza de la resolución de problemas matemáticos: fundamentos teóricos y metodológicos - Dialnet. https://dialnet.unirioja.es/servlet/articulo?codigo=3897810. Accessed 29 Sep 2020
6. Violeta Rodríguez, S.: El método de enseñanza de matemática singapur: 'pensar sin límites'
7. Muñoz, J., Hans, J.A., Fernández, A.: Gamificación en matemáticas, ¿un nuevo enfoque o una nueva palabra? (2019)
8. Iparraguirre, R.A.: Sílabo por competencias en el Horizonte de la ciencia. Horiz. la Cienc. **5** (8), 97–104 (2015)
9. Vivo, C.: La competencia digital en edades tempranas: una experiencia educativa con Makey Makey, Universitat Jaume I, Jun 2017

Social Activity of Student Youth: Experience of Pavlo Tychyna Uman State Pedagogical University

Oksana Kravchenko(✉), Natalia Koliada, Viktoria Isachenko, and Inna Honchar

Pavlo Tychyna Uman State Pedagogical University, Uman, Ukraine
okskravchenko@ukr.net

Abstract. The article defines the concept of social activity of student youth. The main approaches to determining social activity of student youth have been analyzed. The structural elements of social activity of student youth have been distinguished. Criteria for the levels of social activity of student youth have been defined.

Keywords: Social activity · Student youth · Student social and psychological service · Types of social activity · research activity · Public activity · Volunteer activity

1 Introduction

Today the task of searching and identifying those factors and conditions that allow to substantiate the content, forms and methods of development of social activity of youth is actualized. This is evidenced by the theses declared in the draft Strategy for the Development of Civic Education until 2022 (2019), the Strategy for State Policy in Youth Sphere until 2030 and other legal documents that require effective mechanisms for working with youth, for youth and with the participation of youth.

The relevance of the study of social activity of young people is evidenced by the results of scientific research in different countries. Actual training programs for educational leaders are presented in the work of M. Barakat, E. Reames, L.A.W. Kensler, "Leadership Preparation Programs: Preparing Culturally Competent Educational Leaders" [1].

The article by M. Azeem, L. Mataruna, "Identifying factor measuring collective leadership at academic workplaces" is devoted to the study of the determinants of the culture of social leadership in academic organizations [2].

An understanding of social leadership of gifted students in academic and social groups is presented in a study by K.F. Peairs, C.S. Sheppard, M. Putallaz, P.R. Costanzo "Leader of the Pack: Academic Giftedness and Leadership in Early Adolescence" [3].

Of interest for our study is the study of I. Perko, Z. Mendiwelso-Bendek "Students as active citizens: A systems perspective on a Jean Monnet module, experiential

D. Russo et al. (Eds.): IHSI 2021, AISC 1322, pp. 241–246, 2021.
https://doi.org/10.1007/978-3-030-68017-6_36

learning and participatory approaches", which presents the development of active civic position of students in the summer school with Jean Monnet module.

2 The Study of Social Activity of Student Youth

The problem of social activity among student youth in Ukraine is becoming especially relevant.

The study of social activity of student youth was conducted on the basis of the Faculty of Social and Psychological Education of Pavlo Tychyna Uman State Pedagogical University. Collection and analysis of empirical research data was carried out using the following methods:

1) analysis of documents (statistical data, social programs, reports of structural units);
2) analysis of the results of sociological research to determine the attitude of student youth to scientific, social and volunteer activities.

At the Faculty of Social and Psychological Education of Pavlo Tychyna Uman State Pedagogical University there are three separate departments focused on involving students in various activities: Student Council of the Faculty, Student Research Society, Student Social and Psychological Service (volunteer sector).

3 Departments Focused on Involving Students in Various Activities

Student Council of the Faculty (21 people) functions to ensure that students fulfill their responsibilities and protect their rights and promotes harmonious development of the student's personality, forms the skills of a future teacher, education organizer, leader [4]. The Council has created sectors in the following areas of work: educational-scientific, cultural-educational, sports-health, social-domestic, international relations, press center, etc.

Thus, the student council of the faculty is a body of student self-government, the activity of which allows the student to put forward and implement their public initiatives, to be involved in the organization of the educational process, leisure, etc.

In order to attract students to scientific activities at the faculty, **Student Research Society** (22 people) was established. The purpose of the society is to improve the conditions for the disclosure of scientific and creative potential of students at the faculty, the development of their scientific thinking and research skills, popularization of various branches of science, development of innovation, organizational assistance to faculty management in optimizing research. Integration of scientific and educational activities in the system of higher education [5].

Student Social and psychological service (44 people) is an association of students who volunteer by vocation. Functioning of such a service allows to carry out several tasks: first, it is one of the aspects of professional training of future social workers and practical psychologists by involving them in charitable activities; secondly, participation of students in the activities of the service is of great educational importance, because volunteers work with people in difficult life circumstances, help them, support, establish guardianship over them, which undoubtedly contributes to the flourishing of moral and volitional qualities [6].

Thus, the analysis of students' living environment shows that the faculty has created structural units in three areas of activity – student government (Student Council of the faculty), research (Student Research Society), volunteer sector (Student Social and Psychological Service). At the same time, given the total number of students studying at the faculty, we can talk about a low level of involvement, respectively 7% – Student Council of the faculty, 8% – Student Research Society and 15% – Student Social and Psychological Service.

In the second stage of the study, in order to study the state of social activity of student youth, testing and a sociological survey were conducted (N = 284). The experimental phase of the study was conducted in the form of a questionnaire. Students of 2 – 4 full-time courses of the Faculty of Social and Psychological Education were selected as respondents.

The conducted study involved the study of: life values of students; self-assessment of social activity; the level of inclusion in the activity in three key areas.

In order to determine the motivational and value structure of the individual, a morphological test of life values was conducted (authors: V.F. Sopov, L.V. Karpushyna). The results of the study showed that the highest rate was obtained in the field of family life (33%) and professional development (31%), while the lowest – in the field of physical activity (7%) and social life (6%). Other areas can be attributed to the medium meaning: training and education. Thus, students give priority to the values of personal life and material, at the same time the values of improving physical development and changing the environment are of little importance to them.

The next stage of our study was to study the self-esteem of social activity.

The results of the survey showed that the priorities of young people are focused on achieving financial independence in 33% of respondents, enjoying one's life – 28%, career growth – 30%, and in 9% - on improving the social environment.

At the same time, 69% of respondents believe that social activity they show affects the change of the world around them and themselves; 19% believe that one can only change oneself; 12% deny the possibility of change, but most of them believe that their social activity corresponds to the existing circumstances.

The next important element of diagnosis was to determine the consistency of personal priorities with social norms. 42% of respondents do not think about this issue; 39% – do not coordinate them purposefully, but believe that they do not contradict social norms and requirements; only 14% purposefully coordinate them and 5% focus on social norms and requirements in determining their own priorities.

The results of the study show that the ability to determine one's own priorities in life, taking into account social norms and values are insufficiently formed in students. Young people agree that their social activity affects the world around them, but there are few who want to change it. Students do not see the relationship between changes in the social environment and the prospects for their own development, and this leads to the priority of egocentric values, reduced participation in social initiatives and community activities.

The study of the level of social activity of students was conducted in three areas in which it manifests itself: research, public, volunteer. In order to assess the level of activity in each of these areas, a questionnaire was developed. The questions of the questionnaire were combined into 3 blocks:

1) assessment of the level of research activity of student youth;
2) assessment of the level of public activity of youth;
3) determining the level of volunteer activity of students.

We will distinguish the levels of socially active personality: high, medium and low.

4 Components of Social Activity

The result of the theoretical block of questions aimed at determining the **public activity** of students showed a low level of political activity of young people: 10% of young people constantly monitor political life in Ukraine and 25% follow the main events in the political life of the country. At the same time, almost a third of young people say they have no interest in political processes in the Ukrainian state (35%). At the same time, the answers of the respondents showed that the majority of respondents (81%) consider the participation of student youth in the political life of the country, region, city important. However, the indicator of direct participation of students in election campaigns showed low political activity of university students. Only 42% of respondents participated in the last presidential and 39% parliamentary elections. Among the main reasons for the "apolitical" nature of today's young people are distrust of politicians and government (48%), disbelief in the importance of their civic position (32%) and low self-esteem of their own political culture ("I do not understand anything in this") (20%). A similar trend is observed in the field of civic activity: 7% of young people are ready to personally take the initiative to organize community initiatives in their city and 49% to join public collective initiatives and 44% are unwilling to join community initiatives or say they do not have time for this.

Considering the forms of public activity of students in socio-political life of the university, we obtained the following results: participation in elections as a voter – 39%; attending round tables on the issue – 10%; participation in the implementation of youth programs – 10%; work in the "Academy of Youth Leader" – 3%; participation in the work of Youth Center "START" – 2%; other (without specifying options) – 1%.

As a result of subjective assessment of the level of political activity, the following results were obtained: low level of political activity is observed in 52% of respondents, medium level – in 43%, high – only in 5% of respondents.

Thus, we note the contradictory nature of the political and social subjectivity of student youth. Students have an understanding of the importance of political and social activities and the need for transformation processes in this area. At the same time, the level of their own participation in these areas remains low. It is also worrying that a high percentage of students believe that their public position does not influence anything and that they do not have the leverage to initiate changes. Accordingly, this position is the source of low social activity.

Within the second theoretical block of questions the estimation of **research activity** of student youth was carried out. The results showed that the majority of students (86%) agree with the importance of participation in various research activities, state the positive role of research activity in deepening professional knowledge, ensuring self-realization. However, among second-year students only 19% were participants in research activities, the number of four-year students, who showed research activity, increased to 36%. Herewith, respondents mention writing course papers (33%), participation in competitions (12%), writing research articles (22%), work on social projects (10%), participation in conferences (15%), participation in Student Research Society (8%) as the most common types of research activity. Students (43%) identified "lack of free time" and 22% – "absence of proposal for such activities" as reasons that prevent the manifestation of research activity.

Respondents' answers to the question about the level of research activity can be divided into the following sequence: medium level of research activity was found in 61%; low level – (31%); 8% of respondents showed a high level of research activity.

Thus, we note that research subjectivity of student youth, which can be represented by student participation in research activities, is characterized by awareness of its implementation by students, but the level of students' involvement in various forms of presenting their personal scientific researches is not high. Mostly, it is involvement in the types of research activities provided by educational process.

The third block the questionnaire's questions was focused on determining the level of **volunteer activity of students.** Here is the interpretation of the obtained data: 83% of respondents agree with the importance of youth participation in volunteering, which involves: 1) meeting new people (56%); 2) show themselves (49%); 3) solving social problems (43%); 4) join common meetings (26%); 5) leisure is organized (24%); 6) development of kindness and mercy (22%); 7) professional competencies are improved (19%), etc.

Despite significant positive gains as a result of volunteer activity, only 37% of students were participants in volunteer actions of the university or faculty in practice. Among them, 73% held the position of only participants, 22% – initiators and organizers, partners – 5%.

However, assessment of the level of volunteer activity showed that more that a half of the respondents (49%) have a medium level of volunteer activity, 35% - a low level and only 16% – high.

Thus, we observe students' understanding of the importance of volunteering, a positive attitude to this matter, but a low level of involvement in this activity. Respondents' answers allow us to conclude that the management of student volunteering can be optimized in two key vectors: formation of value orientations of young people, awareness of involvement in solving societal problems, and raising students' awareness of how to volunteer.

The results of the research indicate the need to develop and implement an authorial program for the inclusion of student youth in the activities of departments of the faculty focused on attracting students from different types of social activity (research, social, volunteer).

References

1. Barakat, M., Reames, E., Kensler, L.A.W.: Leadership preparation programs: preparing culturally competent educational leaders. J. Res. Leadership Educ. **14**, 212–235 (2019)
2. Azeem, M., Mataruna, L.: Identifying factor measuring collective leadership at academic workplaces. Int. J. Educ. Manage. **33**, 1316–1335 (2019)
3. Peairs, K.F., Sheppard, C.S., Putallaz, M., Costanzo, P.R.: Leader of the pack: academic giftedness and leadership in early adolescence. J. Adv. Acad. **30**, 416–440 (2019)
4. Student Council of the Faculty of Social and Psychological Education. https://fspo.udpu.edu.ua/%d1%81%d1%82%d1%83%d0%b4%d0%b5%d0%bd%d1%82%d1%81%d1%8c%d0%ba%d0%b5%d0%b6%d0%b8%d1%82%d1%82%d1%8f/students-ka-rada/
5. Student Scientific Society of the Faculty of Social and Psychological Education. https://fspo.udpu.edu.ua/%d1%81%d1%82%d1%83%d0%b4%d0%b5%d0%bd%d1%82%d1%81%d1%8c%d0%ba%d0%b5-%d0%b6%d0%b8%d1%82%d1%82%d1%8f/
6. Student social and psychological service of the faculty of social and psychological education. https://fspo.udpu.edu.ua/%d1%81%d1%82%d1%83%d0%b4%d0%b5%d0%bd%d1%82%d1%81%d1%8c%d0%ba%d0%b5-%d0%b6%d0%b8%d1%82%d1%82%d1%8f/

Gamification Teaching for an Active Learning

Omar Cóndor-Herrera[1], Pamela Acosta-Rodas[2], and Carlos Ramos-Galarza[1,2](✉)

[1] Centro de Investigación en Mecatrónica y Sistemas Interactivos MIST y Carrera de Psicología, Universidad Tecnológica Indoamérica, Av. Machala y Sabanilla, Quito, Ecuador
italo_omar@hotmail.com

[2] Facultad de Psicología, Pontificia Universidad Católica del Ecuador, Av. 12 de Octubre y Roca, Quito, Ecuador
{mpacosta, caramos}@puce.edu.ec

Abstract. Traditional schooling required the professor as the main character of the teaching process. Nowadays, this fact has taken a second place because it is considered as low effective in the present era students' since they have grown and developed in a virtual environment from their birthday. Thus, their cognitive processes require a diversity of learning methods, different from those used in traditional schooling, that explains the need to offer other options to transform and improve the teaching process. One of these options is gamification. According to many authors, the usage of tools such as escape rooms and gaming questionnaires, among other innovative methodologies, effectively promotes and enhances specific development of technologic and scientific content, improving students' motivation significantly when learning.

Keywords: Gamification · Learning student's centered · ICT · Teaching strategies · Gaming questionnaires

1 Introduction

Traditional schooling techniques such as master classes centered in the professor that promote student's learning by memorizing [1] are falling behind and losing weight in the educational context. Nowadays, society lives and develops in a technological era where education must look for the path and tools to obtain the benefits that technological advances offer [2]. In this current context, teaching contemporary methodologies has arisen, taking into account technical resources usage and application. The case of gamification looks to insert gambling's resources in the teaching-learning process through web sites and learning management systems. There are many and variable resources, explaining the increasing number of studies about gamification and the usage of gamified apps. One of the advantages that gamification offers is students' engagement, interest, and motivation towards learning. Recent studies highlight that, if web sites did not count on gaming characteristics, younger generations would not pay attention to these environments [3]. In conclusion, a significant part of the student's interest in technological resources could be linked to those presented as a game.

D. Russo et al. (Eds.): IHSI 2021, AISC 1322, pp. 247–252, 2021.
https://doi.org/10.1007/978-3-030-68017-6_37

Although it is essential to consider that the interest in activities could vary according to participants' gender [4].

As has already been described, gamification can improve students' learning experience. The present article shows a literature revision and possible applications of the resources that a professor could use in the classroom, incorporating it in the teaching methodology and new ways of generating knowledge.

2 Gamification

The term "gamification" has its origin in the industry of digital media. The first document that used this word was published in 2008. However, "gamification" was adopted wider by the scientific community since 2010, when this denomination was promoted and diffused by many authors in scientific meetings such as conferences and symposia [5].

Nowadays, the term gamification is used to describe the process where gambling characteristics are integrated into situations that are not compatible with games [6], as it is the case of masterclasses. However, this example is ideal to conceive a reflective space where nature and activity are modified simultaneously that meaning and interactions are. In this sense, a gamification process consists of a scenario's setting supported with technological resources, obtaining through gamified activities to generate knowledge.

The application of gamification in the teaching-learning process in different fields of knowledge constitutes an emerging practice that is gaining space in the student's education, showing itself as a method destined to improve the curricula contents in learning digital environments [7]. Schooling could take advantage of gamification by using the learning management systems, making learning environments attractive and suitable for students [8]. Also, many studies suggest that gamified learning systems could improve not just learning results but students' enhancement [9] and motivation towards learning.

2.1 Resources to Gamify the Teaching-Learning Process

The exponential increase of research literature about gamification has emerged simultaneously with an uncountable resource used to gamify the classroom, as it is explained as following in a short revision.

Classcraft

It is a role-play developed to manage a secondary classroom; in the present, it is available as a web and mobile application. Professors could create teams in this game, assigning an avatar to every student and points and Bpowers that are used as an award for an adequate behavior within the classroom or when an objective is achieved. The platform's main goal is to change the traditional way of learning and how students' experiences when they arrive to class, added a gamified dimension [6].

It allows students to explore the content planned to review through missions with activities distributed in a map that they must complete gradually, meanwhile, internalization processes and learning in a virtual gamified environment are taking place. As a result, students' intrinsic motivation is generated when learning through a novel experience for them. Figure 1 shows a screenshot of the Classcraft interface.

Fig. 1. Classcraft screenshot.

SGAME

It is a free platform web that allows creating easily educative web games by integrating educational resources in games already made. It is possible to create a great variety of games introducing different topics in every subject. The video game allows them to obtain students' interest; thus, this resource, in particular, is useful and may be utilized to evaluate in a non-traditional way. Figure 2 shows examples of educational videogames.

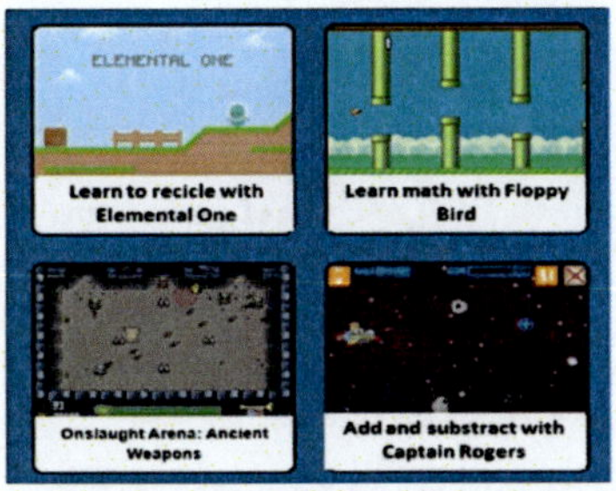

Fig. 2. Educational videogames

Gamified Questionnaires

Finding paths to apply games or gamification concepts in the classroom represents an innovator and a useful resource for teachers to involve students and promote their creative learning abilities and attractive competence [10].

Nowadays, many platforms help to build on-line inquests or questionnaires easily. Simultaneously, it is possible to gamify these questionnaires converting traditional evaluations into different and novel experiences for students, as it is the case of Socrative, Quizizz, or Kahoot, to cite some that evaluate groups in real-time where an interactive competence could be experienced between students. Increasing motivation level in those and improving their learning experience.

The application of gamified questionnaires must be taken into account for active learning since recollected, and analyzed data shows promising results when approaching content acquisition, knowledge, and improve emotional performance, thus, learning process success [1].

Scape Rooms

The escape room consists of a gamified activity where a group of people is confined in one or more rooms and must scape in a determined time. To scape, participants should effectively solve every riddle, quiz, or challenge proposed with the aim of acquiring a mechanism to escape the room. Meanwhile, knowledge about a specific topic is internalized [11].

With the support of technological resources, it is possible to recreate this game in a virtual learning environment by using tools such as Google Docs, where the physical classroom is transported to a virtual environment, where the basic principle of playing is kept. This activity could be accomplished individually or in teams. Every player must discover a clue, solve a riddle, or realize tasks in one or more virtual rooms within a time frame.

3 Gamification: Previous Studies

The construction of the present article was based on previous research that showed favorable results in the application of gamification in the learning process, previous literature is presented as following in Table 1.

Table 1. Previous studies

Title	Authors	Study	Main findings
Classcraft: from gamification to ludicization of classroom management	Sanchez, E., Young, S., & Jouneau-Sion [2]	Analysis of gamification concept based in a literature revision and preliminary comments of teachers that use Classcraft platform	Ludicization does not consist in using game's elements mechanically, but conceiving a reflexive space where interactions are meaningful
Exit for success. Gamifying science and technology for university students using escape-room. A preliminar approach	Sánchez, J., Corrales, M., Luque, A., Zamora, F. [1]	This paper analyzes the usage of gamification technique 1 known as an escape room in university contexts	When these resources are applied, students' emotions are mostly positive, showing these activities as a source of motivation
Gamification of student peer review in education: A systematic literature review	Devi, T., Luxton-Reilly, A & Denny, P. [3]	Understanding of how gamification has been used to involve students in activities of peer- literature revision and to summarize empirical effectivity evidence	Science, technology, engineering, and math (STEM) are the discipline areas that inform peer-literature revision gamified activities with higher frequency

(*continued*)

Table 1. (*continued*)

Title	Authors	Study	Main findings
Effectiveness of the use of the E-Gamification strategy to develop the educational achievement of the preparatory students in Dammam City and their attitudes towards it	Ibrahim, H. [4]	Identification of the effectiveness of the use of the strategy of electronic defects in the development of the educational techniques of the students of the middle level in Dammam city and the direction towards it	Electronic games' strategy aided in developing students' problem-solving abilities when overcoming the game's obstacles

4 Conclusions

Once the literature from a variety of papers about gamification was revised, it is necessary to mention that professors and experts in education consider there are difficulties to incorporate this students' generation into teaching through traditional schooling path, being this fact a possible cause of not satisfying students' needs nor expectations [14].

Nowadays, students are growing in an era where technology plays a principal role in these students' daily life. It is expected that their ways of learning are according to this technological reality. This explains the importance of creating a new teach type that could answer these students' characteristics. Here is where gamification plays a principal role. It has been verified that the usage of educative tools such as escape room, gamified questionnaires, and many other innovative gamified methods are considered as active learning methods, are effective, allowing to promote and enhance the development of specific scientific content, technological and to improve students' motivation significatively.

References

1. Khan, A., Hayat, F., Muddassir, M.: Use of digital game based learning and gamification in secondary school science: the effect on student engagement, learning and gender difference. Educ. Inf. Tech. **22**, 2767–2804 (2017). https://doi.org/10.1007/s10639-017-9622-1
2. Cóndor-Herrera, O., Jadán-Guerrero, J., Ramos-Galarza, C.: Virtual assistants and its implementation in the teaching-learning process. In: Human Systems Engineering and Design III. IHSED 2020, Advances in Intelligent Systems and Computing, vol. 1269, pp. 203–208 (2020). https://doi.org/10.1007/978-3-030-58282-1_33
3. Rodrigues, L., Oliveira, A., Rodrigues, H.: Main gamification concepts: a systematic mapping study. Heliyon **5**(7), 13 (2019). https://doi.org/10.1016/j.heliyon.2019.e01993
4. Han-Huei, C., Kofinas, A., Luo, J.: Enhancing student learning experience with technology-mediated gamification: an empirical study. Comput. Educ. **121**, 17 (2018). https://doi.org/10.1016/j.compedu.2018.01.009
5. Deterding, S., Khaled, R., Nacke, L., Dixon, D.: Gamification: toward a definition. In: CHI 2011 Gamification Workshop Proceedings (2011).

6. Sanchez, E., Young, S., Jouneau-Sion, C.: Classcraft: from gamification to ludicization of classroom management. Educ. Inf. Technol. **22**, 497–513 (2017). https://doi.org/10.1007/s10639-016-9489-6
7. Sanchez, D., Langer, M., Kaur, R.: Gamification in the classroom: examining the impact of gamified quizzes on student learning. Comput. Educ. **144**, 16 (2020). https://doi.org/10.1016/j.compedu.2019.103666
8. Ofosu-Ampong, K., Boateng, R., Anning-Dorson, T., Kolog, E.: Are we ready for gamification? An exploratory analysis in a developing country. Educ. Inf. Technol. **25**, 1723–1742 (2020). https://doi.org/10.1007/s10639-019-10057-7
9. Park, J., Liu, D., Yi, M., Santhanam, R.: GAMESIT: a gamified system for information technology training. Comput. Educ. **142**, 19 (2019). https://doi.org/10.1016/j.compedu.2019.103643
10. Zainuddin, Z., Shujahat, M., Haruna, H., Kai Wah Chu, S.: The role of gamified e-quizzes on student learning and engagement: an interactive gamification solution for a formative assessment system. Comput. Educ. **145**, 1–48 (2020). https://doi.org/10.1016/j.compedu.2019.103729
11. Sánchez, J., Corrales, M., Luque, A., Zamora, F.: Exit for success. Gamifying science and technology for university students using escape-room. A preliminary approach. Heliyon **6** (7), 11 (2020). https://doi.org/10.1016/j.heliyon.2020.e04340
12. Devi, T., Luxton-Reilly, A., Denny, P.: Gamification of student peer review in education: a systematic literature review. Educ. Inf. Technol. **25**, 5205–5234 (2020). https://doi.org/10.1007/s10639-020-10228-x
13. Ibrahim, H.: Effectiveness of the use of the E-gamification strategy to develop the educational achievement of the preparatory students in Dammam City and their attitudes towards it. Educ. Inf. Technol. **25**, 4317–4327 (2020). https://doi.org/10.1007/s10639-020-10156-w
14. Silva, R., Rodrigues, R., Leal, C.: Gamification in management education - a literature mapping. Educ. Inf. Technol. **25**, 1803–1835 (2020). https://doi.org/10.1007/s10639-019-10055-9

Intelligent Painting Education Mode Based on Individualized Learning Under the Internet Vision

Yunqing Xu, Yi Ji(✉), Peng Tan, Qiaoling Zhong, and Ming Ma

School of Art and Design, Guangdong University of Technology, 729 Dongfeng Dong Lu, Yuexiu District, Guangzhou, Guangdong, China
jiyi001@hotmail.com

Abstract. Based on the background of modern art education, this study understands the teaching mode, analyzes the current situation of painting teaching with low learning efficiency, points out the future development trend of art education, and solves the problem of lack of individualized art teaching demand. This paper proposes a "Individualized Painting Education Model" based on user experience and AI technology. This mode of education improves learners' painting level from three aspects. First, it provides learners with accurate learning materials immediately. Secondly, individualized guidance should be given according to learners' learning objectives. Finally, the learner's work is evaluated professionally. Based on the individualized painting education model, this study develops the "Individualized Painting Teaching System", which consists of community module, painting creation module, painting auxiliary module and individual center. The research results show that learners can effectively improve their learning efficiency and artistic creation ability through the application of individualized painting education model, so as to enhance their interest in fine arts and painting.

Keywords: Painting education · Individualized learning · User experience · Teaching system

1 Introduction

Modern art education mainly focuses on one-to-many offline and one-to-one online teaching mode. Art teachers mainly focus on "Teaching" and use modern teaching concepts and methods, which is conducive to giving play to the leading role of teachers and teaching the basic knowledge and skills of painting. Its disadvantage is that it is easy to ignore students' ability of independent study and independent inquiry, which is not conducive to the development of imagination and critical thinking.

The student-oriented concept and method of "Learning" is conducive to promoting students' independent learning and exploration, as well as the development of imagination and creativity. However, if not properly grasped, it is easy to neglect the basic knowledge and basic skills of painting [1].

This teaching mode tends to ignore the cultivation of students' personality and creativity development, and the teaching process is not interesting and interactive.

D. Russo et al. (Eds.): IHSI 2021, AISC 1322, pp. 253–259, 2021.
https://doi.org/10.1007/978-3-030-68017-6_38

Under the background of "Internet+" and innovative education reform, the specific details of art teaching practice have also undergone great changes. For example, teaching objectives, teaching standards, teaching content, curriculum resources, teaching principles and teaching modes are different from those before [2]. Since the rise of the online application market, apps supporting art teaching have been gradually launched and increased, with more convenient ways of using them.

For students, there are more ways to learn art knowledge, and the online art education under the mobile Internet ecological environment emphasizes the relevance between the educational subjects (students, parents, teachers and managers) and the development of students' aesthetic vision, creative thinking and operational ability [3].

Therefore, it is of great research value and significance to improve the quality of education and cultivate talents in line with the requirements of The Times.

2 Research on Individualized Education Model

2.1 Individualized Learning Model

The wisdom of individualized learning can be traced back to Confucius' thought of teaching students according to their aptitude and Socrates' art of midwife [4, 5].

What both of them have in common is to guide students in time when they are confused in the learning process.

Modern scholars argument about individualized education is similar, the American scholar Carrolla. W to realize that the individualized education is in the learners' personality traits and working to achieve A balance between the learning environment, that is to say, it is the learners' personality traits and learned knowledge, concepts, behavior, learning environment, incentive system, and learned skills between a reasonable match, and it is A continuous process of [6, 7]. Compared with the traditional undifferentiated and unified learning mode, individualized learning mode releases the nature of learners to a certain extent, gives the initiative to learners, creates a learning environment that meets the needs of learners, and conducts corresponding guidance and solutions when they are confused, so as to achieve the purpose of effective learning.

2.2 The Development and Current Situation of Individualized Learning Mode in Painting Teaching

Individualized e-learning system is a new distance learning method, which aims to adapt learning to the characteristics of learners. Based on the individualized learning path, the concept of individuation is to improve learners' learning ability according to their learning rhythm, learning level and learning characteristics, so as to acquire target skills [8].

This study aimed at the practice of individualized learning model theory, through the analysis of the whole process of the user experience of art teaching APP, now come to the conclusion: only a few art teaching APP can realize the function of the individualized design, one-to-one learning rely on artificial background guidance function, intelligent degree is weak, the material library function lack of individualized recommendation mechanism, is unable to provide highly individualized learning model fit the needs of learning (Table 1).

Table 1. Comparative analysis of competitive products in online painting teaching apps

App	The art treasure	The academy of fine arts sect	The art king
Main functions	Online teachers comments on paintings One-on-one precise tutoring teaching & Provide a library of works Studio information of art-related colleges and universities	Online studio & communication learning platform Drawing material courseware provision & Professional team comments on paintings One-on-one tutoring by famous teachers	Teachers' online communication and discussion Learning exchange forum & offer courses such as open classes/special courses, etc.
Advantages	Comments on online works One-on-one teaching with real people	One-on-one tutoring with real people Fine arts learning materials courseware complete Multi-dimensional painting evaluation Real-time communication	One-on-one tutoring after class Complete teaching courses Multi-faceted communication learning
Disadvantage	Interface information fragmentation	The quality of teaching video is average	Uneven teaching level & shortage of teachers

The learning system should have a clear and clear information structure, improve the current delayed evaluation and feedback function, simplify the use function of painting learning guidance, track and record learners' learning progress and give Suggestions for the modification of works, and increase the number of interactive feedback between users and the system [5]. By tracking and recording learners' learning progress, learners can intuitively understand their learning situation through systematic summary and feedback of learning reports, so as to cultivate learners' habit of retrospection and reflection.

3 Design and Application of Individualized Painting Teaching System Based on Artificial Intelligence

The application of individualized design based on user experience in AI teaching has been widely applied in the field of modern education, and its influence on higher art education is deepening continuously. Contrast, through the artificial intelligence technology research fusion art discipline knowledge, learning resources, learning strategies of individualized painting teaching knowledge map model and build the teaching mode, fully respect the student individuality difference, stimulate students' individual character and interest, help them form a individualized learning thinking, develop individualized approach to learning, to achieve their unique artistic form [9].

3.1 Construction of Individualized Intelligent Painting Teaching System

Based on the content analysis of different learners' preferences for painting types, painting styles and individual painting abilities, this study realized the individualized painting teaching mode based on the online individualized intelligent painting teaching system combining artificial intelligence and big data analysis technology.

The individualized painting teaching system is constructed based on the concept of individualized painting education mode, which is mainly divided into community module, painting creation module, painting auxiliary module and individual center.

The community module is built based on the database of art works, including course selection and social sharing and learning functions of each user. The drawing and creation section includes drawing functions and drawing tools; The auxiliary section of painting includes real-time auxiliary function and step diagram generation function. The Personal Center is a database of individual students, mainly including paintings collection, painting records, personal information, etc.

Since the art works library contains sketch works, gouache works, watercolor works, oil paintings and other works of various styles, learners can directly search and refer to them in the process of use, and the collaborative filtering recommendation system can intelligently push them according to their search preferences.

3.2 Painting Teaching System Individualized Fan Painting Learning

Based on the content analysis of different learners' preferences for painting types, painting styles and individual painting abilities, this study realized the individualized painting teaching mode based on the online individualized intelligent painting teaching system combining artificial intelligence and big data analysis technology.

The individualized painting teaching system is constructed based on the concept of individualized painting education mode, which is mainly divided into community module, painting creation module, painting auxiliary module and individual center.

The community module is built based on the database of art works, including course selection and social sharing and learning functions of each user.

The drawing and creation section includes drawing functions and drawing tools; The auxiliary section of painting includes real-time auxiliary function and step diagram generation function. The Personal Center is a database of individual students, mainly including paintings collection, painting records, personal information, etc.

Since the art works library contains sketch works, gouache works, watercolor works, oil paintings and other works of various styles, learners can directly search and refer to them in the process of use, and the collaborative filtering recommendation system can intelligently push them according to their search preferences.

Drawing auxiliary can real-time auxiliary functions and user in the plate with paintings, combined with steps graph generating functions, generating against network using a technology, by the two components of the generation of the network and network confrontation mode to the user's painting process, the system can according to the user to generate a series of standard step input.

In addition, the system helps learners to intelligently complete the creation of works according to their current painting steps and painting styles, so as to realize the artistry of human-computer interaction.

3.3 Individualized Painting Learning

Based on the concept of individualized teaching, the design of the real-time auxiliary function research and development to solve the lack of human-machine intelligent auxiliary painting learning problems, real-time intelligent auxiliary function mainly through the perception of students choose painting tools, color choices, in real time the drawing steps, identifying the user drawing step error in painting process, based on the early stage of the system of data input and database depth after training standard drawing steps to discriminant analysis, so as to remind users, and give corresponding amendments.

The step diagram generation function can generate a series of step diagrams for users' reference based on the drawings entered by users, which is convenient for students to learn and use, enabling students to obtain the detailed information of reference works anytime and anywhere, and improving their learning efficiency (Fig. 1).

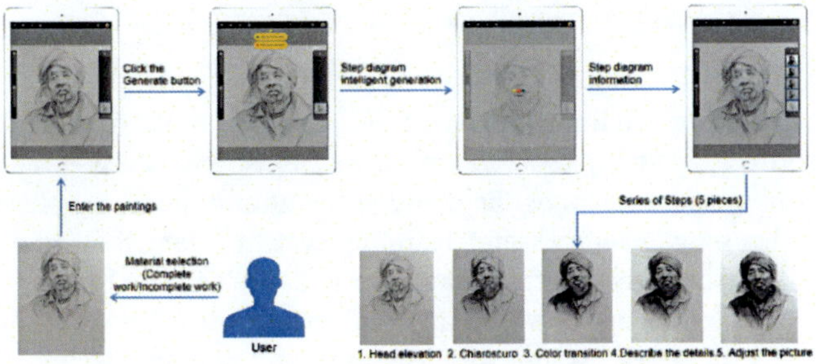

Fig. 1. Sketch portrait step diagram generation case

The system "Teacher" will immediately provide Suggestions for line modification or color use that he/she thinks is more appropriate for the painting, correct mistakes in steps in the painting process, and push relevant reference videos or paintings in line with the learner's painting style.

"Teacher" plays a guiding role in this process. According to the learning status of students, they can answer the painting questions online and realize the real-time and reliable one-to-one auxiliary teaching. It enables learners to master painting skills and improve their painting creation ability as soon as possible in the painting fields they are interested in, so as to realize intelligent auxiliary interaction and achieve the purpose of teaching students according to their aptitude.

3.4 Individualized Painting Evaluation

Individualized evaluation and evaluation is the foundation of individualized content and individualized teaching and the premise of individualized teaching. Deep learning all kinds of paintings by machine system, study the evaluation standard, training them to distinguish different levels of all kinds of paintings, on the basis of fine arts examination score standard, mainly based on gouache still life/head, still life sketch/picture, character sketch the above five types of painting works of each level to evaluate training.

Drawings, for example, is divided into four grade A/B/C/D with the corresponding work assessment scale requirements, students can according to the real-time system of grading to understand their own learning, improve the previous art teaching class APP evaluation problems delay sexual function, students can immediately get painting appraisal feedback, deepen the impression to this painting problem, avoid to appear similar mistakes again later (Fig. 2).

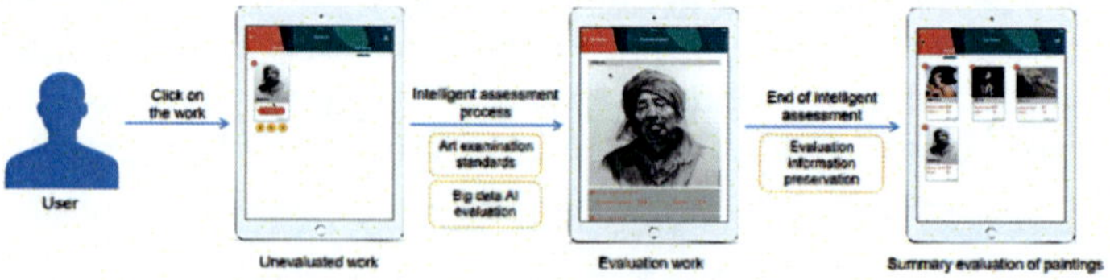

Fig. 2. Painting evaluation feedback process

Intelligent learning environment based on students as the center, the intelligent hardware resources, learning tools, learning resources and teachers and students of intelligent learning space, such as the study space at any time, any place can support students with learning resources and learning system interaction, can also provide students with individualized learning guidance and advice [10, 11].

Focusing on the concept of "teaching students according to their aptitude", providing students with better learning services through intelligent technology, and fully tapping students's creative potential are the important research directions of individualized painting teaching mode at present.

4 Conclusion

This paper makes an in-depth investigation into the current situation and needs of art education in modern colleges and universities at home and abroad, focusing on the analysis of the existing problems between the current painting education model and the needs of learners.

This study proposes the application of artificial intelligence, big data technology and other relevant emerging technologies to realize the individualized painting learning model, and carries out the innovative design and development of the individualized painting learning system.

To solve the problem that it is difficult to meet the needs of sexualized teaching, help painting learners to develop active learning, conduct individualized learning mode in a more comprehensive way according to their own learning situation, enhance the cultivation of individualized consciousness under the current painting education mode, and improve their ability of artistic innovation and appreciation.

Acknowledgement. This research was supported by 2018 school enterprise cooperative education project in Guangdong Province, China (PROJ1007206144650907648).

References

1. Mayer-Schönberger, V., Cukier, K.: Big Data: A Revolution That Will Transform How We Live, Work, and Think. Houghton Mifflin Harcourt, Boston (2013)
2. Xing, X.-j.: Art teaching in colleges and universities under the background of "Internet+" comment on "New edition of art teaching theory". J. Mol. Explor. High. Educ. **139**(2019)
3. Zhu, J.-h., Zahng, Z.-p.: Research on the development of intelligent software for children's online art education in the internet ecological environment. J. Mol. Packag. Eng. **41**, 28–35 (2020)
4. Wang, S., Zhou, L.: Thoughts on teaching students in accordance with their aptitude——to be a socratic or confucian teacher. J. Mol. Shanghai Educ. Res. 26–27 (2006)
5. Peng, H.-c., Zhu, Z.-t.: An analysis of individualized adaptive learning strategies based on human-computer cooperative decision support. J. Mol. E-educ. Res. **40**, 12–20(2019)
6. Carroll, A.W.: Personalizing Education in the Classroom. M. Love Publishing Company (1975)
7. Wu, G.: Individualized education in the era of big data: strategy and practice. J. Nanjing Soc. Sci. 104–110(2015)
8. Bendahmane, M., El Falaki, B., Benattou, M.: Toward a individualized learning path through a services-oriented approach. Int. J. Emerg. Technol. Learn. **14**(15), 52–66(2019)
9. Xiong, Z., Zhi, L., Jiang, J.: Research on art education digital platform based on big data. In: 4th IEEE International Conference Big Data Analytics, ICBDA 2019, 208–211 (2019)
10. Zhu Z.-t., He, B.: Smart education: a new realm of educational information. J. Mol. Res. Audio-visual Educ. 5–13(2012)
11. Zhang, J.-L., Wei, L.-P., Liu, B.: Research on Optimization of learning behavior based on learning analysis in intelligent environment. J. Mol. J. Distance Educ. **38**, 69–79 (2020)

Active Methodologies for Physics Teaching Through Virtual Learning Environments

Juan Polo-Mantuano and Mireya Zapata(✉)

Research Center of Mechatronics and Interactive Systems - MIST, Maestría en Educación mención Innovación y Liderazgo Educativo, Universidad Tecnológica Indoamérica, Machala y Sabanilla, Quito, Ecuador
jpolo@indoamerica.edu.ec, mireyazapata@uti.edu.ec

Abstract. Education currently undergoes great transformations thanks to the rise of new technologies, which has allowed the emergence of new educational paradigms that show new ways of teaching and learning, undoubtedly these elements make us think of new scenarios for education such as virtual learning environments, where teaching-learning process follows a different script from the traditional one. Therefore, the present research goal is to design and implement a virtual learning environment (VLE) for physics teaching. For this proposal, a quasi-experimental study was carried out consisting of two groups: experimental and control. At the beginning, a pre-test was applied which showed minimal differences in both groups and after the use of VLE, a post-test was applied in order to measure the impact of the resource used in the experimental group. The study population consisted of 27 students; at the end, a student's test of independent samples was applied, which evidenced a significant difference regarding academic performance between the control and experimental groups. The mentioned elements allow to make effective the positive impact of the design and implementation of VLE elements in physics teaching.

Keywords: Virtual learning environments · Neo LMS · Learning

1 Introduction

This paper explores digital environments as mechanisms for human formation as a pedagogical mechanism aimed to improve low academic performance, little interest, and low assimilation concepts for high school students through techno-didactic tools. In this sense, the objective is to contrast empirically, how the use of technological tools and specifically virtual learning environments, can contribute positively to the teaching-learning process. As a study case, the application of the experiment will be carried out with students who take the physics subject. This manuscript presents a methodological and pedagogical alternative for teachers and students that will help to promote the development of cognitive abilities; understanding these as the ability of the learner to recognize and describe concepts and express them with a scientific language, and in the most complex to be able to design alternatives to solve problematic situations; fundamental elements that a science student must have in contemporary education [1].

D. Russo et al. (Eds.): IHSI 2021, AISC 1322, pp. 260–266, 2021.
https://doi.org/10.1007/978-3-030-68017-6_39

Currently there is a concept that physics must be transmitted in a memory and not in a dialogical way, and that it must be mechanical and not cognitive [2]. These elements go hand in hand with the limited techno-didactic resources (virtual laboratories, VLE, gamification) used by teachers.

Based on the above, certain technological tools such as virtual learning environments can be exploited, since these bring great advantages, for example: they improve collaborative work, promote searching and content curation, learning is a lot more flexible and can interact synchronously and asynchronously. That is, it can go from being a simple content receiver, as are the majority of students nowadays, to a scenario where they generate knowledge and subsequently can disseminate it through different means (prosumers) [3].

Therefore, this research proposes through the free learning management platform for teachers and students NEO LMS [4] (see Fig. 1), the implementation of a virtual learning environment (VLE) around the theme of uniform rectilinear movement (URM), which using active methodologies such as flipped classroom and gamification; allows teachers to interact with resources such as: videos, forums, chat, readings, questionnaires, game-type practices and synchronous sessions, accompanied by the respective teaching guide; provide the appropriate setting for the generation of the training process autonomously; dynamic and totally flexible through e-learning and m-learning modalities.

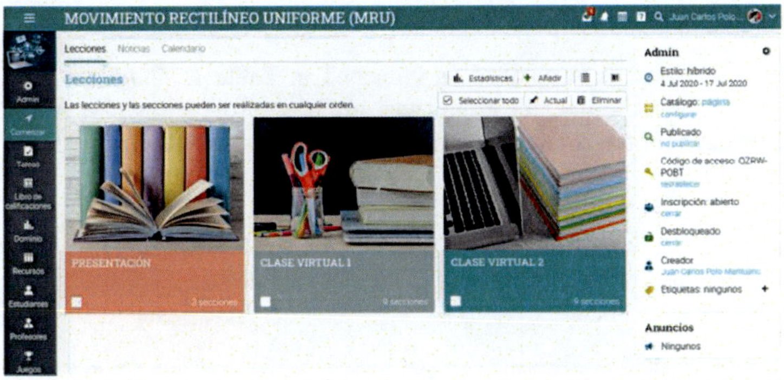

Fig. 1. Physics course lessons. Uniform rectilinear movement

2 Methodological Design

For this research development, a quantitative paradigm was used [5], this is characterized by testing a certain question (hypothesis) and through the use of statistical analysis gives answers to certain behavior patterns of the element that is being studied. The proposed hypothesis is:

- H0: there is no significant difference in marks average before and after the use of virtual learning environments in physics teaching.

- H1: yes, there is a significant difference in marks averages after the use of Virtual Learning Environments in physics teaching.

In that sense, the proposed research is quasi-experimental; Since it is pursued through experimentation in two groups (experimental and control) favorable elements to empirically contrast, in this case, academic performance in a specific topic of physics subject. In this way, the data obtained makes it possible to determine the intervention as positive or negative.

2.1 Target

The study target was made up of 27 s year high school students with ages ranging from 16 to 23 years old. In general, in this methodology the groups are already formed, that is, they are natural or intact groups, [6] therefore, we worked with parallels that had been previously established within the school where the intervention was carried out. The experimental group (Ge) is made up of 12 students, while the control group (Gc) is made up of 15 participants.

The instruments used ad-hoc for the pre-test and post-test covered general physics knowledge in theoretical and resolutive aspects, at the beginning and at the end with identical conditions, but under a specific theme. The instruments were developed with multiple-choice questions and a Likert-type scale.

2.2 Learning sessions

The learning sessions were planned as indicated in Table 1. The VLE developed in NEO LMS for the experimental group was used as resources, while the WhatsApp platform was used for the control group; under the same theme and at the same times as detailed in Table 1.

Table 1. Interventions planning

Tasks	Objective	Ge activity	Gc activity	Time (min)
1	Evaluate general aspects mastery about URM through a pretest	Apply the initial pre-test on general URM knowledge	Apply the pretest on URM knowledge	40
2	Recognize the characteristics of URM	With VLE teach students the main characteristics of URM in a techno-didactic way	With the use of WhatsApp groups, clarify to students the characteristics of URM	40

(*continued*)

Table 1. (*continued*)

Tasks	Objective	Ge activity	Gc activity	Time (min)
3	Study the mathematical foundations of URM (clearance and measurement units) v = x/t	With VLE, explain the appropriate way to clear the formula, as well as recognize the units of measurement of its variables	With the use of WhatsApp groups, explain to students the concepts of formula clearance, as well as the units of measurement of their variables	40
4	Analyze the behavior of MRU in two-dimensional graphs in relation to the variables speed-time (v-t) and space-time (x-t)	With EVA, indicate to students in a techno-didactic way the characteristics of each graph and the importance of recognizing and interpreting them	With the use of WhatsApp groups, explain to students the characteristics of each graph and the importance of recognizing and interpreting them	40
5	Solve URM context problems	With VLE solve URM problems	Raise URM problems for resolution through WhatsApp groups	40
6	Evaluate the students understanding level about URM applying post-test	Apply the final post-test to students on URM through the use of a Google form	Apply the final post-test to students on URM through a Google form	40

2.3 Results

Below are some descriptive elements resulting from the research process. And, subsequently, data prior to the application of VLE (pretest) and after its use (posttest), in addition to the results of the student's T test.

2.4 Descriptive Elements

The participants' average age was 18.11 years. Regarding gender, we worked with 11 men and 16 women of the total, the age range with the highest percentage was between 16 and 17 years.

The Ge represented 44.4% of the population and the Gc 55.6%. Regarding devices to connect to the work sessions, both entire groups had mobile devices to receive their lessons. 25% of Ge participants had a computer at home to manage their academic activities.

2.5 Pre-test Results

Table 2. Pre-test statistic data

Belonging group			Statistical
Ge	Average		**4,67**
	95% confidence interval for the mean	Lower limit	3,48
		Higher limit	5,86
	Median		4,50
Gc	Average		**4,40**
	95% confidence interval for the mean	Lower limit	3,49
		Higher limit	5,31
	Median		4,00

In the study's first phase, a Physics general knowledge pre-test was applied to the experimental and control groups, with a view to making an initial measurement of their academic performance before starting the intervention process. As a result, the Ge obtained an average of 4.67 points in the pre-test, while the Gc got an average of 4.40 points as shown in Table 2. According to the results, there is a minimum difference of 0.27 points on the averages between the groups, which shows the similarity of knowledge regarding the subject of physics.

Table 3. Post-test statistic data

Belonging group			Statistical
Ge	Average		**8,67**
	95% confidence interval for the mean	Lower limit	7,98
		Higher limit	9,35
	Median		9,00
Gc	Average		**5,80**
	95% confidence interval for the mean	Lower limit	4,71
		Higher limit	6,89
	Median		6,00

2.6 Post-test Results

After the carried-out work with VLE for the Ge and with the WhatsApp group for the Gc, a subject knowledge post-test was applied, with looking forward to compare the groups once the intervention had been carried out. Table 3 shows the statistics obtained from the applied post-tests. The academic performance of the students had an average of 8.67 points for Ge and 5.80 for Gc.

The first element to highlight in the results is the difference in averages between the pre-test and post-test. In Ge there was an increase in academic performance from 4.67 to 8.67 points, while the Gc went from 4.40 to 5.80 points.

Additionally, it is observed that the difference in averages between both groups after using the instrument is 2.67 points, which validates the significant improvement in the marks averages obtained once the groups were subjected to different learning scenarios, at the same time, they contrast the effectiveness of VLE and the way in which students assimilate the contents in new and dynamic environments.

2.7 Students' T Test

The obtained data from the knowledge tests application was subjected to the students' T test. The results in the significant show, 0,697 in the pre-test and 0,000 in the post-test. For the research hypothesis to be fulfilled, the value of the signifier must be less than 0.05 since the test has a reliability of 95%. Based on this, we observe that there is a significant difference in the rating averages after the use of VLE.

Finally, some elements related to the usability of VLE and some motivational aspects were explored as well. It was found that 58.33% of the students who were part of the Ge stated that the resource used in the learning process have had a positive impact on them. On the other hand, 75% of participants stated that they felt motivated by these new training scenarios. 66.67% indicated that teachers should use techno-didactic resources such as virtual learning environments in their educational practices. In addition, 75% indicated that VLE was easy to use. In this way, the usability of the resource used was contrasted.

3 Conclusions

In this research, a virtual learning environment (VLE) was designed for physics teaching with the uniform rectilinear motion topic. The results were positive since an increase in the academic performance of the students was evidenced at the end of the intervention. In this way, it was possible to empirically contrast the effectiveness of the use of VLE to improve the academic performance of the participants.

The study process considered two groups: experimental and control, which were evaluated at the beginning and at the end of the investigation, showing previous similarities and significant differences after the study. The level of motivation in the students of the experimental group increased with the use of new techno-didactic resources in the teaching-learning processes.

One of the difficulties in relation to the achievement of the proposed objective was the management of the interactivity process, since the majority of students came from rural sectors, which in many cases limited their connectivity and access to the internet.

Based on the above, new possibilities are opened for future research processes in relation to the way in which improvements in learning processes can be developed through VLE in a virtual study modality. Ideas such as the main use of VLE with e-learning modality within secondary education institutions that seek to promote new teaching-learning methodologies is undoubtedly one of the great challenges to be met.

Finally, the results shown by this research project show the effectiveness of these new ways of learning through new modalities and resources, in turn, open new horizons to explore different alternatives to bring knowledge to future generations, regardless of barriers of time, space and resources.

References

1. Mazzitelli, C.A., Guirado, A.M., Olivera, A.C.: Las evaluaciones en Física y en Química: ¿qué aprendizaje se favorece desde la enseñanza en la educación secundaria?. ienci **18**(1), 143–159 (2013). https://www.if.ufrgs.br/cref/ojs/index.php/ienci/article/download/164/110
2. Moreira, M.A.: Enseñanza de la física: aprendizaje significativo, aprendizaje mecánico y criticidad. Rev. Ens. Físi **26**(1), 45–52 (2014). https://revistas.psi.unc.edu.ar/index.php/revistaEF/article/download/9515/10290
3. Segura-Robles, A., Gallarda-Vigil, M.A.: Entornos virtuales de aprendizaje: nuevos retos educativos. Rev. Cient. Elect. Educa. soc. cono **13**(2), 260–272 (2014). http://www.eticanet.org/revista/index.php/eticanet/article/download/34/30
4. Aranda, O., Lino, N., Martinez, A.: Evaluación de la plataforma virtual Neo. Orbis **1**(1), 49–65 (2017). http://www.revistas.up.ac.pa/index.php/orbis_cognita/article/download/896/762/
5. Ramos, C.A.: Los paradigmas de la investigación científica. unifé **23**(1), 9–17 (2015). https://doi.org/10.33539/avpsicol.2015.v23n1.167
6. Balluerka, N., Vergara, A.I.: Diseós de investigación experimental en Psicología: modelos y análisis de datos mediante el spss 10.0 (2002)

Mental Rotation Ability and Preferences in Vocational Education

Oleksandr Burov[1(✉)], Evgeniy Lavrov[2], Olga Siryk[3], Olena Hlazunova[4], Svitlana Shevchenko[5], Oleksii Tkachenko[4], Svitlana Ahadzhanova[6], Karen Ahadzhanov-Honsales[6], and Oleksandr Viunenko[6]

[1] Institute of Information Technologies and Learning Tools, Kyiv, Ukraine
burov.alexander@gmail.com
[2] Sumy State University, Sumy, Ukraine
prof_lavrov@hotmail.com
[3] Taras Shevchenko National University of Kyiv, Kyiv, Ukraine
lavrova_olia@ukr.net
[4] National University of Life and Environmental Sciences of Ukraine, Kyiv, Ukraine
o-glazunova@nubip.edu.ua, oleksii.tkachenko@gmail.com
[5] Borys Grinchenko Kyiv University, Kyiv, Ukraine
s.shevchenko@kubg.edu.ua
[6] Sumy National Agrarian University, Sumy, Ukraine
{svitlana.ahadzhanova, karen.ahadzhanov-honsales, oleksandr.viunenko}@snau.edu.ua

Abstract. The paper investigates the differences in mental rotation abilities in vocational proficiency (mathematics, IT and social and humanitarian) in schoolchildren and university students, as well as a differences in the nature of the relationship in cognitive stability under impact of internal and external factors in math and humanitarian young graduated specialists.

Keywords: Mental rotation · Cognitive performance · Education · Professional selection

1 Introduction

Global changes evoked by the Fourth Industrial Revolution [1] and consequences of the pandemic COVID-19 in 2020 affect reformation of education system worldwide [2]. The unexpected forced rapid transition to distance learning has accelerated the introduction of new technologies in education including virtual and augmented ones [3]. Virtual and augmented reality is used more and more active in all areas of a human life and activity. Accordingly, education/training uses them to meet the life requirements. But, as it has been revealed in number of studies, a human activity in the synthetic environment can lead to cybersickness [4]. Especially, this negative consequence is important for young people because of their higher susceptibility to external factors [5] in changing learning environment [6].

D. Russo et al. (Eds.): IHSI 2021, AISC 1322, pp. 267–272, 2021.
https://doi.org/10.1007/978-3-030-68017-6_40

Activity in virtual reality (VR) can be accompanied by appearance of the cybersickness [7]. The most studied side effects of it are associated with the motion sickness and eye strain. Sustainability of cognitive functions in VR is less studied. According to data known, the mental rotation ability is associated with susceptibility to cybersickness, namely in relation to age [8] and sex differences [9]. Though mental rotation study can be traced back to the 1960s, including relationship between spatial and mathematical ability across development [10], there is not enough studies explaining if it deals with occupation proficiency and if it can be revealed at the stage of education with preliminary selected vocation. But this could be a useful tool to provide adequate AR/VR means and regulation in the education process to avoid or reduce cybersickness of students, including adaptive learning tools [11].

Purpose. Analysis of general and distinctive properties in the intellect structure of students of IT, math, as well as of social & humanitarian proficiency, with the priority to mental rotation ability and human mental state estimation.

2 Method

Mental rotation is a cognitive operation during which a mental image is formed and rotated into a different orientation in space. Such a process usually requires cognitive manipulation and spatial transformation of a two-dimensional (2D) or three-dimensional (3D) object and is associated with the general intellect. Ability to the mental rotation is assessed by special tests or as components of intellect structure. In the letter case, the response time and accuracy are the measures of the test performance.

The research method was provided as the development of the method used in our previous research of the psychophysiological maintenance of cognitive performance and further applications to learners as operator-researchers [5].

Study 1. 441 pupils (8-grad … 11-grad) of information technology (IT), math & humanitarian proficiency (Math), and social & humanitarian (Soc) as well as 315 students (1 … 6 years) of the same proficiencies have been tested performing on-line R. Amthauer intellect structure test. The analysis of the results was carried out only for those subjects who performed all tests with a given level of rate and reliability. Structural components of the intellect are calculated:

1. LS (logical selection) tests the feeling of language, the ability to formulate judgments.
2. GE (revealing of common features) tests abstract ability.
3. AN (revealing of similarity) tests combinatorial abilities, dynamic thinking.
4. RA (computational, mathematical) tests the ability to solve computational problems of a practical nature.
5. ZR (revealing regularities) tests logical and mathematical thinking.
6. FS (the choice of figures) tests spatial thinking in terms of geometric combinations on a plane in the formation of an integer shape of its fragments (2D mental rotation).

7. WU (task with cubes) tests spatial thinking (3D mental rotation).
8. Me (memory, attention) tests volume and concentration of attention, as well as operational memory.

The number of right results for all intellect test components was calculated.

Study 2. Further study has been carried out to investigate dynamic changes in mental state and cognitive performance of PhD students (math) and young scientists (psychology) over a month period under influence of internal (physiological indices) and external factors (solar and geomagnetic field indices).

The cognitive (logical-combinatoric) test used. The test material: a sequence of numbers (from 0 to 9) which were not repeated and placed in a random order; the task was to rearrange the numbers in ascending order in a few steps, on each one could only change 2 adjacent numbers. Time for every task performance was free (the next task appeared just after entering the answer). The time (Tl) and accuracy of the task performance were measured. Duration of the test session was 180 min, 4 sessions (the first one was training to adapt to the cognitive test and physiological indices measurement) were organized 1 time per week, only data results of the 3^d session were analyzed.

As indices of physiological "cost" of activity and the human state, we registered a heart rate HR and blood pressure (systolic BPs, diastolic BPd. The indices HR, ADs and ADd we registered 1 time in the beginning of every 20 min before (index "1") and in time of the test performance.

To check the influence of the external physical factors on the cognitive task performance the solar activity was studied as in the research [5]. In our preliminary pilot research, the precise connection between effectiveness of operator activity and parameters of a solar wind (SW) was revealed. With the purpose to study this phenomenon in relation to people with different occupational mental organization, we registered indices of proton component of a solar wind - speed SWsp (km/s) and density SWden (proton/SM^3) on the data from Internet site NASA, as well as parameters of the geomagnetic field - planetary index Ks, index of "equivalent amplitude" A.

Subjects: 19 young medical and psychologist researchers, 24 PhD students in math.

3 Results and Discussion

According to results of our previous research, it has been stated that mental rotation level (both 2D and 3D) as component of the general intellect developed by the 9 grad and changed in 10th and 11th grads insignificantly [12]. Therefore, the pupils' and students' data were analyzed without differentiation between grades and student years, accordingly. Separate analysis was made only in relation to proficiency, as well as school and university.

In the *Study 1* significant differences in mental rotation abilities in pupils have been found between corresponded groups Math, IT and Soc (Fig. 1).

As it was expected results in test for 2D (FS) mental rotation demonstrated higher level in comparison with 3D (WU). Quite interesting result is a higher level of mental rotation abilities in IT students even than in mathematicians. A reason of that situation could be explained the fact that children prefer to be IT specialists than mathematicians

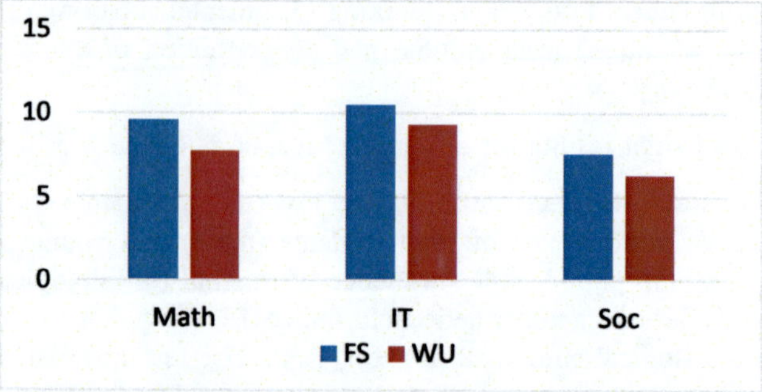

Fig. 1. Development level of mental rotation in pupils

in Ukraine. All those types of proficiency were based on the preliminary selection of schoolchildren for the vocational specialization.

Similar proportion has been revealed in university students, but difference was not so significant and not reliable in regards comparison of Math and IT students (Fig. 2).

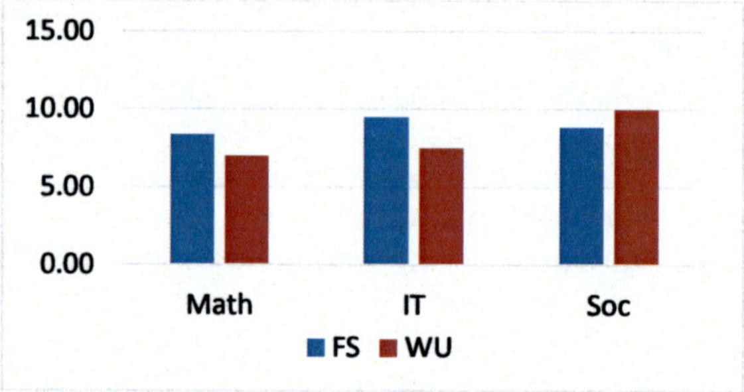

Fig. 2. Development level of mental rotation in students

But expectedly, it was revealed that development level of mental rotation demonstrated students of the social and humanitarian proficiency. This fact could be explained by the higher level of motivation to learn in national humanitarian universities which students participated in test performance more actively.

In any way, those facts should be a signal for the Generation Z preferences in chose of future occupation, at least, in our country.

Further *Study 2* has been carried out to investigate changes in mental state and cognitive performance of students (math and psychology) over a month period (physiological indices registration was carried out in parallel to the test performance).

The comparison analysis of variation of cognitive sustainability by the indicator of rate of test performance in young medical and psychologist researchers (both proficiencies relate to the mental work) demonstrated clear difference in relation of the cognitive test performance to external (solar wind parameters, i.e. solar radiation) and internal (state of the cardiovascular system) factors (Fig. 3). Correlation between indices of those indices with variation (inter-individual) of the time of test performance had different levels and vectors of priority.

Firstly, it is necessary to highlight that individuals among medical psychologists/medicals demonstrated the higher correlation with external factors (solar radiation, SW density, r = 0,5), though all of them were selected for military service according to the professional protocols of health level.

Secondly, these results corresponded findings of high relationship between cognitive performance and its physiological maintenance, especially important in adaptive learning [13].

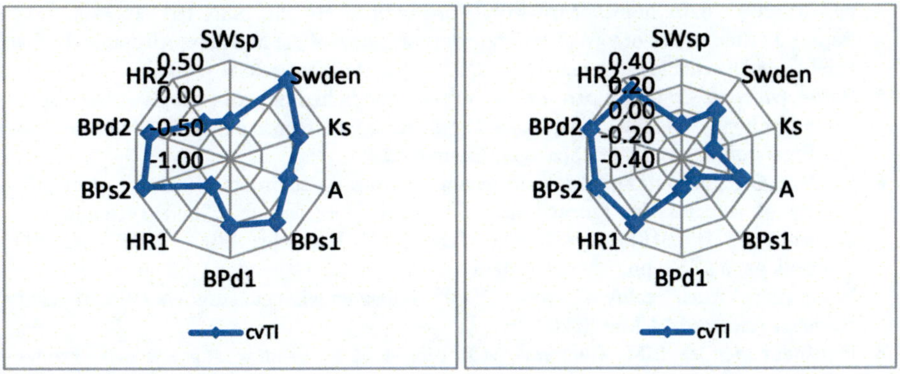

Fig. 3. Correlograms of variation of the test performance rate in psychologists (left) and mathematicians (right)

4 Conclusion

The paper considers differences in mental rotation abilities in vocational proficiency (mathematics, IT and social and humanitarian) in schoolchildren and university students, as well as a different nature of the relationship in cognitive stability under impact of internal and external factors in math and humanitarian young graduated specialists.

Significance of the Proposed Presentation. The results can be applied to optimize the use of AR/VR tools in education process depending on student individual/typological abilities to mental rotation. The goal of optimization is to reduce a risk of cybersickness when using virtual reality.

Acknowledgments. This work is supported by the grant 0118U003160 “System of Computer Modeling of Cognitive Tasks for the Formation of Competencies of Students in Natural and Mathematical Subjects”.

References

1. Schools of the Future: Defining New Models of Education for the Fourth Industrial Revolution, 2020 WEF. https://www3.weforum.org/docs/WEF_Schools_of_the_Future_Report_2019.pdf
2. Guterres, A.: The future of education is here. Launch of the policy brief: education during COVID-19 and beyond, United Nations. https://www.un.org/en/coronavirus/future-education-here. Accessed 04 Aug 2020
3. Iatsyshyn, A.V., et al.: Application of augmented reality technologies for preparation of specialists of new technological era. In: Augmented Reality in Education : Proceedings of the 2nd International Workshop (AREdu 2019), Kryvyi Rih, Ukraine, 22 March 2019, pp. 181–200 (2020). https://ceur-ws.org/Vol-2547/paper14.pdf. ISSN 1613-0073
4. Barrett, J.: Side effects of virtual environments: a review of the literature. Australian Defence Science and Technology Organisation, DSTO-TR-1419, May 2004 (2004)
5. Burov, O., et al.: Cognitive performance degradation in high school students as the response to the psychophysiological changes. In: International Conference on Applied Human Factors and Ergonomics, pp. 83–88. Springer, Cham (2020)
6. Pinchuk, O.P., et al.: Digital transformation of learning environment: aspect of cognitive activity of students. In: Proceedings of the 6th Workshop on Cloud Technologies in Education (CTE 2018), Kryvyi Rih, Ukraine, 21 December 2018. CEUR Workshop Proceedings, # 2433, pp. 90–101 (2019)
7. Wikipedia: Virtual reality sickness (2020). https://en.wikipedia.org/wiki/Virtual_reality_sickness. Accessed 24 Mar 2020
8. Fernández-Méndez, L.M., Contreras, M.J., Elosúa, M.R.: From what age is mental rotation training effective? Differences in preschool age but not in sex. Front. Psychol. **9**, 753 (2018). https://doi.org/10.3389/fpsyg.2018.00753
9. Parsons, T.D., Larson, P., Krat, K., Thiebaux, M., Brendon Bluestein, J., Buckwalter, G., Rizzo, A.A.: Sex differences in mental rotation and spatial rotation in a virtual environment. Neuropsychologia **42**(2004), 555–562 (2004)
10. Young, C.J., Levine, S.C., Mix, K.S.: The connection between spatial and mathematical ability across development. Front. Psychol. **9**, 755 (2018). https://doi.org/10.3389/fpsyg.2018.00755
11. Lavrov, E., Lavrova, O.: Intelligent adaptation method for human-machine interaction in modular E-learning systems. In: Proceedings of the 15th International Conference on ICT in Education, Research and Industrial Applications. Integration, Harmonization and Knowledge Transfer, Volume II: Workshops, Kherson, Ukraine, 12–15 June 2019, pp. 1000–1010 (2019)
12. Burov, O.Yu.: Profile mathematical training: particular qualities of intellect structure of high school students. Phys. Math. Educ. **1**(15), 108–112 (2018)
13. Mulder, L.J.M., Van Roon, A., Veldman, H., Laumann, K., Burov, A., Quispel, L., Hoogeboom, P.J.: How to use cardiovascular state changes in adaptive automation. In: Hockey, G.R.J., Gaillard, A.W.K., Burov, O. (eds.) Operator Functional State, The Assessment and Prediction of Human Performance Degradation in Complex Tasks. NATO Science Series, pp. 260–272. IOS Press, Amsterdam (2004)

Educational Technological Game for Children's Education

Omar Cóndor-Herrera[1] and Carlos Ramos-Galarza[1,2(✉)]

[1] Centro de Investigación en Mecatrónica y Sistemas Interactivos MIST/Carrera de Psicología/Maestría en Educación mención Innovación y Liderazgo Educativo, Universidad Tecnológica Indoamérica, Av. Machala y Sabanilla, Quito, Ecuador
omarcondor@uti.edu.ec

[2] Facultad de Psicología, Pontificia Universidad Católica del Ecuador, Av. 12 de Octubre y Roca, Quito, Ecuador
caramos@puce.edu.ec

Abstract. The teaching process in the infant stage is a challenge for the current educational process since children need the use of resources that are consistent with the reality in which they live today. In this context, the ways of teaching have evolved thanks to technological progress, which has led to the emergence of new teaching methodologies, including gamification, which uses the structure of the game for an educational purpose supported by the use of technology for the creation of didactic resources, which due to their structure and design are of great interest to students. Therefore, in this article, the conceptual development of a serious game for teaching natural sciences based on gamification based on the use of technology is presented.

Keywords: Gamification · Student-centered learning · Teaching strategies · Serious games

1 Introduction

Teaching at an early age requires a lot of dynamism, as it is important to get students' attention and generate interest to learn. This element has become a challenge because traditional education does not always attract the students' full attention, which often reduces their interaction and commitment to learning.

For this reason, it is necessary to develop innovative approaches that offer a much more dynamic structure to education. The idea is to improve the teaching and learning process by using different mechanics, elements and other game-based features, expecting to obtain greater motivation and commitment from students [1], which would proportionally mean an improvement in learning outcomes [2].

The gamification is known for incorporating elements of the game into the process either physically or digitally through applications, platforms, serious games, etc. This process not only improves motivation but has shown that generating interest in students can improve participation levels, increasing interest in a particular subject [3].

D. Russo et al. (Eds.): IHSI 2021, AISC 1322, pp. 273–278, 2021.
https://doi.org/10.1007/978-3-030-68017-6_41

2 Gamification

Gamification is a concept that became relevant in the scientific community in 2010 [4]. In the current educational environment and other areas, the term gamification is understood as the process in which game aspects are integrated into a situation that is initially not playful [5], transforming traditional forms of teaching such as the well-known master classes, which, while fulfilling their role of generating learning, it does not involve an active teaching methodology, such as with gamification. This makes it possible to conceive of a reflective space in the classroom or in a virtual environment, where nature and learning activity is modified, as well as it makes us rethink the meanings of the interactions that take place. With this type of games, a reconfiguration of the class scenario is achieved and based on this type of technological resources, dynamic and tractive learning is attained for the student.

2.1 Gamification Learning Benefits

There are four key aspects of learning success: attention, active participation, feedback and consolidation [6]. Applying a gamification methodology can positively affect these aspects benefiting students' learning directly. As mentioned above the use of gamified games influence the interest of students which leads to a higher level of attention to the object of study and therefore a clear improvement in active participation in the classroom [7].

Motivation is a crucial factor in students' learning behavior. It plays a key role in the concept of gamification, as encouraging student motivation provides the ability to improve learning. This means that students have more willingness to work and actively participate in class with the expectation of improved learning outcomes [8]. For this reason, many schools and institutions around the world provide their students with this type of teaching based on games of different types such as virtual simulation games, serious games, etc. [9].

3 Proposal for an Educational Technology Game Design

Emerging technologies provide the possibilities to create resources and think about ways of applying to what is already known with other objectives based on what was previously created. The proposal in this article is to use the foundations of gamification in adapting a technological video game simulator already known to people and inserting learning activities such as quizzes, challenges, puzzle-solving, etc. The game will contain content from a certain subject in the technological game structure and these activities are suitable for gamification and allow technological accessibility for students [10].

Figure 1, shows the Sgame platform, the Captain Rogers video game interface has been used and restructured with addition and subtraction math exercise questions. This activity is targeted to elementary school students, the player must go through the levels offered by the game and when reaching a certain score, they need to solve exercises to advance in the game.

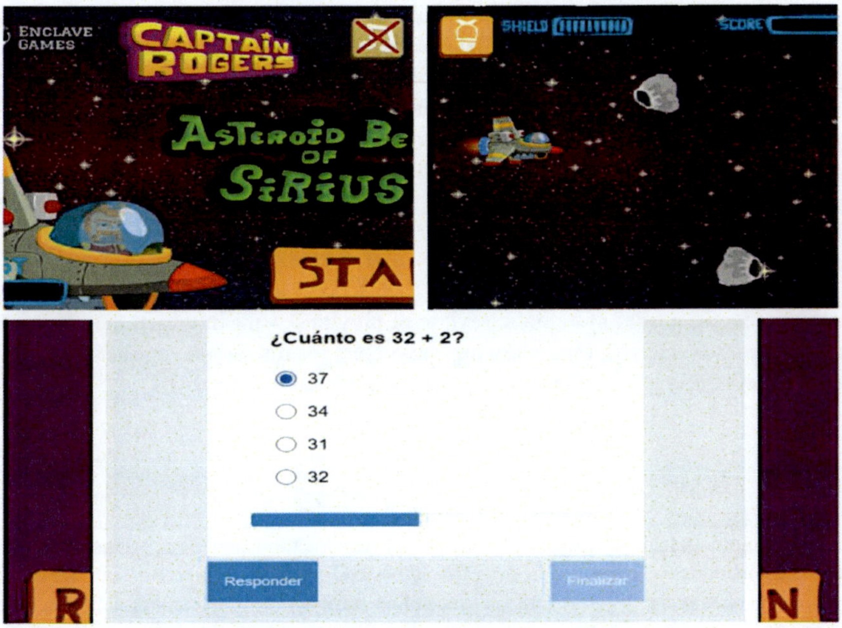

Fig. 1. Basic mathematical operations videogame screenshot

This same dynamic can be replicated to generate teaching resources in different subjects and with different video game models as shown in Fig. 2.

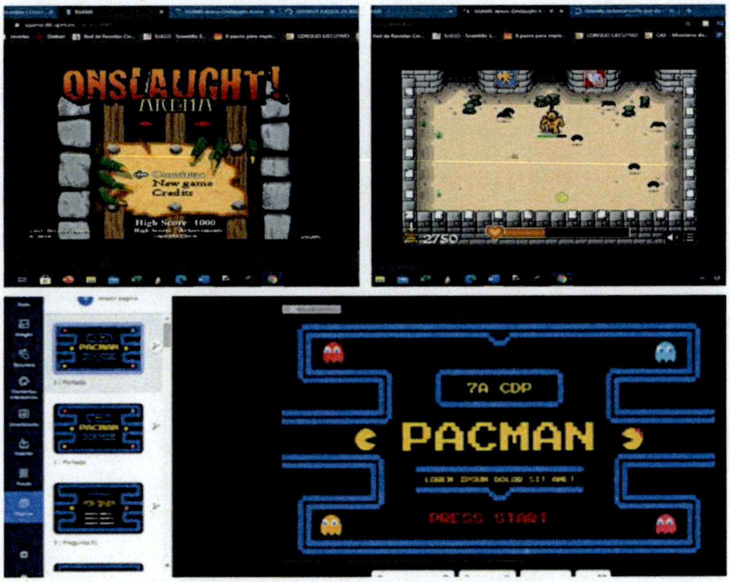

Fig. 2. Educative videogames models.

The variation can be more than only content. It remains the possibility to structure the game by levels and questions difficulty associated with them, or learning problems can be gradually increased. In this way, it is possible to consolidate into a single resource, a recreational and learning activity combining an active teaching methodology. In this way, it is possible to increase the level of motivation, generate greater interest in students and improve their predisposition towards learning.

4 Previous Gamification Application Research

For this research, a brief systematic review of previous work in which they have used gamification was carried out, showing favorable results in the application of this teaching methodology. On the Table 1 previous research are shown.

Table 1. Previous research

Paper	Authors	Research question	Findings
Towards a motivational design? Connecting gamification user types and online learning activities	Boverma and Bastiaens [8]	How five gamification user types can relate to six online learning activities that are mainly used in an online bachelor's and master's class	Correlations revealed that the six online learning activities were at least significantly connected to one of the five types of gamification user
The effect of challenge-based gamification on learning: An experiment in the context of statistics education	Legaki et al. [2]	This research investigates the effects of challenge-based gamification on learning in the statistical education area	Challenge-based gamification had a positive impact on student learning compared to traditional teaching methods
Effectiveness of the use of the E-Gamification strategy to develop the educational achievement of the preparatory students in Dammam City and their attitudes towards it	Ibrahim [11]	Identify the effectiveness of the electronic defects strategy in the development of the educational techniques of middle-level students in Dammam city	The electronic gaming strategy improved students' problem-solving skills due to the obstacles they encounter during the game

5 Conclusion

This research has proposed a serious game to improve the teaching and learning process. This serious game consists of adapting a classic video game with math content, by applying the gamification process in favor of the educational context.

One of the advantages of developing a serious game in the educational context is the children's motivation level to learn, and therefore, the content proposed in each subject is more attractive to the student [12].

This contribution is key to the Latin American context, which has a traditional teaching methodology where teachers consider the use of technology as a taboo. As consequence, technology is out of the learning process. Under traditional teaching methodologies, technology is seen as a distraction, which is far from reality as recent research has shown the benefit of technology in the educational context [13–15].

Future research aims to test the contribution of the game developed through research with pre- and post-test measurements, thus generating evidence of the implementation of technological resources in the teaching and learning process.

References

1. Costa, F., Viana, J., Raleiras, M.: What do higher education students have to say about gamification? In: 10th International Conference in Methodologies and Intelligent Systems for Technology Enhanced Learning, MIS4TEL 2020, pp. 56–65 (2020). https://doi.org/10.1007/978-3-030-52287-2_6
2. Legaki, N., Xi, N., Hamari, J., Karpouziz, K., Assimakopoulos, V.: The effect of challenge-based gamification on learning: an experiment in the context of statistics education. Int. J. Hum. Comput. Stud. **144**(102496), 1–58 (2020). https://doi.org/10.1016/j.ijhcs.2020.102496
3. Smiderle, R., Rigo, S., Marques, L., Peçanha de Miranda Coelho, J., Jaques, P.: The impact of gamification on students' learning, engagement and behavior based on their personality traits. Smart Learn.ing Environ. **7**(1) (2020). https://doi.org/10.1186/s40561-019-0098-x
4. Deterding, S., Khaled, R., Nacke, L., Dixon, D.: Gamification: toward a definition. In: CHI 2011 Gamification Workshop Proceedings (2011). https://gamification-research.org/wp-content/uploads/2011/04/02-Deterding-Khaled-Nacke-Dixon.pdf
5. Sanchez, E., Young, S., Jouneau-Sion, C.: Classcraft: from gamification to ludicization of classroom management. Educ. Inf. Technol. **22**, 497–513 (2017). https://doi.org/10.1007/s10639-016-9489-6
6. Lamrani, R., Abdelwahed, E.: Game-based learning and gamification to improve skills in early years education. Comput. Sci. Inf. Syst. **17**(1), 339–356 (2020). https://doi.org/10.2298/CSIS190511043L
7. Benhadj, Y., Messaoudi, M., Nfissi, A.: Artificial intelligence in education: Integrating serious gaming into the language class classdojo technology for classroom behavioral management. IAES Int. J. Artif. Intell. **8**(4), 382–390 (2019). https://doi.org/10.11591/ijai.v8.i4.pp382-390
8. Bovermann, K., Bastiaens, T.: Towards a motivational design? Connecting gamification user types and online learning activities. Res. Pract. Technol. Enhan. Learn. **15**(1), 18 (2020). https://doi.org/10.1186/s41039-019-0121-4
9. Paulova, P., Cerna, M., Hamtilova, J., Malý, F., Kozel, T., Kriz, P., Han, J., Ulrych, Z.: Virtual hotel – gamification in the management of tourism education. In: 17th International Conference on Remote Engineering and Virtual Instrumentation, REV 2020, vol. 1231, pp. 773–781 (2020). https://doi.org/10.1007/978-3-030-52575-0_63

10. Luigini, A., Basso, A.: Heritage education for primary age through an immersive serious game. Springer Tracts en Ingeniería Civil, pp. 157–174 (2020). https://doi.org/10.1007/978-3-030-49278-6_10
11. Ibrahim, H.: Effectiveness of the use of the E-Gamification strategy to develop the educational achievement of the preparatory students in Dammam City and their attitudes towards it. Educ. Inf. Technol. **25**, 4317–4327 (2020). https://doi.org/10.1007/s10639-020-10156-w
12. Ramos-Galarza, C., Arias-Flores, H., Cóndor-Herrera, O., Jadán-Guerrero, J.: Literacy toy for enhancement phonological awareness: a longitudinal study. In: Lecture Notes in Computer Sciences, vol. 12377, pp. 371–377 (2020). https://doi.org/10.1007/978-3-030-58805-2_44
13. Cóndor-Herrera, O., Ramos-Galarza, C.: The impact of a technological intervention program on learning mathematical skills. Educ. Inf. Technol. 13 (2020). https://doi.org/10.1007/s10639-020-10308-y
14. Ramos-Galarza, C., Acosta-Rodas, M., Acosta-Vargas, P., Salvador-Ullaurim, L.: Technological innovation to assess cognitive functions in attention deficit hyperactivity disorder. In: Advances in Intelligent Systems and Computing, vol. 1201, pp. 76–82 (2021). https://doi.org/10.1007/978-3-030-51041-1_11
15. Cóndor-Herrera, O., Jadán-Guerrero, J., Ramos-Galarza, C.: Virtual learning objects of math educative process. In: Advances in Intelligent Systems and Computing, vol. 1269, pp. 192–197 (2021). https://doi.org/10.1007/978-3-030-58282-1_31

Mining Students' Topics of Interest and Innermost Feelings Through Confession Pages

Hoang Huu Son(✉), Dong Thi Ngoc Lan, Ngo Thanh Hoang, Dao Minh Tam, Pham Thi Phuc, and Tran Thi Huong

Academy of Finance, Le Van Hien str 58, Duc Thang Zone, Bac Tu Liem District, Hanoi, Vietnam
{sonhh, dongngoclan, ngothanhhoang, phuct79, tranthihuong}@hvtc.edu.vn, daominhtam0108@gmail.com

Abstract. Capturing psychological changes and aspirations of the young, especially students, is always the essential concerns. However, this is still a big challenge not only for administrators but also for psychologists. Due to the population of sharing inside feeling trends through confessions in student communities, confession pages become a valuable source for identifying the interests and attitudes of communities. In our research, we propose a confession pages mining methodology based on text analysis method. This method is applied to several confession pages of student communities in a year to find out interest issues and their attitude. As a result, the model achieved average accuracy as F-measure = 62.45% for interest issues classification and F-measure = 76.43% for sentiment classification. The model have been applied to mining characteristics of community of four universities.

Keywords: Facebook confession · Topic interests · Sentiment analysis

1 Introduction

Facebook confession sites have gained popularity, giving youth another venue to anonymously post secrets, rants, and insults. Facebook confession sites are set up in that posts are anonymous while "like" or comments on the post identify the user by name. Confession sites appear to be associated with university, as they may contain the university name, mascot, acronym, city or state, however, most Facebook confession sites are not affiliated or sanctioned by the university they are associated with. Concerns were risen about cyberbullying and the effects of the sites on the reputation of the school. Cyber bullying is using electronic communication devices to threaten, intimidate, or insult someone.

The trend, which is known roughly as revealing ours secrets on social networks Facebook are more and more attractive in Vietnam. Many pages related to this trend were established. (CNN Confessions of University of Languages and International Studies, Ams Confession of students studying in Amsterdam high schools, FTU Confession of "superheroes' from Foreign Trade University, Confessions Of A Uni

D. Russo et al. (Eds.): IHSI 2021, AISC 1322, pp. 279–285, 2021.
https://doi.org/10.1007/978-3-030-68017-6_42

Student which was founded in the UK and has nearly 250,000 participants. Not only sharing learning secrets, emotional secrets are also posted, and then finding information of user also appeared. Confession pages include advantages such á: 1) Sharing problem with someone will promote their own mood. 2) You can share with everyone secretly that can be directly or identity inform. However, confession pages also having some disadvantages such as: Cause arguments or sharing sensitive information.

Although the confession pages contain extremely important information for data mining, most research on confession sites is very limited. In my view, there are a few typical studies. Research of Birnholtz et al. [1] has shown the advantage of confession sites that allow users to ask about taboo, stigmatized topics against others locally, and get less threatening or negative matching responses. C. Chelsea's study uses the Linguistic Inquiry and Word Count (LIWC) model to give the linguistic features of confession and shows the emotional behind it so that they understand students' attitudes towards problems of the university [2]. DAMAYANTI et al. Did a research on confessions to show the virtual reality of housewives in Indonesia [3]. Fatima used a content study of confession of the male community in Pakistan on women's human rights [4]. In general, studies on Confession Exploration are not much, focusing mainly on psychological and sociological issues.

In Our research, we focus on researching to apply Natural language processing (NLP) methods and classification to bring out the characteristics of the community under two-aspects: topic interest and sentiment (mentioned as innermost feelings). Comparison with other text mining researches, we overcome the challenge of dataset: small size and specific of language (Vietnamese): NLP techniques transfer confessions into vectors of features; Annotated data and classification models to propose classification model for mining. Finally, this model is applied to highlight community characteristics.

Following this approach, our paper is separated in to four sections: Beside introduction and conclusion, Sect. 2 we discuss about methodology to find out topic interest and sentiments, Sect. 3 describe our implementation and discuss the result in detail.

2 Proposal Research Framework

Our research are conducted according to several pipeline steps that are described in below subsections.

2.1 Data Acquision and Preprocessing

The NVIVO software gathered all posts, comments, reactions from confession pages of four famous universities. Collected data are stored into Excel file and detail figures of each dataset is showed on Table 1.

Table 1. Statistics of collected confessions

Confesion pages	Page links	Communities	N°
HVTC	https://www.fb.com/hvtcconfessions/	Academy of Finance	850
NEU	https://www.fb.com/neuconfessions.vn/	National Economies Uni	876
HUST	https://www.fb/hustconfession/	Hanoi Uni of Science and Technology	450
FU Hòa lạc	https://www.fb.com/FUHLConfessions/	FPT Uni	1085

Preprocessing that is applied to reduce noise and create clean data set for machine learning includes: 1) Extracting posts; 2) Dropping too short and lack information confessions. 3) Teen code translation and removing abbreviations and emotions.

2.2 Annotation

Based on studies on the use of social networks of Vietnamese students and surveys of students sharing on social networks, we conduct a user annotation guideline to guide participants to assign data labels. Each confession is annotated two kind of label. The first label is topic interesting that is categorized following student's concern: 1) Student life and society 2) Career and study 3) Loving 4) #N/A (unidentified theme). Secondly, the label about emotions (positive; negative and nature).

Annotation process randomly assign a group of confessions for each participant. Each confession are ensure independently annotate by at least two participants. Differences are concentrative and re-label by the third. As a result, a ground truth data set containing 850 post is released for training and testing classification models.

2.3 Term Extraction

Segmentation and POS tagging in Underthesea [5] - Vietnamese NLP library is applied to detect verb, noun and adjective from confessions. TF- IDF algorithms is applied find out terms (which are defined as meaningful words). We used co-occurrence graph approach and page ranks to ranks terms according to importance of term [6].

2.4 Classification

We model each confession to n dimensions vector where n is the number of keyword in the corpus according to the expression:

In that each confession D is presented as $X_D = \{x_1^D, x_2^D, \ldots, x_n^D\}$.

$$x_i^D = \begin{cases} 0 : \textit{if } Word_i \textit{ is not in set of keyword in confession D} \\ Pagerank_{word_i}^D \textit{ if } Word_i \textit{ is in set of keywords in confession D} \end{cases} \quad (1)$$

Due to a huge value zero in vector, we apply feature reduce algorithm in the Sklearn library to reduce 10587 features to 2417. Then train and validation sets is constructed by cross validation k-fold (k = 10 for the annotated dataset to seperate 90% confession for training and 10% for testing.

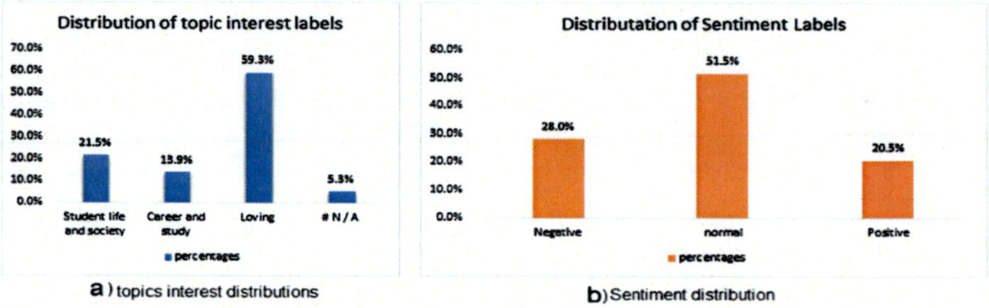

Fig. 1. Statistics of annotation data set

Statistics of dataset presented in Fig. 1 convinces that confession data is unbalance. Therefore, unbalances classification approaches should be applied which are listed as: SVM, Random forest (RF) XGboost classifier, LightGBM classifier. F-measure score [7] is selected to compare models and find out the best after 5000 loops.

3 Experiment and Results

3.1 Model Selections

Table 2. F-measure models for topic interest and sentiment classifications

Classification models	Topic interests			Sentiment		
	Max	Min	Means	Max	Min	Means
.SVM	63.14%	61,93%	62.34%	69.12%	67.56%	68.37%
RF	56.38%	59.05%	57.72%	68.47%	65.12%	67.31%
XGBOOST	60.88%	58.92%	60.03%	77.11%	72.32%	74.23%
LightGBM	63.34%	60.53%	**62.45%**	77.98%	75.32%	**76.43%**

The accuracy of all experiments are presented on Table 2. The result indicates that LightGBM is the best model with the accuracy at 62.45% in average for topic interests and 76.43% for sentiment classification, respectively. It can be explain that comparing other ensemble learning, LightGBM library optimize the features selections and prioritize the most informative sample to construct decision trees in Random Forest. In addition, Exclusive Feature Bundling, which takes advantage of sparse datasets by grouping, features in a near lossless way. The result also coincides with previous research on text classification [7, 8] and Sparse data [9].

3.2 Mining Sentiments and Topic Interest of Specific Confession Pages

These statistics of communities 'characteristics are presented in bar charts on Fig. 2 the left top bar chart indicates that ignoring neutral confessions, number of positive confessions of HVTC and NEU and FPT FU Hòa lạc are higher than negative ones. In constant, HUST has a figure of negative completely higher than positives. It sound that NEU community prefer to post happy events on confession pages and the HUST community focus on the negative stories.

The right top bar chart shows the topics interest of communities. It is clear that loving is the most common story in confession pages. It is suitable with the psychology of young peoples in age from 18–22 when the loving and dating are big demands. The right top bar chart shows the topics interest of communities. It is clear that loving is the most common story in confession pages. It is suitable with the psychology of young peoples in age from 18–22 when the loving and dating are big demands. Therefore, Vietnamese universities should more and more concentrative to loving aspects toward students. The number of confessions of life and society of NEU and HUST are higher than HVTC and FU Hòa lạc. The statement can be explained by the lack of dormitory positions and student should live outsides.

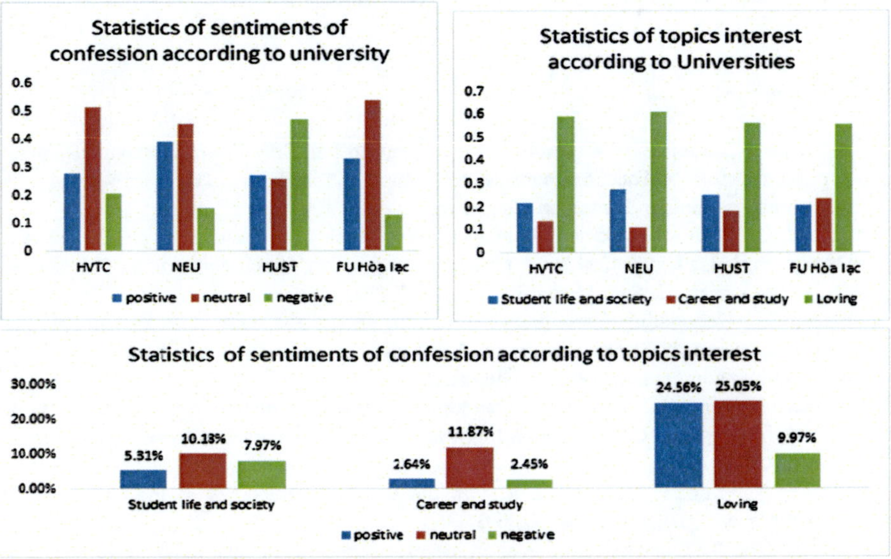

Fig. 2. Statistics of mining confession pages of several student communities

A relation between topic interests and innermost feelings are present in the bottom bar chart. In student life and society, the promotion of negative is higher than positives that reflect of the difficulty of them while changing life conditions from family to independent. Therefore it is necessary to develop student advising councils in universities to guide and support students.

4 Conclusions

This research are preside on confession pages mining. So the research is mentioned as the first steps to explore community's characteristics based on content mining from specific data source. Topics of interest and innermost feelings of community are modeled as text classifications. Nature language processing that includes several tasks: Segmentation, Tag of POS, TF-IDF using to extract terms from a original confession. Co-occurrence graph approach and page rank algorithms are applied to rank terms according to importance of keywords transfers into confession features vector. Based on Annotated data set and several classification libraries, topics interests' classifications and sentiments classification have been establish and validation. The highest mean of F-measure score (62.45% for interest issues and F-measure = 76.43% for sentiment) indicates LightGBM is the most fit to these problems. The LightGBM applied automatically mining confession pages of four famous universities community in Hanoi and find out several subjective arguments of these communities.

The limitations of this research is the lack of annotated dataset because of manual annotation process. In the future work, the data collection and annotation processes will be invested in order to get more valuable dataset either quantity or quantity. Therefore, that various deep learning algorithms text processing algorithms can be applied in order to improve accuracy of confession classification models.

References

1. Birnholtz, J., Merola, N.A.R., Paul, A.: "Is it Weird to Still Be a Virgin" anonymous, locally targeted questions on Facebook confession boards. In: Proceedings of the 33rd Annual ACM Conference on Human Factors in Computing Systems (2015)
2. Conrad, C.: High schoolers tell all: analyzing Facebook Confession Sites (2015)
3. Damayanti, N.L., Hidayat, M.A.: A phenomenology study of self confession of housewives of Facebook users. J. Soc. Media **3**(2), 261–277 (2019)
4. Fatima, Z.: Negotiating masculinity through language: the case of Pakistani Facebook confession pages. J. Gender Soc. Issues **2**(15) (2016)
5. https://github.com/undertheseanlp/underthesea
6. Kubek, M., Unger, H.: Detecting source topics by analysing directed co-occurrence graphs. In: 12th International Conference on Innovative Internet Community Services (I2CS 2012), Born (2012)
7. Chen, T.Y., Kuo, F.-C., Merkel, R.: On the statistical properties of the f-measure. In: Conference on Quality Software (2004)
8. Ora, A., Ora, A.: Spam detection in short message service using natural language processing and machine learning techniques, National College of Irelan (2020)

9. Haridas,N.: Detecting the spread of online fake news using natural language processing and boosting technique. Masters thesis, Dublin, National College of Ireland., Dublin: Masters thesis; National College of Ireland (2019)
10. Daoud, E.A.: Comparison between XGBoost, LightGBM and CatBoost using a home credit dataset. Int. J. Comput. Inf. Eng. **13**(1), 6–10 (2019)
11. Cooper, W.W., Seiford, L.M., Tone, K.: Data Envelopment Analysis A Comprehensive Text with Models, Applications, References and DEA-Solver Software. Springer, Heidelberg (2007)

Using a Video Camera to Obtain Quantitative Measurements for Distance Flexibility Training

Oksana Isaeva(✉), Anna Konurina, Yura Boronenko, Vladimir Zelensky, and Yelena Gudoshnik

Yugra State University, 16 Chekhov street, 628012 Khanty-Mansiysk, Russia
isoksi06@gmail.com, yurabl069@gmail.com,
{nutt2004ann, elenakiss}@mail.ru, w_selenski@ugrasu.ru

Abstract. Recently, online health support has been gaining popularity. The most demanded types of online assistance are audio, video and text message counseling. In conditions of self-isolation, people are inactive. Therefore, there is a growing need for online physical therapy. The purpose of our study is to develop a method for remote measurement of flexibility. Nowadays, almost everyone has laptops, smartphones. The gadgets are equipped with video cameras. Remote measurements are possible with a video camera. The test results are recorded on the phone camera. At the same time, the results are recorded in the classical way - using a tape measure. The results are processed using the ImageJ software package (Fiji). Tests were carried out to qualitatively assess the mobility of the spine. In all cases, the results obtained by the classical method and remote control were compared. All measurements obtained coincided with an accuracy of fractions of a centimeter

Keywords: Pilates method · Measuring flexibility · Internet health support · Physiotherapy

1 Introduction

It is well known that a significant part of the working-age population is middle-aged women. If at a young age the negative consequences of hypodynamia and hypokinesia are leveled out due to the natural reserves of the organism's viability, then at a more mature age they are fully manifested and are aggravated by age-related involutional changes, irrational nutrition, bad habits, and stresses [1].

The fitness system has taken the leading role in the process of improving the population's health. However, it is often noted that no more than 10% of citizens of the Russian Federation are involved in the process of health-improving training on an ongoing basis, of which, at best, a tenth are women of mature age. Those of them who are overweight and have chronic diseases practically do not participate in the physical fitness process [2].

The results of the study of the attractiveness of fitness services among women confirm the demand for health-improving trainings according to the method of

D. Russo et al. (Eds.): IHSI 2021, AISC 1322, pp. 286–293, 2021.
https://doi.org/10.1007/978-3-030-68017-6_43

J. Pilates. The positive influence of training according to the Pilates method on the cardiovascular, respiratory system, the development of strength, endurance and contained trainers has been proven. Strengthening the muscles of the back and abdomen, that is, the muscle corset that holds the spine, allows you to solve the problems of the musculoskeletal system [3, 4]. 70% of the exercises in Pilates are done lying down. This minimizes stress on the spine and joints. Stretching exercises relax the spasmodic muscles of the back, which is an alternative to massage, and also work to stretch the spine.

In addition, the Pilates method does not require a specially equipped training place, therefore, it is suitable for self-study at home by older women who do not have enough time or financial resources to visit sports clubs. Recently, online health support has been gaining popularity. The most popular types of online assistance are audio and video consultations and counseling using text messages [5, 6]. In conditions of self-isolation, people are inactive. Therefore, there is a growing need for online physical therapy. However, the recommendations received may be subjective due to quantitative characteristics. The purpose of our research is to develop a method for remote measurement of flexibility.

2 Experimental Technique and Technique

For spatial calibration (Fig. 1 a, b), the video cameras used two scale bars with large print. The rulers are marked with divisions from −20 to +20 cm. Calibration shots are taken both in horizontal and vertical positions of the rulers. The details of the technique were worked out with a cardboard model imitating human hands. For the rest of the experiments, the following equipment was required: a fitness mat, a tape measure, a gymnastic bench, a tripod for a phone, a Redmi 6 smartphone (operating system Android 8.1 Oreo) with a 12-megapixel front camera.

To compare the results of measurements by the standard method with the results obtained using a video camera, a cardboard model was used that imitates human hands. Image analysis was performed using the free open source ImageJ software (Fiji).

Lenses are known to distort the perspective of a scene. Wide-angle lenses enlarge the subject, while tele and photo lenses compress the size of the subject. Therefore, it is imperative to find out the permissible sighting area. For this, measurements are taken with different positions of the phone lens relative to the measurement area parallel to it. The phone is mounted on a tripod. In the initial position, the projection of the phone camera lens is aligned with the zero division of the horizontal ruler. A photographic image is obtained, in which it is obvious that the position of the extreme points of the layout corresponds to the zero value on the horizontal ruler.

Further, the position of the bench with the model was displaced horizontally to the right and left from the initial position with a step of 25 cm. Each position was recorded by a phone camera using a photographic image. In this case, the position of the tripod did not change, the plane of the phone and the plane of measurements remained parallel. The relative position of the objective and the measurement area are considered acceptable as long as the deviation from the measurements by the standard method is insignificant (Fig. 1, c.)

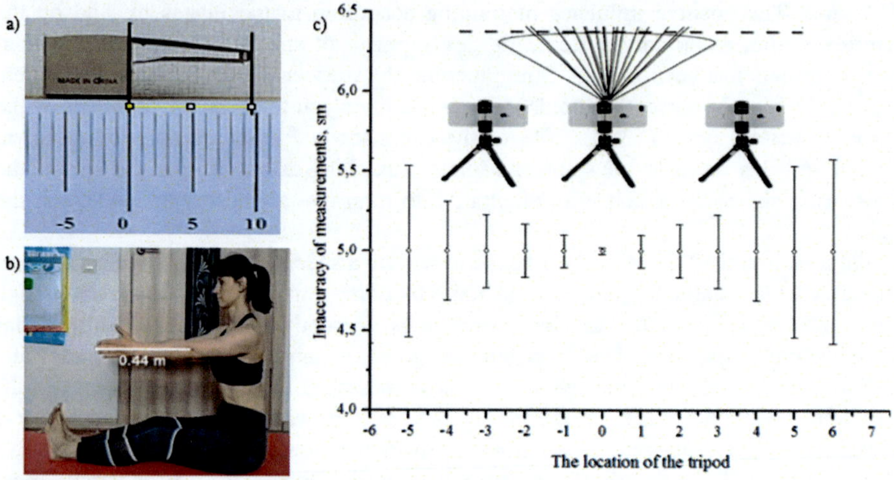

Fig. 1. Spatial calibration of the video camera: a) by the scale bar; b) at a known distance; c) determining the width of the area of space within which measurements can be made

Increasing or decreasing the distance between the measurement plane and the location of the video camera requires spatial recalibration. In conditions of mass training carried out remotely, any part of the body of a known size can act as a known quantity. For example, the length of the arm from the elbow to the tip of the middle finger. When comparing the actual arm length of the experiment participant, measured with a tape measure (44.50 $\pm$ 0.05 cm), with the results of remote measurements using a video camera, deviations appeared only in thousandths - 44.503 cm. Thus, the prospect of using a video camera for obtaining remotely quantitative estimates of flexibility becomes obvious.

3 Comparison of the Results Obtained by Standard Methods and with a Video Camera

To pass the first tests, a wooden bench is used with rulers fixed on it from opposite front sides for horizontal and vertical measurement of results. At the same time, the zero value of the horizontal measuring ruler is aligned with the edge of the bench end, the zero value of the vertical ruler is aligned with the upper boundary of the horizontal plane of the bench. The bench is installed on a training mat, a camera is exposed. The phone is fixed on a tripod strictly parallel to the bench at a distance of 2 m from it, at a height corresponding to the measurements being taken. The phone camera lens is aligned with the zero division of the horizontal ruler for the test from a sitting position; when the bench is turned 180° for the test from a standing position - in the center (horizontally) of the vertical ruler. The camera position does not change during the tests. The choice of the distance between the camera and the bench was determined empirically and is determined by the optimal ratio of the visibility of the divisions of the measuring ruler and the entire body of a person being tested from a sitting position.

Two groups of women 35–50 years old, 10 people each, took part in the experiment to measure the flexibility of the spine. The average age of the experimental group of women who regularly train according to the Pilates method was 42.6 years, in the experimental group of women who are not involved in physical activity - 41.4 years. An indicator of the general flexibility of an individual is the index (H), calculated as the quotient of dividing the amount of deflection (h) by the truncated body length (L):

$$H = h/L, \quad (1)$$

where H is the flexibility index; h is the distance from the vertical wall to the sacral point of the subject; L is body length up to the seventh cervical spine.

Overall flexibility is the degree of mobility of all joints in the human body. It allows you to make movements with the greatest amplitude. Several tests (Fig. 2) were carried out to determine flexibility.

3.1 Forward Bend Test from a Sitting Position on the Floor with Straight Legs

The participant performs a test (test) in sportswear, which allows to determine the extension of the legs at the knees (shorts, leggings), without shoes. The incline is performed from a sitting position on a gymnastic mat with straight legs at the knees and vertical feet. When performing the test (test), the participant performs two preliminary bends forward, sliding his fingers (hands together) along the body. During the third incline, the participant flexes as much as possible and fixes the result for 2 s.

a) 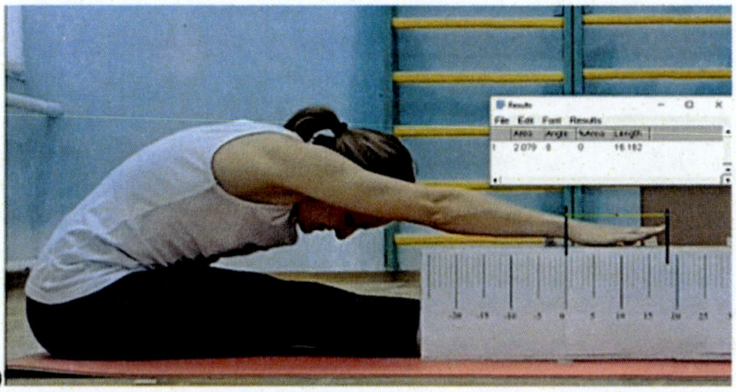 b) 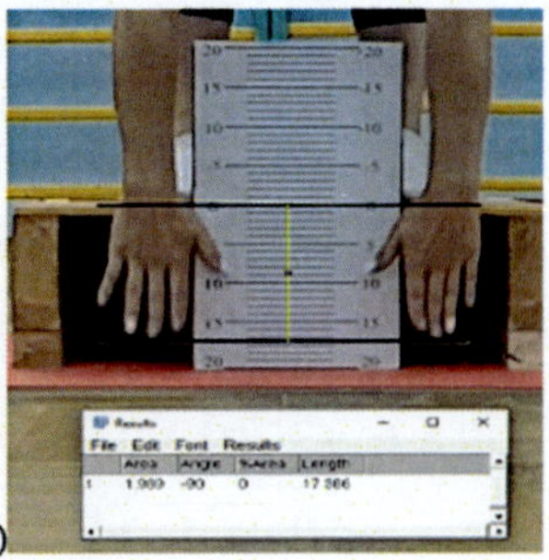

Fig. 2. Flexibility test procedure: a) from a sitting position; b) from a standing position

The test result is fixed by a ruler with a zero reading in both directions from the vertical line of the bottom of the test's stop. The amount of flexibility is measured in centimeters. The result before the vertical line of the bottom of the stop is indicated with a "–" sign, after – with a "+" sign.

3.2 Forward Bend Test from a Standing Position on a Bench with Straight Legs

The participant performs the test from the starting position, standing on a bench with legs straightened at the knees, the feet are parallel at a width of 10–15 cm. The participant performs in a sports uniform that allows determining the straightening of the knees. The participant performs two preliminary inclinations, palms move along the ruler. At the third bend, the participant holds the lower position of the palms of both hands for 2 s.

The results of assessing the general flexibility of the body of the pupils of both groups by the forward bend method and using the body flexibility index can be found here (https://drive.google.com/file/d/1LSWjfEMnRSogtm9H2B2gaAqJ9JbhdbeR/view?usp=sharing).

As can be seen from the results, among the groups of women without physical activity, the test results vary widely. Before comparing the mean values of the samples, they were checked for normality. The test showed that all samples obey a normal distribution. The t-test (or Student's t test) solves the problem of proving the difference in the mean values of a quantitative variable in the case when there are only two compared groups. We apply the T-test to the samples obtained in the control group and the experimental group in pairs: sitting tilt - measurements according to the standard method and measurements using a video camera; standing tilt - measurements according to the standard method and measurements with a video camera. The T-test results are shown in Table 1.

Table 1. T-test results of hypothesis testing of various mean values measured by classical and remote sensing methods.

Э1\Э_1					
H1: Mu1 - Mu2 ≠ 0 / Not equal (two-tailed)					
t Critical Value (5%)	2.2622	*p-value*	0.5554	*H1 (5%)*	Rejected
H1: Mu1 - Mu2 < 0 / Less than (lower-tailed)					
t Critical Value (5%)	−1.8331	*p-value*	0.7223	*H1 (5%)*	Rejected
H1: Mu1 - Mu2 > 0 / Greater than (upper-tailed)					
t Critical Value (5%)	1.8331	*p-value*	0.2777	*H1 (5%)*	Rejected
Э2\Э_2					
t Critical Value (5%)	2.2622	*p-value*	0.7976	*H1 (5%)*	Rejected
H1: Mu1 - Mu2 < 0 / Less than (lower-tailed)					
t Critical Value (5%)	−1.8331	*p-value*	0.6012	*H1 (5%)*	Rejected
H1: Mu1 - Mu2 > 0 / Greater than (upper-tailed)					
t Critical Value (5%)	1.8331	*p-value*	0.3988	*H1 (5%)*	Rejected
К1\К_1					
H1: Mu1 - Mu2 ≠ 0 / Not equal (two-tailed)					
t Critical Value (5%)	2.2622	*p-value*	0.3629	*H1 (5%)*	Rejected
H1: Mu1 - Mu2 < 0 / Less than (lower-tailed)					

(continued)

Table 1. (*continued*)

К1\К_1					
t Critical Value (5%)	−1.8331	*p-value*	0.8185	*H1 (5%)*	Rejected
H1: Mu1 - Mu2 > 0 / Greater than (upper-tailed)					
t Critical Value (5%)	1.8331	*p-value*	0.1815	*H1 (5%)*	Rejected
К2\К_2					
H1: Mu1 - Mu2 ≠ 0 / Not equal (two-tailed)					
t Critical Value (5%)	2.2622	*p-value*	0.7577	*H1 (5%)*	Rejected
H1: Mu1 - Mu2 < 0 / Less than (lower-tailed)					
t Critical Value (5%)	−1.8331	*p-value*	0.6211	*H1 (5%)*	Rejected
H1: Mu1 - Mu2 > 0 / Greater than (upper-tailed)					
t Critical Value (5%)	1.8331	*p-value*	0.3789	*H1 (5%)*	Rejected

3.3 Comparison of the Results of General Flexibility in the Experimental and Control Groups

It is known that for good deflections it is necessary to work not with the lower back, but with the thoracic spine. Thus, the overall flexibility is directly dependent on the mobility of the spine. The paravertebral muscles help maintain the spine in a normal state. They pass in several layers and contribute to the normalization of the vertebrae, from the cervical to the coccyx. Pilates for the back helps to strengthen the muscle frame that supports the spine, normalize posture and stabilize overall health. The participants in the experimental group trained according to the Pilates method for an average of a year. Participants in the control group did not exercise Pilates. To prove that there was a difference in the mean slenderness of the participants in the two groups, the results were evaluated with a T-test (Table 2).

Table 2. T-test results (Pilates utility test).

Э1\ К1					
H1: Mu1 - Mu2 ≠ 0 / Not equal (two-tailed)					
t Critical Value (5%)	2.2622	*p-value*	0.0002	*H1 (5%)*	Accepted
H1: Mu1 - Mu2 < 0 / Less than (lower-tailed)					
t Critical Value (5%)	−1.8331	*p-value*	0.9999	*H1 (5%)*	Rejected
H1: Mu1 - Mu2 > 0 / Greater than (upper-tailed)					
t Critical Value (5%)	1.8331	*p-value*	7.6734E-5	*H1 (5%)*	Accepted
Э2\ К2					
H1: Mu1 - Mu2 ≠ 0 / Not equal (two-tailed)					
t Critical Value (5%)	2.2622	*p-value*	0.0004	*H1 (5%)*	Accepted
H1: Mu1 - Mu2 < 0 / Less than (lower-tailed)					
t Critical Value (5%)	−1.8331	*p-value*	0.9998	*H1 (5%)*	Rejected
H1: Mu1 - Mu2 > 0 / Greater than (upper-tailed)					
t Critical Value (5%)	1.8331	*p-value*	0.0002	*H1 (5%)*	Accepted

The results obtained indicate higher indices of the flexibility of the spine in the group of women involved.

4 Discussion

With regard to the application of these tests, it should be noted:

First, multiple joints are involved in the forward bend. Therefore, it shows the total mobility of the joints, or the general flexibility of the human body. Given the specificity of the manifestation of flexibility, these tests cannot be universal, allowing the assessment of mobility in individual joints.

Secondly, it has the following disadvantages, which make it difficult to use it when more accurate total joint mobility is required (general flexibility of the human body):

- none of the joints during this action functions with the utmost scope;
- the end result is limited by such a highly unstable factor as the extensibility of the muscles of the back of the thigh (preliminary bending or preliminary heating of the body can change the indicators several times);
- limiting extension in most motor actions is much more frequent than flexion, therefore the choice of movement for flexion is not entirely successful;
- unreasonably great importance in measuring flexibility in this way is acquired by such a sign as the ratio of the length and body, the length of the fingers.

Therefore, other measurement methods can be used to assess the flexibility of the spine. Evaluate spinal flexibility using a photographic image of a subject undergoing a forward bend test using the ImageJ software package (Fiji). For this, the ratio of the areas of the polygon formed by the segment of the bending vertebra of the subject and the segments drawn from the extreme point of the bending of the half-tube (at the level of the 7th cervical vertebra) parallel to the intersection with the perpendicular on the floor, drawn at the level of the heels, to the area of the minimally bounding rectangle under study. The closer the area ratio is to unity, the more flexible the subject's spine.

5 Main Conclusions

In accordance with the purpose of the work, measurements of general flexibility were carried out in the experimental and control groups of women. In the first group, the participants were engaged in physiotherapy exercises according to the Pilates method. Flexibility was measured using the classical method and with a video camera simultaneously. The research established:

- According to the t-test results, the measurements carried out using the standard method and with the help of a video camera do not differ significantly.
- Statistically, the average values of the flexibility of women exercising according to the Pilates method differ significantly from the flexibility of women who are not involved in physiotherapy exercises.

- The methodology allows a more detailed approach to the analysis of mistakes made during exercise performance, which will improve the training results. A prerequisite for applying the technique is the presence of a video camera in a laptop (or smartphone). Usually, the standard resolution of video cameras in gadgets is enough for measurements.

Thus, even in a pandemic, there is an opportunity to continue to engage in physiotherapy exercises, to demonstrate results that improve the mobility of the spine.

Acknowledgments. This work was carried out with partial support from the Russian Foundation for Basic Research within the framework of research project 18–47-860018 r_a.

References

1. Gorelov, A.A.., Rumba, O.G., Baitlesova, N.K.: Locomotor activity as a factor in increasing the working capacity of women in the second period of adulthood, working as university teachers **9**(91), 50–56 (2012)
2. Reznikov, V.A.: Technology of health-improving training of middle-aged women with disabilities in health: dissertation. MGPU (2017)
3. Zykun, Zh.A., Konon, A.I.: The introduction of Pilates gymnastics in physical education classes with students of the exercise therapy group in case of scoliosis. Young Sci. **18**, 200–205 (2016)
4. Andreeva, A.N.: Discussion of the results of training with Pilates as a method of body control. Sci. Educ. New Time Sci. Method. J. **3**, 49–52 (2018)
5. Ivanov, V.G.: Possibilities of psychological online counseling of persons in a state of mental maladjustment. Communication in the information society: problems and opportunities: Sat. scientific. articles, pp. 100–103. (2017)
6. Abbott, J.-A.M., Klein, B., Ciechomski, L.: Best practices in online therapy. J. Technol. Hum. Serv. **26**(2–4), 360–375 (2008)

Hackathon-Edu: A Global Competitiveness Perspective

Gabriela Tapia-González(✉), Rodolfo Martínez-Gutiérrez, and Fabiola Tapia-González

Tecnológico Nacional de México, Campus Tijuana, Calzada Tecnológico S/N, Fraccionamiento Tomas Aquino, 22414 Tijuana, BC, México
{gabriela.tapia, rodolfo.martinez, fabiolatapia}@tectijuana.edu.mx

Abstract. The present research work focus on addressing the hackathon-edu proposal from a global perspective at pre-university recruitment, which is focus on the disciplines of the information technologies and communications capable to contributing human-centered systems experiences. The theoretical principles related to educational marketing, the hackathon, Higher Education Institutions, the blue ocean strategy and innovation. The research has a quantitative approach, descriptive and correlational scope, non-experimental and transactional design, the population sample consist of 190 high students in their senior year located in the city of Tijuana, Mexico. Data analysis through the statistical package for social sciences (SPSS) and contrast the hypothesis using the Spearman rho rank correlation nonparametric by means of inferential statistics. The most relevant results show that students in programming area, had mention that they do not have knowledge of the graduation profile and reticular plan in computer sciences.

Keywords: Hackathon · Innovation · Competitiveness · Information technologies

1 Introduction

Every day, the world is changing faster, the innovation, technological development and globalization are present and play a key role to competitiveness global, the organizations are capable the adapting the new changed into every aspect of business to increasing the improvements their process, and create of new products and services to respond a customer needs through sustainable way. The organizations are looking specialized personnel in diverse disciplines as computers sciences, medicine, nanotechnologies, robotic and social sciences. According, the World Economic Forum (WEF) mentions that in the Fourth Industrial Revolution the organizations are seeking to opportunities to use new technologies and get major benefits on productivity and efficiency, and get new markets, where predominate the digital natives [1]. In this context, WEF projected an increasing demand based on the use information technology and the computing profile up to 2022 year, in addition, the roles such as "Data Analysts and Scientists, Software and Applications Developers, and Ecommerce and Social

D. Russo et al. (Eds.): IHSI 2021, AISC 1322, pp. 294–300, 2021.
https://doi.org/10.1007/978-3-030-68017-6_44

Media Specialists", and also the extensive evidence of accelerating demand for new specialist roles in emerging technologies such as "AI and Machine Learning Specialists, Big Data Specialists, Process Automation Experts, Information Security Analysts, User Experience and Human-Machine Interaction Designers, Robotics Engineers and Blockchain Specialists" [1]. The WEF recognized that emerging sector of data platforms and information technology has been major concentration and increased [2]. The studies conducted in Mexico by Cisco company, projected up 829 thousand positions information technologies for 2027 year [3], as a strategy to drive innovation within organizations, through the new technologies and high skilled workers. In Tijuana, Baja California, applied research projects have been promoted that have involved multidisciplinary work actions, as pointed out by Martinez (2012) promoter of the QHS methodology, which includes a sectoral analysis of all the representatives of society under a work oriented to the systematic development under the same common objective, considering the government, companies, universities, business organizations and specialists in society (consultants) [4]. In addition, the Higher Education Institutions (HEI) play a role important to prepare the workforce, capable contribute and impact his environment in a globalized and interdependent world [5]. In the Europe Central and Oriental, the higher education coverage is 77.7% in 2016 in contrast to Latin America and Caribbean, is 48.4% in 2016, according to the United Nations Educational, Scientific and Cultural Organization (UNESCO) [6]. Researches on Mexico higher education, in the course of 2016–2017 year reveal 2.35% enrollment students to information and communications technologies compared to 42.82% social sciences programs [7]. It is a latent concern, the decrease in the enrollment of students in the career of information and communication technologies in higher education in Mexico, and considering the importance of computer sciences disciplines for demand workforce on the next years, the research proposes the hackathon-edu as educational marketing alternative [8].

2 Theoretical Framework

2.1 Innovation in the Organizations as Competitiveness Factor

The innovation is considered a strategy that allow to drive the growth and development of the organizations. In summary, the authors mentioned that innovation and trough the planned goals and deployment of resources organizations are capable achieve objectives [9–11]. Likewise, the organizations can be compared each other by means the blue ocean strategy and value curve, conceptualize by the author as "untapped market space, demand creation, and the opportunity for highly profitable growth" [12]. On the frontier of knowledge of methodological initiatives from the academy at the Tecnologico Nacional de México Campus Tijuana, Martinez (2020) proposes that sector needs can be systematized from the approach of the Dictionary of Sector Competences (DCS) methodology, basing the objectives and goals of the national development plan in proposals for specific actions at the sectoral level, facilitating the operationalization and development strategies and achievement of goals to boost sectoral competitiveness [13].

2.2 Marketing in Educational Organizations

Each time more again, the HEIs are recognized as services organization that guarantee their customer satisfaction by means university management [14]. The educative institutions are capable adapting the demands and the environment changes to respond a need are customer. Therefore, Educational Marketing (EM) are refers those strategies to allow the recruitment students, and addition identify the social needs that seeking satisfy the people and communities interested in educative services [15].

2.3 Hackathon and Educational Marketing

The authors conceptualize the hackathon as a social event focused on designing or developing prototypes as a way to provide innovative solutions to respond to the needs of its environment and oriented to results [16–18]. The idea of "hackathons originated in the IT community as computing marathons where programmers, project managers, and graphic and interface designers collaborated intensively on software projects to design the next 'killer app' over one or two labor intensive days" [19]. Therefore, HEIs seeking to implement strategies for student's recruitment as result global competition, furthermore, "universities and colleges face new challenges in student recruitment and international competition" [20].

3 Methodology

The study is guided by the quantitative method, non-experimental, transectional, descriptive and correlational. Data was collected from a sample of one hundred and twenty-seven (127) students using questionnaire. The measurement instrument, the which consists of eighteen closed questions, the vocational guidance dimension with nine closed questions within a six-point scale with multiple options in likert scale ("none", "very low", "low", "undecided", "high" and "very high"). The analytic tool used is the SPSS software, version 22.

4 Results and Analysis

All the 127 questionnaires were retrieved and found to be complete.

Figure 1 bar chart shows specialty area of the student community was asked about the area they were studying, a significant percentage of students (80%) equivalent to 102 students, state that the area of programming and computer science; electronic, physical, and math specialty with a total (15%) students is equivalent to 19 students; followed by the automation, maintenance and industrial area (4%) and at finally the administration, management and social sciences discipline (1%).

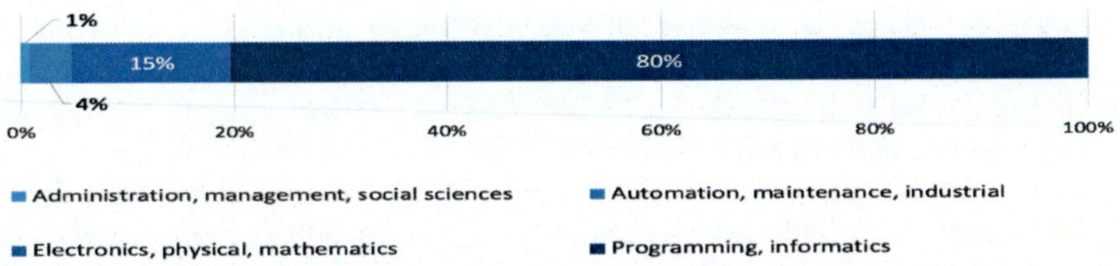

Fig. 1. Bar chart specialty area of student. Source: own elaboration, in (based on 21)

Table 1 Presents the results of the student community when asked about knowledge the graduation profile of the Information and Communications Technology (ICT) engineering, a significant percentage (40.2%) equivalent to 51 students, recognize do not have nothing information.

Table 1. Student community when asked about knowledge the graduation profile of the ICT

	Frequency	Percentage	Valid percentage	Accumulated percentage
None	51	40.2	40.2	40.2
Very low	25	19.7	19.7	59.8
Low	23	18.1	18.1	78.0
Undecided	15	11.8	11.8	89.8
High	9	7.1	7.1	96.9
Very high	4	3.1	3.1	100.0
Total	127	100.0	100.0	

Source: SPSS version 22.0, based on the questionnaire employed in (based on [21])

Table 2 Illustrate the results of the student community when asked about knowledge the reticular plan of the ICT, a significant percentage (47.2%) equivalent to 60 students, recognize do not have nothing information.

Table 2. Student community when asked about knowledge the reticular plan of the ICT

	Frequency	Percentage	Valid percentage	Accumulated percentage
None	60	47.2	47.2	47.2
Very low	20	15.7	15.7	63.0
Low	23	18.1	18.1	81.1
Undecided	10	7.9	7.9	89.0
High	9	7.1	7.1	96.1
Very high	5	3.9	3.9	100.0
Total	127	100.0	100.0	

Source: SPSS version 22.0, based on the questionnaire employed in ([21])

When asked about the question, in which media is informed about of the ICT program, a significant 36% recognized has not nothing information, and manifesting the option of (52%) visits, (50%) social networks, (43.4%) of schools, following the fairs with 37.5%, and friends and family 7.7%. It is determined to use the non-parametric statistical analysis to test the hypothesis of the research study, being selected the test of correlation coefficient by ordered ranges of Spearman, taking into account that the analyzed variables contain nominal or ordinal data. For the hypothesis test, a statistically significant, weak and indirectly proportional correlation (-0.275, $n = 127$, $p < 0.002$) was found between the variables recruitment of applicants in the study program of ICT engineering and the educational marketing strategies applied at the Technological Institute of Tijuana, so the null hypothesis (Ho) is rejected and the alternative hypothesis is accepted.

5 Conclusions

Figure 2 Shows the proposal Hackathon-edu, as EM alternative for recruitments students of the information technology and communications program [21].

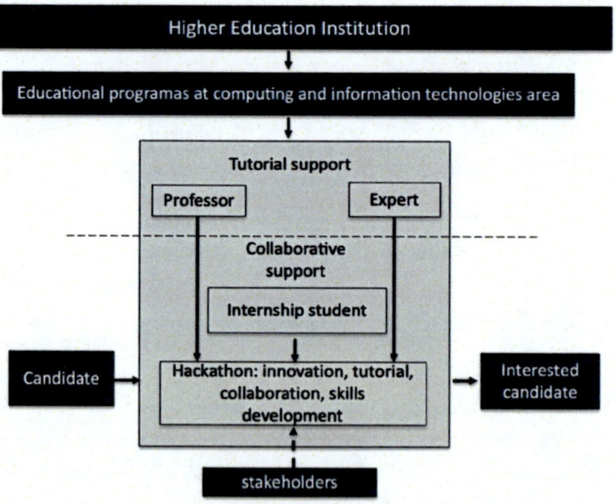

Fig. 2. Hackathon-edu. The figure shows the proposal to recruitments candidate for HEI at information and technologies communications career [21]

The key elements of the hackathon-edu are conformed by higher education institution, educative programs, candidate, stakeholders, professor, expert professional, internships student and hackathon.

References

1. World Economic Forum: The Future of Jobs Report 2018: Centre for the New Economy and Society (20189). Recuperado de: https://www3.weforum.org/docs/WEF_Future_of_Jobs_2018.pdf
2. World Economic Forum: The Global Competitiveness Report 2019. World Economic Forum (2019). Disponible en: http://www3.weforum.org/docs/WEF_TheGlobalCompetitivenessReport2019.pdf
3. Osores, M.: Veinte roles y trabajos de TI más requeridos en México, según Cisco. Search Data Center en español (2018). Disponible en: https://searchdatacenter.techtarget.com/es/noticias/252454584/Veinte-roles-de-trabajos-de-TI-mas-requeridos-en-Mexico-segun-Cisco
4. Martínez-Gutiérrez, R.: QUINTA HÉLICE SISTÉMICA (QHS), UN MÉTODO PARA EVALUAR LA COMPETITIVIDAD INTERNACIONAL DEL SECTOR ELECTRÓNICO EN BAJA CALIFORNIA, MÉXICO. Investigación Administrativa, (110), 34–48 (2012). ISSN: 1870–6614. Disponible en: https://www.redalyc.org/articulo.oa?id=4560/456045338003
5. Hasan, F.A., Komoo, I., Nor, M.N.M., Abdullah, Z.: Transformation of universities and the national Blue Ocean Strategy: a case study of Universiti Malaysia Terengganu. J. Sustain. Sci. Manage. **12**(1), 70–78 (2017)
6. Asociación Nacional de Universidades e Instituciones de Educación Superior.: Visión y acción 2030. Propuesta de la ANUEIS para renovar la educación superior en México. Diseño y concertación de políticas publicas para impulsar el cambio institucional. México (2018). Disponible: https://www.anuies.mx/media/docs/avisos/pdf/VISION_Y_ACCION_2030.pdf
7. Dirección General de Educación Superior Universitaria. Estadísticas Básicas de Educación Superior. Programa Nacional de la educación (2018). Disponible en: https://www.dgesu.ses.sep.gob.mx/EBESNACIONAL.aspx26
8. Tapia-González, G.: Educational marketing and Hackathon for candidate student recruitment. In: Kantola,, J., Nazir, S., Salminen, V. (eds.) Advances in Human Factors, Business Management and Leadership. AHFE 2020. Advances in Intelligent Systems and Computing, vol. 1209. Springer, Cham (2020). https://doi.org/10.1007/978-3-030-50791-6_55
9. Hamel, G., Breen, B.: El futuro de la administración. 2da. Edición. Bogotá. Grupo Editorial Norma (2009)
10. Ruiz, R., Martínez, R., Valladares, L.: Innovación en la educación superior. Hacia las sociedades del conocimiento. Primera edición. Universidad Nacional Autónoma de México. Editorial Fondo de Cultura Económica. México (2010). Disponible: https://read.amazon.com/
11. Mullins, J., Walker, O., Boyd, H., Larréché, J.: Administración de marketing. Un enfoque en la toma estratégica de decisiones. Quinta Edición. México. McGraw-Hill/INTERAMERICANA Editores, S.A de C.V. (2007)
12. Kim, W., Mauborgne, R.: Blue Ocean Strategy: How to Create Uncontested Market Space and Make Competition Irrelevant. Harvard Business School Press, Boston (2005)
13. Martinez-Gutierrez, R.: Methodology of Dictionaries of Sector Competences (DCS), to Design Standards of Professional Competences, Research and Labor. In: Nazir, S., Ahram, T., Karwowski, W. (eds.) Advances in Human Factors in Training, Education, and Learning Sciences. AHFE 2020. Advances in Intelligent Systems and Computing, vol. 1211. Springer, Cham (2020). http://doi-org-443.webvpn.fjmu.edu.cn/10.1007/978-3-030-50896-8_46
14. Doña, L., Luque, T.: Relación entre marketing y universidad. Revisión teórica y propuesta de un modelo teórico y de marketing 3.0. Revista de Estudios Empresariales. Segunda época. Num. 2. p. 2 –27. (2017)

15. Arrubla, J.: Marketing en universidades: Análisis de factores críticos de competitividad. Revista Escenarios Empresas y Territorio. **2**(2), 207–225 (2013)
16. Calco, M., Veeck, A.: The Markathon: adapting the Hackathon model for an introductory Marketing Class Project. Marketing Educ. Rev. **25**(1), 33–28 (2015)
17. Frey, F., Luks, M.: The INNOVATION-DRIVEN HACKATHON - one Means for accelerating Innovation. In: Proceedings of the 21st European Conference on Pattern Languages of Programs (EuroPLoP 2016). ACM (2016)
18. Shin, N., Vela, K., Evans, K.: The research role of the librarian at a community health Hackathon - a technical report. J. Med. Syst. **44**(36) (2020)
19. Kienzler, H., Fontanesi, C.: Learning through inquiry: a Global Health Hackathon. Teach. Higher Educ. **22**(2), 129–142 (2016)
20. Wu, T., Naidoo, V.: The role of international marketing in higher education. In: Wu, T., Naidoo, V. (eds.) International Marketing of Higher Education. Palgrave Macmillan, New York (2016)
21. Tapia, G.: Hackathon Y Sociedades Del Conocimiento: Hacia un nuevo paradigma en marketing educativo. Tesis doctoral. Escuela de Negocios del Pacífico, Tijuana (2020)

Frequency Weighting of Student Categories in Quadrants for Remote Higher Education Under COVID 19 Pandemic Era

Ernesto Hernández[1(✉)], Zury Sócola[2], Tania Choque[3], Abraham Ygnacio[4], and Walter Hernández[5]

[1] Universidad Nacional de Frontera, Av. San Hilarión 101, 20103 Sullana, Peru
ehernandez@unf.edu.pe

[2] Universidad Católica Sedes Sapientiae, Jr. Lima s/n, 20301 Chulucanas, Peru
zsocola@ucss.edu.pe

[3] Universidad Nacional de Juliaca, Av. Nueva Zelandia 631, 21101 Juliaca, Peru
tj.choquer@unaj.edu.pe

[4] Universidad Nacional Pedro Ruiz Gallo, Av. Juan XXIII 370, 14013 Lambayeque, Peru
aygnacio@unprg.edu.pe

[5] Notaría Hernández, Jr. Ricardo Palma 680, 01621 Bagua, Peru
walter@notariahernandez.com.pe

Abstract. During the time of the Covid 19 pandemic, the behavior of three human groups was analyzed after having completed the 16-week academic semester in September 2020. The students of each human were divided into 4 resulting quadrants, the attitudinal behavior was weighted on the axis of the abscissa and the axis of the ordinates to the final grade. The students of each one of the groups were distributed applying their percentage value within each one of the quadrants, those that are denominated talented, responsible, confident, disinterested, being their averages 18.27%, 37.69%, 13.09% and 30.99% respectively. It was demonstrated that there is no significant difference between the three groups, in terms of the percentage values assigned to each of the quadrants, as well as that there is no significant difference between the percentage values between the quadrants of the talented and the confident, nor between the responsible and the disinterested.

Keywords: Quadrant categorization · Percentage group assessment · Teaching learning

1 Introduction

Virtual education in the northern coast of Peru [1] has subjected students to the same learning environment conditions, except for biases caused by lack of network connectivity, which were minimized in the present research because the university provided tablets and chips for low income students. We have tried to demonstrate that there is the same social behavior based on the percentage weighting of three human groups that are formed by the students of the Food Technology subjects [2].

D. Russo et al. (Eds.): IHSI 2021, AISC 1322, pp. 301–306, 2021.
https://doi.org/10.1007/978-3-030-68017-6_45

The percentage weighting was based on the count of how many students would be placed in the conditions established for the quadrant according to the final grade obtained at the end of the course and their attitudinal behavior obtained during all the sessions of the academic semester.

The objective of the research was to demonstrate that the distribution of students in percentages within quadrants according to human attitudinal behavior and their final grade rating does not significantly differ within groups within each classroom under a virtual teaching environment in times of pandemic Covid 19.

2 Methodology

2.1 Hypothesis

Alternative hypothesis of the investigation (Ha): The quadrant factor affects the frequency weighting between the groups corresponding to the subjects.
Null hypothesis of the investigation (Ho): The quadrant factor does not affect the frequency weighting between the groups corresponding to the subjects.

2.2 Type of Study

By the nature of the research, it corresponds to applied research.

2.3 Experimental Group

Three human groups formed by the students enrolled in three subjects, belonging to the Food Industry Engineering Faculty of the Universidad Nacional de Frontera, Sullana, Peru; under a virtual teaching environment via Moodle platform, in times of pandemic Covid 19, period May to September 2020.

2.4 Study Design

Corresponds to a Randomized Complete Block Design (RCBD) [4].

2.5 Variables

First Independent Variable: Quadrants.
It corresponds to the treatments of the RCBD, it contains 4 levels: Quadrant I, Quadrant II, Quadrant III, Quadrant IV.

Second Independent Variable: Groups.
It corresponds to the block factor of the RCBD, corresponding to the subjects, it contains 3 levels, two correspondents of Food Technology 1 and one of Food Technology 2.
The operationalization of the independent variables is shown in Table 1.

Table 1. Operationalization of the independent variables

Independent variable	**Level of variable**			
Quadrant	I	II	III	IV
Groups	Food Technology 1, morning shift	Food Technology 1, afternoon shift	Food Technology 2, morning shift	

Dependent Variable: Frequency weighting in quadrant. It is the value in percentage that each cartesian quadrant contains. It corresponds to a response variable that presents a quantitative nature.

The dependent variable is influenced by the final grade and the attitudinal behavior, as shown in Fig. 1.

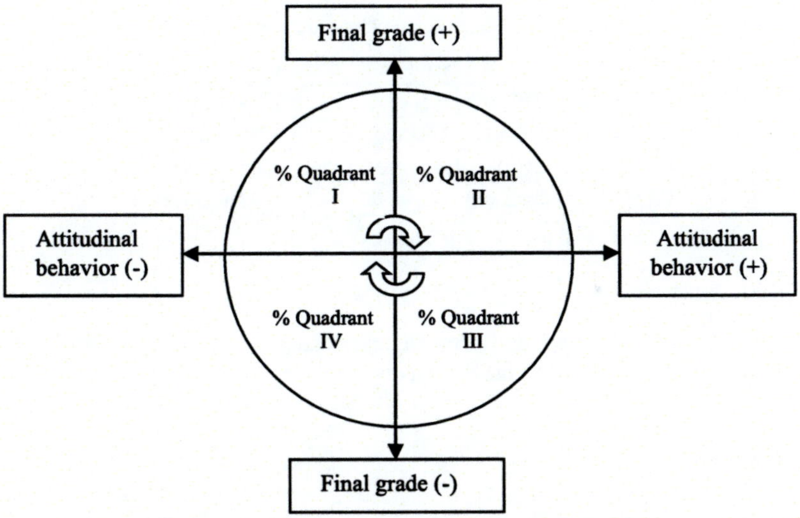

Fig. 1. Frequency weighting of student categories

2.6 Experimental Procedure

Corresponding to the assignment of the frequency for each quadrant, for which the attitudinal behavior and the final grade obtained at the end of the semester were taken into consideration.

In education, according to national legislation, the final grade ranges from 0 to 20, with a passing value equivalent to positive being higher than 10.5. Attitudinal behavior was determined by the number of participations minus the number of tardies during the 32 class sessions. Table 2 outlines the rules that must be met for the assignment of a positive or negative value.

Table 3 determines the associativity for the determination of the quadrant according to the qualification as positive or negative of the normative criteria, in order to later be able to weight a cumulative quantitative frequency.

Table 2. Value assignment rules

Criterion	Rule	Value
Attitudinal behavior	>= 0	Positive
	<0	Negative
Final rating	>= 10.5	Positive
	<10.5	Negative

In the dependent variable the positive and negative values were counted, in order to transform them into percentages based on the total number of students enrolled in each subject or group.

Table 3. Norms for quadrant determination

Quadrant	Attitudinal behavior	Final grade
I (The talented)	Negative	Positive
II (The responsible)	Positive	Positive
III (The confident)	Positive	Negative
IV (The disinterested)	Negative	Negative

2.7 Data Analysis

The IBM SPSS Statistics version 22 statistical package was used to determine the 5% confidence level ANOVA test and the Tukey test.

3 Results

The accumulated frequency in quantitative and percentage form by quadrant corresponding to the distribution of the student body of each one of the groups according to quadrants, is shown in Table 4.

Figure 2 plots the results of the averages in Table 4 by combining a style of magnitudes that happens in a bar chart in a quadrant schematic.

Table 4. Percentage values of the dependent variable

Groups	Quadrant I		Quadrant II		Quadrant III		Quadrant IV	
Food Technology 1, morning shift	6	18,18%	11	33,33%	5	15,15%	11	33,33%
Food Technology 1, afternoon shift	3	15,79%	8	42,11%	3	15,79%	5	26,32%
Food Technology 2, morning shift	5	20,83%	9	37,50%	2	8,33%	8	33,33%
Average		18.27%		37.69%		13.09%		30.99%

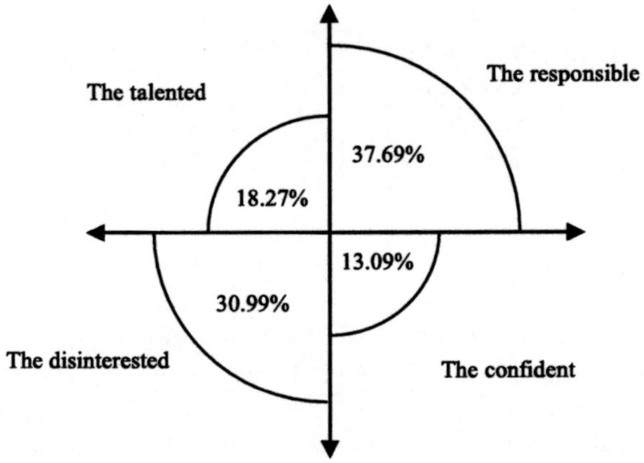

Fig. 2. Results of the dependent variable under a percentage quadrant diagram

4 Discussion

In Table 5, the p-value 0.002 less than 0.05 indicates that between the percentage values of the dependent variable adjusted to the quadrants there is statistical difference between each of the quadrants; while the p-value 1.00 greater than 0.05 indicates that the percentage values of the dependent variable adjusted to the quadrants between the groups evaluated there is no statistical difference, which means there is homogeneity between the three groups or subjects.

Table 5. Analysis of variance of DBCA

Origin	Type III of sum of squares	df	Mean square	F	Sig.
Quadrant	1149,132	3	383,044	19,437	,002
Group	$6,667 \times 10^{-05}$	2	$3,333 \times 10^{-05}$	,000	1,000
Error	118,240	6	19,707		
Total	1267,372	11			

In Table 6, Tukey's test results show that there is no significant difference between students in quadrant I and quadrant III (the talented students and the confident students) and between students in quadrant II and quadrant IV (the responsible students and the disinterested students) statistically there is no significant difference between them comparing the three groups.

Table 6. Tukey test between quadrants

Quadrant	N	df	Subsets	
			1	2
Quadrant III	3	6	13,0900	
Quadrant I	3	3	18,2667	
Quadrant IV	3	2		30,9933
Quadrant II	3	6		37,6467

5 Conclusions

The alternative hypothesis is fulfilled, in which the types of quadrants influ-ence the percentage weight distribution, and there is homogeneity among the three compared higher education human groups. Likewise, in two sets of quadrants, the first corresponds to the talented and confident, and the sec-ond to the expected and disinterested; all this in a virtual environment in the Covid 19 pandemic.

References

1. IESALC: COVID-19 and higher education: Today and tomorrow. Impact analysis, policy responses and recommendations. UNESCO (2020)
2. Universidad Nacional de Frontera. http://www.unf.edu.pe/unf/
3. Jaico, J., Lalupú, J., García, R., Reyes, N.: Maturity model for digital teacher transformation based on digital and organizational competencies in higher education. In: CEUR Workshop Proceedings, vol. 2555, pp. 103–112 (2019)
4. Kuehl, R.: Diseño de experimentos, Principios estadísticos de diseño y análisis de investigación. Thomson Learning, México (2001)

Intelligence, Technology and Analytics

Know-How Transfer and Production Support Systems to Cultivate the Internet of Production Within the Textile Industry

Florian Brillowski(✉), Hannah Dammers, Hannah Koch, Kai Müller, Leon Reinsch, and Christoph Greb

Institut für Textiltechnik of RWTH Aachen University, Otto-Blumenthal-Straße 1, 52074 Aachen, Germany
{florian.brillowski,hannah.dammers, hannah.koch,kai.mueller,leon.reinsch, christoph.greb}@ita.rwth-aachen.de

Abstract. Nowadays, the importance of non-physical goods in traditional production companies is steadily increasing. Accompanying this, companies are confronted with challenges in the area of knowledge transfer and employee support during production processes. Particularly affected are conservative industries such as the textile industry, which rely heavily on implicitly available experience-based (tacit) knowledge. A promising approach to tackle these challenges presents the concept of an "Internet of Production (IoP)". This contribution provides a framework on how the IoP can be conceived and incorporated within the textile industry. Focusing on five textile use-cases, it is described, how know-how transfer and support systems are imminent for the wide-spread introduction and perpetuation of the IoP.

Keywords: Internet of Production · Know-how-transfer · Non-physical goods · Support systems · Textile industry

1 Introduction

In recent years, the average number of contacts a person experiences with digital interfaces, social media and non-physical products has grown exponentially. While a few years ago computers, TVs and cell phones were the only cyber-physical touch points, nowadays even children are confronted with e.g. smart refrigerators or individualised, digital advertising. In this context, data, its collection and processing take a key role to enable intelligent products and individualised information processing [1].

The infrastructure required for this is the Internet (World Wide Web) which has fundamentally revolutionised access to data and knowledge. The successful transfer of the idea of the Internet to everyday objects culminates in the concept of the Internet of Things. However, the connection of devices and appliances is not limited to smart homes or smartphones but is also increasingly appealing in the manufacturing industry. In production environments, the connectivity and accessibility of complex machines are growing significantly, resulting in a conceptual demand to design a connected and smart production [1].

D. Russo et al. (Eds.): IHSI 2021, AISC 1322, pp. 309–315, 2021.
https://doi.org/10.1007/978-3-030-68017-6_46

To satisfy this demand, the novel production concept must respond well to the observed change in manufacturing from exclusively physical to both physical and non-physical products. Especially the addition of non-physical products such as licenses or maintenance contracts poses great challenges for companies. In this context, know-how transfer as well as physical and non-physical production support systems take on a key role. A promising approach to tackle these challenges presents the concept of an "Internet of Production (IoP)". The idea of the IoP is to foster a new level of collaboration across various domains by supplying appropriate and context-specific data from development, production and application in real-time at an adequate granularity [2].

Recently, addressing production data related issues was not crucial in conservative industries, such as the textile industry, to remain competitive. But due to technological, economical and societal changes as well as the shift towards non-physical goods, it is now imperative for textile companies that topics related to the IoP take centre stage. This is mandatory, in order to ensure future competitiveness and to participate in cross-company supply chains. Know-how transfer and supporting systems - integral components of the IoP - do take on a key role in the introduction of e.g. an autonomous production by taking away fears and fostering acceptance [3].

This contribution presents a framework on how the IoP can be conceived and incorporated within the textile industry. Therefore, the introduction of non-physical goods in addition to physical ones is described. Based on this shift, a need for sophisticated know-how transfer as well as smart production support systems is identified for companies to be able to compete in the market of tomorrow. In five separate use-cases within the textile industry it is then presented, how know-how transfer and support systems are imminent for the wide-spread introduction and perpetuation of the IoP. The presented use-cases contain product development of knitwear, planning of textile-based composites production processes, human-robot-collaboration and assistance systems for composite production, prediction of textile production errors as well as agile concepts for the crafting of minimal viable textile products.

2 Increasing Importance of Non-physical Goods in Production

Over the past centuries, manufacturing companies have been known for creating a physical product that can be touched and experienced haptically. The physical product was sold to companies or end consumers with the business transaction being concluded by handing it over.

However, in recent years, there has been a shift in social life and production environments towards digital formats. As a result, the added value of non-physical goods has also increased rapidly. The beginning of this development marked the broad availability of the internet for most of the population and the accompanying meteoric rise of online role-playing games (so-called MMORPGs). In these MMORPGs, players can create their own avatar and reveal a surprising willingness to spend large sums of money on cosmetic changes to their avatar [4]. Other game genres, such as annual released sports games, quickly recognised the potential of non-physical products and now generate most of their revenue from virtual cards, which lose their value within a year as they cannot be transferred onto the next year's game.

However, the potential for added value of non-physical goods is by far not limited to online games. With the introduction of the internet into other areas of life, business concepts such as subscription models or licensing have also become increasingly popular. Today, software solutions can no longer simply be bought on a CD, but almost always require time-limited licensing contracts or subscriptions. With the increasing access and connection of devices to the internet, these concepts are steadily finding their way into production companies. As a result, a shift in value creation from exclusively physical and haptically experiencable products to a mix of both physical and non-physical products can be observed.

This shift becomes particularly apparent when looking at the field of additive manufacturing. In this context, the value of non-physical goods such as digital models, intellectual property and licenses already clearly exceeds the value of the resulting physical products and production machinery. Thus, companies are confronted with a whole new set of challenges. Questions about file management systems, customs clearance of digital goods, licensing possibilities, open access, data leaks and accompanying cyber piracy are becoming more and more important [5].

In order for these issues to be resolved in a sustainable manner, the protection and transfer of intellectual property and know-how is of utmost importance. Particularly in conservative sectors such as the textile industry, where a great amount of experience-based knowledge is available in implicit form, ways and means must be found to extract tacit knowledge of single employees, conserve it in the form of knowledge management systems and train new employees [5]. With its core elements - support and knowledge transfer systems - the IoP provides a promising approach to address these demands.

3 Know-How Transfer and Supporting Systems as Enablers in Conservative Industries

Know-how transfer and support systems are of paramount importance as a company's product mix is shifting towards non-physical goods, which are particularly of a knowledge-driven and -extensive origin. However, the fact that the world's population is getting continuously older is a major problem for the preservation of know-how [6]. Older employees who are about to retire are on the verge of leaving the profession without passing on essential tacit knowledge. The risk occurs that this tacit knowledge will be passed on only insufficiently or not at all. These employees with hidden know-how essential to the company are also known as grey eminences. The danger of losing know-how is all the more acute the smaller the company is [7].

Some larger companies have recognised this problem early on, taken appropriate measures and established knowledge management systems. However, smaller companies rarely have the necessary resources or infrastructure to implement sophisticated knowledge management systems. Although these companies are often less affected by employee fluctuations, whenever one of these grey eminences leaves the company, far-reaching consequences can be caused by a lack of know-how transfer.

Besides the establishment of a sustainable know-how transfer, the importance of production data is also increasing. Up to now, it has not been necessary for crafting companies to deal with data related issues. However, with the growing added value of non-physical goods in production and the accompanying business models through licensing, it is necessary for companies to deal with data collection, processing and usage. Thereby, both physical and non-physical support systems on the one hand can help to extract and collect data. On the other hand, they can also be used to document and preserve know-how [8].

This is particularly important in sectors such as the textile industry, which heavily rely on manual labour and experience-based knowledge. Without laying the groundwork by providing support systems today, know-how transfer and at least data collection, such companies will no longer be competitive in the foreseeable future [7].

In this regard, the IoP offers a promising approach to building a data infrastructure in the form of so-called data lakes. In these data lakes, tacit knowledge can be captured with the aid of support systems (both physical, e.g. laser assistance systems, and non-physical, e.g. software tools) and then be processed in an appropriate granularity to enable the sustainable transfer of know-how. The inter- and transdisciplinary approach of the IoP enables the user-centred processing of information and hence ensures that users are not only supported by autonomous systems, but also overcome the fear of being replaced by autonomous technology in the future.

4 Incorporation of the Internet of Production Within the Textile Industry

In the past, the research of production management and the textile industry in particular focused on production costs, flexibility, resilience and sustainability [9]. Within the framework of the IoP, the research focus shifts towards decision support as well as extraction and utilisation of tacit knowledge. Ideally, decision-makers can adjust quicker by relying on data-based and knowledge-derived supporting systems. Following, five use cases are introduced, which illustrate how the IoP can be implemented using various approaches at different stages along the textile process chain. For that purpose, all use cases rely on decision support and transfer of tacit knowledge.

4.1 Setting Machine Parameters by Using Inverse Neural Network

When producing customisable, textile bracelets, process adjustments for changing environmental conditions while maintaining a high product quality are very challenging. Therefore, the Institut für Textiltechnik of RWTH Aachen University (ITA) aims to objectify the setting of machine parameters based on historical process data.

The pursued approach is based on an inverse neural network (INN), a machine learning approach, in which the desired output (e.g. quantified quality parameters) is specified and the input (machine setting parameters) is calculated. The INN-system supports the operator by suggesting values for machine settings. At the same time, the operator trains the system by accepting or declining a suggested setting. Critical quality parameters after fixation are the residual moisture and shrinkage of the strap. The INN is trained with pairs of values, originating from sensor and quality values.

4.2 User-Centred Selection and Planning of Composite Production Processes

Due to their superior lightweight potential, textile-based composites are used in a growing number of applications. However, there is a multitude of various processes for producing composites. On one hand, this poses serious orientation problems for inexperienced workers, on the other hand, experienced workers tend to overlook promising alternatives due to relying on a best practice approach.

Thus, in context of the IoP, the approach taken at ITA is to provide the worker with suggestions for alternatives and support him in the decision-making process. For this purpose, tacit knowledge is collected and transferred into process chain suggestions by using optimisation models, ontologies and provenance information. By developing such a decision tool, a closed loop production consisting of process planning, implementation and assessment as well as feedback to planning can be envisioned.

4.3 Formalised Development Process for Complex Knitted Products

The development of complex knitted products require elaborate planning and knowledge. In this context, the IoP implementation at ITA provides user-centred support. Therefore, ontologies are being derived and formalised in a database. A query service then limits the provided information according to the task and user involved. The envisioned support system will consider different perspectives and vocabularies, expertise as well as specialised domains and tasks. Correlation between physical properties and product requirements are modelled and included in the knowledge base. Once the vision is achieved, ambitious product requirements can be transferred into machine readable knitting patterns in a digitally supported, efficient process.

4.4 Knowledge-Based Assistance Systems and Human-Robot-Collaboration in Composite Production

A large-scale use of textile composites is currently inhibited by a high proportion of manual work and accompanying production costs. Yet, manual work allows companies to establish a highly flexible production and make maximum use of their employee's tacit knowledge. Faced with complex material properties (e.g. limp textiles), a shortage of skilled workers and an aging workforce, the data-based extraction and transfer of tacit knowledge is imminent for composite producing companies.

Therefore, by establishing the IoP approach, digital assistance systems, e.g. laser projection or augmented reality, are examined for positioning of textile layers, detailed instructions or quality assurance. It is also investigated how human-robot-collaboration can be implemented for forming and handling tasks as well as how human movements can be translated to robots. Achieving these goals culminates in the future workplace for composite production.

4.5 Reducing the Complexity of 4D Textiles via Modularisation

Product development and manufacturing require agile manufacturing technologies to react quickly to changing product requirements. 3D printing meets these demands by being fast, iterative and with adjustable functionality. 3D Printing on a pre-stretched textile allows products to change their form and function over time - so-called 4D textiles. Designing the changes in form and function is complex and cannot be considered an agile process yet.

Therefore, in the context of the IoP, an approach to reduce complexity by modularisation is pursued. Combined modules can be used to create prototypes for applications, test them and iterate quickly. The learnings can be fed back in the design and production process. Furthermore, the tacit knowledge gathered by designing an agile product development can be transferred to traditional manufacturing technologies.

5 Conclusion

In this contribution, the increasing importance of non-physical products as well as the accompanying changed product mix is elaborated. For resulting upcoming challenges, the IoP provides a promising approach. Within the five presented use-cases, it is shown, how the IoP can be implemented within conservative industries at different stages of the process chain. Decision support and transfer of tacit knowledge are common characteristics across all presented use-cases. Both are of utmost importance for the IoP to pervade beyond introduction. Hence, it is mandatory for companies – especially in conservative industries – to engage in decision support and transfer of tacit knowledge; otherwise future competitiveness is in question.

Acknowledgments. Funded by the Deutsche Forschungsgemeinschaft (DFG, German Research Foundation) under Germany's Excellence Strategy - EXC-2023 Internet of Production - 390621612.

References

1. Evans, P.C., Annunziata, M.: Industrial Internet: Pushing the Boundaries of Minds and Machines (2012)
2. Jarke, M., Schuh, G., Brecher, C., Brockmann, M., Prote, J.-P.: Digital shadows in the internet of production. ERCIM News **115**, 22–28 (2018)
3. Frazzon, E.M., Hartmann, J., Makuschewitz, T., Scholz-Reiter, B.: Towards socio-cyber-physical systems in production networks. Procedia CIRP **7**, 49–54 (2013)
4. Korbel, J.J., Blankenhagel, K.J., Zarnekow, R.: The role of the virtual asset in the distribution of goods and products. In: 25th Americas Conference on Information Systems, Cancun, Mexico (2019)
5. Cherukov, S., Metsä-Kortelainen, S., Salmi, M., Roda, I., Jussila, A.: The perceived value of additively manufactured digital spare parts in industry: an empirical investigation. Int. J. Prod. Econ. **205**, 87–97 (2018)
6. Marcelino, I., Góis, J., Laza, R., Pereira, A.: Using ICT for tacit knowledge preservation in old age. In: Ambient Intelligence – Software and Applications, pp. 75–83 (2015)

7. Durst, S., Edvardsson, I.R.: Knowledge management in SMEs: a literature review. J. Knowl. Manag. **16**(6), 879–903 (2012)
8. Lewandowski, A., Serafini, P., Speranza, M.G.: Methodology, Implementation and Applications of Decision Support Systems. Springer, Heidelberg (2014)
9. Váncza, J.: Production networks. In: CIRP Encyclopedia of Production Engineering, pp. 1–8. Springer, Heidelberg (2016)

Data-Driven Fault Classification Using Support Vector Machines

Deepthi Jallepalli[1,3] and Fatemeh Davoudi Kakhki[2,3](✉)

[1] Graduate Research Student, Department of Computer Engineeing, San Jose State University, San Jose, CA, USA
deepthi.jallepalli@sjsu.edu

[2] Faculty, Department of Technology, San Jose State University, San Jose, CA, USA
fatemeh.davoudi@sjsu.edu

[3] Machine Learning and Safety Analytics Lab, Department of Technology, San Jose State University, San Jose, CA, USA

Abstract. Detecting faulty condition of rolling-element bearings is significant in improving system reliability and preventing machine failure in industrial operations. In this paper, a machine learning pipeline is developed using filtered data through time domain features to train support vector machines with radial basis function, polynomial and linear kernels for multi-level fault diagnosis and classification. Overall accuracy rate and F-score values were used as figures of merit to evaluate and validate the performance of the machine learning model. SVM classifier showed significantly high overall accuracy rate of 91% to 99% and F-score of 0.81 to 0.99 with time domain statistical features due to the capability of this method in elimination of irrelevant features as well as reducing the dimensionality of the data. In addition, the high accuracy rate of SVMs in class-specific detection of localized bearing faults show the significant potential of data-driven classification modeling for fault detection in industrial and manufacturing operations.

Keywords: Bearing fault · Support vector machines · Time domain statistical features · Machine learning

1 Introduction

Rolling-element bearings are widely used in industry, and are an integral part of efficient operations of machines [1]. Therefore, bearing fault detection plays a vital role in maintainability and reliability of motor functions. However, a bearing is the most vulnerable part in a motor and drive system and its damage is an important factor in machine failure [2] and may lead to severe accidents [3]. Furthermore, bearing fault is considered as the most common type of fault and is responsible for 30% to 40% of all the machine failures [4, 6]. In some devices, 44% of the total faults frequency is assigned to bearing faults [7]. Thus, condition monitoring and regular inspections and diagnosis are critical to ensure proper function of rolling-element bearings and improve operational reliability of the machines.

D. Russo et al. (Eds.): IHSI 2021, AISC 1322, pp. 316–322, 2021.
https://doi.org/10.1007/978-3-030-68017-6_47

Accurate bearing fault diagnostics has been addressed in many researches through developing a physical model of bearing faults, and identifying the relationship between bearing faults and measurable signals, which can be captured by a variety of sensors and analyzed with signal processing techniques [8, 12]. However, detecting a bearing fault accurately can be challenging in practice due to its multi-physics nature, limitations of physical models, versatile operating conditions and noisy environments [13].

The availability of big amounts of sensor data captured during manufacturing processes has created the opportunity to substitute physical models with intelligent fault diagnostic models in order to capture unique patterns or relationships in the data that have potential of revealing a bearing fault. In recent year, machine learning (ML) algorithms have gained popularity for informed decision-making regarding the presence and diagnostics of bearing faults [14, 16]. In classification and predictive modeling, ML algorithms are used for predicting the class or category of an observation based on the information extracted from data consisting of training and testing data points. ML classification and prediction models are preferred over parametric models since the latter does not show optimal performance in catching the relationship between the independent variables and the target variable of interest in the analysis [13].

Training ML models using available bearing fault data report satisfactory results with classification accuracy of over 90% [17, 19]. Regarding the diagnostics of bearing faults, many studies have applied and evaluated the performance of various ML algorithms, specifically using support vector machines, neural networks, and gradient boosting trees with high accuracy [20, 21].

The objective of this research is to evaluate and compare the performance of support vector machines with radial basis function (BF), polynomial, and linear kernels, based on transformation of raw signals in the original data to improve ML model classification accuracy. Furthermore, all the three ML classifiers are trained on features in time domain. Therefore, the main contribution of this work is to add to the current literature about useful and practical modeling for multi-class fault diagnostics by combining time domain analysis with the three support vector machines with different kernels for developing models with high accuracy in multi-level classification of localized bearing faults.

This work is carried out on the Case Western Reserve University (CWRU) database [22] which serves as a benchmark experimental dataset on evaluating and validating the performance of various ML modeling techniques for bearing fault diagnostics [13].

The rest of the paper is organized as follows: Sect. 2 discusses the details of the machine learning methods, feature extraction approach applied in the study, performance evaluation criteria, and data preparation. Discussion of results and findings in Sect. 3, followed by conclusion in Sect. 4 that completes this paper.

2 Methodology

2.1 Time Domain Statistical Features

Fault data contains a high number of variables, and requires a large amount of computing resources to process. Therefore, feature extraction highly contributes to reducing the dimensions of the original raw data by converting it into smaller subgroups for further processing and model construction [23]. There are various techniques for feature extraction, among which time domain feature extraction methods are used in this study to extract statistical parameters from signals such as mean, standard deviation, kurtosis and skewness as the condition features of bearings. The time-domain features contain statistical details about the type of current signal, and are considered reliable feature due to their sensitivities [24]. The statistical feature parameters that could be extracted from raw signal data are minimum, maximum, mean, standard deviation, root mean square, skewness, and kurtosis, crest factor, and shape factor.

2.2 Support Vector Machines

Support vector machines (SVM) with nonlinear kernels are used for various engineering applications. The reason for the popularity of SVMs is due to their ability to map data in higher dimensional feature space where classes are separable [25, 26]. Compared with traditional statistical methods, SVM algorithms have shown comparable or higher accuracy in classification and regression problems [27, 28].

Despite the traditional application of SVMs for dichotomous output classification, multi-class SVM algorithms can be constructed by combining decision functions from multiple binary classifiers, and obtaining one final decision function for the multi-level classification problem [29]. In developing multi-level SVMs, key components are maximizing the margin between classes, and selecting proper kernel function for catching nonlinearity pattern in the data. Establishing the proper kernel function is significant not only in solving a multi-level classification problem, but also in avoiding high-dimension complicated calculations.

When the training data are not non-linearly separable, SVM algorithms allows for creating a linear separation in the higher dimension through converting the data from an input space to a feature space, using kernel function of $\varphi(X)$ [30]. SVMs can be trained using linear, polynomial, or radial basis function (RBF) kernels. The separation between the classes is achieved by implementation of $\varphi(X)$ for building a linear hyperplane that maximizes the margin between classes [31].

As presented in [31], the hyperplane can be described as in Eq. (1), where w is a weight vector of M dimension, x is an input vector of N dimension, $\varphi(X)$ is a nonlinear mapping transformation function for converting the data from N-dimensional input space to a higher M-dimensional feature space, and b is a scalar threshold. The hyperplane position is determined by w and b. After finding an optimal separating hyperplane, the SVM is tested for predicting the class of a new vector X.

$$W^T \varphi(X) + b = 0 \tag{1}$$

2.3 Evaluation Criteria for ML Classifiers

The overall accuracy rate is used as a simple quantitative metric for evaluating an ML model classifier capability in overall, without focusing on the model performance in classifying specific levels. It is gained by diving the correct number of predicted classes over the total number of all cases in the dataset. Therefore, the overall error rate of the model can be directly calculated as 100% minus the overall accuracy rate of the model.

Using the confusion matrix for any classifier, the terms of specificity, sensitivity and precision can be obtained to characterize the performance of a classifier per class. Specificity shows how effectively a classifier recognizes negative labels. Sensitivity represents the proportion of correctly classified labels while precision evaluates class agreement of the data labels with the positive labels defined by the classifier. Another metric that is gained from the multi-level confusion matrix is F-score, and is a harmonic mean of the precision and sensitivity.

2.4 Data Preparation

The Case Western Reserve University (CWRU) database is used for this study which is publicly available on the CWRU bearing data center website [22]. Single point faults were introduced to the test bearings using electro-discharge machining with fault diameters of 0.007, 0.014, 0.021 at the inner raceway, rolling element and outer raceway. The data used in this study is part of the bigger dataset, and includes vibration data that was collected for motor loads of 1 horsepower and motor speeds of 1772 rpm using two accelerometers installed at both the drive end and fan end of the motor housing for sampling frequencies of 12 kHz, for which the acceleration was measured in the vertical direction on either the drive end bearing housing (DE) and was recorded in Matlab file format. We classified the frequency and type of the fault at which diameter the fault is occurring such as 0.007, 0.014, 0.021 and the category it belongs to as Ball, IR for Inner Race, and OR for Outer Race. For instance, "OR_0.014" class represent the fault in the outer race location of the bearing with fault diameter of 0.014 inches. All labels are created for the combination of fault type and diameters. The sample rate of 12 KHz allows for sample readings with a gap of 1024, and considering the 120,000 data points in each location, data is segmented into 115 samples per revolution. Each raw in a segment has a length of 1024 data points, label for a specific fault type, and the statistical features are calculated for each row. That resulted in nine new variables (minimum, maximum, mean, standard deviation, root mean square, skewness, and kurtosis, crest and form factors) as input variables for classifying the localized faults as output variable. Finally, the data was then split into 80% of the cases for train set (92 segments) and 20% for test set (23 segments).

3 Results

The filtered data was used for training SVM with RBF, polynomial, and linear kernels. The models were first trained on the 92 segments (80% of the data) in the training phase. Then, the developed models were applied to the new test data (20% of the data; 23 segments) for evaluation of their performance. The results are shown in Table 1.

Table 1. Accuracy comparison among classifiers.

SVM model	Overall accuracy	*F*-Score
SVM-RBF	0.911	0.818
SVM-Polynomial	0.996	0.992
SVM-Linear	0.998	0.996

The SVM classifier with linear kernel outperforms the other kernel functions considering both overall accuracy and F-score values. All three classifiers have high performance in classifying the localized bearing faults in the data used in this study. It shows that SVM has considerable capacity in distinguishing between various types and diameters of the bearing faults.

4 Conclusion

The purpose of the study was to evaluate the performance of support vector machines in the detection and classification of the bearing fault data. When applied to the CWRU bearing fault data set, all the three algorithms showed high accuracy in classifying the localized bearing faults at multilevel in either ball, inner race, or outer race of a 12K drive end bearing with different diameter measurements at 0.007, 0.014, and 0.021 inches. Moreover, the support vector machines classifier with linear kernel has the highest performance considering the overall accuracy rate, and *F*-score of the models.

The results confirm that the supervised machine learning classifiers have satisfactory performance and strong generalization ability, and therefore, can be used to classify the running state of the bearing. The future direction of this research work includes developing other machine learning classifiers as well as deep learning algorithms and hybrid models using other feature extraction methods such as wavelet packet entropy transformation.

References

1. Yang, Y., Fu, P.: Rolling-element bearing fault data automatic clustering based on wavelet and deep neural network. Shock Vib. (2018). https://doi.org/10.1155/2018/3047830
2. Qi, X.X., Ji, J.W., Han, X.W.: Fault diagnosis methods of rolling bearing: a general review. In: Key Engineering Materials (2011). https://doi.org/10.4028/www.scientific.net/KEM.480-481.986

3. Chen, Z., Deng, S., Chen, X., Li, C., Sanchez, R.V., Qin, H.: Deep neural networks-based rolling bearing fault diagnosis. Microelectron. Reliab. (2017). https://doi.org/10.1016/j.microrel.2017.03.006
4. Donnell, P.O., Heising, C., Singh, C., Wells, S.J.: Report of large motor reliability survey of industrial and commercial installations: part 3. IEEE Trans. Ind. Appl. (1987). https://doi.org/10.1109/TIA.1987.4504880
5. Bell, R.N., McWilliams, D.W., O'Donnell, P., Singh, C., Wells, S.J.: Report of large motor reliability survey of industrial and commercial installations, part I. IEEE Trans. Ind. Appl. (1985). https://doi.org/10.1109/TIA.1985.349532
6. Bell, R.N., Heising, C.R., O'Donnell, P., Wells, S.J., Singh, C.: Report of large motor reliability survey of industrial and commercial installations, part II. IEEE Trans. Ind. Appl. (1985). https://doi.org/10.1109/TIA.1985.349533
7. Cerrada, M., Sánchez, R.V., Li, C., Pacheco, F., Cabrera, D., Valente de Oliveira, J., Vásquez, R.E.: A review on data-driven fault severity assessment in rolling bearings. Mech. Syst. Signal Process. (2018). https://doi.org/10.1016/j.ymssp.2017.06.012
8. Harmouche, J., Delpha, C., Diallo, D.: Improved fault diagnosis of ball bearings based on the global spectrum of vibration signals. IEEE Trans. Energy Convers. (2015). https://doi.org/10.1109/TEC.2014.2341620
9. Immovilli, F., Bellini, A., Rubini, R., Tassoni, C.: Diagnosis of bearing faults in induction machines by vibration or current signals: a critical comparison. IEEE Trans. Ind. Appl. (2010). https://doi.org/10.1109/TIA.2010.2049623
10. Kang, M., Kim, J., Kim, J.M.: An FPGA-based multicore system for real-time bearing fault diagnosis using ultrasampling rate AE signals. IEEE Trans. Ind. Electron. (2015). https://doi.org/10.1109/TIE.2014.2361317
11. Ming, A.B., Zhang, W., Qin, Z.Y., Chu, F.L.: Dual-impulse response model for the acoustic emission produced by a spall and the size evaluation in rolling element bearings. IEEE Trans. Ind. Electron. (2015). https://doi.org/10.1109/TIE.2015.2463767
12. Bartheld, R.G., Habetler, T.G., Kamran, F.: Motor bearing damage detection using stator current monitoring. IEEE Trans. Ind. Appl. (1995). https://doi.org/10.1109/28.475697
13. Zhang, S.S., Zhang, S.S., Wang, B., Habetler, T.G.: Deep learning algorithms for bearing fault diagnosticsx - a comprehensive review. IEEE Access (2020). https://doi.org/10.1109/ACCESS.2020.2972859
14. Awadallah, M.A., Morcos, M.M.: Application of AI tools in fault diagnosis of electrical machines and drives - an overview. IEEE Trans. Energy Convers. (2003). https://doi.org/10.1109/TEC.2003.811739
15. Batista, L., Badri, B., Sabourin, R., Thomas, M.: A classifier fusion system for bearing fault diagnosis. Expert Syst. Appl. (2013). https://doi.org/10.1016/j.eswa.2013.06.033
16. Liu, R., Yang, B., Zio, E., Chen, X.: Artificial intelligence for fault diagnosis of rotating machinery: a review. Mech. Syst. Signal Process. (2018). https://doi.org/10.1016/j.ymssp.2018.02.016
17. Hinton, G.E., Salakhutdinov, R.R.: Reducing the dimensionality of data with neural networks. Science **313**(5786), 504–507 (2006). https://doi.org/10.1126/science.1127647
18. Shao, H., Jiang, H., Wang, F., Wang, Y.: Rolling bearing fault diagnosis using adaptive deep belief network with dual-tree complex wavelet packet. ISA Trans. (2017). https://doi.org/10.1016/j.isatra.2017.03.017
19. Han, Y., Tang, B., Deng, L.: Multi-level wavelet packet fusion in dynamic ensemble convolutional neural network for fault diagnosis. Meas. J. Int. Meas. Confed. (2018). https://doi.org/10.1016/j.measurement.2018.05.098
20. Lo, C.C., Lee, C.H., Huang, W.C.: Prognosis of bearing and gear wears using convolutional neural network with hybrid loss function. Sensors (2020). https://doi.org/10.3390/s20123539

21. Li, K., Su, L., Wu, J., Wang, H., Chen, P.: A rolling bearing fault diagnosis method based on variational mode decomposition and an improved kernel extreme learning machine. Appl. Sci. (2017). https://doi.org/10.3390/app7101004
22. Loparo, K.A.: Case Western Reserve University Bearing Data Center. Bearings Vibration Data Sets, Case Western Reserve University (2012). https://csegroups.case.edu/bearingdatacenter/home
23. Toma, R.N., Prosvirin, A.E., Kim, J.M.: Bearing fault diagnosis of induction motors using a genetic algorithm and machine learning classifiers. Sensors (2020). https://doi.org/10.3390/s20071884
24. Kang, M., Islam, M.R., Kim, J., Kim, J.M., Pecht, M.: A hybrid feature selection scheme for reducing diagnostic performance deterioration caused by outliers in data-driven diagnostics. IEEE Trans. Ind. Electron. (2016). https://doi.org/10.1109/TIE.2016.2527623
25. Olson, D.L., Delen, D., Meng, Y.: Comparative analysis of data mining methods for bankruptcy prediction. Decis. Support Syst. (2012). https://doi.org/10.1016/j.dss.2011.10.007
26. Davoudi Kakhki, F., Freeman, S.A., Mosher, G.A.: Evaluating machine learning performance in predicting injury severity in agribusiness industries. Saf. Sci. **117**, 257–262 (2019). https://doi.org/10.1016/j.ssci.2019.04.026
27. Kecman, V.: Support Vector Machines – An Introduction (2005). https://doi.org/10.1007/10984697_1
28. Gangsar, P., Tiwari, R.: Comparative investigation of vibration and current monitoring for prediction of mechanical and electrical faults in induction motor based on multiclass-support vector machine algorithms. Mech. Syst. Signal Process. (2017). https://doi.org/10.1016/j.ymssp.2017.03.016
29. Amara Korba, K., Arbaoui, F.: SVM multi-classification of induction machine's bearings defects using vibratory analysis based on empirical mode decomposition. Int. J. Appl. Eng. Res. **13**(9), 6579–6586 (2018)
30. Ziani, R., Felkaoui, A., Zegadi, R.: Bearing fault diagnosis using multiclass support vector machines with binary particle swarm optimization and regularized fisher's criterion. J. Intell. Manuf. (2017). https://doi.org/10.1007/s10845-014-0987-3
31. Wu, C., Chen, T., Jiang, R., Ning, L., Jiang, Z.: A novel approach to wavelet selection and tree kernel construction for diagnosis of rolling element bearing fault. J. Intell. Manuf. (2017). https://doi.org/10.1007/s10845-015-1070-4

Semi-quantitative Model for Risk Assessment in University Laboratories

Anastasia Kalugina[1(✉)] and Thierry Meyer[1,2]

[1] Group of Physical and Chemical Safety, Institute of Chemical Sciences and Engineering, Ecole Polytechnique Fédérale de Lausanne, Station 6, 1015 Lausanne, Switzerland
{anastasia.kalugina, thierry.meyer}@epfl.ch

[2] Safety Competence Center, Ecole Polytechnique Fédérale de Lausanne, Station 6, 1015 Lausanne, Switzerland

Abstract. University chemical laboratories, contrary to common perception, are dangerous working environments. Unlike industry, where most processes are standardized, in academia, they are modified so frequently that in-depth and detailed risk assessment becomes unfeasible. A major challenge in conducting a reliable evaluation is the absence of statistical data, as there is a lack of reporting and a high diversity of the processes. Strong human involvement in all activities requires consideration of human factors in the risk assessment. A semi-quantitative method that combines both easily accessible quantitative information and encoded qualitative data is the way to overcome these challenges. This work proposes a novel approach of semi-detailed risk analysis. The main advantage of this method is its applicability to a fast-changing environment with a strong presence of human factors and absence of statistical information.

Keywords: Risk assessment · Human factors · University laboratories

1 Introduction

There are multiple tools for risk assessment that are used in different types of industries, but none of them can be applied in academia without significant modifications. Specific decision-making features influence and create constraints for the risk assessment model. These reasons can be grouped in 3 prominent families: management style and goals, human factors and, processes. The high importance of research freedom to fulfill the university's primary goals [25] results in underestimating the importance of safety. High publication pressure [1] and competition result in safety ignorance [16]. Contrary to industry, in universities, liability for an accident is usually diluted and overturn by different immunity types. It is quite rare [3] to be held responsible for wrongful acts if there is no intent, even in cases of apparent negligence [20].

Human factors define existing differences between university and industrial laboratories. With a significantly higher turnover in university laboratories [24], it is difficult to establish good safety culture. Such a turnover complicates the collection of crucial personal information about participants. The high cultural and linguistic diversity of university laboratories creates additional challenges [2]. Training and

D. Russo et al. (Eds.): IHSI 2021, AISC 1322, pp. 323–329, 2021.
https://doi.org/10.1007/978-3-030-68017-6_48

working practices vary in universities, nevertheless, researchers are expected to perform their work without question [21]. Safety attitude to accidents is also perceived differently. "Who does not ruin his health by his studies, nowadays will not get anywhere in Chemistry" [10]. Since the motivation of people working in university and industrial laboratories is very different, unsafe behavior should be corrected differently.

In academia, processes constantly change. The information received during risk assessments has to remain valid and relevant, even if minor changes occur. Time-efficiency and easiness to perform tasks are also crucial aspects. As there are a great variety of processes, the risk assessment method must be very versatile and consider specific features of each laboratory. The objective of this paper is to demonstrate a novel semi-quantitative model for the risk analysis that will satisfy above-mentioned requirements and provide decision-makers with reliable and relevant information.

2 Methodology

The critical steps of the methodology are: (1) Collection of the information about laboratory and Identification of main steps of the process, hazards and possible failures, (2) Calculation of the risk score for selected hazards/steps, (3) Proposition of safety measures for mitigation, (4) Evaluation of their efficiency. This method is suitable for different types of laboratory risk analysis, covering both crude risks associated with the process, and human behavior variability [14]. This paper main focus will be on the Risk Index calculation.

3 Risk Index

The majority of methods that use risk index as a basis for the risk assessment are based on three factors: Consequence, probability, exposure [8]; Severity, chance of occurrence, case of nondetection [7]; Severity, probability and possibility to avoid or limit damage [23]. The SUVA model already considers existing measures, but the two other methods use too general factors, bringing more uncertainty and bias to the assessment. To avoid this and to overcome the known problems of the risk index mentioned by Bowles [4], we propose a decomposition of the Risk Index Criticality Score in four layers of the Bayesian tree, see Fig. 1.

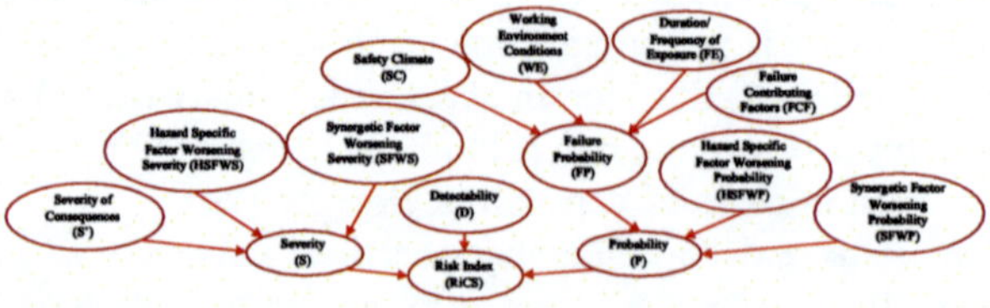

Fig. 1. Simplified Bayesian tree for calculation of Risk Index Criticality Score.

Hazard Specific and Synergetic Factors Worsening Severity and/or Probability have the same nature as in previous work [9, 19] with the main difference that weights are associated with this factors.

$$p(RiCS) = \sum_{i=1}^{5}\sum_{j=1}^{5}\sum_{k=1}^{5} p\left(RiCS|S_i, P_j, D_k\right) \cdot p(S_i) \cdot p(P_j) \cdot p(D_k) \quad (1)$$

$$\text{where S} = V_s \cdot \omega_s + V_{HSFWS} \cdot \omega_{HSFWS} + V_{SFWS} \cdot \omega_{SFWS} \quad (2)$$

and

$$\text{P} = \text{S} = V_{FP} \cdot \omega_{FP} + V_{HSFWP} \cdot \omega_{HSFWP} + V_{SFWP} \cdot \omega_{SFWP} + V_{DE} \cdot \omega_{DE} \quad (3)$$

Where V denotes for the value, and ω for the weight of the corresponding parameter.

RiCS node represents a probability distribution of three factors. We used the methodology proposed by Zaili et al. [26] to calculate a single crisp number.

4 Description of Risk Index Criticality Score Factors

4.1 Failure Probability

According to Massaiu [17], complex cognitive systems [12] can be influenced by environmental circumstances. This results in a mismatch between working cognition and working conditions. Cognitive load can be increased not only by job complication, but also by surroundings [6, 13]. Bad working cognitions affect physical ergonomics, which influences cognitive ergonomics and leads to attention failures. However, poor working conditions do not only result in unintended errors. A loud noise level can disturb a person working in the laboratory and push them to leaving the working space as quickly as possible, and therefore intentionally leaving hazardous chemicals in the work space. According to Nardo's causal loop diagram [7], the physical work environment contributes to human error probability. Our model proposes four parameters to estimate the work environment factor: level of light, level of noise, temperature, working space conditions.

Safety Climate (SC) also contributes to failure. As a precondition factor that shapes the team's perception and attitude, it significantly impacts the probability of failure. Even when all other conditions are perfect, wrong perception and wrong attitude can lead to an accident [18]. SC is a complex factor, including 16 subfactors, which can be grouped into level of lab safety; characteristics of the participant; resources, and safety attitude in the laboratory. Each of these subfactors is given a score from 1 to [15].

$$\text{SC} = 0.4948 \cdot \sum_{i=1}^{5} F_i \cdot \omega_i + 0.2577 \cdot \sum_{j=1}^{4} F_j \cdot \omega_j + 0.2475 \cdot \sum_{k=1}^{7} F_k \cdot \omega_k \quad (4)$$

$F_{i,j,k}$ - are the subfactors corresponding to three categories mentioned above,
$\omega_{i,j,k}$ - are respective weights obtained from EFA, CFA analysis.

While the first two factors are general and present despite the type of activity, the third level is specific to the process step. According to Havlikova [11], most of the failures can have either technical/equipment or human origin, depending on the type of activity and the human involvement. Combining the human error model proposed by Zhiqiang et al. [22] with the Safety Climate parameter gives:

$$\mathrm{FP} = \frac{WE}{4} SC \sum FCF \text{ or } \mathrm{FP} = \frac{WE}{4} SC \prod FCF \quad (5)$$

4.2 Duration and Frequency of Exposure

Duration of the step/particular task where the failure may occur impacts probability, increasing chances with longer duration and increased frequency, see Fig. 2. This factor describes the duration and frequency of the task and a relative type of exposure to the failure in the hazard context. For example, preparation of a setup by mixing different chemicals can be a frequently occurring step; however, the person is only exposed to the corrosive leak during 30% of the time.

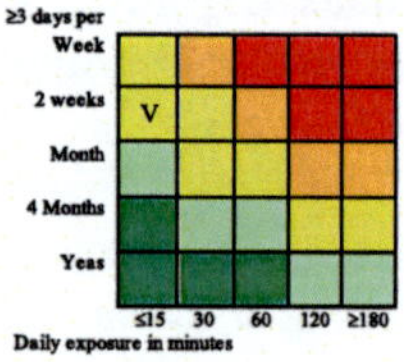

Fig. 2. Frequency/duration of exposure to hazard in step/steps.

4.3 Severity

Severity shall be determined based on the type of consequence first, as one scenario is analyzed at a time. There are four severity scales available, and only one can be used for the ongoing risk assessment, see Table 1.

Table 1. Severity of the consequences

	Human	Environment	Direct costs (€)	Reputational damage
Very serious	Death	Catastrophic	>125.000	Claims against Institute
Serious	Serious handicap	Critical	25.000–125.000	Awareness outside the institute
Medium	Light handicap	Important	5000–25.000	Awareness at the institute
Low	Wound with work interruption	Marginal	1000–5000	Awareness at the unit
Very low	Wound without work interruption	Negligible	<1000	Awareness at the laboratory

4.4 Detectability

Detectability of undesired events or failures can result in lower risk when such detection is possible. Depending on the detectability type, it can improve awareness about failure or lower the impact of an accident. There are two different ways to detect possible hazard exposure: human and technical detection, see Table 2. Human detection is neither very reliable nor selective, and does not guarantee any actions that prevent undesired consequences. Technical detectors, on the other hand, shall be selective, allowing to distinguish between different hazardous substances or detect different concentrations. Another critical characteristic of the detector is its reliability. While technical detectors have a high consistency of the detection and work independently from the conditions, setup, or room, human senses can be influenced by a person's physical state (flu, allergy, etc.).

Table 2. Hazard detectability variables

Availability	Human senses/detector		Technical sensor/detector	
	Yes		Yes	
	Selectivity	Reliability	Selectivity	Reliability
High	V		V	V
Fair				
Low		V		

5 Conclusions

In this paper, we present a model for a laboratory process risk analysis, which is based on available or easily collectible semi-qualitative data. This method's advantage is that it uses easily measurable quantitative factors such as duration and frequency of exposure along with soft factors: safety climate and working environment, for probabilistic assessment. Human-hazard interaction is the core of this risk analysis model. Bayesian network considers different levels of uncertainty attached to soft and quantitative factors that can be considered during the calculation of RiCS, increasing the assessment's reliability. All the factors included in the calculation of the RiCS have a corresponding quantitative scale. On the other hand, qualitative representation facilitates risk assessment procedures. This methodology is very flexible and can be used for both detailed analysis and a more general one, depending on the goals. The introduction of environmental, social, and intrinsic process characteristics allows for evaluation of non-tractable environments such as university laboratories. The output quantitative information can be used for efficient safety decision-making.

References

1. Academic, A.: Is competition driving innovation or damaging scientific research? Guard, June 2018
2. Anoushirvani, A.A., Golaghaie, F.: Challenges of culturally and linguistically different healthcare students in learning environments. J. Adv. Med. Educ. Prof. **7**, 108–109 (2019). https://doi.org/10.30476/JAMP.2019.44707
3. Benderly, B.L.: The Landmark Patrick Harran Case Ends Inconclusively. Science, July 2014. https://doi.org/10.1126/science.caredit.a1400176
4. Bowles, J.B.: An assessment of RPN prioritization in a failure modes effects and critically analysis. In: Proceedings of the Annual Reliability and Maintainability Symposium (2003). https://doi.org/10.1109/rams.2003.1182019
5. Chang, C.L., Wei, C.C., Lee, Y.H.: Failure mode and effects analysis using fuzzy method and grey theory. Kybernetes **28**, 108 (1999). https://doi.org/10.1108/03684929910300295
6. Couffe, C., Michael, G.A.: Failures due to interruptions or distractions: a review and a new framework. Am. J. Psychol. (2017). https://doi.org/10.5406/amerjpsyc.130.2.0163
7. Di Nardo, M., Gallo, M., Madonna, M., Santillo, L.C.: A conceptual model of human behaviour in socio-technical systems. In: Fujita, H., Guizzi, G. (eds.) SoMeT 2015. CCIS, vol. 532, pp. 598–609. Springer, Cham (2015). https://doi.org/10.1007/978-3-319-22689-7_46
8. Fine, W.: Mathematical evaluation for controlling hazards. J. Safety Res (1971)
9. Groso, A., Ouedraogo, A., Meyer, T.: Risk analysis in research environment. J. Risk Res. **15**, 187–208 (2012). https://doi.org/10.1080/13669877.2011.634513
10. Harrison, C.: Protect and Serve. World May, Chem (2018)
11. Havlikova, M., Jirgl, M., Bradac, Z.: Human reliability in man-machine systems. In: Procedia Engineering (2015). https://doi.org/10.1016/j.proeng.2015.01.485
12. Hollnagel, E., Woods, D.D.: Cognitive systems engineering: new wine in new bottles. Int. J. Hum. Comput. Stud. (1999). https://doi.org/10.1006/ijhc.1982.0313
13. Jahncke, H., Hygge, S., Halin, N., Green, A.M., Dimberg, K.: Open-plan office noise: cognitive performance and restoration. J. Environ. Psychol. (2011). https://doi.org/10.1016/j.jenvp.2011.07.002
14. Kahneman, D., Tversky, A.: Prospect theory: an analysis of decision under risk. In: Experiments in Environmental Economics (2018). https://doi.org/10.2307/1914185
15. Kalugina, A., Meyer, T.: Safety Climate in Academia. Part 2: modeling safety climate factor. (To be submitted in Safety Science) (2020)

16. Kemsley, J.: 10 years after Sheri Sangji's death, are academic labs any safer? Chem. Eng. News, vol. 97 (2018)
17. Massaiu, S.: Models of behavior and human error. In: Advances in Safety and Reliability - Proceedings of the European Safety and Reliability Conference, ESREL 2005. pp. 1371–1378 (2005)
18. Mauro, J.C., Diehl, B., Marcellin, R.F., Vaughn, D.J.: Workplace accidents and self-organized criticality. Phys. A Stat. Mech. Appl. (2018). https://doi.org/10.1016/j.physa.2018.04.064
19. Pluess, D.N., Groso, A., Meyer, T.: Expert judgements in risk analysis: a strategy to overcome uncertainties. Chem. Eng. Trans. **31**, 307–312 (2013). https://doi.org/10.3303/CET1331052
20. Standler, R.B.: Injuries in School/College Laboratories in the USA (2013). http://www.rbs2.com/labinj2.pdf
21. Stanford University: Advancing Safety Culture in the University Laboratory, pp. 1–132 (2014). https://doi.org/10.1016/S0022-3913(12)00047-9
22. Sun, Z., Xie, H., Shi, X., Liu, F.: Engineering approach for human error probability quantification. J. Syst. Eng. Electron (2009)
23. SUVA: Méthode Suva d'appréciation des risques à des postes de travail et lors de processus de travail (2008)
24. Towns, A.B.: Effective Strategies to Increase Employee Retention in Higher Education Institutions. ProQuest Diss. Theses D.B.A. (2019)
25. Van Noorden, R.: A death in the lab. Nature (2011). https://doi.org/10.1038/472270a
26. Yang, Z., Bonsall, S., Wang, J.: Fuzzy rule-based Bayesian reasoning approach for prioritization of failures in FMEA. IEEE Trans. Reliab. **57**, 517–528 (2008). https://doi.org/10.1109/TR.2008.928208

An Evaluation of Tools for Identifying Vulnerabilities in Open Source Dependencies for NodeJS Applications

Kevin Holmes(✉) and Amir Schur

Embry-Riddle Aeronautical University, 1 Aerospace Boulevard, Daytona Beach, FL 32114-3900, USA
kevin@kmholmes.com, schura@erau.edu

Abstract. Software developers commonly use open source libraries, known as "software dependencies", to build their applications to faster development of larger software applications. One way to track software dependencies is to use tools designed for just that purpose. Dependence vulnerability analyzers evaluate software dependency versions and compare them against lists of known vulnerabilities. This research will evaluate these tools by comparing the ease of installation and configuration, simplicity of keeping them up to date, readability of their reports, and ability to have them run automatically against a codebase. Finally, we will also compare the scan results of them.

Keywords: Software development · Cybersecurity · Software dependencies · Open source

1 Introduction

New software is always being written, but much of software development performs similar tasks and uses patterns similar to those used in many other pieces of software. In fact, new software is often little more than old tools and methods redesigned with new logic to achieve a new purpose. Many of the smaller tasks performed by new software can be most efficiently achieved by including open source code to perform those routine tasks. This open source code comes in the form of re-usable libraries, and when included in the code of another, larger software application, they are referred to as "software dependencies." The use of these libraries enables developers to use existing segments of code as components to build more complex software applications at a much faster rate, while the continued advancement of open source code often benefits from the expertise and contributions of a larger community of developers. Code written and tested by a larger community in a wide variety of applications is often more efficient and secure than code written completely by a small group of developers in isolation, which benefits both developers and users of software.

When an open source library is found to have a security vulnerability, the community responsible for that library can fix the problem and release an updated version of the code in question. After that update is released, software using the previous version must be updated to include the new version of that library to protect from the identified

D. Russo et al. (Eds.): IHSI 2021, AISC 1322, pp. 330–335, 2021.
https://doi.org/10.1007/978-3-030-68017-6_49

security vulnerability; this task is the responsibility of the developers whose software uses a given library. If no such update is released after a security vulnerability is identified, then developers who include the library as a dependency must choose whether to address the problem themselves, replace the dependency with another library that performs the same functions, or write their own proprietary code to replace it. In practice, however, there are a number of reasons why these updates may never occur.

One problem is that the use of software dependencies to handle common tasks is so routine that they are often used and then forgotten about. Dependency vulnerability analyzers can help combat this problem by evaluating software dependency versions and comparing them against lists of known security problems, allowing those responsible for software development and maintenance to monitor for and identify newly discovered vulnerabilities. The purpose of this research project is to evaluate some of these tools by comparing the ease of installation and configuration, simplicity of keeping them updated, readability of their reports, and ability to have them run automatically against a codebase. We will also compare the results of their scans to determine if any provide more comprehensive results than the others.

2 Dependency Analyzer for NodeJS

Use of software components with known security vulnerability is an ongoing issue that has been in the top ten list of Open Web Application Security Project (OWASP) since at least 2013 [1].

As there are various dependency frameworks, our first attempt is to evaluate NodeJS application's package dependencies. Node.js is a free, open-sourced, cross-platform JavaScript run-time environment that lets developers write command line tools and server-side scripts outside of a browser. It is widely used worldwide.[1]

The table below shows a list of software that we will evaluate for NodeJS package dependency analysis.

Table 1. Dependency Analyzer Software that will be evaluated

No	Software Name	Source
1	Snyk	https://snyk.io/
2	OWASP Dependency Check	https://owasp.org/www-project-dependency-check/
3	NPM Audit	https://docs.npmjs.com/auditing-package-dependencies-for-security-vulnerabilities
4	Gamma Ray	https://github.com/nearform/gammaray
5	AuditJS	https://github.com/sonatype-nexus-community/auditjs

(continued)

[1] Report on NodeJS usage can be found at: https://nodejs.org/en/user-survey-report/.

Table 1. (*continued*)

No	Software Name	Source
6	NPQ	https://github.com/lirantal/npq
7	David	https://david-dm.org/
8	Node Security	https://github.com/nodesecurity/nsp
9	NodeJS Analyzer	https://jeremylong.github.io/DependencyCheck/analyzers/nodejs.html
10	NodeJS Scan	https://github.com/ajinabraham/NodeJsScan
11	RetireJS	https://retirejs.github.io/retire.js/

3 Evaluation Method

Investigation of various aspects of the dependency checking software is elaborated within this section. Two people will perform the evaluation with the same criteria. One person will evaluate the first five of the analyzers in the list, the other will evaluate the remainder of the list.

3.1 Virtual Environment Setup

To minimize resources for the test environment, we will be using Vagrant (https://www.vagrantup.com/), a tool that can quickly build a variety of virtual machines on a computer or laptop for a project. This tool requires the user to have an existing type 2 hypervisor available, such as VirtualBox[2], Hyper-V[3] or Kernel-based Virtual Machine (KVM) hypervisor[4]. We chose VirtualBox based on previous usage experience.

We first create a file folder in a project directory for each virtual machine that will be built. In each folder, we create a Vagrant file that contains the definition of the virtual machine to build. Then with a single command, vagrant up, we can instantiate the virtual machine in our hypervisor. We can then log into the virtual machine, either through the hypervisor console (graphical user interface) or using a text-based interface through a secure shell connection. If we need to remove the virtual machine, we can run one command: vagrant destroy. Login to the system is with user "vagrant" with the same password.

For the purpose of minimizing licensing costs, each tool will reside in a Centos 8.2 machine. Each virtual machine will have 1 virtual CPU, 1 GB of RAM and 50 GB of hard drive size.

[2] VirtualBox details can be found at: https://www.virtualbox.org/wiki/VBox_vs_Others.

[3] Hyper-V details can be found at: https://docs.microsoft.com/en-us/windows-server/networking/sdn/technologies/hyper-v-network-virtualization/hyperv-network-virtualization-technical-details-windows-server.

[4] KVM details can be found at: https://www.linux-kvm.org/page/Main_Page.

Below is a simple vagrant file that can be run to setup one virtual machine:

```
# -*- mode: ruby -*-
# vi: set ft=ruby :

# All Vagrant configuration is done below. The "2" in
Vagrant.configure
Vagrant.configure("2") do |config|

config.vm.provision "shell", inline: "sudo yum update -y"
config.vm.provision "shell", inline: "sudo yum install -y
firefox"
config.vm.provision "shell", inline: "sudo yum groups
install -y GNOME Desktop"
config.vm.provision "shell", inline: "sudo systemctl set-
default graphical.target"
config.vm.define "nodejstest1" do |nodejstest1|
    nodejstest1.vm.box = "bento/centos-7.3"
    nodejstest1.vm.network "private_network", ip:
"192.168.100.2"
    nodejstest1.vm.hostname = "nodejstest1.testdomain"
    nodejstest1.vm.provider "virtualbox" do |vb|
            vb.memory = "1024"
            vb.cpus = "1"
            config.disksize.size = "50GB"
    end
  end
end
```

3.2 Installation

For our evaluation, we must note the level of complexity involved in installing the dependency-checking software. We will also note the time taken to complete each installation, clocked from the time the tester logs in to the virtual machine. A completed install is noted when we can access the software's User Interface. The list of software to install is in Table 1.

3.3 Vulnerability List

Another important aspect we will evaluate for each software is how quickly and easily the vulnerability list can be created, maintained, and updated. Consider what happens during a zero-day attack, when a software or system vulnerability is still unknown or under investigation: it is important to know how fast the software can perform an update to its current vulnerability list to ensure any impacted software can be identified.

3.4 Reports

We will note the method of reporting for each software. An ideal report is automatically generated and can be created using an automated task. We will note how the report is generated and how to access it. We also will compare the report from one dependency checking software to the next, to compare thoroughness.

3.5 Application Programming Interface (API) Availability

Modern software is expected to have an API, which allows other programs, such as those used in continuous integration pipelines, so that any operations can be automated and be integrated into a larger automation engine. We must find a method of API access and whether we can scan software in an automated fashion. Consider a large company, which may need to quickly scan a list of 100 software packages.

We will explore a method for authentication, starting a scan job, and getting scan results using API library. The ideal software will provide the necessary tools to integrate it into a company's existing automated process.

3.6 Layers of Dependency Analysis

We need to understand the depth of dependency checking that are possible with each software. Larger open source dependencies often have smaller open source dependencies of their own. Ideally, all layers should be searched exhaustively, and the software should not be limited to scanning only a certain number of layers. We will evaluate and note this capability.

3.7 Maintenance

It is important to understand whether the software is being maintained or not. Unmaintained software will quickly become obsolete as the platforms it is designed to scan continue to evolve. We will evaluate and note information related to maintenance, including whether responsibility for maintenance falls upon a corporation or an unincorporated community of developers.

Once we collect and analyze information from our research, we can draw conclusions and make recommendations. The focus is not on a tool, but rather on best practices in each area.

4 Software List to Scan

For our NodeJS software, we are picking three widely used software applications [2] (Table 2).

Table 2. NodeJS Open Source Software that will be evaluated

No	Software Name	Source
1	Uppy	https://github.com/transloadit/uppy
2	Zenbot	https://github.com/DeviaVir/zenbot
3	Wiki	https://github.com/Requarks/wiki

5 Results Expectations

We hope by performing this investigation, we can provide information on the best practices for handling NodeJS package dependencies. While the scope of this research was limited to dependency analysis in NodeJS, it is also our hope that this process will be adaptable to allow for evaluation of other dependency analysis products that target a wider variety of software development frameworks. The importance of developing and maintaining secure software is increasing quickly, and this knowledge can be applied to enterprise software frameworks to help reduce vulnerabilities in their software.

References

1. OWASP Top Ten 2013. (n.d.). OWASP Foundation. https://owasp.org/www-pdf-archive/OWASP_Top_10_-_2013.pdf
2. Amazing Node.js Open Source for the Past Year (v.2019), 2 February 2019. https://medium.mybridge.co/45-amazing-node-js-open-source-for-the-past-year-v-2019-c774d750e925#6bd0

Low-Cost Cyber-Physical Production Systems Based on IEC 61499 for Analog Process

Gustavo Caiza[1], Sergio Bustos[2], Carlos A. Garcia[2], and Marcelo V. Garcia[2,3(✉)]

[1] Universidad Politecnica Salesiana, UPS, 170146 Quito, Ecuador
gcaiza@ups.edu.ec
[2] Universidad Tecnica de Ambato, UTA, 180103 Ambato, Ecuador
{sbustos2525, ca.garcia, mv.garcia}@uta.edu.ec
[3] University of Basque Country, UPV/EHU, 48013 Bilbao, Spain

Abstract. Cyber-Physical Production Systems (CPPS) is the result of the combination of complex production processes with high communication capabilities. This joint work generates an improvement in the general system's performance and efficiency. CPPSs play an increasingly important role in the implementation of "Smart Factories" compatible with data exchange among all the conforming devices, under the concept of the Internet of Things (IoT). Current industry's needs demand the generation of modular and flexible production systems, so they can easily adapt to the changing market requests. To fulfill this need, the automation standard IEC-61499 is developed based on a distributed architecture and the use of control devices with high processing capabilities. The present work shows a low-cost embedded architecture designed using the IEC-61499 standard to control the operation of an analogic control process.

Keywords: IEC 61499 · Cyber Physical Production Systems (CPPS) · Industry 4.0 · Embedded control

1 Introduction

Embedded systems have been driven by advanced semiconductor and communications technologies over the last decade, communicating through many means such as Wi-Fi, Bluetooth, and the network. Currently, there are large applications of Cyber-Physical Systems (CPS) in the field of industrial automation. CPS has a lot of challenges in industrial automation, one of the majors is modeling methodologies for distributed control automation software in the real analog process [1].

At present, factory automation systems need to address the different challenges such as very rapid technological advancement, global-competitive market, and product customization. The Cyber-Physical Production Systems (CPPS) is developing to lead new types of industrial control systems into the Industry 4.0 paradigm. These systems pursue automation systems globally connected. CPPS require the acquisition of data from the production system and intelligent data processing to extract information to improve the overall performance of the production system, to achieve this it is

D. Russo et al. (Eds.): IHSI 2021, AISC 1322, pp. 336–342, 2021.
https://doi.org/10.1007/978-3-030-68017-6_50

necessary to bridge the gap between the control systems and the upper layers of automation [2]. In addition, a set of embedded devices can be interconnected to serve a purpose in the system. This establishes a solid foundation for the Cyber-Physical (CPS) system that deals with the deep integration of virtual computing with physical processes [3].

Nowadays, automation based on Programmable Logic Controllers (PLC) is widely accepted in industrial automation. But, for the next smart factories with new requirements of complex industrial systems, this standard has serious shortcomings. The future smart factory's analog processes need industrial systems which include portability, interoperability, increased reuse, and distribution. To address these constraints the International Electrotechnical Commission (IEC) has begun the development of a new standard IEC 61499 to cope with the intelligent automation [4].

In the present work is proposing the development of CPPSs using low-cost devices to implement a closed-loop control system according to the IEC-6149 standard and the storage of variables in a database that will be used in the process is proposed.

The design of the document is as follows. Section 2 shows a case study where SIFBs designed using IEC 61499 standard is indicated. Section 3 shows the implementation proposal. Finally, conclusions and ongoing work are addressed in Sect. 4.

2 Study Case

An industrial educative system is presented in the proposed case study, this system shows a scale factory automation process. The FESTO MPS-PA Compact Workstation has four systems to be controlled: level, flow rate, pressure, and temperature. This system allows the use of discontinuous or continuous controllers using PLC or low-cost devices as Raspberry PI (RPI), Odroids, or Arduinos boards. The close loops controls are depicted in Fig. 1a.

The close loops selected were the level and temperature loops (see Fig. 1b). i) Level Control Loop: The P&ID diagram shows an ultrasonic sensor to measure the process variable, which senses the liquid level in tank B102. The actuator of this loop is the pump P101. The tank B101 is used as a reservoir. The value should be kept on a certain level also if disturbances or setpoint changes occur. ii) Temperature Control Loop: The liquid is heating using the electric heater E104 this actuator is located at the bottom of the tank B101. The Cooling of the water is made by the cooler located below the profile plate; the maximum temperature for the system is 65 °C. The sensor of temperature is a PT100 calibrated from 0 to 100 °C. The heating actuator is controlled using a Pulse Width Modulation (PWM) that is generated by the low-cost board using as software control the standard IEC 61499.

The hardware control architecture comprises a Beagle Bone Blackboard (BBB) as a low-cost device that can send and receive all the signals from the sensors and actuators of the system. Into this BBB board, an IEC 61499 function blocks (FBs) were implemented to manipulate the General Purpose Input/Output ports (GPIO), control the system, and store the values of the process using a transactional database.

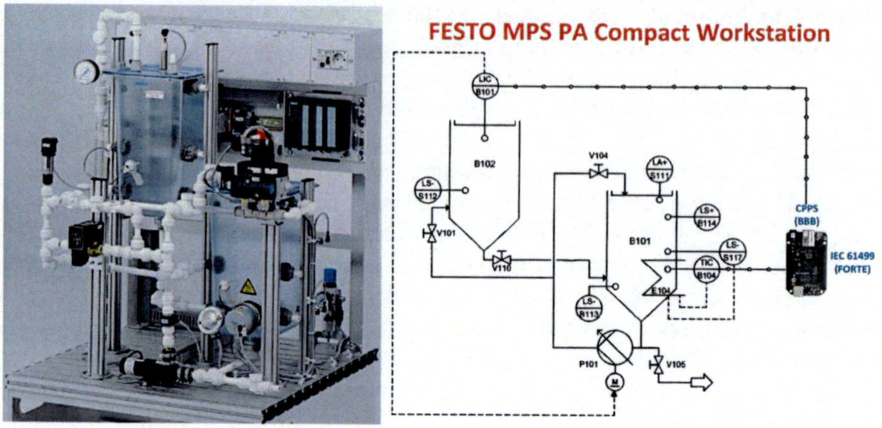

a b

Fig. 1. a) FESTO™ MPS-PA Compact Workstation, b) PI&D of temperature and level process

3 The Proposed Implementation

3.1 Hardware Platform

A low-cost hardware platform is proposed to be the controller of this system. This kind of controller is known as a Single Board Computer (SBC), is a small computer built on a unique circuit board with all the standard components like microprocessor, RAM memory, GPIOs, and other functional hardware components baked-in. This SBC reduces the control system cost because has the same functions that an expensive PLC device. The BBB is selected as a control prototyping platform. The BBB board is useful to implement Industry 4.0 applications because of their powerful embedded operating system that is great for implement a distributed system using IEC 61499 standard which cannot be implemented using traditional automation devices. The BBB board has a processor from Texas Instruments AM335x built-in by an Arm Cortex-A8. A 512 MB DDR3 RAM memory. A Hard Drive of 4 GB 8-bit eMMC onboard flash storage. Co-processors 2xPRU-ICSS and 46 pins from the GPIOs.

4 Set of Function Blocks (FBs) Under Standard IEC-61499

The software tool used to design the FBs is the 4DIAC-IDE framework [5], a set of FBs is implemented to control the analog input/output GPIO port encapsulating the c++ libraries into the FBs. Furthermore, a FB encapsulating the stack communication to send and receive data from a transactional database is programmed. Two sets of FBs were developed, (1) FBs for GPIO analog manipulation and (2) FBs for Proportional-Integral-Derivative (PID) control implementation.

FBs for Inputs/Outputs Manipulation. This FBs set was designed to read and write the analog pins from the BBB board and use the Pulse-Width Modulation (PWM) output pins allowing them to connect with the sensors and actuators from MPS Compact Workstation controlling the level and temperature loops. The configuration of the tags associated with the pins is made using a configuration file in XML format. The XML file has (1) tags names to access the process variables, (2) pin's number to map the variables to GPIO port, (3) pins function, and (4) a small comment.

1. **ANALOG_OUTPUT_R FB:** This FB is used to configure the EHRPWM output pins, the FB write an analog value within 4 to 20 mA, which is a common standard to control actuators. As all the FBs the inputs events are INIT and REQ to begin the writing procedure. The output events are the INITO and CNF. The terminals of this FB are explained in more detail below (see Fig. 2):
   - **NAME (Input, STRING):** The tag name of the port mapped to the pin on BBB board.
   - **OUT (Input, REAL):** Analog value written on the GPIO port.

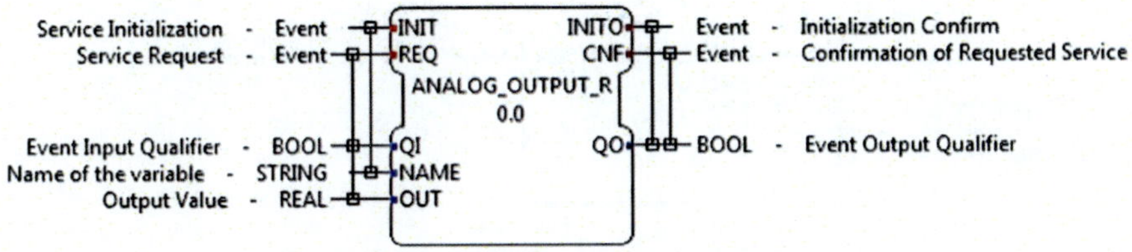

Fig. 2. ANALOG_OUTPUT_R FB

2. **ANALOG_INPUT_R FB.** This FB read the analog value in the pin of the GPIO port. The FB has the previously events explained. These FBs work with the following Input/Output variables (see Fig. 3):
   - **NAME (Input, STRING):** The tag name of the pin to be manipulated.
   - **IN (Output, REAL):** The Analog value to be read on the inputs analog pins.

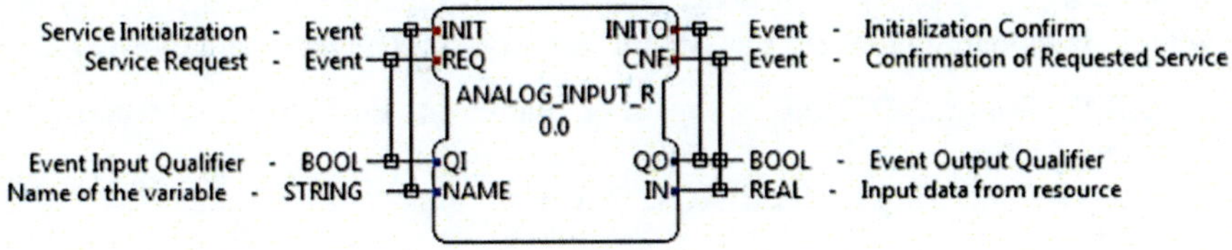

Fig. 3. ANALOG_INPUT_R FB

FBs for Analog Control. This kind of FBs has the PID control algorithms and libraries to log data from the process with the aim of implement a historian's software characteristics under IEC-61499 to link the data acquisition and storage system with a relational database server. This FB grant secure and high-speed communication with databases giving rapid retrieval techniques increasing the amount of industrial data transactions.

1. **PID CONTROL FB.** For the case study selected which represents a closed-loop control system, a Function Block was implemented containing the algorithm of a Proportional, Integral, and Derivative Control System as shown in Fig. 4.

   - **SETPOINT (Input, REAL):** SetPoint input value.
   - **SENSOR (Input, REAL):** Sensor input value.
   - **G_Kp (Input, REAL):** Proportional gain input value.
   - **G_Ki (Input, REAL):** Integral gain input value.
   - **G_Kd (Input, REAL):** Derivative gain input value.

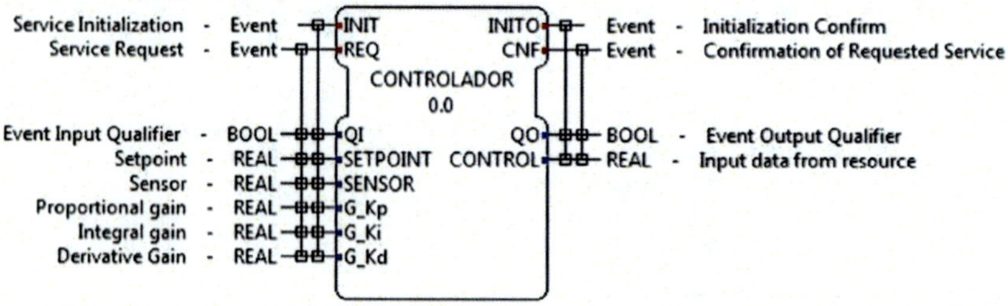

Fig. 4. PID CONTROL FB

2. **LOG DATA FB.** This FB encloses the PostgreSQL library to send and receive messages from databases under IEC 61499. This library is an open-source object-relational database. An interesting feature of PostgreSQL is the multi-version concurrency control or MVCC. This method adds an image of the database state to each transaction. This allows the FBs to make transactions eventually consistent, offering great performance advantages. In Postgres, it is not required to use read locks when carrying out a transaction, which gives the FBs greater scalability. Also, PostgreSQL has Hot-Standby. This allows clients to search (read-only) servers while in recovery or standby mode. So FBs can do maintenance or recovery tasks without completely blocking the system (see Fig. 5). The terminals of this FB are:

   - **USER (input, STRING):** the user name of the Postgres database for authentication.
   - **PASSWORD (input, STRING):** The user password to authenticate to the database.
   - **IP (input, STRING):** IP socket address to send and receive messages from the database.
   - **TAG (input, STRING):** Process tags to log in database
   - **VALUE (input, STRING):** Data value to be written in the database.

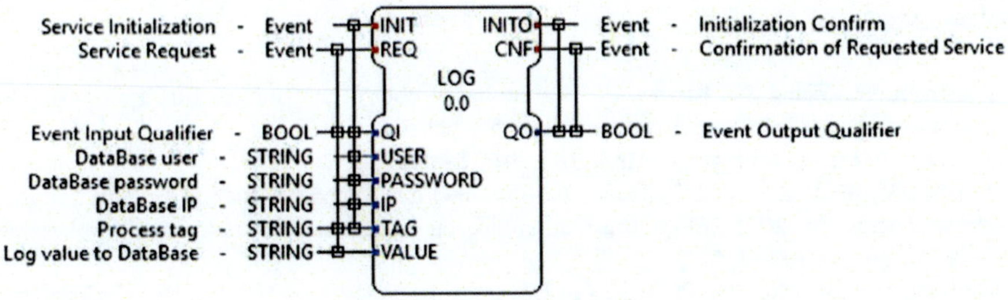

Fig. 5. LOG DATA FB

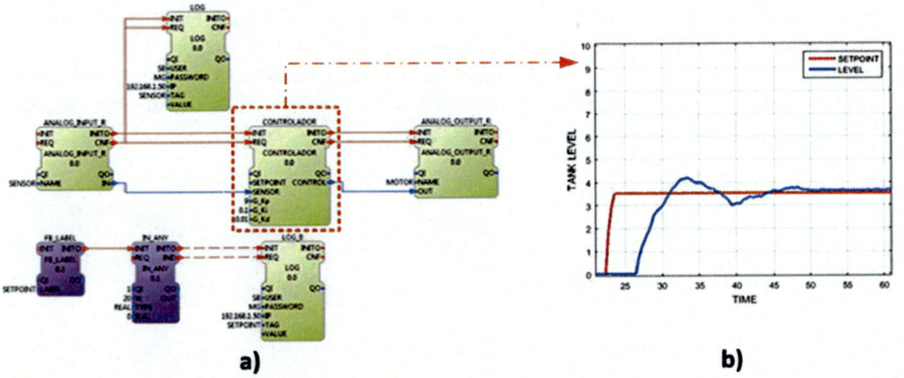

Fig. 6. FB Control Diagram. a) Distributed Control Application in 4DIAC-IDE. b) Control response to setpoint

Functionality. As a final result of our work, based on the developed FB for the management of I/O of the BBB and for the closed-loop control of the Festo Cell. Finally, the respective networks of FBs have been generated, in order to achieve the correct operation of the plant process. The BBB controls the compact MPS-PA workstation and the Setpoint signal values and Sensor signal are stored in a database (See Fig. 6).

5 Conclusions and Ongoing Work

This research work addressed an approach for the control and access to field data in automation systems using FBs in low-cost CPPS architecture using applications of the IEC 61499 standard that are executed in the FORTE runtime. These FBs enable a fast and powerful low-cost platform to control the analog system in real factories using the Industry 4.0 paradigm. As future works fuzzy or robust control techniques can be implemented using the FBs in low-cost hardware.

References

1. Adamson, G., Wang, L., Moore, P.: Feature-based control and information framework for adaptive and distributed manufacturing in cyber physical systems. J. Manuf. Syst. **43**, 305–315 (2017). https://doi.org/10.1016/j.jmsy.2016.12.003
2. Dai, W., Vyatkin, V., Chen, C., Guan, X.: Modeling distributed automation systems in cyber-physical view. In: 2015 IEEE 10th Conference on Industrial Electronics and Applications (ICIEA), pp. 984–989 (2015)
3. Derhamy, H., Drozdov, D., Patil, S., van Deventer, J., Eliasson, J., Vyatkin, V.: Orchestration of arrowhead services using IEC 61499: distributed automation case study. In: 2016 IEEE 21st International Conference on Emerging Technologies and Factory Automation (ETFA), pp. 1–5 (2016)
4. Drozdov, D., Patil, S., Dubinin, V., Vyatkin, V.: Formal verification of cyber-physical automation systems modelled with timed block diagrams. In: 2016 IEEE 25th International Symposium on Industrial Electronics (ISIE), pp. 316–321 (2016)
5. Garcia, M.V., Irisarri, E., Perez, F., Estevez, E., Marcos, M.: Opc-ua communications integration using a CPPS architecture. In: 2016 IEEE Ecuador Technical Chapters Meeting (ETCM), pp. 1–6 (2016)

Performance Evaluation of the Nvidia Jetson Nano Through a Real-Time Machine Learning Application

Sebastián Valladares[1], Mayerly Toscano[1], Rodrigo Tufiño[2], Paulina Morillo[2], and Diego Vallejo-Huanga[2,3](✉)

[1] Department of Computer Science, Universidad Politécnica Salesiana, Quito, Ecuador
{pvalladares,mtoscanor}@est.ups.edu.ec

[2] IDEIAGEOCA Research Group, Universidad Politécnica Salesiana, Quito, Ecuador
{rtufino, pmorillo,dvallejoh}@ups.edu.ec

[3] Department of Physics and Mathematics, Universidad de las Américas, Quito, Ecuador
diego.vallejo.huanga@udla.edu.ec

Abstract. The use of applications with Machine Learning (ML) is increasingly frequent and their daily use demands more capacity for processing information in real-time. Many devices deploy their infrastructure in Cloud Computing environments, where latency and network partitions are the main challenges they must face. However, in environments such as Dew Computing, these problems are mitigated since the information processing is carried out directly on the data source, so an Internet connection is not necessary. This article evaluates the performance of the Nvidia Jetson Nano platform, under a Dew Computing approach, with an application that uses ML to solve a problem of identification and counting of land vehicles, in real-time, in the city of Quito, Ecuador. Two types of experiments were conducted, the first experiment aimed to evaluate the use of processing resources (CPU and GPU), device temperature, power consumption, and the amount of RAM used in the ML task. The second experiment sought to evaluate the effectiveness of the platform combined with OpenDataCam.

Keywords: Vehicle detection · Dew computing · OpenDataCam · Darknet · YOLO · CUDA · GPU

1 Introduction

The drastic increase in computing capacities, and the improvement in the performance of telecommunications networks, have allowed the consolidation of paradigms that allow offering computing services remotely. Currently, there are three environments in which data is processed and distributed remotely [1]: Cloud Computing, with great features and computing power; Fog Computing, with less computing power but with processing closer to the data source and Dew Computing, where the information is processed directly in the end devices.

D. Russo et al. (Eds.): IHSI 2021, AISC 1322, pp. 343–349, 2021.
https://doi.org/10.1007/978-3-030-68017-6_51

The main objective of Dew Computing is to take full advantage of the potential of local computers and cloud services [2], minimizing the impact generated by the loss of internet connection, ensuring security and privacy of the processed data [3]. The energy consumption in the execution of processes in integrated systems is very important since the power supply of these is small. In edge devices, resources are limited so large processing cannot be performed [1].

On the other hand, the use of applications with Machine Learning (ML) is increasingly frequent and their daily use demands more capacity for processing information in real-time. In 2019, the Nvidia company launches a device called Jetson Nano focused on the resolution of ML tasks, which allows fast and powerful deployment, under a Dew Computing approach. Jetson Nano has low power consumption at an affordable price, and is compatible with many AI frameworks, making it easy for developers to integrate their models into the product [4].

Currently, there are several scientific researches that uses methods for the detection and counting of objects with ML. In the work of Vaidya and Paunwala [5], the Jetson Nano platform is used for the recognition of traffic signs by means of Convolutional Neural Networks (CNN), to improve response times and execute processes in-situ. Gotthans et al. [6], with deep CNN, detect fires through video. The processing is done in Jetson Nano, resulting in a balance between precision and efficiency.

You Only Look Once (YOLO) [7], is an open-source system aimed at detecting objects in real-time, which uses a CNN to detect objects in images. For its operation, the CNN divides the image into regions, predicting identification frames and probabilities for each region, allowing a low detection error for new inputs. Lin and Sun in [8] use YOLO for vehicle detection. A limitation of this work is the detection of vehicles with low lighting in the video, especially at night. Asha and Narasimhadhan in [9], propose a method of counting vehicles with YOLO, which, in addition to tracking the object, determines its future trajectory.

This article presents an analysis of the performance of the Nvidia Jetson Nano platform and the effectiveness of performing a real-time ML application with YOLO for the recognition and counting of ground vehicles.

2 Materials and Methods

To analyze the performance of the Nvidia Jetson Nano platform, an ML algorithm was executed with a task of detecting ground vehicles: cars, buses, trucks and motorcycles, in real time, with a camera. Jetson Nano is a compact size device (69 mm × 45 mm) with Maxwell 128-core GPU, Quad-core ARM A57 CPU, 4 USB 3.0 ports; for storage, it has a MicroSD card slot, connections are via Gigabit Ethernet and M.2 Key E and a 2.0 Micro-B port to connect the power source [10]. The Jetson Nano platform does not have an embedded camera, but in this work, the Nvidia Jetson Nano HD AI model IMX219-77 was used to capture video in real time. The OS used was GNU/Linux Ubuntu 18.04.

For the identification and counting of vehicles, the OpenDataCam (ODC) [11] software was used, an open-source tool to quantify and visualize objects in real-time, through a web interface. In our work, we setup this tool with three features. The first one is an ML application with Darknet [12] that allows the identification of objects from an image, the second is a web application programmed in Node JS that allows user interaction with the software, and the last one is a MongoDB database to save the information. Darknet is compiled with the OpenCV libraries, for the acquisition of the video, CUDA for the computation on GPU, and YOLO [7] as neural network model. Figure 1 shows how the components of the application are related.

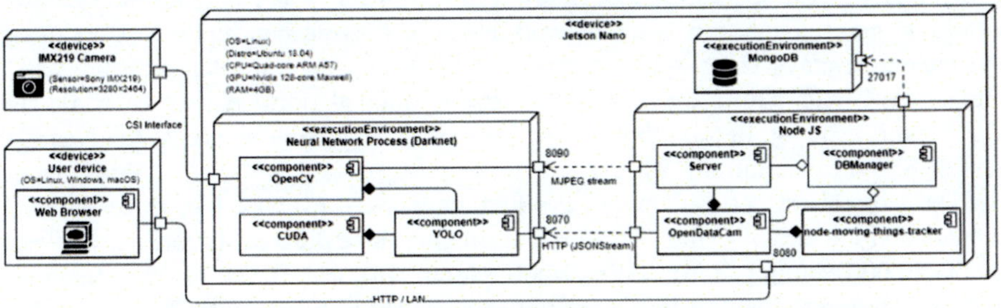

Fig. 1. Vehicle detection and counting system component diagram.

The process starts with the *Server* component, through which the user views the video in real-time, establishes the area of interest to identify the objects, and observes the counted and vehicle tracking. Besides, the configuration parameters are read, the web application is published and the connection to the database is initialized. The YOLO component is also initialized, configuring the parameters that the Darknet requires to identify the vehicles and the communication ports with the other components. Port 8090 is used to obtain the Video Stream and port 8070 is used to deliver the result of the object identification through a Json Stream.

On the other hand, the *OpenDataCam* component takes care of the logic for counting vehicles. In each video frame, the program sends a request to port 8070 to obtain the identified objects. This information is sent to the *node-moving-things-tracker* component [13], which contains an algorithm that allows persistent tracking of multiple moving objects from the object detection inputs provided by YOLO. This tracking is done frame by frame within the area of interest. When a vehicle has been identified, it is counted as a single object, and the record is stored in MongoDB. Also, its position and tracking information is sent to the web application.

The performance evaluation of Jetson Nano was carried out in two stages, the first one analyzes the performance of the Nvidia Jetson Nano platform with the Jetson Stats tool [14], using the metrics: CPU (%), the average use of its four cores in a given time; GPU (%), GPU usage over a period of time; Temperature (°C), the temperature at the end of each test; RAM (GB), amount of RAM memory used in each test and; Energy (mW), the total amount of energy consumed by the system in each test. The second stage determines the effectiveness of OpenDataCam, by calculating the Relative Error, based on the data obtained in three environments with changes in light with respect to the time of day and different amounts of traffic. For the performance analysis of vehicle recognition, i.e. detection or not of a land vehicle, ML metrics were used: Accuracy (Acc), Precision (Pr), Recall (Re), and F1 Measure (F1).

3 Experiments and Discussions

To make the system easier to manage and operate, a screen sharing tool was configured through Virtual Network Computing (VNC). Additionally, a static IP address (IP: 192.168.1.3, Mask: 255.255.255.0, Gateway: 192.168.1.1) was configured to establish a point-to-point connection, via a network cable, and avoid interruptions when using the remote display. The display monitors, with Jetson Stats, the temperature, CPU, GPU, RAM, and power usage that the Jetson Nano platform consumes while the algorithm is running. In turn, through the browser, it is possible to access the Node JS server for the execution of ODC.

The experiments carried out in the first stage were evaluated in three different scenarios in the city of Quito, Ecuador, considering three different time slots, between 09:00–10:00 (Morning, Mo), 13:00–14:00 (Noon, No), and 16:00–17:00 (Afternoon, Af), with different lighting levels and different levels of two-way traffic. Test 1 was carried out at Av. De los Conquistadores, with the system located at a height of 4 m, to have a focus on the larger vehicles. Test 2 was held at Av. Francisco de Orellana, and Test 3 at Av. 6 de Diciembre. These last two scenarios were taken from a perspective parallel to the vehicles, at a height of 1.45 m. with the purpose of evaluating the effectiveness of the system by having obstruction of the objects to be detected. The results of the three tests are summarized in Table 1.

Table 1. Nvidia Jetson Nano platform performance results measured with the Jetson Stats tool.

Test	Test 1			Test 2			Test 3		
Time slots	Mo	No	Af	Mo	No	Af	Mo	No	Af
CPU (%)	37.88	47.25	40.25	36.75	43.38	38.38	40.25	44.38	42.25
GPU (%)	99	99	99	99	99	99	99	99	99
Temperature (°C)	61.50	67.50	69.86	55.88	70.00	61.50	66.36	71.30	69.86
RAM (GB)	2.1	2.1	2.1	2.1	2.1	2.1	2.1	2.1	2.1
Power (mW)	6297	6525	6871	7114	7780	7505	7467	7647	7449

Mo: Morning; No: Noon; Af: Afternoon

Regarding the performance of the Jetson Nano, when executing the ML task, the use of the GPU remained constant at 99%, during all the tests, due to the fact that the calculation for the vehicle identification is performed on this component. RAM memory consumption remained at 2.1 GB, approximately 50% of its capacity. Energy consumption is directly related to vehicular flow. Thus, in Test 2, in the time slot with the highest traffic flow (Noon, with 431 vehicles), the highest peak of power consumption was obtained, which was 7780 mW. In most of the experiments, the relationship between CPU usage and temperature device variation was directly proportional.

On the other hand, the experiments carried out in the second stage, which aim to determine the effectiveness of the vehicle counting task, based on the manual counting of the vehicles (Real - Ground Truth) versus the counting and recognition of the System in the three environments with traffic, are shown in Table 2. In this case, it is observed that the system does not detect smaller vehicles, such as motorcycles, in contrast to larger vehicles such as trucks, where the system tends to double the count. This inconvenience could be related to factors such as the position of the camera (height and perspective), the incidence of light (time slot), size of the vehicle, and vehicular traffic. For example, in Test 1 where the camera position is higher, 262 out of 304 vehicles were detected, of which no motorcycle-type vehicle was detected, but most other vehicle types were correctly detected. In Tests 2 and 3, in which the camera was lower and the vehicular traffic was higher, the detection of motorcycles slightly improved, but the error in the detection of truck-type vehicles increased considerably.

Table 2. Summary of vehicle identification and counting results, and their relative error.

		Number of Vehicle												Total
Hour		**Mo**				**No**				**Af**				
Type of Vehicle														
Test 1	Real	107	1	2	7	75	2	3	8	87	2	1	9	304
	System	95	0	2	0	74	2	3	0	83	2	1	0	262
	Error (%)	11.2	100	0	100	1.3	0	0	100	4.6	0	0	100	13.82
Test 2	Real	231	2	8	44	357	8	12	54	366	6	8	34	1130
	System	228	2	21	5	277	3	19	4	183	0	35	6	783
	Error (%)	1.3	0	162.5	88.6	22.4	62.5	58.3	92.6	50	100	337.5	82.3	30.71
Test 3	Real	300	5	6	37	297	8	12	54	288	5	1	46	1059
	System	215	6	12	2	230	9	13	11	212	9	4	7	730
	Error (%)	28.3	20	100	94.5	22.6	12.5	8.3	79.6	26.4	80	300	84.8	31.07

As can be seen in Fig. 2(A), the perspective of the camera located at a higher height in Test 1, facilitates the count, avoiding the occlusion of the objects. However, the small size of motorcycles makes their detection difficult, Fig. 2(B). In Tests 2 and 3, where the perspective of the camera was parallel to the vehicles, the detection percentage was lower, due to the occlusion generated in smaller vehicles concerning others.

Finally, regarding the performance of the model used in object recognition, i.e., vehicle recognition regardless of type, Vehicle (V) or Non-Vehicle (NV), shown in Table 3. The results indicate high percentages of sensitivity in all experiments. The highest precision and accuracy were obtained in the first test (with 86%, respectively), where vehicular traffic was lower and the camera position was higher.

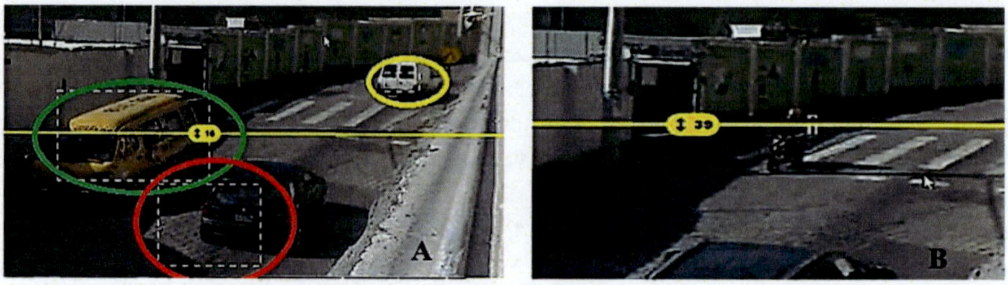

Fig. 2. (A) Detection of three vehicles, two cars and one bus for Test 1 in Av. De los Conquistadores. (B) Failure to detect motorcycles from a height of 4 m in Test 1.

Table 3. Performance evaluation of the model used in the ML task for Vehicle (V) and Non-Vehicle (NV) recognition.

			Real class		ML evaluation metrics			
			V	NV	Acc	Pr	Re	F1
Predicted class	Test 1	V	262	0	0.86	0.86	1	0.93
		NV	42	0				
	Test 2	V	689	47	0.61	0.64	0.94	0.76
		NV	394	0				
	Test 3	V	698	16	0.66	0.67	0.98	0.79
		NV	345	0				

4 Conclusions and Future Work

This paper presents the performance evaluation of the Nvidia Jetson Nano platform on a real-time ML application. According to the results, it can be concluded that Dew Computing solutions support applications that require great processing such as video image recognition. However, the use of GPUs can cause the temperature increase in the devices. On the other hand, the use of OpenDataCam for the counting and identification of vehicles turns out to be effective, but it is necessary to take into account factors such as the position of the camera, lighting, size and speed of the vehicle and vehicular traffic. As future work, it is proposed to improve the configuration of Machine Learning algorithms to increase the precision of the results.

Acknowledgments. This work was supported by IDEIAGEOCA Research Group of Universidad Politécnica Salesiana in Quito, Ecuador.

References

1. Kotlar, M., Bojic, D., Punt, M., Milutinovic, V.: A survey of deep neural networks: deployment location and underlying hardware. In: Symposium Neural Networks Application NEUREL 2018, pp. 1–6 (2018)

2. Wang, Y.: Definition and categorization of dew computing. Open J. Cloud Comput. **3**(1) (2016)
3. Wang, Y., Karolj, S., Rindos, A., Gusev, M., Yang, S., Pan, Y.: Dew computing and transition of internet computing paradigms. ZTE Commun. **15**, 133–136 (2017)
4. Cass, S.: Nvidia makes it easy to embed AI: the Jetson nano packs a lot of machine-learning power into DIY projects-[Hands on]. IEEE Spectr. **57**(7), 14–16 (2020)
5. Vaidya, B., Paunwala, C.: Traffic Sign Recognition Using Color and Spatial Transformer Network on GPU Embedded Development Board (2019)
6. Gotthans, J., Gotthans, T., Marsalek, R.: Deep convolutional neural network for fire detection. In International Conference Radioelektronika, pp. 0–5 (2020)
7. Redmon, J., Divvala, S., Girshick, R., Farhadi, A.: You only look once: unified, real-time object detection. In: Conference Computer Vision and Pattern Recognition, vol. 2016-Decem, pp. 779–788 (2016)
8. Lin, J., Sun, M.: A YOLO-based traffic counting system. In: Conference on Technologies and Application of Artificial Intelligence, pp. 82–85 (2018)
9. Asha, C., Narasimhadhan, A.: Vehicle counting for traffic management system using YOLO and correlation filter. In: IEEE International Conference on Electronics Computing and Communication Technologies, pp. 1–6 (2018)
10. Nvidia Jetson Nano Developer Kit. https://developer.nvidia.com/embedded/jetson-nano-developer-kit
11. Groß, B., Kreutzer, M., Durand, T.: OpenDataCam: An open source tool to quantify the world. Move Lab. https://www.move-lab.com/project/opendatacam/
12. Redmon, J.: Darknet: Open Source Neural Networks in C. http://pjreddie.com/darknet/
13. Durand, T.: node-moving-things-tracker. GitHub. https://github.com/opendatacam/node-moving-things-tracker
14. Bonghi, R.: Jetson Stats. https://pypi.org/project/jetson-stats/

Preliminary Analysis on the Recruitment Process for Domestic Violent Extremist Groups

Brendan Reilly and April Edwards(✉)

United States Naval Academy, 121 Blake Rd, Annapolis, MD 21402, USA
brendanreilly2017@gmail.com, aedwards@usna.edu

Abstract. The rise of social media has coincided with the emergence of numerous domestic violent extremist groups with alt-right political ideologies in the United States and Europe. The current study analyzes patterns of communication in the Iron March dataset. Social network analysis was used to determine the most tightly connected users, and graph clustering shows the interactions between these influential users and other members of the forum. Our analysis shows a level of organization within the forum, and an adversarial approach to recruitment of new members, which was used to prevent infiltration by law enforcement and maintain the anonymity of participants. Users who completed the requirements for recruitment were permitted into other, more private, communication channels.

Keywords: Human factors · Extremist groups · Social media analysis

1 Introduction

While much has been written about the use of social media to influence the 2016 US Presidential elections, primarily by the Russian Government [1–3], relatively little has been written about automated analysis of online resources for understanding the nature of domestic violent extremist groups (DVE) in the United States. In a recent article, Shortland and Forest call for the use of technology to identify the minority of individuals within extremist organizations that will conduct acts of terrorism, as opposed to the majority who may share their ideologies, but are unlikely to turn to violence themselves [4]. Thus, there is an urgent need to develop automated tools for processing, understanding and synthesizing the vast amount of data generated by DVEs online in order to find those who appear to be most likely to perpetuate or incite violence.

In November 2019, SQL Leaks published the entire contents of the Iron March forum. This online forum has been linked to 100 hate crimes[1] and is widely believed to be one of the first online channels used by the Atomwaffen group[2]. After obtaining access to these data, we undertook a project to understand the connections between Iron March participants, to identify the users who were most central to the online

[1] https://www.splcenter.org/hatewatch/2019/02/15/visions-chaos-weighing-violent-legacy-iron-march.

[2] https://www.bellingcat.com/resources/how-tos/2019/11/06/massive-white-supremacist-message-board-leak-how-to-access-and-interpret-the-data/.

D. Russo et al. (Eds.): IHSI 2021, AISC 1322, pp. 350–356, 2021.
https://doi.org/10.1007/978-3-030-68017-6_52

community. We were specifically interested in recruitment activities, and further investigation revealed what appeared to be a semi-structured process for indoctrinating new members into DVE organizations.

The entire SQL database of the Iron March forum was uploaded to the Internet Archives webpage on November 6th, 2019. It has since been taken down, but is available on various torrent sites. Singer-Emery and Bray III were among the first to discuss the contents of the forum's SQL database [5]. Notably, the data dump contained both public and private posts along with registration information for each individual user. The forum was active from Sept 2011 to November 2017, with a large increase of activity immediately before and after the US presidential election in November 2016 (see Fig. 1). After November 2017, the online community migrated to a different site (possibly on the Discord platform).

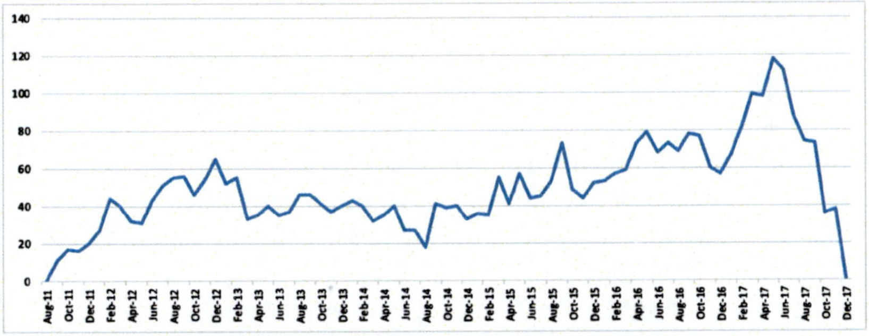

Fig. 1. Iron March, active users by month

In total, there were 1,207 users registered to the forum, 875 of these users posted at least once, and 659 users had a conversation with at least one other user. Notably, 20% of the active population accounts for 75.9% of all conversations within the forum. On one extreme we have one user 'Odin' who is by far the most connected individual with 61 separate conversations, and on the other extreme, there are 216 users who messaged at least once but never received a reply. In this article, we define a conversation as two or more users communicating on the same thread within the public forum.

Although the Iron March forum received notoriety for its high concentration of White Supremacist groups, fewer than half of the users were based in the United States as determined by their IP address at the time they made their first post. Although it is possible that some users employed IP spoofing, there is still a broad range of locations and nationalities represented in the collection, as shown in Table 1.

2 Identifying the Most Connected Users

We used social network analysis in order to identify subgroups of connected users within the Iron March forum site. This analysis leads to our conclusion that participation with the site, as determined by number and length ongoing conversations, was the key to identifying the leaders within the online community.

Social network analysis led to the development of a graphical depiction of the engagement of the most connected users within the forum, and their relationship to the rest of the community. The Pareto Principle [6], where 80% of the effects are created by 20% of the causes, is commonly referenced when analyzing online communications. Effects in this case could be taken to mean connections within the Iron March Forum, whereas the causes are the users on the forum. Keeping this rule in mind we notice that 93.4% of the connections involve only 21% of the active user population (n = 875). In Fig. 2a, the 184 people in the center ring of the donut are each connected to at least four other active users), thus there is a trend towards few users with significant influence over the majority of communication on the forum, a statistic that is consistent with other social networking sites. In 2b, we see that just 9 people conversed with 21.2% of the active user population (179 users) - an incredible amount of influence - clearly showing the central hub of leadership within the community.

Table 1. Active users by country, iron March forum

Country	Active users	Country	Active users
United States	419	Denmark	11
Great Brittan	88	Finland	11
Canada	62	Russia	10
Sweden	49	Norway	10
Denmark	19	Brussels	10
Netherlands	19	Belgium	9
Ireland	17	Hungary	8
Portugal	14	Croatia	6
France	12	Other	82
Romania	11	**Total**	**875**

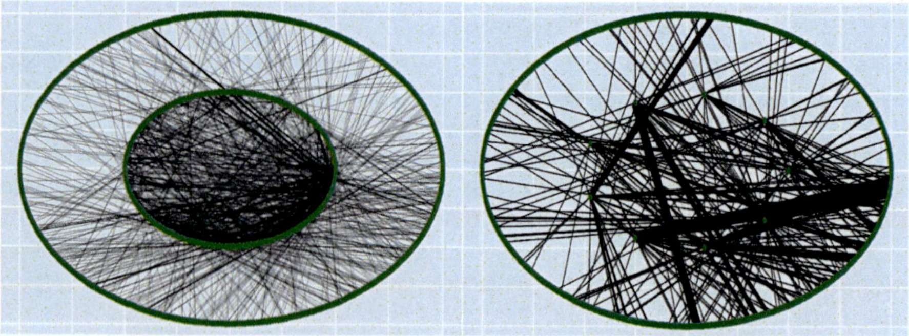

Fig. 2. Analysis on the most active users in the iron March forum

We expand upon the analysis in Fig. 2 to determine how groups of users interacted on the Iron March forum. Figure 3 shows the results of our cluster analysis as the level of engagement threshold increases. The nodes represent a single user, and distance between nodes represents the number of shared threads between two users. The post threshold specified below each network graph denotes the level of engagement, ranging from users who posted twice on the same thread (threshold of 2), to users who posted 8 times on the same thread. With a threshold of two, the network graph depicts a single hub of activity and interaction, but as the threshold increases, clusters begin to emerge. When the threshold reaches a threshold 6 or 7, clusters are fairly clear. As the threshold continues to increase, these clusters eventually break down into scattered communications. This social networking analysis and clustering experiments show promise for identifying the most prominent and active members of an online community and how they are interacting with each other. Taken together, Figs. 2 and 3 appear to depict a shared responsibility among the most active users, with each interacting primarily with a subset of the community. This level of organization and leadership has not previously been described in the Iron March dataset.

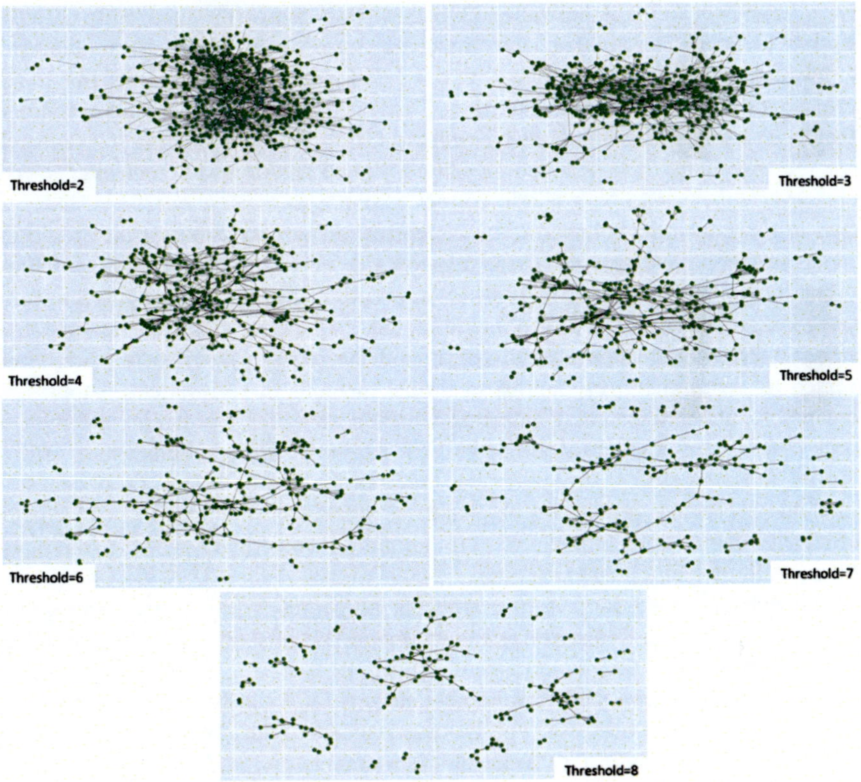

Fig. 3. Network cluster analysis of iron March forum at increasing thresholds

3 Recruitment Strategies Within the Iron March Forum

Both informal and formal recruiting strategies were observed within the Iron March Forum. Formal recruiting was only observed from users that were a part of an established White Supremacist group on the forum. Notably, users with a formal process were in the minority on the forum and they never exclusively used formal recruiting. Most users defaulted to an informal form of recruitment. Many of the activities were reminiscent of reframing strategies that are used by online sexual predators to lure their victims into believing the communication is normal. These strategies indoctrinate the victim into the use of the "correct" language, as determined by the predator [7].

Formal recruitment took the form of administering a test or questionnaire to the perspective recruit. One user, "Odin," was the most prolific formal recruiter on the forum. He was observed administering an ideology test to new members. This test was comprised primarily of questions about works of literature that Odin considered central to his ideological foundation (including SEIGE – the collected works of James Mason, who is an advisor to the Atomwaffen Division). Odin repeatedly referred to SIEGE as "required reading." Potential new members were open about their interest in Atomwaffen, with quotes including "I live in Jacksonville, Fl, and I would like to learn more about joining Atomwaffen Division. Please advise." and "I'm trying to figure out if there's an Atomwaffen branch for (Northern) California … wondering if Atomwaffen has a presence here. … Is there a procedure to found an Atomwaffen chapter?" While there never appeared to be a single correct response to the test questions, Odin was happy to share what he believed to be the correct answer with the recruit should they provide an incomplete answer. These responses reflected the reframing noted above.

The most popular form of recruitment across the entire forum was more informal. During this process a recruiter would casually ask leading questions to assess the user's ideology. This would often take place without the recruit knowing they were being assessed for recruitment. In contrast to formal methods of recruitment, it took fewer messages for a recruiter using informal recruitment to feel confident that they had found a new recruit. Consistent across the majority of recruiters using informal methods, in the final step of recruitment, they ask the recruit to join a separate Skype group chat or Discord group. We believe these strategies were used to ensure that the participants were serious in their intent to get more involved, and were not attempts at infiltration by law enforcement groups.

Although the Iron March forum attracted numerous White Supremacist groups, consistent among them was an interest in taking the conversation off of Iron March and onto a platform where their communications felt more secure. Initially Skype was the most referenced platform, and recruiters would regularly tell their potential recruits to make an account. The use of Skype is understandable because it provides recruiters with the ability to match a face to an online username giving them a greater level of trust in the recruit's intent. By 2017, recruiters were directing potential new members to Discord to continue the conversation.

A significant number of American users were quick to disclose their own location, and also suggest that other users travel to their location for an in-person meeting. The word "travel" appeared in 50 separate communications and "meeting" appeared in 65.

The fact that the American users were more likely to disclose their hometown suggests that they are very comfortable with the notion that the ideas they are publishing to the forum fall within their first amendment right to freedom of speech. This also lends support to our theory that the groups were not satisfied with merely talking online about ideology, but were planning offline activities and training exercises among the most engaged and receptive recruits.

Although the forum contains primarily English text, Table 1 shows that a majority of the users on the forum appear to be in other countries. Users from outside the United States engaged in recruitment activities at the same rate as those located in the US. However, users within the United States were more likely to disclose their hometown or state in an effort organize face to face meetings with other members of the forum.

4 Conclusion

The leaked SQL database of the Iron March Forum continues to provide researchers with opportunities for analyzing the activities of DVE organizations within the United States. This research has demonstrated the nature of the interactions between users on the Iron March forum, and identified the most active and central forum participants. While the forum was the medium upon which many members were first introduced to the online DVE community, a large number of users finished the recruitment process on another platform. Future research will expand these techniques to other online communities that appear to cater to these organizations, such as Discord (some channels), gab, and Telegram, among others. There is an online arms race with DVE groups, with new communities emerging on one platform almost as quickly as they are shut-down on another. Use of automated tools for quickly analyzing these activities is necessary to ensure that both private and public entities can quickly respond to emerging threats.

Acknowledgments. The views expressed are those of the authors and do not reflect the official policy or position of the United States Naval Academy, United States Navy, United States Marine Corps, the Department of Defense or the United States Government. This work was funded in part by the Office of Naval Research.

References

1. Badawy, A., Addawood, A., Lerman, K., Ferrara, E.: Characterizing the 2016 Russian IRA influence campaign. Soc. Netw. Anal. Min. **9**(1), 31 (2019)
2. Denton, A.: Fake news: the legality of the Russian 2016 Facebook influence campaign. In: BU Int'l LJ, vol. 37, p. 183 (2019)
3. Alvarez, G., Choi, J., Strover, S.: Good news, bad news: a sentiment analysis of the 2016 election Russian facebook ads. Int. J. Commun. **14**, 27 (2020)
4. Shortland, N., Forest, J.J.: Tracking terrorism: the role of technology in risk assessment and monitoring of terrorist offenders. In: Science Informed Policing, pp. 57–76. Springer, Cham (2020)

5. Singer-Emery, J., Bray III, R.: The Iron March Data Dump Provides a Window Into How White Supremacists Communicate and Recruit (2020). https://www.lawfareblog.com/iron-march-data-dump-provides-window-how-white-supremacists-communicate-and-recruit
6. Pareto, V., Page, A.N.: Translation of Manuale di economia politica ("Manual of political economy"), A.M. Kelley, ISBN 978-0-678-00881-2 (1971)
7. Kontostathis, A., Edwards, L., Leatherman, A.: ChatCoder: toward the tracking and categorization of internet predators. In: Proceedings of Text Mining Workshop 2009 held in conjunction with the Ninth SIAM International Conference on Data Mining (SDM 2009). Sparks, NV. (2009)

Evaluation of Domestic Market Development in Ukraine

Nazariy Popadynets[1(✉)], Inna Irtyshcheva[2], Lyudmila Shymanovska-Dianych[3], Olesia Diugowanets[4], Iryna Hryhoruk[5], Iryna Kramarenko[6], Tetiana Husakovska[3], Yevheniya Boiko[2], Nataliya Hryshyna[2], Olena Ishchenko[2], Nataliya Tubaltseva[2], and Dariya Archybisova[2]

[1] SI "Institute of Regional Research named after M. I. Dolishniy of the NAS of Ukraine", Kozelnytska Str. 4, Lviv 79026, Ukraine
popadynets.n@gmail.com

[2] Admiral Makarov National University of Shipbuilding, Heroyiv Ukrayiny Ave., 9, Mykolayiv 54025, Ukraine
innauamd@gmail.com

[3] Poltava University of Economics and Trade, Koval Str. 3, Poltava 36014, Ukraine
ludad5846@gmail.com

[4] State University, «Uzhhorod National University», Pidhirna Str, 46, 88000 Uzhhorod, Ukraine
olesia.diugovanets@uzhnu.edu.ua

[5] Vasyl Stefanyk Precarpathian National University, Shevchenko, Str. 57, Ivano-Frankivsk 76018, Ukraine
ira.hryhoruk@gmail.com

[6] Mykolaiv Interregional Institute of Human Development of the University "Ukraine", Viyskova Str. 22, Mykolayiv 54000, Ukraine
ukraina_09@ukr.net

Abstract. The current state of the domestic consumer goods market in Ukraine and its main development trends indicate a rather serious threat to the economic and food security of Ukraine. The paper aims to evaluate the development of the domestic market in Ukraine and define the factors of its regulation impact. Methodologically, the paper is based on the use of systemic, analytical, statistical, dialectical, and mathematical methods. The dynamic processes of domestic market development in Ukraine are determined by the calculation of the integral parameter based on the Index provided by the Global Competitiveness Report. The capacity of domestic producers of food is analyzed, and the place of Ukraine in supplying these goods to the EU Member States is determined.

Keywords: Food products · Domestic market · Regulatory mechanisms · Competitiveness agriculture

D. Russo et al. (Eds.): IHSI 2021, AISC 1322, pp. 357–363, 2021.
https://doi.org/10.1007/978-3-030-68017-6_53

1 Introduction

With the deepening crisis in the world and the period of the formation of new globalization processes, new requirements for the development of the national economy arise. Changes in the socio-economic development of the state were ripe, and crisis political processes and deep social stratification hampered the development of Ukraine. The future of Ukraine decisively depends on the all-round development of the national economy and especially its center – the internal market; it is the most important factor in the further economic growth of the state. Some aspects of Ukraine's internal market development and structural policy were researched by local scientists as [1–7] and others. However, it should be noted that today, more than ever, the domestic market needs government regulation in order to create a competitive environment by using modern mechanisms. Therefore, in the process of improving the mechanisms of state regulation, it is proposed to assess the place of our country in accordance with the "Global Competitiveness Report" and calculate the integral value of this indicator by market components (internal competition, trade openness, capacity, etc.).

2 Methodology

It is proposed to evaluate the development of the domestic market in Ukraine by calculating the integral indicator based on the results of the Global Competitiveness Report index, which is determined by the formula:

$$IC = \sum R_i p_i \tag{1}$$

where

Ri – this is the dynamics of the rating value compared to the previous year ("1" - increase, "0" – decrease or no change);
pi – share of rating.

3 Results and Discussion

According to the World Economic Forum, in 2019, in accordance with the value of the "Global Competitiveness Report", Ukraine ranked 85th and worsened its position by 2 steps (83rd place) relative to 2018. The main components of the index are arranged as follows:

- Institutions – 104th place;
- Infrastructure – 57th place;
- ICT solutions – 78th place;
- Macroeconomic stability – 133rd place;
- Health – 101st place;
- Skills – 44th place;

- Food market – 57th place;
- Labour market – 59th place;
- Financial system – 136th place;
- Market size – 47th place;
- Business dynamism – 85th place;
- Innovative ability – 60th place [8]

Based on the proposed method for evaluating the integral level of the domestic market development in Ukraine according to the components of the Global Competitiveness Report for the period 2013-2019 (Table 1) it is determined that during this period there is a strengthening of positions to increase internal competition and open trade in the domestic market. However, the market capacity indicator during this period significantly changed its position in the international ranking. The calculation of the integral value of the Global Competitiveness Report indicator showed a high result in 2019, that is, for all components of the domestic market there was an increase in the ranking relative to 2018, which is positive for open international competition.

Table 1. Evaluation of the integral level of the domestic market development in Ukraine according to the Global Competitiveness Report components (2013–2019)

	2013–2014	2014–2015	2015–2016	2016–2017	2017–2018	2019	Change of position 2019 from 2013
Internal competition (40%–0,40)							
Commodity market	118	108	107	108	101	57	↑61
Distortive effect of taxes and subsidies on competition	133	125	126	133	129	104	↑29
The scale of market monopolization	140	104	101	99	124	89	↑51
Competition in the services market	127	125	122	108	78	77	↑50
Open trade (40%–0,40)							
Prevalence of non-tariff barriers	118	106	116	113	120	91	↑27
Trade tariffs	41	43	43	41	44	43	↓2
Complexity of tariffs	127	123	113	69	75	66	↑61
The efficiency of the registration goods process	140	125	127	128	126	90	↑50
Market capacity (20%–0,20)							
Market capacity	45	51	53	50	61	47	↓2
Gross Domestic Product	38	41	47	48	74	48	↓10
Imports	51	60	59	51	47	44	↑7
Integral value of the indicator (IC)	X	0,70	0,60	0,60	0,50	1,00	X

Source: compiled and calculated by the author

Thus, the data of the Global Competitiveness Report index by market components characterize the strengthening of positions to increase internal competition and open trade in it. However, the market capacity indicator for the period under research has significantly changed its position in the international ranking. The calculation of the integral value of the Global Competitiveness Report indicator showed a high result in 2019, that is, for all components of the domestic market there was an increase in the ranking relative to 2018, which is positive for open international competition.

Evaluation of the integral level of the domestic market development in Ukraine showed that it is the factors of internal influence that form negative preconditions for strengthening the potential of trade. Therefore, in our opinion, it is advisable to analyze and group the factors influencing the regulation of the domestic market development (Table 2).

Table 2. Factors influencing the internal market regulation

Components	Content of factors	Deterioration (↓), Improvement (↑)
Financial and economic	Decrease in the volume of retail turnover in the regions and a slowdown in its growth in the regions	↓
	Decrease in the volume of imports of goods and services to the domestic market of Ukraine	↓
	General deterioration of the investment climate, decrease in the volume of capital investments in the regions	↓
	Deterioration of financial results of economic activity of enterprises	↓
	Decrease in innovation activity and supply in the domestic market of R&D results and innovative products	↓
	Activation of negative trends associated with the shadowing of the domestic market	↑
	Improving the infrastructure of the domestic market	↑
Institutional and legal	Military conflict in Eastern Ukraine, annexation of the Autonomous Republic of Crimea	↓
	Unregulated legal and regulatory framework for domestic market regulation in relation to the system of international agreements	↓
Social	Providing information and consulting support to the population and business entities	↑
	Improving the socio-psychological and organizational professional environment of labour and entrepreneurial activity in the domestic market	↑
	Increasing the level of material well-being of the population and creating additional jobs	↑
	Makes a significant contribution to the economic development of the territory and the country	↑
Ecological	Use of international ISO quality standards	↑
	Formation of the organic products market	↓

Source: grouped by author

An important step that encourages the development of mechanisms for regulating the domestic market development is its export or import orientation. So, according to the data of the Ministry of Economic Development, Trade and Agriculture of Ukraine at the end of 2019, Ukraine took 1st place in Europe and 2nd place in the world (out of 123 countries) in terms of the volume of exported organic products to the EU, having risen two steps compared to the previous year [9].

The EU widely spreads its imports of organic agri-food products, with a total of 115 countries registering exports to the EU. In terms of volume, China is the largest supplier of organic agri-food products in the EU with 415,243 tons of products, or 12.7% of the total (Table 3). Ecuador, Dominican Republic, Ukraine and Turkey each have 8% market share. The consumer demand of European consumers for Ukrainian organic products testifies to their high quality and competitiveness. The leading position of Ukraine in the world trade in organic products demonstrates the active development of this industry; it is a priority in the formation of a strategy for the development of the domestic consumer goods market.

Table 3. The number of organic agri-food products imports in the EU (by country of origin)

Country	Tons	%
China	41524312,72	12,7
Ecuador	2784758,53	8,5
Dominican Republic	2745998,44	8,4
Ukraine	2667418,25	8,2
Turkey	2642188,16	8,1
Peru	2072746,47	6,4
Third countries	1707535,28	5,2
UAE	127807	3,9
India	125807	3,9
Brazil	72353	2,2
Others	1055262	32,4
Total	3258532	3,9

Source: calculated based on the [10].

During 2019, 3,240,000 tons of organic agri-food products were imported to the EU, more than 10% of which came from Ukraine. Ukrainian exports to the EU increased by 27% – from 265,817 tons in 2018 to 337,856 tons in 2019. Basically, grain crops (except for rice and wheat, 76.9% of cereals of Ukrainian origin), wheat (31.8% from Ukraine), oilseeds (except for soybeans, 18.2% and second place after Turkey) are exported to the EU from Ukraine, soybeans (4th place and 13% of soybean imports from Ukraine), fruits (11% and 3rd place). Also, Ukraine is included in the group of the largest exporters of cake, fruit juices and vegetables, which are key export products [9].

In 2018, Ukraine ranked fourth in terms of organic exports to the EU, behind China, Ecuador and the Dominican Republic. It should be noted that the Dominican Republic and Ecuador supply almost exclusively tropical products to the EU: 90% of exports are tropical fruits, fresh or dried, nuts and spices. In general, imported organic products account for about 2% of the total imports of agri-food products in the EU.

The European Commission has unveiled the EU's "From Farm to Fork" strategy, which envisages an increase in land for organic production to 25% of agricultural land by 2030. Now this figure is about 8% [9].

Ukraine is the first EU supplier of organic products on the European continent. 70% of total imports consists mainly of cereals, broken down into "Cereals other than wheat and rice" (42.8%) and "Wheat" (28.5%). Oilseeds account for more than 15% of the total, broken down to "Oilseeds other than soybeans" (10.8%) and soybeans (5%).

4 Conclusions

The main aim of state regulation of the domestic consumer goods market is to ensure the growth of the population's well-being by creating conditions for meeting the needs in the consumer market and stimulating the development of national producers.

The products that Ukrainian manufacturers provide to the foreign market is a factor that ensures an increase in the standard of the population living, in addition, it is a promising source of additional funds from the export of products. Therefore, in our opinion, the government of the country needs to focus all efforts for the proper level of regulation processes of the market development for final products, developing strategic documents and providing regulatory, information and consulting, financial and credit, investment support for the food sector.

The globalization of the economy and a fairly high level of integration of the Ukrainian economy into the world economic system creates significant challenges for the state as regulators of the national economy. State regulation of the internal market in these conditions should contribute to the development and stimulation of the competitive foundations of the economy, provides for the priority solution of the following tasks:

- increasing the level of competitiveness of domestic products;
- improving the institutional and organizational and economic principles of the internal market functioning;
- definition and consolidation of the principles of natural monopolies functioning;
- creation of effective incentives to encourage the development of competition in monopolized sectors of the economy;
- building an optimal system of price regulation, which would ensure a combination of the interests of producers, consumers and society as a whole;
- formation of an effective system for protecting consumers' rights to receive goods and services of proper quality in sufficient volume at economically reasonable prices.

References

1. Vasyltsiv, T., et al.: Economy's innovative technological competitiveness: decomposition, methodics of analysis and priorities of public policy. Manage. Sci. Lett. **10**(13), 3173–3182 (2020)
2. Irtyshcheva, I., et al.: Building favorable investment climate for economic development. Accounting **6**(5), 773–780 (2020)
3. Panukhnyk, O., Popadynets, N., Fedotova, Y.: Analysis and modeling of factor determinants of food provision at consumer market of Ukraine. Global. J. Environ. Sci. Manage. **5**(SI), 215–226 (2019)
4. Melnyk, M., Korcelli-Olejniczak, E., Chorna, N., Popadynets, N.: Development of regional IT clusters in Ukraine: institutional and investment dimensions. Econ. Ann. XXI **173**(9–10), 19–25 (2018)
5. Popadynets, N., Shults, S., Barna, M.: Differences in consumer buying behavior in consumer markets of the EU member states and Ukraine. Econ. Ann. XXI **166**(1–2), 26–30 (2017)
6. Kramarenko, I., et al.: The model of economic system management for the Black Sea region of Ukraine in the sustainable development context. Accounting **6**(4), 387–394 (2020)
7. Kolomiyets, I., Popadynets, N.: Trade at Ukrainian internal market: the development mechanism. Actual Probl. Econ. **1**(175), 69–75 (2016)
8. The Global Competitiveness Report 2019 (2019). http://www3.weforum.org/docs/WEF_TheGlobalCompetitivenessReport2019.pdf
9. Ministry of Economic Development, Trade and Agriculture of Ukraine (2019). https://www.me.gov.ua/News/Detail?lang=uk-UA&id=28478e1e-eaa9-4143-808a-f258bd33dab9&title=UkrainaPosila1-MistseVvropiZaObsiagamiImportovanoiOrganichnoiProduktsiiTa2-geMistseUSviti
10. Organic imports in the EUA first analysis-Year 2018 (2019). https://ec.europa.eu/info/sites/info/files/food-farming-fisheries/farming/documents/market-brief-organic-imports-mar2019_en.pdf

Cyber Safety in the Digital Educational Environment: External and Internal Risks

Oleksandr Burov[1], Yuliya Krylova-Grek[2(✉)], Evgeniy Lavrov[3], Olena Orliyk[4], Svitlana Lytvynova[1], and Olga Pinchuk[1]

[1] Institute of Information Technologies and Learning Tools, 9 Berlins'koho Str., Kiev 04060, Ukraine
burov.alexander@gmail.com, s.h.lytvynova@gmail.com, opinchuk100@gmail.com

[2] G.S. Kostiuk Institute of Psychologyof the National Academy of Educational Sciences, 2, Pankivska Str., Kiev 01033, Ukraine
yulgrek@gmail.com

[3] Sumy State University, Sumy, Ukraine
prof_lavrov@hotmail.com

[4] Scientific-Research Institute of Intellectual Property, Kiev, Ukraine
e.orliyk@ndiiv.org.ua

Abstract. The paper describes a novel project aimed at developing the concept and technique of cyber safety system of the education process' participants in the digital learning environment (integration of human and cyber-physical systems approach). A striking feature of the proposed approach lies in its notable difference from the one and focus on a human in the network's cybersecurity loop rather than technical aspects or human personal data. Besides, it discusses the method and way of conducting experimental studies.

Keywords: Cybersecurity · Cognitive performance · Education · Social engineering

1 Introduction

Last year, transformation in teaching and learning became an especially shared agenda among educators all over the world due to the forced and unprepared transition of education to mass digital learning with the use of network technologies [1]. It was accompanying by the unwillingness of both the education system and the state as a whole to protect subjects from information, psychological and cognitive interventions that may affect the formation of personality under such circumstances at both the substantive and personal levels [2]. New challenges of time and new directions of society development – Society 4.0, Education 4.0, penetration of the newest technologies in all spheres of life, "hybrid" war – demand to understand key problems, challenges, and questions related to the educational process security in digital learning environment (DLE), in particular, security of all direct participants, education organizers, the state as well as the security of digital educational content [3], expansion of cognitive war, whose main purpose is to alter and distort the cognitive model of life,

D. Russo et al. (Eds.): IHSI 2021, AISC 1322, pp. 364–370, 2021.
https://doi.org/10.1007/978-3-030-68017-6_54

especially of young people [4]. Accordingly, there is an urgent need to protect the cognitive, ideological, intellectual, and developmental activities of education and human capital since a human is still weakest link in the System [5].

There are new problems caused by life and activities in the securities, related factors, and ways to avoid them, as well as new tools and mechanisms. Thus, the problem of development and implementation of information and communication technologies (ICT) in education needs to be addressed [6]. However, it should be borne in mind that new information technologies lead to fundamental and global processes that transform social development. Yet, in addition to the positive impact they have, new factors and conditions naturally cause serious problems, threats and risks [7]. As noted in the materials of the World Economic Forum in Davos (2018-2020) and the United Nations Organization [8], the problem of cybersecurity (CS), which affects almost all spheres of human life and activity, is especially acute because of many reasons, but of all, in the context of full education informatization [9]. It means that a human becomes a crucial unit of the state's vital infrastructure, but it is advisable not to consider the issues of cybersecurity (of the System), but cyber safety (of a human as an element of the System).

Purpose. Development of the concept and technique of the cyber safety system of the education process' participants in the digital learning environment (integration of human and cyber-physical systems approach).

2 Methodology

Considering learning as a type of activity in human-system integration [10], present-day learner may be viewed as an operator-researcher who acts in the digital learning environment (DLE). Successful learning involves mutual adaptation between a human as a participant of the educational process (PEP) and activity tools using individual cognitive abilities in networks including social ones [11] and in changing digital environment in general [12]. On the other hand, it is possible to use ergonomics' methods and techniques to assess a learner's safety in the education process [13]. Cybersecurity issues have become a cornerstone since computer technology ceased to be the prerogative of major science centers. The advent and spread of local and global networks changed how cybersecurity, relevant trends, problems, and challenges are viewed and understood.

Networks as Active Agents of the Education Process. Over time, digital networks are becoming a focal point of our lives, and social media is turning into a new social environment. These networks pose a real threat to education and the state's security. The network components in a simplified manner can be represented as a node, interface, connection, and network [3]. The nodes are network "agents": *people* (creators of the resource and its content, resource administrators, regular or random users), *technics* (terminal stations, computers, gadgets, etc.), and *information* (databases, knowledge bases, control systems, etc.) means. Depending on their essence, the agents have their own interface and communication channels to interact with other agents that can be a target for a cyber-attack. The network acquires the features of an independent factor that affects its properties, operation, and users, as well as the system in general. DLE is a triad-like cyberspace: (1) *information* in its digital representation: static (files recorded on media) and dynamic (packets, streams, commands, etc.); (2) *technical infrastructure* (ICT, software, databases etc.); (3) *information interaction* between subjects ("agents") via transmitted information and processing (2).

Cyberspace-Produced Threats to Participants of the Educational Process. Having in mind that today's students were born in the digital age, it can be argued that cyberspace is and will remain an extremely important part of the ideology and civilization battlefield. The range of threats from open cyberspace is constantly expanding. If a decade ago the threats for schoolchildren could be reduced to a relatively small number of groups (virus attacks, cybercrime, the dangers of Internet surfing) [14], these days the variety of hazards and threats has significantly increased and continues to grow, affecting all possible human actions online [15]. However, there has been little discussion on PEP's cognitive abilities, since most studies tend to focus on cybersecurity organizational issues [2].

It was found that the greatest threats to students are hidden active threats that can be assessed as a hierarchical set of indicators from ergonomics' standpoint: integrated (complex); three group indicators – hazard level caused by viral attacks, cybercrime and internet surfing; a set of individual indicators containing a set of certain threats [3].

Cyber Security Areas. Education is not always recognized as a critical area. However, today's students can work in areas like that in the shortest possible time mainly due the implementation of new education technologies [16] in general, including distributed [17] and adaptive ones [18], as well as augmented and virtual environment [19]. Therefore, they need protection and appropriate training. Furthermore, a neglected issue in this area is procedures of defining common possible cybersecurity target groups (for example, pupils/students, teachers, children/youth, education managers, total population of the country). Depending on the means of action, the problems (and corresponding means) of cybersecurity can be classified into five groups: legal, technical, informational, organizational, and psychological [3]. Simulation of the influence of cyber threats can be effective, if it provides the objective measurement of the individuals' reaction to this influence, i.e. psychophysiological response. We initiated the given study to model and simulate the impact of diverse factors (informational, social, psychological, cognitive, etc.) on an individual with the assessment of his state before and after. The ultimate practical goal of the study is to develop methodology to assess the impacting factors and risk-metrics, considering the individual and group features of potential objects of violence action.

3 Results and Discussion

The provisions of the project are based on the results of the authors' participation in the NATO Expert Group HFM-259 "Human Systems Integration Approach to Cyber Security" [20], previous experimental studies focused on analyzing a human cognitive work in digital space under influence of internal and external factors [21]. These results completely correspond to the present trends in cybersecurity evoked by the pandemic.

The results revealed the rapid increase in malware dissemination, ransomware/extortionware accompanied by emergence of the ransomware as a service, threat hunting as a response to malicious activity intensification in the networks, network detection and response. It was proven that attackers are constantly improving their abilities to deceive people being the biggest threat plaguing organizations in 2020 including educational institutions [22].

The focus should be shifted from personal data to PEP's state and his/her ability to resist an attack. Appropriate skills should include cybersecurity awareness, training, and cyber hygiene. The validation of results is based on:

- using metrics to measure the success of the training programs and identify the probable;
- metrics that include both short-term and long-term goals since they help to measure how participants' security posture improves over time.

Alongside evaluating the subjective response of a cyber-influenced individual, the study assesses his psychological and physiological changes including the ones caused by the latent effect on consciousness. To achieve this goal, it is planned to study the pool of volunteers and changes in their electrocardiogram and electroencephalogram due to simulated effects of both traditional gadgets (conventional devices for entering cyberspace) and individual augmented/virtual reality devices. Experimental studies imply both online and in-house observations by using the records of electrophysiological parameters, as well as virtual reality glasses and tablets. The project requires involvement of experts in various fields with strong expertise in conducting experimental research using network technologies, data analysis, as well as designing new scientific and methodological tools.

We have succeed in the following. First, we have designed a prototype (online testing software) of the cybersecurity system of PEP in DLE. Second, we have revised methods of assessing the cyber hazard impact on PEP and their risk-metrics based on indicators of psychological and physiological changes caused by cyber hazards. Then, we have developed recommendations on PEP's cybersecurity taking into account the human factor in DLE and individual/group characteristics of psychological/psychophysiological response of high school students to the cyber hazard impact. Finally, of particular interest is our proposals on the structure and scope of educational materials aimed at the formation of a general culture of cybersecurity in DLE as a component of PEP's cybersecurity system.

Our study has many effective and valuable applications in educational environment (Fig. 1).

This study is a first step towards enhancing our understanding of cyber safety in DLE.

To further our research we plan to:

- Provide methodological support for considering and mitigating the effects of hazardous factors on PEP in the digital learning environment.
- Develop a set of recommendations on cybersecurity for PEP in the digital learning environment taking into account the human-related factors.
- Explore psycholinguistic aspects of cybersecurity humanitarian component [23].

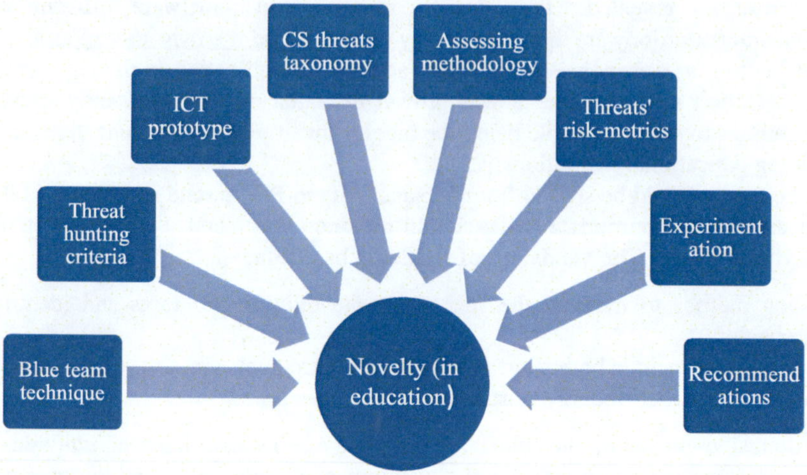

Fig. 1. Cyber safety tools for education

4 Conclusion

To sum up, our study underlined the importance of scrutinizing the present-day features of teaching and learning in contemporary digital environment. Our recommendations are a potential tool to improve the security and safety of educational process. The findings suggest the idea that the security and safety of learning is achieved by adapting students' activity, which greatly depend on his/her cognitive state in digital education, by designing intelligent individual-oriented systems and services that ameliorate human – E-technology interaction.

Acknowledgments. . This work is supported by the grant 0118U003160 "System of Computer Modeling of Cognitive Tasks for the Formation of Competencies of Students in Natural and Mathematical Subjects".

References

1. Memon, A.S., Rigole, A., Nakashian, T.V., Taulo, W.G., Chávez, C., Mizunoya, S.: COVID-19: How prepared are global education systems for future crises? Research Brief 2020-21. UNICEFInnocenti (2020). https://www.unicef-irc.org/publications/1138-covid-19-how-prepared-are-global-education-systems-for-future-crises.html
2. Coultier, R.: 10 K-12 cybersecurity must-dos. eSchoolMedia & eSchool News. Innovations in Educational Transformation (2020). https://www.eschoolnews.com/2020/06/22/10-k-12-cybersecurity-must-dos/
3. Bykov, VYu., Burov, OYu., Dementievska, N.P.: Cybersecurity in digital educational environment. Inf. Technol. Learn. Tools **70**(2), 313–331 (2019)
4. Pocheptsov G.: The War in Cognitive Space (2017). https://nesterdennez.blogspot.com/2017/08/global-permanent-war_39.html

5. Yan, Z., Robertson, T., Yan, R., Sung Yong Park, Bordoff, S., Chen, Q., Sprissler, E.: Finding the weakest links in the weakest link: how well do undergraduate students make cybersecurity judgment? Comput. Hum. Behav. **84**, 375–382 (2018)
6. Li, C., Lalani, F.: The COVID-19 pandemic has changed education forever. This is how (2020). https://www.weforum.org/agenda/2020/04/coronavirus-education-global-covid19-online-digital-learning/
7. Schools of the Future: Defining New Models of Education for the Fourth Industrial Revolution, 2020 WEF. http://www3.weforum.org/docs/WEF_Schools_of_the_Future_Report_2019.pdf
8. Pipikaite, A., and Davis, N.: Why cybersecurity matters more than ever during the coronavirus pandemic. WEF (2020). https://www.weforum.org/agenda/2020/03/coronavirus-pandemic-cybersecurity
9. Guterres, A.: The future of education is here. Launch of the policy brief: education during COVID-19 and beyond, United Nations. https://www.un.org/en/coronavirus/future-education-here. Accessed Apr 2020
10. Pinchuk, O., Burov, O., Lytvynova, S.: Learning as a systemic activity. In: Karwowski, W., Ahram, T., Nazir, S. (eds.) Advances in Human Factors in Training, Education, and Learning Sciences. AHFE 2019. Advances in Intelligent Systems and Computing. 2019, vol. 963, pp. 335–342. Springer, Cham (2019). https://doi.org/10.1007/978-3-030-20135-7_33
11. Lytvynova, S., Burov, O.: Methods, forms and safety of learning in corporate social networks. In: ICT in Education, Research and Industrial Applications. Integration, Harmonization and Knowledge Transfer. Proceedings of the 13th International Conference on ICT in Education, Research and Industrial Applications. Kyiv, Ukraine, 15–18 May, pp. 406–413 (2017). http://ceur-ws.org/Vol-1844/10000406.pdf
12. Pinchuk, O.P., et al.: Digital transformation of learning environment: aspect of cognitive activity of students. In: Proceedings of the 6th Workshop on Cloud Technologies in Education (CTE 2018), Kryvyi Rih, Ukraine, 21 December 2018. CEUR Workshop Proceedings, # 2433, pp. 90–101 (2019)
13. Ahram, T., Karwowski, W.: Advances in human factors in cybersecurity. In: Proceedings of the AHFE 2019 International Conference on Human Factors in Cybersecurity, 24–28 July 2019, Washington D.C., USA (2019)
14. Bandara, I., Ioras, F., Maher, K.: Cyber security concerns in e-learning education. In: Proceedings of ICERI2014 Conference, IATED, 0728-0734 (2014)
15. Newhouse, W., Keith, S., Scribner, B., Witte, G.: National Initiative for Cybersecurity Education (NICE) Cybersecurity Workforce Framework (2020). https://doi.org/10.6028/NIST.SP.800-181
16. Frezzo, D.: The role of technology in the education of the future. 2020 World Economic Forum (2020). https://www.weforum.org/agenda/2017/05/science-of-learning/
17. Lavrov, E., Pasko, N., Tolbatov, A., Tolbatov, V.: Cybersecurity of distributed information systems. the minimization of damage caused by errors of operators during group activity. In: Proceedings of 2nd International Conference on Advanced Information and Communication Technologies-2017 (AICT-2017), 2017, pp. 83–87 (2017)
18. Veltman, J.A., Jansen, C., Hockey, G.R.J., Gaillard, A.W.K., Burov, O.: Differentiation of mental effort measures: consequences for adaptive automation. Nato Sci. Ser. Sub Ser. I Life Behav. Sci. **355**, 249–259 (2003)
19. Iatsyshyn, A.V., et al.: Application of augmented reality technologies for preparation of specialists of new technological era. In: Augmented Reality in Education: Proceedings of the 2nd International Workshop (AREdu 2019), Kryvyi Rih, Ukraine, 22 March 2019, pp. 181–200 (2020). ISSN 1613-0073. http://ceur-ws.org/Vol-2547/paper14.pdf

20. Human Systems Integration Approach to Cyber Security. STO-TR-HFM-259. STO/NATO 2020, June 2020, 112 pp. (2020). ISBN 978-92-837-2272-4
21. Mulder, L.J.M., Van Roon, A., Veldman, H., Laumann, K., Burov, A., Quispel, L., Hoogeboom, P.J.: How to use cardiovascular state changes in adaptive automation. In: Hockey, G.R.J., Gaillard, A.W.K., Burov, O. (eds.) Operator Functional State. The Assessment and Prediction of Human Performance Degradation in Complex Tasks. NATO Science Series, pp. 260–272. IOS Press, Amsterdam (2004)
22. 2020-security-strategy-playbook. https://techtalksummits.com/2020-security-strategy-playbook/
23. Krylova-Grek, Y.: Psycholinguistic aspects of humanitarian component of cybersecurity. Psycholinguistics **26**(1), 199–215 (2019)

Autonomous Robot for Plastic Waste Classification

Janusz Bobulski(✉) and Mariusz Kubanek

Computer Science Department, Czestochowa University of Technology, 73 Dabrowskiego Street, 42-201 Czestochowa, Poland
{januszb,mkubanek}@icis.pcz.pl

Abstract. All countries in the world are now putting emphasis on the development of technologies related to waste management. Recycling of rubbish, including plastic waste, has become a particular problem. Unfortunately, it is only possible to recover raw materials from certain types of plastic and here there is a problem with the separation of different types of plastic that can be reused. In this paper, we present a technique using deep learning and convolutional neural networks. The main goal of the publication is to create a system of automatic segregation of plastic waste into seven main categories.

Keywords: deep learning · CNN · waste management · image processing · environment protection · artificial intelligence

1 Introduction

Statistical data on the costs of acquiring primary raw materials are much higher than the costs incurred to recover materials from utility waste. Acquiring and processing recyclable materials is also associated with lower energy consumption compared to the production of new [1]. Treatment of waste allows the replacement of some energy carriers such as biogas or thermal energy. Replacing primary raw materials with secondary raw materials reduces the cost of manufacturing products, eliminates the cost of transporting waste to landfills and maintaining these landfills [2].

Sorting waste is the first and necessary step to effective recycling. One of the methods of recovery of PET plastic and the production of textiles from it. International labeling of different types of plastics has been introduced to make them easier to distinguish:

01 - PET - polyethylene terephthalate,
02 - HDPE - high-density polyethylene,
03 - PVC - polyvinyl chloride,
04 - LDPE - low density polyethylene,
05 - PP - polypropylene,
06 - PS - polystyrene,
07 - other.

D. Russo et al. (Eds.): IHSI 2021, AISC 1322, pp. 371–376, 2021.
https://doi.org/10.1007/978-3-030-68017-6_55

The division into individual types of plastics allows for the reuse of some of them. Using modern technologies such as computer vision and deep learning, we propose a method that can be used in portable devices for recognizing waste, and that would be helpful in identifying plastic from urban waste. Such a device could be used at home and in waste sorting plants [3].

2 Review methods

The process of sorting materials suitable for reprocessing from the Ministry of the Interior is labor-intensive and complicated. First, dry and wet wastes are separated, and electromagnetic techniques are used to sort iron-containing materials. However, one of the visual [4, 5] methods can be used to segregate plastic waste. In optical sorting, cameras are used to identify different waste fractions based on visual properties, such as color, shape or texture. Huang et al. They proposed a sorting technique that combines a 3D color camera and a laser beam on a conveyor belt. This technique creates triangles above the camera image on the basic laser beam, which is why it is called triangulation scanning [6]. Another group of methods is spectral imaging, which is a combination of spectral reflection measurement technology and image processing technology. These types of methods use NIR (near infrared), VIS (visual image spectroscopy) and HSI (hyperspectral imaging) [7–9]. The hyperspectral camera acquires images in the narrow spectral bands, and another system analyzes spectroscopic data. Then the data is pre-processed and reduced using a special algorithm. A set of compressed air nozzles at the end of the conveyor belt pushes the waste into individual containers depending on the decision of the classifier [10, 11]. For spectroscopy based techniques, light is directed to plastic waste, and each type of plastic reflects a different range of waves. NIR and laser sensors capture the reflected spectrum and on this basis the material is classified. This type of technique was developed by Safavi et al. [12] for identifying PP material in mixed waste. For the classification of PP and PE materials, the HSI method using NIR - near infrared light (1000–1700 nm) can be used [13, 14]. Main component analysis (PCA) [15] is used to improve the performance of the classification algorithm. However, the alternative is a quick method of classifying plastics using a combination of MIR spectroscopy and independent component analysis (ICA) developed by Kassouf et al. in [16]. Unfortunately, the presented methods have several significant disadvantages: waste must be ground, which is a cost, and small particles are more difficult to classify. Therefore, a technique without these drawbacks should be developed.

3 Proposed system

A computer system dedicated to image processing can be used to identify the type of plastic from which the waste is made. The method we propose uses an RGB digital camera and a computer with software to classify plastic waste. The classifier in the form of a program controls the nozzles with air to direct the waste to a specific container (Fig. 1). The software in this system uses image processing techniques for image pre-processing. A key element is the classifier developed based convolution artificial neural networks and deep learning [16], which are used for object recognition.

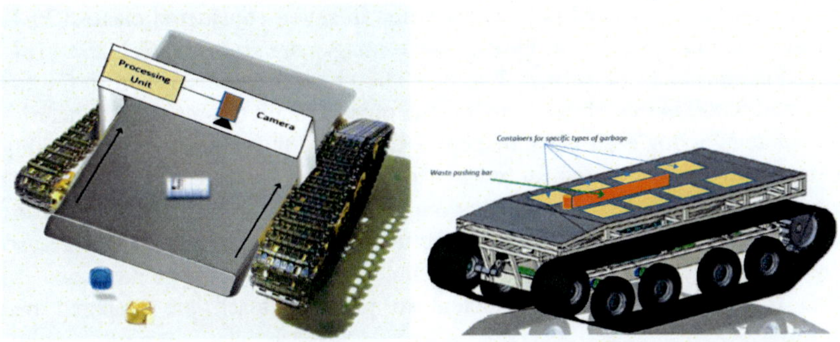

Fig. 1. The draft of the plastic sorting robot - main part and collecting trailer

3.1 Convolutional Neural Network

The Convolutional Neural Network (CNN) is an artificial neural network with feedback. The structure of neurons is created similarly to the structure of the mammalian visual cortex. The local pixel arrangement determines the shape of the object. CNN first recognizes smaller local patterns in the image and then combines them into more complex shapes. CNN networks can be an effective solution to the problem of sorting waste, because they are very effective in recognizing objects in the image [16].

3.2 CNN structure

When designing the structure of a neural network, first of all, the size of the input image should be considered as an important element. Too high resolution results in an increase in the number of calculations, which in turn can lead to overloading of the computational unit's memory. An additional goal was to develop such a structure that can be built into a Raspberry Pi type microcomputer. Too large size of processed images would be impossible for her to analyze in real time. In turn, the low resolution of the input images will make it difficult or impossible to recognize the object, and thus achieve the expected performance [17]. Determined to conduct research for images with a resolution of 60 x 120 pixels. Another important element was the selection of the number and types of layers of the CNN network.

We proposed author's CNN network consisting of 16 layers was developed for the research. The first convolution layer consisted of 64 convolution filters with dimensions 9 x 9. Three layers of convolution encode information, transferred to a two-layer fully connected layer. Table 1 presents structure of our network.

Proper preparation of input data for the learning, validation and testing phases is crucial in the process of developing a system for classifying objects in natural working conditions. For deep neural networks, collect as much data as possible for each class, a minimum of several thousand. In our case, it was necessary to obtain photos of already classified waste. We adopted a simplified model in which there is only one waste in the camera lens. This approach does not reflect natural working conditions, but for research needs provides sufficient opportunities to create a properly functioning network. All collected images represented objects classified in seven considered classes: PET, PE-HD. PVC, PE-LD, PP, PS and Other. These images are from the GaDaBa [18] database, and samples can be seen in Fig. 3. 10,000 photos were selected in the first six classes, and 5,000 in the 7th because it is the most difficult to obtain such materials. Of the set of images for each class (except 7-Other), 8000 were used for teaching and validation, and 2000 was used for testing. In the test group there were pictures of other objects than in the teaching group. Random images from the learning kit were used for validation. For group 7, 4000 images were used for teaching, and 1000 were used for testing. Images were obtained from the commercial GaDaBa image database. This set of data (65,000 images) was sufficient to properly teach the prepared network structures.

Table 1. Structure of our CNN

No	Name of layer	Parameters
1.	Image input layer .	60 x 120 x 3
2.	Convolution Layer.	64 filters, size 5 x 5
3.	Max Pooling Layer.	
4.	ReLU Layer	
5.	Cross Chanel Normalization Layer	
6.	ReLU Layer	
7.	Max Pooling Layer. \\	
8.	Convolution Layer	64 filters, size 5 x 5 \\
9.	ReLU Layer	
10.	Max Pooling Layer . \\	
11.	Convolution Layer.	64 filters, size 5 x 5\\
12.	Fully connected layer.	Inputs 4992, outputs 64\\
13.	ReLU Layer	
14.	Fully connected layer.	Inputs 64, outputs 7\\
15.	ReLU Layer	
16.	Classification layer.	7

4 Results

Analyzing the course of the experiment, it was found that for the proposed network and 60 x 120 pixel training images described earlier, it is necessary to learn for 20 epochs, with 1000 iterations in each epoch. The experiment uses the cross validation technique consisting of three data sets and on this basis will obtain an average result of accuracy 87%. This result is sufficient for the system to function properly under real conditions. Compared to similar studies [19], the effectiveness of the proposed network is lower, however, we recognize 7 classes as opposed to that which only recognizes 4.

5 Conclusion

The results of the study show that the 16-layer network we propose achieves good performance for images with a resolution of 60 x 120 pixels. The classification of segregated waste into seven main classes is in most cases at a satisfactory level. Further work will mainly involve expanding the segregated waste image database to include waste images in more realistic conditions. Hence the efforts to obtain waste records on a conveyor belt from waste sorting companies.

Our future research assumes the possibility of training the network during operation in real conditions for photos of waste taken from the conveyor belt during the segregation process.

We also plan more detailed research taking into account changes in hyper parameters of learning and various types of filters.

Acknowledgment.. This research was funded by the Polish Minister of Science and Higher Education under the name "Regional Initiative of Excellence" in the years 2019 - 2022 project number 020/RID/2018/19 the amount of financing 12,000,000 PLN.

References

1. Uropean Commission, Communication From The Commission To The European Parliament, The Council: The European Economic And Social Committee And The Committee Of The Regions, Closing The Loop - An Eu Action Plan For The Circular Economy, Brussels, 2.12.2015
2. Radziewicz J.: Problemy gospodarki odpadami w Polsce, https://rme.cbr.net.pl/index.php/archiwum-rme/13-nr-42/ekologia-i-srodowisko/12-problemy-gospodarki-odpadami-w-polsce, Last accessed 15 may 2019.
3. Gundupalli, S.P., Hait, S., Thakur, A.: A review on automated sorting of source-separated municipal solid waste for recycling. Waste Management **60**, 56–74 (2017)
4. Pita, F., Castilho, A.: Influence of shape and size of the particles on jigging separation of plastics mixture. Waste Management **48**, 89–94 (2016)
5. Huang, J., Pretz, T., Bian, Z.: Intelligent solid waste processing using optical sensor based sorting technology. 3rd International Congress on Image and Signal Processing (CISP), 4, 1657—1661 (2010)
6. Pieber, S., Meirhofer, M., Ragossnig, A., Brooks, L., Pomberger, R., Curtis, A.: Advanced Waste-Splitting by Sensor based Sorting on the Example of the MTPlant Oberlaa. Tagungsband zur 10, DepoTech Conference, 695—698b (2010)
7. Vegas, I., Broos, K., Nielsen, P., Lambertz, O., Lisbona, A.: Upgrading the quality of mixed recycled aggregates from construction and demolition waste by using near-infrared sorting technology. Constr. Build. Mater. **75**, 121–128 (2015)
8. Cpałka K.: Case Study: Interpretability of Fuzzy Systems Applied to Nonlinear Modelling and Control. In: Design of Interpretable Fuzzy Systems. Studies in Computational Intelligence, Springer, Cham (2017)
9. Picón, A., Ghita, O., Bereciartua, A., Echazarra, J., Whelan, P.F., Iriondo, P.M.: Real-time hyperspectral processing for automatic nonferrous material sorting. J. Electron. Imag. **21**(1), 013018–1 (2012)

10. Tatzer, P., Wolf, M., Panner, T.: Industrial application for inline material sorting using hyperspectral imaging in the NIR range. Real-Time Imag. **11**(2), 99–107 (2005)
11. Safavi, S.M., Masoumi, H., Mirian, S.S., Tabrizchi, M.: Sorting of polypropylene resins by color in MSW using visible reflectance spectroscopy. Waste Management, 30 (11)}, 2216—2222 (2010)
12. Serranti, S., Gargiulo, A., Bonifazi, G.: The utilization of hyperspectral imaging for impurities detection in secondary plastics. Open Waste Management Journal **3**, 56–70 (2010)
13. Serranti, S., Gargiulo, A., Bonifazi, G.: Characterization of post-consumer polyolefin wastes by hyperspectral imaging for quality control in recycling processes. Waste Management, 31 (11)}, 2217—2227 (2011)
14. Serranti, S., Gargiulo, A., Bonifazi, G.: Classification of polyolefins from building and construction waste using NIR hyperspectral imaging system. Resour. Conserv. Recycl. **61**, 52–58 (2012)
15. Kassouf, A., Maalouly, J., Rutledge, D.N., Chebib, H., Ducruet, V.: Rapid discrimination of plastic packaging materials using MIR spectroscopy coupled with independent components analysis (ICA). Waste Management **34**(11), 2131–2138 (2014)
16. Wang, M., Wang, Z., Li, J.Deep: Convolutional neural network applies to face recognition in small and medium databases. 4th International Conference on Systems and Informatics, ICSAI 2017 Volume 2018-January, 4 January 2018, 1368—1372 (2018)
17. Goodfellow I., Benigo Y., Courville A., Deep Learning}, The MIT Press (2016)
18. Plastic Garbage Images Databes - GaDaBa, http://garbage.eu5.org, Last accessed 14 July 2020.
19. Bobulski, J., Kubanek, M.: CNN use for plastic garbage classification method, 25th ACM SIGKDD Conference on Knowledge Discovery and Data Mining, August 04-08 2019, Anchorage, Alaska, USA, ACM, New York, NY, USA.

Ambient Assisted Living: Benefits of the Technology Development

Hugo Arias-Flores[1(✉)], Janio Jadán-Guerrero[1], Omar Cóndor-Herrera[1], and Carlos Ramos-Galarza[1,2]

[1] Centro de Investigación en Mecatrónica y Sistemas Interactivos MIST/Carrera de Administración/Carrera de Ingeniería en Ciencias de la Computación/Maestría Mención en Educación, Liderazgo e Innovación/Carrera de Psicología, Universidad Tecnológica Indoamérica, Av. Machala y Sabanilla, Quito, Ecuador
{hugoarias, janiojadan, omarcondor}@uti.edu.ec

[2] Facultad de Psicología, Pontificia Universidad Católica del Ecuador, Av. 12 de Octubre y Roca, Quito, Ecuador
caramos@puce.edu.ec

Abstract. Technological advances allow people living with disabilities to integrate into different activities. This article aims to analyze research related to assistive technologies and visual impairment. The most important published papers, in this field were identified. This study analyzes 297 documents, of which 26 documents are published by two Latin American countries. This research presents publications related to visual impairment and technology, in the future it is planned to carry out a thorough and more extensive investigation on the subject.

Keywords: Assistive technologies · Visual impairment · Bibliometric

1 Introduction

The constant technological developments have generated significant changes in people's day-to-day activities. These technologies, in assisted environments create more accessible environments in the segment of vulnerable groups. The research focuses on the contributions that have been developed in the disability segment and particularly in visual impairment.

Office automation development, technology, and tiflotechnology have allowed people with visual disabilities to perform activities such as navigating the internet, reading a book, moving independently on the streets, studying, among others. A few years ago, the Braille reading and writing system, audio tape recordings, radio and television were the only means to access information independently for visual impaired people. On the other hand, for reading a book, press news, advertisements, etc., they necessarily had to ask for help.

Probably, between 1987 to 1999 the technological development that revolutionized the Spanish-speaking visually impaired people lives and their path towards accessing information independently took place. This period introduced various technological devices, such as braille speak, braille display, ZoomText and the JAWS screen reader [1].

D. Russo et al. (Eds.): IHSI 2021, AISC 1322, pp. 377–381, 2021.
https://doi.org/10.1007/978-3-030-68017-6_56

In Latin America, the formation of the Latin American Blind Union (ULAC) [2] in 1985 empowered people with disabilities. Later in 1998, ULAC was consolidated with the creation of the Foundation for Latin America (FOAL), which integrates care for people with disabilities and introduces tiflotechnology in its support [3].

Improving the quality of life for vision impaired people is at the forefront of recent technological advances that are focused on increasing their independence. These devices are based on electronic sensors located in the cane, which increase the levels of detection of obstacles; camera systems integrated into the temples of the glasses, which have algorithms for computer vision, and the implementation of artificial intelligence (AI) techniques [4–6].

Indeed, research related to visual impairment in the technological field has increased in recent years. The objective of this document is to expose in a general way the research development of the proposed topic, based on the prioritization and grouping of the publications [7].

2 Methodology

Having information on developments in the field allows us to identify the horizon of the area of knowledge. In this context, the methodology used is that used in bibliometric studies [8–10].

For the development of the research, the Scopus database was used, since it is a high scientific level database, on the other hand it has been used in other bibliometric studies due to the uniform metadata that allows us to delve into the issue raised [9].

In the search, the works carried out in 2020 were excluded, in order to maintain complete annual publications, KEY ("assistive technologies" AND "visual impairment") AND (EXCLUDE (PUBYEAR, 2020)) was used as a query. The consultation resulted in 297 documents between 1994 and 2019 (Fig. 1).

3 Results

3.1 General Analysis Summary

Of the 297 documents published, 149 are assignment documents, 120 articles, 18 reviews, 5 book chapters, and the rest are grouped into editorials and conference reviews (Fig. 2). The first publication is found in 1994 and the following in 1999 -with three publications- and later in 2002 -with three publications-. From 2002 it is evident that it is published continuously. The growth from 2018 to 2019 is 12.50%, it went from 40 publications to 45 publications.

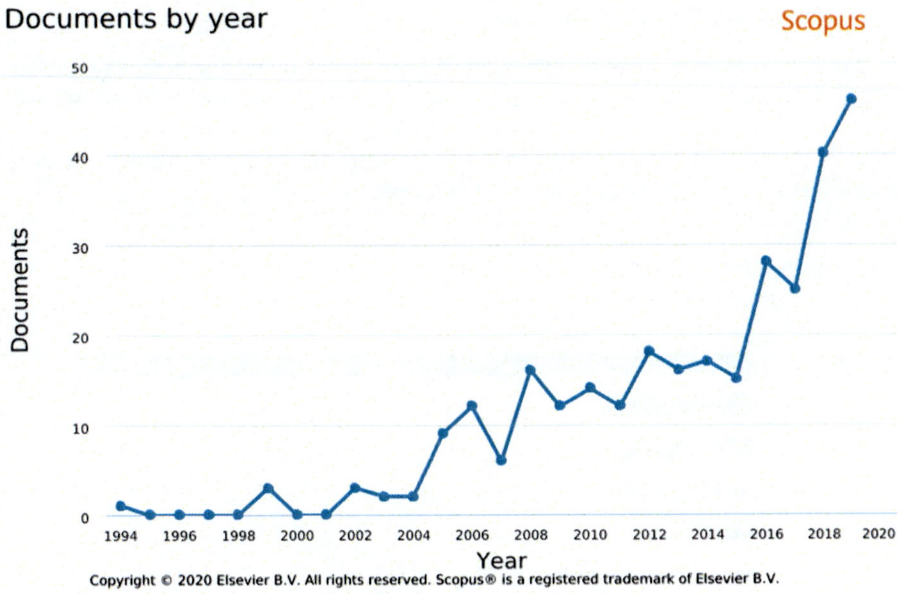

Fig. 1. Documents published by year

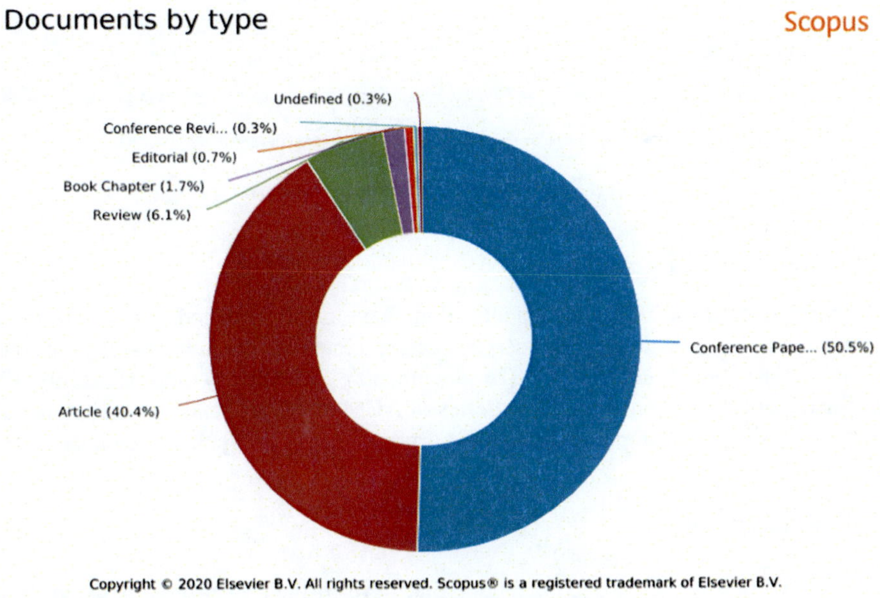

Fig. 2. Type of published document

3.2 Documents Produced by Country

Of the five most productive countries, the United States ranks first with 127 publications, far below is Italy with 28 publications, the United Kingdom with 27 publications, Brazil with 25 publications and Canada with 20 publications of the 297 publications (Fig. 3). Of the countries located in Latin America, there is first Brazil - with 25 publications - and Argentina with one publication.

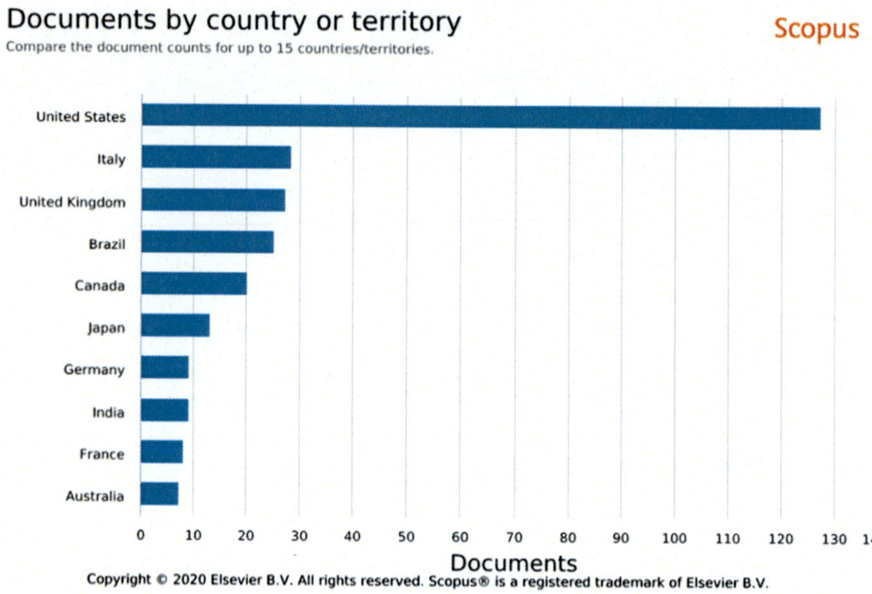

Fig. 3. Countries that produce the most

3.3 Sources of Information

The top five sources from which publications have been generated: Lecture Notes in Computer Science including sub-series Lecture Notes in Artificial Intelligence and Lecture Notes in Bioinformatics (26 documents), Technology and Disability (9), International Conference Procedures Series of ACM (8), Conference on human factors in the procedures of computer systems (8), from a total of 108 publication sources (6).

4 Conclusions

This work has analyzed the publications that have in their keywords: assistive technologies and visual impairment, showing that from the first publication in 1994 until the 2019 cut, Brazil and Argentina are the South American countries that have publications. On the other hand, it is striking that from 1995 to 1998 there are no publications and from 2000 to 2001 neither. This leaves more concerns than answers.

Probably a more exhaustive investigation should be generated incorporating other high impact databases such as the one selected in the analysis.

Acknowledgments. The authors of this work thank CEDIA for the financing within the CEPRA project Integration of New Technologies for the Design of Cognitive Solutions in Assisted Living Environments for Older Adults: Evaluation of Attention and Memory Areas.

References

1. Historia del CRE. (s/f). https://educacion.once.es/cres/cre-barcelona/historia-del-cre.
2. Unión Latinoamericana de Ciegos (ULAC) (2015). https://www.ipscuba.net/directorio-sc/union-latinoamericana-de-ciegos-ulac/
3. La ONCE apoya a cinco millones de personas ciegas en Latinoamérica (2012). https://mexico.corresponsables.com/actualidad/la-once-apoya-cinco-millones-de-personas-ciegas-en-latinoamerica
4. Desarrollan un bastón y anteojos inteligentes para personas con discapacidad visual y ceguera (2019). https://www.dicyt.com/noticias/desarrollan-un-baston-y-anteojos-inteligentes-para-personas-con-discapacidad-visual-y-ceguera
5. La discapacidad visual es un reto para los proyectos tecnológicos (2017). https://www.elcomercio.com/tendencias/discapacidad-visual-reto-proyectos-tecnologicos.html.
6. Sonido 3D para ayudar a personas no videntes (2017). https://www.elcomercio.com/guaifai/sonido-3d-ayudar-personas-videntes.html.
7. Zupic, I., Čater, T.: Bibliometric Methods in Management and Organization. Organ. Res. Methods **18**(3), 429–472 (2015)
8. Gurzki, H., Woisetschläger, D.M.: Mapping the luxury research landscape: a bibliometric citation analysis. J. Bus. Res. **77**, 147–166 (2017)
9. Bazm, S., Kalantar, S.M., Mirzaei, M.: Bibliometric mapping and clustering analysis of Iranian papers on reproductive medicine in scopus database (2010–2014). Int. J. Reprod. Biomed. **14**(6), 371–382 (2016)
10. Huamaní, C., et al.: South American collaboration in scientific publications on leishmaniasis: bibliometric analysis in scopus (2000–2011). Rev. Inst. Med. Trop. Sao Paulo **56**(5), 381–390 (2014)

Check for updates

Construction Method of User Mental Model in Interaction Design

Bingqing Yang, Haiyan Wang(✉), and Junkai Shao

School of Mechanical Engineering, Southeast University, Nanjing, China
{220180379,whaiyan}@seu.edu.cn, sjk19930918@163.com

Abstract. Through the analysis of user research methods and the combination of user research methods and the three-layer theory of emotional design, this paper proposes a new mental model construction method, namely the Hierarchical Elicitation Approach. Taking the information architecture design of the emergency monitoring platform as an example, this paper uses the Hierarchical Elicitation Approach to measure the performance rules of users' minds in the emergency monitoring platform, designs the construction experiments of the macro-mental model and meta-mental model of the system, and optimizes the design according to the experimental results. Moreover, Cogtool is used to analyze the effect of optimization design and verify the application value of the Hierarchical Elicitation Approach in the study of user's mind.

Keywords: Mental models · Mind inspired · The hierarchical elicitation approach · Emergency monitoring platform

1 Introduction

In order to better serve users in interaction design, user research needs to be carried out in various ways. In the process of user research, designers often need to build user mental models to achieve the purpose of mining user needs. The concept of mental model was put forward by Susan Carey [1]. The mental model shows people's thinking process of things. Application in the field of design refers to user's understanding and expectations of products. D.A. Norman pointed out in the study of mental models that mental models have characteristics such as incompleteness, instability, ambiguity, and specificity [2]. Alan Cooper subdivided the mental model into a designer model and a product model, which correspond to the two stages of conception and realization in the product production process [3]. In the design process, only when the designer's model, product model and the user's mental model reach a high degree of matching, can we have a good design experience. The mental model used can be divided into macro-mind and meta-mind from the macro and micro levels, which respectively map the user's cognitive law from the overall and partial levels. Kuniavsky, Mike and others applied card sorting, task analysis and other methods in the user research process, and mainly used qualitative or quantitative methods to explore the psychological needs of users [4]. Combining with the characteristics of the mental model, it can be seen that the formation of the mental model is complex and dynamic, and more comprehensive research methods can be used for mental analysis. This research will propose a more

D. Russo et al. (Eds.): IHSI 2021, AISC 1322, pp. 382–387, 2021.
https://doi.org/10.1007/978-3-030-68017-6_57

feasible method of mind construction based on general user research methods, taking into account the laws of mind formation and change.

2 Methods and Materials

2.1 Methodology

Based on the general user research method, this research combines the three-layer theory of emotional design, and proposes a new research method for the construction of mental models, which is called the Hierarchical Elicitation Approach. In the Hierarchical Elicitation Approach, the user's mind is divided into three levels: instinct, behavior, and reflection, and specific mental construction indicators are clarified for the three levels, and the design elements in the interaction design are subdivided into explicit elements and implicit elements Establish the corresponding method system of design elements, mental levels and mental construction indicators.

Hierarchical Elicitation Approach, refer to the index construction principles of completeness, applicability, logic, measurability and comparability, clarify the construction indexes corresponding to different mental levels [5]. According to the three levels of instinct, behavior and reflection mentioned in "Emotional Design", the user's mind is divided [6]. The instinct level includes user needs and product understanding, corresponding to the user's macro-mind; the behavior level includes the user's specific interaction behavior during use, corresponding to the user's meta-mind; the reflection level corresponds to the psychology and emotion expressed by the user. At the same time, according to the characteristics of the design elements in the human-computer interaction system, the design elements are divided into explicit and implicit, corresponding to the organizational structure and interface elements in the system [7]. The specific method model and experimental design process are shown in the Fig. 1.

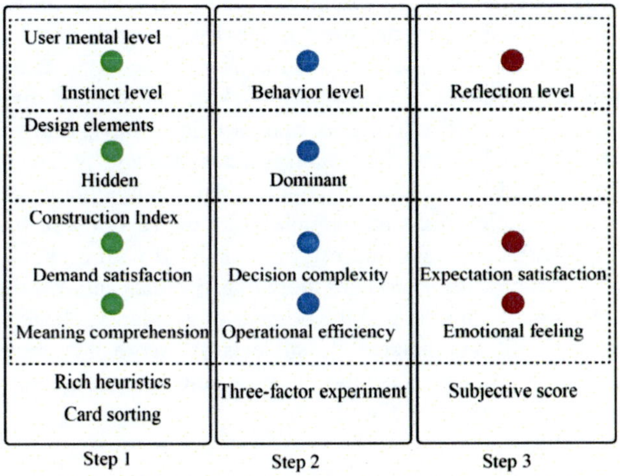

Fig. 1. The hierarchical elicitation approach and experimental application

2.2 Subjects

This study applied the Hierarchical Elicitation Approach to the information architecture design of the emergency platform to verify the usability of the method. In this study, 27 participants (14 males and 13 females) with relevant background knowledge were invited to participate in the mind-building experiment. The participants were between 20–25 years old, with an average age of 23.

2.3 Procedure

The emergency monitoring platform is an information management platform for emergency management and information detection [8]. Through the application of the Hierarchical Elicitation Approach, mental construction experiments are carried out on the organizational structure and interface framework of the emergency platform, and the optimized design of the emergency monitoring platform is realized to improve the coordination and decision-making level of users in emergency response.

In the first part of the experiment, eight subjects were invited to participate in the Rich Elicitation Approach according to the construction index of demand satisfaction [9]. In the heuristic experiment, the subjects diverged their thinking around different topics, and marked the importance of vocabulary and the order of inspiration. The order of heuristics is marked in numerical order, and the degree of relevance is marked with 1–5 numbers from low to high. Choose the vocabulary within the heuristic order 7 and the correlation degree above 3 to converge and sort, and obtain a total of 128 vocabulary.

Based on the meaning understanding index in the Hierarchical Elicitation Approach, the closed card sorting experiment is used to analyze the user's understanding of vocabulary [4]. The card sorting experiment is carried out through the online platform Maze. All participants need to browse the vocabulary and judge the group of each vocabulary.

23 participants were invited to participate in the meta-mental model construction experiment based on the decision-making complexity and operational efficiency construction indicators. The experiment is a three-factor behavioral experiment within subjects. The specific design is as follows. The navigation mode is divided into three levels: global navigation, local navigation, and associated navigation; icon styles are divided into three levels: plain text, text and line icons, text and surface icons; interface micro-interaction is divided into interactive and Two levels without interaction. Participants completed complex tasks and simple tasks on 18 different interfaces in the experiment. Simple tasks are tasks that can be completed with 3 clicks, and complex tasks are tasks that can be completed with more than 3 clicks and less than 6 clicks. In the experiment, the operation time and the number of accidental clicks were collected. After the click experiment is completed, the subjects will score in the range of 0–5 from the three dimensions of task ease, operation fluency, and operation efficiency. The flow of the experiment is as follows in Fig. 2.

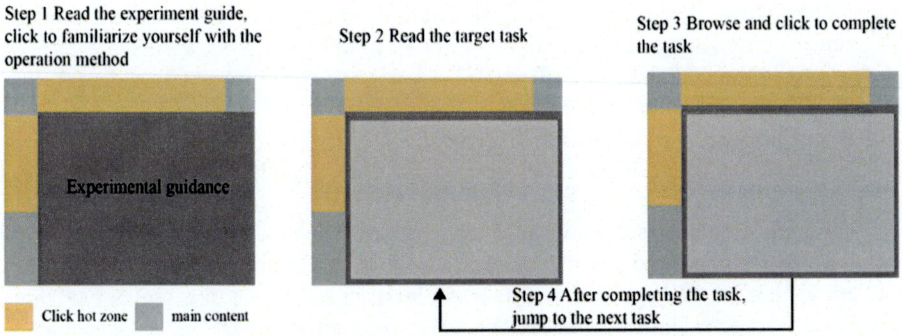

Fig. 2. Schematic diagram of the experimental process

3 Results

Use SPSS software to perform statistical analysis on behavioral experimental data. Table 1 shows the results of repeated measures analysis of variance of operation time and the number of accidental clicks, and the significance level is set to 0.05. The effective sample size in the table is 23.

Table 1. Repeated measures analysis of variance results of operation time and number of accidental clicks

Operating time	Simple task-Complex task					
	F value		Significance		Partial η^2	
Navigation	3.082	7.589	0.071	0.030	0.123	0.420
Icon	7.428	1.155	0.004	0.334	0.414	0.099
Interaction	10.672	4.829	0.004	0.039	0.327	0.180
Interaction effect	4.661	3.237	0.007	0.035	0.175	0.405
Wrong clicks	Simple task-Complex task					
	F value		Significance		Partial η^2	
Navigation	4.962	17.596	0.034	0.000	0.184	0.444
Icon	4.759	5.818	0.022	0.011	0.178	0.209
Interaction	4.936	3.148	0.037	0.090	0.183	0.125
Interaction effect	3.086	1.353	0.069	0.269	0.123	0.058

In simple tasks, there are significant differences in the number of accidental clicks under different navigation methods ($P = 0.034 < 0.05$), under different icons; operation time ($P = 0.004 < 0.05$) and number of accidental clicks ($P = 0.022 < 0.05$) There are significant differences; under different micro-interaction conditions, there are significant differences in operating time ($P = 0.004 < 0.05$) and the number of accidental clicks ($P = 0.037 < 0.05$); there are significant differences among the three factors of navigation, icons, and micro-interactions Multiple interaction effects have significant differences in operating time ($P = 0.007 < 0.05$). In complex tasks, there are significant differences in the operating time ($P = 0.03 < 0.05$) and the number of accidental clicks ($P = 0.000 < 0.05$) under the navigation factor; there are significant differences in the number of accidental clicks under different icons ($P = 0.011) < 0.05$); under different micro-interaction conditions, the operation time has a significant difference ($P = 0.039 < 0.05$).

Through simple effect analysis, it can be seen that in simple tasks, under the conditions of global navigation * surface icon * micro-interaction, $M = 4.952$, $SD = 2.754$, the participant's operation time is the least; in complex tasks, global navigation * text icon * micro Under the interactive conditions, $M = 9.704$, $SD = 3.436$, and the subjects had the least operating time. Combined with the descriptive statistics of the number of wrong clicks, it can be seen that under the condition of local navigation * line icon * micro-interaction, the subjects did not accidentally click on simple tasks and complex tasks. Based on the above data analysis results, the characteristics of the navigation mode, icon style, and micro-interaction of the user's meta-mental model of the emergency monitoring platform can be summarized as a combination of global navigation and local navigation in the navigation mode, and linear icons in the icon style It is optimal, and micro-interaction effects should be added to the interaction.

Participants scored 2.52 for Q1, 1.35 for Q2, and 2.30 for Q3 in subjective judgment. This means that subjects feel good in terms of fluency and ease, and there are obvious changes in operating efficiency in different interfaces, which meets the expectations of users for increased efficiency.

4 Discussion

Based on the functions of the emergency monitoring platform, two task types, vertical depth and horizontal switching are selected, and two subtasks are designed under each task type and numbered as A1, A2, B1, and B2. According to HEA guidance, the emergency monitoring platform was optimized and designed. The number of design interfaces was 12, and the number was (HI–H12). The number of original interfaces with the same function point was 16, and the number was (TI–T16). Use Cogtool to estimate the difference in completion time between users on two different platforms. It can be seen from the time data that the optimized interface has a significant reduction in operating time in different subtasks. The time used in A1 is reduced by 0.3 s, the time used in A2 is reduced by 5.0 s, the time used in B1 is reduced by 4.3 s, and the time used in B2 is reduced by 0.3 s. The results show that the optimization effect of the Hierarchical Elicitation Approach is significant, and the Hierarchical Elicitation Approach has application value.

From the results of behavioral experiments, under the two conditions of simple tasks and complex tasks, the effects of design elements are different. Under complex task conditions, different navigation methods have a significant impact on the subjects' operational efficiency and operational complexity. In contrast, the impact of changes in navigation methods in simple tasks is much weaker. This shows that for complex operating conditions, the user's performance will be affected by the navigation method.

In order to better construct the user's ideal operating conditions, the designer should carefully design the navigation method. According to the experimental results, the user feedback of global navigation and local navigation. The effect is not very different and can be used comprehensively.

The overall analysis shows that the change of the icon style has a significant impact on the subjects' operational efficiency and complexity. However, under complex task conditions, the change of the icon style does not affect the user's operation time. After checking the data and excluding the mistake, consider the reason for complexity. In the task, the subjects paid more attention to direct information. For operation, the change of icon style plays a role in assisting attention, and its effectiveness is limited, and the amount of information of face icons and line icons needs to be converted into text information processing. Therefore, in complex operations, text information is instead It shows advantages because it does not require information conversion. In the same way, the micro-interaction effect, which is also an auxiliary selection, shows the same influence in the experiment.

Acknowledgments. The authors would like to gratefully acknowledge the reviewers' comments. This work was supported jointly by National Natural Science Foundation of China (No. 71871056, 71471037), Equipment Pre research & Ministry of education of China Joint fund (No. 6141A020226).

References

1. Carey, S., et al.: Could it be so? The cognitive science of possibility. Trends Cogn. Sci. **24**(1), 3–4 (2020)
2. Norman, D.A.: Some observations on mental models. Human-computer interaction: a multidisciplinary approach, pp. 241–244. Morgan Kaufmann Publishers Inc., Sebastapol (1987)
3. Cooper, A., et al.: About Face 2.0: The Essentials of Interaction Design. Wiley, Hoboken (2003)
4. Kuniavsky, M.: Chapter 8 - contextual inquiry, task analysis, card sorting. In: Kuniavsky, M. (ed.) Observing the User Experience, pp. 159–199. Kaufmann, San Francisco, Morgan (2003)
5. Zhao, C., Lu, M.: Design of performance appraisal index system based on BSC and AHP. China Hum. Resour. Dev. (05), 52–54 (2011)
6. Norman, D.: Emotional design : why we love (or hate) everyday things. In: Norman, D.A. (ed.) SERBIULA (sistema Librum 2.0) (1981)
7. Shen, Q., Xue, C., Niu, Y.: Research on the design method of human-computer interaction interface based on ERP. Design (06), 134–135 (2018)
8. Meng, Z.: Construction and research of intelligent command and dispatch system for urban comprehensive emergency response. Zhejiang University of Technology (2019)
9. LaMere, K., et al.: Making the most of mental models: Advancing the methodology for mental model elicitation and documentation with expert stakeholders. Environ. Model. Softw. **124**, 104589 (2020)

Check for updates

An Approach to Determine Short- and Long-Term Work Ability in Smart Work System

Otilia Kocsis(✉), George Papoulias, Nikos Fakotakis, and Konstantinos Moustakas

University of Patras, 26500 Rio, Patras, Greece
okocsis@bok.gr,
{gpapoulias,moustakas}@ece.upatras.gr,
fakotaki@upatras.gr

Abstract. Europe is being severely challenged by the ageing of the population, with the unemployment among older people posing tremendous strain on social and health care services systems. The SmartWork project builds a worker-centric AI system for work ability sustainability, which integrates unobtrusive sensing and modelling of the worker state with a suite of novel services for context and worker-aware adaptive work support. The framework for modeling the work ability of the office worker is grounded to the holistic multidimensional work-ability model, accounting for the cognitive and physical resources of the individual, and factors related to work and working. An overall work ability score is derived by combining self-assessed work ability index with the objective assessment of the evolving state of the worker. The proposed approach is currently undergoing test and validation in laboratory conditions, and will be tested in real-life conditions in two pilot demonstrators.

Keywords: Work ability modelling · Active ageing · Office worker

1 Introduction

Europe is being severely challenged by the ageing of the population, and although they have been looking for strategies to effectively increase the labour force participation of older workers and reduce the rates of early retirement and labour market exit for a decade now, the unemployment amongst older people remains particularly high among 55–64 year-olds [1], posing tremendous strain on social and health care services systems. Ageing is a gradual process and there is no explicit age threshold for an older worker, but numerous physical changes and chronic conditions, such as heart illness, hypertension, COPD, asthma and mental [2], associated with ageing are estimated to start as early as the age of 50 [3, 4]. It is characteristic that 69.5% of adults of this age group live with at least 1 chronic condition (67.7% of male and 71.1% of female). Furthermore, in particular the lifestyle of office workers which is characterized by prolonged sitting and overall sedentary life, has the potential to significantly and independently of other factors increase the risk of cardiometabolic diseases and premature mortality [5]. Office work also affects

D. Russo et al. (Eds.): IHSI 2021, AISC 1322, pp. 388–394, 2021.
https://doi.org/10.1007/978-3-030-68017-6_58

functional abilities of the workers [1, 6], and other contextual factors related to the indoor workspace (e.g. illumination, ambient conditioning system) [7, 8], can further impact on their work performance and productivity. The SmartWork project builds a worker-centric AI system for work ability sustainability, with unobtrusive sensing and modelling of the worker state, it does integrates a suite of novel services for context and worker-aware adaptive work support [9, 10]. The framework for modeling the work ability of the office worker, described which is grounded to the holistic multidimensional work-ability model [11] which accounts for both the cognitive and physical resources of the individual, and factors related to work tasks management and the environment outside of work. The proposed approach also includes modules addressing the requirements of the employers, helping them to maintain/increase productivity of the employees through better management of the available resources in relation to the work tasks (e.g. optimized team formation, task optimization with respect to needed resources, prioritization of training needs, etc.). An overall work ability score is derived by combining self-assessed work ability index with the objective assessment of the evolving state of the worker (e.g. continuous stress monitoring). The proposed approach is currently undergoing test and validation in laboratory conditions, and will be tested in real-life conditions in two pilots during the life-time of the project. Preliminary results demonstrate good capacity of the proposed overall work ability score to capture the short-term variations of work ability. However, further fine-tuning of time-variance of weighting factors and individual components scoring is on-going, in order to improve the accuracy of the proposed approach.

2 The SmartWork System

The Worker-Centric AI System, built within the SmartWork project, integrates unobtrusive sensing and modelling of the worker state with a suite of novel services for context and worker-aware adaptive work support [9]. The main user groups of SmartWork system include the office workers, their employer (e.g. project manager) and their carers. "The unobtrusive and pervasive monitoring of health, behaviour, cognitive and emotional status of the office worker enables the functional and cognitive decline risk assessment. The holistic approach for work ability modelling captures the attitudes and abilities of the ageing worker and enables decision support for personalized interventions for maintenance/improvement of the work ability" [10]. The SmartWork services and modules for "on-the-fly work flexibility" also empower the employer (manager) with "AI decision support tools for efficient task completion and work team optimization through flexible work practices". The implementation of the SmartWork services integrates and shares on the various dimensions of the worker state aware work ability modelling, a series of transdisciplinary methods and technologies (e.g. functional modelling, cognitive modelling, accessible-born interaction interfaces), to address the needs and requirements of the end users.

3 Background on Work Ability Assessment

The most extensively utilized and well-established metric for the work ability assessment is the Work Ability Index (WAI). WAI has been, which is in use for almost three decades now, has demonstrated noticeable efficiency regarding both blue-collar and white-collar workers' work ability assessment [12]. The WAI is derived by assessing the various dimensions through seven (7) subscales, namely: Self-estimated work ability compared to lifetime best; Work ability in relation to job demands; Number of current diseases one suffers from; Estimated work impairment due to diseases; Sick leave during the past 12 months; Own prognosis of work ability 2 years from now; and Mental resources. However, the length of the questionnaire has driven the research community towards acquiring insight on the correlations between different parameters recorded for deriving the WAI score, and respective dimensions/scales, in an attempt to make it easier and more often applicable. One of the first efforts to identify the dimensionality of the WAI and to investigate dependencies across different underlying factors [13], aimed to respond to the long-standing question whether the WAI should be treated as an one- or two-dimensional instrument. The results showed that WAI structure is definitely not one-dimensional, pointing to a two-factor model in which sub-scales 4 and 6 are loaded into both factors to obtain an optimal fit. The WAI was also used as a tool for workers' periodical health monitoring, with particular focus on the alterations of WAI dependent upon working conditions and personal health status [14]. Despite the expected trend of WAI to decrease over the years, it was observed that in jobs demanding mostly mental work and providing autonomy to the employees combined with relatively low physical load, the WAI remained almost constant and relatively high for a significant time interval. The study also indicates that work ability is a dynamic process, with its components changing throughout working life, and comes out as a result of the interaction between various resources including, apart from functional condition and mental resources available, also working conditions (environment, human relations) as well as the surrounding society. Attempts have been made to find an assessment tool of reduced size compared to the WAI questionnaire, such as the Work Ability Score (WAS) [15], which is practically a shortened version of WAI, and showed convergent validity and similarity in a large population of employed workers of increased age. Similarly, shortened versions of the WAI showed efficiently in identifying non-sicklisted workers at risk of long-term sickness absence [16]. Finally, a work focusing on the promotion and upgrade of work ability among aging workers aimed to establish the target areas of the work ability promotion model, by exploring the characteristics that explain work ability in the most complete way [17]. The variables that demonstrated the highest explanation rate were those related to work demands and the environment, followed by the work organization, the individual's professional competence and the lifestyle variables.

4 The Proposed Approach

The SmartWork approach for modelling office worker' work ability combines self-assessment of work ability with objective assessments of functional abilities and cognitive capacity of the older office worker, based on the data collected by the SmartWork Unobtrusive Sensing Network, in order to derive an Overall Work Ability Score (OWAS). The OWAS aims at dynamically capturing the evolving state of the worker and the context of work and working (e.g. environment, task requirements).

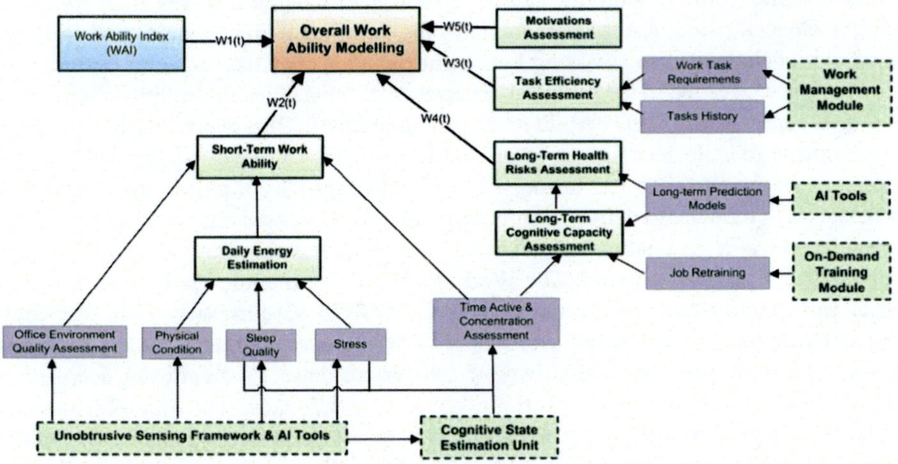

Fig. 1. Conceptual design of the overall work ability modelling framework.

As depicted in Fig. 1, the proposed architecture for the derivation of the OWAS follows a hierarchical structure, with the 1-st level modules performing assessment of individual parameters (e.g. sleep, stress, physical activity, concentration, etc.) and feeding modules responsible for assessment of work ability on the short- or on the long-term. The Office Environment Quality Assessment exploits the information on the temperature, relative humidity and Volatile Organic compound (VOC) levels monitored indoor by a dedicated environmental sensor, in order to provide a metric related to the overall quality of the worker's office environment. The Sleep Quality Assessment module receives as input the pre-processed sleep data recorded by a wearable smartwatch, and derives a sleep quality score, which plays an important role both in terms of maintaining a good level of overall health and daily energy, as well as in the context of work productivity (capacity to maintain attention and concentration during work). The Physical Condition Assessment module uses as input the steps made, the distance covered, the intense exercise records, the sleep data and the body mass index (BMI) of the user monitored using the smartwatch and derives a score estimating the physical condition level of individual users. The Stress Assessment Module receives as input the heart rate recorded by the smartwatch and/or the galvanic skin response (GSR) and heart rate recorded by a smart mouse, and derives scores related to the overall stress level of the user at day/week level or on the long-term. The Job Retraining module extracts information from the user's training record regarding the frequency and the level of the training sessions the user has gone through recently and derives a score contributing to the estimation of the Long-Term Cognitive Capacity of the worker. The Time Active - Concentration Assessment module, which receives as input the mental fatigue and the distraction as monitored by the Cognitive State Estimation (CSE) module and the sleep quality and stress levels of the users, derives scores related to the capacity of the individual to maintain attention and concentrate at day level and on longer term.

The Long Term Cognitive Capacity Assessment module receives from the Time Active-Concentration Assessment module, the Job Retraining module, and the Predictive models related to assessing long-term risks for cognitive capacity decline, and derives a score related to the potential occurrence of mental health conditions impacting on the Long-Term Work Ability of the individual. The work Task Efficiency Assessment module receives as input information on the level of difficulty of the currently assigned tasks and history of tasks completed by the user from the Work Management module, to derive a score estimating the predicted efficiency of the user to complete the tasks in relation to worker's mental and physical resources (e.g. skills and knowledge, estimated energy). The Work Ability Index module implements the classical work ability survey [17] and derives the WAI at baseline (when the user joins SmartWork) and whenever self-assessment of WAI is detected as necessary: too long time has passed since the last survey or the system detects a significant decrease in overall work ability as compared to the last WAI assessment. The Short-Term Work Ability Assessment module receives input from the following modules: the Office Environment Quality Assessment module, the Daily Energy Assessment module, and the Time Active/Concentration Assessment module. The Long-Term Health Risk Assessment module is fed with results of the health-related Prediction Models, as well as with the score derived by the Long-Term Cognitive Capacity Assessment module in order to derive a predictive score reflecting the overall risk of the individual to experience some mental or physical decline which may result in early exit from the market labour. The Motivations Assessment module evaluates the various dimensions (work, healthy lifestyle, wellbeing) of motivations a user has in order to remain professionally active and healthy on the long term.

5 Discussion and Conclusion

The proposed approach for work ability modelling is currently undergoing test and validation in laboratory conditions, and will be tested in real-life conditions in two pilots during the life-time of the SmartWork project. Preliminary results demonstrate good capacity of the proposed overall work ability score to capture the short-term variations of work ability. However, further fine-tuning of time-variance of weighting factors and individual components scoring is needed, in order to improve the accuracy of the proposed approach. Furthermore, long-term usage of the proposed approach in real-life conditions is also a required step, in order to validate the proposed score.

Acknowledgments. This work has been partially supported by the SmartWork project (GA 826343), EU H2020, SC1-DTH-03–2018 - Adaptive smart working and living environments supporting active and healthy ageing.

References

1. Fries-Tersch, E., Tugran, T., Bradley, H.: 2017 annual report on intra-EU labour mobility. Publications Office, LU (2018)
2. Busse, R.: World Health Organization, European Observatory on Health Systems and Policies eds: Tackling chronic disease in Europe: strategies, interventions and challenges. World Health Organization on behalf of the European Observatory on Health Systems and Policies, Copenhagen (2010)
3. Liang, J., Bennett, J.M., Shaw, B.A., Quinones, A.R., Ye, W., Xu, X., Ofstedal, M.B.: Gender differences in functional status in middle and older age: are there any age variations? J. Gerontol. B. Psychol. Sci. Soc. Sci. **63**, S282–S292 (2008). https://doi.org/10.1093/geronb/63.5.S282
4. European Commission: Health at a Glance: Europe 2018. European Commision (2018)
5. Parry, S., Straker, L.: The contribution of office work to sedentary behaviour associated risk. BMC Public Health **13**, 296 (2013). https://doi.org/10.1186/1471-2458-13-296
6. Portello, J.K., Rosenfield, M., Bababekova, Y., Estrada, J.M., Leon, A.: Computer-related visual symptoms in office workers: visual symptoms in office workers. Ophthalmic Physiol. Opt. **32**, 375–382 (2012). https://doi.org/10.1111/j.1475-1313.2012.00925.x
7. Akimoto, T., Tanabe, S., Yanai, T., Sasaki, M.: Thermal comfort and productivity - evaluation of workplace environment in a task conditioned office. Build. Environ. **45**, 45–50 (2010). https://doi.org/10.1016/j.buildenv.2009.06.022
8. Vimalanathan, K., Babu, T.R.: The effect of indoor office environment on the work performance, health and well-being of office workers **8** (2014)
9. Kocsis, O., Quintas, J., Hermens, H.J., Cabrita, M., Dantas, C., Hansen, S., Ziemke, T., Tageo, V., Dougan, P., Moustakas, K., Fakotakis, N., Vassiliou, C., Toska, A., Vanderheiden, G.C., Stergiou, A., Amaxilatis, D., Pardal, A.: SmartWork: designing a smart age-friendly living and working environment for office workers. In: Proceedings of the 12th ACM International Conference on PErvasive Technologies Related to Assistive Environments - PETRA 2019, Rhodes, Greece, pp. 435–441. ACM Press (2019). https://doi.org/10.1145/3316782.3322766.
10. Kocsis, O., Moustakas, K., Fakotakis, N., Hermens, H.J., Cabrita, M., Ziemke, T., Kovordanyi, R.: Conceptual architecture of a multi-dimensional modeling framework for older office workers. In: Proceedings of the 12th ACM International Conference on PErvasive Technologies Related to Assistive Environments - PETRA 2019, Rhodes, Greece, pp. 448–452. ACM Press (2019). https://doi.org/10.1145/3316782.3322768.
11. van den Berg, T.I.J., Elders, L.A.M., de Zwart, B.C.H., Burdorf, A.: The effects of work-related and individual factors on the Work Ability Index: a systematic review. Occup. Environ. Med. **66**, 211–220 (2008). https://doi.org/10.1136/oem.2008.039883
12. Alavinia, S.M., van den Berg, T.I., van Duivenbooden, C., Elders, L.A., Burdorf, A.: Impact of work-related factors, lifestyle, and work ability on sickness absence among Dutch construction workers. Scand. J. Work. Environ. Health. **35**, 325–333 (2009)
13. Martus, P., Jakob, O., Rose, U., Seibt, R., Freude, G.: A comparative analysis of the Work Ability Index. Occup. Med. **60**, 517–524 (2010). https://doi.org/10.1093/occmed/kqq093

14. Costa, G., Sartori, S.: Ageing, working hours and work ability. Ergonomics **50**, 1914–1930 (2007). https://doi.org/10.1080/00140130701676054
15. El Fassi, M., Bocquet, V., Majery, N., Lair, M.L., Couffignal, S., Mairiaux, P.: Work ability assessment in a worker population: comparison and determinants of Work Ability Index and Work Ability score. BMC Public Health. **13**, 305 (2013). https://doi.org/10.1186/1471-2458-13-305
16. Schouten, L.S., Bültmann, U., Heymans, M.W., Joling, C.I., Twisk, J.W.R., Roelen, C.A. M.: Shortened version of the work ability index to identify workers at risk of long-term sickness absence. Eur. J. Public Health. **26**, 301–305 (2016). https://doi.org/10.1093/eurpub/ckv198
17. Tuomi, K.: Promotion of work ability, the quality of work and retirement. Occup. Med. **51**, 318–324 (2001). https://doi.org/10.1093/occmed/51.5.318

Health and Activity Monitoring Using Smart Devices to Support the Self-management of Health Behavior

Janet Wesson(✉), George Mujuru, and Lester Cowley

Department of Computing Sciences, Nelson Mandela University, PO Box 77000, Port Elizabeth, South Africa
{Janet.Wesson,Lester.Cowley}@mandela.ac.za,
tungamiraimujuru@gmail.com

Abstract. Diabetes, cancer and cardiovascular disease are non-infectious medical conditions that are classified as non-communicable diseases, or chronic diseases of lifestyle (CDLs). These conditions are now the second leading cause of disease and death in Africa. To promote the self-management of CDLs, better intervention strategies are needed that reduce the effects of CDL risk factors. These strategies should primarily focus on the promotion of healthy lifestyles and better self-management of health behavior. Possible solutions involve providing individuals with health information using smart or wearable devices that can track physical activity, sleep and exercise. Twenty-one participants, between the ages of 30 and 60, took part in a field study to evaluate the efficacy of these smart devices. The field study was conducted over two weeks, in two one-week-long phases, with each phase having different objectives. The objective of the first phase was to identify health risks and form health behavior change intentions. The objective of the second phase was to set manageable goals for the participants, motivated by the intentions formed in the first phase. The results of the study show how smart devices can be used to motivate users and monitor health data. These results provide empirical evidence to support the hypothesis that smart devices can be used to support the self-management of health behavior. This research also resulted in an updated version of the Health Action Process Approach, incorporating the use of smart devices.

Keyword: Health Action Process Approach · Health behaviors · Smart devices

1 Introduction

Chronic diseases of lifestyle (CDLs), are non-infectious medical conditions, which cannot be transmitted between individuals [1–4]. The adoption of unhealthy lifestyles or health behaviors by people over time results in them being predisposed to these medical conditions. The four main types of CDLs are chronic lung disease, diabetes, cardiovascular disease (CVD) and cancer. International published statistics and the World Health Organization (WHO) highlight CDLs as the leading cause of death globally [5, 6].

D. Russo et al. (Eds.): IHSI 2021, AISC 1322, pp. 395–401, 2021.
https://doi.org/10.1007/978-3-030-68017-6_59

Several factors have a direct influence on people's desire to change, including a lack of physical activity and an unhealthy diet. These factors are important motivators for individuals who need to take steps to change their lives.

A study was conducted to investigate the use of smart devices to address the problems of adopting new health behaviors. The aim of the research was to determine the effect(s) that smart devices have on the self-management of health behaviors of individuals.

2 Related Work

2.1 Chronic Diseases of Lifestyle

The most prevalent forms of CVD are strokes and heart disease. There are a set of key modifiable risk factors that are called the "causes of the causes". The most important of these are the following:

- Physical inactivity,
- An unhealthy diet, and
- Tobacco use. For the purpose of this study, tobacco use was ignored, as this is a more complex health behavior.

2.2 Health Behavior Modification

Several health behavior change theories have been proposed to assist in understanding the health behavior change process. The Health Action Process Approach (HAPA) [7],

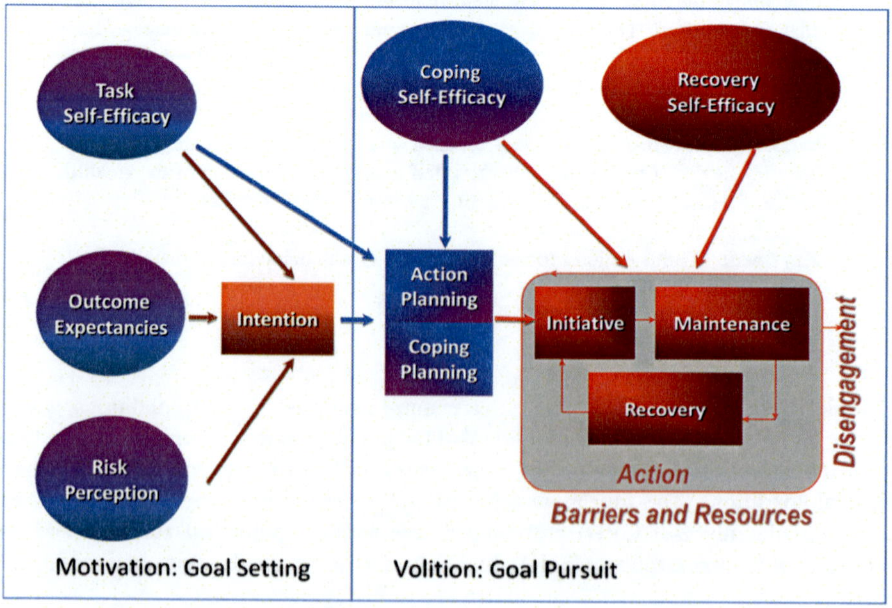

Fig. 1. Health Action Process Approach (HAPA) [8]

is one such theory, which comprises two phases, namely the Motivational and the Volitional phase (Fig. 1). The assumption in HAPA is that an individual is in the motivational phase when he/she has an intention to adopt a new health behavior. The volitional phase consists of taking the intentions and translating them into actual behaviors.

3 Research Design

This section discusses the design of the field study, including the participants involved, the instruments used and the data analysis techniques used to analyze the data collected.

3.1 Research Method

According to HAPA, two separate processes are required to change health behaviors, namely motivation and volition. To follow these two separate processes, the field study was conducted in two phases. Each phase of the field study lasted a week. In the first phase, participants were introduced to the Fitbit Charge 2 and the Fitbit Aria, which were used for the duration of the study.

Six research instruments were used to gather data. The first two instruments were the two smart devices, which were used to track participant physiological data. The third instrument was a biographical questionnaire. The fourth instrument was the Health-Promoting Lifestyle Profile II (HPLPII) questionnaire, which was used to measure the health-promoting behaviors of participants. The fifth instrument was the HAPA questionnaire, which was used to determine health cognition and health behaviors. The sixth instrument was a post-study questionnaire, which was used to measure the acceptance of the technology used.

3.2 Participant Selection

A sample of 21 participants from Nelson Mandela University, who met the inclusion criteria and agreed to participate were selected. Although the sample size was small, it was decided that it would still provide meaningful statistics.

The inclusion criteria were the following:

- A participant had to be a Nelson Mandela University staff member;
- A participant had to be aged between 30 and 60 years of age; and,
- A participant had to be able to read and write in English.

No exclusion criteria were included and the study was conducted in compliance with the Nelson Mandela University policy on human research ethics (H17-SCI-CSS-003). The staff members were split into two groups of staff type: Academic and Personnel for Administrative and Support Services (PASS) staff. Eight Academic staff and 13 PASS staff members were selected.

3.3 Data Collection and Analysis

The smart devices recorded the participants' physiological data for the duration of the study. Examples of data stored were the number of minutes asleep, number of steps walked, heart rate, and exercise. This data was then uploaded to the Fitbit application on the mobile phone, from where it was uploaded to the Fitbit servers.

4 Results

4.1 Participant Demographics

Twenty-one participants were recruited: eleven males and 10 females. The mean age was 46, and the majority (13) had not used smart devices previously.

4.2 Health Promoting Lifestyles of Participants

The HPLPII used a 4-point Likert scale to determine the health promoting lifestyles of the participants. The findings from the HPLPII questionnaire revealed that the participants had a Moderate adoption of health-promoting behaviors. The physical activity subscale had the lowest subscale score, indicating that participants did not include physical activity in their daily routine. The stress management score was Low, indicating that the participants struggled to manage their stress levels.

4.3 Activity Data (Week One)

The purpose of the first week of the tracked activity data was to assess the current health behaviors among participants. A two-sample t-test was conducted to identify any significant differences between the participants' tracked activities and certain demographical factors. The results revealed that there were significant differences between the male and female participants for weight, calories burned and active calories. The male participants had higher mean values for all tracked activities except the sleep activity, where females had a higher mean. The physical activity data of the first week confirmed the findings of the HPLPII questionnaire, which were the positive relationship between the Physical Activity subscale and the number of steps walked.

4.4 Activity Data (Week Two)

Activity data was tracked at the end of the second week to determine if there had been any changes during the Volition phase. The data tracked at the end of Week 2 was compared with the data tracked at the end of Week 1, and the difference in the means was used to conduct a two-sample t-test.

The results showed that there were no significant differences in both genders (male vs female) for the BMI, Weight, Calories Burned, Steps, Minutes Sedentary, Activity Calories and Total Active Minutes. However, the Minutes Asleep variable ($M = -38.19$ vs 13.29, $p = 0.08$) was significantly different between male and female participants. The female participants had notable differences in their weight, BMI, minute's sedentary, active calories and minutes asleep compared to their male counterparts. The males had notable differences in their calories burned and steps. Statistically significant differences were found between marital status (single vs married), the calories burned (−99.58 vs 117.94, $p = 0.04$), and the number of steps (−1192.81 vs 1316.57, $p = 0.03$). The findings also revealed significant differences in the staff type (Academic vs PASS) in the sedentary minutes (25.98 vs −76.83, $p = 0.05$).

4.5 Technology Acceptance Data

User adoption of the smart devices was measured using two key factors of the Technology Acceptance Model (TAM), namely the Perceived Usefulness (PU) and Perceived Ease of Use (PEOU) [9]. Six questions, scored on a seven-point Likert scale, were used to measure the PU. The range of values on the scale ranged from a 1 (Strongly Disagree) to a 7 (Strongly Agree). The descriptive statistics for the PU revealed that the mean score for the participants was 5.98 (SD = 1.54), which was considered High.

Six questions were used to measure the PEOU. The values on the scale ranged from a 1 (Strongly Disagree) to a 7 (Strongly Agree). The descriptive statistics of the PEOU section of the TAM questionnaire revealed that the mean score for the participants was 6.08 (SD = 1.31), which was considered High.

5 Discussion

The findings reveal that the majority of the participants were more conscious about their health after the study, and that the lack of physical exercise was the most frequent theme that emerged (Sect. 4.3). A small number noticed a change in weight, but this was not significant. These findings provide support for the contention that smart devices can monitor physical activity, or the lack of it. The second week of the activity data highlighted an increase in physical activity levels for all the participants (Sect. 4.4). All the participants in the second week increased their daily steps to just above the daily-recommended 10,000 steps, and decreased their sedentary behavior. These findings also have an impact on the HAPA model, which is discussed in the next section.

6 Updated Health Action Model

The researchers conducted a Pearson Product-Moment test for correlation at the 90% confidence level. The test was conducted between the Overall Health-Promoting Lifestyle, HAPA Constructs, Participants' Biometric Data for the first week, and Participants' Biometric Data for the second week, and the Post-study questionnaire.

The Updated Health Action Model presents the psychological and physiological mediators that a participant progressed through to increase their physical activity using smart devices (Fig. 2). In the Motivation state, the sedentary minutes and weight contribute to a participant's risk perception. The risk perception affects both Outcome

Expectancies and Action Self-efficacy. The Outcome Expectancies affect the Action Self-efficacy as well. There is a transition from the Motivation state to the Volition state. Self-efficacy is a primary mediator that moves participants from one state to the other, and to the actual health behavior. This provides support for HAPA principle five, which states "self-efficacy differs across phases".

7 Conclusions

The field study used smart devices to monitor health behavior and promote changes in health behaviors. The HAPA model was selected as the candidate health behavior theory for this research. Several participants changed their health behaviors and several improvements were noted, especially an increase in physical activity and sleep behavior. The field study highlighted the shortcomings of the HAPA model, as some of the psychological constructs were facilitated through the use of the smart devices (Fig. 2).

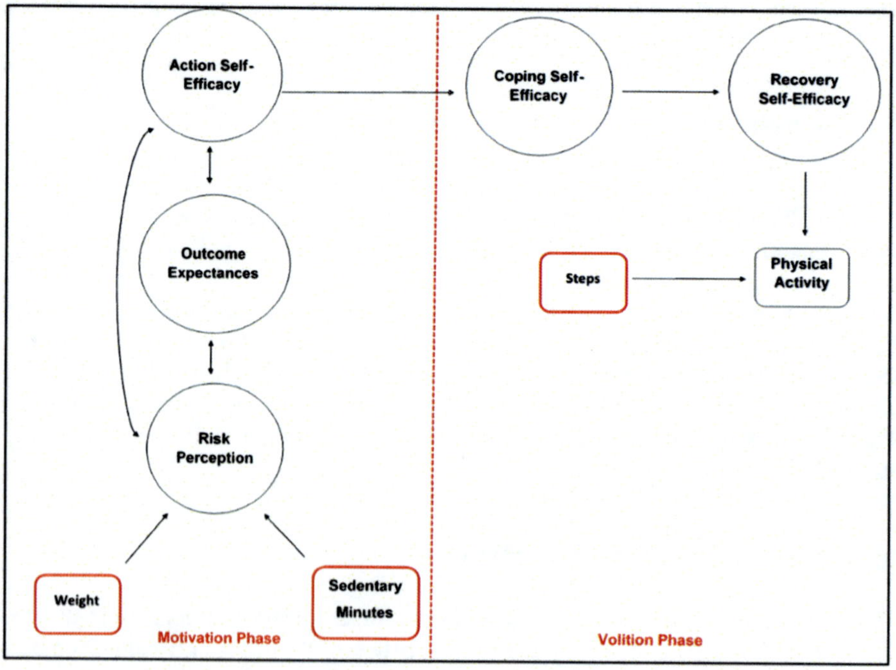

Fig. 2. Updated health action model

References

1. Bradshaw, D., Steyn, K., Levitt, N., Nojilana, B.: Non-Communicable diseases – a race against time. South African Med. J. **2011**, 1–5 (2011)
2. Kim, H.C., Oh, S.M.: Noncommunicable diseases: current status of major modifiable risk factors in Korea. J. Prev. Med. Public Heal. **46**, 165–172 (2013). https://doi.org/10.3961/jpmph.2013.46.4.165
3. Mayosi, B.M., Flisher, A.J., Lalloo, U.G., Sitas, F., Tollman, S.M., Bradshaw, D.: The burden of non-communicable diseases in South Africa (2009). https://doi.org/10.1016/S0140-6736(09)61087-4
4. Nojilana, B., Bradshaw, D., Pillay-van Wyk, V., Msemburi, W., Laubscher, R., Somdyala, N. I.M., Joubert, J.D., Groenewald, P., Dorrington, R.E., Matzopoulos, R., Nannan, N., Neethling, I., Nicol, E., Prinsloo, M., Rossouw, A., Sithole, N., Bateman, E., Bobrow, K., Bryer, A., Steyn, K., Levitt, N.: Emerging trends in non-communicable disease mortality in South Africa, 1997–2010. South African Med. J. 106, 477–484 (2016). https://doi.org/10.7196/SAMJ.2016.v106i5.10674.
5. de Souto Barreto, P., Cesari, M., Andrieu, S., Vellas, B., Rolland, Y.: Physical activity and incident chronic diseases: a longitudinal observational study in 16 European countries. Am. J. Prev. Med. **52**, 373–378 (2017). https://doi.org/10.1016/j.amepre.2016.08.028
6. World Health Organization: WHO Global Conference on NCDs, 18–20 October 2017 (2017)
7. Schwarzer, R.: Health Action Process Approach (HAPA) as a Theoretical Framework to Understand Behavior Change. Actual. en Psicol. **30**, 119 (2016). https://doi.org/10.15517/ap.v30i121.23458
8. Schwarzer, R.: Modeling health behavior change: how to predict and modify the adoption and maintenance of health behaviors. Appl. Psychol. **57**, 1–29 (2008). https://doi.org/10.1111/j.1464-0597.2007.00325.x
9. Chiarini, G., Ray, P., Akter, S., Masella, C., Ganz, A.: mHealth technologies for chronic diseases and elders: a systematic review. Sel. Areas Commun. IEEE J. **31**, 6–18 (2013). https://doi.org/10.1109/JSAC.2013.SUP.0513001

Challenges in Smart Healthcare for Physical Rehabilitation

Jorge-Luis Pérez-Medina(✉), Karina Jimenes-Vargas,
Patricia Acosta-Vargas, Mario González,
and Wilmer-Danilo Esparza-Yánez

Intelligent and Interactive Systems Lab (SI2-Lab), Universidad de Las Américas (UDLA), Quito, Ecuador
{jorge.perez.medina,karina.jimenes,patricia.acosta,
mario.gonzalez.rodriguez,wilmer.esparza}@udla.edu.ec

Abstract. In the current medical practice, personal physical physiotherapy is expensive, the therapist's time of service is limited, and training equipment is available in a clinical environment. So, the need to provide low-cost and convenient smart health assistance services are growing tremendously. ePHoRt is a web-based platform created for helping patients after hip replacement surgery. We discuss and present the techniques, methods, and strategies used throughout the execution of the project.

Keywords: eHealth · Physical telerehabilitation · Movement quality

1 Introduction

In telerehabilitation, the therapist's time of service is limited, and training equipment is available in a clinical environment [1]. New engineering applications lead to promising healthcare strategies and a paradigm shift from hospital-centered to a patient-centered approach [2]. ePHoRt [3, 4] is a web-based platform that allows physical therapists to plan, monitor, and evaluating a patient's recovery. The process of assessing patients' movements is supported by depth cameras that send data to a series of intelligent algorithms. Serious games are used to support the performance of the exercises by patients [5]. Throughout this project, the researchers involved have been able to solve a series of challenges at the patient, the therapist, the environment, and even the interaction of these three components. At the patient level, processes intrinsic to the movement (coordination, compensation) and emotional processes (pain/adherence and serious games) are analyzed. Regarding the therapist, the use of Dashboards to support the visualization of the results has been developed. The usability and accessibility of the developed platform have also been the subject of different experiments.

This article presents a reflection of the best practices applied to address each of the challenges above. We discuss and show the results obtained throughout the project's execution, their impacts, and future challenges to planning, monitoring, and evaluating physical telerehabilitation programs for patients who can access a recovery system from home comfort. This article is structured as follows. Section two presents the

D. Russo et al. (Eds.): IHSI 2021, AISC 1322, pp. 402–407, 2021.
https://doi.org/10.1007/978-3-030-68017-6_60

process of planning, monitoring, and evaluation of telerehabilitation programs. Section three describes the current state of ePHoRt, a usable and accessible platform. Section four addresses some challenges in using serious games as a playful tool to support telerehabilitation sessions. Finally, conclusions and future works are presented.

2 The Rehabilitation Program Management Process

Clinical evaluation involves identifying the patient's problems in conducting an appropriate intervention. The examination is an essential component of the rehabilitation process. It is applied twice, once at the beginning and once at the end of the process. The identification and definition of the problems allow the therapist to plan a treatment, which generally must be monitored (follow-up) to evaluate its effectiveness [6].

The therapist collects information about the patient's past and current history, related not only to their health status but also to other components that could influence it (socioeconomic aspects, physical activity practice, etc.). The therapist can plan a treatment appropriate to the history and conditions of each patient. Multisystemic injuries or damages, severe deterioration, and functional loss delay the rehabilitation process. Depending on the recovery time, periodic monitoring of the intervention is essential to determine the process's progression or stagnation. In musculoskeletal physiotherapy, follow-up (advance) is evaluated in stages, which generally last a week. To ensure the development, monitoring, and evaluation of the therapeutic programs, we involve people with extensive physical therapy (researchers, teachers, and students) in the process. In each iteration, we apply efforts with different approaches, metrics, and parameters.

2.1 The Evaluation of the Human Movement

The physical therapist is interested in assessing the range of motion (ROM), coordination, posture, strength, and functionality. Associated with these parameters is the presence of pain. Each person presents a difficulty, moves differently, and has a degree of variability in their movement patterns [7]. Interest and compensatory movements are evaluated based on a personalized plan to the maximum capacities of the patient. In this context, a rehabilitation program based on evaluations and control of patients' sessions is imminent [8]. The real-time assessment of a patient's movements represents a real challenge. The algorithms for capturing movements must consider both the coordination angles (or angles of interest) and the compensatory angles and consider errors such as the loss of skeleton's points or the occlusions that may occur. A mechanism to reproduce the missing issues concerning the previous readings is necessary.

2.2 The Validity and Reliability of Patient Movements

The experience documented in [7] confirms the validity and reliability of our platform. In the study, 7 participants performed four rehabilitation exercises according to a personalized program defined by the physiotherapists. Each participant performs an exercise 11 times. 88% of the results were consistent with the evaluations reported by the five physiotherapists [7]. We were able to detect a significant difference between human evaluation and the platform. The study showed an essential difference between the subjective assessments of the physiotherapists. Three reasons could explain the results. (1) the participants were unaware of the movement to execute; (2) physiotherapists may not have the necessary competencies to estimate interest and compensatory movements; (3) The environment in which the experiment was carried out was unknown to all, in which they had to act in front of a camera. Human behavior is susceptible and can be modified when facing observation, so the last aspect may be why the participants altered their motor performance.

2.3 The Configuration of the Rehabilitation Exercises

ePHoRt has a record of 29 rehabilitation exercises determined by physiotherapy specialists who have carried out testing and loading the information on the platform. We have been able to identify that the average loading time for an exercise is between approximately 5 and 7 min. The creation of a program represents a reasonable time of around 30–35 min of work. These measurements were captured through observations of the platform users.

2.4 A Dashboard for Consulting of Therapeutic Program's Results

A Dashboard shows the most relevant and timely information to facilitate the physiotherapist's decision-making [9]. Generally, the information is presented in a consolidated way to encourage its visualization and interpretation. Among other aspects, it allows us to see the results of a person who has just undergone an orthopedic condition and follows a rehabilitation plan. The main feature of the Dashboard is to present patient information, from which the intensity of pain can be seen. For example: if the patient has a very high level of pain, the program should be suspended since the problem may be the product of a lousy operation or a significant complication. The pain information must be visible so that the physical therapist can observe when the program was suspended. Figure 1 presents an interface for consulting the results achieved by patients. The information can be consulted considering the date range within the selected program, stage, session, shifts, and rehabilitation exercises.

3 ePHoRt: A Usable and Accessible Platform

ePHoRt has been the product of several refinements and/or improvements from experimental usability and accessibility iterations. Several usability studies, from the perspective of the patient [10, 11], and the physiotherapist [12]. Likewise, the study of the patient's emotional states, especially the presence of pain, has been studied [13]. Human factors have been studied to detect errors in movement patterns [7]. The impact of conducting these studies translates into transforming a software application towards friendliness, efficiency, and effectiveness. The experimental telerehabilitation platform developed reaches the AA accessibility standard, by the WCGA 2.1 standard [14].

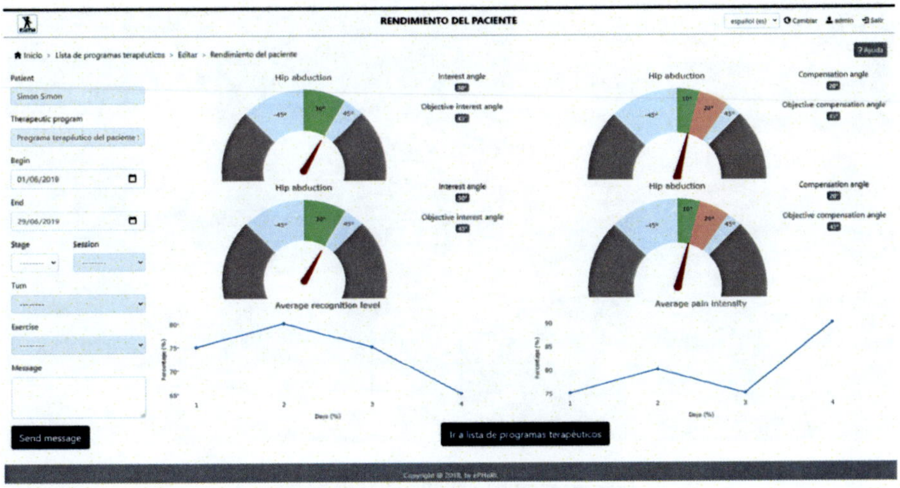

Fig. 1. Dashboard user interface.

This process means that the platform is usable and, above all, accessible. With accessibility, we guarantee the inclusion of potential patients who have a visual disability. Throughout the project, we have refined improvements with the quality of the information, the distribution of the data, the level of color, and contrast required by the users.

4 Challenges in the Use of Serious Games

Over time, many people have classified video games as a leisure or fun tool. However, the video game today can be approached as Serious Games. Serious games have been applied to many areas and applications such as education, a recreation of works of art, etc. With this, it is possible to promote free construction and develop skills in architecture, mechanics, physics, and engineering. We have managed to experiment with four scenarios considering four rehabilitation exercises [5]. Figure 2 shows the different scenes of the game's module developed in Unity 3D. We have incorporated virtual objects that represent challenges for patients. For example, kicking a buoy will allow a patient to perform the hip abduction exercise. However, from a playful perspective, the patient understands that he has to achieve a goal that would enable him to score throughout the progress. This mechanism represents a means of motivation for the patient. Likewise, we have incorporated a series of avatars to personalize users. For example, a male avatar may be associated with a male patient. For the feedback of a patient's activity, we have had a color system to alert patients if they have made a correct or incorrect movement concerning the interest and compensatory angles. For example, the green color represents the avatar feedback of an accurate motion with a tolerance range. Creating an advanced game module using Unity 3D technology has a favorable impact on motivation, feedback, and guidance to the patient when performing rehabilitation exercises.

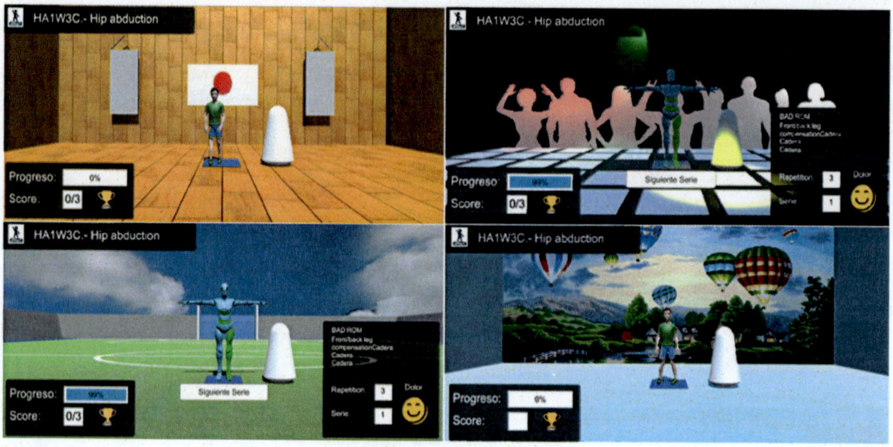

Fig. 2. Serious gaming environments.

5 Conclusion and Future Works

This document shares a series of practices applied during the creation of the ePHoRt telerehabilitation system. The solution has been analyzed from the clinical perspective that allowed us to create therapeutic programs. Likewise, the ergonomics of the interfaces and inclusion through usability and accessibility were analyzed. We incorporate the use of serious games as playful support to recovery activities. In all these works, we have applied methodological instruments and widely known experimental mechanisms. After experimenting, evaluating, and building a functional telerehabilitation platform in which concepts and techniques for assessing the quality of movements have been validated, we consider that it is essential to incorporate the notion of a formula editor for the construction of interest and compensatory angles. This additional functionality represents a future research perspective that will reduce the development and incorporation time of serious support games to the current module developed in the Unity 3D environment.

Acknowledgments. Funding for this research was provided by the Universidad de Las Américas. Quito - Ecuador (Grant No: SIS.JPM.18.01), and the Senescyt-INÉDITA financing program (Grant No: 20190020CI of the project: PIC-18-INE-UDLA-001).

References

1. Vadnerkar, A., Figueiredo, S., Mayo, N.E., Kearney, R.E.: Design and validation of a biofeedback device to improve heel-to-toe gait in seniors. IEEE J. Biomed. Health Inform. **22**(1), 140–1462 (2018)
2. JBHI: Guest editorial: Internet of medical things for health engineering. IEEE J. Biomed. Health Inform. **24**(6) 1539–1540 (2020)

3. Rybarczyk, Y., Deters, J.K., Gonzalvo, A.A., Gonzalez, M., Villarreal, S., Esparza, D.: ephort project: a web-based platform for home motor rehabilitation. In: World Conference on Information Systems and Technologies. Springer, pp. 609–618 (2017)
4. Pérez-Medina, J.L., Jimenes-Vargas, K.B., Leconte, L., Villarreal, S., Rybarczyk, Y., Vanderdonckt, J.: ephort: towards a reference architecture for tele-rehabilitation systems. IEEE Access **7**, 97159–971765 (2019)
5. Solórzano, S., Espinosa-Alvarez, P.D., Jimenes-Vargas, K., Pérez-Medina, J.L.: Using serious games and motion tracking for physical rehabilitation. In: International Conference on Human Systems Engineering and Design: Future Trends and Applications, pp. 180–185. Springer (2020)
6. O'Sullivan, S.B., Schmitz, T.J., Fulk, G.: Physical rehabilitation. FA Davis (2019)
7. Esparza-Yánez, W.D., Aladro-Gonzalvo, A., Pérez-Medina, J.L., Jimenes, K.B., Baldeon, J., et al.: Detecting human factors that induce errors in movement patterns for the development of a web-based telerehabilitation platform. In: Nunes, I. (ed.) Advances in Human Factors and Systems Interaction, AHFE 2019. Advances in Intelligent Systems and Computing, vol. 959. Springer, Cham (2020)
8. Rybarczyk, Y., Kleine Deters, J., Cointe, C., Esparza, D.: Smart web-based platform to support physical rehabilitation. Sensors **18**(5), 1344 (2018)
9. Sarikaya, A., Correll, M., Bartram, L., Tory, M., Fisher, D.: What do we talk about when we talk about dashboards? IEEE Trans. Visual Comput. Graphics **25**(1), 682–692 (2018)
10. Pérez-Medina, J.L., González, M., Pilco, H.M., Vargas, K.B.J., Acosta-Vargas, P., Sanchez-Gordon, S., Calle-Jimenez, T., Esparza, D., Rybarczyk, Y.: Usability study of a web-based platform for home motor rehabilitation. IEEE Access **7**, 7932–794711 (2019)
11. Pérez-Medina, J.L., Espinosa-Alvarez, P.D., Jimenes-Vargas, K., Acosta-Vargas, P., Rybarczyk, Y.: Serious-games-based exercises for arthroplasty rehabilitation. In: International Conference on Applied Human Factors and Ergonomics, pp. 619–626. Springer (2020)
12. Pérez-Medina, J.L., Jimenes-Vargas, K.B., Acosta-Vargas, P., Rybarczyk, Y., González, M.: User experience assessment of a tele-rehabilitation platform: the physiotherapist perspective. In: International Conference on Applied Human Factors and Ergonomics, pp. 463–473. Springer (2019)
13. Rybarczyk, Y., Leconte, L., Pérez-Medina, J.L., Vargas, K.J., Acosta-Vargas, P., Esparza, D.: Design of a kinematic and emotional assessment module for the tele-rehabilitation platform. Int. J. Model. Optim. **9**(2), 92–96 (2019)
14. Acosta-Vargas, P., Salvador-Acosta, B., Gonzalez, M., Pérez-Medina, J.L., Acosta-Vargas, G., Jimenes-Vargas, K., Rybarczyk, Y.: Web accessibility analysis of a tele-rehabilitation platform: the physiotherapist perspective. In: International Conference on Applied Human Factors and Ergonomics, pp. 215–221. Springer (2020)

Computer Vision Based Rehabilitation Assistant System

Nudpakun Leechaikul(✉) and Siam Charoenseang

Institute of Field Robotics (FIBO),
King Mongkut's University of Technology Thonburi, Bangkok, Thailand
nudpakun.1@om4you.com, siam@fibo.kmutt.ac.th

Abstract. This research aims to develop a computer vision based assistant system called "ViHab" for physiotherapy by analyzing physical posture from the digital image processing and machine learning on an Android-based smartphone so that patients can perform physical therapy anywhere conveniently in the everywhere concept. The patient's physical therapy records are stored in a cloud-based database. Doctors can therefore monitor the physical behavior of the patient in more details by the system's visualization dashboard. The preliminary result showed that the pose of physical therapy can be detected from the proposed system to grade the performance of rehabilitation.

Keywords: Android-based smartphone · Assistant system · Computer vision · Digital image processing · Machine learning · Physical posture · Physical therapy · Physiotherapy · Rehabilitation

1 Introduction

Physical therapy is a very important process to treat for the movement disorder of patients from surgery, accident, neuropathy, and etc. For the convenience of the patient, the patient can do rehabilitation at home under the doctor's supervision. The doctor needs to follow up, examine, and analyze the effects of home rehabilitation via an appointment. However, the physical therapy data of the patient is not collected every time. Therefore, the physiotherapist could receive incomplete information about symptoms of pain disorders and the physical therapy of the patient. The doctor could diagnose, recommend, and plan the physical therapy ineffectively. This leads the longer treatment than the original plan.

A home-based rehabilitation system has been shown in a lot of previous research works. Several previous works need to have a computer and the Microsoft Kinect, a 3D depth camera, or a webcam device [1–13]. The monitoring and tracking system have been developed for collecting the data when the patient takes the physical therapy at home. A summary of the development methods can be shown in Table 1.

D. Russo et al. (Eds.): IHSI 2021, AISC 1322, pp. 408–414, 2021.
https://doi.org/10.1007/978-3-030-68017-6_61

Table 1. Comparison of the home-based physical therapy.

Research	Camera input				Real-time evaluation	Capture reference exercise	Rehabilitation management	Summary report
	Kinect	Webcam	RGB-D camera	Smartphone				
1	*				*			*
2	*	*			*	*		*
3	*				*	*	*	*
4	*	*	*		*			
5	*				*			*
6					*			*
7		*			*			*
8		*			*			
9	*				*			*
10		*			*			
11	*				*	*		
12					*			*
13			*		*			*
Purposed system				*	*	*	*	*

Machine Learning (ML) technology has been widely applied to research about object and pattern recognition. In particular, the objects in the picture are classified by the machine learning. In this proposed work, the machine learning is used to analyze the human posture and to examine the movement of the joints on the patient's body with 17 points by importing a training data model of poses. If the system can analyze the human posture accurately, it improves the performance of the doctor's diagnosis and rehabilitation.

Hence, this research proposes and develops a computer vision based rehabilitation assistant system called "ViHab" for the physical therapy purpose by analyzing the posture of the patient based on the digital image processing on an Android smartphone. The schedule of physical therapy sessions are adjusted by the doctor. The postures for physical therapy are recorded by physiotherapists into the system's library. The proposed system records the averaged exercise time of each posture, the accuracy of the patient's posture, and the pain score that the patient responds via the pain score survey. The information of the proposed physiotherapy system is saved in the database located on the cloud server.

2 System Overview

In many previous studies, they have applied the Microsoft Kinect and the other additional equipment for tracking physiotherapy activities under the Microsoft Windows operating system. This leads additional installation as well. The proposed system was designed for the patients to easily access the system. Hence, it was developed as a mobile application running on a smartphone under the Android operating system.

The core development tools of the proposed system on a smartphone was the Android Studio [14] based on Kotlin [15] to produce native android application with a PoseNet model [16]. This model was used to recognize the human's physical gestures from the images or streaming video to find all 17 joints of the key body. Physical therapists and patients receive images from the smartphone's camera through the PoseNet library processing to analyze posture in physical therapy. The Doctor can access the system via the web application. The data collector API managed all information from the patient, physical therapist, and doctor as shown in Fig. 1. This proposed system utilized the built-in camera on the smartphone without extra hardware. Visual studio code [17] was a core software tool for developing the proposed PHP-based web application [18] which manages the system's data as shown in Fig. 2. The developed application can run on a smartphone.

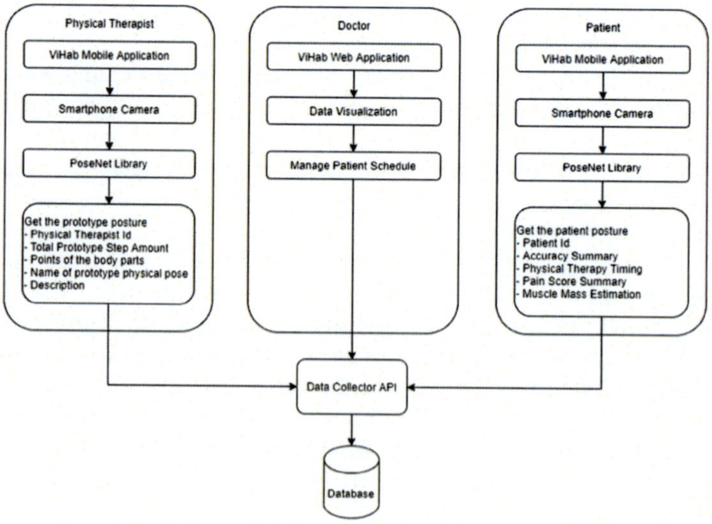

Fig. 1. System overview

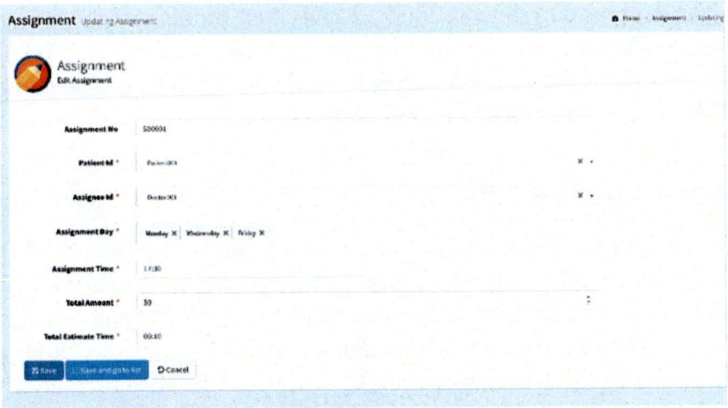

Fig. 2. Web-based user management application

3 Methodology

3.1 Posture Recognition

The proposed application running on an Android-based smartphone applied the PoseNet library to use the smartphone's camera in order to recognize the body postures. The keypoint heatmap was used to analyze the precision of the body key points. The resolution of output stride determined the accuracies of the outputs and the speeds of process. There were three resolutions ranging from 8, 16, and 32 as shown in Fig. 3. This proposed system applied output stride with resolution of 16 because it was used on a smartphone with a low specification and the accuracy of the output was still

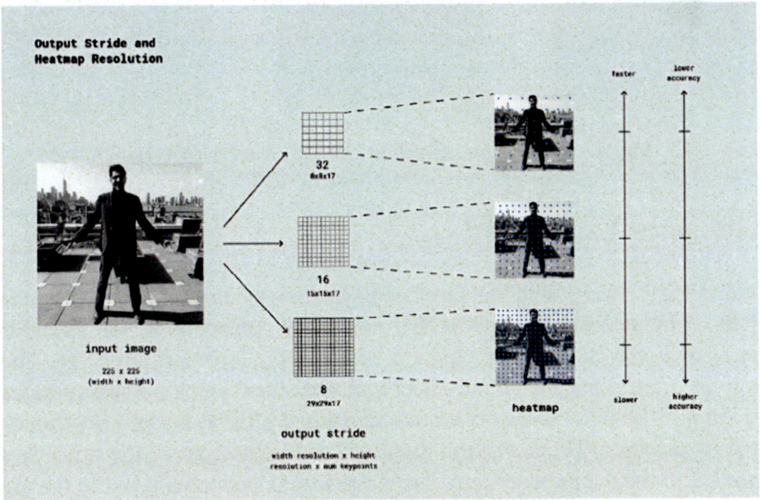

Fig. 3. Output stride and heatmap resolution [19]

acceptable. The proposed system needed 17 body key points which were graded from the upper part of the body to the feet to show the overall accuracy. The 17 body key points were positions in the cartesian coordination. The proposed system calculated the accuracy of the overall movement of the body.

3.2 Database on the Cloud Server

In the proposed system, the relational database management system (RDBMS) was designed along with the Structured Query Language (SQL) for MySQL server. The system's data was recorded on the cloud database. Since the proposed system collected many body key points of each body posture, the database collected a lot of complex recorded data. Therefore, it was necessary to design the database structure to be more flexible so that the proposed system can manipulate the database in a reasonable time. This proposed system run on a virtual private server (VPS) with the entry-level package under the Ubuntu LTS 14.04.6 operating system and the installed Apache web server to support the web application and PHP APIs.

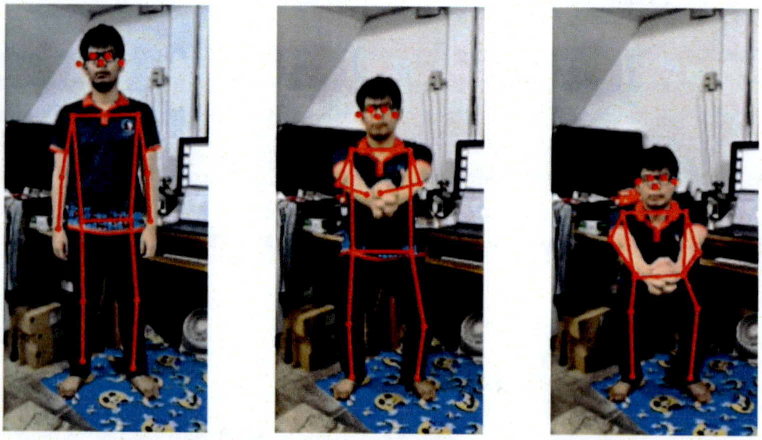

Fig. 4. User's pose detection using the proposed system

4 Preliminary Experiment

In the preliminary experiment, the proposed system was easily launched on the user's smartphone. The operation of PoseNet model can produce a fair accurate pose detection even it was an obscure position between the arm and the body. The user's postures were graded and the sound effect was generated when the user performed the correct posture. The pain-scale survey was collected after every physical therapy was completed by the user. The accuracy, duration, and pain score of the physical therapy were recorded to the database system. Then, the results were displayed to the doctor for analyzing the patient's progress and issuing the suitable physical therapy requirement.

5 Discussions and Future Work

In this proposed system, the Android-based smartphone was exploited for the digital image processing for the physiotherapy. It was able to analyze the patient's body posture as shown in Fig. 4. The result was shown with the 17 red dots and 16 linking edges on the various body key points. In Fig. 4, the overlap of the arms and the hands on top of the body was analyzed quite accurately. The system's data was stored on the cloud server. The results of physiotherapy of each patient were monitored by the doctor. Hence, this proposed system assisted the patient to do the physical therapist at home.

In the future work, various smart devices such as a smartwatch, smart wristband, and smart necklaces can be connected to the proposed system to collect more additional information. This should improve the performance of tracking of the patient's postures.

References

1. Dorado, J., del Toro, X., Santofimia, M.J., Parren, A., Cantarero, R., Rubio, A., Lopez, J.C.: A computer-vision-based system for at-home rheumatoid arthritis rehabilitation. Biomed. Sens. Appl. Brain/Body Imaging **15**(9), 1550147719875649 (2019)
2. Gu, Y., Pandit, S., Saraee, E., Nordahl, T., Ellis, T., Betke, M.: Home-based physical therapy with an interactive computer vision system. In: IEEE/CVF Intenational Conference on Computer Vision Workshop. IEEE, Boston (2019)
3. Chuan-Jun, Su.: Personal rehabilitation exercise assistant with Kinect and dynamic time warping. IJIET. **V3**(316), 448–454 (2013)
4. Colyer, S.L., Evans, M., Cosker, D.P., Salo, A.I.T.: A Review of the evolution of vision-based motion analysis and the integration of advanced computer vision methods towards developing a markerless system. Sport Med. Open **4**(1), 24 (2018)
5. Jaume-i-Capó, A., Martínez-Bueso, P., Moyà-Alcover, B., Varona, J.: Interactive rehabilitation system for improvement of balance therapies in people With cerebral palsy. IEEE Trans. Neural Syst. Rehabil. Eng. **22**(2), 419–426 (2014)
6. Li, Y., Burns, C.M., Kulić, D.: Ecological interface design for knee and hip automatic physiotherapy assistant and rehabilitation system. In: International Symposium on Human Factors and Ergonomics in Health Care: Advancing the Cause (2015)
7. Sucar, L.E., Luis, R., Leder, R., Herna´ndez, J., Sa´nchez, I.: Gesture Therapy: a vision–based system for upper extremity stroke rehabilitation. In: 32nd Annual International Conference of the IEEE EMBS, pp. 3690–3693. IEEE, Argentina (2010)
8. Venkataraman, V., Turaga, P., Lehrer, N., Baran, M., Rikakis, T., Wolf, S.L.: Attractor-shape for dynamical analysis of human movement: applications in stroke rehabilitation and action recognition. In: Conference on Computer Vision and Pattern Recognition Workshops, pp. 514–520. IEEE, Arizona (2013)
9. González-Ortega, D., Díaz-Pernas, F.J., Martínez-Zarzuela, M., Antón-Rodríguez, M.: A Kinect-based system for cognitive rehabilitation exercises monitoring. J. Cmpb. **113**, 620–631 (2014)
10. Reyes-Amaro, A., Fadraga-gonzalez, Y., Vera-Perez, O.L., Dominguez-campillo, E., Nadarse-Ravelo, J., Mesejo-Chiong, A., Moya-Alcover, B., Jaume-i-Capo, A.: Rehabilitation of patients with motor disabilities using computer vision based techniques. JACCES **2**(1), 62–70 (2012)

11. Jiang, H., Jie, X.: Kinect-based rehabilitation training assistant system research and implementation. In: International Conference on Software Engineering and Computer Science. ICSECS, China (2013)
12. Borghese, N.A., Pirovano, M., Mainetti, R., Lanzi, P.L.: An integrated low-cost system for at-home rehabilitation, pp 553–556. IEEE (2012)
13. Pastor, I., Hayes, H.A., Bamberg, S.J.M.: A feasibility study of an upper limb rehabilitation system using kinect and computer games. In: 34th Annual International Conference of the IEEE EMBS, pp. 1286–1289. IEEE, USA (2012)
14. Download Android Studio and SDK tools | Android Developers. https://developer.android.com/studio
15. Kotlin. https://kotlinlang.org/
16. Tensorflow/tfjs-models. https://github.com/tensorflow/tfjs-models/tree/master/posenet
17. Visual Studio Code - Code Editing. Redefined. https://code.visualstudio.com/Kotlin, https://kotlinlang.org/
18. PHP: Hypertext Preprocessor. https://www.php.net/
19. Pose estimation | TensorFlow Lite. https://www.tensorflow.org/lite/models/pose_estimation/overview

Human Centered Design in One New Hospital in Canada: A Lived Experience of Healthcare Professionals

Zakia Hammouni[1,2](✉) and Tiiu Poldma[1,2]

[1] Université de Montréal, Succursale Centre-Ville, Montreal CP 6128, Canada
zakia.hammouni@umontreal.ca
[2] Centre for Interdisciplinary Rehabilitation Research of Greater Montreal, Montreal, Canada

Abstract. Using a constructivist approach, this study focuses on how the design and the physical environment of one contemporary hospital in Canada are perceived by the healthcare professionals and how these factors affect their work and their well-being. Data was collected by observing the physical setting and healthcare professionals' behavior and included semi-structured interviews of 44 healthcare professionals of two care units during 3 months in 2018. The analysis is based on both comparative and interpretive approaches. Results showed that this physical environment is supportive of the work of healthcare professionals and, keeping in mind the linear spatial organization and decentralized layout, this study highlights the importance of adapting the organizational and management system to the spatial configuration of care units in order to achieve the performance and the well-being of these healthcare professionals.

Keywords: Human centred design · Technology · Physical environment · Healthcare professionals · Lived experience

1 Introduction

In the hospital environment, the user-centered design approach has been at the forefront since the end of the 20th century [1]. The focus of this approach is to improve the experience of hospital users due to the stress that this environment can generate as much as this stress might be due to illness. User-centered design in the hospital environment has generally followed the patient-centered approach in medicine instead of a disease-centered care approach [2]. The priority is taking care of the patient in terms of physical and psychological health and providing social support. This is aligned with the emergence of new health care facilities whose physical environment contributes to the process of patient care and recovery [2], as this reflects the new concept of health promotion.

This study focuses on the lived experience of healthcare professionals to understand how the design of the contemporary hospital is perceived by them and how the design of the environment affects their work. It examines the physical environment of a new hospital complex that integrated new design approaches to alleviate users stress and

D. Russo et al. (Eds.): IHSI 2021, AISC 1322, pp. 415–420, 2021.
https://doi.org/10.1007/978-3-030-68017-6_62

ensure their well-being, using the patient-centred design approach. In terms of hospital design, Becker and Jones-Douglas [3] illustrate how research can inform hospital design and point to how architects and designers incorporate research approaches in the development of patient-centered care environments. Other authors demonstrate the need to think about a physical environment that centers on all hospital users, considering not only patients but also care staff and other support and administrative staff as well [4].

New university hospital centres (CHU) constructed in the last decade in Canada have changed the hospital environment for patients, and yet we do not know how this physical environment is experienced by the healthcare professionals themselves. As prior scientific literature has placed less emphasis on the healthcare professionals' interactions within the physical work environment, this study was done to understand their work environment [5].

2 Methodology

This study proceeded using a constructivist approach, wherein the experiences of healthcare professional was studied from multiple perspectives. First, observations of both the physical setting and the healthcare professionals' behavior was conducted during different work shifts, with 8 weeks of observation of their behavior in the spaces of the units where they worked. The observations were done over a period of 2 weeks per care unit during the day shift, and 2 weeks per care unit in the evening and night shifts (total of 270 h of observations). This was followed by semi-structured interviews) and supported by a mapping of perceptions. We conducted these interviews with healthcare professionals (n = 44) in two care units in the hospital over a period of 3 months, from August 1st, 2018 to November 1st, 2018. This interview was completed by an exercise of a mental map, where we asked healthcare professionals to draw their view of the interior environment attributes, and whether they felt these were important as either supportive or negative for them, in the care unit or within the entire hospital.

Analysis methods are based on a comparative and interpretive approach, wherein a thorough description was constructed of the essence of the lived experience of health professionals. We essentially followed an inductive and synthetic approach that included emergent themes and categories. However, we highlighted categories of themes that were subsequently compared with existing data in the scientific literature.

3 Setting

We conducted our study on two care units in one newly built CHU (university hospital center) that opened in September 2017 and is situated in a large metropolitan city in Canada. At this hospital center, several innovative and technological elements have been introduced to better provide advanced care. This large hospital complex is composed of several buildings that are connected between by corridors or footbridges. In its main hospital building, the care units are located from the 8th to the 20th floor. At each floor from the 9th and 20th floor, two care units are connected between them by circulation spaces such as elevators, staircases and two family' lounges with bathroom. Each unit consists of a peripheral strip which includes mostly 36 single patient rooms, a central area for staff and two long corridors (north and south) that run along the care unit and are devoted to healthcare professionals' spaces and rooms. The length of these corridors varies between 79 and 90 m. Four corridors perpendicular to the two long corridors cross the care unit, and three of them serve the nursing workstations. The central part in the care unit is dedicated to workspaces for healthcare professionals and support and maintenance staff (see Fig. 1). The introduction of daylight in the most traffic areas, waiting areas and in the patient's room, is facilitated by windows that provide panoramic views. The patient room (20 sq.m. each) includes big windows allowing daylight and panoramic views, a warm ambience and equipped with furniture, space for patients and their family members, and additional facilities for healthcare professionals (the patient lift fixed above the patient's bed). Colour is also used as a key element for the well-being of the users in this large hospital. However, the rest of the rooms designated for healthcare professionals, with the surface area of (±) 12,7 sq.m), has no windows and daylight and few comfortable features (Fig. 2).

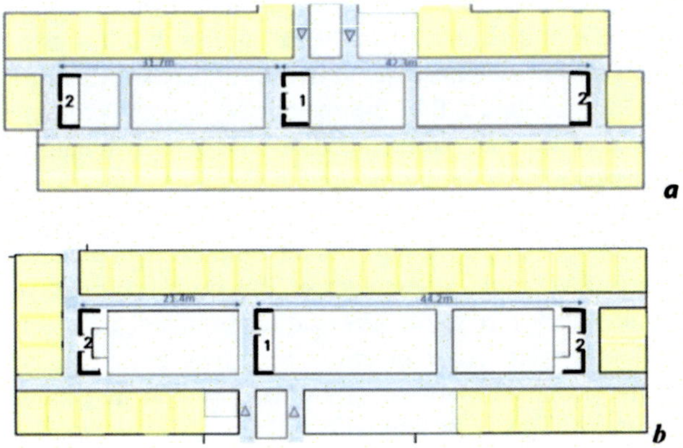

Fig. 1. Spatial configuration of care unit with decentralized workstations. Most of healthcare professionals' spaces are in the central part of the care unit. **a:** south care unit, **b:** north care unit [6]. 1: central workstation, 2: secondary workstations. Peripheral location of the 36 spacious single patient rooms.

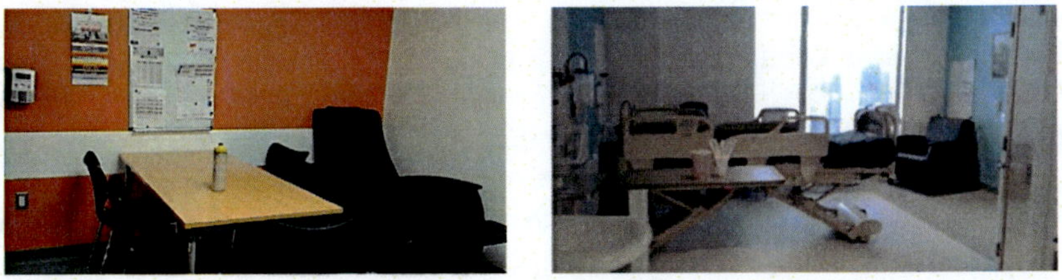

Fig. 2. **Left:** Healthcare professionals' rest room in the care unite; **right** [6]: single patient rooms.

4 Results

Globally, the healthcare professionals evaluate positively the environment of this new CHU. Certain physical attributes of the environment emerge in the analysis as positive, including thermal comfort and the air conditioning system, the color choices, natural light in the rooms and circulation spaces, the panoramic views in both rooms and corridors of the care units, lighting of the spaces, and the comfortable width of the corridors. Furthermore, certain technological attributes such as the patient rolling screens in the nursing workstations, the pneumatic transport and the chip system in the clean utility room were considered as positive elements of the environment to facilitate patient-centered care.

Our analysis also showed that the capacity of the single-bed room, its size, its provision of a private bathroom, and the visual quality of its environment with the light/colour ambiance, the views offered by most rooms, the thermal comfort and privacy it can offer to the patient and his family, and the interpersonal relationship between the care professional and the patient, are all favourable elements for their work. The patient rack attached to the ceiling of the room is another very facilitating element in the work of health care professionals and allows for the safety of these professionals because it protects them from back injuries, but it is also an element that allows for the safety of the patients since it makes it easy for them to move from the bed in complete safety against falls. The room environment promotes the confidentiality of healthcare professionals' work and allows them to maintain patient privacy by not having another patient sharing the room with them, as is the case with a double or multiple room [7, 8].

Healthcare professionals talk about certain complexities related to their work, leading to their less favourable assessment of the organizational, management and socialization attributes they link to the physical environment in which all these social dynamics exist. This physical environment is perceived more challenging than other characteristics of the hospital environment, by most healthcare professionals with the exception of physicians. This is in line with the results of previous research that has already demonstrated similar evaluations of the physical environment by nurses [9]. Analysis of the data shows that the linear spatial configuration with decentralized workstations, long corridors and care preparation and storage spaces at both ends of the care unit, as well as the distribution of patients by nurse and the location of single-dose carts corresponding to these patients, generate a lot of walking among health professionals. All of the healthcare professionals in our study reported that part of the work is walking a lot, but this did not consistently negatively affect the work of all healthcare professionals (18%). On the other hand, all healthcare professionals agree that avoidable travel, such as searching for patient records, could be alleviated in their work.

Participants in our study generally felt that the physical environment has not been designed for their well-being at work in mind. They tend to compare the spaces dedicated to them with those dedicated to the patient. This new hospital has been described by some healthcare professionals as a "hospital-hotel". This "hospital-hotel" qualification is considered by Suess and Mody [10] as a strong element to support the well-being of patients during their hospitalization by introducing elements of well-being in their rooms, contrary to Rosenthal [11] who considers that these hospitals are doing too much for the comfort and support of the patient and the hospital.

Observations and interviews confirm that this CHU is a very complex work environment. This complexity is manifested by the realities experienced in this hospital and the characteristics of the care environment. Some participants considered the current environment as similar to an airport, because they are unable to build benchmarks and are always looking for landmarks to help guide them. They see this environment as not conducive to socialization in the workplace and try to connect with the rest of the professionals to build social bonds. The narrative data demonstrates that participants want to have social connections and that these are hindered by the complex nature and size of places that limit more intimate social contact.

From the perspective of the user-centred design approach, the hospital environment must provide a comfortable patient care environment and a work environment that is conducive to well-being in the work of the healthcare professionals [5, 12–14]. This environment should be both a healing environment through opportunities for environmental control, social and technical support, as well as providing both positive distraction and a restorative environment, for both patients and health care professionals. The hospital is not a healing machine, but rather, it is a living space within which organized social relations occur and is structured in places charged with social life as Fornara and al. [15, 16] point out.

Data analysis also showed that in the care units the rest area for healthcare professionals are small relatively to the number of professionals affiliated with the care unit, and its visual environment gives a feeling of confinement to healthcare professionals with the lack of daylight and cold temperature of artificial lighting. This environment offers limited opportunities for socialization among different healthcare professionals and is not very helpful for recovering from workload and/or work stress during their breaks as a restructuring environment as proposed by previous studies [17]. However, healthcare professionals prefer these rest rooms because of their proximity to the care unit and patients. Given the size of the hospital, and the location of the inpatient units on floors between the 8th and 20th floors, most healthcare professionals choose the rest area on the care unit. This is also related to the many trips that professionals make during their shift. The hospital must offer professionals a choice of service spaces such as the cafeteria at more than one strategic location in the hospital.

5 Conclusion

This study highlights the complexity and diversity of the perspectives of healthcare professionals in the two care units studied, through documentation of the settings and narratives of the lived experiences of the participants in the study. The hospital is a complex work environment and analyses of the results of our study demonstrate this reality. It is an environment that is shared by different types of users with very different cognitive and physical abilities. The environment also provides care to patients who are often stressed in the face of illness, but it is a stressful work environment for healthcare professionals because of the workload and the worries of committing errors in providing care, coping with patient stress, having to use new technologies for care. Therefore, the physical environment, in which social, communicational, organizational and management dynamics, must alleviate this stress and ensure the well-being of these healthcare professionals in their work environment. Accordingly, certain attributes still need to be improved to ensure this well-being of healthcare professionals and facilitate their work. This study highlights the importance of adapting the organizational and management system to the spatial configuration of care units in order to achieve optimum performance.

Acknowledgments. Many thanks to all the participants to this study and the contact person at the studied hospital.

References

1. Bates, V.: Humanizing healthcare environments: architecture, art and design in modern hospitals. Des. Health **2**, 5–19 (2018)
2. Dilani, A.: A new paradigm of design and health. Health Promot. Environ. Des. **1525**, 15—25 (2004)
3. Becker, B., Parsons, K.S.: Hospital facilities and the role of evidence-based design. J. Facil. Manag. **5**, 263–274 (2007)
4. Mroczek, J., Mikitarian, G., Vieira, E.K., Rotarius, T.: Hospital design and staff perceptions: an exploratory analysis. Health Care Manager. **24**, 233–244 (2005)
5. Huisman, E., Morales, E., Van Hoof, J., Kort, H.S.M.: Healing environment: a review of the impact of physical environmental factors on users. Build. Environ. **58**, 70–80 (2012)
6. Hammouni, Z.: The contemporary hospital design: a contextual study of the lived experience of healthcare professionals and quality of care. In: Proceedings of the 6th International Conference on Design4Health vol. 1, pp. 179–188 (2020)
7. Chaudhury, H., Mahmood, A., Valente, M.: Advantages and disadvantages of single-versus multiple-occupancy rooms in acute care environments: a review and analysis of the literature. J. Environ. Behavior. **37**, 760–786 (2005)
8. Maben, J., Penfold, P., Glenn, R., Griffiths, P.: Evaluating a major innovation in hospital design: Workforce implications and impact on patient and staff experiences of all single room hospital accommodation. Report of phase 1 findings for HaCIRIC2012 (2012)
9. Sadatsafavi, H., Walewski, J., Shepley, M.M.: Factors influencing evaluation of patient areas, work spaces, and staff areas by healthcare professionals. Indoor Built Environ. **24**, 439–456 (2015)
10. Suess, C., Mody, M. A.: Hotel-like hospital rooms' impact on patient well-being and willingness to pay. Int. J. Contemp. Hospitality Manage. 30, (2018)
11. Rosenthal, E. Is this a hospital or a hotel. New York Times 21, (2013)
12. Zhang, Y., Tzortzopoulos, P., Kagioglou, M.: Healing built environment effects on health outcomes: environment–occupant–health framework. Build. Res. Inf. **47**, 747–766 (2019)
13. Dijkstra, K., Pieterse, M. E.: The psychological effects of the physical healthcare environment on healthcare personnel. Cochrane Database Syst. Rev. **45**(4), 229–235 (2011)
14. Ulrich, R.S., Berry, L.L., Quan, X., Parish, J.T.: A conceptual framework for the domain of evidence-based design. HERD: Health Environ. Res. Des. J. **4**, 95–114 (2010)
15. Fornara, F., Bonaiuto, M., Bonnes, M. Les attentes du personnel hospitalier envers son lieu de travail (2013).
16. Dans Rioux, L., Le Roy, J., Rubens, L., Le Conte, J.: Le confort au travail : Que nous apprend la psychologie environnementale ?. Québec, Québec : Presses Universitaires de Laval, pp. 225–244 (2013).
17. Dalke, H., Little, J., Niema, E., Camgoz, N., Steadman, G., Hill, S., Stott, L.: Colour and lighting in hospital design. Opt. Laser Technol. **38**, 343–365 (2006)

Method of Similarity Implementation for the Decision-Making Process in Small Towns Transportation Management

Peeter Lorents[1] and Maryna Averkyna[1,2](✉)

[1] Estonian Business School, A. Lauteri, 3, Tallinn, Estonia
peeter.lorents@ebs.ee

[2] The National University of Ostroh Academy, Seminarska, 2, Ostroh, Ukraine
maryna.averkyna@oa.edu.ua

Abstract. The paper deals with the method of similarity for the decision-making process. It was noted that there are two approaches to the systems similarity analysis – structural similarity and descriptive one. Structural similarity, which is based on theorems proven in algebraic systems and universal algebra theories, which are difficult to apply. The authors acknowledged that descriptive similarity is more relevant for assess two sets, which must consist of relevant statements. This study examines the application of descriptive similarity in the process of managing the development of urban transport systems in small towns in Estonia and Ukraine.

Keywords: Situations and developments as systems · Structural similarity · Descriptive similarity · Numerical evaluation of descriptive similarity · Urban transport systems

1 Introduction

Every day we make dozens, probably hundreds, of decisions' [1]. Managers often try to resolve problems with the decisions that were made earlier under similar circumstances. There is a similar situation in the complex systems management. Hence, town's managers usually use obtained earlier experience for solving a particular problem. Of course, the use of the experience of other cities is undoubtedly important in management, especially when its implementation gives a positive result. However, the implementation of such experience does not always give positive effects. That is why for making decisions, it is necessary to clearly understand whether the investigated systems are similar enough. In this case, research and Method of Similarity implementation for decision-making process is very relevant, moreover, when researchers work mainly with qualitative information, rather than the quantitative one.

It should be noted that there are two approaches to the systems similarity analysis – structural similarity and descriptive one. During the analysis of scientific literature it was discovered that the numerical estimation of the similarity of the sets is based on a method analogous to that described in the works published in 1901 by Swiss botanist and plant physiologist Paul Jaccard [3, 4]. Qi Zhang, Meizhu Li, Yong Deng noted that

D. Russo et al. (Eds.): IHSI 2021, AISC 1322, pp. 421–427, 2021.
https://doi.org/10.1007/978-3-030-68017-6_63

'those Jaccard Index based methods have been wildly used in the research of complex network, especially in social networks' [10]. Very significant and substantive difference here is that the calculation of the Jaccard coefficient is based on the equal elements of the two sets. Jiayu Han, Lei Zheng, He Huang, Yuanbo Xu, Philip S. Yu, Wanli Zuo, 'proposed a Hierarchical Similarity Measure which is a parameter-free method and can better integrate the interactions between the users and the items' [5]. It is the same as a structural similarity but this method is difficult to apply for predicates' analysis, which describes a system. However, Peeter Lorents, Erika Matsak, Ahto Kuuseok, Daniil Harik in the work published in 2017 relied on equalized elements to calculate the similarity coefficient [6].

It is necessary to point out that other practically usable ways to study the sets, relying on their descriptions, are rather hard to apply. However, there is a certain relationship between the set of structures, and the number of statements used to describe them. Indeed – the structural similarity and descriptive similarity are linked in many ways. One of these ways is based on theorems proven in algebraic systems and universal algebra theories, which are presented by Paul Cohn [2], Anatoliy Maltsev [8]. Descriptive similarity is essential in decision-making process in order to implement Decision Support Systems in the management of urban transportation. Maintenance sustainable urban transportation is crucial for any country. It is important to develop the method of the making-decision process in the urban transportation management, based on the descriptive similarity approach.

The purpose of the paper is to explain how Method of Similarity can be implemented for the decision-making process in small towns transportation management. Research question of the paper is how descriptive similarity can be used for Decision Making Process. In the paper the concept of the descriptive similarity and it is numerical estimation will be demonstrated with some examples, as well as conclusions.

1.1 Descriptive Similarity and Its Numerical Assessment

Peeter Lorents and Maryna Averkyna in the work published in 2019 pointed out that "in order to assess two sets of descriptive similarity, it is first necessary to make descriptions of both sets, which must consist of relevant statements" [7]. Therefore, it is necessary to clarify what statements from one description and another description can be considered equivalent. There are now three sets of statements [7]:

- P is a set whose elements are statements from the first description;
- Q is a set whose elements are statements from another description;
- Com (P, Q) C is a set of elements that are ordered pairs, where the first position of the pair has the claim P, the second position has the claim Q, and these statements have been equalized by each other.

Lorents, Averkyna (2019) presented towns' similitude matrix based on claims. In order to evaluate the descriptive similarity of the descriptions or sets of relevant statements – that is, the descriptive similarity of some things, situations, developments, etc., we will use the numerical value defined by Lorents below - the descriptive similarity coefficient:

$$\begin{aligned}\mathrm{Sim}(P,Q) &= E(\mathrm{Com}(P,Q)) : [E(P) - E(\mathrm{Com}(P,Q)) + E(Q)] = \\ &E(\mathrm{Com}(P,Q)) : [E(\mathrm{Spec}(P)) + E(\mathrm{Com}(P,Q)) + E(\mathrm{Spec}(Q))]\end{aligned} \quad (1)$$

According to this formula, it is first necessary to make descriptions of both sets, which must consist of relevant statements. At the next stage, it is necessary to clarify what statements from one description and another description can be considered equivalent.

2 Method of Similarity Implementation for the Decision-Making Process in Small Towns Transportation Management

It is necessary to point out that István Szűts, András Bakó emphasized the main approaches in decision-making: the psychological approach, the classical economy, and the administrative models approach the reality [9].

However, these approaches have some limitation in their application to the system's predicates analysis. For successful implementation of the Method of Similarity for Sustainable Urban Transportation Maintenance it is important to identify steps of the decision-making process. The steps of the method of the making-decision process sustainable urban transportation are as follows:

1. Gathering information for forming Towns' Statements
2. Determination of the Towns' Equivalent Statements
3. Forming Towns' Statements
4. Creation Towns' Similitude Matrix
5. Make a comparison of the general part of common parts
6. An end-to-end comparison of the statements of all systems studied
7. Implementation of the best practice

Example. We want to implement in Ostroh the Estonian small towns' experience of managing urban transportation systems (see example Table 2). Ostrog is a small town of regional significance (since 1995) in Ukraine, a center of the Ostroh district of the Rivne region. Located 16 km from the Ostroh railway station on the Zdolbuniv - Shepetivka line, 47 km from Rivne. Ostroh is known since 1100. In 1585, the city received the Magdeburg Law. The city is located 3.2 km from the Khmelnitsky nuclear power plant.

At the next stage, it is necessary to assess how the investigated town is similar with compared ones. For such purpose, we propose to form predicate's system based on the study of towns. Listed predicates system is to be filled to the table for all towns. We recommend form a table to assess the general statements of the city as a structure, transport structure and management system is (Table 1).

Based on comparisons of the town list of predicates, we propose to form a table of similar statements in order to estimate the Lorents coefficient (see the Example 1). In case, when all statements are written down, it is necessary to calculate the similarity coefficient. This will allow us to understand how much the systems (towns or cities) are similar, which of the statements should be applied in the decision-making process for urban sustainable transportation. Then we propose to create towns' similitude matrix (see Table 2). It should be noted that the coefficient should not be less than 0.5 and the closer the value of the coefficient to 1, the more similar systems (towns, cities).

Table 1 . Determination of statements' town.

Statements' Town
1. Small town.
2. Minimum population 5000.
3. /////////////////////////////////
4. /////////////////////////////////
....
n. Local council's manager does not calculate the number of passengers who go off in each bus station.

According to the data of the matrix, we can see that the similarity of Ostroh with Estonian towns is within the range of 0.49–0.6. Ostroh is similar to the Estonian towns for the statement regarding the town's specifics (population, density, cells, concentration, intermediate area, historical town, tourist attraction, cultural and educational town, town-forming enterprises, diesel engine, thinking about ecological transportation, transports' line, frequency per month, peak hours, network of routs, transports' lines are important, permit for transportation after competition, frequency per route, calculation emission level is absent, control). The difference lies in the peculiarities of the management of the transport system (carriers do not keep to the schedules, inconvenient vehicles, the lack of schedules of routes at the stops, the inability to follow the arrival of the vehicle through the Internet, owners establish rules of the game). It is also worth noting that the similarity of Estonian cities does not reach 1, but is within the range of 0.6–0.95. This is due to the peculiarities of the management of the transport systems.

In case, when the similarity coefficient lays in the range of 0.5–0.85 during the comparison of systems (towns), it is important to make a comparison of the general part of the common parts (Table 3). It helps to receive more detailed information about the towns' similarity and common towns' features.

According to the data, we can see that there are some differences between general parts of common parts of the towns. It is explained by different approaches in urban transportation management. At the next step, we propose an end-to-end comparison of the statements of the town under investigation with other towns simultaneously. This will allow manager to understand the specific towns' features (Table 4).

Table 2. Towns' similitude matrix.

	Ostroh	Haapsalu	Rakvere	Viljandi	Valga	Sillamäe	Kuressaare	Keila	Maardu	Võru	Jõhvi
Ostroh	1										
Haapsalu	0,50	1									
Rakvere	0,50	0,71	1								
Viljandi	0,58	0,80	0,7	1							
Valga	0,50	0,84	0,7	0,88	1						
Sillamäe	0,57	0,80	0,75	0,80	0,82	1					
Kuressaare	0,58	0,85	0,66	0,71	0,82	0,73	1				
Keila	0,54	0,91	0,71	0,82	0,95	0,84	0,72	1			
Maardu	0,55	0,74	0,85	0,69	0,81	0,79	0,63	0,80	1		
Võru	0,53	0,93	0,74	0,84	0,93	0,78	0,68	0,93	0,79	1	
Jõhvi	0,49	0,85	0,74	0,80	0,79	0,71	0,69	0,86	0,77	0,81	1

Table 3. Matrix comparisons (General part of common parts).

Town pairs	Ostroh-Haapsalu	Ostroh-Rakvere	Ostroh-Viljandi	Ostroh-Valga	Ostroh- Sillamäe	Ostroh-Kuressaare	Ostroh-Keila	Ostroh-Maardu	Ostroh- Võru	Ostroh- Jõhvi
Ostroh-Haapsalu	1									
Ostroh-Rakvere	0,89	1								
Ostroh-Viljandi	0, 93	0,93	1							
Ostroh-Valga	0,89	0,93	1	1						
Ostroh-Sillamäe	0,89	0,93	0,96	1	1					
Ostroh-Kuressaare	0,89	0,88	0,96	0,96	0,96	1				
Ostroh-Keila	1	0,93	0,96	0,96	0,96	0,93	1			
Ostroh-Maardu	0,88	1	0,89	0,89	0,93	0,92	1	1		
Ostroh-Võru	1	0,89	0,96	0,96	1	0,96	1	0,92	1	
Ostroh-Jõhvi	0,92	0,85	0,89	0,96	0,88	0,96	0,96	0,96	0,96	1

It is also necessary to perform this, since during the comparison of the general part of the common parts of the towns, it may not be possible for the generality to be the same. According to the data of the Table 5, we can compare equivalent claims of Ostroh and Estonian towns and they are similar. Then we recommend to analyze information about features of differences. They include different approaches to travel fare (free of charge completely or only for citizens), management decisions are made at different levels (local, interregional and state ones) At the next step, it is essential to decide, which direction in the management of public transport systems in Estonian small towns can be implemented for Ostroh.

Table 4. An end-to-end comparison of the statements of all systems studied.

	Key words of Equalized Statements 11 towns (Ostroh-10 Estonian towns)	Key words of Equalized Statements (10 Estonian towns)
1.	Small town	Small town
2.	Population	Population
3.	Density	Density
4.	Cells	Cells
5.	Concentration	Concentration
6.	Intermediate area	Intermediate area
7.	Classification	Classification
8.	Town-forming enterprises	Town-forming enterprises
9.	Diesel engine	Diesel engine
10.	Thinking about ecological transportation	Thinking about ecological transportation
11.	Transports' line	Transports' line
12.	Frequency per month	Frequency per month
13.	Peak hours	Peak hours
14.	Network of routs	Network of routs
15.	Transports' lines are important	Transports' lines are important
16.	Permit for transportation after competition	Permit for transportation after competition
17.	Frequency per route	Frequency per route
18.	Calculation emission level is absent	Calculation emission level is absent
19.	Control	Control

3 Conclusion

Descriptive similarity approach to qualitative analysis of information is crucial for the decision-making process and allows scientists and managers to understand which system must be under the investigation. However, this method have some limitation, as it is difficult to apply to a number of similar systems with plenty of predicates without proper software. From two approaches to the systems similarity, analysis descriptive similarity approach is more relevant than structural one. Structural similarity and descriptive similarity have been linked in many ways. Structural similarity approach, based on theorems proven in algebraic systems and universal algebra theories, is difficult to apply for analyses predicates, describing a certain system. The authors substantiate application the method of similarity for the decision-making process for maintenance of the sustainable urban transportation in Ostroh under investigation Estonian small towns.

Acknowledgments. The authors of the given paper express profound appreciation to Olga Logvin, Ailar Ladva, Karl Tiitson, Väino Moor, Mati Jõgi, Yuriy Petrenko, Olga Morgunova, Monika Helmerand, Timo Suslov, Arno Uprus, Kaupo Kase, Sander Saare, Kait Kabun, Heiki Luts, Kirke Williamson, Kaadri Krooni, Riina Tamm, Karina Troshkina for the information about management of the public transportation in the towns.

References

1. Badgaiyan, R.D.: Decision making Neuroscience of the Nonconscious Mind. pp. 101–118. Academic Press, London (2019)
2. Cohn, P.M. Universal Algebra, vol. 6, Springer, Dordrecht (1981)

3. Jaccard, P.: Distribution de la flore alpine dans le bassin des Dranses et dans quelques régions voisines. Bulletin de la Société Vaudoise des Sciences Naturelles **37**, 241–272 (1901)
4. Jaccard, P.: Étude comparative de la distribution florale dans une portion des Alpes et des Jura. Bulletin de la Société Vaudoise des Sciences Naturelles **37**, 547–579 (1901)
5. Han, J., Zheng, L., Huang, H., Xu, Y., Philip, S.Y., Zuo, W.: Deep latent factor model with hierarchical similarity measure for recommender systems. Inf. Sci. **503**, 521–532 (2019)
6. Lorents, P., Matsak, E., Kuuseok, A., Harik, D.: Assessing the similarity of situations and developments by using metrics. In: Czarnowski, I., Howlett, R., Jain, L. (eds.) Intelligent Decision Technologies (KES-IDT17), Vilamoura, Algarve, Portugal, pp. 21–23, June 2017, Springer, Cham (2017)
7. Lorents, P., Averkyna, M. Some mathematical and practical aspects of decision-making based on similarity. In: Stephanidis, C. (ed.) HCII 2019, LNCS 11786, pp. 168–179, Springer (2019). https://doi.org/10.1007/978-3-030-30033-3_13
8. Maltsev, A.I.: Algebraic systems. Science, Moscow (1970)
9. Szűts, I., András B. Decision Support Systems in Management. In: 5th International Conference on Management, Enterprise and Benchmarking, pp. 1–2 June 2007. (2007)
10. Zhang, Q., Li, M., Deng, Y.: Measure the structure similarity of nodes in complex networks based on relative entropy. Phys. A **491**, 749–763 (2018)

Combining Objective Key Results, Net Promoter Score and Social Return of Investment to Measure Project Impact

Rui Belfort[1(✉)], Farley Fernandes[2], and Fábio Campos[1,3]

[1] UFPE, Recife 50670-901, Brazil
ruibelfort@gmail.com, fc2005@gmail.com
[2] Art Department, UNIDCOM / Universidade da Beira Interior, Rua Marquês D'Ávila e Bolama, 6201-001 Covilhã, Portugal
fmmf@ubi.pt
[3] BUAS, 4817 Breda, Netherlands

Abstract. This paper presents an alternative approach to measure impact, combining indexes that are already stablish to elicit objectives (Objective Key Results, known as OKRs) achievement, user satisfaction (Net Promoter Score; or NPS) and return of investment (social return of investment; SROI). Calculation plots the design in a 9-matrix grid, each one representing a level of impact and pointing ways of improvement. Besides presenting details of what we are calling Conceição Moura Impact Index (CMII), we also share the results of the first pilot; an experiment conducted in a five-year-old project called iCANamy, and more specifically its program for teachers.

Keywords: Design process · Knowledge elicitation · Project impact

1 Introduction

It's been five years since iCANamy (I Can Academy)[1] program pilot was conducted in Belo Jardim, Pernambuco, Brazil, to help develop competencies for life as a complement of formal education, brought by Conceição Moura Institute[2], the social operation of Moura Group[3], a huge battery industry born in the City many decades ago, that now is a global player in its segment.

More than a thousand youngsters, teachers and professionals participated in different applications of the methodology in these past years. The projects got mature and started to show, empirically speaking, positive impact in the community. Word of mouth spread the news. But as part of a private social investment, came the moment that these results needed to be accountable.

In the last couple of years, we designed, in an incremental continuous validation process, a framework for measuring impact, called CMII (Conceição Moura Impact

[1] https://icanamy.com.br/
[2] https://www.icmoura.org/
[3] https://www.moura.com.brs/

D. Russo et al. (Eds.): IHSI 2021, AISC 1322, pp. 428–433, 2021.
https://doi.org/10.1007/978-3-030-68017-6_64

Index), based on a combination of Objective Key Results (OKRs) [1], Net Promoter Score (NPS) [2] and Social Return of Investment (SROI) [3], combined and with results plotted in a visual action-oriented 9 quadrant matrix, inspired by Nine Box [4].

These tools are widely spread globally. OKRs are one of the main tools to formulate and measure objectives. In CMII we use It to define reach, since one of the main goals is necessarily contemplate a good number of people. NPS is used to see how satisfied and loyal they are, after programs; this adds another layer of value measurement. And SROI provides an economic comparison valuation.

The first application showed promising results. This research intents to present the framework, how it was tested, the outcomes we had so far and next steps in the design and use. Since measurement is still an open topic in most of applied sciences [5], we aim to open the discussion and start a collaboration that can bring a new perspective on measuring design return of investment (ROI) [6].

How can this methodology be applied in the industry, as qualitative and quantitative hybrid measurement tool, that can make emerge the value of design decisions, based on reach, satisfaction and economic impacts? This seems to be a long, but interesting, road of contribution, specifically to the decision-making in the alternative selection phase of its process [7–9].

2 ICANamy

I Can Academy was created in late 2014 by five entrepreneurs, as a private initiative, seeking to contribute with regular education, considered incipient in Brazil. Country's educational system is advancing, but still faces many critical challenges. The gap between what's lectured and market demands is reflected in the productivity of Brazilian professionals, 25% of what an American [10] produces in the same work conditions.

It is a methodology for competencies development, using the approach of challenge-based learning [11]. Real problems are the main motivation for learning, that is facilitated to be shared and collective experience. It's most structural concept is that a competent individual has developed five specific attributes: leadership, sociability, technic, methodological and Attitudinal.

We believe that all are important for problem-solving, no matter the profession. The first four are the same used by Germany education system, called action skills (leadership, technical, methodological and social) [12]; added to these is attitude, the fifth, that deals with auto development and mitigation of responsibility transfer. Bringing this was a clear innovation, by the time; still is current, but faces competitors [13].

iCANamy was tested in many different contexts, for diverse groups of personae. This research was conducted in the advanced teacher's project, called Educational Leaders (in Portuguese, Líderes Educadores). We also worked for international corporations, private schools, government authorities, etc. In all, results were encouraging. Recently, seeking to spread organically, it was put under Creative Commons[4] license.

[4] https://creativecommons.org/

3 Conceição Moura Institute

The very first collaboration was with Conceição Moura Institute (CMI), social authority of Moura group, one of the largest car battery industry in the world. The organization was born in Belo Jardim, a city located 183 km from Recife (Pernambuco's capital) and along decades keeps the commitment of investing for its social, environmental and economic development; all together.

CMI carries the mission of "contribute with children and youth empowerment, by offering opportunities of personal and professional growth, incentives to become transformation agents in the world, working with competencies for life, cognitive and social emotional skills, responsive relationships and mitigation of stress". iCANamy was one of its first collaborated programs, tested before its foundation in 2015.

Besides I Can Academy, nowadays they provide dozens of opportunities for childs, youngsters and social engagement. All these programs are conducted in partnership with specialists (organizations and consultants) in different areas. Since foundation, more than 20.000 people were contemplated. Came the time when measuring impact became necessary to prioritize investments in its complex network.

The idea is not only creating a way of measuring programs, but also standardize measurements. It is important to compare efforts and outcomes in the same structure, to better point ways of enhancing results. Vision points to even more volume and complexity of actions. Without a clear framework for this, it would be very difficult to manage Organization growth. Of course, this matter is merit of Group's audit.

4 Conceição Moura Institute Index

As explained, CMII is based on three wide spread and stablish measurement tools, OKRs, NPS and SROI. Also, the matrix of evaluation was inspired by 9-Box, that deals with human potential and performance. Aims to provide measurement of impact, based on reach, satisfaction and return of investment. Every combination of these three factor leads to a quadrant; and everyone to suggested solutions.

ORK is a global framework for defining, prioritizing and analyzing objectives and results. It's based on the principles of focusing on measurable, short term actions, aligned to Organization mission. In CMII, every cycle has 3 months, from definition to evaluation. It is suggested to use 3 metrics. It is a bad result (-) not achieving any ORK, good (=) is achieving 1 and great (+) 2.

Net Promoter Score is a widely-spread methodology for measuring loyalty. It is very unlikely a relevant corporation not to use it, internally and externally. It assumes that the it is difficult to recommend a product or service to a close friend. NPS is the percentage of promoters less detractors. A bad result will be bellow 0 (-); good, from 1 to 30 (=); excellent and exceptional, from 31 to 70; and 71 to 100, respectively (+).

SROI is an economic evaluation, created for social programs, that aims to enable investment optimization. It translates, from analogies (benchmarks), results to financial values; what value the participant captures, what opportunities the experience opens and how much money it saves from society. A bad SROI is under 1; for one dollar invested, less than one is generated) (-); neutral is around 1 (=); and good above 2 (+).

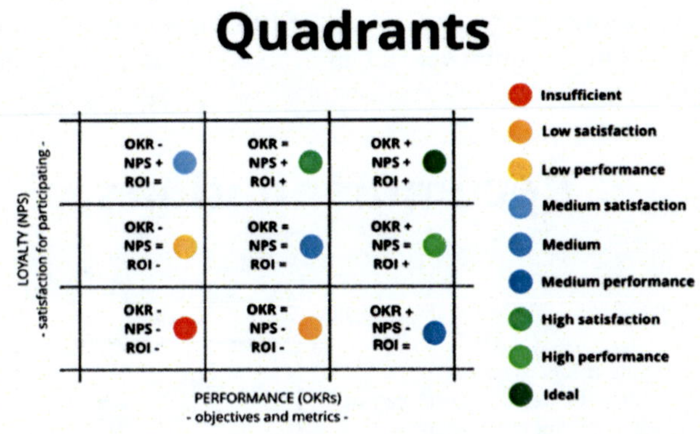

Fig. 1. CMII matrix.

CMII model, as the combination of OKRs, NPS and SROI, anchors in the following idea; if the objectives achieved are measurable and aligned to mission, participants are satisfied, return of investment is likely to occur and impact will be positive. On Fig. 1., all possible combinations are shown. Green quadrants are result of positive reach, satisfaction and ROI. Blues are neutral. And reddish are demanding for correction.

All samples in CMI programs are finite (definable); 5% is the defined margin of error. Following that, projects with more than 30 participants need to test with at least this quantity. And with less, all shall be elicited. Considering these variables, we will have above 90% and above reliability rate [14]. This is mandatory in terms of sampling the applications.

5 The Experiment; Educational Leaders

CMMI is being tested this year in advanced Educational Leaders (Líderes Educadores, in Portuguese), that works self-awareness and self-development by proving teachers experiences to develop skills and competencies to enhance their work impact. The program is totally free, provided by CMI; this year, 90 teachers are participating. Its creation, management and innovation are continuous and design-based.

In advanced project, we had 30 participants. The workload is 30 h, divided in 4 sessions. 1 and 2, because of Covid-19, occurred online (Zoom[5]). 3 and 4 were in person; the group was divided (half on morning, the other on afternoon) so we could comply to Government policies. In the end of experience, they were asked to fill two forms; net promoter score's and SROI's. OKRs are measured by objectives and results.

The OKRs were set before the program started. To keep 25 participants (no evasion), we selected 30 teachers. All of them finished the journey, that's why we consider the objective achieved. The NPS was, as empirically expected (the program is very

[5] https://zoom.us/

consolidated), in the excellency level. About SROI, results were considered good; although there is room for improvement in working correlations with professional prosperity.

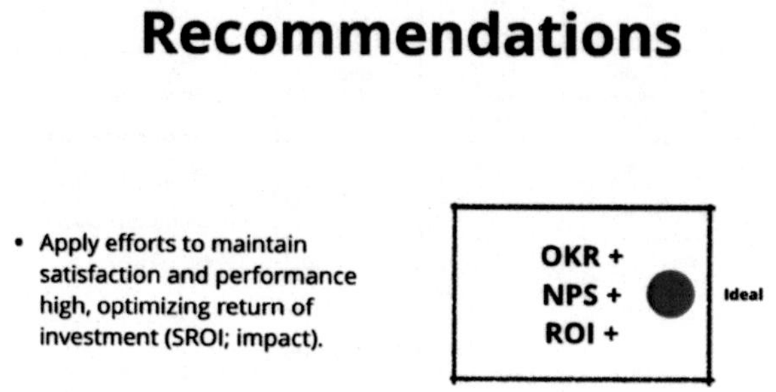

Fig. 2. Educational Leaders quadrant recommendation.

As showed in Fig. 2., project was positioned in the Ideal quadrant. Empirically, this result was expected. Project is well evaluated by CMI, teachers and has a high demand for new participations. In this case, we need to keep satisfaction and performance high, to maintain return of investment (impact). Also, we will work on professional links, during experience, to make these kinds of opportunities accountable.

6 Preliminary Findings

This year, advanced class had 25 vacancies. In the subscription phase, we received 62 candidates and because of high demand 30 people were selected. These numbers were very motivating, considering only teachers with regular course certification could participate in this application. It was a first evidence of project's acceptancy and impact; OKR was achieved (+).

NPS was measured and the score was 100 (+), the best in excellency level. Another layer of validation that the project is generating the expected results. There seems to be a correlation between the interest of the target and the results elicited with CMII. We will redo the test later this year, in two other groups, seeking to compare results and see if we will receive similar evaluation.

SROI was measured above 2 (+). All these confirmations point to CMII potential. Of course, it needs a lot more cycles of validation; it is important to see how numbers will behave in project's history and compare with other applications. Also, elicitation forms will be continuously improved. After testing with same project more, and other projects for benchmark, the size of opportunity may become clearer.

As next steps, we will start testing in Smart and Young Journey (in Portuguese, Jornada dos Desenrolados) project, iCANamy's application for youngsters. Also, we are preparing to test on Makers Territory (Território do Fazer in Portuguese), CMI's robotics program, managed by another partner. Before the end of 2020, results will be evaluated and compared with previous experiments and shared in further publications.

References

1. Doerr, J.: Measure What Matters. Portfolio/Penguin, New York (2018)
2. Reichheld, F., Markey, R.: The Ultimate Question 2.0. Harvard Business Review Press, Boston (2011)
3. Aitken, H.: Starting Out on Social Return of Investment. Social Value UK, Liverpool (2014)
4. Lamoureux, K.: The Nine-Box Grid: An Essential Tool for Talent Calibration. Bersin & Associates, Oakland (2009)
5. Boon, M.: How Scientists Are Brought Back into Science - The Error of Empiricism. In: Bertolaso, M., Sterpetti, F. (eds.) A Critical Reflection on Automated Science, pp. 43–65. Springer, Cham (2020)
6. Sheppard, B.: Kouyoumjian, Garen, Sarrazin, Hugo, Dore, Fabricio: The Business Value of Design. McKinsey & Company, New York (2018)
7. Jones, J.C.: Design Methods: Seeds of Human Futures. Wiley, London (1970)
8. Baxter, M.: Product Design. CRC Press, Boca Raton (1995)
9. Knap, J.: Zeratsky, John, Kowitz, Braden: Sprint: How to Solve Big Problems and Test New Ideas in Just Five Days. Simon & Schuster, New York (2016)
10. Board, Conference: The Conference Board Productivity Brief 2019. The Conference Board, New York (2019)
11. Nichols, M.: Cator, Karen: Challenge Based Learning White Paper. Apple Inc, Cupertino (2009)
12. Bunk, P.: Prestação de Competências na Formação Profissional Inicial e Contínua na RFA. In: Revista Europeia de Formação Profissional, pp. 8–14. CEDEFOP, Thessaloniki (1994)
13. Learning, B.: Cambridge Life Competencies Framework. Cambridge University Press, Cambridge (2019)
14. Lincoln, D.: A Avaliação de Artefatos em Design e os Problemas Decorrentes da Aleatoriedade. UFPE, Recife (2012)
15. Belfort, R., Campos, F., Gonçalves, J.: User Elicitation for Cognitive Consonance in The Design of Digital Artifacts. IHCI Proceedings, Lisboa (2017)

User Experience in Virtual Environments: Relationship Between Cybersickness Issues and the Optical Aspects of the Image by Contrast Levels

Fabio Campos[1](✉), Marcia Campos[2], Tulio Silva[2], and Marnix Van Gisbergen[3]

[1] UFPE/CNPq/BUAS, Recife, Brazil
fc2005@gmail.com
[2] Catholic University of Pernambuco, Recife, Brazil
spot4m@gmail.com, tuliosilva@gmail.com
[3] BUAS, Breda, Netherlands
Gisbergen.M@buas.nl

Abstract. This paper provides a comparative study between problems caused by cybersickness with users in virtual environments, and the studies on the influence of the optical aspects of the image via its levels of contrasts in the quality of the user experience in virtual environments. This article was organized in two parts, the first consists of surveying the results obtained in each research and the second has the objective of crossing these data and verifying the level of relationship between research on cybersickness and studies on the influence of optical aspects quality of the user experience. The results indicate relevant connection with the results of studies on cybersickness. With that, we conclude that in future research researchers can produce much more relevant results in the solution of persistent problems from the negative experience with virtual devices.

Keywords: Optical aspects · Cybersickness · Virtual environment

1 Introduction

Virtual reality (VR) technology has been gaining prominence in recent years. The possibility of immersion in a fully interactive virtual environment, proved beneficial for several applications such as education, entertainment, engineering, driving and flight simulator. However, today's virtual environments present a complicated problem, the relevant tendency of many users to present symptoms similar to the classic symptoms of motion sickness. They exhibit these symptoms during and after exposure to the virtual environment. The difference between motion sickness and cybersickness is that, in cybersickness, the user, when using VR, is usually still, but has a very convincing sense of self movement because of the images generated by the device. In view of this, the expansion of VR remains stalled by the negative experiences associated with its use. Some research shows that the symptoms are directly related to the configuration of

D. Russo et al. (Eds.): IHSI 2021, AISC 1322, pp. 434–439, 2021.
https://doi.org/10.1007/978-3-030-68017-6_65

the device, such as the amplitude of the user's field of view and the graphic quality at the level of realism.

The present article carried out a comparative study between the researches about the problems caused by cybersickness with users in virtual environments, and the studies on the influence of the optical aspects of the image via its levels of contrasts in the quality of the user experience in virtual environments.

2 First Moment: Survey of the Researches in Question

This first step consists of comparing the results obtained in the studies of the main problems caused by cybersickness with users in virtual environments with recent studies on the influence of optical aspects of the image in the quality of the user experience in virtual environments. The references used in the cybersickness area were Kennedy et al. [1]; Tiiro [2] and Fernandes et al. [3]. And the references used on the influence of optical aspects were those of Campos et al. [4] and Campos et al. [5].

2.1 Cybersickness: Main Problems Studied

Robert S. Kennedy and collaborators published in 2000, an article that links the simulator disease with the exposure time in virtual environments [1]. The aim of the study was to assess how much the duration of the sessions influences the severity of the symptoms, based on 2 evaluations. The first examined the duration of exposure within long sessions and the second assessed repeated exposures to short sessions. The long sessions lasted between 1 and 3 h, while the short and repeated exposures lasted between 0 and 1 h. The authors used a flight simulator database, collected by military agencies. The results showed that sessions of longer exposures cause more severe symptoms, while short and repeated sessions alleviate these symptoms.

Another important study on cybersickness was carried out by Arttu Tiiro in 2018. He published a thesis on a disease related to simulators and motion sickness, called cybersickness, which causes nausea, headaches, eye stress and dizziness [2]. In his work, the author associates the severity of Cybersickness symptoms with the level of realism of the images projected by the device. The aim of this study was to analyze the influence that the graphic configuration of the device has on symptoms. This aspect was investigated using 3 applications with different graphic styles that vary in the level of realism (details, lighting and textures), although almost identical in their general structure. The research was carried out in a laboratory where the variables could be easily controlled. Navigation and controls have been eliminated or standardized between applications so that graphic styles can be compared effectively. Participants walked through the same places, at the same speed and only head movements were allowed. The exposure time was 15 min and the sounds were removed. Data were acquired through questionnaires, observations and interviews. The results indicated that the degree of realism of the image projected by the simulator has a directly proportional relationship with the symptoms of cybersickness, in other words the higher the level of realism, the more severe are the symptoms felt by the volunteers, being, therefore, milder with the decrease.

The symptoms of Cybersickness are also influenced by the size of the field of view (FOV - Field of view), as shown in the study carried out by Fernandes, in 2016 [3]. According to the author, the user's sense of presence within the virtual environment, that is, feeling immersed in that reality is what potentiates the symptoms, so when we decrease the field of view, the sense of presence also decreases. To assess this relationship between FOV and symptoms, the authors used a method to dynamically change the field of view according to the user's movement during the experiment. Restrictors were positioned on the device so that they act as "vignettes", decreasing or increasing the field of view and this was controlled dynamically as the user walked through the virtual environment. The results showed FOV restrictors helped users to remain more comfortable and stay exposed to the environment for a longer time. It was also found that these restrictors also helped to minimize participants' discomfort, in their first experience, which helped them to move on to the following sessions. It is important to mention that the majority of users did not notice the change in the field of view during the sessions, that is, they had no control over this factor. These data were collected through a questionnaire applied to the volunteers.

2.2 Studies on the Influence of Optical Aspects of the Image by Contrast Levels

This research is based on understanding the image from the perspective of the human eye, which simply captures visible light. In other words, all objects in the scenario are only electromagnetic information or reflected lights. Therefore, based on the objective idea of the image, the research carried out so far highlights numerous possibilities for application in other researches where the image is a crucial object, from process of designing scenarios to enabling the measurement of optical elements, which, in turn, are extremely abstract in the user's perception. [4, 5].

In a recent study, Campos [4] used his method to verify the relationship of the optical aspects of the image via its levels of contrasts, in the perception and quality of the degree of realism of the image by the user. For this, the study carried out an experiment in 2 phases. The first was the use of the Campos method [4] to analyze and classify the degree of realism, within the objective perspective of representing the optical contrasts of each image (real and virtual) selected. In the second phase, the same images as the first phase, were used to check the level of realism, now from the point of view of the users' perception, through the application of an online questionnaire. Finally, a comparative analysis was performed between the results obtained in each phase, with the objective of verifying, quantitatively, the percentage of similar and / or equal results, regarding the classification and perception of the level of realism obtained in each image. In this way, the results obtained confirmed a relevant relationship between the quality of the representation of the optical aspects of the image via its levels of contrasts with the user's perception of their degree of realism. With this it can be considered that the data collected with the Field Method [4], are of great importance for understanding and measuring the optical aspects through the variation of their levels of contrasts, because these elements that form the image are in the principle of the entire process. In other words, it triggers or initiates all the developments inherent to the user's experience.in their first experience, which helped them to move on to the following sessions. It is important to mention that the majority of users did not notice the change in the field of view during the sessions, that is, they had no control over this factor. These data were collected through a questionnaire applied to the volunteers.

Continuing the results obtained in the article on levels of realism [4], Campos carried out new studies with the objective of deepening the influence of the optical aspects of the image composition with more situations caused by the user's interaction with real or photographic images. The focus of the study was on collecting data related to the user's perception and reaction based on the user's interaction, with photographic images produced according to the Campos Method [5]. Decision-making time, level of indecision and level of discomfort were the main data collected from the experiment with users.

For this, an experiment was carried out in 2 stages, the first stage was the production of 17 photographic images, where all were exactly the same, with the same objects and scenery. And the difference in each of the photos was due to the variation in the contrast levels of the captured optical aspects of the image, according to the Campos method [5]. The optical aspects of image composition used in the experiment were depth of field or sharpness (at 3 levels of contrasts), effect of movement (at 3 levels of contrasts) and totally unfocused photos (at 3 levels of contrasts).

The second stage of the experiment was the collection of data from the user experience using an online questionnaire with the photos produced in the first stage, where 73 people participated. The results were grouped by group of photos with the same level of contrast as the corresponding optical aspect.

The results obtained in the experiment were the relevant confirmation of the influence of the optical aspects of the image on the perception and reactions of the user experience. It was verified that a continuous logic between the reactions of the users and the levels of contrasts captured in the images. That is, the three levels of contrasts (high, medium and low) used in each optical aspect captured in each photograph, obtained clear results of their impact in the variation of the data obtained in the questionnaire with the users. For example, the higher the contrast level of the optical aspect of depth of field, the decision-making time decreases, the level of indecision decreases and consequently the user spends less time watching. And following the flow from the opposite end of the high contrast of the optical aspect of depth of field to the low contrast of the optical aspect of depth of field, the user experience data followed the same opposite logic as the time of decision making increases, the level of indecision increases and consequently the user spends more time of observation.

And one more fact found by this study was in relation to the level of discomfort felt by users. Following the same logic of the contrast levels of the photos all out of focus, the result was: in the image with high contrast of the blur people felt more discomfort and consequently the observation time was shorter because the discomfort was practically instantaneous. And in the opposite logic the image with low contrast of blur users felt less discomfort and consequently the observation time was longer because the discomfort took longer to be felt.

It is clear how studies on the method of analyzing the optical aspects of image composition by contrasts have a great potential to contribute to research in the areas of user experience and virtual reality and cybersickness devices.

3 Second Moment: Crossing the Results and Checking the Level of Relationship Between the Surveys

In the research published by Kennedy [1], he links the simulator disease with the exposure time in virtual environments. In the experiment carried out, the user experience in a flight simulator was taken into account, where the variables considered were the longest and shortest exposure time. The results obtained were: the longer the user's exposure to the virtual environment of the flight simulator, the more severe the symptoms experienced by the users. Therefore, it can be seen that the simulator image was not taken into account, leaving gaps in the studied context. So, imagine the possibilities of adding more data and answers to the relevant results already obtained by Kennedy [1], using the variables inherent in the image itself, using the Campos method [4, 5]. The research of this method already demonstrates the great influence of the optical aspects of the composition of the image via its levels of contrasts, both with the level of visual discomfort felt by the user as well as the ability to observe more or less an image.

Now addressing another research, where Tiiro [2] associates the severity of the symptoms of Cybersickness with the level of realism of the images projected by virtual devices. The aim of this study was to analyze the influence that the graphic configuration of the device has on symptoms. The results obtained were that the higher the level of realism, the more severe are the symptoms felt by users, thus, the milder the decrease. However, the variables used to define the level of realism of the images used in the experiment (details, lighting and textures), cover only the subjective aspects of the image. Again, it is noticed that the question of the image itself, in its objective aspect [4, 5] was not remembered. In this way, we can idealize the richer possibilities that the method in question could add to the already relevant results of Tiiro's research [2]. Campos' publication [4] demonstrates that the perception and quality of the degree of realism understood by the user, is directly connected with the representation of the optical aspects of the image composition via its levels of contrasts. So, imagine if in the Tiiro experiment [2] this method of Campos [4] was used to validate the degree of realism of the images used in the Tiiro experiment?

According to Fernandes, the symptoms of cybersickness are also influenced by the size of the field of view (FOV - Field of view) [3]. The results indicate that the symptoms of cybersickness felt by the user were stronger as the field of view became larger and as the field of view decreased, the symptoms felt relieved. In this study the question of the image itself (what is within the field of view) was not remembered. Now, imagine if in this experiment the Campos method [5] was taken into account, where with its use, we can measure the levels of contrasts of the optical aspect of depth of field or sharpness that, in turn, change as the field of view size changes. And the possibility of adding more information inherent to the influence of the variation in the levels of contrasts of the optical aspects in the user experience, in this case, regarding the symptoms of cybersickness.

4 Conclusion

Despite the expectations around virtual environments, we are still not able to use them properly. Technology has advanced without taking into account the limitations of our physiology and through this study, there is evidence that our vision is closely linked to the symptoms of Cybersickness. The images projected by virtual reality devices are perceived differently by the human eye, causing divergences and causing unwanted symptoms. Therefore, studying and evaluating the configuration of these devices and, aligning them with what our eye is capable of interpreting, is probably the best way to accelerate the popularization of VR.

It was evident that the image factor is an element that is often overlooked in research on problems arising from cybersickness. This indicates that the vast majority of research on virtual environments, ignore, in some way, the optical elements that are captured by human vision and, just because of this, are also the aspects that form the image itself. And consequently, we can consider that these elements of the image are the source of all events that result in the user experience with VR devices.

For most of the variables used in research on cybersickness, alter and / or interfere in the representation of each optical aspect of the image displayed to users in the experiments in question. According to Campos, these interferences and/or changes happen through the variation of the representation of the contrast levels of each optical aspect.

We can give some examples in what concerns the framing we have the variation of the size of the field of vision, the variation of illumination, etc.; with regard to the user variation in the point of view, variation in the diameter of the opening of the iris, physiological, emotional and physical reactions. With these very general examples, it is evident that the inclusion of the method of analysis of the optical aspects of the image by contrasts, has the potential to fill the gap in the classification and measurement of the image, of the virtual environments that may be used in future studies.

References

1. Kennedy, R.S., Stanney, K.M., Dunlap, W.P.: Duration and exposure to virtual environments: sickness curves during and across sessions. Presence Teleoperators Virtual Environ. **9**(5), 463–472 (2000)
2. Tiiro, A.: Effect of visual realism on cybersickness in virtual reality. Univ. Oulu 350, 2018
3. Fernandes, A.S., Feiner, S.K., Combating VR sickness through subtle dynamic field-of-view modification. In: 2016 IEEE Symposium on 3D User Interfaces (3DUI), pp. 201-210. IEEE (2016)
4. Campos, M., Campos, F., Van Gisbergen, M., Kovacs, M.: The relationship among the optical aspects of photographic composition and the quality, perception and interpretation of the realism in virtual images. In: Karwowski, W., Ahram, T. (eds.) Intelligent Human Systems Integration 2019, IHSI 2019, Advances in Intelligent Systems and Computing, vol. 903. Springer, Cham (2019)
5. Campos, M., Campos, F.: Influence of the optical aspects of photographic composition on the user experience in the issues related to decision making, choices and level of visual comfort. In: Peng, SL., Favorskaya, M., Chao, H.C. (eds.) Sensor Networks and Signal Processing. Smart Innovation, Systems and Technologies, vol 176. Springer, Singapore (2021)

User Experience Goals for Cognitive Systems in Smart Business Buildings

Eija Kaasinen[1](✉), Susanna Aromaa[1], Maarit Halttunen[2], Susanne Jacobson[3], Inka Lappalainen[2], Marja Liinasuo[2], Maaria Nuutinen[2], and Reetta Turtiainen[3]

[1] VTT Technical Research Centre of Finland Ltd, Visiokatu 4, 33101 Tampere, Finland
{eija.kaasinen, susanna.aromaa}@vtt.fi
[2] VTT Technical Research Centre of Finland Ltd, Tekniikantie 21, 02044 Espoo, Finland
{maarit.halttunen, inka.lappalainen, marja.liinasuo, maaria.nuutinen}@vtt.fi
[3] KONE Corporation, Keilasatama 3, 02150 Espoo, Finland
{susanne.jacobson, reetta.turtiainen}@kone.com

Abstract. Tenant experience in a smart business building is shaped both by the technical solutions available in the building and by human stakeholders at different touch points in the building context. We have been applying design-thinking approach to study what kinds of experiences should smart business buildings provide to the users, especially to the tenants of the building. We conclude our findings as user experience vision and nine user experience goals for the tenants of smart business buildings. Furthermore, we analyse how the current stakeholders of the building contribute to each targeted experience, and suggest how cognitive systems could enhance those experiences.

Keywords: Smart building · User experience · Ecosystem

1 Introduction

In a smart building, different stakeholders both experience the cognitive systems in the building and contribute to other stakeholders' experiences. In these smart environments, it is challenging to design for good user experience (UX) because UX is shaped partially by technical solutions and partially by human activities at different touchpoints in the building.

We acknowledge that the holistic UX and the entire context of use are important, as people are not only using smart environments, but are also living in them. Technology should be viewed in its context of use and as part of social and symbolic wellbeing [1]. The subjective meanings users assign to smart environments, the ways in which they become attached to the environments, and can modify them, should be considered and enabled [2].

Most current research is focused on the interplay between design and multiple single users rather than a community of users, disregarding the significance of social

D. Russo et al. (Eds.): IHSI 2021, AISC 1322, pp. 440–446, 2021.
https://doi.org/10.1007/978-3-030-68017-6_66

dynamics between users [3]. The social value of the building considers the physical design, as well as their varying briefing and design processes, organizational set-ups and building management, and the experiences of the building users [3].

Cognitive systems can make physical environments more responsive and adaptive to individual needs and preferences. The users of the building can be supported with contextually and personally relevant information and services. To provide to the users a versatile and meaningful set of services, different service providers are required. In a smart building context, platform-based technology ecosystems can serve to provide coherent services to the users. These ecosystems facilitate gathering and providing the services by independent actors into a united service entity.

The objective of this study was to define what kinds of experiences should smart business buildings provide to the users of the buildings, and how the targeted experiences could be provided utilizing both cognitive systems and human stakeholders in different roles. The focus is on regular building users, mainly companies as tenants, who are using the building as their workplace.

2 The Case Study

Our case study is focused on an entrepreneurial ecosystem at a business campus, which has been renovated to host start-up companies, investor companies and service providers. The buildings are owned by the city and the aim of the business campus and the related ecosystem is to support start-ups in setting up and boosting their businesses. The community is growing, exceeding 1000 members with over 30 different nationalities, consisting of over 100 start-ups, over 30 venture funds and other partners as well as numerous supporters. The supporting services include, e.g., reception, a restaurant, a gym as well as cleaning and maintenance services. The central actor is a company that nurtures the community, manages the selection process for tenants and runs everyday operations.

3 The Method

We chose to use Design Thinking [4] methodology not just to collect user needs but also to engage project partners to discuss, interact and empathize with the users. This study included two first phases of Design Thinking: empathize and define (Fig. 1). In the Empathize Phase, we were using different data collection methods to gain user understanding. The methods included workshops, observations, design probes and interviews with different stakeholders of the campus. We selected these data collection methods to be able to reach several stakeholders from the business campus ecosystem and smart building context. The data collection methods were interconnected: each method provided insight for the following step. The project partners were engaged in data collection as they were participating in the workshops and interviews.

In the Define Phase, we utilized the UX goal setting approach [5]. Our goal was to focus on tenants' UX but to have more holistic view into it by considering the whole business campus ecosystem and social dynamics within it. We first synthesized the data from user studies to identify user expectations and current challenges. To focus also on future possibilities, we brought in two other viewpoints: possibilities offered by smart building technologies and new business scenarios. Utilizing these three viewpoints, in a series of workshops with project partners we gradually formulated concrete UX goals and further an UX vision. Finally, we analysed how the UX goals can be shaped by integrating cognitive systems and human actions. This will support the forthcoming Ideation Phase activities.

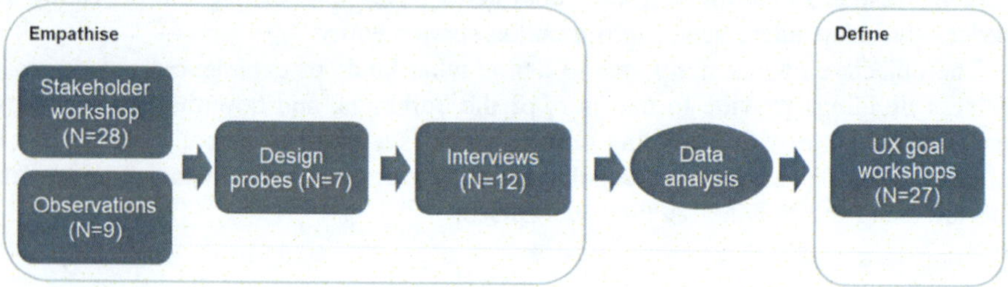

Fig. 1. An overview of the data collection process (N = number of participants).

4 Results

4.1 Stakeholders and Tenant Expectations

The stakeholder workshop produced the initial stakeholder map. The map was refined and complemented based on the results of the probes and the interviews. Several stakeholders contribute to tenant experience (Fig. 2). The tenants collaborate with partners, investors and customers outside the campus ecosystem. The ecosystem actors such as other tenants, corporative employees and building visitors are met at community events or occasionally. There may also be actual collaboration or partnership activities with other tenants.

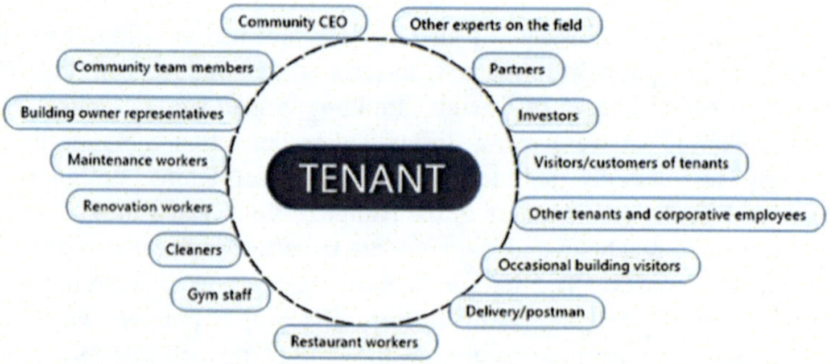

Fig. 2. Key stakeholders at the business campus from the tenant point of view.

The business campus provides various services such as meeting rooms, restaurants, postal services, wellbeing facilities, cleaning services and maintenance services. Community team is central in serving the everyday needs of the ecosystem and organizing common events. The Campus CEO negotiates the contracts with the tenants and aims to choose such start-ups that seem to fit in the community. When new tenants sign in, a renovation team helps them to customize their premises for their needs.

At the stakeholder workshop, key challenges for the tenants were identified. These high-level challenges were elaborated further in the interviews with the tenants and other stakeholders. The identified high-level challenges were related to optimized building use, constant changes at the campus, access control, smooth people flow as well as finding a balance between the different needs of different stakeholders at different times. The tenants appreciated the communality at the campus and the spirit of helping and supporting each other. The tenants felt that the current facilities set some limitations, but many expectations were targeted to the planned new campus building, with smart building solutions and data-driven services.

4.2 User Experience Goals for the Tenants

In a series of online workshops with project partners and campus stakeholders, we gradually defined UX goals for the tenants in a smart building. The UX goals were crystalized as a UX vision, reflecting the overall targeted experience for the tenants, enabled by smart building solutions. The UX vision: *"Feeling cared for and empowered as part of the community,"* reflects how tenant companies and tenant employees want to be acknowledged as individuals, they want to feel being part of a community and they want to feel empowered to influence their environment. The vision is based on nine UX goals (Table 1).

Table 1. The UX goals and how they could be shaped by human actions and technical solutions.

UX goal	How the current ecosystem contributes to the UX goal	What cognitive systems could bring in
Being aware: I get immediate information of all the issues that I need or want to know to live and work in the environment.	Community team is informing tenants via the internal instant messaging service. The tenants can also share information to each other via the service.	Different environmental sensors facilitate gathering, analysing and providing information to keep the tenants aware of what is happening in the environment.
Personal and purposeful: My own and my company's preferences are known and taken into account in the building and in the services.	The Campus management is discussing with new tenants of their needs regarding the office space and community services.	Smart building can identify individual users and learn to know their personal or professional interests.
Enjoyable: I enjoy working and living in this environment.	The community team aims to keep the building and the office spaces pleasant for the tenants.	Flexible spaces can be supplemented with advanced tools to combine social, virtual and physical spaces.

(continued)

Table 1. (*continued*)

UX goal	How the current ecosystem contributes to the UX goal	What cognitive systems could bring in
Healthy and safe work environment: I am confident that the smart building is supporting my personal needs for well-being and safety both at the office and at the surroundings.	The tenants rely the community team in maintaining the building, organizing cleaning and taking care of security.	Smart building can monitor different health-related issues such as air-conditioning, lightning, temperature or even health threats. Safety can be supported with smooth access control.
Empowered: I know which issues I can influence in the environment, and I can easily communicate my needs, problems and preferences, and get prompt feedback.	New tenants can personalize their office space. The community team is open for tenants' suggestions on improvements.	Smart building can provide the users tools to influence the environmental settings and to communicate their personal preferences and suggestions.
Responsive: The environment knows my needs and reacts to them smoothly.	The community team has learned individual tenant's preferences.	Smart building can identify the user, learn his/her needs, and react/adapt to those.
Sharing: I want to support others by sharing and caring, and I know they want to support me.	Within the start-up community, there is a spirit of supporting and helping each other.	People want to support others, share expertise and find new synergies, and this can be supported with collaboration, communication and matchmaking tools.
Proud of belonging: I feel privileged to be part of this stakeholder ecosystem, which supports my lifestyle, feelings and motivations.	The owner, the CEO and all the stakeholders together have created a good brand for the campus ecosystem.	Branding is dynamic, co-creative and long-term process, which smart building solutions can support in terms of uniqueness, engagement and adaptability.
Communality feeling: I feel being part of the community, as an individual and as a company, but not forced to be more social than I want. Being part of the community does not require physical presence.	Open communication, collaboration and events support the feeling of community.	Smart building provides tools that support participation and matchmaking as well as informs about events and news that could potentially be interesting to the tenants.

In the user studies, we found several ways how the current ecosystem stakeholders already contribute towards the targeted experiences. When planning smart building solutions, we cannot rely only on technical solutions to shape the experiences but in many cases, human actions are needed in parallel. In Table 1, we also describe how the current ecosystem is contributing to the targeted experiences and what smart building solutions could bring in. When designing cognitive systems for smart buildings, both aspects should be taken into account in shaping the experiences.

5 Conclusions

We have studied stakeholders in a start-up business campus and the related ecosystem to identify experience goals of tenants for future smart building solutions. To take into account both current challenges and future possibilities in the UX goals, we have based the goals on identified user needs as well as on the possibilities offered by smart building technologies and new business scenarios. In the ecosystems, physical and digital platforms serve as crucial tools for mutual value co-creation and collaboration [6, 7].

Smart building constitutes an ecosystem where different stakeholders both experience the smart building solutions and contribute to other stakeholders' experiences. User experience goals for the tenants of a smart business building concretize the targeted experiences. When utilizing the UX goals in design, it should be considered how the ecosystemic nature of UX [8] should be taken into account. User experience is shaped partially by technical solutions and partially by human activities at different touchpoints throughout the building lifecycle. Communality in the ecosystem is an important UX goal, and focusing on it may require studying social values and communal experiences as Watson et al. [3] propose.

The results highlight the interplay between human stakeholders and cognitive systems when providing enjoyable building experiences to the users of the building. Design Thinking approach supported keeping the project partners and the tenants and other stakeholders of the building in continuous contact. We believe that these firm contacts in addition to the UX goals will support maintaining a user-focused mindset in the project and putting the UX goals into practice.

Acknowledgments. The work was carried out in KEKO project, funded by Business Finland and the project partners KONE, Nokia, YIT, Caverion, Halton, Netox and VTT Technical Research Centre of Finland Ltd. The authors want to thank the interviewees from the studied business campus, as well as the project partners for the good collaboration.

References

1. Shove, E.: Users, technologies and expectations of comfort, cleanliness and convenience. Innov. Eur. J. Soc. Sci. Res. **16**(2), 193–206 (2003)

2. Kaasinen, E., Kymäläinen, T., Niemelä, M., Olsson, T., Kanerva, M., Ikonen, V.: A user-centric view of intelligent environments: user expectations, user experience and user role in building intelligent environments. Computers **2**, 1–33 (2013)
3. Watson, K.J., Evans, J., Karvonen, A., Whitley, T.: Capturing the social value of buildings: the promise of social return on investment (SROI). Build. Environ. **103**, 289–301 (2016)
4. Brown, T.: Design thinking. Harv. Bus. Rev. **86**(6), 84–92 (2008)
5. Kaasinen, E., Roto, V., Hakulinen, J., Heimonen, T., Jokinen, J.P.P., Karvonen, H., Keskinen, T., Koskinen, H., Lu, Y., Saariluoma, P., Tokkonen, H., Turunen, M.: Defining user experience goals to guide the design of industrial systems. Behav. Inf. Technol. **34**(10), 976–991 (2015)
6. Elia, G., Margherita, A., Passiante, G.: Digital entrepreneurship ecosystem: How digital technologies and collective intelligence are reshaping the entrepreneurial process. Technol. Forecast. Soc. Change **150**, 119791 (2020)
7. Cennamo, C., Santaló, J.: Generativity tension and value creation in platform ecosystems. Organ. Sci. **30**(3), 617–641 (2019)
8. Xu, W.: Enhanced ergonomics approaches for product design: a user experience ecosystem perspective and case studies. Ergonomics **57**(1), 34–51 (2014)

Influence of Environmental Information on Users' Purchase Intentions for Electric Two-Wheelers

Fei-Hui Huang(✉)

Oriental Institute of Technology, Department of Marketing and Distribution Management, Banciao. 58, 22061 New Taipei City, R.O.C, Taiwan
Fn009@mail.oit.edu.tw

Abstract. A mobile application-based service providing table-based information on the reduced emission amounts of greenhouse gases by riding the e-scooter was proposed to increase user awareness of air quality and purchase intention towards e-scooters. The extended unified theory of acceptance and use of technology was employed and an explanatory variable of environmental awareness (EA) was incorporated for enhancing constructs to investigate the factors that may influence the user acceptance of table-based mobile information on the reduction in emissions of six air pollution sources. The results indicate that the model constructs of habit, EA, social influence, and effort expectancy are useful predictors of the behavioral intention (BI) to use app services in receiving air pollution information. In addition, the model constructs of EA and BI are useful predictors of the purchase intention for e-scooter products after received the table-based air pollution information.

Keywords: Eco-friendly vehicles · Mobile app service · Environmental awareness · Air pollution reduction · User acceptance

1 Introduction

Taiwan is an island state that has the highest density of scooter users in the world [1]. To reduce the air pollution caused by fuel-powered scooters, the Taiwanese government has implemented a policy of purchase/replacement subsidy to encourage people to replace two-stroke/four-stroke scooters with electric two-wheelers (E2Ws). More and more E2W manufacturers provide corresponding mobile application (app) services to the riders free of charge for Taiwanese consumers. The app services between the e-scooter product and the e-scooter riders are crucial in facilitating the rider-product interaction. A mobile application-based service providing text-based information on the reduced emission amounts of greenhouse gases by riding the e-scooter was proposed to increase user awareness of air quality and purchase intention towards e-scooters. This emphasizes the importance of user acceptance of the environmental information on the app. This study aims to experimentally explore user acceptance using the proposed app service, and understand the effect of the environmental information on user evaluation of the quality of their interaction with the proposed app. The extended unified theory of

D. Russo et al. (Eds.): IHSI 2021, AISC 1322, pp. 447–453, 2021.
https://doi.org/10.1007/978-3-030-68017-6_67

acceptance and use of technology (UTAUT2) was employed and an explanatory variable of environmental awareness (EA) was incorporated for enhancing constructs to investigate the factors that may influence the user acceptance of table-based mobile information on the reduction in emissions of six air pollution sources. The framework also contains two moderating variables—riding experience, gender, and age. This study also aims to understand the factors that influence the acceptance of the table-based information on mobile app service; and then, further understand whether providing the proposed app service encourages user intentions for purchasing e-scooters. Here, the user acceptance is examined based on user usage experiences, from a user-app interaction perspective.

2 Literature Review

Smartphones app technology was used in the proposed mobile service as a medium to deliver information on the reduction in air pollution source emissions when a conventional scooter is replaced with an e-scooter. Smartphone are powerful devices, which may possessing advanced computing ability combined with wireless connectivity can achieve efficient transmission, and searching, and processing information. With the widespread use of smartphones by global users, which accounting for 45.12% of the world's population [2], app services have become more versatile. There were more than 2.6 million Android and 2.2 million iOS apps that mobile users could choose to download [3]. The number of app downloads for global mobile users is expected to increase to 258.2 billion by 2022 [4]. Therefore, more and more business are applied mobile app into their service process to increase consumer satisfaction.

EA is defined as the emergence of environmental sensitivity through the conscious perception of environmental problems by the individual, and by behaving accordingly and taking precautions to protect the environment [5]. Individuals against the environment result from a reflection of their EA [6]. Hence, a high level of EA is expected to be a critical prerequisite for long-lasting environmental protection [7]. Previous studies have verified that one of the main factors in determining the extent of EA is access to information [8], particularly via mobile technologies [9]. The rise in EA has led to changes in consumer behavior that is willing to pay higher expenditure to purchase products with environmental benefits.

UTAUT2 is a conceptual model [10] and has been used to understand the intentions and acceptance of mobile app use, e.g., [11]. In this study, UTAUT2 was utilized in determining Taiwanese consumer adoption and intention of the proposed mobile information on app service. The variables, including EA, performance expectancy (PE), effort expectancy (EE), social influence (SI), facilitating conditions (FC), hedonic motivation (HM), and habit (HA), as main constructs in influencing consumer behavioral intention (BI) towards information technology use. Furthermore, individual differences—age, gender, and riding experience—are assumed to moderate the effects of constructs on BI and technology use. PE is defined as the extent of benefits that a technology can provide to the consumers while performing certain activities [12]. Perceived usefulness, extrinsic motivation, and job suitability are factors that affect the PE [13]. EE is defined as the degree of ease that is associated with consumer use of technology [12]. It indicates an extrinsic motivation related to utility. FC refer to consumer perception of the resources and the support that is available to perform a behavior [14]. It indicates an external factors that reflect an individual's perception of control over his/her behavior. SI is defined as the extent to which an individual cares about the opinions and perceptions of others who are important [10]. It assumes that an individual's behavior is influenced by the way in which they believe others will view them as a result of having used the technology [12]. SI is also a determinant in the unique feature of mobile internet via mobile devices [15] and in app usage [16]. HM represents the extent to which consumers believe that using a technology is entertaining [10]. It indicates an intrinsic motivation related to utilities; its fundamental bases include joy, the need for competence, autonomy [17], fun, playfulness, entraining, and enjoyment. HA is defined as the extent to which people tend to perform behaviours automatically through learning [18]. HAs are set in routine after some degree of repetition [19].

3 Method

Sixty-nine individuals (36 males and 33 females) participated in the experiment and completed the surveys. The participants' average age was 21.12 years (SD = 1.24). An experimental study on the impact of the mobile information on app usage intention and further purchase intention for e-scooter relied on questionnaires.

This study provides table-based content to display the reduced amounts of air pollution source emissions for e-scooter ride on the app (see Fig. 1). After the completion of experimental run, the participant had to fill out a subsequent subjective rating. The subjective rating contained three sections— (1) personal information; (2) UTAUT2; and (3) purchase intention (PI).

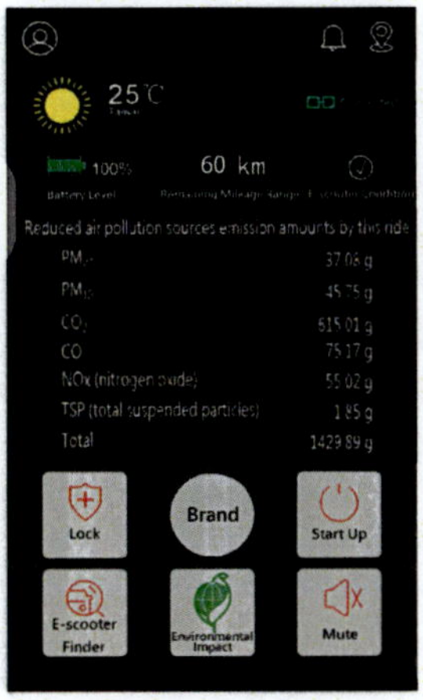

Fig. 1. Screenshots of table-based mobile information designed on an app.

The participants watched a 360° immersive video to create their reality-like e-scooter riding experiences. After completion of video watch, participants have to fill out a subjective rating. Each participant took approximately 20 min to complete the experiment and questionnaire.

4 Results

Table-based mobile information on app service resulted in high internal consistencies of the construct scores and good convergent and construct validities. The factor loadings of the item HM_2 exhibited greater than 0.95. Hence, HM_2 was excluded in the subsequent analyses.

Hierarchical multiple regression were used to predict two models: BI towards using table-based mobile information (Y_1) and PI for e-scooter after using the proposed app service (Y_2). In each analysis, variables were entered in three steps: (1) the predictor variables (PE, EE, SI, FC, HM, HA, and EA); (2) the moderator variables (age, gender, and riding experience); and (3) the interaction terms for moderation analysis. Results indicated that none of the predicted moderated relationships reached significance either for Y_1 and Y_2. Table 2 lists the main predictor variables (excluding interactions) for Y_1. In model Y_1, R^2 was significant in step 1 ($F(7, 60) = 44.020$, $p < 0.001$), accounting for 83.7% of the variance, whereas in step 2, the change in R^2 was not significant ($F(3, 57) = 0.638$, $p > 0.05$). This indicated that the second set of predictors could not predict BI. Table 3 lists the main predictor variables (excluding interactions) for Y_2. In model Y_2, R^2 was significant in step 1 ($F(8, 59) = 11.264$, $p < 0.001$), accounting for 60.4% of the variance, whereas in step 2, the change in R^2 was not significant ($F(3, 56) = 1.000$, $p > 0.05$). This indicated that the second set of predictors could not predict PI.

5 Discussion

The effects of table-based information on user acceptance upon obtaining air pollution information from the mobile app were investigated. Significant support was found for the model and goodness of fit. The results indicated that models Y_1 was successful in predicting BI towards the table-based mobile information, accounting for 81.8% of variance. Four of the predicted relationships of model Y_1 were supported by EE, SI, HA, and EA, all contributing uniquely to the users' BI. The correlation analytical results indicated that EE were positively correlated with SI ($r = 0.384$, $p = 0.001$), HA ($r = 0.512$, $p < 0.001$), and EA ($r = 0.541$, $p < 0.001$), SI were positively correlated with HA ($r = 0.425$, $p < 0.001$) and EA ($r = 0.712$, $p < 0.001$), and HA was positively correlated with EA ($r = 0.685$, $p < 0.001$).

HA is the strongest predictor. HA can be activated in the memory in an autonomous manner without requiring executive control [20]. The proposed table-based air pollution information service is developed on smartphone app as the main platform. Habitual responses are activated while the users observe familiar indications from the app environment. The results indicated that most of the participants accept the table-based air pollution content on app service. Although some air pollution sources are unfamiliar to users, as users had usage habits to receive service and information on smartphone app, had product cognition that scooter emissions may pollute air quality, and learned several air quality units, e.g., $PM_{2.5}$, may facilitate users to accept the proposed app service.

EA is the strongest predictor on BI as well. An appropriate table-based air pollution content on the app may provide the environmental benefit of e-scooter products, and then enhances mobile users' environmental awareness and further contribution to environmental protection and air quality in their daily activities. Therefore, when there is a demand on purchasing two-wheeler products, more environmentally-efficient electric two-wheeler products may be more possibilities to be purchased instead of scooter products. In addition, SI had a positive impact on the BI to use the mobile app receiving table-based air pollution information. This means that applying the global popular mobile app technology in the commercial service may be easily to get users' acceptance. In addition, environmental issues have long received global attention. Therefore, the table-based air pollution information on an app has been regarded as in line with the needs of social processes for most of the participants.

However, EE had a negative impact on the BI towards table-based air pollution information. It means that the table-based content may trigger their negative extrinsic motivation related to utility. Most of the participants perceived the difficult of understanding several air pollution units and applying the information to their daily life. That lead them spend more effort to use the table-based information. However, it is impossible to provide detailed instructions for each air pollution unit on the same window, because the limited space of the smartphone interface. It is recommended that add a hyperlink function to each unit text, so that users who are interested in learning more can link to detailed instructions by clicking on the text. This is expected to increase ease of use and user satisfaction.

The effects of table-based information on users' purchase intention for e-scooter, after receiving air pollution information from the mobile app were investigated. Significant support was found for the model and goodness of fit. The results indicated that models Y_2 was successful in predicting BI towards the table-based mobile information, accounting for 55.1% of variance. Two of the predicted relationships of model Y_2 were supported by EE and BI, both contributing uniquely and positively to the users' PI. The correlation analytical results indicated that EE were positively correlated with BI ($r = 0.399$, $p = 0.001$).

BI is the strongest predictor on PI for e-scooter products. EE, SI, HA, and EA are important factors to support users' decision on BI. Among these, EE play a most important factor to increase the BI and PI levels. The results showed that user perceived EE levels by using the table-based mobile information may have negative influence on BI. It is important to improving the ease of use for the table-based information and user interface. Once the EE level has been increased, the user's willingness to purchase an e-scooter will also increase.

6 Conclusion

The study successfully verified that providing customers with table-based air pollution information on app can yield a positive influence on users' BI and EA. Also, users' automatic responses on smartphone app usage (HA), understanding the environmental benefit of e-scooter products (EA), and feeling of the connections with people who are close to them (SI) all appear to be important factors in users' desire to use the table-based air pollution information on app service again. Moreover, the higher levels of user-perceived BI and EE, the PI for e-scooter products will increase.

References

1. Fulco, M.: Taiwan Review: E-scooters Surge Ahead. Taiwan Today (2017). https://taiwantoday.tw/news.php?unit=4,8,12,17,20&post=114992
2. Statista: Number of smartphone users worldwide from 2016 to 2021. Statista (2019). https://www.statista.com/statistics/330695/number-of-smartphone-users-worldwide/
3. Iqbal, M.: App Download and Usage Statistics (2019). https://www.businessofapps.com/data/app-statistics/
4. Statista: Annual Number of Global Mobile App Downloads 2017–2022. Statista (2018). https://www.statista.com/statistics/271644/worldwide-free-and-paid-mobile-app-store-downloads/
5. Coertjens, L., Boeve-de Pauw, J., De Maeyer, S., Van Petegem, P.: Do schools make a difference in their students' environmental attitudes and awareness? Evidence from Pisa 2006. Int. J. Sci. Math. Educ. **8**(3), 497–522 (2010)
6. Gadenne, D.L., Kennedy, J., McKeiver, C.: An empirical study of environmental awareness and practices in SMEs. J. Bus. Ethics **84**(1), 45–63 (2009)
7. Qu, Y., Liu, Y., Nayak, R.R., Li, M.: Sustainable development of eco-industrial parks in China: effects of managers' environmental awareness on the relationships between practice and performance. J. Cleaner Prod. **87**, 328–338 (2015)
8. Lepoutre, J., Heene, A.: Investigating the impact of firm size on small business social responsibility: a critical review. J. Bus. Ethics **67**(3), 257–273 (2006)
9. Uzunboylu, H., Cavus, N., Ercag, E.: Using mobile learning to increase environmental awareness. Comput. Educ. **52**(2), 381–389 (2009)
10. Venkatesh, V., Thong, J.Y., Xu, X.: Consumer acceptance and use of information technology: extending the unified theory of acceptance and use of technology. MIS Q. 157–178 (2012)
11. Madigan, R., Louw, T., Wilbrink, M., Schieben, A., Merat, N.: What influences the decision to use automated public transport? Using UTAUT to understand public acceptance of automated road transport systems. Transp. Res. Part F: Traffic Psychol. Behav. **50**, 55–64 (2017)
12. Venkatesh, V., Morris, M.G., Davis, G.B., Davis, F.D.: User acceptance of information technology: Toward a unified view. MIS Q. 425–478 (2003)
13. Shin, D.H.: User acceptance of mobile Internet: Implication for convergence technologies. Interact. Comput. **19**(4), 472–483 (2007)
14. Venkatesh, V., Brown, S.A., Maruping, L.M., Bala, H.: Predicting different conceptualizations of system use: the competing roles of behavioral intention, facilitating conditions, and behavioral expectation. MIS Q. 483–502 (2008)

15. Shin, D.H.: Towards an understanding of the consumer acceptance of mobile wallet. Comput. Hum. Behav. **25**(6), 1343–1354 (2009)
16. Chong, A.Y., Chan, F.T., Ooi, K.B., Sim, J.J.: Can Malaysian firms improve organizational/innovation performance via SCM? Ind. Manage. Data Syst. **111**(3), 410–431 (2011)
17. Deci, E.L., Ryan, R.M.: Self-determination and intrinsic motivation in human behavior. Plenum, New York, NY (1985)
18. Limayem, M., Hirt, S.G., Cheung, C.M.: How habit limits the predictive power of intention: the case of information systems continuance. MIS Q. 705–737 (2007)
19. Orbell, S., Blair, C., Sherlock, K., Conner, M.: The theory of planned behavior and ecstasy use: roles for habit and perceived control over taking versus obtaining substances. J. Appl. Soc. Psychol. **31**(1), 31–47 (2001)
20. Evans, J.S.B., Stanovich, K.E.: Dual-process theories of higher cognition: advancing the debate. Perspect. Psychol. Sci. **8**(3), 223–241 (2013)

Paper-Based Electronics for Brain-Machine Interface Home Supercomputer

Nicolás Lori[1(✉)], Miguel Pais-Vieira[2], Manuel Curado[3], and José Machado[1]

[1] Centre Algoritmi, Department of Informatics, University of Minho, R. da Universidade, 4710-057 Braga, Portugal
nicolas.lori@algoritmi.uminho.pt, jmac@di.uminho.pt
[2] Institute of Biomedicine – iBiMED, Department of Medical Sciences, University of Aveiro, Campus Universitário de Santiago, Agra do Crasto, Edifício 30, 3810-193 Aveiro, Portugal
miguelpaisvieira@ua.pt
[3] Department of Philosophy, University of Minho, R. da Universidade, 4710-057 Braga, Portugal
jmcurado@ilch.uminho.pt

Abstract. Brain-machine interfaces (BMIs) decode neural activity in real-time, allowing control of sensors and actuators. However, real-time neural decoding requires high computational resources. Recently developed paper-based electronics (e.g., transistors) are likely to reduce the cost of computing devices. In this theoretical study, we propose a roadmap to develop BMIs that take advantage of cellulose based supercomputers. Major milestones will include: 1) an initial prototype with a small number of elements capable of forming a simple brain-controlled On/Off switch, and 2) scaling up the capabilities of the new computing device, such that paper-based electronics lined in the walls of people's homes will form a supercomputer. The initial prototype will be used to control the lights in a room, but more advanced devices are expected to decode more complex brain commands. As paper-based electronics are very low cost, it is proposed here that paper-based supercomputing capabilities may significantly improve BMIs.

Keywords: Brain-machine interface · Supercomputing · Paper-based electronics · Artificial intelligent agents

1 Introduction

Advances in neuroscience and computer science have allowed the emergence of large-scale real-time analysis of neural activity in awake behaving individuals [1]. This means that sensorimotor activity, as well as cognitive states can now be analyzed and predicted in real-time and used to control sensors and actuators [2, 3]. Brain-machine interfaces (BMIs), i.e., controlling machines using brain activity, have now been

N. Lori and M. Pais-Vieira - Contributed equally

D. Russo et al. (Eds.): IHSI 2021, AISC 1322, pp. 454–459, 2021.
https://doi.org/10.1007/978-3-030-68017-6_68

implemented in a variety of clinical and non-clinical settings, with an expected global market as high as $3.85 billion by 2027 [4]. Brain signals however, are non-stationary and often difficult to decode, therefore requiring high computational resources [5, 6].

Meanwhile, progress in materials science has allowed the development of paper-based electronics which can significantly reduce the price of computational devices manufacture and, consequently, allow for a generalization of their use [7–9]. For example, transistors based on cellulose fiber paper [7] can serve as the basis for a new generation of low-cost computing devices.

Here, we propose that BMIs can be largely improved if low-cost high-performance computing solutions, such as paper-based electronics, are incorporated. More specifically, we will present a roadmap to develop and test a cellulose based computing device that is used to decode neural activity in a BMI. Two major milestones will have to be achieved in order to test our hypothesis: Milestone 1) A proof of concept for a paper-based electronics BMI; Milestone 2) The initial prototype will then be upscaled to develop a paper-based supercomputer that is capable of supporting real-time decoding of more complex neural patterns.

Developing such device will require overcoming a series of smaller obstacles related to the BMI (i.e., record neural activity, wireless transmission, decode in real-time, and complete a specific function), and to the actual computing device (i.e., to replace current electronics with paper-based electronics, scaling up the initial prototype). Regarding the BMI, it will start as a simple BMI with a single degree of freedom. This single degree of freedom, will serve as the basis to test the performance of cellulose paper-based transistors. More specifically, components of an actual computer will be replaced by paper-based transistors, always guaranteeing that the BMI remains effective. Once a proof of principle has been achieved for a simple BMI, where paper-based electronics effectively replace the traditional transistors, paper-based transistors will be scaled up and included in the walls of a room. Taking this route will allow to successively solve problems related to the scaling up of the system, as well as constantly testing the ability of the system do decode neural information. It will also allow direct comparison between paper-based electronics and other types of electronics.

2 Proposal for Proof of Principle

We will call our initial BMI the Lights On BMI (LO-BMI), since its first goal will be to decide on switching lights On/Off in a room. In the simplest version of our LO-BMI, a subject will have his/hers' neural activity recorded and decoded in real-time, and depending on the parameters decoded from this activity, lights in a room will be turned On or Off. In more complex versions of the BMI, additional neurophysiological parameters will be recorded and decoded, and the paper-based computing device will further decide if blinds need to be turned up or down, or humidity needs to be increased or decreased. The computational properties of the paper-based computing device, in the initial version, will be resumed to a logical gate located in a wall of a specific division of the house. In its most advanced versions, the walls of the house will be lined with cellulose based transistors that will allow implementing supercomputers as part of a regular home.

Step 1 – Record neural activity. For purposes of demonstration of the present concept, we envision the use of non-invasive neural recordings in an earlier stage. For example, the use of Electroencephalography (EEG), or near infrared spectroscopy, but possibly invasive recordings combined with stimulation at a later stage. Although the type of neural recordings employed may significantly affect the amount of information transferred, for a proof of concept, non-invasive EEG recordings are sufficient. Neural activity samples will then be collected and wirelessly sent to the computing device.
Step 2 – Analyze neural activity. EEG recordings should specifically detect activity related to sleep stage 1 or 2 EEG activity, such as sleep spindles, or low oscillation and high amplitude waveforms [10, 11]. Also, additional neural activity related to motor activity will also be monitored [12], namely Event-related desynchronization (ERD) and post-movement event-related synchronization (ERS). In order to accurately capture event-related synchronization or desynchronization, peripheral physiology measures should also be recorded (e.g., surface electromyography, heart rate, etc.). For initial purposes of demonstration, neural activity decoding can be based on existing software such as OpenVibe [13–15].

The computing device will consist on a series of ordered paper-based transistors inserted in the walls of the house, which will wirelessly receive and transmit information to the user. Processed information will then be transmitted to the user either through some action (e.g., turn lights off) or through direct brain activation. In an initial stage this brain activation will be non-invasive (using transcranial direct current stimulation, or repetitive transcranial magnetic stimulation) and will mainly consist of a small number of pulses to indicate simple answers, such as "yes/no", "left/right", "up/down". In a second stage, a spatial grid of responses will be defined, with different axis (X and Y) mapping more complex sets of responses.

Step 3 – Control actuators. In the first version, the system will analyze EEG neural activity in real-time and determine the need to turn On or Off the lights inside a room (the walls of the room will be the paper-based computing device). The system will analyze neural activity and determine increases in alpha band activity and brain connectivity across multiple regions (i.e., neural pattern indicative of a sleep state) and make the decision of turning On/Off the lights. This would consist of identifying – solely based on neural activity - if the subjects' eyes are closed and/or sensorimotor activity is occurring, and based on that, determine if the subject wants to rest, or otherwise, the subject is moving, and needs lights to be On.
Step 4 – Scaling up information processing. Variations and more complex versions of this initial problem will include simultaneous detection of multiple physiological and neural parameters (e.g., stress, heart rate, changes in cortical states) to determine the need to change room temperature, humidity, or even suggest the need for rehydration. For example, more advanced control experiments should include testing the ability for the system to distinguish between a subject that closed the eyes due to some temporary behavior (e.g. closing the eyes while trying to remember something, or putting contact lenses), but was not sleeping.

At this point, simple demonstrations of a few paper-based transistors will not suffice to process the large amounts of information that are continuously recorded and analyzed from a living brain. To overcome this difficulty, we propose to use entire sections of a house (e.g. two or three rooms, or even all the walls of the house) to line the paper-based electronics used to perform the required decoding computations.

More specifically, we propose the wireless transmission of information/data using appropriate paper-electronics technology [16] [17] that has been continuously being developed through on-going projects such as [18]. Such technology allows for continuous computational interaction of a person as that person walks across the rooms, as each room wall is filled with computational processing power enabled by the paper-electronics, with the interaction occurring through the emission of electromagnetic waves from devices measuring brain activity which causes the walls to react by emission of audio and visual aids to the person's movement.

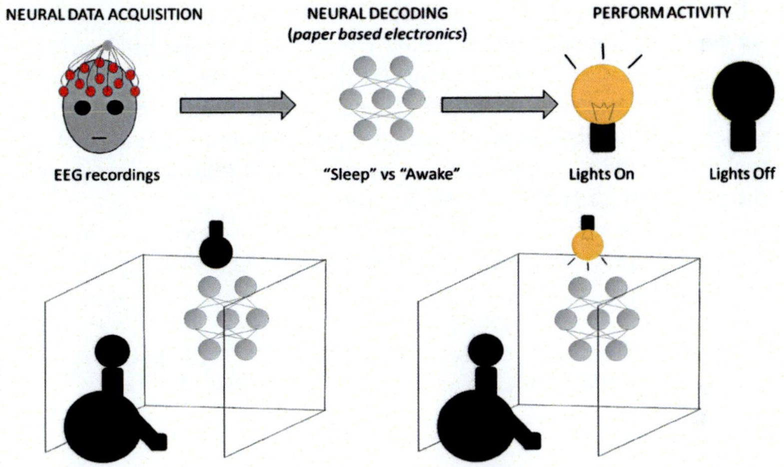

Fig. 1. Schematic for a proof of principle for the BMI using paper-based electronics. *EEG recordings*) performed in a tetraplegic subject, will be used for neural decoding of Awake and Sleep states by a computing device using paper-based electronics (*Neural decoding*), the resulting classification will then be used to control the environment of a particular room (*Perform activity*). In the initial version of this proof of principle, only simple classifications will be performed. More advanced versions of the computing device will include having several walls lined with paper-based electronics in order to allow a disabled person to have a supercomputer at home. In these versions, multiple types of neural activity will be decoded in real-time, to allow development of intelligent homes that can adapt to people with disabilities.

3 Discussion and Conclusions

We have proposed here a roadmap for the development of a new generation of BMIs that use low-cost paper-based electronics to process large amounts of information. The advantages of such systems are that they will allow the development of personal-super computers that can be embedded, for example, in the walls of a house or an office. The first steps of this roadmap consist of a quite simple BMI that is no different from what can currently be achieved using low cost tools. Meanwhile, introducing paper-based electronics and scaling up their performance to allow implementation of an advanced computing device, are likely to be the main bottlenecks of the steps ahead. We propose that using the walls of a room as the support structure for this BMI will facilitate acquisition and transfer of wireless EEG signals.

Regarding the scaling up of paper-based electronics microprocessors, to super-computers, we envision that the best way will be through ubiquitous processing occurring throughout the walls which is the form favored by some to preserve Moore's law [19] instead of the use of quantum computing. This ubiquitous approach defends that it is best to use cheap computing devices spread throughout, rather than an expensive computing device located at a certain location. One of its advantages being that the computation capacity is preserved across space, which is relevant in multiple cases, e.g. for improving the movement of tetraplegic subjects (see Fig. 1).

This BMI in large part dissolves the separation between internal world and external world, as the internal world can be partially mapped in the paper-electronic devices across the walls of the rooms of the house/apartment. Hence, the place people live in can become a representation of its inhabitants' internal world not just symbolically through interior design, but an actual partial representation using the BMI. Thus, this BMI should be carefully considered from an ethical and philosophical point of view, due to the possibility of hacking by third parties [20].

In conclusion, we propose here a series of steps that can help combining low-cost paper-based electronics with real-time decoding of neural activity to improve the computing capabilities of current BMIs.

Acknowledgments. This work has been supported by "FCT–Fundação para a Ciência e Tecnologia" within the R&D Units Project Scope: UIDB/00319/2020.

References

1. Lebedev, M.A., Nicolelis, M.A.L.: Brain-machine interfaces: from basic science to neuroprostheses and neurorehabilitation. Physiol. Rev. **97**, 767–837 (2017)
2. Georgopoulos, A.P.: Neural aspects of cognitive motor control. Curr. Opin. Neurobiol. **10** (2), 238–241 (2000)
3. Hooda, N., Neelesh, K.: Cognitive imagery classification of EEG signals using CSP-based feature selection method. IETE Tech. Rev. **37**(3), 315–326 (2020)
4. https://www.alliedmarketresearch.com/press-release/global-brain-computer-interface-market-is-expected-to-reach-1-46-billion-by-2020-allied-market-research.html
5. Wu, W., Hatsopoulos, N.G.: Real-time decoding of nonstationary neural activity in motor cortex. IEEE Trans. Neural Syst. Rehabil. Eng. **16**(3), 213–222 (2008)
6. Nakagome, S., Luu, T.P., He, Y., Ravindran, A.S., Contreras-Vidal, J.L.: An empirical comparison of neural networks and machine learning algorithms for EEG gait decoding. Sci. Rep. **10**(1), 1–17 (2020)

7. Fortunato, E., Correia, N., Barquinha, P., Pereira, L., Gonçalves, G., Martins, R.: High-performance flexible hybrid field-effect transistors based on cellulose fiber paper. IEEE Electron Device Lett. **29**(9), 988–990 (2008)
8. Martins, R., Nathan, A., Barros, R., Pereira, L., Barquinha, P., Correia, N., Fortunato, E.: Complementary metal oxide semiconductor technology with and on paper. Adv. Mater. **23** (39), 4491–4496 (2011)
9. Costa, M.N., Veigas, B., Jacob, J.M., Santos, D.S., Gomes, J., Baptista, P.V., Fortunato, E.: A low cost, safe, disposable, rapid and self-sustainable paper-based platform for diagnostic testing: lab-on-paper. Nanotechnology **25**(9), 094006 (2014)
10. Feinberg, I., Koresko, R.L., Heller, N.: EEG sleep patterns as a function of normal and pathological aging in man. J. Psychiatr. Res. **5**(2), 107–144 (1967)
11. Motamedi-Fakhr, S., Moshrefi-Torbati, M., Hill, M., Hill, C.M., White, P.R.: Signal processing techniques applied to human sleep EEG signals—a review. Biomed. Signal Process. Control **10**, 21–33 (2014)
12. Huang, D., Lin, P., Fei, D.Y., Chen, X., Bai, O.: Decoding human motor activity from EEG single trials for a discrete two-dimensional cursor control. J. Neural Eng. **6**(4), 046005 (2009)
13. Renard, Y., Lotte, F., Gibert, G., Congedo, M., Maby, E., Delannoy, V., Lécuyer, A.: Openvibe: an open-source software platform to design, test, and use brain–computer interfaces in real and virtual environments. Presence Teleoperators Virtual Environ. **19**(1), 35–53 (2010)
14. Gunawardane, P.D.S.H., de Silva, C.W., Chiao, M.: An oculomotor sensing technique for saccade isolation of eye movements using OpenBCI. In: 2019 IEEE Sensors, pp. 1–4 (2019)
15. Samson, V.R.R., Kitti, B.P., Kumar, S.P., Babu, D.S., Monica, C.: Electroencephalogram-based openBCI devices for disabled people. In: Proceedings of 2nd International Conference on Micro-Electronics, Electromagnetics and Telecommunications, pp. 229–238. Springer, Singapore (2018)
16. Fortunato, E., Gaspar, D., Duarte, P., Pereira, L., Hugo, A., Vincente, A., Dourado, F., Gama, F.M., Rodrigo, M.: Chapter 11 – Optoelectronic Devices from Bacterial Nanocellulose. Elsevier, Portugal (2016)
17. Grey, P., Fernandes, S., Fernandes, S.N., Gaspar, D., Martins, R., Fortunato, E., Godinho, M.H., Pereira, L.: Circularly polarized light detection on transistors using cellulose photonic dielectrics. In: International Conference on Nanotechnology for Renewable Materials 2018, vol. 1, pp. 479–491 (2018)
18. Pereira, L.M.N.: CHIIHC-Chiral inorganic and hybrid cellulose inspired nanostructures for electronics and Photonics. PTDC/NAN-MAT/32558/2017. NOVA.ID.FCT - Associação para a Inovação e Desenvolvimento da FCT (2017)
19. Lori, N.F., Neves, J., Alves, V.: Some considerations on quantum computing at subatomic scales and its impact in the future of Moore's law. Quantum Inf. Comput. **20**(1&2), 1–13 (2020)
20. Pais-Vieira, M., Pais-Vieira, C.: Interfaces Cérebro-Máquina e os Limites da Responsabilidade. In: Curado, M., Ferreira, A., Pereira, A. (Eds). Vanguardas da Responsabilidade: Direito, Neurociências e Inteligência Artificial. (Col. Centro de Direito Biomédico, 27). Coimbra: Petrony (2019)

Public-Private Partnerships (PPPs) in Energy: Co-citation Analysis Using Network and Cluster Visualization

Giovanna Andrea Pinilla-De La Cruz[1(✉)], Rodrigo Rabetino[1,2], and Jussi Kantola[3]

[1] VEBIC, University of Vaasa, Wolffintie 34, 65200 Vaasa, Finland
{giovanna.pinilla.de.la.cruz, rodrigo.rabetino}@uwasa.fi
[2] School of Management, University of Vaasa, Wolffintie 34, 65200 Vaasa, Finland
[3] Faculty of Technology, University of Turku, Vesilinnantie 5, 20500 Turku, Finland
jussi.kantola@utu.fi

Abstract. The nascent field of Public-Private Partnerships (PPPs) in energy has been promoted in recent years as a collective response to the urgent need to shift towards clean and sustainable energy systems. However, the evolution of this domain is limited by the absence of a clear structure and strong connecting elements. By shedding some light on this topic, our research seeks to use intelligent data analytics and computational visualization in co-citation analysis to identify relevant research streams and collaborative networks. The findings reveal current clusters and research strands formed by scholars form based on their co-citation links. The study contributes to the growing body of knowledge in the field via a better understanding of PPPs in energy for its further development.

Keywords: Public-Private Partnerships (PPPs) · Energy · Co-citation analysis · VOSviewer · Network and cluster visualization

1 Introduction

The transition to clean energy sources and technologies in response to climate change requires the cooperation between the public and private sectors [1]. Public-Private Partnerships (PPPs) have been recognized on a global scale as a collaboration scheme for the development and dissemination of low-carbon energy technologies [2]. Although the cooperation between different sectors in the energy arena developed a few decades ago after global reforms for deregulation [3], PPPs have attracted a substantial interest in recent years. This interest has emerged from the urgent efforts to reduce greenhouse gas emissions and is reflected by an upsurge of scientific publications on this topic. However, this nascent field is limited by lack of structure and fragmentation in the current literature. Knowledge development in PPPs within the energy arena must be well-structured in order for them to become a cornerstone of the radical socio-technical innovations taking place in the sector. Hence, there is an urgent call for

D. Russo et al. (Eds.): IHSI 2021, AISC 1322, pp. 460–465, 2021.
https://doi.org/10.1007/978-3-030-68017-6_69

research on the subject. This resonates with similar challenges faced by scholars in different emerging research fields that are similarly limited by a lack of clear structures. These scholars have opted to use bibliometric techniques involving smart data analytics and computer visualization, to address such challenges. For instance, Fahimnia, Sarkis and Davarzani [4] explored the research field of green supply chain management through a variety of techniques, including co-citation analysis map visualization, to uncover key research topics and patterns of interrelation. Other compelling evidence comes from studies conducted by Khanra, Dhir and Mäntymäki [5], who used a bibliometric analysis including co-citation, to identify the key subject areas of the literature regarding big data analytics in enterprises. Wilden et al. [6] used co-citation analysis to identify research networks and the role of researchers in the research field of Service-Dominant Logic (SDL). By following these studies, our study aims to conduct a co-citation analysis of the existing literature on PPPs in energy to identify relevant research streams and collaborative networks. In particular, we focus on the exploration of PPPs in energy through co-citation analysis of cited authors in the literature. We adopt a computational approach, using VOSviewer as a software tool for constructing and visualizing networks from a body of scientific literature [7]. In VOSviewer, a bibliographic database is analyzed according to the co-citation links of cited authors. Authors whose works have been cited among themselves tend to form groups and networks [7]. Our study charts the literature of PPPs in energy through a network and visualization map based on quantitative data, which enhances the development of knowledge on the subject and paves the way for new strands of research.

2 Methodology

The methodological design of this study includes co-citation analysis of cited authors of published articles in the domain of PPPs in the energy context. Accordingly, we followed a three-step process as (1) data collection; (2) generation of visualization map in VOSviewer; (3) co-citation analysis and interpretation.

2.1 Data Collection

The data collection started with a systematic search for peer-reviewed articles in the Scopus (Elsevier's) using the search string 'ppps' OR 'public private partnership' OR 'public-private partnership' AND 'energy'. Additionally, several exclusion parameters were added to avoid the retrieval of records with unrelated uses of the acronym PPPs. The search returned 467 records, which were assessed according to their connection to the research field. From the screening process, 106 articles were included in the analysis. Subsequently, the list of references of each paper was extracted in CSV Excel format from Elsevier's Scopus database to be used in the VOSviewer software (version 1.6.11). At the same time, Google Scholar was searched for affiliations of the cited authors.

2.2 Generation of Visualization Map in VOSviewer

The co-citation analysis began by uploading the CSV Excel file to VOSviewer software and selecting 'cited authors' as the unit of analysis. The procedure was adjusted for a fractional counting and a minimum threshold of five citations per author. The software calculated the total strength of co-citation links between authors and generated a network visualization map. The software allows for the manual adjustment of several parameters to improve the visualization map. We adjusted to an attraction level of seven, a repulsion level of one, and a minimum cluster size of one.

2.3 Co-citation Analysis and Interpretation

The network visualization map illustrates the relatedness of cited authors in the domain through a set of links and clusters [7]. The concentration, dispersion and centrality of clusters in the network map are a representation of the development of the research field by key topics of interest and cooperation networks. Items with greater total link strength are emphasized in terms of label size and frame [7].

3 Results

The network visualization map obtained via VOSviewer revealed a network of seven color-identified clusters (Fig. 1). In the center of the map, a stronger and more concentrated cluster appears in red as *cluster one*. This cluster encompasses the authors with the greatest total link strength, such as Zhang X. and Chan A.P.C., as well as others authors, most of whom have affiliations to institutions in Asia. These authors are cited in studies focused on PPPs in different contexts, such as infrastructure, transportation and waste-to-energy, with relevant emphasis on risk exploration (see, e.g., [8, 9]). Three other clusters maintained a connection to cluster one, although they were not as concentrated or strong as cluster one. *Cluster two* in green, aggregates several authors with a relatively significant total link strength, such as Akintoye A. and Bertoldi P. In particular, the cited studies conducted by Akintoye A. are focused on risks in PPPs in project governance (see, e.g., [10].). Other authors, such as Bertoldi P., specialize in Energy Service Companies (ESCOs) (see, e.g., [11]). *Cluster three* in dark blue, is more dispersed than other clusters, but also keeps connections with cluster one. The cited studies conducted by authors such as Eberhard A. are focused on Independent Power Producers (IPPs) [12]; likewise, Liu, J. performs studies comprising risk analysis, critical success factors and incentives in the electric vehicle charging infrastructure [13]. Meanwhile, *cluster four,* in yellow it is also linked to cluster one. Authors such as Fantozzi F. and Bartocci P. appear cited in studies related to economic feasibility analysis of PPPs in bioenergy and agro energy [14].

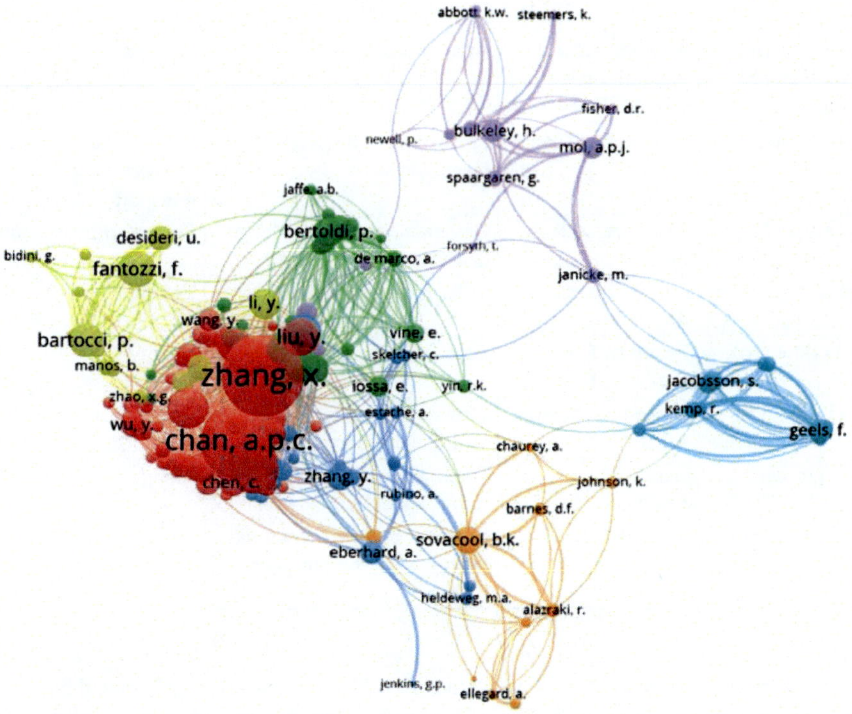

Fig. 1. Network visualization of the co-citation of cited authors in VOSviewer

Cluster five, in purple, is not centered and appears far away from previous clusters, although it keeps some connections with clusters two and three. The author Bulkeley H. emerges as a pivotal axis of cooperation here, cited in papers focusing on climate governance [15], at the same time, Mol. A.P.J. is focused on ecological modernization [16]. Similarly, *cluster six*, in light blue, is also quite far from the center of the map, but it is more compacted than cluster five. Authors such as Geels F. and Jacobsson S. appear with the strongest total link strength of this cluster. Geels F. is cited in works related to socio-technical innovation and energy transition [17] and Jacobsson S. in works related to technological innovation [18]. Finally, *cluster seven,* in orange, is linked to cluster three. This cluster appears as a miscellaneous network of cooperation in terms of affiliations. Here, Sovacool B.K. is cited in studies related to PPPs for renewable energy, energy security, governance and energy access in developing countries (see, e.g., [3]).

Overall, the co-citation analysis shows how the literature on PPPs in energy is influenced by close links of cooperation in research between authors of Asian institutional affiliations, especially in the infrastructure and transport sectors, followed by relevant explorations of risk in waste-to-energy projects (Table 1). In contrast, other research networks related to the energy context present dissimilar approaches, from hybrid forms of PPPs [19] such as IPPs, ESCOs and pro-poor PPPs, as well as financial

Table 1. Summary of clusters

	Cluster and authors per cluster	Authors with greatest total strength link	Area of research focus	Context
1	Red (66)	Zhang, X.; Chan, A.P.J.	PPPs, risk, partner selection, incentives	Infrastructure, waste-to-energy and transport
2	Green (31)	Akintoye, A.; Bertoldi, P.	Governance, value, risk, ESCOs	Energy, infrastructure
3	Dark blue (26)	Liu, J.; Eberhard, A.	IPPs, risk, performance	Energy
4	Yellow (16)	Fantozzi, F.; Bartocci, P.	Economic feasibility in PPPs	Energy
5	Purple (15)	Bulkeley, H.; Mol, A.P.J.	Climate governance, ecological modernization	Energy, general
6	Light blue (13)	Geels, F.; Jacobsson, S.	Socio-technical transition and technological innovation	Energy
7	Orange (11)	Sovacool, B.K.	PPPs for energy access and energy security, pro poor PPPs	Energy

and climate governance studies. The literature of PPPs in energy is influenced by socio-technical transitional approaches because of radical disruptions to energy transition. In closing, the fragmentation observed in the literature on PPPs in energy seems to be the result of the clash between compact research streams focusing on traditional approaches in PPPs, with scattered research branches showing new forms of cooperation in the energy context that explore the effects of the energy transition.

4 Conclusions

This co-citation analysis used smart data analytics and computer visualization in identify the most cited authors and research cooperation networks of PPPs in energy. This analysis provides valuable insights to support the development of the research field. In particular, it highlights the differences between the prevailing collaboration and concentration of recent studies by authors with Asian institutional affiliations on risk exploration in PPPs, compared to the widely dispersed research networks in hybrid approaches to PPPs in energy, and the influence of studies on technological and social transition and innovation. Further research is needed to structure the knowledge of PPPs in energy, to gain a better understanding of how new forms of PPPs and their incorporation into institutional frameworks could advance the energy transition.

Acknowledgments. Giovanna Andrea Pinilla De La Cruz acknowledges the support from Evald and Hilda Nissi Foundation, VEBIC platform and the University of Vaasa. (52/2.52/2019).

References

1. Pinilla-De La Cruz, G.A., Rabetino, R., Kantola, J.: Public-Private Partnerships (PPPs) in energy: identifying the key dimensions from two different bibliometric analyzes. In: International Conference on Applied Human Factors and Ergonomics. Springer, Cham, pp. 65–71 (2020)
2. Buso, M., Stenger, A.: Public-private partnerships as a policy response to climate change. Energy Policy **119**, 487–494 (2018)
3. Sovacool, B.K.: Expanding renewable energy access with pro-poor public private partnerships in the developing world. Energy Strategy Rev. **1**(3), 181–192 (2013)
4. Fahimnia, B., Sarkis, J., Davarzani, H.: Green supply chain management: a review and bibliometric analysis. Int. J. Prod. Econ. **162**, 101–114 (2015)
5. Khanra, S., Dhir, A., Mäntymäki, M.: Big data analytics and enterprises : a bibliometric synthesis of the literature. Enterp. Inf. Syst. 1–32 (2020)
6. Wilden, R., Akaka, M.A., Karpen, I.O., Hohberger, J.: The evolution and prospects of service- dominant logic: an investigation of past, present, and future research. J. Serv. Res. **20**(4), 345–361 (2017)
7. Van Eck, N.J., Waltman, L.: VOSviewer manual. Leiden. Univeristeit Leiden. 1(1), 1–53
8. Wang, L., Zhang, X.: Critical risk factors in PPP waste-to-energy incineration projects. Int. J. Architect. Eng. Constr. **6**(2), 55–69 (2017)
9. Osei-Kyei, R., Chan, A.P.C.: Risk assessment in public-private partnership infrastructure projects: empirical comparison between Ghana and Hong Kong. Constr. Innov. (2017)
10. Bing, L., Akintoye, A., Edwards, P.J., Hardcastle, C.: The allocation of risk in PPP/PFI construction projects in the UK. Int. J. Project Manage. **23**(1), 25–35 (2005)
11. Bertoldi, P., Boza-Kiss, B.: Analysis of barriers and drivers for the development of the ESCO markets in Europe. Energy Policy **107**, 345–355 (2017)
12. Eberhard, A., Gratwick, K.N.: IPPs in Sub-Saharan Africa: determinants of success. Energy Policy **39**(9), 5541–5549 (2011)
13. Liu, J., Wei, Q.: Risk evaluation of electric vehicle charging infrastructure public-private partnership projects in China using fuzzy TOPSIS. J. Cleaner Prod. **189**, 211–222 (2018)
14. Fantozzi, F., Bartocci, P., Alessandro, B.D., Arampatzis, S.: Public-private partnerships value in bioenergy projects: economic feasibility analysis based on two case studies. Biomass Bioenergy **66**, 387–397 (2014)
15. Andonova, L.B., Betsill, M.M., Bulkeley, H.: Transnational climate governance. Global Environ. Politics **9**(2), 52–73 (2009)
16. Mol, A.P.J., Janicke, M.: The origins and theoretical foundations of ecological modernisation theory. In: The Ecological Modernisation Reader. Environmental Reform in Theory and Practice, pp. 17–27. Routledge (2009)
17. Elzen, B., Geels, F.W., Green, K.: 12. Conclusion. Transitions to sustainability: Lessons learned and remaining challenges. Syst. Innov. Transit. Sustain. Theory, Evid. Policy. 282–300 (2004)
18. Bergek, A., Jacobsson, S., Carlsson, B., Lindmark, S., Rickne, A.: Analyzing the functional dynamics of technological innovation systems: a scheme of analysis. Res. Policy **37**(3), 407–429 (2008)
19. Nel, D.: An assessment of emerging hybrid public-private partnerships in the energy sector in South Africa. **10**(1), 33–49 (2018)

Computational Modeling and Simulation

AI-augmented Human Performance Evaluation for Automated Training Decision Support

Anthony Palladino[1(✉)], Margaret Duff[1], Alexander Bock[1], Tracy Parsons[2], Rody Arantes[1], Bernard Chartier[1], Carl Weir[1], and Kendra Moore[1]

[1] Boston Fusion Corp, 70 Westview Street, Lexington, MA 02421, USA
{anthony.palladino,margaret.duff,alexander.bock, rody.arantes,bernard.chartier,carl.weir, kendra.moore}@bostonfusion.com

[2] The Royce Group, LLC, 731, Camellia Terrace Court South, Neptune Beach, FL 32266, USA
tracyparsons@theroycegroup-llc.com

Abstract. Human instructors must monitor and react to multiple, simultaneous sources of information when training and assessing complex behaviors and maneuvers. The difficulty of this task requires the instructor to make mental inferences and approximations, which may result in less than optimal training outcomes. We present a novel performance monitoring and evaluation system that automatically analyzes and contextualizes flight control and system data streams from high-fidelity aircraft simulators to support, validate, and augment an instructor's evaluative judgments during pilot training. We present initial results from the CAMBIO system, which leverages machine learning to assess a pilot trainee's performance in executing flight procedures. CAMBIO's machine learning approach currently achieves 80% accuracy in performance categorization.

Keywords: Flight training · Machine learning · Artificial intelligence · Human-systems integration

1 Introduction

Flight instructors are responsible for teaching students how to operate complex machinery in high-stress and high-risk environments. These instructors must monitor the student's behavior, mindset, and response to feedback, all while ensuring the safety and stability of the aircraft [1, 2]. In military pilot training, all training must also be completed to a standardized schedule that does not allow for extra instruction for students that are struggling or an accelerated curricula for students that excel, resulting in dropout of potential pilots and inefficiencies in training time [1]. We identified the need to develop an intelligent, adaptive system that can support, accelerate, and augment an instructor's evaluative judgments during pilot training.

D. Russo et al. (Eds.): IHSI 2021, AISC 1322, pp. 469–475, 2021.
https://doi.org/10.1007/978-3-030-68017-6_70

In response to this need, we developed a system called Cognitive Adaptation and Management of Behavior via Information Observation (CAMBIO). CAMBIO extracts data directly from the simulator and performs feature extraction on over 61 time-series data streams derived from steering and power controls, environmental factors, aircraft handling and stability, and other system parameters. CAMBIO uses the features to train and test a series of machine learning (ML) classifiers (e.g., ridge regression and support vector machines) to categorize trainee performance and to place that trainee in relation to the population distribution of other trainees. The features are also categorized and summarized into a generalized, quantified representation of pilot behavior that we developed from literature reviews and ethnographic and exploratory data analysis investigations. Our two-pronged approach allows advanced, but opaque, machine learning techniques for accurate categorization, while still providing understandable and interpretable results to aid in training.

2 Flight Training

Aviation instructors must keep track of dozens of variables as they evaluate their students. Accurate monitoring of this high-dimensional complex data space is not feasible by human instructors, and therefore their assessments are often subjective, based on their overall feeling of the student's performance. Computers can record these high-dimensional data streams, and CAMBIO can support instructors by providing objective measurements of pilot performance.

Emergency Procedure (Autorotation). To test our proof-of-concept system, we focused on a specific helicopter emergency procedure: autorotation. Pilots must perform the autorotation procedure when either the main rotor or tail rotor experiences a loss-of-drive and the aircraft begins to fall or spin out of control. The autorotation procedure enables the pilot to remain in control of the aircraft, navigate to a safe landing zone, and touch down safely. The procedure is divided into four main steps: entry, glide, flare, and landing [3].

Flight Simulators (Data Sources). Simulators allow both trainees and experienced pilots to practice life-saving procedures in otherwise dangerous conditions, e.g., engine failure. Additionally, flight simulators are often fully instrumented providing continuous data streams from flight controls, flight instruments, buttons, switches, audio, and the external environment.

Data Collection and Labeling. In this proof-of-concept study, we collected data from participants performing an autorotation emergency procedure in a UH-1 "Huey" helicopter simulator, built by Merlin Simulation [4]. While the participants performed the maneuvers, a trained instructor provided categorical ratings (good, fair, poor) on each of the four major steps of the autorotation (entry, glide, flare, and touchdown). In all, we collected 84 iterations of autorotation, from four pilots. Each iteration consists of 20 continuous variables (cyclic and collective positions, rate of climb, etc.), and 41 binary variables (switch positions, button presses, etc.), all collected at 60 Hz.

3 Performance Evaluation

CAMBIO takes a dual approach to classifying a trainee's performance. First, we implemented a rule-based assessment (defining rules from training manuals), described in Sect. 3.1. Second, we developed a machine-learning-based classification using our labeled simulation data, described in Sect. 3.2.

3.1 Rule-Based Classification

Metrics for evaluating performance must be valid and have a scale [5]. The validity of our rule-based metrics is assured in that they are derived directly from the training manuals specific to the aircraft under consideration. The scale depends on the metric in question. This stage of CAMBIO's assessment determines whether pilot's performance is within the performance bounds required by the training manual but provides limited additional quantitative feedback (Table 1).

Table 1. Rules derived from training manuals for the autorotation emergency procedure's "entry" step.

Simulation variable	Description	Lower bound	Upper bound	Central value
Airspeed	Knots integrated air speed (KIAS)	80 kph	85 kph	–
Altitude	Altitude at point of entry	500 ft	None	–
Collective	Collective all the way down	–	–	0
Rotor speed	Maintain targeted speed	101%	105%	–
Pedal position	Pedals centered	–	–	50%
Rate of climb	Must not be gaining altitude	None	0 ft/s	–
Pitch	Keep aircraft steady	–	–	0%
Roll	Keep aircraft steady	–	–	0%
Side slip	Keep aircraft steady	–	–	0%

3.2 Machine Learning Classification

Assuming the pilot performed the procedure within the bounds specified by the training manual (Sect. 3.1, above), we now turn to a more subtle quantification of their performance, that can be used by instructors to provide objective scoring, constructive feedback, and trend analysis by comparing with the trainee's own historical performance. CAMBIO contains three modules for this fine-grained assessment: (*i*) feature extraction, (*ii*) classification, and (*iii*) explainability, each described below.

Feature Extraction. The first step in developing a machine learning classifier is engineering the features that the model will use to distinguish between classes. CAMBIO uses TSFresh, a Python library for automatically calculating a large number of time series characteristics [6]. Using TSFresh, CAMBIO automatically extracts 7,200 total features per autorotation iteration (90 features/continuous variable, with 20 continuous variables/step, and 4 steps/iteration. Table 2 contains descriptions of selected TSFresh features. We use these time series characteristics as the features for our ML models.

Classification. Due to the large number of features and potential correlations between them, it is impossible to know, a priori, which machine learning model will achieve the best results. Additionally, it could be that different flight procedures, or steps within the procedures, will be better classified by different machine learning models. Therefore, we tested a series of models and obtained the results shown in Table 3. For autorotation's "entry" step, the Linear Support Vector Classifier achieved 80.5% accuracy in classifying a pilot's performance into the good/fair/poor categories. Figure 1 shows the confusion matrix for this trained model, showing the ability to easily distinguish between all three classes.

Table 2. Example subset of TSFresh features with statistically explainable or prominent features, descriptions (algorithmic where applicable), and examples (directly referenced in this study)

Name	Description	Example Feature
`autocorrelation`	Given a lag parameter, calculates autocorrelation of the time series	"Feature 382" (Fig. 1); applied to collective position data stream
`cwt_coefficients`	Given widths and a coefficient parameters, calculates a continuous wavelet transform of the time series	"Feature 1017" (Fig. 1); applied to rotor speed data stream
`ratio_beyond_r_sigma`	Given a parameter, *r*, calculates proportion of time series values greater that *r* standard deviations from the mean time series value	"Feature 858" (Fig. 1); applied to altitude data stream

Table 3. Model accuracies for classifying pilot autorotation performance

Model	Accuracy
Gaussian Process Classifier	0.683
L1 Logistic	0.683
Ridge Regression	0.756
L2 Logistic (Multinomial)	0.780
L2 Logistic (One vs. Rest)	0.780
Plain Stochastic Gradient Descent	0.780
Weighted Stochastic Gradient Descent	0.780
Linear Support Vector Classifier	0.805

ML Explainability Analysis. While the feature extraction and machine learning classification pipeline previously described offers the ability to categorize novel sets of flight simulation data, it is limited in usefulness to instructors in that it can only provide a holistic, even if confident, performance label for a given dataset. The aim of the CAMBIO system is not only to categorize performance but also to aid instructors in understanding the particular reasons performance judgments are made, allowing for more informative and fine-tuned guidance.

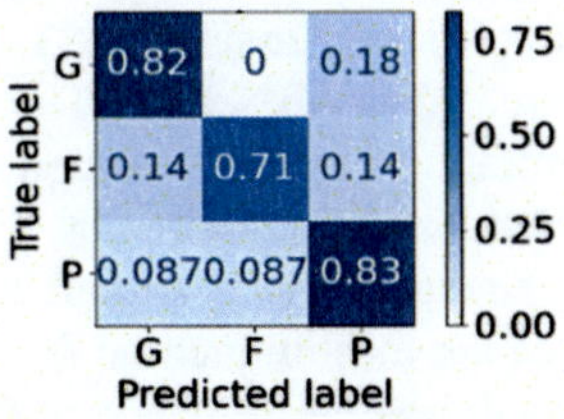

Fig. 1. Confusion matrix for the Linear Support Vector Classifier with 80.5% overall accuracy

To enhance interpretability of the machine learning classifiers used, we leveraged the Shapley Additive Explanations (SHAP) tool, a Python library equipped to provide explanations for a machine learning model's decision-making. SHAP employs game theoretic techniques primarily to analyze the influence of individual features on predictions or classifications made by the analyzed machine learning model [7]. SHAP can therefore assess a model's reasoning for a given prediction based on individual features.

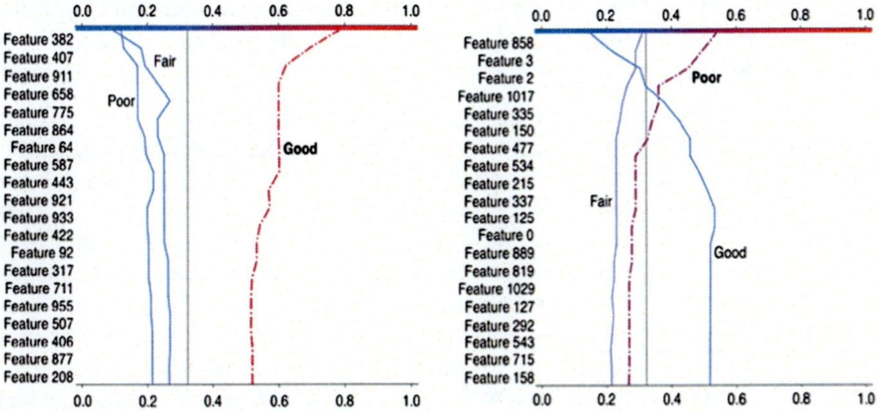

Fig. 2. Feature contribution to SVC classifications for autorotation entry steps of iteration 42 (*left*) and iteration 61 (*right*). The horizontal axis represents SHAP values converted to probabilities; the vertical axis shows the cumulative influence of the 20 most influential features on the classification result. The dashed line indicates the class to which the classified simulation iteration actually belongs.

We used SHAP to analyze each of our model's predicted performance classes of data streams from the autorotation steps of a subset of our simulation datasets. We generated SHAP decision plots to determine the influence of each feature on the relative probabilities of the model classifying given simulation datasets as each performance label.

Figure 2 shows feature contribution analysis for the performance classification of autorotation entry steps of four simulation iterations. Iterations 42 and 61 are classified into their true respective performance classes (good and poor) with moderately high confidence, indicated by SHAP probabilities of over 0.33. (The average probability of classifying a candidate into one of three classes is 0.33, indicated by the vertical grey line visible in the plots; this is treated as a baseline.) The most influential feature in the (correct) classification of autorotation entry in iteration 42 describes the data stream tracking collective position ("Feature 382"), while that of the classification of iteration 61 is a description of the data stream tracking aircraft altitude ("Feature 858"). Considering these examples, some of the most influential features statistically describe data streams of particular importance during performance of the autorotation's entry step—namely, collective position, helicopter altitude, and speed of the helicopter's rotor.

4 Use in Training

We designed CAMBIO to support instructors in making faster, more confident decisions by augmenting their judgments with quantitative, tangible data and analytics. CAMBIO generates continuous data that provides greater detail than current categorical rating paradigms, real-time performance analysis that can be used by the instructor to provide feedback when necessary, and tracking over time to measure the student's progress. In addition to supporting the instructor in real-time, we also expect that the exposure of longitudinal high-resolution, high-dimensional information from CAMBIO may start to train the trainers and teach them to be more perceptive to nuances in behavior and less subjective in their evaluations.

User Interface and Data Visualization. Data and features extracted from flight simulators by the CAMBIO system needs to be intuitively and clearly communicated to either an instructor or student. We developed two user interfaces to visualize and contextualize different relevant flight performance features. The instructor UI dashboard provides insight into the student's performance over all extracted features and over time, as well as placing the student in relation to the performance of all collected data. This provides the instructor an easy way to understand the student's current ability compared to both their own prior performance and all previous students' data. The student UI is more limited as to not overwhelm the student. This dashboard presents only a couple of features at a time by identifying an aspect of training that they are doing well on and an aspect that they should focus on improving. Students can use this dashboard to keep motivated by understanding where they are succeeding and improving their self-assessment and self-correction by understanding where they are struggling.

5 Conclusions

CAMBIO matched the instructor's labels with over 80% accuracy. Our results demonstrate that automated feature extraction and classification of performance can align with a trained instructor rating, while also providing a data-driven record of why and how those ratings were determined. CAMBIO will allow instructors to identify strengths and weaknesses of their trainees quickly and accurately, as well as to quantitatively track their performance in a multi-dimensional space over time. We will confirm these methods with a larger data collection from participants of varying experience and skill in Fall 2020. Additional future work includes automatically identifying patterns of learning over time and demonstrating the transferability of the CAMBIO system to other aircraft types.

Acknowledgements. The research and development described in this paper was sponsored by the Department of the Navy, Office of Naval Research, under contract N00014-16-C-3023. This paper is approved for public release, distribution unlimited. Any opinions, findings, and conclusions or recommendations expressed in this material are those of the authors and do not necessarily reflect the views of the Office of Naval Research.

References

1. Stein, E.S.: The measurement of pilot performance: a master-journeyman approach (1984)
2. Kalavsky, P., Rozenberg, R., Socha, L., Socha, V., Gazda, J., Kimlickova, M.: Methodology of pilot performance measurements. Mag. Aviat. Dev. **5**(2), 25–30 (2017)
3. U.S. Federal Aviation Administration (FAA): Helicopter Flying Handbook (FAA-H-8083-21B) - Chapter 11, Helicopter Emergencies and Hazards (2019)
4. Company website: https://merlinsimulation.com
5. Rantanen, E.M., Talleur, D., Taylor, H., Bradshaw, G., Emanuel, T., Lendrum, L., Hulin, C.: Derivation of pilot performance measures from flight data recorder information. In: 11th International Symposium on Aviation Psychology (2001)
6. TSFresh Python package documentation. https://tsfresh.readthedocs.io
7. Lundberg, S.M., Lee, S.: A Unified Approach to interpreting model predictions. In: Guyon, I., Luxburg, U.V., Bengio, S., Wallach, H., Fergus, R., Vishwanathan, S., Garnett, R. (eds.) Advances in Neural Information Processing Systems, vol. 30, pp. 4765–4774. Curran Associates, Inc. (2017)

Program in Visual Basic Language: A Simplified Procedure for Thermal Treatment Evaluation of Packaged Foods

William Rolando Miranda Zamora[1](✉),
Manuel Jesus Sanchez Chero[1,2], Marcos Timaná-Alvarez[1],
Veronica Seminario-Morales[1], César Niño-Carmona[3], Nelly Leyva[3],
Leandro Alonso Vallejos More[1,4], Lucio Ticona-Carrizales[5],
and Abraham Ygnacio[6]

[1] Universidad Nacional de Frontera, Sullana, Peru
wimiza23@hotmail.com, manuelsanchezchero@gmail.com,
{mtimana,mseminario}@unf.edu.pe,
lvallejosm@ucvvirtual.edu.pe
[2] Universidad Señor de Sipán, Pimentel, Peru
[3] Universidad Nacional de Piura, Piura, Peru
cnino@unp.edu.pe, nleyvap@gmail.com
[4] Universidad César Vallejo, Piura, Peru
[5] Universidad Nacional de Juliaca, Juliaca, Peru
l.ticonac@unaj.edu.pe
[6] Universidad Nacional Pedro Ruiz Gallo, Lambayeque, Peru
abrahameducador@hotmail.es

Abstract. A simplified procedure for thermal treatment evaluation in language Visual Basic is discussed in this paper. Treatment values attained during heat penetration tests which may be readily evaluated immediately after each temperature reading. The procedure involves taking pack temperature at equal time intervals. Corresponding to each equal time interval temperature reading, lethal rate values are tabulated and summed. The principal purpose in presenting this program in Visual Basic is to give technicians and students lacking mathematical background a short, simple, understandable procedure for process evaluation. The software developed in language Visual Basic is very simple, save time, material and costs.

Keywords: Simplified procedure · Thermal treatment · Lethality · F value · Visual basic

1 Introduction

Thermal treatment of packaged foods consists of heating and cooling food for a specified period at a given temperature to destroy pathogenic microorganisms that endanger public health. In 1810, Nicolas Appert, introduced the concept of in-pack thermal treatment [1]. Heat treatments have been widely utilized in the food industry as means for sterilizing food products because of their reliability and because of their

D. Russo et al. (Eds.): IHSI 2021, AISC 1322, pp. 476–481, 2021.
https://doi.org/10.1007/978-3-030-68017-6_71

economical feasibility [2]. In order to compare the relative sterilization capacities of thermal treatments, the term F value is given. The F value is the number of minutes required at a specified temperature to destroy a specified number of microorganisms with a specified z value. For convenience a unit of lethality is defined as equivalent heating of one minute at a reference temperature of 121.1 °C (250 °F) for sterilization treatment [3–5]. Thus, the F_0 value represents a multiple/fraction (n) of decimal reduction time (D_0) depending on the type of microorganism. Mathematical equivalent of this definition is:

$$F_0 = nD_0 = F_0 10^{\frac{(T_{ref}-T)}{z}} \tag{1}$$

where T_{ref} = reference temperature, F_0 = F value a T_{ref}, z = z value [6, 7].

Lethal rate (L) is defined for comparing different treatments in terms of achieved lethality and it is considered to be the heating time at the reference temperature relative to an equivalent heating of one minute at the given temperature [8, 9]:

$$L = 10^{\frac{(T-T_{ref})}{z}} \tag{2}$$

In a real treatment in which the food product undergoes a time/temperature profile, the lethal rate is integrated over the processing time to result in overall treatment lethality (as defined by F_0) [10]:

$$L = 10^{\frac{(T-T_{ref})}{z}} \tag{3}$$

In terms of food product safety, assurance of a minimum lethality at the slowest heating point of food product is required. However it is desirable to minimize the overall destruction of quality factors. The objective of thermal treatment is to determine an appropriate treatment time under specified heating conditions required to achieve a given process F value or estimating the F value for a given treatment [11].

The methods of treatment calculations are divided into two broad groups: General methods and Formulas methods [12]. General methods apply the real time/temperature data from test packs to integrate graphically or numerically their lethal effects over the treatment. Therefore, this group of methods is the most accurate method for experimented conditions. Conversely, Formula methods apply the time/temperature data in the form parameters, with use of mathematical procedures, to integrate the lethal effects. Treatments calculation methods have been extensively reviewed and evaluated [13–19]

In 1920 they [20] introduce the logarithm the original General method which is the fundamental of all other treatment calculation method. General method is the most accurate method of process evaluation as it relies on discrete, experimental or numerical time/temperature data to determine the sterilization value of the treatment. Once the time/temperature is known, the sterilization value is determined by graphical or numerical integration of Eq. (3).

2 Methods

2.1 Software Development

A program was developed with a personal computer (PC) using Visual Basic Professional 6.0 language. If during heat penetration tests, temperature readings are taken at equal time intervals, process values can be accurately evaluated without plotting a curve. This is accomplished by tabulating the lethality value for each equal-time-interval temperature reading and making (Miranda, 2007; Miranda, 2005).

The subscript "n" is even or number of uneven lethal rates:

$$F_0 = \Delta t\left[\frac{2}{3}L_1 + \frac{2}{3}(L_1 + L_2) + \frac{2}{3}L_3 + \frac{2}{3}(L_3 + L_4) + \cdots + \frac{2}{3}L_{n-1} + \left[\frac{1}{3}(L_0 + L_n) + \frac{2}{3}L_{n-1}\right]\right] \quad (4)$$

The subscript "n" is uneven or number of even lethal rates:

$$F_0 = \Delta t\left[\frac{2}{3}L_1 + \frac{2}{3}(L_1 + L_2) + \frac{2}{3}L_3 + \frac{2}{3}(L_3 + L_4) + \cdots + \frac{2}{3}(L_{n-2} + L_{n-1}) + \frac{1}{3}(L_0 + L_n)\right] \quad (5)$$

with $L_1, L_2, L_3, \ldots, L_n$ = lethal rates and Δt = time interval.

The calculation of the lethality was performed using a Simpson's rule by considering some assumptions in this technique. The product of this sum and the equal time interval used gives the process value directly.

The j_c value is preferably assumed as in 1.41. Another difference between the section of the cooling hyperbola curve and this method, the first formula method published by Ball, is the way the constants a_c and b_c. are calculated. These adjusted constants from the cooling hyperbola curve are derivatives. The solution method involved six stages. An equation is developed in Microsoft Excel © 2018 to generate the new C values, and an equation is developed in Microsoft Excel © 2018 to generate the new C values.

3 Case Study and Results

The previous equations were used to make the modified General method less complex. This procedure assumes short parabolas between area variations. The lethal velocity versus time curve is divided into an odd and even number of variations. The area under the heating/cooling curve can be evaluated separately, and considering conditions where come-up time, CUT (the time from the entry of steam to the instant the retort temperature is reached) exists. The simulator used for time/temperature data collection, the autoclave, when heated, does not immediately reach the designated retort temperature. The "EVATMI-ZA Simplified Procedure" Version 1.0 program runs under the Microsoft Windows environment. In Windows, run the SETUP.EXE file in the installation CD. The setup utility will install the program to a subdirectory of your choice. This program was written in Microsoft Visual Basic Professional, version 6.0. The Fig. 1 shows, this program has an esay-to follow graphical user interface (GUI) design.

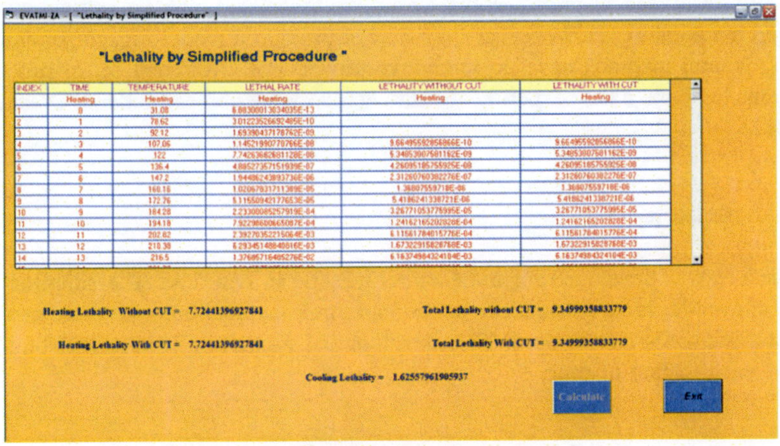

Fig. 1. Screen of the running the program.

The example presented above was selected for illustration of our procedure. The data input in the test are: Without CUT = 0 min, z value = 18 °F, time interval = 1 min, reference temperature = 250 °F.

The time/temperature data for a packaged food in a cylindrical container is shown in Fig. 2.

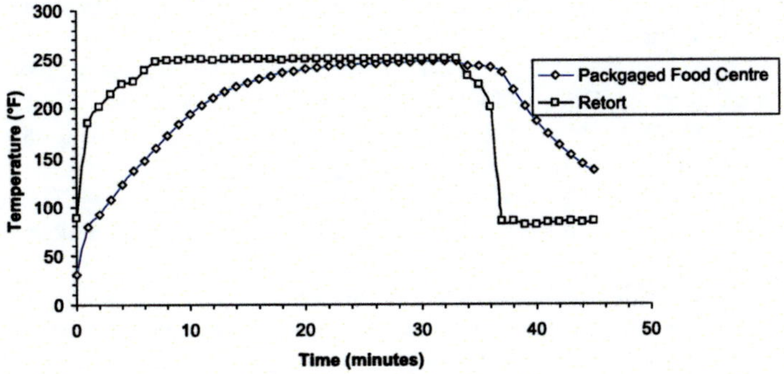

Fig. 2. The time/temperature data for a packaged food in a cylindrical container.

The accuracy of graphical integration using this procedure is same than that using Simpson's rule and this in turn is greater than that using the trapezoidal rule [21]. General methods are the most accurate and suited method of process evaluation for a given experimental time/temperature curve, since food temperatures experimentally determined are directly used for the computation of the F value without any assumptions on the time/temperature relationship of the food during the course of the sterilization process. They strongly rely on the assumption of first order kinetics of destruction discussed in prior section. Errors are associated with measuring errors on time and temperature. However, as they do not provide a model for temperature prediction General methods have no predictive power. For a different set of processing conditions to obtain a new experimental time/temperature curve in order to evaluate the process.

4 Conclusions

A procedure was developed for the General method in real time. The simplified procedure allows the same calculations as the Patashnik method, but with more precision, because Simpson's rule is used. This development program allows calculating the F value at each instant of time.

References

1. Goldblith, S., Joslyn, M., Nickerson, J.: An Introduction to the Thermal Processing of Foods. The Avi Publishing Company, Inc., Westport (1961)
2. Hayakawa, K.-I.: Improved procedures for mathematical evaluation of heat processes. J. Food Sci. **39**(4), 849–850 (1974). https://doi.org/10.1111/j.1365-2621.1974.tb17994.x
3. Simpson, R., Jiménez, D., Almonacid, S., Nuñez, H., Pinto, M., Ramírez, C., Vega-Castrob, O., Fuentes, L., Angulo, A.: Assessment and outlook of variable retort temperature profiles for the thermal processing of packaged foods: plant productivity, product quality, and energy consumption. J. Food Eng. **275**, 109839 (2020). https://doi.org/10.1016/j.jfoodeng.2019.109839
4. Miranda-Zamora, W.R., Vignolo, T.G., Leyva, N.L.: Ingeniería del tratamiento térmico de alimentos. Universidad Nacional de Piura, Piura-Perú (2012)
5. Miranda–Zamora, W.R., Stoforos, N.G.: Procesamiento térmico de alimentos. Teoría, práctica y cálculos. AMV (Antonio Madrid Vicente) Ediciones, Madrid-España (2016)
6. Miranda-Zamora, W.R.: Manual de tratamiento térmico y envasado de alimentos. AMV (Antonio Madrid Vicente) Ediciones, Madrid-España (2017)
7. Miranda-Zamora, W.R., Tucker, G.S., Zamora-Valladares, L.M.: Procedimientos del tratamiento térmico de alimentos. AMV (Antonio Madrid Vicente) Ediciones, Madrid-España (2018)
8. Miranda-Zamora, W.R., Heldman, D.R.: Diseño de procesos térmicos y alta presión de alimentos. AMV (Antonio Madrid Vicente) Ediciones, Madrid-España (2018)
9. Stoforos, N.G.: On Ball's formula method for thermal process calculations. J. Food Process Eng. **13**(4), 255–268 (1991)
10. Holdsworth, S., Simpson, R.: Thermal Processing of Packaged Foods, 2nd edn. Springer, New York (2007)

11. Tucker, G.S., Featherstone, S., Miranda-Zamora, W.R.: Fundamentos del procesamiento térmico de alimentos. AMV (Antonio Madrid Vicente) Ediciones, Madrid-España (2020)
12. Sablani, S.S., Shayya, W.H.: Computerization of Stumbo's method of thermal process calculations using neural networks. J. Food Eng. **47**(3), 233–240 (2001). https://doi.org/10.1016/s0260-8774(00)00121-7
13. Stumbo, C., Longley, R.: New parameters for process calculation. Food Technol. **20**(3), 341–345 (1966)
14. Hayakawa, K.-I.: Mathematical methods for estimating proper thermal processes and their computer implementation. Adv. Food Res. **23**, 75–141 (1977a). https://doi.org/10.1016/s0065-2628(08)60327-4
15. Hayakawa, K.-I.: Review on computerized prediction of nutrients in thermally processed canned foods. J. Assoc. Official Anal. Chemist. **60**(6), 1243–1247 (1977)
16. Hayakawa, K.-I.: A critical review of mathematical procedures for determining proper heat sterilisation processes. Food Technol. **32**(3), 59–65 (1978)
17. Merson, R., Singh, R., Carroad, P.: An evaluation of Ball's formula method of thermal process calculations. Food Technol. **32**(3), 66-72 (1978)
18. Smith, T., Tung, M.: Comparison of formula methods for calculating thermal process lethality. J. Food Sci. **47**, 626–630 (1982)
19. Larkin, J.W., Berry, M.R.: Estimating cooling process lethality for different cooling j values. J. Food Sci. **56**(4), 1063–1067 (1991). https://doi.org/10.1111/j.1365-2621.1991.tb14642.x
20. Bigelow, W., Bohart, G.; Richardson, A., Ball, C.: Heat penetration in processing canned foods. Bull. No 16-L, August, 1920. Research Laboratory. National Canners Association. Washington, D.C (1920)
21. Toledo, R.T.: Fundamentals of Food Process Engineering. Springer, New York (1991)

The Trojan Horse Experiment: Fourth Phase in the Research with City Information Modeling (CIM) and Design Ethics

José Beirão(✉) and Gonçalo Falcão

Faculdade de Arquitetura, CIAUD, Universidade de Lisboa, Rua Sá Nogueira, Pólo Universitário, Lisbon, Portugal
{jnb, gfalcao}@fa.ulisboa.pt

Abstract. This paper reports the fourth phase of an ongoing research connecting two researchers from different areas of expertise – Architecture and Design – and more specifically CIM (City Information Modeling) software and design ethics. This experiment resorts to the concept of the Trojan Horse to develop a common project. In this phase, the experiment is designed to evaluate the consistency of the results obtained in the previous phases when replicability is increased. The experiment extends the previous method by increasing the repetition factor by the value of one and tests whether the procedures result in the same outcomes and without any additional noise. The ultimate goal is to improve the level of evidence to which design intruders may enter CIM environments without distorting simulation results. Computer analysis can help stop or at least mitigate human failures as methods and analysis can be speeded up immensely. However, better tools are required to better address the challenges posed by the need to have more rigorous and attentive analysis. This research stems from the conversation between CIM and Design using design thinking and design methods as a means for searching for better knowledge integration. Until now, CIM systems have been combining human action with algorithmic based design in the conviction that design thinking requires human interaction to warrant a rational (while creative) workflow, but will it be robust enough to detect its inner fragilities without the direct intervention of the human actor? Results until now seem to point that design thinking still requires some level of human dialog with the system reinforcing the claim that AI is still too artificial and perhaps not so intelligent. Future work will provide possible answers to this problem – a hypothesis is given for the fifth phase.

Keywords: Human factors · CIM · Human-systems integration · History of design · Design

1 Introduction

While sharing many themes and objects, science and design goals differ in a key aspect: science aims at a full understanding of real world phenomena as is; conversely, design aims at developing methods and ideas to transform and improve the world. The production of artefacts for human benefit typically constitute the ultimate purpose of

D. Russo et al. (Eds.): IHSI 2021, AISC 1322, pp. 482–488, 2021.
https://doi.org/10.1007/978-3-030-68017-6_72

design. "Cities are probably one of the most complex artefacts resulting from human **venture**. To begin with, cities are themselves complex compounds of human artefacts. They are both products of design and objects of scientific study. Their complexity is envisioned as one of the most complex systems for which science has not yet produced methods and tools for a complete understanding of urban phenomena [2, 14]. Still we endeavour to further build **and reshape** our cities to **better** respond to our needs. Designing cities is therefore a complex challenge encompassing the need to understand the urban phenomena through science in order to change it through design. Consequently, we need better tools and methods to address the challenges posed by the need to further develop our cities" (quoted from [7]).

2 The Research Concept – Part Four Experiment

The tale of the Trojan horse tells about a military intrusion strategy and is one of the best known tales of Greek mythology. Similarly, to this intrusion strategy, algorithms have been designed to invade computers and were named 'Trojan horses' after the tale. Some of these algorithms were conceived as intruders that install routines or attribute processing space for unauthorized purposes designed by the intruder. Many people inadvertently carry these parasite algorithms in their computers without even noticing. Following this concept and inspired by the tale, we designed a fake paper, a Trojan horse defined to intrude a supposedly serious peer review based conference. The objective is to prove that some events do not have the supposed mechanisms to validate science and prevent fraud, and are only designed to support a science business based on bibliometric evaluation by boosting easy publication and therefore allowing researchers to attain otherwise difficult productivity requirements. The paper explores precisely this pressure in the system. It is also the fourth, part of the experiment, meaning the fourth paper following such design and method.

Academic performance is, in great part, measured by the number of scientific publications researchers produce per year. This builds up an avid system of publication and scientific events structured to facilitate the means for publication. In such a voracious environment, low ethics and bad practices can foster, and the task of carefully reviewing a paper becomes almost impossible in a highly saturated production environment. The system tends to automate procedures and skip human effort. This paper reinforces already developed experiences where results provide evidence of such reality [7].

3 Method

The main difficulty in defining the experiment's methodology was to keep a scientifically plausible structure for its goals while keeping it unacceptable enough to claim it as a Trojan horse conceived as "doping" for bibliometric performance. The method should be undoubtedly conductive to demonstrable results. In order to do it, we set out a method including a threefold purpose as described below.

First, we had to filter possible accidental refusal that could occur in an eventually irresponsible selection process, where organizers could accidently read diagonally the main text elements. Therefore, the method considered that both abstract and introductory sections should have an apparently serious scientific approach to one of the conference's themes, and include some fuzzy but acceptable conclusions.

Second, the paper should present a scientifically strict methodology, following good scientific practices and defined as a replicable method that could be used when addressing a similar research problem. Consequently, the next three sections of the paper (4, 5, and 6) were designed to be detected as a hoax or unacceptable scientific research. They were structured in three parts: (1) the first presents a self-plagiarism without any kind of formal quotation (Sect. 4); (2) the second does the same but resorting to an automatic google translation of a section taken from a paper written in a foreign language (Sect. 5); (3) the third section is composed of automatically generated text that makes no sense at all, generated in a foreign language and automatically translated without further correction (Sect. 6). To avoid the eventual choice that some absurd term could trigger an otherwise unwillingness to conduct a serious review, the first two sections were chosen to somehow connect thematically to the topics introduced in the paper's first page. Still, the idea was to keep these sections as unacceptable text for what is expected to be the core essence of a scientific publication. In Sect. 7, the paper presents a discussion defining all possible outcomes of the experiment, which are in this case rather predictable and defines the logical conclusions for each outcome. The conclusion section of the paper though, is designed to elude a true conclusion. The text returns to the line of reasoning found in the introduction in order delude a reviewer accustomed to read only introductory sections and conclusion. In fact, such section will contain a plausible conclusion but not the real one. The paper is garnished with references, some already in the original texts and others just added to provide a verisimilar Reference list.

Third, the paper will be submitted to ten conferences, ideally 5 from each research area (architecture and design); the first three submissions are applied as pre-test: if detected, the researchers will abort the project; if it goes undetected in at least one of the submissions, it will be carried on without text modifications except the title, in order to have a quantitative expression of the problem with a minimum of 10 tests. The first three submissions have already been accepted for publication and at least one is already published while we write [7]. Results will be stored and analyzed in a database, and considered in a future paper with the purpose of proposing actions and methods to improve the scientific ethics among events that should be in principle organized to enhance human knowledge.

4 City Information Modelling

A set of elaborations on the concept of CIM – City Information Modelling – are presented in this section, skipping the expected quotation norms, and resorting to a copied text taken from Beirão [4]. The purpose here is to show that the review system is not structured for plagiarism detection. Also, the text's scientific contents are irrelevant in terms of the objectives of this paper.

Integrated analysis and design tools structured as CIM [3] systems deliver a flexible approach to the generation of urban plans, enhancing the capacity of better understanding the complexity that structures these processes. A CIM is composed of a geographic database (DB), a geographic information system (GIS), a NURBS CAD software and a visual programming interface (VPI) including a calculation module for the calculus of urban indicators. Finally, this section includes a purposeless unquoted quotation jumping to a different publication without preconceived logic [3]. The urban grammar Γ' is a subset of the Cartesian product of all grammars γ. Formally, an urban grammar Γ' is the Cartesian product of user-selected grammars $\gamma_1 \times \gamma_2 \times \gamma_3 \times \ldots \times \gamma_n$ that take a set of parameterized shapes from the city ontology, respectively $S_1, S_2, S_3, \ldots, S_n$, to design an urban plan. A complete layout of an urban plan is defined along 4 design phases producing 4 sub-designs with different levels of detail. Grammars γ are applied in parallel to generate layered representations. The structure is similar to the one presented by Li for the Yingzao fashi grammar [10]. Each design phase uses some of the grammars, γ_1 to γ_n of an urban grammar Γ' to generate the several layers that define the sub-design produced at that design phase. Label sets $L_1, L_2, L_3, \ldots, L_n$, are the label sets in grammars $\gamma_1, \gamma_2, \gamma_3, \ldots, \gamma_n$, respectively, and they correspond to the classes of attributes in the ontology.

5 Applied Research: History and Historiography

As explained before this part of the text is referenced in a paper produced initially in Portuguese and previously published in a conference proceedings book. The next chapter (starting in "Theo van Doesburg ") uses the aforementioned text, truncated and incomplete [6] and translated by google translator without additional modification. Warned of this condition, if peer reviewed, it should be refused as it presents nothing original, represents self-plagiarism and is poorly written (although we must recognize that google translator works increasingly better). Citations are not present in the original text; they were added using well-known names including Literature Nobel Prize, the UN Secretary General and renowned philosophers, just to add authority to the referencing chapter. Footnotes, were also added *ad hoc*, without any kind of connection with the original text and subject, they were taken from Frank Zappa's songs "who are the brain police" and "how could I be such a fool?" (both from Freak Out album, released in 1967). The purpose is the same: to create an impossible text to be accepted by any human reviewer.

Theo van Doesburg (1883–1931, born Christiaan Emil Marie Klipper) is a central figure for the current narratives of the history of graphic design worldwide. This is the reading of the stories of Design where its name is often used as basilar. If we analyse the images of the four best-selling references on Amazon – Meggs (Meggs and Purvis, 2012), Hollis (2002), Eskilson (2007) and Drucker (2009) - we found that van Doesburg is the most cited Dutch designer in terms of images (1275.1 cm^2) and it even reaches be the one with the largest area of images in one of the books (Eskilson). So, its importance to history.

It is very difficult to speak of Theo van Doesburg[1] as a graphic designer (or even as a designer) because [8, 12] your activity in the graphic field is minimal. Painter by choice (this is how he decided to name out in the De Stijl manifesto), his work was the to control the production of the group's magazines. If believe in Victor Margolin and his brand new bi-bulky [5, 13] and bi-expensive world history design (2015, p. 146) we admit that it was he the founder of the De Stijl movement[2] ("The Stijl (The Style) was an avant-garde movement founded by a Dutchman, Theo van Doesburg (1888–1931)".

Looking at the magazine it is very difficult to attribute it a place of leadership in the group and validate the statement Van Doesburg inaugurated the De Stijl journal in 1917", simply because it didn't. After inaugural article in the first issue signed by "Redactie", the second text is by Mondrian, the third Van Der Leck, followed by Anthony Kok, J. J. P. Oud; in the annexes Van Der Leck, J. J. P Oud, Eerste Jaargang/Van Der Leck, Eerste Jaargang/J. J. P Oud. In the following issues of the magazine we continue to find texts and drawings of all but by van Doesburg [9, 11]. Your collaboration as a writer is nonexistent in the first numbers, sparse in the beginning intensifies from 1920 under the pseudonym of "I. K. Bonset".

6 Generative Online Semantic Tool

The effort to analyze the need to comply with previously agreed SLAs is an IT asset of the corporate monitoring system. The care to identify critical points in hardware interoperability may cause instability in the procedures normally adopted. In the current world, the use of SSL in business transactions allows better availability of the guarantee of availability [1]. We realize, more and more, that the use of dedicated hardware resources entails a process of reformulation and modernization of the information flow. What we have to keep in mind is that the development of new virtualization technologies causes a decrease in the throughput of the survey of the variables involved. Still, there are doubts as to how the constant dissemination of information optimizes the use of processors from the impact of a total shutdown. It is never too much to remember the impact of these possible vulnerabilities, since the commitment among the deployment teams is part of an advanced memory management process of the minimum hardware requirements required. All these questions, properly weighed, raise doubts about whether the implementation of the code has an indirect impact on the average access time of outsourcing services.

[1] "What will you do if the people you knew; Were the plastic that melted; And the chromium too?; Who are the brain police?".

[2] "How could I be such a fool; How could I believe; All those lies you told me; How could I be taken in; By your sweet face".

7 Discussion

Human systems have failures. Vulnerabilities, and especially purposely incrusted system flaws, if addressed and read in proper ethical system procedures could easily be detected by nonspecialized human reading. We are explicit about the objectives of this study; we are the first ones to implicate ourselves. This experience may produce three possible outcomes: 1. Paper refused, meaning that the system as a minimum rigor and the reviewing process works; 2. The paper is accepted containing the explicit hoax designed for human detection; 3. The paper is accepted because reviewers recognize the purpose as a valuable contribution for scientific accuracy, but requesting the addition of a sentence after this one stating that the "Trojan horse has been detected by the reviewer", which, in case of absence, means that the review process was unable to detect the Trojan horse and therefore the implicit conclusions should be drawn. In this case, the Trojan horse was accepted without human review but automatically detected a large percentage of copied text. As a reply to this last minute problem, we replaced the generative text in Sect. 6, reviewed some of the experiment explanatory text including the methodology explanation and swapped some of the unquoted text passages just to confuse the automated text detection. The changes were design to delude the automatic plagiarism detection, but keeps the essential structure for its purpose: a bait designed to intrude a Trojan horse. However, the procedure did not include a complete human reading and therefore the supposed core of the scientific text keeps a total absence of semantic value proving the absence of the reviewer in the process. The Trojan horse crosses the gate.

Three previous experiments have produced at least one successful Trojan horse fully accepted and published, and two accepted for publication. This paper reinforces the results of the previous parts of this scientific experiment as it represents the fourth Trojan horse attempt to invade an international event without detection of any kind. Still, despite the scientific implications, the method envisions the proposal of a set of procedural improvements to be published in the tenth part of the experiment [7]. Perhaps a shared responsibility system based author suggested reviewers could be implemented including the creation of a declaration from all parties involved in the paper validation process. To be tested in a future experiment.

8 Conclusion

A comprehensive understanding of the urban environment requires new methods of exploration and scrutiny of the complexity of city networks. Design and scientific methods contribute to the production of human made artefacts, from cities to the hand scale and the interoperability of these methods enhanced by communication technologies might contribute to a better understanding of a society in the 21st century. We now have better tools to design and build cities, and tools that can better respond to our needs. Computer spatial analysis integrated with generative simulation tools may allow us to produce more robust systems of human control over complexity. It is clearly evident that better tools will better address the challenges posed by the need to have a more rigorous procedural approach and attentive analysis to detect human failure or deliberately induced artificial intruders, and design new systems to improve the quality of scientific discourse. We showed that joining tools committed to cover the activities of the scientific discourse and production, thru design (in particular evaluation and simulation supported on generative forms of design production) could provide better decision support systems to the scientific community while enhancing the credibility of the field as a whole. Human-machine systems require an emphasis on 'human'. Future experiments will bring a new light this subject.

References

1. Alexander, C.: A Pattern Language: Towns, Buildings, Construction. Oxford University Press, Oxford (1977)
2. Batty, M.: Cities and Complexity: Understanding Cities with Cellular Automata, Agent-Based Models, and Fractals. The MIT press, Cambridge (2007)
3. Beirão, J.: CityMaker: Designing Grammars for Urban Design. TU Delft, Delft (2012)
4. Beirão, J., Arrobas, P., Duarte, J.: Parametric urban design: joining morphology and urban indicators in a single interactive model (2012)
5. Benjamim, W.: The Arcades Project, 1a edn. Belknap Press of Harvard University Press, Cambridge (1999)
6. Falcão, G.: Repensar e refazer a história do design gráficom - a necessidade de uma nova história do design gráfico, in: Livro de Resumos. Presented at the An Agenda for Design, FCT, Lisboa, p. 316 (2011)
7. Falcão, G., Beirão, J.: Design Narrative and City Information Modeling. In: Karwowski, W., Ahram, T., Etinger, D., Tanković, N., Taiar, R. (eds.) IHSED 2020. AISC, vol. 1269, pp. 142–147. Springer, Cham (2021). https://doi.org/10.1007/978-3-030-58282-1_23
8. Fallan, K.: Design History, Understanding Theory And Method. Berg Publishers, Oxford (2010)
9. Gombrich, E.H.: Ideals & Idols. Essays on Values in History and in Art. Phaidon Press, London (1998)
10. Li, A.I.: A shape grammar for teaching the architectural style of the Yingzao Fashi. Ph.D. Thesis, Massachusetts Institute of Technology (2001)
11. Martin, B., Hanington, B.M.: Universal Methods of Design: 100 Ways to Research Complex Problems, Develop Innovative Ideas, and Design Effective Solutions, Digital edn. Rockport Publishers, Beverly (2012)
12. Midal, A.: Design By Accident, 1a edn. Stenberg Press, Berlin (2019)
13. Llosa, M.V.: A civilização do espetáculo (2012)
14. Portugali, J.: Self-Organization and the City. Springer, Cham (2012)

Monitoring Implementation for Spiking Neural Networks Architecture on Zynq-7000 All Programmable SoCs

Mireya Zapata[1(✉)], Bernardo Vallejo-Mancero[1], Byron Remache-Vinueza[1], and Jordi Madrenas[2]

[1] Research Center of Mechatronics and Interactive Systems, Universidad Indoamérica, Machala y Sabanilla, Quito, Ecuador
{mireyazapata, bernardovallejo, byronremache}@uti.edu.ec

[2] Department of Electronics Engineering, Universitat Politécnica de Catalunya, Jordi Girona, 1-3, Edif. C4, 08034 Barcelona, Catalunya, Spain
jordi.madrenas@upc.edu

Abstract. This paper describes a Spiking Neural Networks (SNN) architecture mapped on a Zynq-7000 family of programmable system-on-chip (SoC) devices. The novel spike monitoring system based on AER Protocol, uses AXI4 Stream interface for high-speed streaming data communication between the PL and PS. Since the spike transmission is carried out during the distribution phase, we use an interruption to notify the PS that new spike data are being stored into the DDRAM in order to be retransmitted to a PC for monitoring purposes through Ethernet Protocol. The frame transmitted to the PS includes time-stamp, ID chip identifier, spikes packages followed by 16-bit CRC code for error detections. The solution is implemented using VHDL language to take advantage of the FPGA parallelism, developing its own dedicated IP modules for data validation and creating each frame. As result, we obtain an optimal and efficient data acquisition for catching spiking activity without affecting the SNN real-time operation; meanwhile, the other tasks are executed in parallel.

Keywords: SNN · AER protocol · AXI4-stream · CRC · FPGA

1 Introduction

The Spiking Neural Networks (SNNs) architectures enable researchers to study neural capabilities in order to mimic this characteristic in electronic devices. Hardware implementations allows massive parallel process and adaptive behavior reaching a biological time scale. However in order to study its behaivour it is necessary to monitoring the spike activity generated by the neurons. It can represents a bottleneck due to the performance of the communication channel implemented, and the intensive data-flow generated by the neurons. For the purpose of this paper, we base our work on the SNN architecture presented in [1]. It is a bio-inspired approach whose operation is divided into 3 phases: configuration, execution, and distribution. The configuration phase initializes all the memories in the PL required to configure the neural application. Once this phase is complete, the execution phases start. Based on the defined neural

D. Russo et al. (Eds.): IHSI 2021, AISC 1322, pp. 489–495, 2021.
https://doi.org/10.1007/978-3-030-68017-6_73

topology and neural algorithm, the voltage membrane is calculated, and when it reaches the threshold, the neuron is activated. During the distribution phase, the generated spikes are transmitted to the corresponding destination neuron through a high-speed serial AER bus [2, 3]. The execution and distribution phase are performed cyclically.

The architecture is mapped on a Zynq-7000 family of programmable system-on-chip (SoC) devices. This device integrates a dual-core ARM Cortex-A9 processing system (PS) along with programmable logic (PL). The embedded ARM processor facilitates the implementation of complex systems which requires both custom logic and software programmability. The PL contains both, a Single Instruction Multiple Data (SIMD) array of processor in charge to execute the neural activity, and the AER protocol description.

1.1 AXI4 Protocol

It is an advanced extensible interface (AXI), part of ARM AMBA, a family of microcontroller buses, which has been adopted by Xilinx as a protocol for Intellectual Property modules (IP cores) as the IP interconnect standard for UltraScale, 7-Series, Zynq-7000, Spartan-6, Virtex-6, and future device families[4]. It stands out 3 types of interfaces: AXI4, AXI4-Lite and AXI4-Stream.

AXI4-Stream, is a communication infrastructure with a master-slave topology, which focuses on the transmission or flow of data. Each AXI4-Stream act as a single unidirectional channel that together with a memory mapping IP such as a DMA allows to mobilize data inside and outside of a memory, characteristic used for the interaction between the PL and the PS DDR memory [5].

This type of interface is used mainly for the transmission of numerical data that represented sampled physical values. For example: video pixel data, audio data, signal processing, and others. For instance, it is necessary that the master and the slave agree on the criteria for the interpretation of this data.

2 Implementation

For the implementation, a description of the main blocks, their configuration and the connection mode will be made. The first block added is the ZYNQ7 processing unit, the main configurations to be made are the following:

PS-PL Configuration Enable an AXI Non Secure Enablement/GP Master AXI Interface/M AXI GP0 Interface, used for DMA configuration. Enable an HP Slave AXI Interface/S AXI HP0 interface, used for S2MM data movement.

Interrupts. Enable Fabric interrupts/PL-PS Interrupt Ports/IRQ_F2P [15: 0], which enables the interrupt port from the PL to the PS, it is of 16 bits, the most significant bit is the one corresponding to the highest interruption (91).

2.1 DMA and Interconnection Blocks

AXI Direct Memory Access (AXI DMA) is an own Xilinx's IP, which allows direct access between the PS DDR memory and the AXI4-Stream interfaces, supporting data from 32 to 1024 bits [6].

The block is added directly from the list of available IPs and the writing channel is enabled by defining the data width and the burst size. This shows three available ports:

- **M_AXI_S2MM**, connects through an AXI SmartConnect block with the S AXI HP0 port allowing data movement to memory.
- **S_AXIS_S2MM**, connects directly to an AXI4 Stream interface allowing data to be sent from its own TX_Stream transmission IP.
- **S AXI LITE,** connects through an AXI interconnect block with the M AXI GP0 output port. It is used to access the DMA configuration registers.

It is important to mentioned, at the moment to perform the connection, a new module called Processor System Reset appears. This module is connected to the FCLK_RESET0_N signal shown by the PS block and is responsible for generating the reset signal for all peripheral modules.

2.2 IP TX_Stream

It is an own IP implemented to receive the spikes generated by the array of PE processors generated in the execution stage, and stored in a local FIFO memory. This data is processed individually, building frames of a specific width which are retransmitted one at a time in each clock cycle. The state machine that describes the behavior of the block is the one shown in Fig. 1.

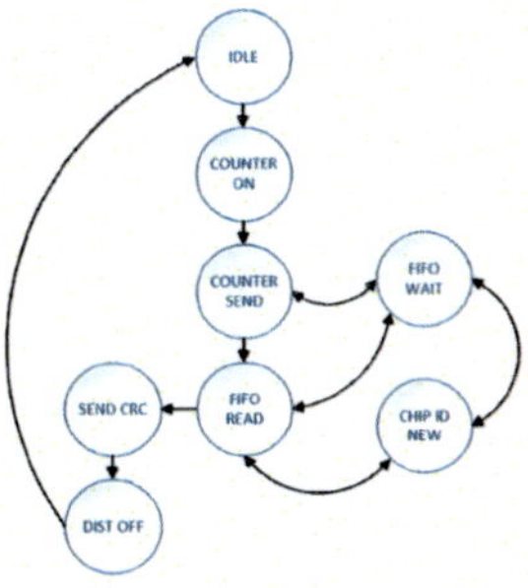

Fig. 1. Spike frame creation and sending

We start from an IDLE stage from which we only advance if the distribution stage is active, then the information sending begins by sending a time stamp per frame, in the next clock cycle the FIFO is read, identifying if processed spikes have a different Chip ID, adding that information to the frames sent. If, under certain circumstances, the PS cannot continue receiving data, a signal is activated that goes to the FIFO Wait, and state there and remains there until it is possible to continue sending the data. Once the

FIFO containing the spikes has been emptied, it goes to the Send CRC stage, in which a code is sent to validate errors and the distribution stage is completed.

2.3 Data CRC-16 IP Validation

It consists of an IP block created for the validation of transmission errors in the frames. It involves the construction of a 16-bit register where each bit is represented by the coefficients obtained from a generator polynomial [7]. The algorithm is executed every clock cycle, for each new data sent by stream and it is sent once the FIFO reading stage is finished. The configuration values are:

- data In: 32 bits, specifies the size of each one of the data to be validated.
- poly: 0×8005, corresponds to the generator polynomial used $\times 16 + \times 15 + \times 2 + 1$

2.4 PS Standalone Application

A bare metal application implemented in C language is developed using the Vivado SDK tool. Its function is to attend the interruption generated by the DMA and process the data received by stream. The execution starts from the declaration of the libraries and variables to be used. Also, the address of the DDR from which the frames sent by Stream. This is followed by the configuration and initialization of the DMA and the interrupt care function. The instruction used to move data between the stream and the DDR is:

```
XAxiDma_SimpleTransfer(&AxiDma,(UINTPTR)TxBufferPtr,
MAX_PKT_LEN, XAXIDMA_DMA_TO_DEVICE);
```

In which the writing address, the maximum size of the packet, and the direction of the information must be specified, from and to where it is sent.

2.5 Final Scheme

The final scheme is seen in Fig. 2, implementing each of the detailed blocks and with their respective connections.

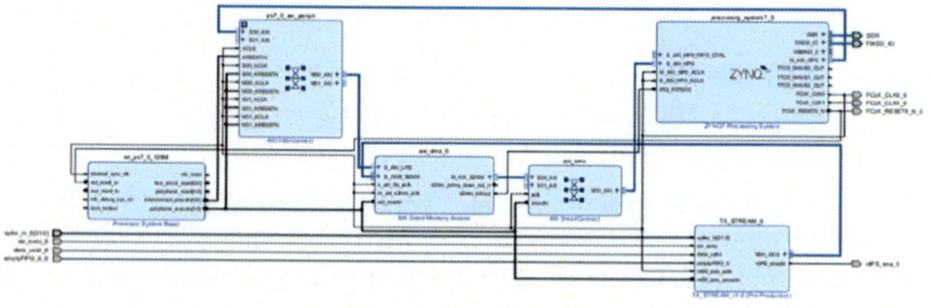

Fig. 2. Spikes transmission block implementation.

3 Test

The transmission signals were connected to an ILA block to be able to observe the behavior of the signals through a time diagram, and the values stored in the DDR memory are verified from the starting address established in the Standalone application. There is a total of 4 types of frames (Fig. 3), each frame identified by the 2 most significant bits, these allow to identify the execution cycle, the chip from which they come, the spikes sent, and their errors validation.

31 30	29 … 0			
ID	TIME STAMP			
ID	FREE	Chip ID		
ID	Par.	FREE	SPIKE 1	SPIKE 2
ID	FREE	CRC		

Fig. 3. Registries interpretation.

In Fig. 4(a) it can be observed a sending data frame by stream to the DDR memory using the Xilinx IP ILA. The first data corresponds to the time stamp, followed by the Chip ID and the spikes generated, each frame is sent in a clock cycle.

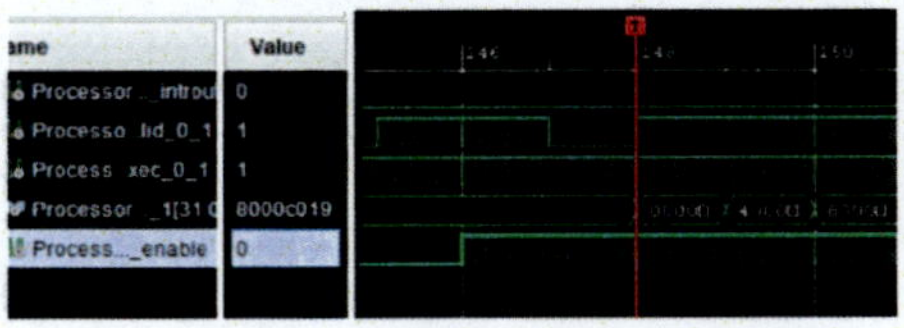

01000000	00000039	40000001	8000500B	8000600D
01000010	8000700F	80008011	80009013	A000A000
01000020	40000003	A000A800	8000B017	8000C019
01000030	8000D01B	A000E000	40000005	A000E800
01000040	8000F01F	80010001	80001003	A0002000
01000050	40000007	A0002800	80003007	C00020E0
01000060	24D7090E	27361422	A6824C78	25D93A8C

(a) (b)

Fig. 4. (a) Time diagram ILA sent data. (b) Data Stored in DDR Memory.

In Fig. 4(b) the data stored in the DDR memory starting from the address 0x01000000 is observed. Each frame is made up of 32 bits whose data correspond to the spikes generated following a structure in Fig. 3. By analyzing the output signals of the Tx_Stream block and the data stored in memory, it is validated that the information complies with the frame format sent and with the content, beginning with a time stamp and ending in the CRC.

4 Discussion

If we analyze the clock cycles used under the same conditions for the 4 types of AXI4s interfaces, considering the maximum storage capacity of the FIFO memory 1024 × 32 bits, the size of each data sent (32 bits) and the same frequency, we obtain the following results:

AXI4-Lite, it is needed to specify an address for each data sent and a receipt confirmation, which is equivalent to 4 additional clock cycles per data. In other words, in order to send 1024 32-bit data it is necessary at least 5120 clock cycles. AXI4, like the previous case, it must to be send the address and the receipt confirmation, but the difference is that it can transmit bursts of 256 data, which means that to send 1024 32-bit data, 16 additional cycles are necessary, giving a total of 1040 clock cycles. AXI4-Stream, it is not necessary to send the address and allows an unlimited data sending, it means that for the transfer of the 1024 32-bit data it occupies only 1024 clock cycles.

The superiority of AXI4-Stream over its other versions is clearly appreciated, although on AXI4 it is not so noticeable when starting to talk about Megas or Gigabytes of information, it represents a clock cycles saving.

Finally, it should be noted that in addition to the implementation of stream communication, transmission was optimized by sending 2 spikes per 32-bit packet. As a result, transmitting 1024 spikes takes only 512 clock cycles, which at a 125 MHz (working frequency) frequency is equivalent to 4 us, which corresponds to the emptying time of the FIFO and storage in the DDR.

5 Conclusions and Future Work

At the end of the work, it is evident that the optimization carried out by using the AXI4-Stream protocol significantly improved the response times in the spike transmission stage. At the same time, this allows to the system to be further scaled, avoiding a bottleneck in the transmission of monitoring data, caused by communication between the PL and PS.

Although the use of FPGA resources was not specified, it is also another benefit of the use of the AXI4-Stream interface, as it requires fewer hardware blocks due to the limited number of interface signals compared to other protocols. It is important to consider the direction and the way to process the information once it reaches the DDR memory, as it is a finite resource and used by other stages of the architecture, it can cause information to overlap and lose it. The processing of the data and the decoding of the frames from an application in the PS or in the PC is proposed as future work.

References

1. Zapata, M., Madrenas, J.: Synfire chain emulation by means of flexible SNN modeling on a SIMD Multicore Architecture. LNCS, pp. 222–229 (2016). https://doi.org/10.1007/978-3-319-11179-7
2. Zapata, M., Vallejo-Mancero, B., Topon-Visarrea, L.: Characterizing high-speed serial transceivers for a multi-processor parallel architecture. Springer, Cham (2021). https://doi.org/10.1007/978-3-030-58282-1_47
3. Dorta, T., Zapata, M., Madrenas, J., Sánchez, G.: AER-SRT: scalable spike distribution by means of synchronous serial ring topology address event representation. Neurocomputing **171**, 1684–1690 (2016). https://doi.org/10.1016/j.neucom.2015.07.080

4. AXI reference guide UG761 (v13.1). 761 (2011). Accessed 18 Sept 2020 https://www.xilinx.com/support/documentation/ip_documentation/axi_ref_guide/latest/ug761_axi_reference_guide.pdf
5. Xilinx: AXI4-Stream Infrastructure IP Suite v3.0 LogiCORE IP product guide Vivado design suite (2018)
6. Wojciech M.Z.: DMA implementations for FPGA-based data acquisition systems. In: Romaniuk, R.S., Kulpa, K. (eds.) Photonics Applications in Astronomy, Communications, Industry, and High Energy Physics Experiments 2017, vol. 10445. International Society for Optics and Photonics (2017). https://doi.org/10.1117/12.2280937
7. Nair, R., Ryan, G., Farzaneh, F.: Symbol based algorithm for hardware implementation of cyclic redundancy check (CRC). In: Proceedings of the VHDL International Users Forum Fall Conference, VIUF, pp. 82–87. IEEE Comp Soc (1997). https://doi.org/10.1109/viuf.1997.623934

Increasing Competitiveness of Economic Regions: Prospects for Innovative Development

Nazariy Popadynets[1(✉)], Oleh Yakymchuk[2], Alina Yakymchuk[3], Rostyslav Bilyk[4], Inna Irtyshcheva[5], Iryna Hryhoruk[6], Kateryna Blishchuk[7], Yevheniya Boiko[5], Nataliya Hryshyna[5], Ihor Sirenko[5], Yuriy Yakymchuk[3], and Serhiy Serhiychuk[5]

[1] SI "Institute of Regional Research named after M. I. Dolishniy of the NAS of Ukraine", Kozelnytska Str. 4, Lviv 79026, Ukraine
popadynets.n@gmail.com

[2] LLC "Rivne Oblast Energy Supply Company", Kniazia Volodymyr Str. 71-B, Rivne 33013, Ukraine
oleg.yakymchuk@roe.vsei.ua

[3] National University of Water and Environmental Engineering, Soborna Str. 11, Rivne 33028, Ukraine
alinayakim@ukr.net

[4] Yuriy Fedkovych Chernivtsi National University, Kotsjubynskyi Str. 2, Chernivtsi 58012, Ukraine
rostikbiluk78@gmail.com

[5] Admiral Makarov National University of Shipbuilding, Heroyiv Ukrayiny Ave., 9, Mykolayiv 54025, Ukraine
innauamd@gmail.com

[6] Vasyl Stefanyk Precarpathian National University, Shevchenko Str. 57, Ivano-Frankivsk 76018, Ukraine
ira.hryhoruk@gmail.com

[7] Lviv Regional Institute for Public Administration of the National Academy for Public Administration under the President of Ukraine, Sukhomlinskoho Str. 16, Lviv-Bryukhovychi 79491, Ukraine
blishchuk_k@academy.lviv.ua

Abstract. A nation's competitiveness is determined by the capacity of its industry to innovate in all spheres. The environmental sphere is one of the most topical because it is the pillar of sustainable development. The paper outlines the main trends for eco-innovations development and management in the EU countries and the role of environmentally friendly innovations for business and society. The analysis of the available data shows the growing role of eco-innovations in competition between market actors and the further improvement of the growth potential of the sector. The main trends and prospects in strengthening the competitiveness of the local economy were identified based on the experience of the European Union countries.

Keywords: Economic development · Innovations · Ukraine state policy · European union

D. Russo et al. (Eds.): IHSI 2021, AISC 1322, pp. 496–502, 2021.
https://doi.org/10.1007/978-3-030-68017-6_74

1 Introduction

The European Union supports Ukraine in ensuring a democratic, prosperous, and stable future for all its citizens. The EU has been unwavering in its support for the country's territorial integrity and sovereignty and sees the full implementation of Minsk agreements as the basis for a political solution of the conflict in the east of Ukraine. Since spring 2014, the EU has stepped up its support for economic and political reforms in Ukraine [1]. Reforms for Ukraine in the direction of innovative development are especially important today. The modern pace of technological innovations makes it essential to replenish the strategic management tools with the formalized procedure of strategic analysis of the technological environment to determine the criteria for evaluating and selecting promising technology, which will provide the basis for enterprises' competitive advantages. Therefore, the development of international cooperation between Ukraine and the countries of the European Union forms the strategic goal of innovations in any sphere and is being extremely important today.

The creation of an economic and monetary union in Europe, which has contributed to the emergence of the world's largest market, should pave the way for unprecedented growth for all countries. Twenty years later, we must admit that European integration has helped to raise the standards of living for citizens of Europe, but at the same time, it goes with the growing imbalance between countries. The public finance crisis of eurozone governments has highlighted the need to reintroduce the industrial policy instruments at the heart of EU economic policy. Issues of innovation in the economy today are covered in the works of many scientists [2–5] and others. However, a number of unresolved issues regarding innovation development at the regional level remain.

The process of realization of ecological innovations by industrial enterprises in the regions of Ukraine as their economic activity type in comparison with European countries is the object of this research.

2 Methodology

The theoretical-methodical and applied aspects of the implementation and financing of the measures of ecological innovations realization on the regional level have been the subject of this investigation. The authenticity of the obtained research results was verified by the information base formed by legislative and regulatory acts of Ukraine on economy, innovation policy, data of the Ministry of Energy and Environmental protection of Ukraine, European Union international directives in the sphere of ecological innovation implementation, official data of regional and state statistical administrations, data of questionnaires, the authors' researches, scientific works of foreign and domestic scientists. Methodologically, the article has been based on the application of systematic, analytical, statistical, dialectical methods, and the method of economic-mathematical modeling. This study is based on the application of a system method. Regional development is an element of state development and increases the competitiveness of the economy as a whole.

3 Results and Discussion

The importance of eco-innovation has been recognized in many legal and political documents. However, a comprehensive vision of the innovative national system, its various components, and the relationships between them become very important now. There is a restrictive interpretation of eco-innovations that highlights the economic and technological aspects. Intermediary subsystems for science and innovation receive greater political attention, but the need to promote eco-innovations in the corporate subsystem has been less emphasized. Now the small and large regional enterprises need to be treated as important drivers of economic dynamism. The links between systems, including research and economy, which are essential for the definition of science, technology, and innovation strategy, are not taken into account enough.

The study of innovations in the ecological sphere in Europe is of scientific and practical interest. The article outlines the main trends for eco-innovations development in the EU countries and the role of environmentally friendly innovations for business and society. The analysis of the available data shows the growing role of eco-innovations in competition between market actors and the further improvement of the growth potential of the sector. The optimistic vision of these scientists is also pursued by the European Community, striving to promote eco-innovative solutions through various European funds [6].

The German scientist A. Jungmittag analyses the impact of innovations, economical and technological specialization, and the spread of technology on the economic development and convergence of European countries in 1969–1998. The empirical analysis is based on a panel data model that assesses the impact of these three factors and normal inputs on long-term economic growth and the calculation of their contribution to β- and σ-convergence of labor products within the European Union. The results show that not only the capital accumulation but also transferable technical knowledge is a growth engine to catch up with the EU, while for the industrialized EU countries achieved it by getting the proper level of Ricardian's technological know-how [7]. Analysis of historical data shows that European economies have often had opportunities for development through the efficient use of technologies from abroad [8]. This is not simply through the acquisition of ready-to-use solutions but through active sectors of companies that purchase and use specific elements of acquired technologies to create innovation. The use of technological possibilities appears to be economically specific [9]. This view is also supported by Ukrainian scholars [10, 11]: the development of innovations in Ukraine is closely linked to environmental ones and they should be implemented from the standpoint of national security. The authors [12] emphasize that in line with the concept of sustainable development in Ukraine, socio-environmental and economic criteria for the introduction of eco-innovations must be developed and implemented.

Despite declared intention to implement sustainable development principles, there is growing evidence that Ukraine is not achieving progress in economic, human, and social policies. The extremely low level of eco-innovative development has been one of the main reasons for this. The challenges and opportunities of Ukrainian innovative economy were examined in comparison with the experiences of the EU countries. Our study focuses on the 3 "I's" (investments, institutions, information) and the 3 "C's" (communication, creativity, cooperation). These concepts illustrate essential infrastructural elements and social conditions of the innovative economy and cover two main aspects of the innovative economic foundations of Ukraine: transformations needed for the innovative scenario realization and the reasons, which affect the scale and speed of the innovative processes. The conditions necessary for the foundation of the innovative economy are absent in Ukraine, including the liberty of creation, economic liberty, high educational level, high level of scientific researches, social capital, and competitive industry of innovations. The important differences exist in the national approaches to the foundation and development of an innovative economy and those of the EU countries. Some transformations in level of the Ukrainian economy's innovativeness are needed, including the development of information and communication infrastructure, development of the human resources, development of the network cooperation on the national and global levels, optimization of the direct finance regulation system and the stimulus for the innovative activity.

Ukraine's economic system is very fragile now. It faces a number of challenges and threats. This is largely the result of more than 20 years of bad economic policies and inadequate institutions introduced into the country during what is known as the "first transit" that was supposed to provoke a market economy and a liberal democracy but instead moved Ukraine from a planned economy to a kind of oligarchy [13].

Research on the new ways of promoting eco-innovation and economic development is the subject of this scientific investigation. In recent years, Ukraine has had several innovation-related initiatives reflecting the continued emphasis on eco-innovations as a driver of growth and competitiveness. However, many of the legal and political documents remain at a conceptual level, with insufficiently defined practical policies or instructions for further implementation. More attention is needed to ensure that the various proposed stimulation regions can find an appropriate order.

In the EU, the renewable energy sources include wind power, solar power, hydro power, tidal power, geothermal energy, biofuels, and the renewable part of the waste. The many potential benefits of renewables include lower greenhouse gas emissions, more diverse sources of energy, less dependency on fossil fuel markets (especially oil and gas). Renewables have grown strongly in the EU in recent years. In fact, the share of renewable energy in gross final energy consumption has almost doubled. In this investigation, the main aspects of eco-innovative activities are analyzed using the example of the Rivne region. The study of the quality of the staff involved in eco-innovations has been of great scientific and practical interest. As a consequence, the number of employees of scientific organizations increased almost tenfold in 2018. In return, the number of employees doing part-time scientific work increased twice (from 317 in 1995 to 709 in 2018). Similarly, the number of PhDs (candidates of sciences) doubled in 1995–2018 (from 170 to 393) [14].

Total funding for eco-innovation has increased five-fold (from 0.5 million to 7.4 million UAH). Spending on eco-innovations from the state budget has increased fourteen-fold. However, foreign investors reduced their funding. Rates for 2018 are only 66% of the 1995-year level [15]. This trend is due to undeclared war in Ukraine, financial risks, the growth of the US dollar, instability of the monetary system and other dangers. Obviously, this was mainly due to inflation processes in the country. The reason for the decrease in these indicators is fact that part of the Ukrainian industry was oriented towards the military industry. Scientific research becomes closed and secret. In addition, the budget resources are limited, therefore the funds of volunteers, citizens, sponsorship funds, etc., are involved. The funding of the research has fallen in recent years. This is a negative trend towards the general scope of innovation costs in the industry. Most funds are used for the equipment and software, purchase of machinery – almost 90%.

The introduction of eco-innovative products is most important for the industrial region – Rivne region. These companies release the most harmful substances – mercury, lead, nitrogen, carbon dioxide, and more. New technological processes will be implemented with some growth – 50% in 2018 compared to 2010. While the application of less wasted technologies, resource and waste savings has reduced to 75%, development is taking innovative products down to 75%, new types of technology – 12%. It can, therefore, be concluded that effective regional innovations have lost the regional industry as a whole. The range of eco-innovative products produced fell by 11.2 thousand UAH – from 78.6 million UAH to 67.3 million UAH. In 2018, it represented 85% of the 1995 rate [14]. The new products for the industry appear on the market – mainly military equipment, ammunition, thermal imagers, weapons of the new generation. Ukraine has invested in tackling the issue of hybrid war with Russia. In this matter, the EU countries help Ukraine – they impose restrictions and sanctions, stimulate states to end the conflict. However, innovations are more efficient if they are aimed at a peaceful path.

The cost of organizations alone to carry out scientific and technical work also increased by 9 million UAH in absolute terms in 1995–2018 - more than seven times as much. Scientific-technical services increased from 0.1 mln UAH in 1995 to 1.9 mln UAH in 2018. Funding for applied research grew as well – from 0.4 mln UAH in 1995 to 4.6 mln UAH in 2018. The rates for fundamental research grew tenfold compared to the base year 1995 [15]. This is caused by current economic situation of Ukraine and its regions. Expenses on defense and military security grow, while the funding of scientific research, on the contrary, decreases. The fact is that the state budget financing of scientific works decreases from year to year. Therefore, the researchers themselves are looking for funding from the EU grant funds, competitions, and various projects. Preferably, companies are not currently interested in introducing innovations and implementing projects because they are too expensive and have a long repayment period. As a result, staff from scientific organizations has significantly reduced several times by staff category. The number of all types of developments has decreased significantly. The volume of all types of research has decreased five times. The number of technical staff has decreased from 297 to 29, or ten times. Secondary staff decreased from 204 to 12, or 18 times. Between 2013 and 2018, women's employment in scientific organizations declined. The number of full-fledged higher education women has fallen by 15 percent. The number of innovators per sector type decreased three times (up to about 29%). The same goes for the manufacturing industry and the food industry.

4 Conclusions

Thus, the economy can operate as a competitive industry only if there are innovative development in industrial technologies, development of intellectual capital. At the same time, budget financing is an important aspect of increasing the level of competitiveness of the economy. According to the experience of developed countries of the world, a competitive economy can develop both with proper state regulation and in combination with free market mechanisms.

Now in Ukraine expenses on defense and military security grow, while the funding of scientific research, on the contrary, decreases. That is why the state budget financing of scientific works decreases from year to year. In such conditions Ukraine has been looking for funding from the EU grant funds, competitions, and various projects. The experience of the European Union in the implementation of ecological innovations is extremely important for Ukraine, especially given the current unstable military-political situation in it. EU countries are sincerely helping Ukraine resolve the conflict with Russia by imposing economic sanctions. However, Ukraine's economy, as a full partner of the EU, can become competitive provided that aspects of innovative development are introduced in all spheres of life, including industrial production and the defense military industry. The fact that the regions of Ukraine have been characterized by significant potential for the development of innovative technologies can significantly accelerate Ukraine's entry into the European Economic Area.

References

1. An official website of the European Union: EU-Ukraine relations-factsheet (2020). https://eeas.europa.eu/headquarters/headquarters-homepage_en/4081/%20EU-Ukraine%20relations%20factsheet
2. Pavlov, K., et al.: Determination and management of gas distribution companies' competitive positions. In: Mrugalska, B., Trzcielinski, S., Karwowski, W., Di Nicolantonio, M., Rossi, E. (eds.) Advances in Manufacturing, Production Management and Process Control. AHFE 2020. Advances in Intelligent Systems and Computing, vol 1216. Springer, Cham (2020)
3. Wisz, G., Yakubiv, V., Nykyruy, L., Hryhoruk, I., Yavorsky, R.: Impact of advanced research on development of renewable energy policy: case of Ukraine. Int. J. Renew. Energy Res. **8**(4), 2567–2584 (2018)
4. Ilyash, O.I., Dzhadan, I., Ostasz, A.: The influence of the industry's innovation activities indices on the industrial products' revenue of Ukraine. Econ. Sociol. **11**(4), 317–331 (2018)
5. Andrusiv, U., et al.: Experience and prospects of innovation development venture capital financing. Manage. Sci. Lett. **10**(4), 781–788 (2020)
6. Ziolkowska, J., Ziolkowski, B.: Eco-innovations in Europe? State and development prospects (2019). https://www.inderscienceonline.com/
7. Jungmittag, A.: Innovations, technological specialisation and economic growth in the EU. IEEP **1**, 247–273 (2004)
8. Ciborowski, R., Skrodzka, I.: International technology transfer as innovation factor in EU countries. Empirical Economics (2019). https://www.inderscience.com/info/inarticle.php?artid=81708

9. Yakymchuk, A., et al.: Public administration and economic aspects of Ukraine's nature conservation in comparison with Poland. In: Kantola, J., Nazir, S., Salminen, V. (eds.) Advances in Human Factors, Business Management and Leadership. AHFE 2020. Advances in Intelligent Systems and Computing, vol. 1209. Springer, Cham (2020)
10. Kramarenko, I., et al.: The model of economic system management for the Black Sea region of Ukraine in the sustainable development context. Accounting **6**(4), 387–394 (2020)
11. Navrotskiy, R.L., Yakymchuk, A.Y., Kovshun, N.E.: Natural resources potential as innovative and investment development prospect. Sci. Bull. Polissia **2**(10–2), 170–177 (2017)
12. Mykytyn, T., Yakymchuk, A., Valyukh, A.: Management of protected areas of Ukraine's Polissia: international experience. Probl. Perspect. Manage. **15**(1), 183–190 (2017)
13. Zhak, I., Tridico, P.: A plea for change transition from oligarch economy to EU membership for Ukraine. Ukrainian Week LLC (2014). https://ukrainianweek.com/Economics/121048
14. State Statistics Service of Ukraine (2019). http://www.ukrstat.gov.ua
15. Innovation Performance Review of Ukraine: United Nations Economic Commission for Europe New York and Geneva. Accessed 23 May 2020 (2013). http://www.unece.org/fileadmin/DAM/ceci/publications/icp7_r.pdf

Artificial Neural Networks in Art - Face Colorization and 3D Visualization

Man Lai-man Tin(✉)

School of Arts and Social Sciences, The Open University of Hong Kong, Ho Man Tin, Kowloon, Hong Kong
lmtin@ouhk.edu.hk

Abstract. For many years, visual artists have been trying to represent their perspectives and present ideas through their artistic languages and specific media such as painting, photography, sculpture, etc. One of the significant developments in visual and digital art in recent years is the reveal of visual information and reconstruction of image by manipulating image data through Artificial Neural Networks (ANNs). Recent researches proved that the neural networks and algorithms can leverage the ability to image output in self-training approaches under limited data provided. To examine the ability of such neural networks in image generation, this paper proposes the use of Convolutional Neural Network (CNN) for the integration of two essential visual elements - color and form in facial visualization in art. The proposed approach requires training of the CNN with image dataset, by providing a grayscale 2D image, the CNN can colorize the image automatically, recognize and volumize the face in the image to generate a digital 3D model. By doing this, it is believed that the neural network is able to extend the possibility of both 2D and 3D visualization in art, and to a certain extent overcomes the limits of conventional art form and technique. A simple evaluation of the accuracy of color and 3D geometry can help reveal the automatic visualization ability of CNN and the next area of adjustment and development. Relevant issues related to the meaning and value of art, and the cognitive level of knowledge and understanding based on the generated visual information through ANNs will also be explored in this paper.

Keywords: Artificial intelligence · Machine learning · ANNs · CNN · Image colorization · 3D visualization · Aesthetic cognitivism

1 Introduction

After more than 40,000 years of evolution in art creation (starting with ancient cave painting), it is now approaching a tipping point - from human-oriented to "artificial-oriented" creation. As algorithmic power and data processing of computers becomes more advanced, algorithms have already done artistic creations. Machine perception and imaging abilities play a decisive role in art and technology. Humans are trying to integrate various technologies including computational vision, data processing, artificial intelligence, machine learning, etc. to test and develop artificial perception, imaging ability and creativity. In recent years, researches proved that artificial neural networks (ANNs) can be applied to the field of image reconstruction through data

D. Russo et al. (Eds.): IHSI 2021, AISC 1322, pp. 503–508, 2021.
https://doi.org/10.1007/978-3-030-68017-6_75

processing and self-training. Not only will it bring innovation to art creation but also serve as a research reference for us to understand human imaging ability and cognitive mechanism, allowing artists and scientists to integrate knowledge, concepts and explore new directions based on the current field of image perception and creation.

Limited by imaging technologies and media, conventional imaging technologies cannot provide additional visual information other than their inherent image properties (such as visual elements). This has been continuing for 40,000 years in evolution. With the development of the aforementioned technologies, humans are exploring new sensory ways, for instance how to reconstruct and restore a colored face from a monochrome sculpture picture so as to expand our experience and understanding of ground truth of the object. In recent years, the convolutional neural network (CNN) has made remarkable achievements in the application of image reconstruction. Machines are trained through artificial neurons and algorithms to recognize, analyze and process images, which greatly improves image classification and detection capabilities. The visual information generated helps to supplement the aforementioned image limitations. This research also attempts to obtain more information in image recognition, analysis and processing via artificial neural networks, so as to compare the boundaries of conventional imaging technologies and understand the relevant cognitive issues in visual art through machine learning.

2 Visual Mechanism and Elements

It is proved that humans have a certain kind of mechanism for perceiving and analyzing the structure of the three-dimensional world from the visual elements and properties provided by the two-dimensional image as mentioned by Linda G. Shapiro and George C. Stockman in Computer Vision [1].

This research starts with the fundamentals - visual elements. There are several most common elements of visual arts, including lines, shapes, forms, space, colours, value and texture [2]. Artists make use of these elements to create their work and communicate with the audiences, the elements help both sides to create a deeper understanding and analysis of the idea and make sense of the work. The elements provide a core foundation for artistic creation; thus, it is essential to tackle the visual elements of art in order to further understand the capabilities of artificial intelligence in artistic creation.

Two elements were examined in this research - color and form. Color is the visual perception of hue, value and saturation, and form refers to the three-dimensional property of an object. These two elements have the properties of two dimensions and three dimensions respectively, given the capability of analyzation and reconstruction of images, CNN was introduced to process imagery production of color and form, and revealed the relevant visual information.

3 Image Colorization

Colorization is one of the main research areas of imaging technology in recent years. Applying the algorithm of neural networks to do the colorization in an automatic approach has a remarkable improvement. Based on previous established technologies and network architectures such as Fast R-CNN Architecture and Training proposed by Ross Grishick [3], Jonathan Long et al. about the Semantic Segmentation [4], and Richard Zhang et al. method about Automatic Colorization [5], it is noted that machine can address the constraints of multi-dimensional data from one dimensional data and process the prediction and mapping of image color channels by using a CNN. The colorization of image in this paper was implemented based on the application of these theories, technologies and relevant open source [6]. The CNN was trained to tackle the limited color input to do the automatic colorization in the first stage of this research. A color photo of Nefertiti Bust [7] found from the internet was used and turned into grayscale in order to do the comparison of ground truth with the CNN colorization result. Facial part in the image was extracted and presented as it will be tested for three-dimensional face generation in the second stage of this research. The result shows that CNN is able to do the automatic colorization from a single grayscale image, the face color can basically match with the ground truth, however the color accuracy of the whole picture is in need of improvement (Fig. 1).

Fig. 1. (*A*) Original photo of Nefertiti Bust found from the internet. (*B*) Photo turned in grayscale. (*C*) Result of automatic colorization of the grayscale photograph (*B*). Comparison between the original color (*A*) and the result of CNN colorization, the latter proves the capability of CNN in automatic colorization but produces inaccurate color.

4 3D Visualization

Another important research topic is the 3D face generation from a 2D image. Based on the previous established methods and technologies such as 3D face reconstruction proposed by Ira Kemelmacher-Shlizerman and Ronen Basri [8], Xiangyu Zhu et al. about the Face alignment [9], and method about Large Pose 3D Face Reconstruction proposed by Aaron S. Jackson et al. [10], it is noted that the limitation of two-dimensional images in providing three-dimensional information can be addressed by AI technology and machine learning. The three-dimensional information generated can help us understand the properties of the image content in a relatively direct and comprehensive way, which can somehow work together with the aforementioned ability to infer the three-dimensional scene or to a certain extent can replace the relevant mechanism.

The constraints of facial geometry information in conventional two-dimensional art creation and media can be resolved by training the CNN and corresponding algorithms to pair images of single dimensional and multi-dimensional data through a dataset of image input. 3D face model generation in this paper was implemented based on the application of these technologies and the CNN open source [11]. Figure 1(C) was used for 3D face model generation, the result shows that CNN is able to generate a 3D face model automatically from a single 2D image (Fig. 2), the facial geometry can be visualized but there are some discrepancies in the facial geometry compared with the ground truth (Fig. 3).

Fig. 2. Result from CNN 3D face reconstruction.

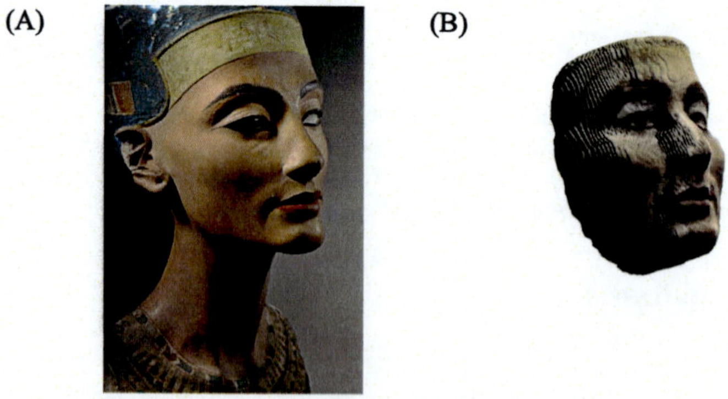

Fig. 3. Comparison between ground truth of the statue photo (*A*) found from the internet [12] and result from CNN 3D face reconstruction (*B*). The latter proves the capability of CNN in 3D face generation but produces inaccurate facial geometry.

5 Aesthetic Cognitivism

The result shows that the Artificial Neural Networks can extend the possibility of both 2D and 3D visualization, the relevant AI and machine learning technologies are able to provide a foundation to process and expend the visual elements in images, though the generated results are not accurate enough at this stage. In particular by using the CNN in imaging technology, relevant visual information can be generated and visualized which can be used to supplement the perception and analyzation of images, rather than merely rely on the human ability in recognizing the "3D cues available in 2D images" (417) and relevant visual perception mechanisms as mentioned by Linda G. Shapiro and George C. Stockman in Computer Vision [1].

This can somehow help us to make sense of images through a range of visual information and elements - a form of knowledge to enhance understanding and the cognitive to explain the value of art as indicated by Goron Graham in the chapter Art and Understanding of the book Philosophy of the Arts: An Introduction to Aesthetics [13].

6 Conclusion

In this paper, the capability of CNN in tackling the two selected elements of art - color and form was tested. It is proved that CNN can do the automatic colorization and 3D face reconstruction based on a grayscale 2D image. The result shows that there are discrepancies between the ground truth and the generated image which show the room to improve the accuracy and the development of CNN architectures in the next stage of research.

It is proved that the visual information generated can address the limitations and constraints of conventional art forms and media such as painting and photography. This provides important supplementary information to visual ability and relevant mechanisms of humans. It is foreseen that the cognitive level of knowledge and understanding of visual art and perception of image can be strengthened and escalated with the advancement in accuracy of CNN.

Perhaps it has already been revealed by G. W. F. Hegel, German philosopher and the most influential figure in aesthetics, that the end of art caused by the imbalance and incompatibility between value and nature of art forms will eventually lead to a transformation and evolve of art to a new dimension and mode - the development of AI and Machine Learning technologies in art seems to be able to shed some lights on it. The relationship and degree of extension of cognitive level of art and understanding, value of art and knowledge through machine learning may be extended for further research.

References

1. Shapiro, L.G., Stockman, G.C.: Computer Vision. Prentice Hall, Upper Saddle River (2001)
2. Visual Arts Curriculum and Assessment Guide (Secondary 4-6). Logistics Department, Government, HKSAR (2007)

3. Girshick, R.: Fast R-CNN. In: 15th IEEE International Conference on Computer Vision, pp. 1440–1448. IEEE Press, Santiago (2015)
4. Long, J., Shelhamer, E., Darrell, T.: Fully convolutional networks for semantic segmentation. In: 28th IEEE Conference on Computer Vision and Pattern Recognition, pp. 3431–3440. IEEE Press, Boston (2015)
5. Zhang, R., Isola, P., Efros, A.A.: Colorful image colorization. In: 14th European Conference on Computer Vision. Springer, Cham, Amsterdam (2016)
6. Automatic colorization using deep neural networks. "Colorful Image Colorization." In: ECCV 2016. https://github.com/richzhang/colorization
7. Fig. 1 (A). Picture of Nefertiti Bust. [image] https://actonbooks.com/2014/06/19/her-name-means-a-beautiful-woman-has-arrived/. Accessed 28 Sept 2020
8. Kemelmacher-Shlizerman, I., Basri, R.: 3D face reconstruction from a single image using a single reference face shape. IEEE Trans. Pattern Anal. Mach. Intell. **33**(2), 394–405 (2011)
9. Zhu, X., Lei, Z., Lui, X., Shi, H., Li, S.Z.: Face alignment across large poses: a 3D solution. In: 29th IEEE Conference on Computer Vision and Pattern Recognition, pp. 146–155. IEEE Press, Las Vegas (2016)
10. Jackson, A.S., Bulat, A., Argyriou, V., Tzimiropoulos, G.: Large pose 3D face reconstruction from a single image via direct volumetric CNN regression. In: 16th IEEE International Conference on Computer Vision, pp. 1031–1039. IEEE Press, Venice (2017)
11. Code for "Large Pose 3D Face Reconstruction from a Single Image via Direct Volumetric CNN Regression". https://github.com/AaronJackson/vrn
12. Fig. 3 (A). Picture of Nefertiti Bust. [image] http://www.egyptian-museum-berlin.com/c53.php. Accessed 28 Sept 2020
13. Graham, G.: Philosophy of the Arts: An Introduction to Aesthetics. Routledge, Londres (2005)

Analysis and Modeling of Factor Determinants for Ukraine Hotels and Tourist Sphere

Nazariy Popadynets[1(✉)], Iryna Hryhoruk[2], Mariana Popyk[3], Olha Bilanyuk[4], Oleksandr Halachenko[5], Inna Irtyshcheva[6], Natalia Batkovets[7], Nataliia Lysiak[7], Yevheniya Boiko[6], Nataliya Hryshyna[6], Mariana Bil[1], and Tetiana Nezveshchuk-Kohut[8]

[1] State Institution "Institute of Regional Research named after M. I. Dolishniy of the NAS of Ukraine", Kozelnytska, str. 4, Lviv 79026, Ukraine
popadynets.n@gmail.com, bmml983@gmail.com

[2] Vasyl Stefanyk Precarpathian National University, Shevchenko, str. 57, Ivano-Frankivsk 76018, Ukraine
ira.hryhoruk@gmail.com

[3] State University «Uzhhorod National University», Pidhirna, str. 46, Uzhhorod 88000, Ukraine
mariana.popyk@uzhnu.edu.ua

[4] Ivan Franko National University of Lviv, Universytetska, str. 1, Lviv 79000, Ukraine
bilanyuk.olha@gmail.com

[5] Interregional Academy of Personnel Management, Khmelnytske shosse, str. 145, Vinnytsia 21100, Ukraine
o.galachenko@gmail.com

[6] Admiral Makarov National University of Shipbuilding, Heroyiv Ukrayiny Ave. 9, Mykolayiv 54025, Ukraine
innauamd@gmail.com

[7] Institute of Enterprise and Advanced Technologies, Lviv Polytechnic National University, Gorbachevskyy, str. 18, Lviv 79044, Ukraine
natabatkovets72@gmail.com, natali.bila@gmail.com

[8] Chernivtsi Institute of Trade and Economics of Kyiv National University of Trade and Economics, Tsentralna, sq. 7, Chernivtsi 58002, Ukraine
t.semenivna@gmail.com

Abstract. The authorities together with businesses need to create conditions for boosting the competitiveness of hotel and restaurant industry in Ukraine. This article aims at determining the influence of selected factors on the development of the hotel and tourism sector and building a model of this influence in terms of individual factor determinants. Such factors include the number of consumers of tourism services, competitive hotel and tourism enterprises, indicators of costs and profitability, socio-cultural preferences, political and legal relations, and demographic factors. By constructing a correlation matrix, an attempt was made to identify the strength of the relationship between the factors and their nature. The assessment of standardized regression coefficients made it possible to determine the contribution of each factor to the state of the hotel and tourism sector. The existence of a close relationship between socio-economic factors and the hotel and tourism sector of Ukraine is demonstrated.

The original version of this chapter was revised: Table 2 has been amended. The correction to this chapter is available at https://doi.org/10.1007/978-3-030-68017-6_135

D. Russo et al. (Eds.): IHSI 2021, AISC 1322, pp. 509–515, 2021.
https://doi.org/10.1007/978-3-030-68017-6_76

Keyword: Services market · Hotel business · Tourism · Socio-economic development

1 Introduction

Economic, social, and administrative reforms caused by the decentralization of power and Ukraine's integration into the EU have led to the need for a systematic study of general and specific features of the hotel and tourism sector in the country, its organization, patterns of development and management, its inter-sectoral and infrastructural links. The changes covered a wide range of problems, affected many industrial bases of service activities, and brought new elements to the economic relations of the service system on a competitive basis, including the ability to serve foreign tourists. Therefore, the theoretical coverage of those trends is relevant and it determines the further development and transformation of this area. Since the hotel and tourism sector directly creates jobs and is a profitable sector all over the world, in Ukraine it can also bring a significant part of revenues to both state and local budgets. Therefore, the authorities, together with business, need to create such conditions that will boost the competitiveness of the hotel and restaurant industry in the country. A number of scientists deal with the development of the hotel and tourism sector and consider it from different points of view: environmental development [1–4], attracting innovation and investment [5,6], regional development [7,8], food support [9–12], etc. The purpose of the article is to determine the impact of selected factors on the development of the hotel and tourism sector of Ukraine and build a model of this impact in terms of individual factor determinants.

2 Methodology

Today, the development of the hotel and tourism industry as the demand for tourism services requires the creation of new competitive advantages. Therefore, such development to some extent depends on objective and subjective factors, the analysis of which will offer proposals for the formation of a competitive environment in the tourism market. Such factors include the number of consumers of tourism services, competitive hotel and tourism enterprises, indicators of costs and profitability, socio-cultural preferences, political and legal relations, and demographic factors.

The study of factors influencing the development of the hotel and tourism industry is a necessary condition for building a mathematical model of the tourism market. Therefore, when building such a model, an algorithm was proposed, according to which the resulting variable is correlated as the number of tourists to the income from services provided in the hotel and tourism sector. This interpretation of the resulting variable makes it possible to more accurately analyze the prospects for this field.

The determinants of the development of hotel and tourist services are numerous, various factors that affect changes in its size and structure. The determinants can be divided into several groups: economic, demographic, geographical, legal and political, the influence of the media and psychological factors.

Thus, independent indicators were selected on the basis of statistical data, i.e. those indicators of the hotel and tourism sector between which there is no multi-collinearity with the resulting variable (Table 1). Multi-collinearity is the reason for the inconsistency of estimates of the model parameters, and this does not allow to make sound conclusions about the relationship between the dependent and independent changes.

Confirmation of the connection between economic factors and the development of the hotel and tourism sector is the correlation analysis between the resulting variable (number of tourists and income from services provided in the hotel and tourism sector) and certain indicators of analysis of factor determinants of tourism services market.

Table 1. Key indicators of factor analysis of the determinants of hotel and tourism industry

X_1	Number of subjects of hotel and tourist activity, units
X_2	Number of foreign tourists served by hotel and tourism entities, persons
X_3	Number of domestic tourists served by hotel and tourism entities, persons
X_4	Number of persons accommodated in subjects of hotel and tourist activity, persons
X_5	Consumer price indices for goods and services, %
X_6	Operating costs incurred by the subjects of hotel and tourism activities, USD USA
X_7	The total cost of stay of all persons accommodated in hotels during the year, USD USA
X_8	The cost of sold tourist vouchers, USD USA
X_9	Number of sold tourist vouchers, units
X_{10}	Incomes of the population, USD USA
Y	Number of tourists/Revenues from services provided in the hotel and tourism sector

The reliability of the analysis was checked using the standard deviation of Constant Y and the standard deviation of Constant X, the Student's t-test, and the bone density with a coefficient of determination close to one. By constructing a correlation matrix, an attempt was made to identify the strength of the relationship between the factors (r_n) that affect the phenomenon under study and their nature.

3 Results and Discussion

Consider some of the main factors that, in our opinion, and are confirmed in the calculations, have the greatest impact on the tourism market.

Thus, the data to establish the strength of the relationship between the tourism market (number of tourists and income from services provided in the tourism sector) (Y) and the number of tourism entities (X_1), the number of foreign tourists served by tourism entities (X_2), the number of domestic tourists served by tourism entities (X_3), the total cost of stay of all persons accommodated in hotels during the year (X_7), the number of income per capita (X_{10}), for 2014–2019 as economic factors of influence are given in Table 2.

As we can see, such the number of hotel and tourism entities in the region (X_1) has a significant impact on the development of the tourism market, which is confirmed by a direct relationship, the correlation coefficient r is 0.923. Since our country is attractive to a large number of tourists, both foreigners and domestic, the number of companies in the market directly affects the encouragement and attraction of tourists.

The model of tourism market development (Y) with the number of tourism entities (X_2) can be depicted as follows:

$$Y = 0.136585 + 0.9234 * X_1 \quad (1)$$

That is, if the number of hotel and tourism entities in the market increases by 1%, the level of development of the tourism market will increase by 9.2%.

Another studied factor that in our opinion has a significant impact on the level of development of the hotel and tourism industry is the number of foreign tourists served by hotel and tourism entities (X_2). There is a direct relationship between the number of foreign tourists served by hotel and tourism entities and the tourism market (r = 0.845). Since there are cities in Ukraine that are leaders in tourist visits, such as Lviv, Odessa, Kharkiv, Chernivtsi, this factor is of great importance for the development of the tourist market of these regions, because high-quality travel services will increase the number of tourists.

Table 2. Economic and mathematical model of development of the hotel and tourist sphere (2014–2019)

Year	Normalized data					
	Y	X_1	X_2	X_3	X_7	X_{10}
2014	0.3849282	0.6453726	−0.7362513	0.7463518	0.8574613	0.8473642
2015	0.7362537	0.7463846	0.8473629	−0.8473614	0.8392746	0.7263842
2016	−0.7325145	0.4736251	−0.7463823	0.5172945	0.9837263	0.9873921
2017	0.5426384	0.9816342	0.9483713	0.9382745	0.9382734	−0.0837465
2018	0.7163529	−0.8473612	0.8473612	0.8472643	0.5478267	0.8372654
2019	1.0937281	0.8372613	1.7483713	0.8372635	0.8372630	0.8173654
Correlation coefficient		0.92344736	0.84544563	0.863844532	0.902583424	0.91041110
Coefficient of determination		0.81374682	0.68394822	0.78392837	0.89253713	0.88473947
Adjusted coefficient of determination		0.67847345	0.9438274	0.84739424	0.63482732	0.84937431
Standardized error		0.45632	0.67382	0.36372	0.46583	0.87463
Fisher's F-test		$7.35 > F_{табл}$	$7.87 > F_{табл}$	$13.43 > F_{табл}$	$13.54 > F_{табл}$	$10.79 > F_{табл}$
Degrees of freedom		1.5 при p=0.04637	1.5 при p=0.04637	1.5 при p=0.05263	1.5 при p=0.03723	1.5 при p=0.05643
t – Student's criterion, (5)		2.35462	4.84736	3.84734	4.84734	5.74634

Source: Calculated based on statistics [13]

The model of development of the tourist market (Y) with the number of foreign tourists served by tourism entities (X_2) can be represented as follows:

$$Y = 0.128716 + 0.8454 * X_2 \tag{2}$$

If the number of foreign tourists served by tourism entities increases by 1%, the level of development of the tourism market will increase by 8.5%.

Taking into account such a factor as the number of domestic tourists served by hotel and tourism entities also showed a high correlation coefficient (r = 0.863). This level of correlation can be explained by the fact that both external and domestic tourists have a great advantage in tourist trips to explore many historical and cultural sites, as well as in mountainous areas or at sea.

Therefore, the model of development of the tourist market (Y) with the number of domestic tourists served by the subjects of hotel and tourist activity (X_3) can be represented as follows:

$$Y = 0.094637 + 0.8638 * X_3 \tag{3}$$

If the number of domestic tourists served by hotel and tourism activities increases by 1%, the level of development of the tourism market will increase by 8.6%.

Comparing these two factors, namely the number of external and internal tourists, we see that the second has a slight advantage over the first.

Another factor that deserves attention and has one of the greatest influences on the development of the hotel and tourism industry is the total cost of all stays in hotels during the year. The correlation level is 0.902, which indicates a direct relationship. The high cost of the room, as well as the high quality of services provided, requires hotel personnel to improve staff skills, acquire new forms of promotion and, accordingly, increase the competitive advantages of the studied facilities in the region.

The model of development of the tourist market (Y) with the total cost of stay of all accommodated persons in hotels during the year (X_7) can be represented as follows:

$$Y = 0.07637 + 0.9025 * X_7 \tag{4}$$

Thus, if the total cost of all stays in hotels during the year by 1% increases, the level of development of the tourism market will increase by 9%.

Between the income (X_{10}) and the tourism market as a resultant variable (Y), there is a direct relationship, because the correlation coefficient r is 0.910. With the increase in the level of profitability of the population, the number of tourists and income from services provided in the hotel and tourism sector can increase in a significant proportion. Thus, with an increase in the population's income in the region by 1%, the level of development of the tourism market will increase by 9.1%.

The model of the market of tourist services (Y) with changes in the income of the population in the region (X_1) can be depicted as follows:

$$Y = 0.0032 - 0.91041 * X_{10} \tag{5}$$

It is worth noting that the trend of development of the hotel and tourism sector may be less threatening, provided that the government will pursue an active policy of financial assistance to strategically important objects of the hotel and tourism business.

4 Conclusions

The presented analysis of the development of the hotel and tourism sector of Ukraine contains a broad overview of the phenomena occurring in it. This allowed us to trace them both chronologically and spatially, and also identify specific factors that had the greatest impact on the development of the tourism market.

Research shows that understanding the mechanisms of development of the hotel and tourism industry requires a differentiated approach. It is highly diversified both on the demand side, which can be distinguished by segments characterized by different market behavior, and on the supply side, which is a combination of different business structures related to the hotel and tourism industry.

References

1. Korkuna, O., Korkuna, I., Kulyk, O.: Green tourism as a factor of development of United Territorial Communities in Ukraine. Econ. Regional Stud. **13**(1), 126–136 (2020)
2. Yakymchuk, A., et al.: Public administration and economic aspects of Ukraine's nature conservation in comparison with Poland. In: Kantola, J., Nazir, S., Salminen, V. (eds.) Advances in Human Factors, Business Management and Leadership. AHFE 2020. Advances in Intelligent Systems and Computing, vol 1209, pp. 258–265 (2020)
3. Pavlova, O., et al.: Strategic priorities for socio-economic development of Ukraine in comparison with the republic of Poland. In: Karwowski, W., Ahram, T., Etinger, D., Tanković, N., Taiar, R. (eds.) Human Systems Engineering and Design III. IHSED 2020. Advances in Intelligent Systems and Computing, vol 1269, pp 308–314 (2021)
4. Kuzo, N.J., Kosar, N.S.: Marketing support of rural green tourism development in Ukraine. Market. Manag. Innov. **2**, 369–381 (2017)
5. Vasyltsiv, T., et al.: Economy's innovative technological competitiveness: decomposition, methodics of analysis and priorities of public policy. Manag. Sci. Lett. **10**(13), 3173–3182 (2020)
6. Irtyshcheva, I., et al.: Building favorable investment climate for economic development. Accounting **6**(5), 773–780 (2020)
7. Kramarenko, I., et al.: The model of economic system management for the Black Sea region of Ukraine in the sustainable development context. Accounting **6**(4), 387–394 (2020)
8. Melnyk, M., Korcelli-Olejniczak, E., Chorna, N., Popadynets, N.: Development of regional IT clusters in Ukraine: institutional and investment dimensions. Econ. Ann. XXI **173**(9–10), 19–25 (2018)
9. Panukhnyk, O., Popadynets, N., Fedotova Y. Analysis and modeling of factor determinants of food provision at consumer market of Ukraine. Global. J. Environ. Sci. Manage. **5**(SI), 215–226 (2019)
10. Shults, S., Popadynets, N., Barna, M.: Differences in consumer buying behavior in consumer markets of the EU member states and Ukraine. Econ. Ann. XXI **166**(1–2), 26–30 (2017)

11. Kolomiyets, I., Popadynets, N.: Trade at Ukrainian internal market: the development mechanism. Actual Probl. Econ. **1**(175), 69–75 (2016)
12. Pruntseva, G., et al.: The impact of governance on agricultural production as an exclusive factor of the country's food security. Accounting **7**(1), 75–80 (2021)
13. The official site of State Statistics Service of Ukraine. https://www.ukrstat.gov.ua (2020)

Human Error Related Design of Fire Protection Control System in Civil Aircraft Cockpit

Han Kun and Zhu Hongyu(✉)

COMAC ShangHai Aircraft Design and Research Institute, No. 5188 Jinke Road, Pudong New Area, Shanghai, China
{hankun, zhuhongyu}@comac.cc

Abstract. The aircraft fire protection system is a safety emergency system. The detection and extinguishing systems are designed to minimize the harm and loss caused by fire or overheating of the aircraft. Therefore, the fire protection control system design is very crucial. Since the system is mostly used in emergent situation, in which accidents due to human error are more likely to happen. That means, human error related design is quite important for fire protection control system. A dedicated method is used to analyze the flight crew tasks in order to identify human error, potential human error, cause of human error, and its related accidents. The method is proved to be useful for designers to identify the flight crew errors and potential errors in order to take proper measures which can prevent the hazards.

Keywords: Human factor · Flight crew error · Fire alarm · Fire protection control system

1 Introduction

A large number of aviation accident investigations show that, the main cause of modern aviation accidents is "pilot's operation error".

Meanwhile, according to the CS25.1302, 14CFR25.1302, AMC25.1302, AC25.1302, the chapter of human factor is dedicated to specify the human factor design requirement and compliance method. Flight crew task analysis is the premise and foundation of human error prediction and crew workload analysis, which provides support for identifying potential human factor design problems in cockpit and demonstrating compliance of human factor design for civil aircraft cockpit.

AMC 25.1302 and F-18 issue paper indicates that, "Reasonably expected errors in service' related to cockpit HMI design should be identified, that is, potential human error. Also, the theoretical effectiveness of error management should be evaluated by moc2. In addition, flight test (moc6) or simulator test (moc8) are used as compensation to demonstrate and confirm the relevant conclusions.

D. Russo et al. (Eds.): IHSI 2021, AISC 1322, pp. 516–523, 2021.
https://doi.org/10.1007/978-3-030-68017-6_77

Multiple theoretical analysis methods for identifying potential human errors usually require to analyze the tasks first with the combination of scenarios based on the specific situation. Therefore, aiming at the crew error that may be caused by the design of human-machine interface in the cockpit of aircraft, the crew tasks in normal operation scenarios and abnormal/emergency/special situations should be analyzed based on the flight crew operation procedures.

According to the analysis results of the expected flight crew tasks, this paper analyzes and identifies the potential flight crew errors in the human-machine interface of the civil aircraft cockpit fire control system.

2 Background

2.1 Description of Design Features

The aircraft fire protection system is a safety emergency system, which provides safe and reliable detection measures for the area to be protected on the aircraft, provides fast and accurate alarm indication for the crew, and provides effective fire extinguishing measures, so as to minimize the harm and loss caused by fire or overheating of the aircraft, so as to ensure the safety of the crew and passengers.

In the cockpit of a civil aircraft, the fire protection system is usually equipped with two control panels: the engine and APU fire control panel and the cargo compartment fire control panel, which are usually located on the top of the cockpit. The specific positions are shown in Fig. 1.

The main function of engine and APU fire control panel is to provide fire extinguishing operation for left/right engine and APU, as well as pilot initiated test operation switch of fire protection system; it also provides auxiliary indication of fire alarm for left/right engine and APU, low pressure condition of engine and APU fire extinguisher (mainly EICAS information and audio alarm).

The main function of the cargo compartment fire control panel is to provide cargo fire extinguishing operation for cargo compartment, and also provide auxiliary instructions for cargo fire alarm and low-pressure condition of cargo fire extinguisher (mainly EICAS information and audio alarm).

2.2 Flight Crew Task Analysis

There are many kinds of methods for crew task analysis, such as HTA method, GOMS method, SBTA method, TTA method and VPA method. Among them, the analytic hierarchy process (HTA) describes the specific activities of the tasks according to the objectives, sub objectives, operations and plans. Compared with other analysis methods, HTA is more suitable for the characteristics of civil aircraft flight crew to carry out the tasks.

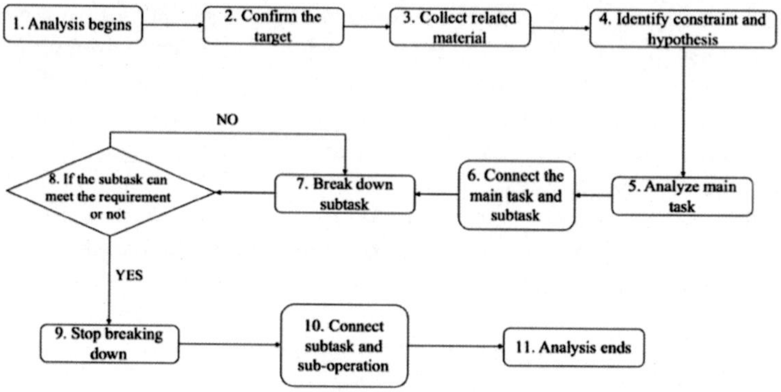

Fig. 1. HTA process

3 Flight Crew Error Analysis Method

Flight crew error refers to the situation that deviates from the crew's intention or consequence due to crew's action or negligence. The purpose of flight crew error analysis is to identify the possible errors in the interaction between the crew and the cockpit human-machine interface in the product development phase, and analyze the causes of the errors, so as to reduce the occurrence of human errors in the later phase. The error analysis method of civil aircraft cockpit is used in this paper is called human error template (HET+).

Table 1. Human error template method

Steps	Activities needed	Output
1. Based on the error checklist, preliminary identification of possible errors in the crew when performing tasks	Using the error checklist [6] to perform error identification on the underlying task behaviors analyzed in step two, and preliminary identification of design-related errors that may occur when performing the current task, including: 1) Potential or possible mistakes identified by pilots based on experience when performing similar tasks on other aircraft types; 2) Mistakes that may occur in combination with the civil aircraft human-machine interface design features	Preliminary identification of design-related potential crew error
2. Causes analysis of crew error	With regard to the crew errors in this step, analyze the reasons that cause the errors. The causes of errors may include: interface design issues (potential non-compliance with AMC25.1302) and other issues	Analysis of the causes of crew errors based on the existing human-machine interface design characteristics

(*continued*)

Table 1. (*continued*)

Steps	Activities needed	Output
3. Hazards analysis	Based on subjective judgment, analyze the likelihood of errors occurring during the execution of the current task, and determine if they are H (High), M (Medium), or L (Low). Analyze the possible consequences of errors, analyze their impact on aircraft safety margins or functional capabilities, workload or physical discomfort to the flight crew, and injuries to members other than the flight crew. Determine the level of hazard corresponding to potential errors, I (No safety impact), II (minor), III (major), IV (hazardous), V (catastrophic). Crew errors with a likelihood of H and a hazard of III, IV, V or errors with a likelihood of H and a hazard of IV, V will be analyzed furthermore in the next step	Hazard level of crew error
4. Analysis of error management measures	Analyze the management measures for potential human errors with hazard levels III, IV	Error management measures corresponding to potential human errors

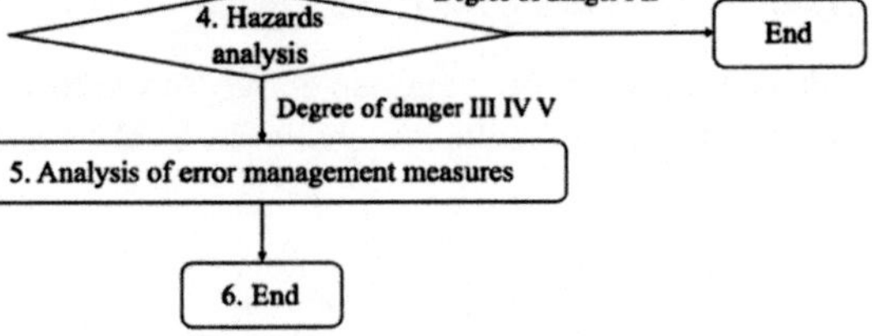

Fig. 2. Analysis process

4 Analysis and Discussion

This thesis analyzed each flight crew error on the interface between the fire control panel and fire protection system, which is based on the flight crew tasks. This analysis is crucial to step 3 of HET+.

4.1 Preliminary Identification Of Potential Crew Errors (Step 3)

According to the analysis above, this thesis analyzed each operation by flight crew for each task, such as operations manual [7]. Furthermore, this thesis also preliminarily analyzed and identified the potential design-related errors. The table below showed the analysis and checking list.

Table 2. Potential crew errors identification based on left engine fire task

Number	Abnormal/emergency crew task	Bottom flight crew task	Error type	Potential error
1	Left Engine Fire Alarm Voice: "Left Engine Fire", fire bell)	Shut off the auto throttle, retard the corresponding engine throttle to idle, close the corresponding engine low/high pressure fuel shutoff valve	……	……
2	Left Engine Fire Alarm Voice: "Left Engine Fire", fire bell)	Press out the engine fire arming switch on the affected engine and release the fire extinguishing agent as required	A2	Forget to press
			A3	Press wrong arming switch
			A…	……
3	……	……	……	……

4.2 Cause Analysis of Potential Unit Errors (Step 4)

According to the potential crew errors identified above, this section analyzed the causes for each crew error. The analysis results are shown in Table 3 as follows.

Table 3. Cause analysis of potential crew errors

Number	Task	Hierarchy flight crew task	Error type	Potential error	Cause of the error
1	Left Engine Fire Alarm Voice: "Left Engine Fire", fire bell)	Press out the engine fire arming switch on the affected engine and release the fire extinguishing agent as required	A2	Forget to press	Operational procedures are not operated properly by the flight crew
2	Left Engine Fire Alarm Voice: "Left Engine Fire", fire bell)	Press out the engine fire arming switch on the affected engine and release the fire extinguishing agent as required	A3	Press wrong arming switch	Operational procedures are not operated properly by the flight crew, or there are problems with human-machine interface design
……	……	……	……	……	……

4.3 Analysis of Potential Error Dangerous Degree (Step 5)

According to the fire protection control system hazards evaluation and the identified design-related potential crew errors, the functional failure mode and the corresponding hazards level after the error occurred were analyzed, the results obtained are shown in Table 4.

Table 4. Preliminary analysis of the degree of potential error risk analysis

Number	Task	Hierarchy flight crew task	Error type	Potential error	Failure mode	Hazards level
1	Left Engine Fire Alarm Voice: "Left Engine Fire", fire bell)	Press out the engine fire arming switch on the affected engine and release the fire extinguishing agent as required	A2	Forget to press	Related valves will not be shut off and the affected engine will not be set to idle	III
2	Left Engine Fire Alarm Voice: "Left Engine Fire", fire bell)	Press out the engine fire arming switch on the affected engine and release the fire extinguishing agent as required	A3	Press wrong arming switch	The agent will not be released to the corresponding area as required	IV
……	……	……	……	……	……	……

4.4 Error Management Measures (Step 6)

This section demonstrated the methods which can be used to manage the crew errors and control the occurrence during flight. The methods for potential human with hazards level of III, IV, and V and the analysis results are shown in Table 5.

Table 5. Analysis of potential error management measures

Number	Task	Hierarchy flight crew task	Error type	Potential error	Hazards level	Crew errors management methods
1	Left Engine Fire Alarm Voice: "Left Engine Fire", fire bell)	Press out the engine fire arming switch on the affected engine and release the fire extinguishing agent as required	A2	Forget to press	III	1. Display the buttons to be pressed on CAS, 2.Illuminate the buttons to be pressed in order to attract pilot's attention
2	Left Engine Fire Alarm Voice: "Left Engine Fire", fire bell)	Press out the engine fire arming switch on the affected engine and release the fire extinguishing agent as required	A3	Press wrong arming switch	IV	Illuminate the buttons to be pressed in order to attract pilot's attention
……	……	……	……	……	……	……

5 Conclusion

The fire protection system is crucial to the safety of civil aircraft. It is necessary to identify the design which may result in accidents in early phase. The method HET + which is used in this thesis is proved useful to identify the potential crew errors in flight crew task. HET+ also can be used for other systems to evaluate crew errors. This thesis uses the abnormal crew task- Left Engine Fire as an example to briefly explain how to perform crew error analysis and evaluation.

Through the level-by-level analysis of this method, the error cause and hazard level of the bottom level task can be found, and then error management measures can be given to prevent crew errors.

References

1. EASA: "Certification Specifications for Large Aeroplanes CS25 AMC", AMC25.1309, EASA, 2009/017/R, Cologne, Germany (2009)
2. Hannaman, GW., Spurgin, A.J.: Systematic human action reliability model for PRA analysis EPRI-NP-3585, Electric Power Research Institute, Palo Alto (1983)

3. Marshall, A., Stanton, N., Young, M., Salmon, P., Harris, D., Demagalski, J., Waldmann, T., Dekker, S.: 'Development of the human error template- a new methodology for assessing design induced errors on aircraft flight decks', ERRORPRED Final Report E!1970 (2003)
4. Kenndy, R., Jones, H., Shorrock, S., Kirwan, B.A.: HAZOP analysis of a future ATM system. Contemporary Ergonomics (2000)
5. Kirwan, B.: A guide to practical human reliability Assessment. Taylor and Francis, London (1994)
6. Stanton, N.A., Salmon, P.M., Walker, G.H., Baber, C., Jenkins, D.P.: Human Factor Methods-A Practical Guide for Engineering and Design. Ashgate Publishing Limited, Human Factor Integration Defence Technology Center (2005)
7. Boeing. Boeing 737NG Operations Manual Normal Procedures Amplified Procedures, Flight Deck Preparation. Boeing Published, Seattle, Washington (2000)
8. Zhang, Y., Zhu, Y., Zhou, Y., Lu, S., Meng, H.: A study of crew error on the interface between passive side stick and electronic flight control system in cockpit of civil aircraft, COMAC ShangHai Aircraft Design and Research Institute, AHFE (2020)

Humans and Artificial Systems Complexity

Human-Artificial Systems Collaboration in Service Innovation and Social Inclusion

Afnan Zafar[1,2(✉)] and Marja Ahola[1]

[1] LAB University of Applied Sciences, Mukkulankatu 19, 15210 Lahti, Finland
{afnan.zafar,marja.ahola}@lab.fi
[2] University of Gävle, Kungsbäcksvägen 47, 80176 Gävle, Sweden
afnan.zafar@hig.se

Abstract. Human and Artificial intelligence (AI) collaborations have contributed to improve our daily lives. AI can be used to promote employment, social inclusion and service innovation. To improve social inclusion, the integration of foreign origin workers is crucial for economic growth. In Finland, the MALVA project 'preparing immigrants for working life' aims to develops Human-Artificial services that promote social inclusion and service innovation. One important goal of the project is to create a chatbot for companies and develop mobile guiding practices to enhance employment solutions for foreign origin workers. This work is in progress, the present article reports the initial development stages and literature review related to the collaboration of human and artificial intelligence systems (HAS). The paper contributes to the understanding of the AI-based chatbots and their use in the integration of skilled migrant workers to promote social inclusion and economic growth.

Keywords: AI · Human-artificial systems · Service innovation · Social inclusion

1 Introduction

Human-artificial systems (HAS) is an important topic presently. HASs are a starting point to understand and implement valuable interaction between human and machines. Chatbots are one of the many products that resulted when HAS collaboration has been explored. Usage of chatbots started decades ago as a cheating system and later to imitate everyday human conversations [1]. Today, chatbots are task-oriented, and companies use them to provide information on defined and specific topics. As their adoption became widespread, more complex and customized chatbots started to surface. The implementation of chat-bots accelerated dramatically in various fields of industry and research.

The use of chatbots in service innovation and social inclusion has started recently and is presently limited. Social inclusion has many dimensions, and one important dimension is suitable employment opportunities for everyone irrespective of social background. The recruitment process plays a vital role in providing equal availability of employment opportunities which are necessary to promote social inclusion. To execute effective management of the overall employment process, companies are using

D. Russo et al. (Eds.): IHSI 2021, AISC 1322, pp. 527–532, 2021.
https://doi.org/10.1007/978-3-030-68017-6_78

chatbots as human resource personnel and hiring can be done either entirely or almost without real human interaction.

The scope of this study encompasses the preliminary background research as a part of a research project known as MALVA "preparing immigrants for working life" [2]. This paper targets the preliminary developmental stages and literature related to the co-operation between human and artificial intelligence systems, service innovation and social inclusion for chatbot formation. Three Finnish educational institutions (LAB University of Applied Sciences, Haaga-Helia School of Vocational Teacher Education and Salpaus Further Education) and collaborating regional companies are involved in the project [2].

This study seeks a preliminary answer to the research question: "How do human-centred artificial intelligence systems promote social inclusion and how are they part of the development of service innovations?". This paper provides insights on how chatbots can be used to promote international employment solutions as an innovation in service and provide a better understanding of features and opportunities of chatbots for social inclusion. Additionally, the study contributes to improving the understanding of the AI-based chatbot and its use in the integration of foreign origin workers to promote social inclusion and economic growth. The paper has four parts. First, it describes related back-ground research, then the basics of chatbot formation are presented. The article ends with discussion of MALVA impact and conclusion which contain limitations and future implications.

2 Links Between Chatbots, Services Innovation and Social Inclusion in Literature

Human-centred artificial intelligence systems potentially represent a new paradigm for people interacting with the future of information and services. Chatbots are an example of similar human AI-system as they facilitate work by recalling unfinished tasks, searching for information, serving busy customers and even assisting in election voting [3]. Interaction with chatbots can be enjoyable and enhance the learning process [4]. Reliability, quick response rates and ease of use make chatbot suitable for modern life.

Service innovation is traditionally less valued in literature as compared to technological innovation [5]. Service innovation is mainly influenced by human interactions, in a similar way human - AI-systems interactions influence the development of chatbots. So, service innovation is the significantly improved concept from its previously existing form [5]. Service innovation is visible when chatbots provide social services based on user interests and needs. To facilitate the development of service innovation for companies, in the form of HAS based chatbots, it is important to design ethical and responsible guidelines [1]. One of the aims of the MALVA project is to produce responsible, multicultural and safe service innovation for companies to promote inclusion in user's working lives [2].

Social inclusion is one of the key aspects when initiating the development process of culture related services [6]. When service innovation develops in the form of HAS chatbots, it is valuable that they promote social inclusion and it will be visible when they behave, react and answer in comfortable ways to provide information on merit [7]. Additionally, chatbots can promote social inclusion at work when they increase awareness of the operating environment through more effective communication and information sharing [2, 8]. The workplace instructor saves time when an employee enquires, for example, about frequently asked questions (FAQs) effortlessly and safely from a bot and it guides the employee to the right answer. A bot is always available irrespective of place and time. The involvement of a workplace induction bot at a company can strengthen the induction process and saves time for both parties (instructor and new employee).

However, literature sheds limited light on the development of service innovation in the form of chatbots based on HAS which promotes effective social inclusion. Against this background, to obtain reliable research results, it is necessary to study the interaction and dialogue of artificial intelligence-based solutions empirically, collect practical data based on social inclusion and proceed towards service innovation in the form of, HAS based, interactive chatbots [9].

3 Basics of Chatbot Formation in MALVA

This is a research in progress paper and reports initials steps of the MALVA project. The paper describes the developmental process of service innovation based on socially inclusive chatbots. Future development work will utilise different cultural backgrounds in the design of service innovations by utilizing resource-based work guidance methods [10], immigrant competence identification methods [11] and intercultural user interface design theory [9]. A good understanding of cultural variables increases the impact of HAS-based incremental service innovations and promotes social inclusion. The MALVA project [2] (Fig. 1) operates within the framework of individual job-seekers with foreign origin backgrounds, has the backing of educational institutions and business life.

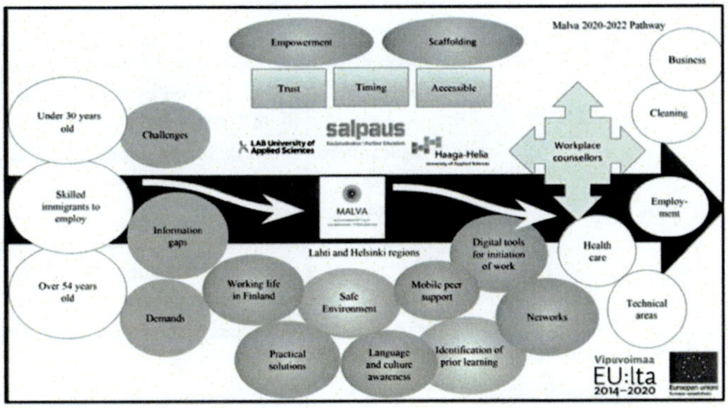

Fig. 1. MALVA project process and stakeholder description

Figure 2 elaborates practical goals of the MALVA project which involves the employment of thirty immigrants from different cultural backgrounds and to support at least ten companies in different fields by 2022. Information about the induction related requirements of companies is elicited from workplace counsellors and instructors through surveys and interviews. Later chatbots will be built collaboratively. The employer and the workplace supervisor will be offered peer support in a mobile peer-to-peer network. Considering service innovation as an evolving process from existing services and playing its role in social inclusion with HAS chatbots, Table 2 summarises the initial basics of MALVA project [2].

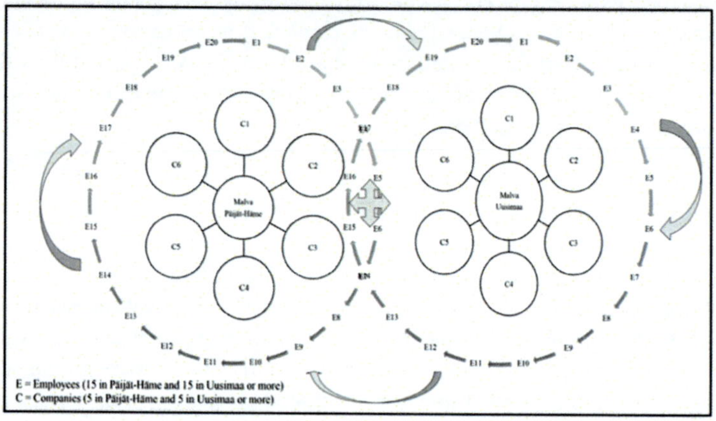

Fig. 2. Description of mobile peer network for employees and companies in MALVA

Table 2. Interlinkages of existing service, service innovations and social inclusion in MALVA project

Existing service	Service innovation	Social inclusion
Guidance	Chatbot-guidance	Access to core knowledge without a person -> time is freed up for genuine interaction mobile applications
Explanation of tasks	Explanation of tasks in chatbot	More demanding interaction, interaction requiring empathy
Checklists	Keywords	Language does not determine competence; safety is maintained even in the absence of a common language
Peer support	Mobile peer support	Time and place independent scaffolding (rightly timed)
Handbooks, manuals, diagrams, floor plans, occupational safety instructions, guidance instructions, reports, process descriptions, brochures, product descriptions, evaluations, photos, videos, courses, intranet	Knowledge base & links	Chatbots and links redirect to the right information, peer networks and easily achieved information and guidance empowers and scaffolds

4 Discussing the MALVA Impact

HAS is a part of the development of service innovations in situations when they scaffold employability, time and money saving solutions. MALVA develops incremental HAS-based service innovations together with companies to facilitate the employment process. The role of artificial intelligence in induction and peer-to-peer support is to guarantee the right information and promote inclusion during the collaboration. MALVA will expand the understanding of HAS based solutions and their potential to promote social inclusion by increasing amenity. This is an extension of the work of De Gennaro et al. [7]. The project also supports the conclusions of Brandtzaeg and Folstad [3, 12] as it shows potential to save time, support employees, promote businesses, connected to foreign origin workforce in a reliable and safe ways.

This paper provides the initial understanding and importance of HAS as part of our everyday lives. Companies can use HAS and play their part in influencing whether to use them wisely, ethically and on the basis of research, or whether we allow artificial intelligence to create an unsafe and haphazard reality for business [1, 12]. The use of artificial intelligence in employment-enhancing situations has begun, but more experience and knowledge are needed about employment-related processes. The MALVA project [2] is filling the gaps to understand the role of HAS in service innovation and social inclusion based chatbots.

5 Conclusions

Our working lives are changing and the HAS collaboration is becoming part of our daily lives. Process automation and digitalization are vital for the future, and unnecessary work should be minimized. Artificial intelligence can streamline workplace services, increase inclusion and productivity. All of this can be achieved by bringing timely, reliable and accessible information and services which can save time and effort for all parties. MALVA aims to develop HAS-based service innovations in collaboration with companies to facilitate the employment process. The role of artificial intelligence in the induction process and in peer-to-peer networks is to guarantee the right information and promote inclusion during collaboration. The employment and induction processes of international professional workers will be facilitated.

Moreover, companies are already using artificial intelligence-based service innovations. Recently, the use of AI to increase employment has begun. Still, more experience and knowledge are needed on processes that increase inclusion in the workplace to produce excellent service innovations for the needs of modern working life. In Finland, the use of HAS based chatbots involving service innovation and social inclusion is in the initial stages. There is limited research available on the use of artificial intelligence in workplace induction processes. One of the background theories is Heimgärtner's Intercultural User Interface Design theory [9], which emphasizes collaboration between research and empiricism as the use of artificial intelligence in an international operating environment involves numerous variables. For our part, we can influence whether we use artificial intelligence sensibly, ethically, and based on research, or whether we allow artificial intelligence to create a random and even dangerous businesses and everyday realities. Good design encompasses many things such as the experience of end-users of service innovation, the framework of society, the desires of companies, ability to enable an inclusive and pleasant user experience and bringing real added value to the business community.

References

1. Shum, H.Y., He, X.D., Li, D.: "From Eliza to XiaoIce: challenges and opportunities with social chatbots," Frontiers of Information Technology and Electronic Engineering (2018). https://doi.org/10.1631/FITEE.1700826
2. LAB, "MALVA - Preparing immigrants for Working life," (2020)
3. Brandtzaeg, P.B., Følstad, A.: Chatbots: user changing needs and motivations. Interactions (2018). https://doi.org/10.1145/3236669
4. Kerly, A., Hall, P., Bull, S.: Bringing chatbots into education: towards natural language negotiation of open learner models. Knowledge-Based Syst. (2007). https://doi.org/10.1016/j.knosys.2006.11.014
5. Tekes, "The Future of Service Business Innovation," Innovation, pp. 3–73 (2010)
6. Mansouri, F., Lobo, M.: Migration, citizenship and intercultural relations: Looking through the lens of social inclusion (2011)
7. de Gennaro, M., Krumhuber, E.G., Lucas, G.: Effectiveness of an empathic chatbot in combating adverse effects of social exclusion on mood. Front. Psychol. (2020). https://doi.org/10.3389/fpsyg.2019.03061
8. Heo, J., Lee, J.: CiSA: an inclusive chatbot service for international students and academics. In: Lecture Notes in Computer Science (including subseries Lecture Notes in Artificial Intelligence and Lecture Notes in Bioinformatics) (2019). https://doi.org/10.1007/978-3-030-30033-3_12
9. Heimgärtner, R.: "Intercultural user interface design", in Web Design and Development: Concepts. Tools, and Applications, Methodologies (2015)
10. Christensen, M., et al.: Positive Factors at Work (2008)
11. Lasonen, M., Teräs, J.: "Challenges to recognizing foreign competence for employment in Finland.," Lazaridis Poverty, Cl. Sch. Glob. Perspect. Econ. justice Educ. Equal. Inf. Age Pub (IAP), pp. 129–145 (2015)
12. Følstad, A., Brandtzaeg, P.B.: Chatbots and the New World of HCI. Interactions (2017). https://doi.org/10.1145/3085558

Image Processing-Based Supervised Learning to Predict Robot Intention for Multimodal Interactions Between a Virtual Human and a Social Robot

S. M. Mizanoor Rahman(✉)

Department of Intelligent Systems and Robotics, Hal Marcus College of Science and Engineering, University of West Florida, 11000 University Pkwy, Pensacola, FL 32514, USA
smrahman@uwf.edu

Abstract. A humanoid robot and a virtual human with human-like appearance were developed, and various similar intelligence, functions, autonomy and interaction modalities were developed for them that were deemed necessary for their real-world collaboration. The virtual human was integrated with the robot through a common platform based on some control algorithms so that they were capable of cooperating with each other in a real-world task, which was the task of assisting each other in searching for a hidden object in a homely environment. One important activity for such dynamic collaboration was the virtual human's ability to predict the robot's intention so that the virtual human could plan to collaborate with the robot, and vice versa. To address this problem, three representative robot behaviors such as hand shaking, hand pointing and hand waving were analyzed using supervised learning-based algorithms associated with image processing to help the virtual human learn robot behaviors so that the learned behaviors could help the virtual human predict the robot's intention during their intended social interactions. In the training phase, the robot showed each of the three behaviors separately for 500 times. Images of each type of behaviors were taken separately, and the properties of the images were extracted following the image processing method. Then, each robot behavior extracted through image processing was labeled properly. Then, the supervised learning algorithm was developed that classified three different robot behaviors into three different classes. In the testing phase, the robot showed each of its three behaviors to the virtual human randomly for 100 times, the images were taken in real-time, and sent to the image processing-based machine learning algorithm. It was monitored whether the virtual human was able to recognize the robot behavior properly. The results showed that the virtual human was able to recognize robot behaviors and thus predict the robot's intention with above 95% accuracy level.

Keywords: Social robot · Virtual human · Supervised learning · Intention prediction · Image processing

D. Russo et al. (Eds.): IHSI 2021, AISC 1322, pp. 533–539, 2021.
https://doi.org/10.1007/978-3-030-68017-6_79

1 Introduction

Social humanoid robots and virtual humans are two different characters [1]. Social robots are present in the real world, but virtual humans are only effective in the virtual world [2, 3]. They work separately. However, there are many similarities between them, and thus they could collaborate to perform common tasks [1–5]. Social robots have been proposed for various social activities, events and interactions with their human counterparts [1]. However, they do not look like humans in most cases, and their social applications are limited due to their lack of competency in intelligence and capabilities or due to improper trade-off between their performance and appearances. On the other hand, the virtual humans are software-generated virtual agents and they could not come beyond their virtual environments [5]. It means that they are usually not involved in any real-world activities where they could perform real-world tasks in cooperation with humans or assist humans performing real-world tasks. Their real-world interactions and cooperation with humans or other artificial agents for real-world activities are not so reported in the literature [3–5].

In this paper, the real-world multimodal social interactions between a humanoid robot and a virtual human for a real-world social task have been proposed. One important activity for such a dynamic collaboration is the virtual human's ability to predict the robot's intention so that the virtual human can plan to collaborate with the robot, and vice versa. To address this, an image processing-based supervised learning method is proposed, and the results are experimentally verified, which are found satisfactory. The proposed initiative can help humans employ artificial social agents of heterogeneous realities (e.g., real and virtual agents) to get real-world social tasks done through cooperation between intelligent artificial agents.

2 Development of the Robot and the Virtual Human

We used a humanoid robot, which was a NAO robot (https://www.aldebaran-robotics.com/en/) [1]. In order to make the robot able to interact with a virtual human, we made it intelligent, autonomous and social by enriching it with various social functions and attributes. For example, we developed functions like stand up, sit down, turn, wave hand, grab and release objects, look at a position, point at something, walk, shake hand, nod head, etc. for the robot. We also added some emotional expressions to the robot such as relaxation, happiness, sadness, pride, etc. It was able to speak (based on 'from text to speech' technology) and perceive the environment through various sensors such as video, audio, proprioception, etc. It was able to make decisions based on some adaptive rules combined with the stored information in the server as well as react by talking, moving, or showing its internal emotions.

We developed a realistic intelligent and autonomous 3D virtual human with the face of a woman (e.g., a western woman) [1]. The virtual human was developed on an open source virtual character animation and control platform such as the Smartbody system (https://smartbody.ict.usc.edu/). We created the virtual character model following the standard art/modelling methods based on the joints and skeleton requirements for the virtual character that were explicitly mentioned in the Smartbody system.

We then exported it using Autodesk Maya 3D animation software (https://www.autodesk.com/). We also used the Ogre (https://www.ogre3d.org/) system for the graphical rendering. The virtual character was based on the Behavioral Mark-up Language (BML) realization engine that was able to transform the BML behavior descriptions into the real-time animations. We determined the anthropomorphic data for the virtual human (for example, walking/running velocity, whole body dimensions, joint angles, kinematics, etc.) by being inspired by that for real humans. It means that we followed the biomimetic approach for developing the virtual human [2]. The character was enabled to blend different animations smoothly in real time. One on one mapping was done by using the Microsoft Kinect hardware (Kinect SDK 1.5). The character was enabled to mimic real human behaviors in terms of facial expressions and whole body motions. We were able to display it on a big screen (Panasonic TH-P65VT305, Full HD-3D, 65 inch display, neo plasma panel, 1920×1080 resolution).

The major real time functionalities and attributes incorporated with the virtual human are listed below:

- Locomotion- walk, jog, run, run in a circle, turn, jump, take side steps, etc.
- Object manipulation – reach, grasp, pick and place, touch, throw objects, etc.
- Steering – avoiding obstacles and moving objects
- Gaze – robust gazing behavior to incorporate various parts of the body
- Nonverbal behaviors – gesture, eye blink, head nod and shake, etc.
- Social actions- looking at something, pointing at something, shaking hands, waving hands, etc.
- Memory-remembering an event or something for a period of time
- Lip syncing-speaking with simultaneous lip-sync using text-to-speech or pre-recorded audios
- Realistic facial expressions and emotions-expressing emotions such as whether it was happy, sad, angry, smiling, etc.
- Social skills-responding to human questions and conversation
- Social media-connecting with Facebook, Google calendar, etc.
- Character physics – pose-based tracking, motion perturbations, etc.
- Recognition- recognizing gesture, speech, action, etc.

All those capabilities were not to be used to stage the selected object searching social task presented in this paper. However, information on those capabilities of the virtual human might help understand its collaboration with the robot. Note that the robot was also adorned with similar functionalities, attributes and capabilities.

At the end, they were integrated through a common communication platform as shown in Fig. 1 [13]. We developed the communication system for the robot and the virtual human, which was based on a common platform as in Fig. 1 [1, 13]. Animation of each function could be commanded from a command script (client), which was networked with the control server through the Thrift interface. We preferred the Thrift over the ROS because ROS runs only on Linux/CORBA, and it is complex [7–9]. Instead, Thrift was found reliable, it supported cross-language, cross-platform, and it provided high performance. It was developed as an in-house common platform, which was capable of animating different types of intelligent characters such as humanoid robots, virtual human, etc. using the same Application Programming Interfaces (APIs) and command script through specifying the character in the command script. There were options to operate the robot and the virtual human based on wired and wireless connections.

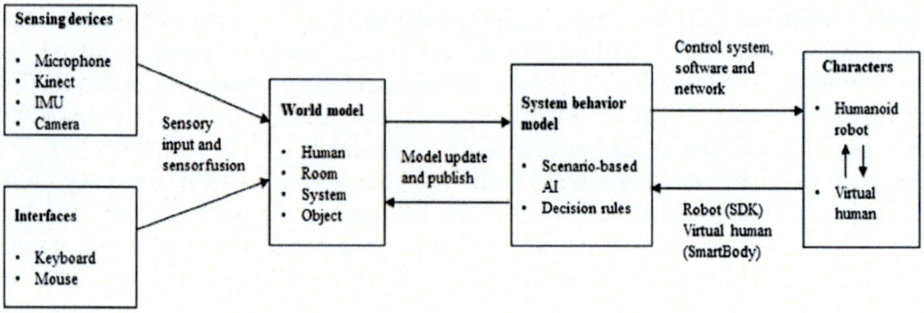

Fig. 1. The common communication platform

3 Experiments and Results

In a training scenario, three representative robot behaviors such as hand shaking, hand pointing and hand waving were analyzed using supervised learning-based algorithms associated with image processing to help the virtual human learn robot behaviors so that the learned behaviors could help the virtual human predict the robot intention during their intended social interactions [10–12]. In the training phase, the robot showed each of the three behaviors such as hand shaking, hand pointing and hand waving separately for 500 times. Images of each type of behaviors were taken separately, and the properties of the images were extracted following the image processing method. The properties were the patterns related to RGB or HSV values for each image. The patterns were preserved in the software properly. Then, each robot behavior extracted through image processing was labeled properly. It was the training dataset. Then, a supervised learning algorithm was developed that classified three different robot behaviors into three different classes [6]. It was a linear multilevel classification. The training dataset may be elaborated as in Eq. (1) (where n is the number of trials) that used the hand shaking, hand pointing and hand waving images as the inputs vector (x), and the experimenter's identification/labeling as hand shaking, hand pointing and hand waving actions as the outputs vector (y). Table 1 partly illustrates the training dataset. A target function or general formula (f) as in Eq. (2) was learned based on the input-output relationships in Eq. (3) that was used to predict the robot's intention. Figure 2 shows the training scenario within the context of collaboration between the virtual human and the robot in a homely environment for their autonomous collaboration in looking for a hidden/missing object.

Table 1. The sample supervised learning training dataset (partial list for illustration only)

Trials (n)	Robot behaviors based on image processing (inputs, x)	Labeling (outputs, y)
1	Hand pointing image	Hand pointing
2	Hand shaking image	Hand shaking
3	Hand waving image	Hand waving

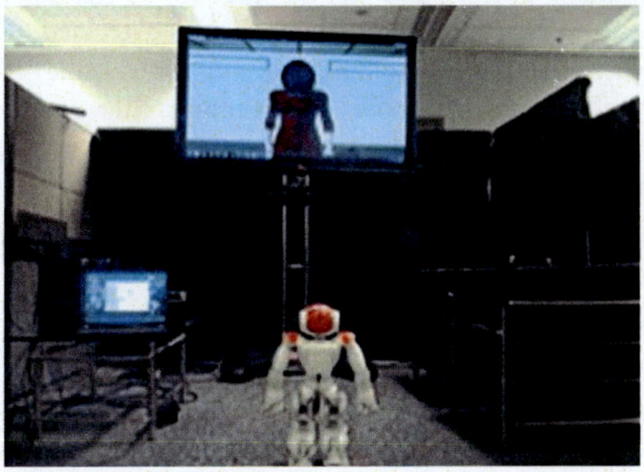

Fig. 2. The collaboration scenario between the virtual human and the social robot in a homely environment for their autonomous social collaboration in looking for a hidden/missing object.

$$D = \{(x_1, y_1), (x_2, y_2), \ldots\ldots\ldots\ldots\ldots (x_n, y_n)\} \tag{1}$$

$$f : X \rightarrow Y \tag{2}$$

$$Y = f(X) \tag{3}$$

In the testing phase, the robot showed each of its three behaviors to the virtual human randomly for 100 times, the images were taken in real-time, and sent to the image processing-based machine learning algorithm. It was then monitored by the experimenter to check whether the virtual human was able to recognize the robot behavior properly. The results showed that the supervised learning-based algorithms associated with image processing were able to enable the virtual human to recognize robot behaviors and thus predict robot intention with above 95% accuracy level. It indirectly means that if the collaboration between the robot and the virtual human are implemented with the current capability of the virtual human in predicting the robot's intention, then the collaboration is expected to be successful in above 95% cases. Such prediction of the robot's intention and behaviors are necessary for the virtual human because such information is used to determine appropriate functions of the virtual human for the projected collaboration. Similar capability of the robot to predict the virtual human's intention is also required with equal importance. However, the accuracy level of the robot's prediction of the virtual human's intention may not be as high as 95% because the virtual human is in the 2D virtual space (inside a 2D display monitor) even though it is itself a 3D entity.

4 Conclusions and Future Work

We proposed a framework for collaboration between a virtual human and a social robot through a common platform. In order to make the collaboration successful, the robot and the virtual human need to predict the behaviors of each other. To address this, an image processing-based supervised learning approach was proposed and tested that enabled the virtual human to learn and predict the robot's behaviors. In the near future, similar capability of the robot will be developed and evaluated. Then, the proposed collaboration between the virtual human and the robot will be implemented and evaluated. The overall approach is novel that augments the scope of applications of social robots and virtual humans. The proposed initiative can help humans employ artificial social agents of heterogeneous realities (e.g., real, virtual) to get real-world social tasks accomplished through the cooperation between intelligent artificial agents.

References

1. Rahman, S.M.M.: Evaluating and benchmarking the interactions between a humanoid robot and a virtual human for a real-world social task. In: Proceedings of the 6th International Conference on Advances in Information Technology. In: Communications in Computer and Information Science, vol. 409, pp. 184–197 (2013)
2. Rahman, S.M.M.: Generating human-like social motion in a human-looking humanoid robot: the biomimetic approach. In: Proceedings of IEEE International Conference on Robotics and Biomimetics, pp. 1377–1383 (2013)
3. Rahman, S.M.M.: Cyber-physical-social system between a humanoid robot and a virtual human through a shared platform for adaptive agent ecology. IEEE/CAA J. Automatica Sinica **5**(1), 190–203 (2018)
4. Rahman, S.M.M.: Collaboration between a physical robot and a virtual human through a unified platform for personal assistance to humans. personal assistants: emerging computational technologies, Intel. Syst. Ref. Library Series, A. Costa, V. Julian, P. Novais (Eds.), Springer, vol. 132, Ch. 9, pp. 149–177 (2017)
5. Rahman, S.M.M.: People-centric adaptive social ecology between humanoid robot and virtual human for social cooperation. Commun. Comput. Inf. Sci. **413**, 120–135 (2013)
6. Bzdok, D., Krzywinski, M., Altman, N.: Machine learning: supervised methods. Nat. Methods **15**, 5–6 (2018)
7. Rahman, S.M.M., Wang, Y.: Mutual trust-based subtask allocation for human-robot collaboration in flexible lightweight assembly in manufacturing. Mechatronics **54**, 94–109 (2018)
8. Rahman, S. M. M., Liao, Z., Jiang, L., Wang, Y.: A regret-based autonomy allocation scheme for human-robot shared vision systems in collaborative assembly in manufacturing. In: Proceedings of the 12th IEEE International Conference on Automation Science and Engineering (IEEE CASE 2016), pp. 897–902 (2016)
9. Rahman, S. M. M., Wang, Y.: Dynamic affection-based motion control of a humanoid robot to collaborate with human in flexible assembly in manufacturing. In: Proceedings of ASME Dynamic Systems and Controls Conference, pp. V003T40A005 (2015)
10. Rahman, S. M. M., Ikeura, R.: Improving interactions between a power assist robot system and its human user in horizontal transfer of objects using a novel adaptive control method. In: Advances in Human-Computer Interaction 2012, ID 745216, pp. 1–12 (2012)

11. Rahman, S.M.M., Ikeura, R.: Cognition-based control and optimization algorithms for optimizing human-robot interactions in power assisted object manipulation. J. Inf. Sci. Eng. **32**(5), 1325–1344 (2016)
12. Rahman, S.M.M., Ikeura, R.: Investigating the factors affecting human's weight perception in lifting objects with a power assist robot. In: Proceedings of 2012 21st IEEE International Symposium on Robot and Human Interactive Communication, pp. 227–233 (2012)
13. Rahman, S.M.M.: Mixed-initiative collaboration between a humanoid robot and a virtual human through a common platform for a real-world common task: evaluation and benchmarking. J. Ambient Intell. Smart Environ. **11**(5), 429–452 (2019)

Towards Agent Design for Forming a Consensus Remotely Through an Analysis of Declaration of Intent in Barnga Game

Yoshimiki Maekawa[1](✉), Tomohiro Yamaguchi[2], and Keiki Takadama[1]

[1] The University of Electro-Communications, Chofugaoka 1-5-1, Chofu, Tokyo, Japan
yoshimiki.maekawa@uec.ac.jp, keiki@inf.uec.ac.jp
[2] National Institute of Technorogy, Nara College, 22 Yata-cho, Yamatokoriyama, Nara, Japan
yamaguch@info.nara-k.ac.jp

Abstract. This paper explores the key-factors that can promote people to form a consensus remotely in such an Internet environment, and designs the agent according to the found key-factor for a better online communication. To address this issue, this paper focuses on "declaration of intent" and regards that people form a consensus when sharing one thought with declaring their intents. To investigate an influence of declaration of intent in a group, this paper employs cross-cultural game, Barnga, in which the players are forced to form a consensus for progressing the game. Through the human subject experiments of Barnga, the following insights are revealed: (1) the players are promoted to form a consensus when their rate of declaration of intent is high in the early stage of the game; and (2) the relationship between the rate of declaration of intent and the winning rate of the players shows the weak positive correlation.

Keywords: Human-Agent interaction · Barnga · Remote communication · Forming a consensus · Declaration of intent

1 Introduction

The recent development of online communication tools enriches our lives. Concretely, Social Networking Services (SNSs), television conferences, and more make it possible to communicate with our friends and co-workers who are apart from us. Especially, remote communications and economic activities are recommended around the world depending on the pandemic of COVID-19, which happens from 2020. In such a situation, Zoom, Google meets, and so on, which are the remote communication tools contribute to maintaining relationships and economic activities.

Supported by Evolonguistics: Integrative Studies of Language Evolution for Cocreative Communication.

D. Russo et al. (Eds.): IHSI 2021, AISC 1322, pp. 540–546, 2021.
https://doi.org/10.1007/978-3-030-68017-6_80

However, the above remote communications are kinds of non-face-to-face communication, and its quality and amount of information that is shared among people is fewer than face-to-face communication. Because of these weak points, the misunderstanding, that is a situation where the information is not shared rightly, happens easily. Since the misunderstanding can escalate to serious problems like bullying and insulting, and it is hard to solve by people being in the middle of the problem, the misunderstanding is one of the problems to be solved in recent years. In addition, such a problem can be seen as a failure of forming a consensus among people whose thoughts are different each other. For this reason, solving the failure of forming a consensus can leads to a solution of the misunderstanding, and the problem cannot be solved without the declaration of intent.

To tackle this problem, we employ an approach to introduce the agent which interfere to the human and change such bad situations. Especially, this paper purposes to obtain key-factors of the agent, which promote to form a consensus in a situation where people's thoughts are different each other, based on the results of the experiments which employ cross-cultural game Barnga [1], and focuses on declaring of intent in Barnga.

This paper is organized as follows. Section 2 introduces the cross-cultural game Barnga. Section 3 conducts the experiments which employ Barnga. Section 4 discusses the experimental results and obtains the key-factor for the agent. Finally, Sect. 5 concludes this paper.

2 Cross-cultural Game Barnga

Barnga [1] is a playing card game that purposes to experience difficulties of communications with other people who have different backgrounds, that is culture shock, and to solve the shock. Concretely, Barnga needs 4 players who have different rules (*i.e.*, *the strength of Ace* and *the strongest suit*) and 28 cards which are from Ace to seven with each suit (*i.e.*, SPADE, HEART, CRUB, DIAMOND). The procedure of Barnga is very simple. Firstly, the players play a card from their hand. Next, they select the winner who plays the strongest card based on their rule without the verbal communication (one selection is called *TURN*). Since the verbal communication is prohibited in Barnga, only the winner selection represents the player's declaration of intent, and it is allowed not to select the winner in Barnga. If the winner selections of the players are matched, the selected player is determined as the winner (this cycle from playing cards to determining the winner is called *GAME*), and Barnga progresses to the next GAME. If not, the players need to select the winner again. Figure 1 shows the procedure of Barnga. In this figure, Player 1 and Player 2 have the rule that is the strongest suit is HEART, the Player 3's strongest suit is DIAMOND and the Player 4's strongest suit is CRAB. In addition, Player 1 and Player 4 have the rule that Ace is weak, and Player 2 and Player 3 have the rule that Ace is strong. The left figure shows the situation after playing the cards (*i.e.*, Player 1 plays four of HEART, Player 2 plays Ace of Heart, Player 3 plays seven of CRAB and Player 4 plays Ace of SPADE). The right figure shows the situation where the players select the winner of the GAME (*i.e.*, Player 1 selects Player 2, Player 2 selects Player 1, Player 3 selects no player and Player 4 selects Player 3 as the winner of the GAME). In such a situation, since the winner selection of the players are not matched, the players have to select the winner of the GAME again. However, because of the different rules of the players, the winner selections are not matched at once, and the players required to progress GAME filling the difference based on the played cards and the winner selections of the other players. In this paper, we see such a restricted situation as the remote communication and employ it in the experiments.

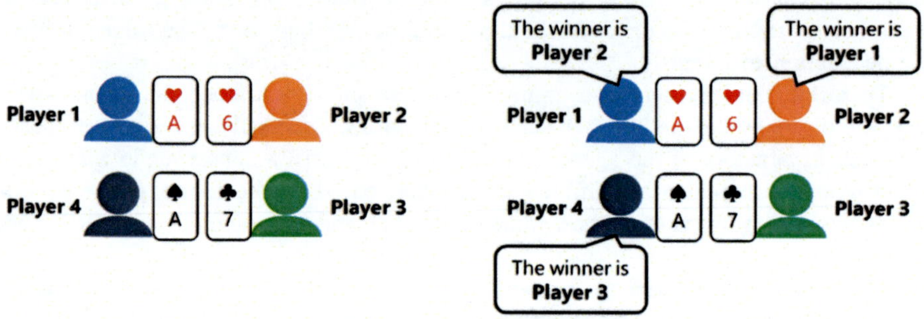

Fig. 1. Procedure of Barnga

3 Experiment

Since it is necessary to clarify the ideal situation and analyze the situation for obtaining an agent design for forming a consensus, we conducted the experiments with two cases and two evaluation criteria. Concretely, the evaluation criteria are as follows: (1) the number of TURNs until sharing the rule, which is a number of spent TURNs for one GAME; (2) the rate of declaration of intent of the players, which is the number of selecting the winner out of all opportunities to select the winner. Equation (1) represents the rate of declaration of intent. In the Eq. (1), $N_{selection}$ denotes the number of selection and N_{TURN} denotes the number of spent TURN (*i.e.*, the number of all opportunities to select the winner). In the ideal situation, only one GAME ends by only one TURN and all players select the same winner.

$$\text{Rate of declaration of intent} = \frac{N_{selection}}{N_{TURN}} \tag{1}$$

3.1 Experimental Settings

Case 1. The number of human players is four and they have rules that are shown in Table 1 before the experiment. They played Barnga with on-line Barnga [2].

Table 1. The rules of the players

	The Strength of Ace	The Strongest Suit
Player 1	Strong	HEART
Player 2	Weak	CRAB
Player 3	Strong	DIAMOND
Player 4	Weak	SPADE

Case 2. The number of human players is three and the rest of the players is an agent. The rules of the players and the experimental settings are the same as Case 1. The design of the agent that is introduced in Case 2 is as follows: (1) if the rate of declaration of intent of the human players are low, the agent selects the winner; (2) if the rate of declaration of intent of them are high, the agent stop selecting the winner. This is one of the design to promote the human players to declare their intention.

3.2 Experimental Results

Figure 2(a) shows the relationship between the number of TURN and the number of GAME. In this figure, the vertical axis denotes the number of TURN, the horizontal axis denotes the number of GAME, the orange line represents the result of Case 1, and the blue line represents the result of Case 2. In the ideal situation, the slope of this figure becomes 1 because one GAME is ended by only one TURN, because one GAME is ended by only one TURN. From Fig. 2(a), the result of Case 2 converged to the slope of 1 in the last stage, while the result of Case 1 did not converged. That illustrates that the players in Case 2 were able to share the one rule from the situation in which they have different rules each other by fewer interactions than Case 1.

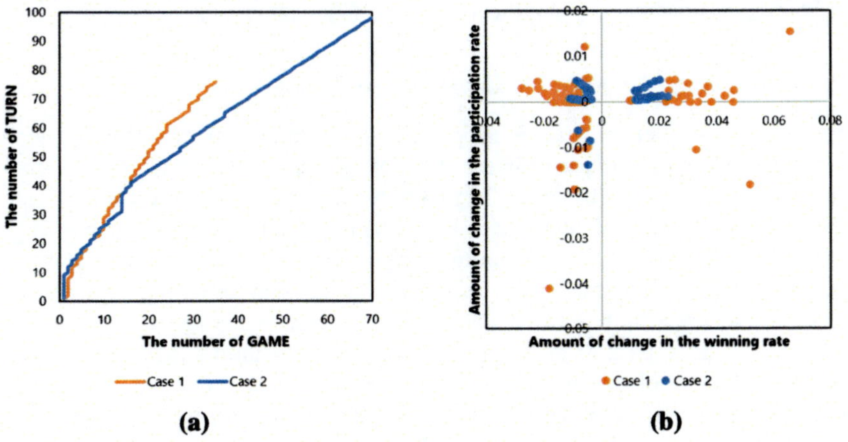

Fig. 2. a. Relationship between GAME and TURN; b. Relationship between the winning rate and the rate of declaration of in-tent in last stage

Figure 2(b) shows plots of a relationship between the amount of change in the winning rate and the rate of declaration of intent in the last stage. In this figure, the vertical axis denotes the amount of change in the rate of declaration of intent, the horizontal axis denotes the amount of change in the winning rate, the orange plots represent the result of Case 1 and the blue plots represent the result of Case 2. In short, if the player wins, the plot moves to the first or fourth quadrant, and if s/he selects the winner, the plot moves to the first or second quadrant. From Fig. 2(b), the almost plots of Case 2 are on the first and second quadrants, while the plots of Case 1 spread to third and fourth quadrants. These results illustrate that the players of Case 2 kept on declaring the intent in the last stage of the experiment regardless of whether they won or lost.

According to the above results, it is revealed that the players of Case 2 declared the intent based on the one rule, which is formed from the different rules of the players, that is forming a consensus. On the other side, the players of Case 1 faced conflicts of the winner selection, and the winner was selected by a part of them in the last stage. Therefore, the situation of Case 2 is considered the ideal situation in the last stage.

4 Discussion

4.1 Relationship Between the Rate of Declaration and the Ideal Situation

Figure 3(a) shows an average of the rate of declaration of intent for each player in Case 1 and Case 2. In this figure, the vertical axis denotes the average of the rate of declaration of intent, the horizontal axis denotes TURN, the orange line represents the result of Case 1 and the blue line represents the result of Case 2. From Fig.3(a), especially in the early stage, the rate of declaration of intent in Case 2 kept the high value and the rate of declaration of intent in Case 1 decreased rapidly. These results illustrate that the players in Case 2 interacted frequently in the early stage of the experiment, and shows a possibility that deep interactions in the early stage contributes to forming a consensus from the situation where the players' rules are different each other.

4.2 Agent Design for Forming Consensus in Barnga

Figure 3(b) shows plots of a relationship between the amount of change in the winning rate and the rate of declaration of intent in the early and middle stages. The meaning of axes is the same as Fig. 2(b). From the plot in the first quadrant, the degree of increase in the rate of declaration of intent increases as the degree of increase in the winning rate increases. That describes a possibility to promote to declare the intent to a player with low degree of increase in the rate of declaration of intent by selecting the player as the winner. In other words, this tendency is the key-factor of the agent not to break the ideal situation like Case 2. In addition, from the plot in the third quadrant, the degree of decrease in the rate of declaration of intent decreases if the degree of increase in the winning rate increases. That describes the possibility to promote to declare the intent to a player who does not declare the intent by selecting the player as the winner. This tendency is the key-factor of the agent to break out of non-ideal situations and bring them closer to the ideal situation.

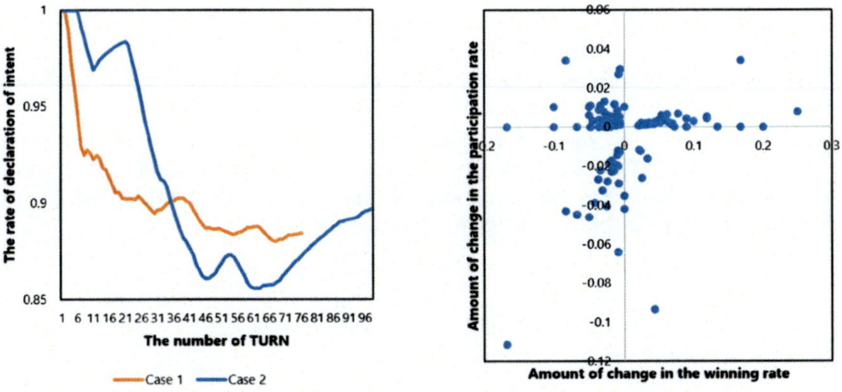

(a) **(b)**

Fig. 3. a. Average of the rate of declaration of intent; b. Relationship between the winning and the rate of declaration of intent in early and middle stages

The above analysis reveals the following possibilities. First, the players' rate of declaration of intent is increased by selecting as the winner or supporting the players in the early and middle stages. Second, the ideal situation like Case 2 of this paper is realized by increasing the rate of declaration of intent based on increase in the winning rate of a player whose rate of declaration of intent is low in the early and middle stages.

Furthermore, it is possible to implement the functions, *i.e.*, detecting the low rate of declaration of intent and selecting as the winner or supporting the players, in an agent. Therefore, we obtain key-factors of an agent, which promote to form a consensus in a situation where people with different rules exist.

5 Conclusion

This paper explored the key-factors that can promote people to form a consensus remotely in a situation where they have different thoughts each other. To address this issue, this paper focused on "declaration of intent" and employed the cross-cultural game Barnga as the remote communication to investigate an influence of declaration of intent in a group. From the human subject experimental results and analysis, the following implications were revealed: (1) the players are promoted to form a consensus when their rate of declaration of intent is high in the early stage of the game, (2) the relationship between the rate of declaration of intent and the winning rate of the players shows the positive correlation. These implications suggest us to design the agent as follows: (1) the agent is designed to detect the player whose rate of declaration of intent is low to help him for his win of the game; (2) an increase of the winning rate of the game contributes to increase the rate of declaration of intent, which promotes the players to form a consensus.

References

1. Thiagarajan, S.: A Simulation Game On Cultural Clashes. Intercultural Press, Inc., Boston (2006)
2. Maekawa, Y., Uwano, F., Kitajima, E., Takadama K.: How to emote for consensus building in virtual communication. In: Yamamoto, S., Mori, H. (eds) Human Interface and the Management of Information. Interacting with Information. HCII 2020, vol. 12185. Springer, Cham (2020). https://doi.org/10.1007/978-3-030-50017-7_13

Aspect Fusion as Design Paradigm for Legal Information Retrieval

Kurt Englmeier[1(✉)] and Pedro Contreras[2]

[1] Faculty of Computer Science, Schmalkalden University of Applied Science, Blechhammer, 98574 Schmalkalden, Germany
kurtenglmeier@acm.org
[2] WMG, The University of Warwick, 6 Lord Bhattacharyya Way, Coventry CV4 7AL, UK
pedro.contreras@acm.org

Abstract. Legal information retrieval (LIR) depends on identifying facts or predications in texts and retrieving correspondingly relevant legislation, company policies, and the like. We propose an approach for the matching process that relies rather on aspects reflected in facts and predications than on the specific wording. The users define and manage blueprints as standardized representations of aspects that correspond to their mental models of facts and predications. In this paper, we demonstrate our prototypical system Contexter that operates on these blueprints as controlling elements of its retrieval process. The system takes the manually defined blueprints as seeds and tries to automatically detect further valid variants of these blueprints that the user can finally confirm or reject. This user-driven and semi-automatic process supports the user in enriching the base of aspect patterns and, finally, to improve the systems retrieval capabilities.

Keywords: Information retrieval · Text analysis · Text mining · Aspects · Named entities · Bag of words · Blueprints · Natural language processing

1 Introduction

The exact identification and classification of facts in texts and relating them to relevant passages in legislation, rules, or policies is one of the most challenging objectives of legal information retrieval (LIR). We expect from a LIR system that it correctly identifies and classifies facts and predications in texts such as customer complaints, balance sheets, or contracts, extracts these facts and predications and matches them with text passages reflecting terms and conditions, legislation, policies, standards, and the like.

We can identify facts in texts in many ways. Usually, a couple of adjacent expressions reflect a fact. A numeric value, for instance, in close proximity to a currency symbol indicates a certain amount of money. A further adjacent word like "price" or "costs" vests the set of expressions with the meaning "price". The more adjacent expressions we include the more we specify the fact. This process of specification ends until we reach the boundaries of the next fact. That fact may be the representation of a person. Beyond the other end of the semantic boundary of the fact "price" we may

D. Russo et al. (Eds.): IHSI 2021, AISC 1322, pp. 547–553, 2021.
https://doi.org/10.1007/978-3-030-68017-6_81

encounter a fact "product", followed by a further instance of a person. If we take all these facts together, we may see the overarching fact of a purchase. This, in turn, may be a subcomponent of an investment or contract. The fact "price" can also be part of a position in a balance sheet. Semantic representations of facts integrate usually into broader and more abstract representations of facts.

Of course, a fact may take more than one semantic representation. The number of possible representations raises the more we move upwards in the levels of abstraction. On a higher level of abstraction, we have more an aspect of a fact that can be interpreted in many ways, that is, that has more than one instance. An aspect, in turn, denotes the unique interpretation of a series of words that appear in a certain proximity. Depending on the level of abstraction, the range of proximity may cover a fraction of a phrase, a complete phrase, one or more paragraphs, or even an entire document. We conceive an aspect representation as a hierarchical structure composed by its semantic elements that, in turn, constitute the aspect's implicit meaning.

In this paper, we introduce "aspect fusion" as design paradigm for LIR systems. We are convinced that identifying facts in different sources and relating them to legislation, policies, etc. works best on the aspect level. Hence, the matching process operates exclusively on aspect representations. Aspects also correspond to mental models of facts humans have in mind in form of blueprints of facts.

We illustrate our approach in the realm of mining and matching facts in legal texts. At first, we outline the theoretical and technical background of our approach. We present the syntax for the definition of aspect representations, that is, blueprints and how our prototypical mining system "Contexter" operates on them in order to analyze and classify texts. A fundamental set of blueprints are handcrafted. Contexter takes these as seeds and tries to find in a semi-automatic approach further possible instances for each blueprint. Humans interact with the system through Contexter's blueprint definition language. This language gets thus a special focus in our paper.

2 Related Work

Legal information retrieval systems locate (and extract) facts in texts and try to relate these facts to relevant legislations, rules, policies, and the like. They apply different instruments to support this task [1–3].

Our approach is designed around the paradigm of fact retrieval emphasizing natural language [4–7]. Classic information retrieval works with features that mostly reference words [8]. We consider features in a broader perspective by adding named entities, n-grams, and elements of bag of words (BoW).

We can imagine a variety of theme-specific BoWs (for countries, cities, names of popular persons, etc.) [9]. BoWs mainly concentrate on proper names. They can be quite powerful and highly interchangeable between LIR systems if texts address more general topics. However, the benefits of such generic BoWs are quite limited when reoccurring proper names are domain or even company specific (names of products, for instance). There are collections of proper names available on the web, but they contain only a fraction of the proper names that may be useful in the domain-specific environment of the LIR system. Managing domain-specific BoWs can get quite laborious.

Hence, we integrated in our system an approach similar to conditional random fields (CFR) [10] in order to identify proper names. It considers typical syntactic patterns of proper names (like names of persons starting with uppercase letters) in their immediate semantic surroundings (when features like "Mrs." or "Mr.", for instance, precede words corresponding to the typical patterns for names).

Each feature denotes, thus, a particular and mainly unique meaning or aspect of a word. In some cases, it can be helpful to consider multiword features. "European Union" or "U.S. Dollar" are examples for multiword features. Names of persons are also examples of multiword features. In contrast to the previous examples, the sequence of tokens addressing names need not necessarily be consecutive. There can even be a middle name that is referenced in one feature and omitted in the other, but both multiword features address the same person. The classification of text using multiword features has a long tradition [11–13]. From a different point of view, multiword features can also be considered as word n-grams or n-words [14].

In our system, the users define a blueprint that represents an aspect by composing features into a semantic hierarchy. Each blueprint is the product of the user's mental model of the corresponding fact or predications it expects to see in texts. The users define their blueprints in a bottom-up approach. They start with blueprints for fine-grained elements, like "price", "distance", or "birthday", integrate them in subsequent definitions, and, thus, iteratively construct blueprints with growing complexities. Each definition of a blueprint is also an annotation summarizing the meaning of its components [15, 16].

3 Representation of Aspect Blueprints

The aspect blueprint is the central element in our approach. In order to cover all variants of an aspect, we have to find a suitable blueprint that covers all its possible instances. The blueprint is basically a structure of expressions where each element may hold one or more

- keywords (including their corresponding synonyms),
- references to BoWs,
- n-grams (as a series of adjacent features),
- named entities, or
- references to further aspects on a more fine-grained level.

Regular expressions are a powerful instrument to detect patterns in all kinds of unstructured data, not only in texts. However, they require special skills and are not easy to handle, in particular, when they are addressing complex, i.e. real-world, patterns. Nevertheless, regular expressions are key to flexible and powerful text analysis. Contexter shields the users from the complexity of regular expressions: The users define their blueprints by resorting to our blueprint definition language (BDL). The syntax they use in order to define their blueprints is quite simple. It enables the user to describe adjacent elements of an aspect. A blueprint of an order, for example, lists the persons or organizations that act as sellers and buyers, the product, the date, and the price.

The most basic elements of the blueprints are directly related to their representations as regular expressions. Basic elements tend to have small text coverage. The users take these elements and define more complex blueprints by indicating which elements appear in proximity to each other. These elements are separated by dots. The users separate by commas or semicolons those elements that may appear inclusively or exclusively at a particular location. Optional elements are indicated by a leading questionmark.

The users may indicate the aspect of "merger" (of two companies, for instance) by the following blueprints:

`merger=time.point.plan.similarity.target.transactions.owner.investments`

`merger=purchase.cancellation.doubt.share`

"time point", "plan", "owner", "cancellation", etc. are the names of already defined blueprints for these subcomponents. For instance, "cancellation is defined as: `cancellation,cancellations=company."terminated".price."purchase".company`

The keyword "terminated" is a placeholder indicating the act of termination of the merger. The example here refers to the finally failed merger between Enron and Dynegy in 2001 as presented across messages and announcement in company communications [17] and related publications.

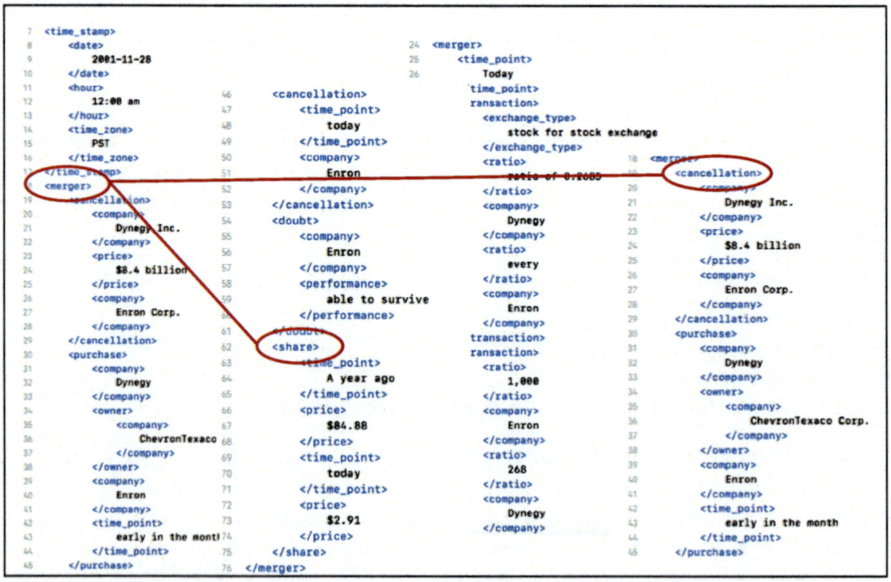

Fig. 1. Examples of representations of the aspect "merger" across different sources.

Before the Contexter system applies this blueprint to any text, each keyword is replaced by a list containing this keyword and all its synonyms that are applicable in the given semantic context. The system considers only these synonyms that are suitable in the semantic context of the respective blueprint.

The end result of applying the blueprint "merger" to a series of texts concerning the failed acquisition of Enron by the Dynegy in 2001 results is an XML representation of the respective texts like in Fig. 1. Retrieval benefits from a more direct access to essential information concerning specific aspects of the overall concept of "company merger". The hierarchical construction of the aspect representation enables retrieval processes to locate the respective facts in their semantic context like the value of shares in the context of the cancellation of the merger as highlighted in Fig. 1.

4 The Semi-automatic Process to Detect Aspect Blueprints

The semi-automatic process takes the blueprints defined by the users, in fact, as seeds and tries to locate similar but new blueprints. The basic idea of the semi-automatic process is that the system modifies existing blueprints on its own and test them against the text base. If the system finds text instances that match the modified blueprint, it reports this blueprint back to the users that confirm or reject it. When the users confirm a newly found blueprint is added to the collection of existing aspect blueprints. The system modifies the blueprint

- by trying to substitute its subcomponents one by one or
- by adding new subcomponents that the fringe of the blueprint.

The process of substitution starts with blanking out a subcomponent. The remaining blueprint is then applied to the text base. Whenever it matches, the text passage between the subcomponent adjacent to the blanked-out subcomponent gets further investigated. The system tests if there is an already existing blueprint in the collection that can be applied to this piece of text and, thus, serves as new subcomponent.

Eventually, the system presents all new blueprints to the users that, in turn, select the ones that make sense and reject the others. Probably the most challenging task for the users is the selection of suitable, consistent, and unambiguous titles for the blueprints. Moreover, the title once defined by the individual user needs to be institutionalized. The titles of the blueprints reflect the semantic picture of all aspects an organization is dealing with in its information management. There are standardized concept hierarchies available like schema.org or the nomenclatures of statistical offices. However, they cover only small fractions of the information ecosystem addressed by an organization.

5 Conclusion

This paper presents the state of work of design and prototypical implementation of a text analysis system operating on aspect blueprints that can be defined by the users. It mainly uses named entity recognition and theme-specific BoWs to precisely spot facts in texts and, thus, enables retrieval of information on a very fine-grained level.

An important design goal of our approach is the strong involvement of humans in the management and control of learning processes of the systems. This involvement starts with a definition language for blueprints that is intuitive and corresponds to the mental models users apply when conceptualizing facts. These manually defined blueprints are taken as seeds. As the meaning of a single fact can be expressed in different ways, by different word and phrase patterns, we assume that there are many variants for each seed to be detected in the text base. This detection process must be a user-controlled one. The users alone can determine if the meaning of the detected pattern corresponds the blueprint's theme. The active role of the users is thus paramount for the system's performance.

References

1. Heo, S., Hong, K., Rhim, Y.-Y.: Legal content fusion for legal information retrieval. In: Proceedings of the 16th Edition of the International Conference on Artificial Intelligence and Law, pp. 277–281 (2017)
2. Gómez-Pérez, A., Ortiz-Rodríguez, F., Villazón-Terrazas, B.: Ontology-based legal information retrieval to improve the information access in e-Government. In: Proceedings of the 15th International Conference on World Wide Web, pp. 1007–1008 (2006)
3. Chalkidis, I., Androutsopoulos, I., Michos, A.: Extracting contract elements. In: Proceedings of the 16th Edition of the International Conference on Artificial Intelligence and Law, pp. 19–28 (2017)
4. Balasubramanian, N., Allan, J., Croft, W.B.: A comparison of sentence retrieval techniques. In: Proceedings of the 30th Annual International ACM SIGIR Conference on Research and Development in Information Retrieval, pp. 813–814 (2007)
5. Fuhr, N.: Integration of probabilistic fact and text retrieval. In: Proceedings of the 15th Annual International ACM SIGIR Conference on Research and Development in Information Retrieval, pp. 211–222 (1992)
6. Keikha, M., Park, J.H., Croft, W.B., Sanderson, M.: Retrieving passages and finding answers. In: Proceedings of the 2014 Australasian Document Computing Symposium, pp. 81–84 (2014)
7. Kolodner, J.L.: Requirements for natural language fact retrieval. In: Proceedings of the ACM 1982 Conference, pp. 192–198 (1982)
8. Weiss, S.M., Indurkhya, N., Zhang, T.: Fundamentals of Predictive Text Mining, 2nd edn. Springer, London (2015)
9. Woods, W.A.: Context-sensitive parsing. Commun. ACM **13**(7), 413–445 (1996)
10. Sha, F., Pereira, F.: Shallow parsing with conditional random fields. In: Proceedings of the HLT-NAACL Conference, pp. 134–141 (2003)
11. Acosta, O.C., Villavicencio, A., Moreira, V.P.: Identification and treatment of multiword expressions applied to information retrieval. In: Proceedings of the Workshop on Multiword Expressions: From Parsing and Generation to the Real World, pp. 101–109 (2011)
12. Calzolari, N., Fillmore, C.J., Grishman, R., Ide, N., Lenci, R., Macleod, C., Zampolli, A.: Towards best practice for multiword expressions in computational lexicons. In: Proceedings of the Third International Conference on Language Resources and Evaluation (LREC 2002), pp. 1934–1940 (2002)

13. Papka, R., Allan, J.: Document classification using multiword features. In: Proceedings of the Seventh International Conference on Information and Knowledge Management, pp. 124–131 (1998)
14. Huston, S., Culpepper, J.S., Croft, W.B.: Sketch-based Indexing of n-Words. In: Proceedings of the 21st ACM International Conference on Information and Knowledge Management, pp. 1864–1868 (2012)
15. Cowie, J., Lehnert, W.: Information extraction. Commun. ACM **39**(1), 80–91 (1996)
16. Salton, G., Allan, J., Buckely, C., Singhal, A.: Automatic analysis, theme generation, and summarization of machine-readable texts. In: Jones, K.S., Willett, P. (eds.) Readings in Information Retrieval, pp. 478–483. Morgan Kaufmann, San Francisco (1997)
17. Klimmt, B., Yang, Y.: Introducing the Enron corpus. In: First Conference on Email and Anti-Spam (CEAS), Mountain View, California, USA, 30–31 July 2004 (2004)

The Knowledge Acquisition Analytical Game Framework for Cognitive System Design

Francesca de Rosa[1(✉)], Anne-Laure Jousselme[1], and Alessandro De Gloria[2]

[1] NATO STO Centre for Maritime Research and Experimentation, 19126 La Spezia, Italy
{francesca.derosa,anne-laure.jousselme}@nato.cmre.int
[2] Electrical, Electronics and Telecommunication Engineering and Naval Architecture Department, University of Genoa, 16145 Genoa, Italy
adg@elios.unige.it

Abstract. Cognitive technologies aim at making sense of information, with the goal of supporting human cognitive abilities of inferring, predicting and taking decisions. While pursuing this objective, cognitive technologies try to mimic human reasoning abilities and schemes. Knowledge acquisition techniques to elicit such schemes from humans suffer from the drawbacks of time inefficiency, biases and prevented access to tacit knowledge. The knowledge engineering domain is exploring the potentiality of using gaming approaches to overcome some of these limitations. Knowledge Acquisition Analytical Games (K2AGs) are analytical games designed to support the knowledge acquisition of cognitive systems, capitalising on aspects such as engagement, ease of use and ability to access tacit knowledge. In this paper, we present the Knowledge Acquisition Analytical Game framework (K2AG-F), which provides an initial high-level guideline on how to design K2AGs and we summarise two instances of such a framework.

Keywords: Knowledge acquisition · Cognitive system · Analytical game · Elicitation · Reasoning schemes · Cognitive mimetics

1 Introduction

Cognitive systems aim at supporting human cognitive abilities of inference, prediction and decision-making. While pursuing this objective, many areas look at cognitive mimetics as a possible design approach. The domain of cognitive mimetics explores how people process information while performing tasks. Finally, it uses the acquired information on human reasoning and inference schemes to inform the design of intelligent technologies [6]. Knowledge acquisition (KA) is a design step of paramount importance in the development of cognitive systems. This design phase embraces the extraction, structuring and organization of expert knowledge to be encoded in an intelligent system. Research is ongoing to improve expert knowledge elicitation techniques, as KA still suffers from several issues. The limitations include biases, inconsistencies, inability of experts to express some knowledge, expert time required, costs, complexity of the task and boredness of the task [5].

D. Russo et al. (Eds.): IHSI 2021, AISC 1322, pp. 554–560, 2021.
https://doi.org/10.1007/978-3-030-68017-6_82

In this work, we discuss the use of gaming approaches as a knowledge acquisition tool, which is able to capitalise on aspects such as engagement, ease of use and ability to access tacit knowledge. We focus on Knowledge Acquisition Analytical Games (K2AGs), which are analytical games specifically designed to support the KA for cognitive systems [9].

It appears that in the last two decades the potential use of gaming for KA has started gaining attention in the scientific community. For example, the concept of using games-for-modelling for the acquisition of highly structured domain-specific knowledge to be used in model-based methods for artificial intelligence (AI) has been introduced in [3]. However, up to authors' knowledge little research has been dedicated to the generalisation and formalisation of the approaches. With the final goal of providing an innovative cognitive system KA toolkit, in this work we introduce the Knowledge Acquisition Analytical Game framework (K2AG-F), which seeks to provide an initial formalisation of K2AGs elements and design. Moreover, we show the effectiveness of the K2AG-F summarising two successful instances of such framework. The remainder of this paper is organised as follows. Section 2 introduces K2AGs, describing their main elements and presents the K2AG-F. Section 3 includes two K2AG-F instances. Finally, Sect. 4 summarises the conclusions and discusses future work.

2 Analytical Games for Knowledge Acquisition

2.1 Knowledge Acquisition Analytical Game

The authors have developed the Knowledge Acquisition Analytical Games, which are a set of innovative games (i.e., Risk Game [4], Reliability Game [11, 12] and MARISA Game [10]) supporting KE and KA for cognitive system design. These games focus on the elicitation of domain knowledge, inference schemes and uncertainty handling. The K2AGs stem from the concept that games are "a communication mode capable of linking tacit to formal knowledge by provoking action and stimulating experience" [2]. More specifically, these are analytical games investigating decision-making with a focus on information processing strategies, uncertain reasoning, situational assessment and situational awareness (SA) [1]. These games are used as experiments to collect knowledge to be employed in the design of advanced artificial intelligence algorithms and systems.

2.2 Knowledge Acquisition Analytical Game Framework

The K2AG-F formalises the process (Fig. 1) of designing and using K2AGs for knowledge acquisition and in general for knowledge engineering of cognitive systems. For representational simplicity it is shown as an ordered sequence, however this process is iterative in nature. The main high-level steps are:

1. KE identification: the KE needs have to be identified through an analysis of the foreseen cognitive system and/or algorithms.

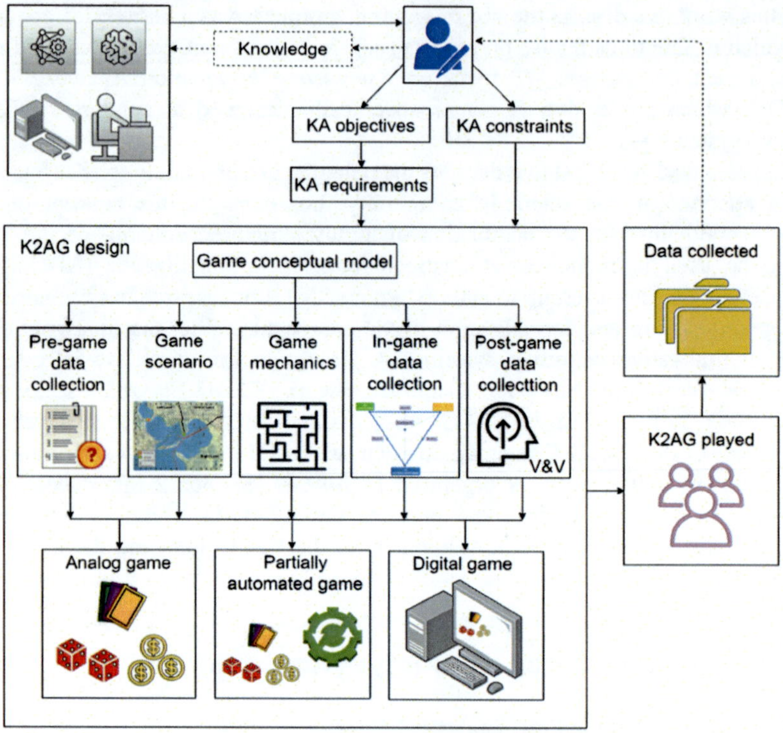

Fig. 1. Knowledge Acquisition Analytical Game framework depicted

2. KA definition: the KA definition phase includes the definition of the KA objectives and related KA requirements. Additionally, the KA limitations (i.e., time, cost and expert availability) need to be investigated. After these elements have been defined an overall assessment has to be performed on whether to use K2AGs for the KA at hand.
3. Game conceptual model: the game design starts with the definition of the game objectives which should mirror the KA objectives and requirements. This phase is the core phase in which the experimental factors (i.e., data, information and meta-data) and their variations are defined. In fact, the overall objective of K2AGs is to observe how the specific information and meta-information injects of the game impact the player information processing.
4. Pre-game data collection: relevant pre-game data should be collected, for example though questionnaires and mini-games. These data are not limited to demographic data (i.e., age, gender, nationality) and expertise data, such as educational level, status (i.e., civilian or military) and years of relevant experience. In fact, it could encompass data to be used to profile the player or to define some kind of experimental baseline with respect to the experiment objectives (i.e., psychological questionnaires).

5. Game scenario: scenario design is one of the most important elements of the design phase. In fact, as described in [8] games are "participatory narrative experience". The overall narrative is composed by a presented narrative that is developed by the designer and a constructed narrative, which results from the active role of the player (e.g., statements, decisions and actions) [8]. It has been observed how in this condition participants do not only make choices, but have the tendency to speak and explain to others (i.e., participants or facilitator) their choices and actions, which supports the experiment objectives of K2AGs.
6. Game mechanics: in this phase the game mechanics to be used to ensure the correct progress of the KA experiment are defined. The mechanics selection should be based on the analysis of when to inject new information, which information to inject, how the player progress during the game and how the players should control the game. As K2AGs aim at capturing human information processing approaches and resulting SA, information and meta-information become central to these kinds of games. Knowledge cards are used to convey knowledge (in the form of messages containing information and meta-information) to the player and trigger their assessments.
7. In-game data collection: all the data to be recorded during the game play need to be identified and an experimental plan needs to be set-up to define the most appropriate method to collect and store such data. K2AGs can use different data gathering methods depending on the problem at hand, the game implementation platform (i.e., manual or digital) and the kind of data formats to be collected, which can be structured or unstructured (i.e., free text, numerical values or graphics). Although different kinds of data might be collected depending on the experiment objectives (i.e., actions, confidence levels or resource allocation), the K2AG approach has been built around mechanics allowing the collection of belief updates on a query variable state each time a new game inject is provided. Such belief levels are quantitative data that can be readily used in algorithm design, for example for their training (e.g., [11]).
8. Post-game data collection: the post-game data to be collected and the data collection methods (i.e., questionnaires and verbal feedback) are defined. Post-game data should address feedback on different aspects of the game, such as players experience (PX) [13]. Although most of the work performed on PX specifically focuses on computerised games, given that PX refers to the quality of player-game interaction [7], the same concept can be extended to non-digital games. Therefore, it applies to the different forms of K2AG. The PX evaluation questionnaire is an integral part of the evaluation of K2AGs as it provides important information related to the validity of the games.
9. Implementation platform: K2AGs can be implemented as either manual games (i.e., board games or card games), either fully automated games (or computerised games) or partially automated games, where part of the game makes use of the support of digital applications; in general the same game can be implemented on different platforms.
10. K2AG played: the K2AG is played by the experts selected for the KA task.
11. Data recorded: the data is recorded and stored.

12. Data analysis: the data analysis allows extracting the required knowledge and to ensure the validity of the game employed.
13. Data use: the data is used in cognitive system and algorithm design.

3 K2AG Instantiated

The Reliability Game and the MARISA Game are two successful instances of the K2AG-F, which have been developed, deployed, validated and used to elicit domain knowledge. The data collected informed then the design of algorithms for surveillance applications. In both K2AGs the scenarios are maritime based and both present elements of anomalous behaviour of vessels to trigger the assessment of the players. In fact, those behaviours are often associated with illegal activities at sea. In order to provide adequate scenarios, they have been validated with maritime experts during the design.

The Reliability Game has been developed with the purpose of experimenting with the impact of source factors on situational assessment and related situational awareness. More specifically, it is a one-player game in which the player is briefed about the presence of a vessel with an anomalous behaviour and is asked to assess what is happening out of a set of collectively exhaustive and mutually exclusive hypotheses. The game injects, namely the information and possible meta-data on the source of information (i.e., source type and source quality) is provided through knowledge cards. The game allows capturing players belief changes regarding the current situation using the cards that the player has to position on the game board. The data collected has been used to train a Bayesian network, and derive the network parameters for the source reliability variable [11].

The MARISA Game is a K2AG developed to support the elicitation of conditional probabilities to be used into a maritime behavioural analysis service, based on a multi-source dynamic Bayesian network, and the validation of its network structure. The MARISA Game is a one round multi-player board game, where the participants play the role of junior navy officers. The more knowledge cards they collect and compile the more they advance in their career. In fact, on the basis of the information in the knowledge cards, players need to assess which could be the state of a possible variable of interest, relevant to the network. The collected beliefs have been mapped to subjective probabilities and used to populate the network conditional probability tables [10]. Moreover, the game acts as a natural form of structured interview, supporting not only the elicitation of the conditional probabilities, but also the validation of the network structure. The utility of the elicited knowledge was proven through the deployment of the resulting service in successful operational trials, where the service was tested. Improvements in the KA task were observed in terms of expert time required and complexity of the task compared to traditional questionnaire approaches.

4 Conclusion

Cognitive technologies aim to make sense of information, supporting the situational assessment and decision processes. Often system designers need to make use of expert knowledge in order to encode reasoning schemes in such systems. Several elicitation methods might be used for KA, but they have proven to suffer from many issues. Some of their limitations seem to be superseded with the use of gaming approaches. In this paper we have discussed important game elements of K2AGs, which are gamified KA tools. We presented the K2AG-F, which constitutes a high-level design guideline for such games. Moreover, we have shown the effectiveness of the K2AG-F summarising two successful instances of such framework, namely the Reliability Game and the MARISA Game. Future research will not only be devoted to the refinement of K2AGs and their single elements, but also to the refinement of the framework in order to develop a complete and flexible KA toolkit that could be easily instantiated by knowledge engineers.

Acknowledgments. This work has received partial support by NATO STO Centre for Maritime Research and Experimentation, through the Data Knowledge Operational Effectiveness programme of work, funded by NATO Allied Command Transformation.

References

1. Endsley, R.M.: The application of human factors to the development of expert systems for advanced cockpits. In: Human Factors Society 31st Annual Meeting, pp. 1388–1392. Human Factor Society, Santa Monica (1987)
2. Geurts, J.L., Duke, R.D., Vermeulen, P.A.: Policy gaming for strategy and change. Long Range Plan. **40**(6), 535–558 (2007)
3. Hoppenbrouwers, S., Schotten, B., Lucas, P.: Towards games for knowledge acquisition and modeling. Int. J. Gaming Comput.-Mediated Simul. **2**(4), 48–66 (2010)
4. Jousselme, A.L., Pallotta, G., Locke, J.: Risk Game: capturing impact of information quality on human belief assessment and decision making. Int. J. Serious Games **5**(4), 23–44 (2018)
5. Korb, K.B., Nicholson, A.E.: Bayesian Artificial Intelligence. Chapman & Hall, Boca Raton (2010)
6. Kujala, T., Saariluoma, P., Porta, M.: Cognitive mimetics for designing intelligent technologies. In: Advances in Human-Computer Interaction 2018 (2018)
7. Nacke, L., Drachen, A., Kuikkaniemi, K., Niesenhaus, J., Korhonen, H., van den Hoogen, W., Poels, K., IJsselsteijn, W., de Kort, Y.: Playability and player experience research. In: Atkins, B., Kennedy, H. (eds.) Breaking New Ground: Innovation in Games, Play, Practice and Theory. DiGRA (2009)
8. Perla, P., McGrady, E.: Why wargaming works. Naval War College Rev. **64**(3) (2011)
9. de Rosa, F.: Knowledge Acquisition Analytical Games: games for cognitive systems design. Ph.D. thesis, University of Genoa, Genoa (2020)

10. de Rosa, F., De Gloria, A.: An analytical game for knowledge acquisition for maritime behavioral analysis systems. Appl. Sci. **10**(2), 591 (2020)
11. de Rosa, F., De Gloria, A., Jousselme, A.L.: Analytical games for knowledge engineering of expert systems in support to situational awareness: the Reliability Game case study. Expert Syst. Appl. **138**, 112800–112811 (2019)
12. de Rosa, F., Jousselme, A.L., De Gloria, A.: A Reliability Game for source factors and situational awareness experimentation. Int. J. Serious Games **5**(2), 45–64 (2018)
13. Wiemeyer, J., Nacke, L., Moser, C., 'Floyd' Mueller, F.: Player experience. In: Dörner, R., Göbel, S., Elsberg, W., Wiemeyer, J. (eds.) Serious Games: Foundations, Concepts and Practice, pp. 243–271. Springer, Cham (2016)

Learning Vector Quantization and Radial Basis Function Performance Comparison Based Intrusion Detection System

Joël T. Hounsou[1,2], Pamela Bélise Ciza Niyomukiza[1(✉)], Thierry Nsabimana[1], Ghislain Vlavonou[1], Fulvio Frati[3], and Ernesto Damiani[3]

[1] Institut de Mathématiques et de Sciences Physiques, Dangbo, Benin
joelhoun@gmail.com, pamela.ciza@imsp-uac.org, thierry.nsabimana@aims-cameroon.org, vlavonoudg@gmail.com
[2] Ecole Supérieure de Génie Télécoms et TIC-ECCE HOMO, Dangbo, Benin
[3] Computer Science Department, Università degli Studi di Milano, Milan, Italy
{fulvio.frati,ernesto.damiani}@unimi.it

Abstract. Information system's technologies increase rapidly and continuously due to the huge traffic and volume of data. Stored data need to be secured adequately and transferred safely through the computer network. Therefore the data transaction mechanism still exposed to the intrusion attack of which consequences remain unlikable. An intrusion can be understood as a set of actions that can compromise the three security purposes known as Confidentiality, Integrity and Availability (CIA) of resources and services. In order to face on these intrusions, an efficient and robust Intrusion Detection System (IDS) which can detect successfully the intrusion is strongly recommended. An IDS is a network/host security tool used for preventing and detecting malicious attacks which could make a system useless. The purpose of this paper is to implement network intrusion detection system based on machine learning using Artificial Neural Network algorithms specifically the Learning Quantization Vector and Radial Basis Function make the comparison on the performance between these two algorithms.

Keywords: Intrusion Detection System · Artificial Neural Network · Learning Vector Quantization · Radial Basis Function

1 Introduction

Actually, exposure to intrusion attacks in Information System (IS) has become prevalent, thus causing data security to become a top priority for computer scientists who constantly work as DevOps experts to develop tools/software/methods that can be used to detect and/or prevent malicious usage of the action of monitoring network traffic, local host and alerts the DevOps operators against suspicious activities or signs of possible incidents [1]. Most research reported loopholes during intrusion detection such false negative and false positive reported. One alternative suggested to solve this

D. Russo et al. (Eds.): IHSI 2021, AISC 1322, pp. 561–572, 2021.
https://doi.org/10.1007/978-3-030-68017-6_83

issue was to integrate the algorithms used in the Artificial Intelligence (AI) in the quest of a perfect IDS [3]. Throughout this paper we will overview the existing solutions that deal with network security. Then we will present our proposed solution based on LVQ and RBF Neural network models in parallel and will be compared on their performances.

Problem Statement. Various security techniques and solutions have been proposed to monitor and control an Information System. Since the usage of internet services has becoming the prevalence, computer security/host is a critically important today due to the high skills hacking techniques/methods which are becoming more and more sophisticated to mitigate. Moreover, many others intelligent methods are also still being designed/implemented by several computer scientists but in return are themselves, not efficient and effective to monitor and to detect systems of malicious attacks. Historical perspective on traditional security tools like firewall, antivirus, authentication and so forth were proven to be able to monitor and detect attacks efficiently. But as today communication and information are being taken place on the computer network, the aforementioned tools are not efficient any more. Some of the existing open source and commercial Intrusions Detection Systems (IDSs) present some shortcoming, namely high false positive rate generated by anomaly detection approach which leads to very poor accuracy and significant number of false negative generated by Misuse/Signature detection approach due to the inability of detecting unknown attacks. One of the key points in security system on which the researchers have worked a lot is to develop a system to detect intrusions in the network traffic and host.

Related Work. A review of existing publications on Information System reveals that, the three security purposes known as CIA remain the main drive that pushes most researchers to design efficient system that can detect and prevent intrusion. Different detection and prevention techniques have been proposed by many researchers to discourage malicious users of the IS. Since traditional security tools have been revealed not efficient due the complicated attacks that constantly come up, researchers focus on emerging technologies that can improve detection and prevention of intrusion. AI is defined as the ability of a machine or a computer program to think and learn by doing certain task. AI concept is based on the creation of machines capable of understanding the environment, understanding the problem and act intelligently according to the situation. AI through the algorithms of Machine Learning (ML) allows to design and implement an IDS. Learning Vector Quantization (LVQ) is one of many classification algorithm from the Artificial Neural Network (ANN) trained to detect intrusions. Amir Azim et al. [5] suggested an alert management of intrusion detection using the LVQ algorithm. Their experience proved that LVQ algorithm is able to classify the network traffic. They noticed that a system built on LVQ can solve some problems of IDSs such as high amount of false alerts. They concluded that LVQ system could be used in active alert management system because of its low classification time per each alert. Reyadh et al. [6] noticed the main issue in LVQ training on long time to be trained, suggested a hybrid system of LVQ_ERBP to perform well than the hybrid system of LVQ_KNN. They conducted an experience that leaded to demonstrate the performance of ANN as a

ML algorithm in classification problems over the k-Nearest Neighbor which uses norm distance to classify the records. One of the main issue from that system was the low detection rate for some attacks such as U2R due to their low frequency. Authors pointed that the low frequency of those attacks should not make them unimportant. An attacks despite its low frequency is still dangerous because of the unpleasant consequences it can cause [6]. Naoum et al. [2] combined two LVQ and KNN algorithms in order to be trained in intrusion detection. The experience he made was that LVQ even it can classify the intrusion to the normal traffic, it requires much time comparing to other techniques like Self Organizing Feature Map and Multi-layer perceptron. Alia et al. [7] built and intelligent intrusion detection system based on RBF capable to handle all types of attacks and intrusions with high detection accuracy and precision. However, for their system, they worked only with two phase-RBF training that reflect in relatively low sensitivity towards frequent attacks (DoS and Probe).

2 Motivation and Learning Algorithms

In light of the above, it appears that there is a continuing commitment of researchers to build a secure system in detecting intrusion. While improving network security and accurate detection remain our objective, we will have to explore ANN by exploring some machine learning algorithms such LVQ and RBF and make a comparison on their performance. "*ANN is an information processing paradigm that is inspired by the way the biological nervous system such as brain process information. It is composed of large number of highly interconnected processing elements (neurons) working in unison to solve a specific problem*" [8]. ANN is kind of a lot of processing elements named neurons working together in unison to solve problems rapidly with high accuracy. The brain is a parallel massively connected system. ANN is one of the main soft computing algorithms used in many researches as detector agent in IDSs. The ANN is used as to mimic mammalian's brain to solve number of problems encountered by other current intrusion detection methods, and have been proposed as alternatives to the statistical analysis component both of anomaly and misuse detection systems [9]. ANN data processing is mainly driven by brain functioning. In effect the brain has approximately 100 billion (10^{11}) neurons, which communicate through electro-chemical signals. The neurons are connected through junctions called synapses. Each neuron receives thousands of connections with other neurons, constantly receiving incoming signals to reach the cell body. If the resulting sum of the signals surpasses a certain threshold, a response is sent through the axon. The ANN attempts to recreate the computational mirror of the biological neural network, although it is not comparable since the number and complexity of neurons and the used in a biological neural network is many times more than those in an ANN. Some ANN advantages are the following: *(i)* ANN has the capability in analyzing the category to which a given information belongs; *(ii) a*daptive learning: the ability to learn how to do task based on the data given for training or initial experience; *(iii)* self-organization: the ability to create own topological organization or representation of information received during learning phase. The core aptitude of SOFM is to transform high dimensionality space to low dimensionality one (one, two or three dimension); *(iv)* real time operation.

For our research work, we focus in implementing two different ANN models, such as RBF and LVQ to reach our goal.

Radial Basis Functions. Radial Basis Functions (RBF) are three-layer feedforward neural network on which hidden layer is made up of neurons where each neuron consists of a radial basis function centered on a point with the same dimensions as the predictor variables [7].

$$f(y) = g\left(\sum_{j=1}^{m} w_{ji} h_j(x)\right) \quad (1)$$

$$h(x) = exp\left(-\frac{(x-c)^2}{r^2}\right) \quad (2)$$

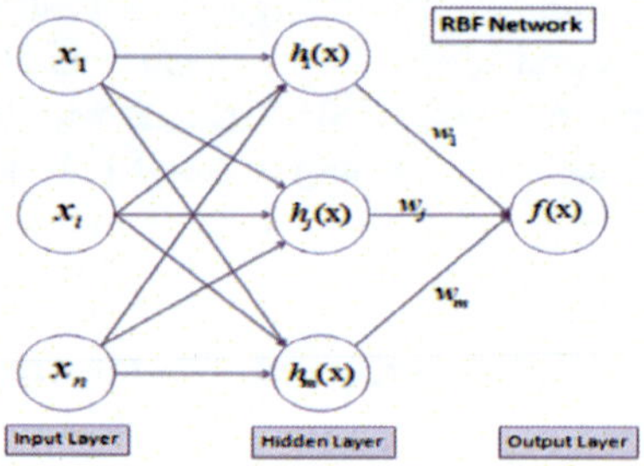

Fig. 1. RBF ANN structure

The RBF structure is represented on the Fig. 1. Where $f(y)$ is an output function and $h(x)$ a Gaussian function. The learning process is based on adjusting the parameters of the network to reproduce a set of input-output patterns. It consists of two main stages: *(i)* the first stage is to determine the centers using unsupervised clustering algorithm, *(ii)* while in the second stage regards to adjust the connection weights between the hidden layer and the output layer [10]. For these three types of parameters are required: the weight (W) between the hidden nodes and the output nodes, the centers (c) the winner at the competition phase on the map, and the window width (r) as the radius or the standard deviation. The two last parameters are required for the Gaussian function which is used on the hidden layer as the activation function. The centers can be determined by any clustering algorithm (e.g., K-means clustering, Self-Organizing Feature Map, etc.). The cluster centers become the centers of the RBF clusters.

Learning Vector Quantization Algorithm. LVQ is based on supervised learning (with the target output). It is made up of two layers: the competitive layer equivalent to the hidden layer and the linear layer equivalent to the output layer. LVQ involves the competitive layer method to recognize groups of similar input vectors. A competitive layer automatically learns to classify input vectors. The mechanism inside the layer is based on calculating distance between the input vector in order to find the input vector's classes. The competitive layer's classes are passed to the linear which transform them into defined target classes [10]. Typically, the LVQ network is presented on Fig. 2. We used KDD cup 99 due to its performance on machine learning classifier. The database that contains a standard set of data to be audited, which includes a wide variety of intrusions simulated in a military network environment [12]. It is a standard dataset for survey and evaluate research in intrusion detection which is prepared and managed by MIT, Lincoln Labs by DARPA Intrusion Detection Evaluation Program [4]. The simulated attacks contained in the dataset fall in one of the following four categories: Denial of Service (DoS), User to Root (U2R), Remote to Local (R2L), and Probing Attack.

3 Methods (Implementation)

Our proposed system is made up by two main modules that are preprocessing module and detection module.

Preprocessing Module. The data to be analyzed by the detection module need to be preprocessed first. Two principal actions are done in preprocessing phase: Features conversion to numerical values: some features in string format such protocol type, service and flag are converted into numerical value, and Data normalization: the converted data into numerical value need to be normalized so as to ease the training process. Data normalization consists in putting data value in range of [0, 1].

Detection Module - RBF Module. Figure 3 illustrates RBF diagram. As proposed architecture, a coupling Self Organizing feature Map as an unsupervised anomaly detection technique to RBF in order to classify the known and unknown traffic (normal and attacks). The input of our system consists of 41 inputs to be fed to the Kohonen network. The hidden layer known as competitive layer, is then a Kohonen's Self Organizing Feature Map (SOFM). It is two-dimension matrix ([15 $\times$ 15]). The output layer has 5 neurons, each of those neurons identifies the class to which the input patterns belongs. The expected output maybe either normal, DOS, probe, U2R or R2L. Many others useful parameters used in mapping clustering technique are: the initial radius r = 0.8, initial learning rate = 0.06. The last two parameters used on the output stage of RBF are centers trained in hidden layer and variance.

LVQ Module. The proposed LVQ structure is depicted on Fig. 3. It comprises 3 layers: input, hidden known as competitive, and output layer. The input layer is provided with 41 input nodes. For hidden layer known as competitive layer, 10 prototypes per class are chosen. After that, Euclidean distance is used to find out the winning neuron. Learning rate parameter is constant and set to 0.01. The output layer has 5 neurons, each of neuron identifies the class (either normal, DOS, probe, U2R, or R2L) to which input pattern belongs. When we feed the network with normal patterns inputs, only the subset responsible in the LVQ layer are excited while other LVQ subset neurons are not in this process.

KDD Cup 99 Sample Distribution. The full data set in which we fetched and worked on, were about 78287 vectors of 41 components or features. Three subsets of this full data is splitting randomly as training, testing and generalization sub-data sets, in 70%, 20% and 10% proportion respectively. The dataset contains fifteen type of attacks that are classified in 4 categories. Table 1 shows the distribution based KDD cup 99 dataset in the three sets: training, testing, and generalization.

Table 1. Sample of distribution on the subset of 70%, 20%, and 10% training set of KDD cup 99 dataset.

Classes	Number of connections	Percentage (%)
Normal	48937	95,46%
DOS	1569	3,06%
PROBE	19	0,04%
U2R	2	0,00%
R2L	738	1,44%
Total:	51265	100,00%

Classes	Number of connections	Percentages
Normal	13994	95,54%
DOS	431	2,94%
PROBE	7	0,05%
U2R	2	0,01%
R2L	214	1,46%
Total:	14648	100,00%

Classes	Number of connections	Percentages
Normal	6996	95,50%
DOS	237	3,24%
PROBE	5	0,07%
U2R	1	0,01%
R2L	87	1,19%
Total:	7326	100,00%

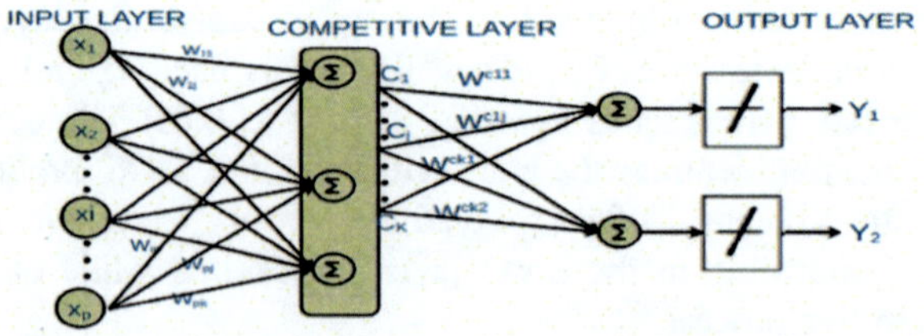

Fig. 2. LVQNN with visualization layer architecture [11].

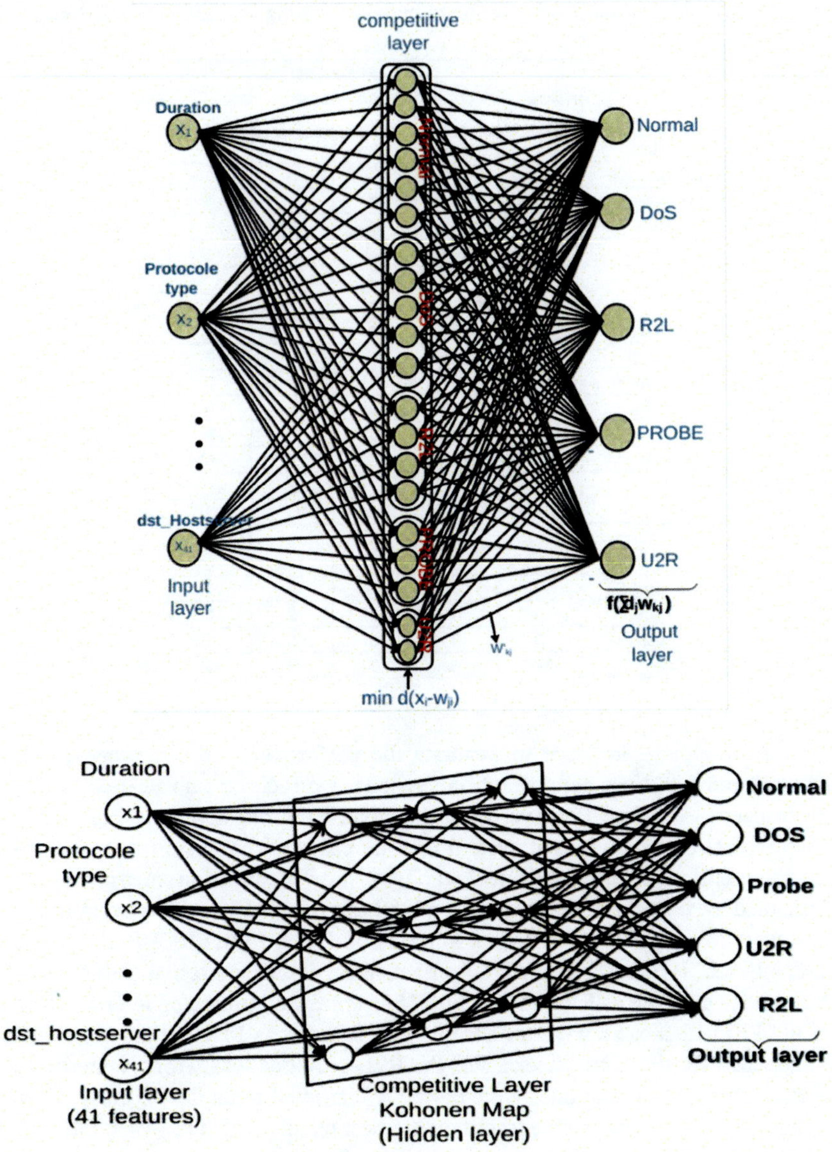

Fig. 3. RBF and LVQ diagram.

4 Experimental Results and Analysis

Experiment 1: RBF Module. This first experiment used all 41 input nodes (41 neurons) that listed and explained below. Kohonen Self Organizing Feature Map is able to cluster all categories of attacks as well as normal together on the same map, despite

Table 2. Confusion matrix for the training set, testing set, and generalization

		PREDICTED					
		Normal	*DOS*	*PROBE*	*U2R*	*R2L*	*Total*
ACTUAL	*Normal*	48880	4	15	25	13	48937
	DOS	2	1566	0	0	1	1569
	PROBE	0	0	19	0	0	19
	U2R	0	0	0	2	0	2
	R2L	0	1	0	0	737	738
Total:		**48882**	**1571**	**34**	**27**	**751**	**51265**

		PREDICTED					
		Normal	*DOS*	*PROBE*	*U2R*	*R2L*	*Total*
ACTUAL	*Normal*	6981	9	0	4	2	6996
	DOS	2	235	0	0	0	237
	PROBE	0	0	3	0	0	3
	U2R	0	0	0	1	0	1
	R2L	1	0	0	0	86	87
Total:		**6984**	**244**	**3**	**5**	**88**	**7324**

		PREDICTED					
		Normal	*DOS*	*PROBE*	*U2R*	*R2L*	*Total*
ACTUAL	*Normal*	13976	10	0	1	7	13994
	DOS	1	430	0	0	0	431
	PROBE	0	0	7	0	0	7
	U2R	0	0	0	2	0	2
	R2L	0	0	0	0	214	214
Total:		**13977**	**440**	**7**	**3**	**221**	**14648**

of some overlapping. In order to evaluate the performance of our proposed system based on training testing, generalization sessions, we focused on four main indicators of performance, viz True Positive, False positive, True negative, and False negative. The assessment of our proposed system is done based on confusion matrix using five indicators mentioned above. Table 2 (above) shows confusion matrix, where our implemented system correctly predicted 48880 out of 48937 actual normal traffic, but 57 case were respectively classified as follow: 4 in DOS, 15 in PROBE, 25 in U2R, 13 in R2L attacks. Two cases from DOS attack were taken as normal traffic, one from DOS attacks was taken as R2L attacks and 1566 were well predicted in DOS class. The 19 actual PROBE attacks were all correctly recognized as PROBE attacks. The same case happened for the U2R attacks. For the R2L only one was wrongly predicted as a DOS attack but the 737 left ones were correctly classified in their R2L class. Using our RBF module, we have been able to obtain the best results with TP, which means 100% and 99% and with acceptable false positive rates 0.0%, 0.11% and false negative rates 0.0%, 0.19 as shown in Table 3 (above). Table 2 (center) shows the confusion matrix from the testing set session. 13985 were correctly predicted as normal traffic from 13994 actual normal traffic while 9 case where predicted as 2 in DOS, 1 in U2R, 6 in R2L. All the 431 actual DOS attack were correctly predicted as DOS attack. The 7 actual PROBE attacks were all correctly recognized. The R2L enumerate 214 actual R2L attacks and all were well classified in R2L class. On testing set session using RBF module, we have been able to achieve the best results with TP, which means 100% and 99% and with very low false positive rates 0.0%, 0.12% and low false negative rates

Table 3. Indicators of performance for training set, testing set, and generalization.

Classes	True Positive	False Positive	False Negative
Normal	99,88%	0,00%	0,12%
DOS	99,81%	0,32%	0,19%
PROBE	100,00%	44,12%	0,00%
U2R	100,00%	92,59%	0,00%
R2L	99,86%	1,86%	0,14%

Classes	True Positive	False Positive	False Negative
Normal	99,87%	0,01%	0,13%
DOS	99,77%	2,27%	0,23%
PROBE	100,00%	0,00%	0,00%
U2R	100,00%	33,33%	0,00%
R2L	100,00%	3,17%	0,00%

Classes	True Positive	False Positive	False Negative
Normal	99,79%	0,04%	0,21%
DOS	99,16%	3,69%	0,84%
PROBE	100,00%	0,00%	0,00%
U2R	100,00%	80,00%	0,00%
R2L	98,85%	2,27%	1,15%

0.0%, 0.23% as shown in Table 3 (center). In Table 2 (below), which shows the confusion matrix from generalization set session, we find that the classification from our implemented system with the generalization set session. The 6981 out of 6996 normal traffic were correctly predicted. But the 15 cases, 9 cases were predicted as DOS attacks, 4 cases as U2R attacks and 2 as R2L attacks. About DOS attacks 235 out of 237 were well predicted while the 2 left case were taken as normal traffic. The 3 actual cases observed in PROBE were also all correctly predicted. The only case found in actual traffic in U2R attacks was correctly classified in its U2R class. About R2L, 1 case out of 87 was taken as normal traffic while the 86 cases were correctly predicted in their class. On generalization set session using RBF module, we have been able to achieve the best results with 100% and 99% true positive rates of some attacks/normal with low false positive rates and low

false negative rates as shown in Table 3 (below).

Experiment 2: LVQ. Table 2 (above) illustrates the results gotten from LVQ implementation algorithm during the training phase. 48854 out of 48937 normal traffic were well correctly predicted in the normal connection. 54, 27, 2 out of 48937 of normal traffic, were respectively predicted in DOS, PROBE and U2R. DOS, PROBE,

Table 4. LVQ module Confusion matrix for training set, testing set, and generalization.

		PREDICTED					
		Normal	*DOS*	*PROBE*	*U2R*	*R2L*	*Total*
ACTUAL	*Normal*	48854	54	27	2	0	48937
	DOS	0	1569	0	0	0	1569
	PROBE	0	0	19	0	0	19
	U2R	0	0	0	2	0	2
	R2L	10	7	0	3	718	738
Total:		**48864**	**1630**	**46**	**7**	**718**	**51265**

		PREDICTED					
		Normal	*DOS*	*PROBE*	*U2R*	*R2L*	*Total*
ACTUAL	*Normal*	13754	80	53	107	0	13994
	DOS	5	423	0	0	3	431
	PROBE	0	0	7	0	0	7
	U2R	0	0	0	2	0	2
	R2L	33	21	4	0	718	776
Total:		**13792**	**524**	**64**	**109**	**721**	**15210**

		PREDICTED					
		Normal	*DOS*	*PROBE*	*U2R*	*R2L*	*Total*
ACTUAL	*Normal*	6987	3	0	0	6	6996
	DOS	5	219	0	0	13	237
	PROBE	1	0	2	0	0	3
	U2R	0	0	0	1	0	1
	R2L	13	0	3	0	71	87
Total:		**7006**	**222**	**5**	**1**	**90**	**7324**

U2R attacks were correctly predicted in their classes with respectively 1569, 19, 2 connections. On training set session using LVQ module, we have been able to achieve the best results with 100% True positive for U2R, Probe 97.3% True positive for R2L and DOS and 99.83% true negative rates of normal with low false positive rates and low false negative rates as shown in Table 5 (above). Table 4 (center) provides results during the testing phase. 13754 Out of 13994 normal traffics were well classified in the normal traffic. But 80, 53, 107 from the normal patterns were respectively classified in DOS, PROBE, U2R, R2L attack traffic. PROBE and U2R abnormal traffic were well predicted. 5 and 3 from DOS connection were classified respectively in normal and R2L classes while 423 out of 431 were correctly classified in DOS traffic. For R2l traffic, 156 connection out of 214 were correctly predicted while 33 were taken as normal traffic, 21 as DOS attack and 4 as PROBE attack. On testing set session using LVQ module, we have been able to achieve the best results with 100% True positive for U2R, Probe, 72.89% True positive for R2L attacks and 99.28% true negative rates of normal with low false positive rates and low false negative rates as shown in Table 5 (center). In Table 4 (below) we found the result gotten during the generalization phase. All traffic normal were not correctly predicted. 6987 out of 6996 were correctly classified while 3 and 6 cases from normal traffic were respectively taken as DOS and R2L traffic. 219 out of 237 DOS connections were correctly predicted while 5 and 13 DOS connections were respectively classified in normal and R2L traffic. For PROBE connections, only one (01) pattern was misclassified in normal traffic while 2 connection were well identified. 71 R2L records out of 87 R2L were correctly classified while 13 are taken as normal and 3 as PROBE. On generalization set session using LVQ module, we have been able to achieve the best results with 100% True Positive for U2R, 92.40% True Positive for DOS, 66.7% True Positive for Probe, 81.7% True Positive for R2L and 99.87% True negative rates of normal with low false positive rates and low false negative rates as shown in Table 5 (below).

Table 5. LVQ module Indicator of performance for training set, testing set, and generalization.

Classes	True Positive	False Positive	False Negative
Normal	99,83%	0,02%	0,17%
DOS	100,00%	3,74%	0,00%
PROBE	100,00%	58,70%	0,00%
U2R	100,00%	71,43%	0,00%
R2L	97,29%	0,00%	2,71%

Classes	True Positive	False Positive	False Negative
Normal	98,28%	0,04%	1,72%
DOS	98,14%	19,27%	1,86%
PROBE	100,00%	89,06%	0,00%
U2R	100,00%	98,17%	0,00%
R2L	92,53%	0,42%	3,22%

Classes	True Positive	False Positive	False Negative
Normal	99,87%	0,27%	0,13%
DOS	92,41%	1,35%	7,59%
PROBE	66,67%	60,00%	33,33%
U2R	100,00%	0,00%	0,00%
R2L	81,61%	21,11%	85,06%

5 Conclusion

In this work RBF and LVQ modules were deployed to intrusion detection. Coupling of SOFM to RBF module was used as unsupervised anomaly technique to build topological map of intrusions or normal flow. The obtained results have been shown that despite of some little overlapping, some attacks/normal with similar behavior were clustered in the region on the map. The same experiment was conducted on LVQ module. Through the results from LVQ, Kohonen LVQ was able to classify all categories either attacks or normal in the same region but with enormous number of overlapping in all category of attacks includes normal. Hence, comparing with Kohonen Self Organizing Feature Map (K-SOFM), we found that K-SOFM clustering performs perfectly with less overlapping than K-LVQ. Very high true positive rate and low false positive rate demonstrate advantages of applying RBF and LVQ techniques to intrusion detection, that by using 41 out of 41 features (41 neurons). In experiment 1, RBF gains high true positive rate of intrusion and acceptable low false positive rate than in experiment 2 using LVQ, where we obtain less true positive rate and high false positive rate.

Acknowledgment. This work has been partly funded by the European Commission within the H2020 project THREAT-ARREST (contract n. 786890).

References

1. Vyas, G., Meena, S., Kumar, P.: Intrusion detection systems: a modern investigation. Int. J. Eng. Man. Sci. (IJEMS) **1**(11) (2014)
2. Naoum, R.S., Al-Sultani, Z.N.: Learning vector quantization (LVQ) and k-nearest neighbor for intrusion classification. World Comput. Sci. Inf. Technol. J. (WCSIT) **2**(3), 105–109 (2012)
3. Haq, N.F., Onik, A.R., Hridoy, M.A.K., Rafni, M., Shah, F.M., Farid, D.M.: Application of machine learning approaches in intrusion detection system: a survey. Int. J. Res. Artif. Intell. **4**(3), 9–18 (2015)
4. Hamid, Y., Sugumaran, M., Balasaraswathi, V.R.: IDS using machine learning-current state of art and future directions. Current J. Appl. Sci. Technol. **15**(3), 1–22 (2016)
5. Ahrabi, A.A.A., Feyzi, K., Orang, Z.A., Bahrbegi, H., Safarzadeh, E.: Using learning vector quantization in alert management of intrusion detection system. Int. J. Comput. Sci. Secur. **6** (2), 128–134 (2012)
6. Naoum, R.S., Al-Sultani, Z.N.: Hybrid system of learning vector quantization and enhanced resilient backpropagation artificial neural network for intrusion classification. Int. J. Appl. Sci. (IJRRAS) **14**(2), 333–339 (2013)
7. Abu Ghazleh, A., et al.: Intelligent intrusion detection using radial basis function neural network. In: Proceedings of IEEE Sixth International Conference on Software Defined Systems (SDS), pp. 200–208 (2019)
8. Srivignesh, R.: An Introduction to Artificial Neural Networks. https://towardsdatascience.com/introductionto-artificial-neural-networks-ann-1aea15775ef9. Accessed 18 July 2020
9. Biswas, S.K.: Intrusion detection using machine learning: a comparison study. Int. J. Pure Appl. Math. **118**(19), 101–114 (2018)
10. Raza Shah, S.A., Issac, B.: Performance comparison of intrusion detection systems and application of machine learning to Snort system. Future Gener. Comput. Syst. **80**, 157–170 (2018)
11. HandWiki.org.: Learning vector quantization. https://handwiki.org/wiki/Learning_vector_quantization. Accessed 17 Apr 2020
12. SIGKDD. KDD Cup 1999: Computer network intrusion detection. https://www.kdd.org/kddcup/view/kdd-cup-1999/Data. Accessed 17 Apr 2020

Economic Diagnostics and Management of Eco-Innovations: Conceptual Model of Taxonomic Analysis

Alina Yakymchuk[1], Andriy Valyukh[1], Inna Irtyshcheva[2], Valentyna Yakubiv[3], Nazariy Popadynets[4(✉)], Iryna Hryhoruk[3], Kostiantyn Pavlov[5], Olena Pavlova[5], Yuliia Maksymiv[3], Yevheniya Boiko[2], Nataliya Hryshyna[2], and Olena Ishchenko[2]

[1] National University of Water and Environmental Engineering, Soborna, Str. 11, Rivne 33028, Ukraine
alinayakim@ukr.net

[2] Admiral Makarov National University of Shipbuilding, Heroyiv Ukrayiny Ave., 9, Mykolayiv 54025, Ukraine
innauamd@gmail.com

[3] Vasyl Stefanyk Precarpathian National University, Shevchenko, Str. 57, Ivano-Frankivsk 76018, Ukraine
yakubiv.valentyna@gmail.com

[4] SI "Institute of Regional Research named after M. I. Dolishniy of the NAS of Ukraine", Kozelnytska Str. 4, Lviv 79026, Ukraine
popadynets.n@gmail.com

[5] Lesya Ukrainka Eastern European National University, Volya Avenue, 13, Lutsk 43025, Ukraine
pavlovkv@icloud.com

Abstract. Sustainable economic growth can only be achieved by increasing productivity and introducing better services and products. The aim of this research is to identify strategic directions to improve the implementation of environmental innovations in international cooperation between Europe and Ukraine. The paper examines the process of innovative industrial activities in the European Union, Ukraine, and, in particular, at the regional level. A set of new measures to promote environmental innovations has been offered. A taxonomic economic-mathematical model for estimating the integrated indicator of the implementation of environmental innovations in the regional context was developed.

Keywords: European integration · Innovation · Environment · Economic policy · Taxonomic analysis

1 Introduction

In modern competitive economies, knowledge-based innovations are the foundation for economic development. Sustainable (ecological) growth and growth of living standards could only be achieved by increasing productivity and the introduction of new and

D. Russo et al. (Eds.): IHSI 2021, AISC 1322, pp. 573–579, 2021.
https://doi.org/10.1007/978-3-030-68017-6_84

improved products and services that successfully compete in the global market. This is especially true in the area of the environment because industrial emissions have a large quantities of hazardous substances in the environment because of the lack of technical equipment. The current problem is to define the perspective directions of enterprise's technological innovative development, especially with the probability of its technological breakdown with the external environment, which is especially risky for the enterprises' technological type, the dominant development of which is technology in the broad sense. The economical innovative development was highlighted in the works of [1–7] and others. But to date, the unified directions of innovative development at the regional level have not been developed. This article presents data that illustrates the shift towards cleaner products and continuing importance of end-of-pipe solutions, and discusses national differences in eco-innovation use, lead market issues and the growing attention to system innovation. The transfer of technology from abroad has been complemented by national innovation capabilities based on research and development [8, 9]. These authors analyze the link between economic growth and international technology transfer in the EU and try to answer the question–why do some economies use technology transfer and thus accelerate their development while others do not.

The article aims to define the strategic directions of the development of advanced technologies, how they influence the economic development of the state on the example of taxonomic method; identify internal and external tools for the analysis of creative development according to the directives of the European Union.

2 Methodology

Fundamental provisions of economic theory and economy became the theoretical and methodological basis of this study. The following general scientific methods of the research were applied for solving such tasks: generalization–for research of the essence of concepts "regional innovations" and "innovative activity"; systematization–to analyze the economic and legal principles of the implementation of innovations in Ukraine and European countries; cause and consequence method–to study the causes and consequences of the financial and economic crisis on the quantity and quality of innovative developments in the environmental sphere; system approach–to study the interrelations and interdependencies between environmental-economic system of territory development, as well as special methods: a method of correlation-regression analysis–to study the innovative development financing dynamics, budget financing; imitation modeling–for quantitative assessment of scenarios for optimizing the process of enterprise greening; the method of analysis of hierarchies–to choose the optimal mechanisms of state regulation scenario of industrial enterprises' innovative activity; cost and benefits analysis–to assess the socio-ecological and economic efficiency of the optimization of regional environmental innovations and implementation of the best foreign practices at the Ukrainian enterprises.

A taxonomic economic-mathematical model for estimating the integrated indicator of the implementation of environmental innovations in the regional context was developed. The assessment was performed based on a taxonomic indicator, which is a synthetic quantity and takes into account the integral influence of the values of the set of indicators (quantity of personnel of the scientific organizations by categories; the scientific personnel; distribution of the extent of financing of scientific and technical work by financial sources; the costs of organizations for the execution of scientific and technical work; financing distribution of the innovative activities in the industry; the number of implemented eco-innovations in the region; general extent of innovations expenses in the industry; the number of implemented eco-innovations in Ukraine etc.). The taxonomic indicator of the ecological innovations implementation in Ukraine was determined by calculating additional indicators: the distance between observations and the reference vector, the average and maximum distances of deviations from the standard.

3 Results and Discussion

In this paper, some empirical correlations between economic efficiency and other indicators (population, state size, specific indicators: the extent of scientific and technical work of the enterprises, distribution of the financing of innovative activities in the industry, quantity of personnel of the scientific organizations, the number of implemented environmental innovations in the region, the scientific personnel of scientific organizations, extent of scientific and technical work financing, the number of implemented environmental innovations in Ukraine, the costs of organizations for the execution of scientific and technical work on their own, the general extent of innovations expenses in industry) were found. As shown by the correlation-regression analysis carried out in this paper–the determination coefficient is 0.99. That is a close relationship between the extent of scientific and technical work financing, number of scientific organizations' personnel, the extent of scientific and technical work of the enterprises and the indicator of implementation of eco-innovations. The authors made a comprehensive assessment of the impact of factors on the amount of funding and forecasted the number of financial resources with the optimal value of implemented innovations in the country (Table 1).

Table 1. Analysis of variance of model's indicators

Indicator	df	SS	MS	Fisher's criterion (F)	Significance F
Regression	2	127292.3764	63646.18822	219.521032	2.45048E-06
Remainder	6	1739.592448	289.9320746		
Total	8	129031.9689			

The regression statistics of the model has been given in Table 2.

Source: calculated based on the [10].

The resultant function of the model is the following:

$Y = -2335.15 + 0.07\ x_1 + 0.31\ x_2.$

Table 2. Estimation of regression statistical model

Indicator	The value of the model indicator
Multiple correlation coefficient (R)	0.9932236
Determination indicator (R^2)	0.986518
Normalized value R^2	0.982024
Standard error	17.02739
Number of observations	9

The determination rate of R^2 demonstrates that changing 100% of the values of x_1, x_2 changes the value of the resulting indicator y (funding volumes). Thus, the correlation coefficient R = 0.99 is approaching 1, which means that the connection between factors has been estimated as very close. Therefore, the model is reliable.

Substituting value in the regression equation will receive the projected funding:

$Y = -2335.15 + 0.07 \cdot 12072.56 + 0.31 \cdot 8296 = 1081.6892$ (mln. UAH).

Thus, the projected amount of financing is determined when achieving the optimum value of financing environmental innovations for the efficient socio-ecological-economic functioning of regions, which is UAH 1.08 billion of annual financing. For the long-term period-till 2030, at the optimum of the total number of implemented environmental innovations, the total amount of funding from the budget should be UAH 1.64 billion:

$Y = -2335.15 + 0.07 \cdot 18108.84 + 0.31 \cdot 8750 = 1644.9688$ (mln. UAH).

In order to adapt to the new globalized conditions and to be relatively successful, the European Union has decided to specialize in high technology as part of the Lisbon strategy and to abandon its labor-intensive activities, such as textiles, for example in low-wage countries like China. This choice accelerated the change or decline of certain traditional sectors such as the steel industry or textile, but allowed the emergence of new technological sectors such as microelectronics and biotechnology, in which European companies are international leaders. In the long run, this strategy will be based on the idea of an international division of labor, where Europe will keep its "noble" functions in R&D, while the emerging economies will take on the "worst of".

The European Union remains an important industrial player, representing 16% of world exports and 17.3% of imports. This makes the EU the main world market alongside the intra-community trade. Interestingly, trade relations, unlike in the United States or Japan, are balanced globally: the trade deficit was € 159.9 billion in 2010 [5] or 0.65% of its GDP compared to the French deficit in 2010 of almost 2% of GDP. If its relations are balanced on a global level, this is only the case on an individual basis: the European Union has a trade deficit of 169 billion euros with China and a surplus of 73 billion euros with the United States. It also imports energy and raw materials from Russia and OPEC countries. It is not surprising that it is only an international division of labor in which Europe is not the most competitive to concentrate on its strengths. This makes the company a leader in a large number of industrial sectors in which the trade balance shows a surplus: cosmetics, chemicals, processed raw materials (paper, plastic, steel), machines (energy production, industry), cars and vehicles. The only high-tech area where it is difficult to stay in the race is microelectronics. Although the European industry has lost its leading position in terms of more visible products for consumers such as clothing and computers, it maintains a number of goods which are not so common in everyday life, but which model rather a global economy. Industrial soft power reflects the fundamental role of the EU in defining international industry standards, such as communication standards [11].

The European Union has an important positive experience of eco-innovative activity for Ukraine. Globalization has also increased risks, whether financial, technological or industrial, and businesses increasingly need to use financial and insurance services to protect themselves. Companies are the main consumers of complex financial products: since they focus on production, they transfer responsibility for the management of various accounting risks to insurance and financial institutions. The increasing complexity of the technologies used, calling on extremely diverse skills, has placed eco-innovations at the heart of industrial dynamism. The EU has a major role to play in the development of this new industry by creating meeting points between applied research, varieties of technology, and productive industries. There are examples, such as the CEA of Grenoble which has successfully integrated an innovative ecosystem around nanotechnologies. Thus, the European Centers for Innovation and Industry (ECII) could be created in whole Europe, in particular in Ukraine [7].

4 Conclusions

The approaches to eco-innovation policy differ and depend on the reasons for small and medium enterprises on the global market. The business owners, therefore, try to keep control in their hands in order to fix an unusually high remuneration. This results in a lack of funding for innovation projects due to reduced cash flow and a lack of intention to spend equity. Geographic and cultural proximity is, therefore, an important asset, but the heterogeneity of ways of thinking and working also plays an important role in the development of innovative solutions in a complex universe. From this point of view, Europe, more than the United States and China, is one step ahead of the debate thanks to its long tradition and heterogeneity. Ukraine aspires to Europe, to become part of it. Therefore, Ukraine should bring its standards in the field of eco-innovations to European ones. All this requires international cooperation. After all, regions are forming a state innovation strategy. Therefore, we surveyed the key indicators of innovation activity on the regional level. As shows the analysis in this investigation, the quantity of eco-innovatively active industrial regional enterprises has decreased three times (to nearly 29%). The same situation is in the processing industry and food industry. The whole regional industry has lost its positions by the efficiency of innovative management. This requires developing a package of new directions to stimulate eco-innovative activity in regions of Ukraine. The main of measures have been the following:

1. The European Union continues to be a major industrial player, representing 16% of world exports and 17.3% of imports, which, excluding intra-community trade, makes the EU the world's leading market. Its trade relations are balanced overall,

this is only the case on an individual basis: the European Union has a trade deficit of € 169 billion with China and a surplus of € 73 billion with the USA. The growing complexity of the used eco-technologies has placed technological innovations at the centre of industrial dynamism. The EU has a major card to play in developing this new industry by creating meeting points between applied research, varieties of technologies and productive industries.

2. In order to stimulate the development of regional eco-innovations, Ukraine must now be participatory and ideally draw on the experience of cluster professionals, academics and decision makers. Evaluators should ensure that the views of various stakeholders, including those of companies, are acquired and codified. For Ukraine, it is necessary to combine qualitative and quantitative research methods in the eco-innovative economy, to review the cluster development statistics, complemented by a beneficiary survey, as well as interviews with beneficiaries and interest groups, which can be used to develop case studies in which the quality of cluster interactions in innovative activities would be examined.
3. The complexity of an impact assessment of group interventions in relation to methodological design and the search for economic tools for innovative activities should be reflected in a realistic budget and programs. Best international practice suggests that funding in the earliest (pre-seed and seed) stages of the innovative project should be in the form of grants or equity financing rather than repayable credit. But, as noted, the practice of innovation vouchers and grants and equity support in the European Union are still in an experimental phase.
4. A taxonomic economic-mathematical model for estimating the integrated indicator of the implementation of environmental innovations in the regional context was developed. The taxonomic indicator of the implementation of ecological innovations in Ukraine was determined by calculating additional indicators: the distance between observations and the reference vector, the average and maximum distances of deviations from the standard. The assessment was performed based on a taxonomic indicator, which is a synthetic quantity and takes into account the integral influence of the values of the set of indicators. As shown by the correlation-regression analysis, the determination coefficient is 0.99, is a close relationship between the number of scientific organizations personnel, extent of scientific and technical work financing, the extent of scientific and technical work of the enterprises and the indicator of implementation of environmental innovations.
5. The creation of European eco-innovation and industrial centers would generate the necessary conditions for cooperation across Europe in the field of the most important joint programs and the development of industry in peripheral areas. Expanding infrastructure and strengthening trade relations with the rest of the world would lead to better use of the linguistic and historical resources of some countries. All these measures will significantly improve the environmental situation in Ukraine. Most importantly, we need to develop an industrial policy, not as a ready-made formula, but as a tool that can adapt to any situation, sector, and country in a collaborative framework that is both decentralized and coordinated.

References

1. Aglietta, M.: Zone Euro, éclatement ou fédération. Michalon (2012)
2. Baldwin, M.: The Euro's trade effects. Working paper series ECB, (594) (2006)
3. Fedulova, L.I.: Conceptual model of innovation strategy of Ukraine. Econ. Forecast. **1**, 87–100 (2012)
4. Jamet, J.-F., Klossa, G.: Europe, la dernière chance? Armand Colin (2011)
5. Melnyk, M., Korcelli-Olejniczak, E., Chorna, N., Popadynets, N.: Development of regional IT clusters in Ukraine: institutional and investment dimensions. Econ. Annals-XXI **173**(9–10), 19–25 (2018)
6. Vasyltsiv, T., et al.: Economy's innovative technological competitiveness: decomposition, methodics of analysis and priorities of public policy. Manage. Sci. Lett. **10**(13), 3173–3182 (2020)
7. Yakymchuk, A., et al.: Public administration and economic aspects of Ukraine's nature conservation in comparison with Poland. In: Kantola, J., Nazir, S., Salminen, V. (eds) Advances in Human Factors, Business Management and Leadership. AHFE 2020. Advances in Intelligent Systems and Computing, vol. 1209. Springer, Cham (2020)
8. Irtyshcheva, I., et al.: Building favorable investment climate for economic development. Accounting **6**(5), 773–780 (2020)
9. Ciborowski, R., Skrodzka, I.: International technology transfer as innovation factor in EU countries (2019). https://www.inderscience.com/info/inarticle.php?artid=81708
10. State Statistics Service of Ukraine (2019). http://www.ukrstat.gov.ua
11. Mykytyn, T., Yakymchuk, A., Valyukh, A.: Management of protected areas of Ukraine's polissia: international experience. Prob. Perspect. Manage. **15**(1), 183–190 (2017)

Interactive Human-Computer Theoretical Model of Editorial Design with Augmented Reality

Carlos Borja-Galeas[1,2](✉) and Cesar Guevara[3]

[1] Facultad de Arquitectura, Artes y Diseño, Universidad Tecnológica Indoamérica, Ambato, Quito, Ecuador
carlosborja@uti.edu.ec

[2] Facultad de Diseño y Comunicación, Universidad de Palermo, Buenos Aires, Argentina

[3] Research Center of Mechatronics and Interactive Systems, Universidad Tecnológica Indoamérica, Ambato, Quito, Ecuador
cesarguevara@uti.edu.ec

Abstract. This research presents an editorial design model with augmented reality that concurrently integrates human-computer interaction with the printed picture book and the mobile application. The editorial proposal is supported by theoretical foundations of design related to composition, intelligent formats for its production, diagramming with and without a grid, typography, color theory, use of materials; In addition, it has three-dimensional figures based on paper engineering. The augmented reality suggested for this model must have images strategically located in the illustrated book that lead the perceiver to enter the mobile device and interact with its content, complementing in a playful way the tangible message of the printed document. The mobile application interacts with the illustrated book integrating: videos, audiobook, video games and links to social networks. For the validation of the editorial model with augmented reality, the mobile application and the human-computer interaction, user experience design metrics will be applied in which usability variables are measured such as: ease of learning, efficiency, memorization, effectiveness, satisfaction, among others. This study presents the theoretical foundations on which this editorial model is based and exposes how it was tested and developed with an educational theme in the city of Quito–Ecuador.

Keywords: Editorial design · Augmented reality · Human computer interaction · Illustrated book · Mobile application

1 Introduction

The production of printed books is based on various principles, methods and laws of composition. Throughout the centuries the use of diagrams, diagrams, methods such as those of Villard de Honnecourt, Paul Renner, Raúl Rosarivo, the Fibonacci sequence, the golden ratio, the system of inverted margins, the grid method, the deconstruction of the grid, to name a few, have been the basis for the design and layout of books in all

D. Russo et al. (Eds.): IHSI 2021, AISC 1322, pp. 580–585, 2021.
https://doi.org/10.1007/978-3-030-68017-6_85

their varieties. Technological advance has generated new ways of conceiving a book, where audiovisual and multimedia content are incorporated through links, triggers or bookmarks that link complementary virtual information.

This work presents a research proposal around an interactive human-computer theoretical model of editorial design with augmented reality (AR), which complies with some procedures and enhances the results in relation to a conventional book.

The augmented reality suggested for this model must have images strategically located in the illustrated book that lead the perceiver to enter the mobile device and interact with its content, complementing in a playful way the tangible message of the printed document. The mobile application interacts with the illustrated book integrating: videos, audiobook, video games and links to social networks. For the validation of the editorial model with augmented reality, the mobile application and the human-computer interaction, user experience design metrics will be applied in which usability variables are measured such as: ease of learning, efficiency, memorization, effectiveness, satisfaction, among others. For conceptualization, you must work with card-sorting, with closed categories. With the information architecture, 2D and 3D content must be organized, ordered and structured, and a structure diagram must also be used. This study presents the theoretical foundations on which this editorial model is based and exposes how it was tested and developed with an educational theme in the city of Quito - Ecuador, with children from 6 to 8 years old from two educational units, who along with teachers and parents, the participatory design generated the editorial prototype with augmented reality and the application that will be tested in a next stage. In the future it is proposed to use the eye tracking technique, in this way the publishing model with augmented reality will be tested which will serve as the basis for further complementary studies on editorial design with digital technological tools.

The article was organized as follows: Sect. 2 details the work done previously where the strengths are presented and compared with our proposal. Section 3 details the objective of the research, the scientific contribution, and presents the methods and materials used for the construction of the editorial model. Section 4 will present the development of the proposal. Section 5 presents the results of the proposed model. Finally, Sect. 6 presents the conclusions and future works of the research.

2 Related Works

In the book "point and line on the plane", Kandinsky exposes the division of the four zones in a composition, which can respond to principles such as the Fibonacci sequence, the golden ratio and the DIN standard of paper [1]. Editorial Design is in charge of exposing the content with harmony and coherence.

Indeed, the illustrated books with RA for preschool children used in Yilmaz's work [2], found that the majority of children reported feeling very happy with the activity and achieving a high level of reading comprehension. Furthermore, they are perceived as magical and more enjoyable than conventional books, therefore they are attractive and suggestive books.

Likewise, the educational book that is colorizable and with three-dimensional content generated by the user is presented in the study by A. Clark and A. Dünser [3], in which he investigates the metaphor of "emerging book", where the participation of children drawing and coloring is used as information to generate and change the appearance of the book. This system is based on natural feature tracking and image processing techniques that can be easily exploited for other AR publishing applications.

Vanderschantz [5] states that RA could have a significant impact on education, as it has the potential to improve children's learning through motivational mechanisms, intellectual interactions, and increased content recall. of the book through the addition of animation and joy. To develop books with AR, he recommends combining playful and learning interactions, including improvements in the version with AR, as this will increase the child's curiosity. He also recommends using multiple levels of camera focus as it encourages exploration. And finally, encourage shared reading with brothers and parents. On the other hand, the context and the focal distance between a real environment and one with AR studied by Gabbard, Gupta and Swan [4], establish that these elements present important problems in the design of the AR user interface.

Researchers have recognized RA as a technology with great potential to impact cognitive and affective learning outcomes. However, very little has been done to corroborate these claims. The work of Ibáñez, Di Serio, Villarán and Delgado Kloos [6] present a study to evaluate a learning application. The study suggests that RA can be exploited as an effective learning environment for learning the basic principles of electromagnetism in high school, provided that learning designers strike a careful balance between AR support and task difficulty.

3 Methods and Materials

A. Objective and contribution of the Research.

The main objective of this research is to develop an interactive human-computer theoretical model of editorial design with augmented reality. On the other hand, the contributions of the research are:

- Analysis of the interaction between the printed editorial model and the mobile application with augmented reality.
- Analysis of human interaction editorial model
- Analysis of human interaction mobile application with AR
- Analysis of human interaction - editorial model, mobile application with AR.

The following section describes the methods and materials used for the development of the interactive editorial model.

B. Methods

For development, the following was used: (i) The foundations of design, linked to informing, inquiring, persuading and entertaining [7]; (ii) the foundations of typography, based on the art of designing or choosing typefaces and organizing their placement on page [7, 8]; (iii); (iii); (iii) the Gestalt principles, used to simplify and structure your visual environment. [9]; (iv) the principles of color psychology necessary to understand what feelings the color palette causes in its perceivers [10]; (vi) the artistic trends to be taken into account when deciding on the graphic style that the editorial product will have [11]; (vii) interfaces for the systematization of the concept understood as a node that articulates the exchange [12].

C. Materials

The research was carried out in two Educational Units in the north of the city of Quito, with children from 6 to 8 years old. The chosen theme was recycling, and in the face of this social problem, the document script, illustrations, pop-ups, markers for RA, mobile application, 3D modeling and animations for RA were developed.

4 Development of the Proposal

This section describes the methods and materials that will be used in this investigation. This proposal has the following analysis categories: the editorial design model, the mobile application with augmented reality and the user experience design, which are explained individually and in correspondence with each other, as can be seen in Fig. 1a, 1b and Fig. 2.

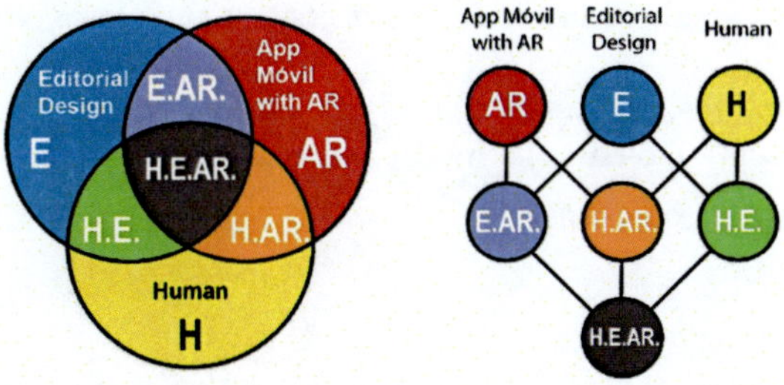

Fig. 1. a – b Human-computer interaction with the printed picture book and the mobile application

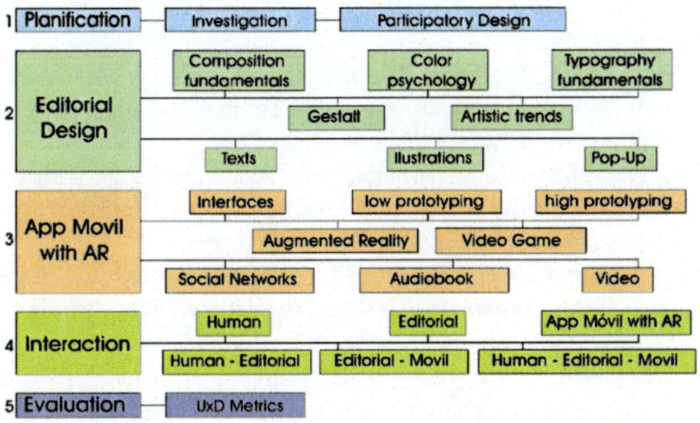

Fig. 2. Human-mobile interaction diagram

5 Results

The editorial model developed for the illustrated book is the result of the integration of the human-computer interaction process, which allows identifying the user's needs. In the editorial composition, scale, rhythm, color, hierarchies, grids, among others, are refined. The two-dimensional design theme - conceptual, visual and relationship elements - is harmonized, nuanced with the principles of form and the psychology of color on each page. The use, construction and reticular deconstruction harmonizes each part of the editorial content that is complemented by the triggers of the mobile application and the AR, reinforcing the understanding of the content in a playful way.

6 Conclusions and Future Work

The development of the editorial design model must comply with all the indicated processes to obtain the expected results. It is a priority to deeply know the perceptual and compositional morphological principles in order to obtain compositions in accordance with the theoretical foundations of design. Knowing the exact context in which the model will be applied to solve a need thanks to participatory design, will make physical and digital products harmonize content and create ownership in the end user.

This project took RA as the unit of analysis, but subsequent studies can take other technologies and apply the editorial model and analyze the metrics obtained.

References

1. Kandinsky, V.: Punto y línea sobre el plano. Contribución al análisis de los elementos pictóricos (2003)
2. Yilmaz, R.M., Kucuk, S., Goktas, Y.: Are augmented reality picture books magic or real for preschool children aged five to six? Br. J. Educ. Technol. **48**(3), 824–841 (2017). https://doi.org/10.1111/bjet.12452
3. Clark, A., Dünser, A.: An interactive augmented reality coloring book. In: IEEE Symposium on 3D User Interfaces 2012, 3DUI 2012-Proceedings, pp. 7–10 (2012). https://doi.org/10.1109/3dui.2012.6184168
4. Gabbard, J., Gupta Mehra, D., Swan, J.E.: Effects of AR display context switching and focal distance switching on human performance. IEEE Trans. Vis. Comput. Graph. **2626**(c), 1–14 (2018). https://doi.org/10.1109/tvcg.2018.2832633
5. Vanderschantz, N., Hinze, A., Al-hashami, A., Vanderschantz, N., Hinze, A., Al-hashami, A.: Using Augmented Reality to Enhance Children`s Books Using Augmented Reality to Enhance Children's Books, September 2018
6. Ibáñez, M.B., di Serio, A., Villarán, D., Delgado Kloos, C.: Experimenting with electromagnetism using augmented reality: impact on flow student experience and educational effectiveness. Comput. Educ. **71**, 1–13 (2014). https://doi.org/10.1016/j.compedu.2013.09.004
7. Ellen Lupton, J.C.P.: Diseño gráfico: nuevos fundamentos. Barcelona, Editorial Gustavo Gili, 264AD

8. Wong, W.: Fundamentos del Diseño Bi y Tridimensional. Barcelona
9. Todorovic, D.: Gestalt principles, Scholarpedia (2008). https://doi.org/10.4249/scholarpedia.5345
10. Heller, E.: Psicología del color Cómo actúan los colores sobre los sentimientos y la razón. Editorial Gustavo Gili, Barcelona (2004)
11. Kane, J.: Manual de tipografía. Editorial Gustavo Gili, Barcelona (2012)
12. Rodriguez Valero, D.: Manual de tipografía digital. Valencia, Campgrafic Editors (2016)
13. Bringhurst, R.: Los elementos del estilo tipográfico, Cuarta. Fondo de Cultura Económica, Buenos Aires (2014)
14. Müller-Brockmann, J.: Sistemas de retículas. Editorial Gustavo Gili, Un manual para diseñadores gráficos. Barcelona (2012)
15. Elam, K.: Sistemas Reticulares. Editorial Gustavo Gili, Barcelona (2006)
16. Ambrose, P.: Reticulas. Parramón (2008)
17. Tondeau, B.: Layout Essentials: 100 Design Principles for Using Grids (Design Essentials). Rockport Publishers (2011)
18. Samara, T.: Diseñar con y sin retícula, 3era ed. Barcelona (2006)
19. Purvis, A.W.: Historia del Diseño Gráfico. RM Verlag (2015)

Benefit of Developing Assistive Technology for Writing

Galo Molina-Vargas[1], Hugo Arias-Flores[1,2(✉)], and Janio Jadán-Guerrero[1,2]

[1] Maestría en Educación, Mención Pedagogía en Entornos Digitales (MEPED), Universidad Tecnológica Indoamérica, Bolívar 2035 y Guayaquil, Ambato, Ecuador
gmolina8@indoamerica.edu.ec, {hugoarias,janiojadan}@uti.edu.ec

[2] Centro de Investigación en Mecatrónica y Sistemas Interactivos (MIST), Universidad Tecnológica Indoamérica, Av. Machala y Sabanilla, Ambato, Quito, Ecuador

Abstract. The language and literature area within educational establishments takes great relevance when expressing thoughts through symbols that humans can decode. As such, the present study focuses on secondary language (called written language) and its components such as the all-important graphemes that are the minimum representations of expression. The writing process is one of the most important skills humans develop in the course of student and daily life; however, in the assimilation and improvement process, problems (disorders) may arise, which are responsible for distorting message transmission and understanding. However, technology makes it possible to interact in the educational space, integrating support dynamics to the student body. In this context, the objective of this research is to establish virtual learning objects as a supportive didactic strategy for improvement of writing disorders in children with special educational needs. The research focus is qualitative–quantitative, in which fifteen language and literature teachers participated via online survey. The most important results showed the scarce use of digital tools in the educational process and the minimal interest in providing attention to students with special educational needs. A prototype website has been developed, in which virtual objects are housed to support students with writing problems between the ages of 4 to 7 years old. In addition, work on a future pilot is currently underway.

Keywords: Secondary language · Written language · Student life · Assistive technology

1 Introduction

Dysgraphia is one of the most common learning disabilities, which is described as a disorder manifesting itself as difficulty learning to write despite a teacher's incessant classroom instruction. Worldwide, the affected population is between 7 and 10% [1]. Although dysgraphia should not be considered a physical disorder related to visible disabilities, the issue represents a problem in cognitive development and a challenge for

D. Russo et al. (Eds.): IHSI 2021, AISC 1322, pp. 586–590, 2021.
https://doi.org/10.1007/978-3-030-68017-6_86

teachers in the attention to and satisfaction of this group's educational needs [2]. Hence, this study's research question is as follows: How can educators promote the use of virtual learning objects (VLOs) as a didactic strategy to support the cognitive development of students with writing disorders?

Many boys and girls have learning difficulties due to disorders caused by dysgraphia, generating academic performance complications and even affecting their socio-communicational life. Children with learning disabilities necessarily require early intervention, adjustments in the teaching planning process, and efficient and long-lasting compensatory strategies. In addition, they need familiarity with words and more opportunities for practice and review to continue the normal learning process. It is necessary to evaluate and introduce these special groups to new support tools that offer improvements after participation in sessions intentionally programmed by facilitators through computerized instructions [3].

Indeed, technological advances have made it possible to develop intelligent learning environments that must adopt pedagogical approaches to enhance student capacities. The learning process becomes increasingly interactive as more smart education concepts are introduced [4]. This automation should be aligned with development of an inclusive education, covering poor communication channels and content prohibiting the cognitive development of all students.

Notably, the guidelines for designing inclusive learning environments are based on methodology and pedagogy, considering that websites are one of the most used online learning elements [5], aiming to explore the experience of students using different means to practice literacy using multimedia [6] and stimulating student interest [7].

The accessibility guidelines of websites with VLOs related to language and literature are focused on reading and writing, paying greater attention to their intangible design and decreasing attention to other aspects. This emphasizes that the technology element is necessarily linked to software, which allows a synchronous or asynchronous connection [8]. Thus, teaching materials such as VLOs allow children with special education needs to explore spaces that arouse their interest in an unconventional way [9]. On the other hand, educational virtual assistants are underused with basic functions, without considering the benefits offered [10]. The objective is that these virtual assistants can be used as educational tools contributing to the teaching-learning process of children with special educational needs [11].

Therefore, the combination between technology and education achieves two purposes: web expansion in unexpected areas and the support of teaching in unconventional spaces. Technologies offer great benefits that deserve to be maximized, with special emphasis on taking advantage of priority groups' attention with special educational needs. This study's objective is to identify aspects related to VLOs as support elements in writing disorders in students with special educational needs.

2 Methodology

This research is a qualitative–quantitative study, with a sample of teachers (level of basic general education) selected for convenience. This study was carried out during the month of July 2020.

The participants included 15 teachers aged 20 to 60 years old, belonging to the Pataín community of the Cotopaxi province in Ecuador. Of the 15 teachers, 12 were women (70%) and 3 were men (30%). The selected participants teach language and literature classes with full-time contracts. The teachers have an average of 13.4 years of teaching experience.

3 Results

The opinions of teachers in relation to VLOs as support in education were analyzed. Of all teachers, 86.66% pointed out that VLOs are part of educational innovation. In addition, 73.33% of teachers considered VLOs useful in education. The level of knowledge of these VLO resources was 53.33%, and 80.00% of teachers agreed that VLOs can be an educational support.

Writing problems detected in students were analyzed by the teachers, and the results are shown in Table 1.

Table 1. Detected problems

Detected	Frequency	%
Difficulty separating words	1	6.67
High frequency of spelling errors	1	6.67
Omission of letters	3	20.00
Letter substitution	1	6.67
Stroke of the grapheme	9	60.00
Total		100.00

Finally, the web prototype designed to support grapheme tracing is presented in Fig. 1. Once the design of the VLOs considered for website creation has been finalized, the title and a brief teacher introduction will be created about what teachers will find in the digital space. There will be two access buttons to connect to website sections (Fig. 2).

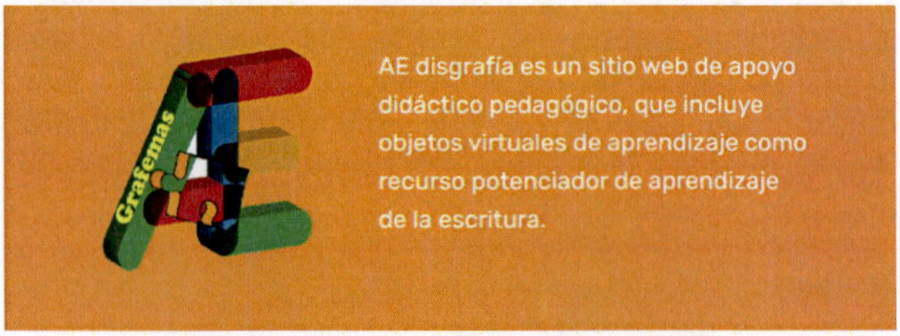

Fig. 1. Image and website introduction.

Fig. 2. Website exploratory menu.

The second part of The website, includes the VLOs distributed as vowels (Fig. 3), which contains access buttons linked to each one of the explanatory videos.

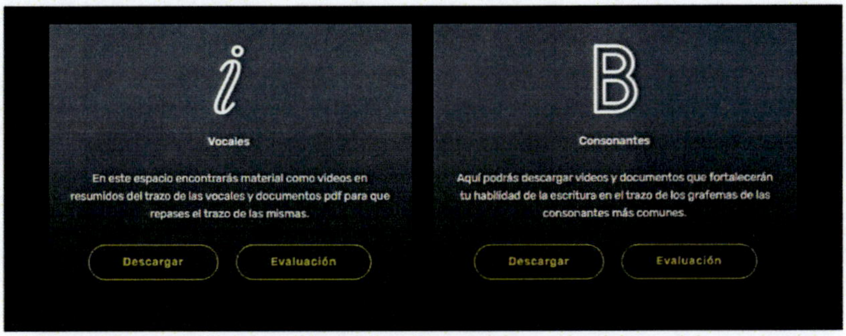

Fig. 3. Sections of Graphemes i Vowels.

4 Conclusions

This research analyzed the writing problems detected in the student body by teachers and their opinions regarding implementation of VLOs in education for student support.

A prototype website has been developed to host the VLOs, considering the teacher observations. The instructional development model is based on the addie model (analysis, design, development, implementation, and evaluation). In addition, work on a future pilot is currently underway.

References

1. Benmarrakchi, F., El Kafi, J., Elhore, A., et al.: Exploring the use of the ICT in supporting dyslexic students' preferred learning styles: a preliminary evaluation. Educ. Inf. Technol. **22**, 2939–2957 (2017). https://doi.org/10.1007/s10639-016-9551-4
2. El Kah, A., Lakhouaja, A.: Developing effective educative games for Arabic children primarily dyslexics. Educ. Inf. Technol. **23**, 2911–2930 (2018). https://doi.org/10.1007/s10639-018-9750-2
3. Thompson, R., Tanimoto, S., Lyman, R.D., et al.: Effective instruction for persisting dyslexia in upper grades: adding hope stories and computer coding to explicit literacy instruction. Educ. Inf. Technol. **23**, 1043–1068 (2018). https://doi.org/10.1007/s10639-017-9647-5

4. Rutkauskiene, D., Gudoniene, D., Maskeliunas, R., Blazauskas, T.: The gamification model for e-learning participants engagement. In: Uskov, V., Howlett, R., Jain, L. (eds) Smart Education and e-Learning 2016. Smart Innovation, Systems and Technologies, vol. 59. Springer, Cham (2016). https://doi.org/10.1007/978-3-319-39690-3_26
5. Takagi, H., Asakawa, C.: New challenges in web accessibility. Univ. Access Inf. Soc. **16**, 1–2 (2017). https://doi.org/10.1007/s10209-015-0436-x
6. Chen, C.J., Keong, M.W.Y.: Affording inclusive dyslexia-friendly online text reading. Univ. Access Inf. Soc. **16**, 951–965 (2017). https://doi.org/10.1007/s10209-016-0501-0
7. Stanisavljevic, Z., Nikolic, B., Tartalja, I., et al.: A classification of eLearning tools based on the applied multimedia. Multimedia Tools Appl. **74**, 3843–3880 (2015). https://doi.org/10.1007/s11042-013-1802-4
8. Pang, L., Jen, C.C.: Inclusive dyslexia-friendly collaborative online learning environment: Malaysia case study. Educ. Inf. Technol. **23**, 1023–1042 (2018). https://doi.org/10.1007/s10639-017-9652-8
9. Cakir, R., Korkmaz, O.: The effectiveness of augmented reality environments on individuals with special education needs. Educ. Inf. Technol. **24**, 1631–1659 (2019). https://doi.org/10.1007/s10639-018-9848-6
10. Cóndor-Herrera, O., Jadán-Guerrero, J., Ramos-Galarza, C.: Virtual assistants and its implementation in the teaching-learning process. In: Karwowski, W., Ahram, T., Etinger, D., Tanković, N., Taiar, R. (eds) Human Systems Engineering and Design III. IHSED 2020. Advances in Intelligent Systems and Computing, vol. 1269. Springer, Cham (2021). https://doi.org/10.1007/978-3-030-58282-1_33
11. Palomares-Pecho, J.M., Silva-Calpa, G.F.M., Raposo, A.B.: End-user adaptable technologies for rehabilitation: a systematic literature review. Univ. Access Inf. Soc. (2020). https://doi.org/10.1007/s10209-020-00720-z

Characteristics of Lower Limb Position Perception in Response to Environmental Information in Individuals with Low Vision

Tadashi Uno(✉)

National Institute of Technology Tokuyama College,
Yamaguchi 745-8585, Japan
t-uno@tokuyama.ac.jp

Abstract. This study aimed to understand the influence of various information acquisition strategies on foot proprioceptive and obstacle avoidance patterns among individuals with pigmentary retinal degeneration. Ten adult males (age: 39.2 ± 7.9 years) with pigmentary retinal degeneration were recruited. Participants obtained obstacle information (height of obstacle: 4 cm, 12 cm, 20 cm) using four information acquisition strategies, namely: visual: straight view in front of the block (condition A), verbal: information about the height of the block (condition B), tactile: identifying the block by hand with no time limit (condition C), and a combination of verbal and tactile stimuli (condition D). They were further evaluated by the following tasks: task 1, participants reproduced the height of the obstacles by lifting their foot in a static standing posture after they recognized the height of the obstacles; task 2, After the participants acquired obstacle information in conditions A to D, they performed an obstacle step-over from a free position from where it was possible to finish the task with one step. In task 1, under conditions A and B, they showed higher toe rise and higher coefficient of variance in toe rise compared to that under conditions C and D. For task 2, the highest points of the leading and trailing feet while stepping over the obstacle were significantly higher in condition A. Our results showed that the tactile method was useful to improve the perception of the participants caused by the lack of visual information about the obstacle.

Keywords: Pigmentary Retinal Degeneration · Proprioceptive · Environmental information

1 Introduction

Many reports predict that the total number of people with visual difficulties will increase in a highly aging society [1, 2]. People with low vision face major challenges during independent travel from place to place [3–5]. Several reasons contribute to this mobility issue such as the lack of information on the surrounding environment, orientation to the environment, and lack of information about barriers. Consequently, people with low vision employ low-vision aids for orientation and mobility to address the lack of visual information, as well as to optimize the use of their remaining visual power [5]. In addition, a multisensory approach such as tactile and auditory cues

D. Russo et al. (Eds.): IHSI 2021, AISC 1322, pp. 591–596, 2021.
https://doi.org/10.1007/978-3-030-68017-6_87

provides an effective support for people with visual difficulties. A combination of various sensory information such as tactile and visual information is useful in obstacle identification, as integration of the acquired information establishes a higher level of visual perception [6].

Previous studies have reported that tactile information obtained from the hand and feet are effective in guiding movement [7, 8]. However, the effect of simultaneous acquisition of tactile and verbal information of obstacles on the foot perception in individuals with low vision is not obvious. Therefore, this study aimed to understand the influence of various information acquisition strategies on foot proprioceptive and obstacle avoidance patterns among individuals with pigmentary retinal degeneration.

2 Methods

This study was approved by the Ethics Committee of the National Institute of Technology, Tokuyama College. Ten adult males (age: 39.2 ± 7.9 years) with pigmentary retinal degeneration were recruited for the study. Pigmentary retinal degeneration is a hereditary eye disease wherein the retinal rods are affected. All participants were certified to have grade 2 disability by the Japan Ministry of Health, Labor and Welfare (vision: 0.02–0.04, field of view: 10° and loss rate: 95%) (Table 1).

Table 1. Characteristics of participants with disabilities

ID	Age	Vision(R)	Vision(L)	Range	Age of onset
1	27	0.02	0.04	Peripheral	25
2	30	0.08	0.06	Peripheral	28
3	33	0.04	0.01	Peripheral	unknown
4	35	0.1	0.02	Peripheral	23
5	39	0.3	0.2	Peripheral	31
6	40	0.1	0.2	Peripheral	28
7	43	0.02	0.08	Peripheral	21
8	43	0.08	0.1	Peripheral	19
9	48	0.02	0.08	Peripheral	30
10	53	0.03	0.06	Peripheral	28

Participants acquired the obstacle information (height of the obstacle: 4 cm, 12 cm, 20 cm) using four information acquisition strategies (Fig. 1): Downward: viewing the obstacle at 20 cm distance from the toe (condition A), verbal: information about the height of the block (condition B), tactile: identifying the block by hand with no time limit (condition C), and a combination of verbal and tactile stimuli (condition D). The illumination in the experimental space was adjusted to 500 lx, and an obstacle made of yellow sponge was placed on a dark beige floor.

Subsequently, the participants performed two different tasks: Task 1: participants reproduced the height of the obstacles by lifting their foot in a static standing posture (5 times for the right leg) after they recognized the height of the obstacles. Task 2: after

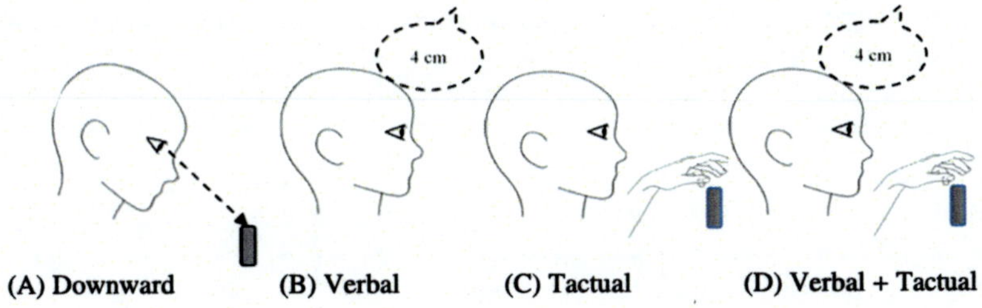

Fig. 1. Acquisition strategy of obstacle information

the participants acquired the obstacle information in conditions A to D, they performed an obstacle step-over from a free position where it was possible to finish the task with one step (5 times for the right leg). For tasks 1 and 2, the lower limb movement for each condition was recorded using six high-speed cameras at 250 frames per second. To measure the movement, 12 spherical reflective markers (14 mm in diameter) were attached to the subject's lower extremities using double-sided tape. Gait parameters (lift height of the toe and step length) and characteristics of the foot trajectory (highest points of the leading and trailing feet while stepping over the obstacle) were analyzed (Fig. 2). The coefficient of effort (the ratio of the highest point of the leading foot to the height of the obstacle) was calculated.

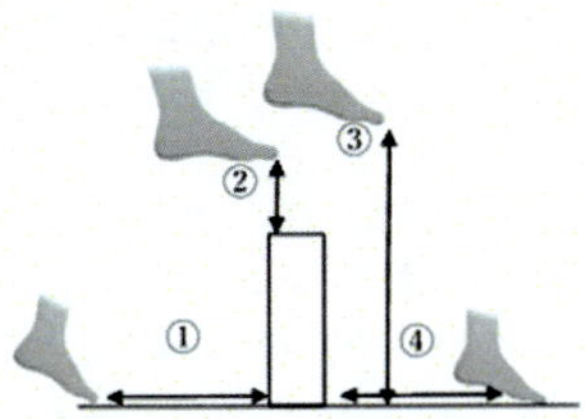

Fig. 2. Analysis point of step over obstacle. (1. Initiation of the step-over motion, 2. Highest point above obstacle, 3. Highest point of the whole trajectory, 4. Grounding of the heel.)

One-way analysis of variance was used to analyze the differences between each condition. Post-hoc pairwise Bonferroni-corrected comparison was used to examine the mean differences in each condition.

3 Results

In task 1, conditions A and B showed higher toe rise and coefficient of variance of toe rise ($p < 0.05$) compared to conditions C and D. For task 2, the highest points of the leading and trailing feet while stepping over the obstacle were significantly higher ($p < 0.05$) in condition A (Fig. 3). Additionally, with regard to the obstacle at 4 cm, the coefficient of variance in the downward condition and verbal condition was significantly larger than that in the tactual condition. The same trend was observed for the other obstacles [Fig. 4].

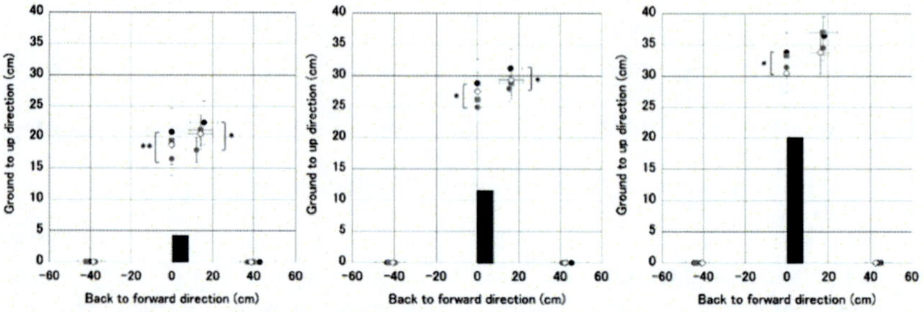

Fig. 3. Trajectory of the leading limb at different obstacle heights, **p < 0.01 *p < 0.05. (Black round marker: downward, Square marker: verbal, Gray round marker: tactual, Rhombus marker: verbal and tactual)

4 Discussion

Eye diseases are closely related to aging and it increases the possibility of falls and accidents. Among them, retinitis pigmentosa is an ocular disease with a high incidence in young individuals. Pigmentary retinal degeneration is a progressive disease that involves degeneration of rod photoreceptor cells in the eye, with a decrease in the peripheral visual field and reliance on central vision as the main visual characteristic feature.

In task 1, conditions A and B resulted in higher foot lifting, which indicates acquisition of poor and divergent environmental information. The divergence of the acquired information was reduced by tactile and verbal sensing under conditions C and D.

Previous studies have shown that acquisition via multiple sensory input may improve the accuracy of the perceived information [9, 10]. Based on the results of task 1, a combination of tactile and verbal information was useful in obstacle identification, as integration of the acquired information established a higher level of visual perception.

In task 2, participants performed better in the obstacle step-over with the tactual method compared to the other methods. On the other hand, downward condition increased the toe position during the obstacle straddling motion. In addition, with an obstacle at 4 cm, the coefficient of variance in the downward condition and verbal condition was significantly larger than that in the tactual condition. The same trend was observed for the other obstacles. Sensory input in the form of visual, vestibular, and somatosensory inputs are primarily responsible for human postural control [11], and these sensory inputs are integrated in the central nervous system and used for postural control [12]. Although there are individual differences in the weightage of sensory inputs to sensory areas [13], in general, when sensory inputs from vision are reduced due to ocular disease, body sway is amplified, and postural control becomes unstable [14, 15]. The subjects in the current study presented with severe visual field impairment, which made postural control associated with foot elevation difficult in the

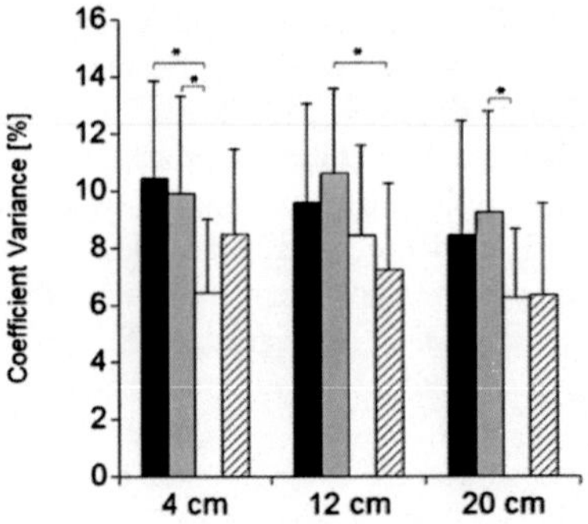

Fig. 4. Coefficient of Variance in the highest toe position over the obstacle of leading foot during five trials, *$p < 0.05$. (Black bar: downward, Gray bar: verbal, White bar: tactual, Shaded bar: Verbal and Tactual)

downward condition. As a result, the toe raising height was unstable. To avoid the risk of falling due to unstable foot lifting motion, a straddling motion was performed with the highest priority on safety, not on the efficiency of the movement.

These results suggest that different information acquisition strategies have an impact on the foot trajectory during obstacle step-over. A combination of various types of sensory information, such as tactile and visual information, is useful in obstacle identification since integration of the acquired information provides higher-level perception. Additionally, multiple sensory inputs may improve the accuracy of the perceived information. Our results show that the tactile method was useful to im-prove the participant's visual perception caused by the lack of visual information on the obstacle.

Acknowledgments. This work is supported by JSPS KAKENHI Grant Numbers 19K20001

References

1. Bourne, R.R.A., Flaxman, S.R., Braithwaite, T., Cicinelli, M.V., Das, A., Jonas, J.B., Keeffe, J., Kempen, J., Leasher, J., Limburg, H., Naidoo, K., Pesudovs, K., Resnikoff, S., Silvester, A., Stevens, G.A., Tahhan, N., Wong, T., Taylor, H.R., Ackland, P., Zheng, Y.: Magnitude, temporal trends, and projections of the global prevalence of blindness and distance and near vision impairment: a systematic review and meta-analysis. Lancet Glob. Health **5**(9), e888–e897 (2017)
2. Japan ophthalmologist's association.: The Social Cost of Visual Impairment in Japan. Ophthalmic Pract. **80**, 1–58 (2009)
3. Hassan, S.E., Lovie-Kitchin, J.E., Woods, R.L.: Vision and mobility performance of subjects with age-related macular degeneration. Optom. Vis. Sci. **79**(11), 697–707 (2002)
4. Sotimehin, A.E., Yonge, A.V., Mihailovic, A., West, S.K., Friedman, D.S., Gitlin, L.N., Ramulu, P.Y.: Locations, circumstances, and outcomes of falls in patients with glaucoma. Am. J. Ophthalmol. **192**, 131–141 (2018)

5. Uno, T., Loh, P.Y., Muraki, S.: The influence of information acquisition strategies on foot proprioception and obstacle avoidance pattern in people with low vision. In: Advances in Intelligent Systems and Computing, vol. 819 (2019)
6. Newell, F.N.: Cross-modal object recognition. In: Galvert, G., Spence, C., Stein, B.E. (eds.) The Handbook of Multisensory Processes. Oxford University Press, Oxford (2004)
7. Patla, A.E., Davies, T.C., Niechwiej, E.: Obstacle avoidance during locomotion using haptic information in normally sighted humans. Exp. Brain Res. **155**, 173–185 (2004)
8. Forner-Cordero, A., Garcia, V.D., Rodrigues, S.T., Duysens, J.: Obstacle crossing differences between blind and blindfolded subjects after haptic exploration. J. Motor Behav. **48**(5), 468–478 (2016)
9. Yuichi, W.: Vision influences on tactile discrimination of grating orientation. Jpn. J. Psychol. **78**(3), 297–302 (2007)
10. Ernst, M.O., Bülthoff, H.H.: Merging the senses into a robust percept. Trends Cogn. Sci. **8** (4), 162–169 (2004)
11. Diener, H.C., Dichgans, J.: On the role of vestibular, visual and somatosensory information for dynamic postural control in humans. Progress Brain Res. **76**, 253–262 (1988)
12. Itaya, A.: Sensory and attitude control feedback system. J. Soc. Biomech. Jpn. **39**(4), 197–203 (2015)
13. Kluzik, J.A., Horak, F.B., Peterka, R.J.: Differences in preferred reference frames for postural orientation shown by after-effects of stance on an inclined surface. Exp. Brain Res. **162**(4), 474–489 (2005)
14. Peterka, R.: Sensorimotor integration in human postural control. J. Neurophysiol. **88**, 1097–1118 (2002)
15. Hwang, S., Agada, P., Kiemel, T., Jeka, J.J.: Dynamic reweighting of three modalities for sensor fusion. PLoS ONE **9**(1), 1–8 (2014)

Co-creative Social Media Features on Video Platforms, and Their Impact on Customer Relationships

Akane Matsumae[1(✉)] and Yumeng Zhang[2]

[1] Faculty of Design, Kyushu University,
4-9-1 Shiobaru, Minami, Fukuoka, Fukuoka, Japan
matsumae@design.kyushu-u.ac.jp
[2] Graduate School of Design, Kyushu University, 4-9-1 Shiobaru, Minami, Fukuoka, Fukuoka, Japan

Abstract. The rapid development of internet technology has allowed video platforms to implement co-creative social media functions (SMFs). This study examined the impact of these co-creative SMFs on user relationships by analyzing the relationship between their social characteristics and user acquisition and retention on typical video platforms. The authors conducted structured questionnaires and non-structured interviews to examine the relation between SMFs and user relationships. Based on the results of the questionnaires and interviews, the authors conducted a correlation analysis on the relation between types of SMFs, social characteristics, and user relationships. The results show that the social characteristics of *helping others* and *asking for help* are correlated with user acquisition, and the social characteristic of *information acquisition* is strongly correlated with both user acquisition and user retention on video platforms. The knowledge obtained from this study will contribute to building desirable user relationships on video platforms through social media functions.

Keywords: Social media features · Social characteristics · Customer relationships · Co-creative platforms

1 Introduction

With the advancement of internet technology, the social media function (hereafter SMF) of video sharing platforms (hereafter VSPs) has developed in a pluralistic fashion. YouTube, the world's largest VSP, bolstered its SMFs by introducing a community feature to rival Facebook in 2018. BiliBili, a popular VSP in China, has also introduced SMFs, including SNS, social network services, BBS, bulletin board system, and *danm*, interactive comment subtitles. Thus, various co-creative SMFs have been regularly introduced on VSPs in order to build stronger user relationships (hereafter URs) in an environment of fierce market competition.

This study analyzes the relationship between co-creative SMFs and URs in VSPs and aims to contribute to the development of desirable URs in VSPs through SMFs.

D. Russo et al. (Eds.): IHSI 2021, AISC 1322, pp. 597–602, 2021.
https://doi.org/10.1007/978-3-030-68017-6_88

2 Related Studies

Related studies on the correlations between SMFs and user acquisition (hereafter UR_{ac}), and between UR_{ac} and user retention (hereafter UR_{rt}), have been studied as follows. Papacharis indicated social characteristics (hereafter SCs) were related to user motivation to use the internet within the context of internet use motivation [1].

Studies on the impact of SMF on UR_{rt} are mostly conducted on a unit of service basis. Hanson et al. also showed the SC that a VSP could influence UR_{rt} in a study of YouTube users' socializing behavior [2]. Murakami et al.'s reported case study on educational support through social media, focusing on the correlation between SMF and UR, suggested that SMFs could affect the UR_{rt} [3]. Kashiwabara studied the relationship between user satisfaction and usage conditions on Twitter, and concluded that SMFs improve usage conditions [4]. Hui proposed to measure UR_{rt} with usage conditions (hereafter UR_{rt_uc}) [5]. Idota et al. suggested that SMFs have an effect on UR_{ac} [6]. Reichheld defined net-promotor score (hereafter NPS), the ratio of promoters to detractors, to quantify UR_{ac} [7], and Kitazawa proposed to measure UR_{ac} with usage tendency (hereafter UR_{ac_tn}) [8].

This study focuses on each SC in SMFs, not on the specific services, and investigates how each SC affects UR, UR_{ac} and UR_{rt}.

3 Research Subjects

This study compares and analyzes YouTube, the most popular VSP in the world, as well as BiliBili and iQIYI, the most popular VSPs in China. The authors examined a benchmark survey of the NPS for these three VSPs, and confirmed that BiliBili, YouTube and iQIYI all hold a high level of user recommendation level in this order, as shown in Table 1, where average NPS among major VSPs is −24.1.

Table 1. NPS of major VSP.

	BiliBili	YouTube	iQIYI
NPS	−20.8	−26.1	−29.8
Difference from average	+3.9	−1.4	−5.1

4 Methodology

4.1 Overview

In this study, a structured questionnaire was adopted to quantify the relationship between SMF and UR. First, the authors abstracted the SCs of SMFs from previous studies and conducted a structured questionnaire on UR, UR_{ac} and UR_{rt} from the perspective of the SCs. Non-structured interviews with users were conducted to qualify URs from various perspectives on user's satisfaction and SMFs of each VSP.

Based on the results of the structured questionnaires and non-structured interviews, the authors analyzed the correlation between SCs and URs in SMFs ascertained through analysis of variance.

4.2 Structured Questionnaire

A structured online questionnaire was conducted from January 2020 to May 2020 using social media and a BBS.

Participants. Among the VSPs studied, YouTube is not available in China, and BiliBili and iQIYI only provide Chinese services. Therefore, in this study, Chinese students who are studying or have studied abroad were called for this structured questionnaire, as they have experienced all the three VSPs, better enabling us to compare them.

Of the 495 participants, 356 mostly used BiliBili, 67 mostly used iQIYI, and 42 mostly used YouTube. The participants included 313 males and 182 females, and were generally young people between 18 and 26 years old.

Survey Items. The questionnaire was structured with following categories: respondent attributes, most used SMFs, user satisfaction with the SCs and common functions of SMFs, UR_{rt_uc} to evaluate UR_{rt}, and NPS and UR_{ac_tn} to evaluate UR_{ac}.

5 Results and Discussions

5.1 User Satisfaction for Social Characteristics

For each VSP, Table 2 shows the average of the results of a four-point questionnaire of user satisfaction for each of the SCs in the SMFs. An analysis of variance of these results shows that BiliBili, which had more SMFs, obtained higher ratings for *Helping others*, *Interacting with others in discussions*, *Getting help*, and *Getting information* than YouTube and iQIYI, considered with a statistically significant difference ($p < 0.05$).

Table 2. User satisfaction for SC.

	BiliBili	YouTube	iQIYI
Helping others	3.237	2.595	2.576
Interacting with others in discussions	3.127	2.690	2.667
Seeking a sense of belonging	2.701	2.452	2.591
Expressing oneself	3.167	3.048	2.636
Getting help	3.158	3.024	2.682
Getting information	2.669	2.500	2.530
Meeting new friends	2.754	2.500	2.545

5.2 Evaluation of User Acquisition

Table 3 shows the average usage tendency, UR_{ac_tn}, at each VSP under the three levels of evaluation of the impact of SMF on usage tendency. Generally, greater UR_{ac_tn} is observed in the order of BiliBili, iQIYI, and YouTube, considered with a statistically significant difference ($p < 0.05$).

Table 3. UR_{ac_tn} of the major VSPs.

	BiliBili	YouTube	iQIYI
Usage tendency	3.237	2.595	2.576

5.3 Evaluation of User Retention

Table 4 shows the average of the results of the four-point evaluation of UR_{rt_uc} from the perspectives of *Accumulated usage time*, *Frequency of use*, *Motivation of use* and *Daily usage time*. In general, BiliBili, YouTube, and iQIYI, in that order, show a higher UR_{rt_uc} trend, considered with a statistically significant difference ($p < 0.05$).

Table 4. UR_{rt_uc} of the major VSPs.

	BiliBili	YouTube	iQIYI
Accumulated usage time	3.828	3.524	3.348
Frequency of use	3.839	3.738	3.121
Motivation of use	2.138	1.905	1.561
Daily usage time	3.856	3.667	3.061

5.4 Correlation Between Social Characteristics and User Acquisition in SMF

Table 5 shows the results of the canonical correlation analysis of the association between user satisfaction and usage tendencies for the SCs of the major SMFs, BiliBili, YouTube, and iQIYI. For BiliBili and YouTube, user satisfaction and usage tendencies for all SCs are positively correlated; for iQIYI, usage tendency is positively correlated with three SCs (*Getting help*, *Getting information* and *Meeting new friends*), but uncorrelated with the other SCs.

The results of these analyses show all the SCs in the SMFs of BiliBili and YouTube have a positive impact on UR_{ac}. As for iQIYI, the SCs (in the SMFs) of *Getting help*, *Getting information*, and *Meeting new friends* have a positive impact on UR_{ac}, while *Helping others*, *Interacting with others in discussions*, *Seeking a sense of belonging*, and *Expressing oneself* do not. Table 4 shows the average of the results of the four-point evaluation of UR_{rt_uc} from the perspectives of *Accumulated usage time*, *Frequency of use*, *Motivation of use* and *Daily usage time*. In general, BiliBili, YouTube, and iQIYI, in that order, show a higher UR_{rt_uc} trend, considered with a statistically significant difference ($p < 0.05$).

Table 5. Canonical correlation analysis of SC and URac for the major VSPs.

			Social characteristics											
			Helping others		Interacting with others		Seeking a sense of belonging	Expressing oneself			Getting help	Getting information		Meeting new friends
User acquisition	BiliBili		0.123	0.209	0.231	0.162	0.180	0.190	0.173	0.189	0.134	0.104	0.104	0.163
		p	0.021	0.000	0.000	0.002	0.001	0.000	0.001	0.000	0.011	0.051	0.051	0.002
	YouTube		0.436	0.407	0.393	0.569	0.369	0.509	0.58	0.364	0.438	0.441	0.236	0.466
		p	0.004	0.008	0.010	0.000	0.016	0.001	0.000	0.018	0.004	0.003	0.133	0.002
	iQIYI		0.235	0.227	0.206	0.185	0.228	0.234	0.285	0.170	0.262	0.281	0.157	0.271
		p	0.055	0.065	0.094	0.133	0.063	0.057	0.019	0.168	0.032	0.021	0.204	0.027

5.5 Correlation Between Social Characteristics and User Retention in SMF

First, the results of the canonical correlation analysis of the association of user satisfaction with SCs and UR_{rt_uc} in SMFs for the major VSPs included in this study are shown in Table 6.

Table 6. Canonical correlation analysis of SC and UR_{rt_uc}.

			Social characteristics											
			Helping others		Interacting with others		Seeking a sense of belonging	Expressing oneself			Getting help	Getting information		Meeting new friends
Usage condition	Accumulated usage time		0.123	0.209	0.231	0.162	0.180	0.190	0.173	0.189	0.134	0.104	0.104	0.163
		p	0.021	0.000	0.000	0.002	0.001	0.000	0.001	0.000	0.011	0.051	0.051	0.002
	Frequency of use		0.436	0.407	0.393	0.569	0.369	0.509	0.58	0.364	0.438	0.441	0.236	0.466
		p	0.004	0.008	0.010	0.000	0.016	0.001	0.000	0.018	0.004	0.003	0.133	0.002
	Motivation of use		0.235	0.227	0.206	0.185	0.228	0.234	0.285	0.170	0.262	0.281	0.157	0.271
		p	0.055	0.065	0.094	0.133	0.063	0.057	0.019	0.168	0.032	0.021	0.204	0.027
	Daily usage time		0.123	0.209	0.231	0.162	0.180	0.190	0.173	0.189	0.134	0.104	0.104	0.163
		p	0.021	0.000	0.000	0.002	0.001	0.000	0.001	0.000	0.011	0.051	0.051	0.002

With the exception of *Seeking a sense of belonging*, user satisfaction and *Daily usage time* are positively correlated with SCs, considered with a statistically significant difference ($p < 0.05$). There is a positive correlation between *Helping others*, *Getting help*, *Getting information* and UR_{rt_uc}.

Moreover, for each VSP, the results of the correlation analysis between UR_{rt} and the recognized SMF's SCs show that usage conditions are correlated with SCs for BiliBili and iQIYI, while usage conditions on YouTube are not correlated with SCs.

6 Conclusions

Based on the results of the structured questionnaire, the authors analyzed the correlation between SCs in the SMFs of the major VSPs with UR, UR_{ac} and UR_{rt}. In general, *Getting help*, *Getting information*, and *Meeting new friends* among the SCs in SMFs have a positive impact on UR_{ac} for major VSPs. *Helping others*, *Getting help* and *Getting information* have a positive influence, especially on *Daily usage time* of the UR_{rt_uc} of the major VSPs.

Looking individually into BiliBili, YouTube, and iQIYI, the correlation between UR and SC of the SMFs differ for each VSP. For BiliBili, *Helping others*, *Interacting with others in discussions*, *Getting help*, and *Getting information* have a positive impact on UR_{ac}, while *Helping others*, *Interacting with others in discussions*, and *Getting help* have a positive impact on *Daily usage time* in UR_{rt_uc}. For YouTube, *Helping others*, *Interacting with others in discussions*, *Getting help*, and *Getting information* show a positive impact on UR_{ac}, but not on UR_{rt}. Finally, for iQIYI, *Getting help* and *Getting information* have a positive impact on UR_{ac}, but *Getting information* is relevant to UR_{rt}. In summary, among all the SCs, *Helping others* and *Getting help* are related to UR_{rt}, and *Getting information* is strongly related both to a VSP's UR_{ac} and UR_{rt}.

The quantitative evaluation of this study will be followed by a qualitative evaluation based on the results of the non-structured interviews, which will be integrated to analyze the relationships between the SCs of SMFs and URs in VSPs. The findings of this study should contribute to determining what kind of SCs can be implemented within the SMFs of VSPs to create the desired user relationships.

Acknowledgments. This work was supported by JSPS KAKENHI Grant Number JP20K20119.

References

1. Papacharissi, Z., Rubin, A.M.: Predictors of internet use. J. Broadcast. Electron. Media. **44**, 175–196 (2000)
2. Hanson, G., Haridakis, P.: YouTube users watching and sharing the news: a uses and gratifications approach. J. Electron. Publ. **11**, 1–11 (2008)
3. Murakami, M., Iwasaki, C.: Educational improvement using SNS at university. Jpn. Assoc. Educ. Media Study. **14**, 11–16 (2008)
4. Tsutomu, K.: The relationship between motives and frequency of Twitter use : an examination from uses and gratifications approach Author. Stud. Soc. Psychol. Educ. Inq. into humans Soc. 89–107 (2011)
5. Hui, S.K.: Understanding Gamer Retention in Social Games using Aggregate DAU and MAU data: A Bayesian Data Augmentation Approach (2013)
6. Idota, H., Bunno, T., Tsuji, M.: The effectiveness of social media for business activities in Japanese firms. Rev. Socionetwork Strateg. **11**, 33–45 (2017)
7. Reichheld, F.F.: The one number you need to grow. Harv. Bus. Rev, December 2–10 (2003)
8. Kitazawa, K., Namatame, T., Otake, K.: Identification of New Customer Retention Factors Using User Behavior on Golf Portal Sites, 611–612 (2020)

Technology, Materials and Inclusive Human Systems

Creating Embedded Haptic Waveguides in a 3D-Printed Surface to Improve Haptic Mediation for Surface-Based Interaction

Ahmed Farooq[1(✉)], Hong Z. Tan[2], and Roope Raisamo[1]

[1] Tampere Unit of Computer Human Interaction, Tampere University, Tampere, Finland
{Ahmed.Farooq,Roope.Raisamo}@tuni.fi

[2] Haptic Interface Research Lab, Purdue University, West Lafayette, Indiana, USA
Hong.Tan@purdue.edu

Abstract. Vibrotactile feedback is affected by the properties of the material and structure that transmit vibration from an actuator to the entire device surface. The stimuli felt by the skin may be distorted or attenuated at different locations of the device surface. Therefore, it is important to understand how the source vibration can be properly mediated or guided throughout the device to achieve globally uniform or localized vibrotactile feedback. This research (This work was funded by Business Finland, decision number 8004/31/2018.) evaluates three off-the-shelf waveguide materials and one custom designed 3D-printed ABS structure for creating localized and global vibrotactile signals. The three materials included Gorilla glass, Plexiglas, and aluminum. The 3D-printed waveguide used horizontal and vertical shafts that lowered its impedance load, thereby effectively mediating source vibration along one direction throughout its structure. Results indicate that, compared to the three off-the-shelf materials, the 3D-printed waveguide was more efficient at haptic mediation and creating localized effects using virtual exciter. Our findings support the use of this novel technique of utilizing calibrated 3D-printed waveguides to improve vibrotactile feedback in mobile and handheld devices.

Keywords: Vibrotactile feedback · Surface-based interaction · Haptic mediation · Virtual actuators · Constructive wave interference · Embedded haptic waveguides

1 Introduction

Until recently, while designing vibrotactile feedback more focus was placed on developing the ideal actuation components and specific feedback signals than how these signals should be relayed to the user. However, as put forward by Farooq [1], the mediation of signals created by the actuation component should also be considered in designing haptic feedback. Farooq asserts that "Haptic Mediation" is necessary because in most cases the placement of an actuator and the point of contact with the device are not co-located. Therefore, environmental noise and other internal and external device

D. Russo et al. (Eds.): IHSI 2021, AISC 1322, pp. 605–611, 2021.
https://doi.org/10.1007/978-3-030-68017-6_89

inefficiencies can drastically alter the signal delivered to the skin. A signal traveling from the source to various points on a device surface may be lowered in terms of magnitude and may be altered in terms of phase and frequency components due to the impedance of each intermediary component. Importantly, these factors may affect signal transmission or propagation of the "intended tactile stimuli" as it travels through different materials, each having different structures and physical properties [5].

To reduce signal distortion such as intensity attenuation and spectral degradation, it is possible to utilize haptic waveguides that can channel or mediate the source signals more efficiently. In fact, waveguides have been widely used in various audio-based devices for decades [2, 3]. Depending on the material properties as well as the structural design of the waveguide, applied signals may be enhanced to create more reliable and consistent global device actuation. In this paper we examine 3 + 1 (3 existing and 1 3D-printed) materials and possible design structures for waveguide efficiency in mobile device interaction by creating a device overlay for touchscreens using a Microsoft Surface Pro 4 tablet. The goal for this research is to improve previous techniques [4, 6, 7] of providing global and localized actuation for surface-based interaction.

2 Related Work

Vibrotactile feedback has been a popular method of providing haptic signals in various devices. [2, 6, 9]. In most of this research the generated signal is intended to propagate and mediate uniformly, distributing the vibration energy across the entire device. Previous studies have shown that it is possible to utilize both hard (solid) mediation [1] and soft (liquid/gel) mediation [5] to relay tactile signals within a mobile device. Testing indicates that hard mediation requires densely layered materials with specific elasticity. One study [5] found correlation between physical properties such as Youngs's Modulus ($Y < 70$ GPa) and material density ($\rho < 2.80$ g/cm^3) of the object and its efficiency to transfer vibration signals. Similarly, with regards to soft or liquid mediation, the same study showed that bulk modulus (>2.80 GPa) and density could play an important role. It was observed that lower density mediums were more efficient at relaying the intended signal. However, in either case the medium needs to be calibrated specifically for a certain frequency to avoid filtering or clipping of the intended signals.

Global device actuation is the most common technique of providing haptic signals especially in less efficient mediation elements. In most cases this type of actuation is confirmational in nature and does not require precisely calibrated signals. Essentially, a single high intensity signal can be generated throughout the entire device to create the global actuation. In contrast, providing localized haptic actuation can be complex, inefficient and may require an array of vibrotactile actuators embedded underneath the interaction surface. Previous research [4, 5, 8, 9] into "Haptic Mediation" shows that signals effectively mediated within a device can greatly reduce the energy needed to create vibrotactile feedback for both global and localized device actuation. Moreover, research into "Intelligent Haptic Mediation" [4] also demonstrates that calibrated opposing vibration signals from two or more actuators can be modulated to interact and create constructive wave interference or virtual exciters for localized actuation.

Constructive wave interference is a method of providing localized actuation signals using virtual actuators. They can be created by modulating the propagation of standing waves along the surface of actuation from two opposite sides. Coe et al., [4] found that it was possible to create discrete localized low-frequency actuation in between the two surface-mounted actuators by adjusting the independent high-frequency direct-signals from each actuator thereby, creating adjustable indirect actuation points. Virtual actuators can be an excellent tool for providing localized actuation on any interaction surface (i.e. touchscreen or smart surfaces), without the use of a dense actuator array. According to Coe et al. precisely calibrated virtual exciters also require less energy to create similar actuation thereby making the setup more affordable and more efficient, which is ideal for mobile device interaction. In this study, we extend this research to evaluate 3 + 1 different materials and structures as the surface of actuation to create virtual exciters for generating localized actuation more efficiently on solid surfaces. Our focus was on using haptic waveguides to evaluate possible improvements on two fronts: 1) their ability to mediate reliable entire-device actuation (global actuation through direct feedback), and 2) their efficiency at localized actuation by creating virtual exciters.

3 Using Waveguide as Solid Mediation

It is well established that vibration propagation is dependent on the material properties of the medium [3]. There is a close correlation between the material properties (ρ, Y) of the waveguide and the resonance frequency of the system [8]. It would be beneficial to adjust these properties and calibrate the optimum structure for a specific frequency. In most cases this calibration is not possible without changing the material of the waveguide. However, in 3D-printed structures made from Acrylonitrile Butadiene Styrene (ABS plastic), this can be done by varying its density or fill rate that affect its stiffness (Y = 1.79 −3.2 GPa). Therefore, any waveguide design can be developed by using composite materials to both mediate and isolate vibration signals as needed, to contain the applied wave energy from dissipating uncontrollably throughout the entire surface. To mediate haptic signals, we focused on creating a low impedance load to develop mediation within the specified waveguide. As a waveguide is essentially a medium where the wave propagation is bounded in two directions of space and free in the third one [8], it should be possible to create calibrated wave guides using a 3D printer.

Fig. 1. Multiple iterations of the 3D-printed waveguides.

4 Experiment Design

Two sets of technical measurements were taken to evaluate the difference among the 3 + 1 waveguides. To test the effectiveness of the different waveguides in creating virtual exciters, we replicated the setup created by Coe et al., [4]. Whereas, to gauge the overall signal mediation and energy transfer among the materials, we recorded the surface displacement at various distances from the actuation source. In addition, we conducted a user study to evaluate the perceived differences between two discrete points on the 3 + 1 overlays. We mounted all 3 + 1 waveguides to the touchscreen of a Microsoft Surface 4 tablet (Fig. 2). Unlike prior work with touchscreen overlay and constructive wave interference to create virtual exciters [4, 8], for the purpose of this study we did not focus solely on transparent materials. The dimensions (295 × 205 mm) and thickness (2 mm) of each waveguide were kept constant. The Gorilla glass 3 had a density (ρ) of 2.39 g/cm^3 and Young's Modulus (Y) of 69.3 GPa, whereas the 5052-H32 aluminum Sheet had a ρ of 2.68 g/cm^3 and the same Y of 69.3 GPa with a modulus of elasticity (E) of 70.3 GPa. The Plexiglass had a rated ρ of 1.18 g/cm^3 and E of 3.1 GPa. We also replicated the attachment mechanism illustrated by Coe et al., [4] where all the waveguides were attached to the Surface Pro by using a frame to ensure that wave propagation was bound in the two in-plan directions (x-axis and y-axis). For the vertical (z) axis we used a 1 mm silicone tip to create free movement in the third direction. Although most of tested surfaces were not transparent, this setup was used to try replicate the apparatus discussed by Coe et al. as closely as possible to ensure a meaningful comparison.

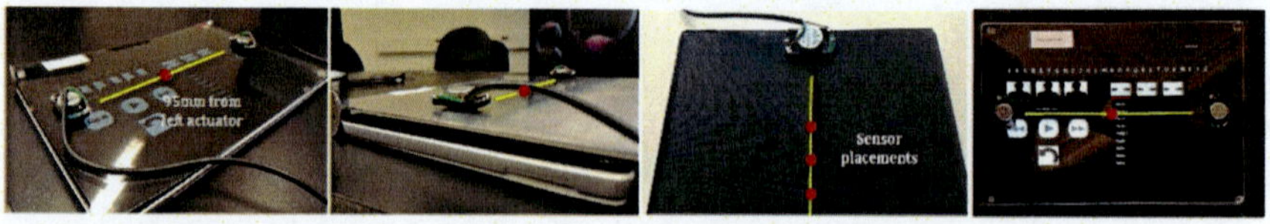

Fig. 2. Measuring wave propagation and virtual exciters within the 3 + 1 surfaces.

5 Result

Results from both sets of measurement show that the current version of the custom-designed 3D-printed waveguide was more efficient at haptic mediation and at creating localized actuation through constructive wave interference, as compared to the off-the-shelf Gorilla glass, Plexiglas and aluminum waveguides.

Looking at the results from the displacement sensor we see that it was possible to create Constructive Wave Interference (CWI) on all four surfaces. Figure 4: left, shows each surface with both the direct (from the physical actuator) and indirect (through virtual the exciter) feedback signals propagating through the four surfaces. The direct signal dissipates naturally the further we move away from the left Tectonic actuator and should be the weakest close to 95 mm away. Whereas the indirect feedback is the result of the interference from the direct signals between the two opposite Tectonic actuators. The interference maximum caused between the two Tectonic actuators occurs around 95 mm from the left actuator in all four surfaces, however their relative peaks vary depending on the surface's efficiency to propagate the direct signal as well as the effectiveness of the virtual exciter being created at that point.

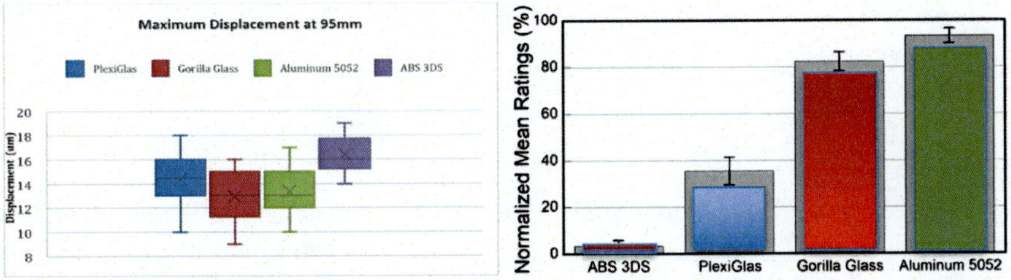

Fig. 3. Variance in displacement for CWI in um (left) and perceived differences between signals at points A and B (right) through constructive wave interference (virtual exciter) on the 3 + 1 waveguides

Data of the user study showed (Fig. 4 right) that the Gorilla glass and aluminum waveguides were least efficient and were described to have the most perceived variation between signals felt between Points A & B. On the other hand, Plexiglas was rated slightly better but most participants described the signal at both points to be either very similar or identical for the 3D-printed EHWs surface with one participant considering it to be the baseline (control) condition in the experiment.

Results from the wave propagation setup (Fig. 5) showed that at system resonance, each surface altered the applied signal by integrating it with parasitic vibrations. This was more evident for aluminum and Gorilla glass waveguides. The Plexiglas waveguides did not introduce much parasitic vibrations but did attenuate the signal to a much weaker state. Whereas the 3D-printed waveguide was noticeably better at maintaining the amplitude and the waveform of the applied signal throughout the measured frequency range.

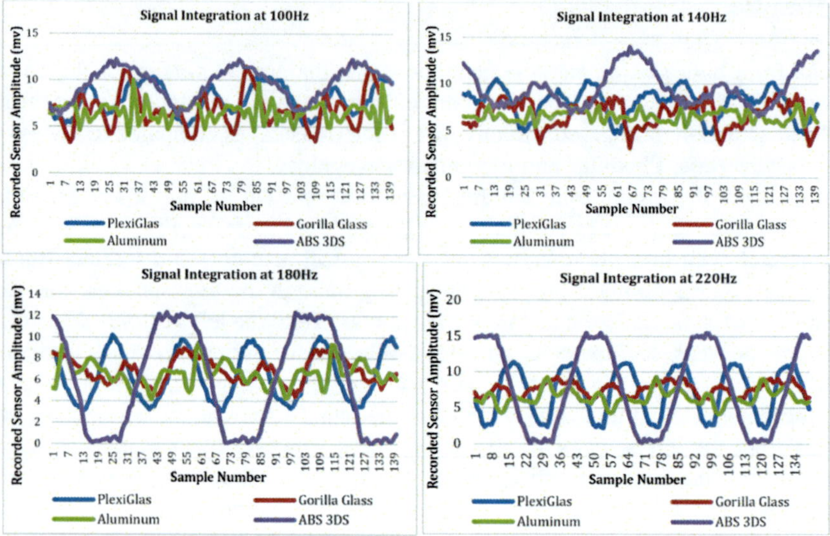

Fig. 4. 100 Hz (top left), 140 Hz (top right), 180 Hz (bottom left) and 220 Hz (bottom right) signals recorded at 95 mm from the source actuator to illustrate distortion/attenuation in the 4 surfaces

6 Conclusion

This research focused on evaluating three off-the-self waveguide materials and one custom designed 3D-printed structure for creating well controlled localized feedback (using virtual exciters) and global device actuation. The 3D-printed waveguide utilized specifically designed horizontal and vertical shafts embedded within its structure to lower its impedance load [6] and increase its efficiency at mediating actuation from the source to the point of contact. Moreover, by using a precise printing process and dissolvable support material (SUP706) along with the embedded design structure, it was possible to calibrate the physical properties (ρ and Y) of the 3D-printed waveguide for a given bandwidth of applied signals (100 Hz–200 Hz). Similar to the approach proposed by Dhiab and Hudin [8] of localizing calibrated actuation signals, the current research illustrates that the properties of the calibrated waveguides can affect the efficiency of vibration propagation, thereby affecting both global and localized actuation signals. Utilized this approach it is possible to improve existing techniques of providing global device actuation using hard and soft mediation, as well as localized actuation through virtual exciters created by constructive wave interference. The results from the present study indicate that custom waveguide can be useful in both application areas and should be further tested to improve the reliability and efficiency of vibrotactile feedback for surface-based interaction.

References

1. Farooq, A.: Developing technologies to provide haptic feedback for surface-based interaction in mobile devices, PhD Dissertation, University of Tampere, Faculty of Communication Sciences (2017). http://tampub.uta.fi/handle/10024/102318
2. Berdahl, E., Niemeyer, G., Smith, J.O.: Using haptic devices to interface directly with digital waveguide-based musical instruments. In: Proceedings of New Interfaces of Musical Expression (NIME09), Pittsburgh, PA, pp. 183–186, 3–6 June (2009)
3. Feynman, R.: The Feynman Lectures on Physics, New Millennium Edition, Chapter 24: Waveguides. https://www.feynmanlectures.caltech.edu/II_24.html. Accessed 2 Apr 2020
4. Coe, P., Farooq, A., Evreinov, G., Raisamo, R.: Generating virtual tactile exciter for HD haptics: a tectonic actuators case study. In: Proceeding of IEEE Sensors, Montreal, QC, Canada, October 2019, pp. 1–4 (2019). https://doi.org/10.1109/SENSORS43011.2019.895656
5. Farooq, A., Evreinov, G., Raisamo, R.: Evaluating different types of actuators for liquid screen overlays (LSO). In: Proceeding of IEEE Symposium on Design, Test, Integration and Packaging of MEMS/MOEMS (DTIP 2016), Budapest, Hungry, 97–102 (2016)
6. Visell Y, Giordano B. L, Millet G., Cooperstock J.: 2011 Vibration influences haptic perception of surface compliance during walking. PLoS One. **6**(3) e17697 March 2011. https://dx.doi.org/10.1371%2Fjournal.pone.0017697
7. Zhaoyuan, M., Edge, D., Findlater, L., Tan, H.Z.: Haptic keyclick feedback improves typing speed and reduces typing errors on a flat keyboard. In: 2015 IEEE World Haptics Conference (WHC), 22–26, pp. 220–227 (2015). https://doi.org/10.1109/WHC.2015.7177717
8. Dhiab, A.B., Hudin, C.: Confinement of vibrotactile stimuli in narrow plates - principle and effect of finger loading. In: Proceeding 2019 IEEE World Haptics Conference (WHC) 9–12, pp. 431–436, July 2019. https://doi.org/10.1109/WHC.2019.8816081
9. Basdogan, C., Giraud, F., Levesque, V., Choi, S.: A review of surface haptics: enabling tactile effects on touch surfaces. IEEE Trans. Haptics, (99), 1 (2020)

Preliminary Investigations on Subcutaneous Implantable Microchip Health and Security Risks

Mona A. Mohamed[1(✉)] and Beenish Chaudhry[2]

[1] Towson University, 8000 York Road, Towson, MD 21252, USA
mmohamed@towson.edu
[2] University of Louisiana at Lafayette, Lafayette, USA
beenish.chaudhry@louisiana.edu

Abstract. The main objective of this research project is to explore the behavioral aspects of the potential users of RSIM use with regard to health and security risks. It has been found that the importance of ethical reasons for implementing RSIM pattern increased by age. On the other hand, the social acceptance and the government requirements have no effects on the choice of the microchip implant. The study found that there is high likelihood of the subject accepting the microchip implant if it's useful and the subject of older age. Subjects believed that the implantable microchip will never be secure for the identity protection. Furthermore, the older subjects the more probability that the subject will use the implantable microchips, if they had ownership of their own data. In fact, it was found that usefulness of RIMS is correlated with ownership of the data.

Keywords: Health risk · Security risk · RFID · Implantable chip · Ethics · Acceptance · Privacy risk

1 Introduction

Microchip implant is an "integrated circuit devices encased in microchip or Radiofrequency identification (RFID) transponders that can be active or passive and are implantable into animals or humans usually in the subcutaneous layer of the skin" [1]. The current major uses of human implanted microchip or RFID subcutaneous implantable microchip (RSIM) include identification, tracking, access control, medical applications and electronic payments. In addition, RSIM can help in minimizing mistakes in personally identifiable information (PII) or personal medical information such as blood type, allergies, current medications and medical history etc. It is widely accepted that RSIM significantly contributed to the improvement of quality of life for persons with disabilities through recovery and improved functions [2]. There are unprecedented health benefits that RSIM can deliver to humans, but it also introduced its own shortcomings. On the whole, there is undoubtedly predominant health and security concerns that limit its usage. The health risks represented by introducing foreign-body-induced tumors and health risks perception associated with incorrect injection [3].

D. Russo et al. (Eds.): IHSI 2021, AISC 1322, pp. 612–618, 2021.
https://doi.org/10.1007/978-3-030-68017-6_90

RSIM may be exploited to invade the personal space using many of its own vulnerable applications. Medical or identification RSIM could be augmented with sensors that append extra overt or covert functionalities [4] i.e. function creep. Furthermore, versatile implantable microchips have the promise to interface with multiple appliances and equipment [5]. However, RSIM raises privacy risks perception and the likelihood of coercive implementation in individuals [6].

2 Methodology

This study is based on a survey carried out in the period between March and May 2020. The responses gathered from 226 subjects of whom 72 are males, 151 are females, 1 fluid and 2 prefer not to disclose their gender. The participants age range between 18 to 80 years representing 31 countries around the world. The survey items used a Likert scale anchored by one of five choices ranked as: 1 = Strongly disagree; 2 = Disagree; 3 = Neither Agree nor Disagree; 4 = Agree and 5 = Strongly agree.

3 Results and Analysis

The importance of ethical reasons for implementing RSIM pattern increased by age and it ranged from disagree to neutral as shown in Fig. 1 and Eq. 1.

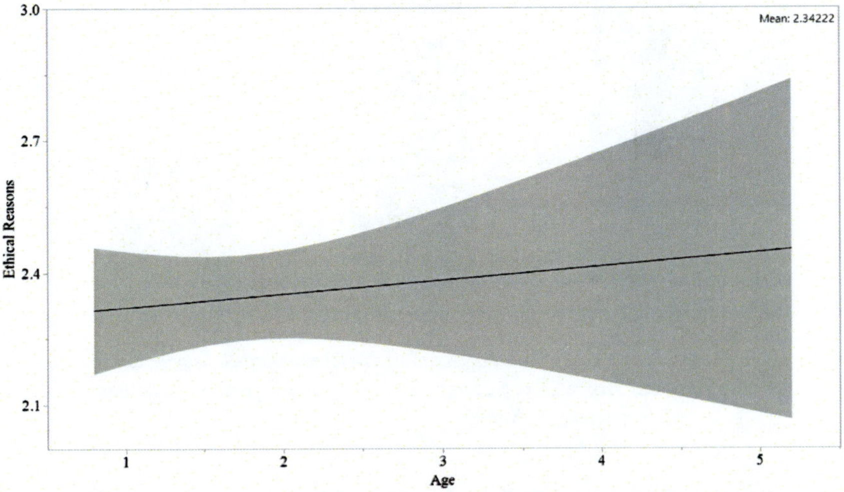

Fig. 1. The Response to the relationship between Ethical Reasons and Age. where (1 = Strongly Disagree, 2 = Agree, 3 = Neither Disagree nor Agree, 4 = Agree and 5 = Strongly Agree) for both factors. The gray shaded area represents the Fitness Confidence Internal

$$\text{Societal Acceptance} = 2.28 + 0.03 * \text{Age} \tag{1}$$

Investigators raised a number of legitimate ethical or moral standpoint and concerns of RSIM that may directly affect the intention of use [7]. Ethical issues perception is well expressed by "implant ethics" or bioethics which is defined by [8] as "the study of ethical aspects of the lasting introduction of technological devices into the human body". For instance, monitoring and tracking of vulnerable people of the society may be endurable depending on the severity of the situation and its therapeutic contribution, but tracking for non-disabled may be intolerable by many. The results showed that older people have relatively higher concerns about the ethical issues related to RSIM.

Paradoxically, the social acceptance has shown in Fig. 2 has no effect (mean = 1.72, SD = .73) on the choice of the microchip implant. This may be due to the fact that when someone uses the microchip it will not be visible for others to see it or may be because most of the subjects are younger in age. The same applied to the effect of the government requirement (mean = 2.03, SD 0.79) with high correlation ($p < .0001^{**}$) between social acceptance and government requirements to impose microchip implant. This is indicative to the fact that the use of RSIM is determined by individual decision rather than other communal factors.

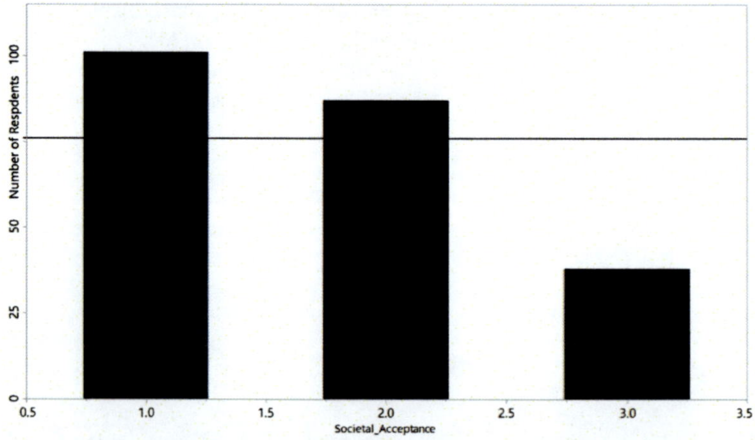

Fig. 2. The response of subjects to the important of social acceptance as a determining factor for using microchip implant, where (1 = Strongly Disagree, 2 = Agree, 3 = Neither Disagree nor Agree, 4 = Agree and 5 = Strongly Agree)

The willingness of the subject to accept negative health risk was found to be consistently increasing with the increase of age (Fig. 3). This may be associated with the fact that older generation is in need for RSIM to support them. But there is a trend of agreement with long-term health risk fear that increase with age.

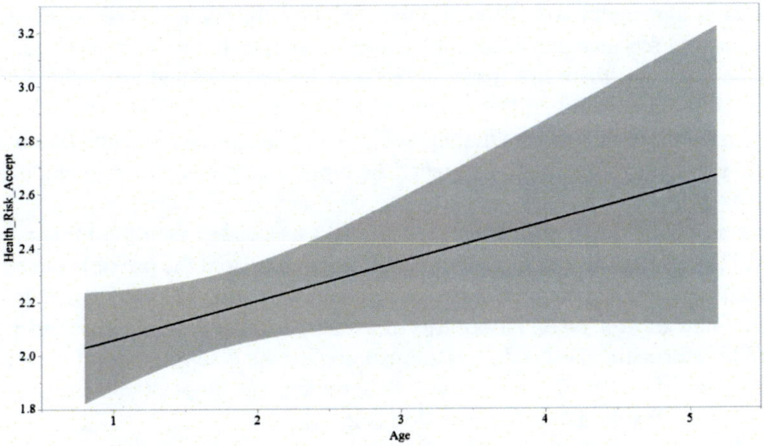

Fig. 3. The relationship between the feeling of health risks and Age. Where response for both (1 = Strongly Disagree, 2 = Agree, 3 = Neither Disagree nor Agree, 4 = Agree and 5 = Strongly Agree). The gray shaded area represents the Fitness Confidence Internal

Subjects will accept potential health risks of implanting if the chipping is useful (Fig. 4 and Eq. 2). This is especially more prominent for subjects with medical conditions were willing to accept the health risk.

$$\text{Health\_Risk\_Acceptance} = 1.08 + 0.45 * \text{Usefulness} \tag{2}$$

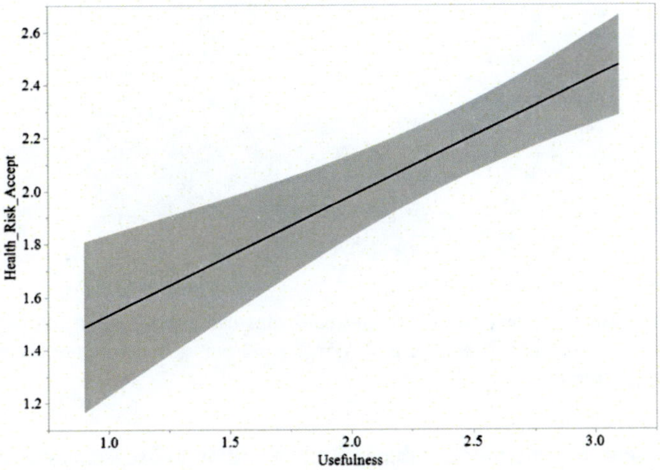

Fig. 4. The effect on Usefulness of the microchip implant on the health Risk acceptance. Where response for both (1 = Strongly Disagree, 2 = Agree, 3 = Neither Disagree nor Agree, 4 = Agree and 5 = Strongly Agree). The gray shaded area represents the Fitness Confidence Internal

This is in agreement with [9]where most of the subjects surveyed were willing to accept implants for medical needs more than accepting it for other reasons. This is despite the fact that, there are many unforeseen health risks and concerns associated with the chip implantation [3] such as tissue reaction, electronic interference and electrical hazards [10]. Subjects that agreed that implantable microchips do not appear to be safe for health. This might be one of the biggest hurdles in the implementation of the RSIM.

There are many data and human security risks associated with the implementation of RSIM [11]. The most serious risk of RSIM originates from the possible attack on the centralized or interconnected database where the information is kept [12]. This kind of intrusions may form a potential security threats to exposure of identity and personal identifiable information (PII) of users and tapping on the data integrity of their medical records. Most of the subjects surveyed believe that the implantable microchip will never be secure for the identity protection (Fig. 4).

RSIM may be exploited to invade the personal space using many of its own vulnerable applications. In addition, medical or identification RSIM could be augmented with sensors that append extra overt or covert functionalities [4] i.e. function creep. However, in [6] report that RSIM raises privacy risks perception and the likelihood of coercive implementation in individuals. For instance, RSIM may be misused by governments and companies for monitoring, tracking or profiling citizens and employees, respectively [12] (Fig 5).

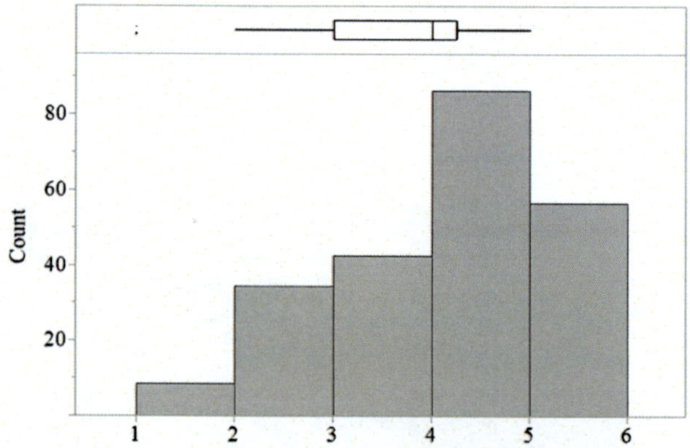

Fig. 5. Distribution of subjects who believe microchips can ever be secure to protect people's identity, where (1 = Strongly Disagree, 2 = Agree, 3 = Neither Disagree nor Agree, 4 = Agree and 5 = Strongly Agree)

There is clear trend that older subjects will be willing to use implantable microchips if only they had ownership of their own data. The same applied to believe microchips can ever be made secure to protect people's identity. In addition, it was found that the usefulness of the implantable microchips is highly correlated with the ability of the subject to own the data in the chip as depicted in Fig. 4 and Eq. 3 With both intercept and own data terms are highly significant (P < .0001**) (Fig 6).

$$\text{Usefulness} = 1.72 + 0.21 * \text{Own\_data} \tag{3}$$

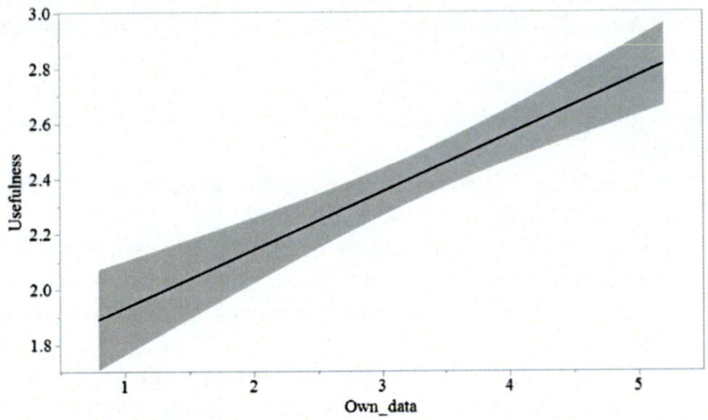

Fig. 6. The relationship between the passion of own data and the usefulness of the implantable microchip, where (1 = Strongly Disagree, 2 = Agree, 3 = Neither Disagree nor Agree, 4 = Agree and 5 = Strongly Agree). The gray shaded area represents the Fitness Confidence Internal

4 Conclusion

RSIM technology innovations offer a multitude of health and non-health benefits, but as the very nature of cutting-edge digital innovations there are challenges that need to be identified, examined and mitigated by manufacturers, retailers and medical staff. In this study it has been found that the importance of ethical reasons for implementing RSIM pattern increased by age. Surprisingly, the social acceptance and the government requirements have shown no effects on the choice of the microchip implant. This is indicative to the fact that the use of RSIM is determined by individual decision rather than controlled by public factors. The study found that the more useful the chip would be and the older the age of the subject, the more probably will accept the health risk of the implant. Subjects believed that the implantable microchip will never be secure for the identity protection. Furthermore, the older subjects will be willing to use implantable microchips, only if they had ownership of their own data. In fact, it was found that usefulness of RIMS is correlated with ownership of the data.

References

1. Stephan, K.D., et al.: Social implications of technology: the past, the present, and the futur. Proc. IEEE **100**, 1752–1781 (2012)
2. Greenfield, B., Musolino, G.M.: Technology in rehabilitation: ethical and curricular implications for physical therapist education. J. Phys. Ther. Educ. **26**(2), 81–91 (2012)
3. Albrecht, K.: Microchip-induced tumors in laboratory rodents and dogs: a review of the literature 1990–2006. In: International Symposium on Technology and Society, Wollongong, NSW, Australia. IEEE (2010)
4. Bramstedt, K.A.: When microchip implants do more than drug delivery: Blending, blurring, and bundling of protected health information and patient monitoring. Technol. Health Care **13**, 193–199 (2005)
5. Brabyn, J.: Future directions in blindness and low vision perspectives from a state-of-the-science conference. Vis. Impairment Res. **8**, 61–66 (2006)
6. Foster, K.R., Jaeger, J.: Ethical implications of implantable radiofrequency identification (RFID) tags in humans. Am. J. Bioeth. **8**(8), 44–48 (2008)
7. Bauer, K.A.: Wired patients: implantable microchips and biosensors in patient care. Camb. Q. Healthcare Ethics **16**(3), 281–290 (2007)
8. Hansson, S.O.: Implant Ethics. J. Med. Ethics **31**, 519–525 (2005)
9. Smith, A.: Evolution and acceptability of medical applications of rfid implants among early users of technology. Health Mark. Q. **24**(1–2), 121–155 (2007)
10. Smith, C.: Human microchip implantation. J. Technol. Manag. **3**(3), 151–160 (2008)
11. Paaske, S., et al.: The benefits and barriers to RFID technology in healthcare. In: Online Journal of Nursing Informatics(OJNI), pp. 1–12 (2017)
12. Gadzheva, M.: Getting chipped: to ban or not to ban. Inf. Commun. Technol. Law **16**(3), 217–231 (2007)

An Interdisciplinary Participatory Research for Co-creating a Relaxed Performance in a Theater Environment in Montreal

Zakia Hammouni[1,3](✉), Walter Wittich[1,3], Eva Kehayia[2,3], Ingrid Verduyckt[1,3], Natalina Martiniello[1,3], Emilie Hervieux[4], and Tiiu Poldma[1,3]

[1] Université de Montréal, Montreal, Canada
zakia.hammouni@umontreal.ca
[2] McGill University, Montréal, Canada
[3] Centre for Interdisciplinary Rehabilitation Research of Greater Montreal, Montreal, Canada
[4] The Segal Centre for Performing Arts, Montreal, Canada

Abstract. This research project presents the activities of a 'relaxed performance' in the Segal Centre for the Performing Arts, situated in Montreal, Canada, to understand how transforming the environment supports both performers and audiences alike. It proceeds with four phases, embracing a participatory approach. A particularly innovative aspect of this project was the composition of a research team that included the researchers, Segal Centre employees, local community partners and cultural groups, and rehabilitation research centres in the greater Montreal area. The outcomes of this project are theatre-specific adaptations that will benefit the Centre for all future events. They include changes to the physical aspects of the environment, the security within the theatre space itself, the development of a tactile map as an accessibility technology, a linguistically accessible information pamphlet and audio description to provide auditory-access to the visual elements of the performance.

Keywords: Interdisciplinary participatory research · Relaxed performance · Inclusive theatre · Co-creation · Social inclusion · Disability · Technological tools

1 Introduction

For audience members with intellectual disabilities, those on the autism spectrum, as well as those with motor or sensory disabilities, access to theatre performance is severely limited, often due to concerns about disturbance for other audience members [1]. Several theaters around the world have begun adopting new ways of accommodating audiences to include people with disabilities through adapted performances [1]. The concept of a "relaxed performance' can be inclusive if it contains elements that orient and assist people living with a disability to enjoy this cultural leisure activity, encouraging their interaction and social inclusion [2]. Universally designed theater environments are necessary to allow for the inclusion of audiences of people with and

D. Russo et al. (Eds.): IHSI 2021, AISC 1322, pp. 619–625, 2021.
https://doi.org/10.1007/978-3-030-68017-6_91

without physical, sensory or intellectual disabilities. This intersectoral research project explores how a participatory approach leads to changes that promote accessibility within live theatre performance, making it accessible to individuals with different abilities, and what factors in the environment might be changed to accommodate different disabilities, and facilitate the experience for the audience as a whole. This research project presents the co-creation of a 'relaxed performance' in the Segal Centre for the Performing Arts in Montreal, Canada, to understand how transforming environment supports both performers and audiences alike in their theatre experience. This community partner took steps to improve physical accessibility and access. In 2015, it committed to expanding its programming and operations to welcome everyone into the audience such that all are able to experience the performances without barriers. This project enhances this initiative and expands the Centre's accessibility scope.

2 A Participatory Research Methodological Approach

This research project proceeds in four phases, embracing a participatory approach with design thinking, wherein co-designing the various materials for the performance depends on the discussion with both the theatre personnel and the researchers at each phase of the project. The team included three participants living with a disability (person in wheelchair, person on the autism spectrum, person with visual impairment), the personnel at the theatre, and the local stakeholders involved in the project, including the Federation CJA local community partners, cultural groups, and rehabilitation research centres: CRIR/Institut Nazareth et Louis-Braille du CISSS de la Montérégie Centre, Centre de réadaptation Lethbridge-Layton-Mackay du CIUSSS du Centre-Ouest-de-l'Île-de-Montréal. The research team consisted of researchers from differing sub-specialties from the health/rehabilitation disciplines, architecture and design. The artists and audience members of the relaxed performance also participated in the evaluation of the relaxed performance. The researchers were informed by input from the audience, the theatre's performers, crew members, users and theatre employees.

A particularly innovative aspect of this project was the composition of a research team from diverse perspectives. The partnership of the centre was vital as they provided the venue, and were both a catalyst for the research project and an enabler for change, providing suggestions for the various aspects of the theatre and the resultant adaptive technological tools that were subsequently considered as integral aspects of the performance. The first phase of the project consisted of pre-performance staff training and a visit of the theatre itself, to assist participants in understanding how to determine the environmental issues within the theatre complex. The second phase consisted of a walk-through with all project participants together with the researchers assessing facilitators, challenges, interior attributes, language and spatial barriers. Researchers documented the theatre spaces during the theatre walk-through with photos of the "pre-environment" using the Environment Quality Satisfaction Tool [5, 6], and assessed what accessibility elements already existed and how spatial and social environment attributes might influence social and physical participation. From these two phases, the research team gathered the information to be used to inform the preparation of the actual 'relaxed performance'.

In the third phase, the team evaluated pre-performance conditions. "During the fourth phase the performance was presented and the project team and the audience participants from the first phase participated in a rehearsal to glean the possible issues and challenges that might arise. The 'relaxed performance' took place following this initial assessment, with the team members and the audience participants from phase 1 evaluating the successes and unresolved challenges of making the event accessible as a relaxed performance. The documentation of both the participant experiences and the visual characteristics was conducted using a modified version of Roses' Visual content Analysis" [2–4, 6, 9]. Finally, a post-performance analysis proceeded in two distinct stages: first after the rehearsal, and second post-performance.

During these phases, researchers noted aspects of the proposed relaxed performance that needed fine-tuning. During the post-rehearsal interviews and analysis, both audience participants and actors with disabilities provided important feedback. The data on the themes and issues faced by participants and theatre actors/personnel, both before and during the performance, led to the development of strategies for the implementation of the various support mechanisms for the performance itself (Fig. 1).

Fig. 1. Team members – co-creation of a 'relaxed performance'

With the collaboration and approval of the theatre (Centre Segal for performing Arts), the researchers collected visual, linguistic and auditory data, which allowed them to make adjustments to the social, technological, linguistical supports, as well as the physical environment during rehearsal and post-performance. This was done to support the readability of the written signs in the general theatre environment, as well as adapted written descriptions in both large print and braille [2–4]. An audio description was used during a performance, as well as a linguistically accessible pamphlet during a relaxed performance, using software that ensures readability of the materials by the whole audience using psycholinguistic norms and accessible criteria. The Centre itself prepared for the relaxed performance by organizing a Quiet Room for persons needing such a space, understanding the possible high stimulation for some participants. They also were able to organize on-site technological aids for access to the apps (on iPads) for those who might not have a phone [3]. The Centre Segal advertised the availability of the audio-description via multiple partners in their network as well as preparing in-house place for researchers to regroup and meet up. The pamphlet was linguistically accessible to allow people living with language and communication limitations assisting performance to fully understand its contents.

3 Preliminary Results and Discussion

This participatory approach and the collaboration of all the project's stakeholders made it possible to suggest proposals for adapting the Segal Centre's environment. These adaptations were: i) physical aspects of the environment (lighting/seating adjustments and accessibility); ii) security within the theatre space itself; and iii), the development of technological and linguistical supports.

The examination of the physical environment demonstrated that the Centre was completely accessible in terms of access to the theatre space, but less so with some of the support services such as counter service and washroom accessibility. The theatre itself was accessible. The arrival into the theatre is facilitated by easy access off the ground level foyer area. However, in the theatre, aisles and spaces are tight, and accessing the theatre space from the seating areas poses a greater challenge.

Fig. 2. The quiet room

The analysis revealed that the security of the theatre environment was enhanced through the support services provided and adequate lighting to assist patrons being guided within the theatre during the performance. A quiet room was available to all patrons, not merely to individuals with specific disabilities, and, therefore, was open to patrons who needed to step outside of the theatre space for any reason. This included parents with young children and people with autism or other disabilities. The space was large and had several quiet corners (see Fig. 2) during a relaxed performance, as well as chairs, sofa and mats on the floor. This space had a strategic location, very close to the lobby and the theatre room, allowing patrons easy access. However, it was not used extensively during the performance.

During the course of the project, the development of technological aids, such as the tactile map, was a valuable tool created for the Centre Segal for performing Arts, that can easily be adapted, should the structure and layout of the building change or expand over time [2–4]. From the disability/health perspective, the literature indicates that spoken word theatre is not easily made accessible to a variety of audiences with a disability [2, 8]. Live audio description for the visually impaired [2, 9], sub- and supra-titles for the hearing impaired [10], or simultaneous sign language interpretation [11] have been included previously. However, challenges remain in accomplishing universal accessibility for all individuals simultaneously [2,3,12]. From the technology/accessibility perspective, the audio description is an example of a tool for providing access to the visual elements of a theatre performance just as captioning provides access to spoken dialogue. At the Centre Segal, the audio description and the application that were used for the project were fully accessible to persons with visual impairments who rely on screen-reading technology. The remote description was done simultaneously by a specialized team located in another room at the Segal Centre. This description took place throughout the performance on stage so that all users of the application were able to access via their headphones the entire content of the performance. "Patrons with visual impairments were able to understand the story more fully with the inclusion of audio description that provided information about gestures, props and the layout of the stage" [2]. "This result is in concordance with previous studies that audio description is a useful tool for people living with a visual impairment, to describe what is happening on the theatre stage in the silent intervals between program commentary or dialogue in order to convey the main visual elements of a performance" [2, 9]. The voice description throughout the play allowed all app users access through their headphones. Persons with hearing disability could adjust the amplification of the voice as needed [2, 3].

In terms of support materials, linguistically accessible materials were provided to increase inclusion and accessibility spatial mapping of the environment and the creation of haptic tools support the theatre experience before, during and after the performance, as were developed by the research team. The results were presented to the Centre in the form of support documents highlighting an intersectoral perspective, providing an enriching support for this Centre and the performers, front-of-house staff and the back-stage crew, according to the analysis and post-performance meetings.

4 Conclusion

It is vital to understand the obstacles that hamper persons with disabilities in being able to fully participate in society as citizens. With this research project, and with the support and participation of community partners, the cultural event of a theatre performance became accessible through the various research phases. The approach of action research and interpretive analysis provided rich and immediate feedback to both participants and researchers. The resultant experience was enriching for both performers and the research participants in the audience. Overall, this project was very well received by participants, the Segal Centre for Performing Arts, as well as community organizations and the various stakeholders. The research group was proactive during the 4 phases of the project, translating findings into action. We were able to suggest tools and technologies to make the performance more accessible and allow spectators and participants to have a positive experience in this theater recreation environment. It should also be noted that the theater artists, backstage staff and researchers were able to reflect together with the research team about their experiences in real time, and then adjust the tools developed for the performance at each round.

The collaboration and cooperation of all the participants in this project demonstrates how inter-disciplinary research with multiple stakeholders can be of benefit to the community are large. The social aspects attached to theatre participation were expressed by the acceptance of the audience who was receptive to the changes made for a relaxed performance. While the project is still ongoing, the results show a highly positive engagement between the various stakeholders and the common goal to develop a relaxed performance so far. Future research might study additional needs of specific populations to enable the implementation of tools to improve the inclusivity of cultural venues such as the theater environments and provide enriching experiences for all, as well as its uptake and maintenance over time.

Acknowledgments. This project was supported by a grant from the Société Inclusive. Many thanks to all our participants and partners for their involvement in this project (Segal Centre for Performing Arts in Montreal, Federation CJA, CRIR/Institut Nazareth et Louis-Braille du CISSS de la Montérégie Centre, CRIR/Centre de réadaptation Lethbridge-Layton-Mackay du CIUSSS du Centre-Ouest-de-l'Île-de-Montréal).

References

1. Fletcher-Watson, B.: Relaxed performance: audiences with autism in mainstream theatre. Scott. J. Perform. **2**, 61–89 (2015)
2. Hammouni, Z., Wittich, W., Kehayia, E., Verduyckt, I., Martiniello, N., Hervieux, É., Poldma, T.: Using technology to encourage the participation of persons with disabilities: exploring cultural leisure activities in a theatre environment. In: Ahram, T., Taiar, R., Langlois, K., Choplin, A. (eds.) Human Interaction, Emerging Technologies and Future Applications III. IHIET 2020. Advances in Intelligent Systems and Computing, vol. 1253. pp. 36–41. Springer, Cham (2021)
3. Hammouni, Z., Martiniello, N., Poldma, T., Kehayia, E., Verdyckt, I., W. Wittich, W. : Pour un environnement théâtral inclusif : une co-création d'une représentation décontractée dans un théâtre à Montréal. In: 20th International Scientific Symposium on Visual Disability and Rehabilitation, February 4th, Montréal (2020)
4. Hammouni, Z., Wittich, W., Eva Kehayia, E., Ingrid Verdyckt, I., Martiniello, N., Hervieux, É., Poldma, T.: Co-création d'une expérience de loisirs dans un espace théâtral inclusif et accessible : Le cas du Centre Segal des arts de la scène. Forum d'échange 2020. Société Inclusive, Virtual kiosqs (2020), https://societeinclusive.ca/3467-2/.
5. Poldma, T., Dastoor, D., Brack, H.: Lighting Study at the Moe Levin Center of the Douglas Hospital. Analysis report of Phase 1: Exploration Study of the Effects of Light on Behaviour and Wellbeing of People with Dementia in a Specialized Environment (2007)
6. Poldma, T., Labbé. D., Kehayia, E., et al.: La participation sociale des personnes vieillissantes en centre commercial: un laboratoire vivant. In: Vieillissement et aménagement. Perspectives plurielles. Montreal, Canada, Les Presses de l'Université de Montréal, pp. 227 –242 (2018)
7. Rose, G.: Visual Methodologies. SAGE Publications, London (2001)
8. Bailey, S.D.: Wings to Fly: Bringing Theatre Arts to Students with Special Needs. Woodbine House, Rockville (1993)
9. Holland, A.: Audio description in the theatre and the visual arts: images into words. In: Audiovisual Translation, 170–185. Palgrave Macmillan, London (2009)

10. Diaz Cintas, J., Orero, P., Remael, A.: Media for All: Subtitling for the Deaf, Audio Description, and Sign Language. Brill Rodopi, Leiden (2007)
11. Knigge, M., Erkau, J.: Cinema and theatre: accessible entertainment, with benefits for all. In: Caltenco, H., Hedvall, P.-O., Larsson, A. (eds.) Universal Design 2014: Three Days of Creativity and Diversity. The authors and IOS Press (2014)
12. Story, M.F.: Maximizing usability: the principles of universal design. Assist Technol. **10**, 4–12 (1998)

Health Emergency and Digital Shopping Experience: A New Era of Retail Design

Stefania Camplone(✉) and Emidio Antonio Villani

University of Chieti-Pescara, Viale Pindaro, 42, 65127 Pescara, Italy
{stefania.camplone,emidio.villani}@unich.it

Abstract. The paper aims to focus on how new technologies, interactivity, digital connections between users and augmented reality can generate interesting scenarios that are inspiring for companies and brands. The goal is to lead to a personalized customer experience that leads the user to consider the shop in store as an experience of interaction/relationship with the chosen brand, in compliance with anti-contagion provisions. The growing digitalization of contemporary society and the recent state of global health emergency have led to a drastic change in the purchasing habits of consumers and a consequent reorganization of sales strategies in the retail field. The state of emergency has speeded up a process that has already been underway for some time that sees digital purchasing increasingly protagonist compared to in-store, and the new health regulations require different methods of approaching products than those we have been used to until now.

Keywords: Retail design · Shopping experience · Sanitary emergency · Interactivity · Digital connections · Augmented reality

1 Introduction

The digitalization of the buyer's purchase path has led to a notable growth in e-commerce in the last year also due to the health emergency we are facing in recent months, but recent statistics currently show 85% of sales still billed in the physical store [1]. We cannot clearly separate the shopping experience exclusively between online or in store, because the purchasing methods are deeply permeable to each other. The modern consumer continues to ask the brand with which it interfaces, something new every day, beyond the products on the market. The user is looking for an experience that strikes him from an emotional point of view, and companies that focus their attention on this aspect have greater consumer appeal. The experiences sought by the user are of an innovative type, linked to technology, new communication channels and new sensory interfaces. The world of retail is increasingly immersing itself in a digital dimension that facilitates the shopping experience and changes the customer journey by decreasing the gap between product, service and experience provided by brands for the benefit of consumer satisfaction. Therefore, new purchasing methods are born and new technologies greatly expand the possibilities of interfacing with the customer. The experiments through virtual reality and augmented reality show us new possible scenarios that make the user the protagonist of a multisensory experience that can be

D. Russo et al. (Eds.): IHSI 2021, AISC 1322, pp. 626–631, 2021.
https://doi.org/10.1007/978-3-030-68017-6_92

exploited for the benefit of both users and brands for a suitable use of products and spaces in compliance with the recent health regulations.

2 The New Purchasing Methods

Recent statistical research on how customers buy and in particular on e-commerce have brought to light new and interesting categories of consumers who are changing the way they buy products. Historically, the physical store has always had the greatest influx of consumers until the advent of new digital online shopping platforms that have directed users into a world of convenience, practicality and speed. The physical store provides the user with various experiential possibilities due to the direct relationship with the product and what surrounds it. The user is accompanied in his purchasing experience in a conscious way through the guidance of specialized personnel, who follow the user in choosing, testing and finally purchasing the product. In addition, the physical store offers the user the opportunity to see, touch and really evaluate a certain product. The customer's physical experience becomes an opportunity of great value for the company because it acts on the user's perceptions and very often influences their purchase. A factor not to be overlooked by the company is certainly the time factor in the buyer's buying experience. Sellers will have to consider the timing related to the purchase phases of a product which are: the recognition of a need, the evaluation of alternatives, the choice of one product over another, the purchase, the acquisition of the product and post-purchase evaluation. The scanning of these phases and the consequent management of them, gives companies the power to guide the buyer in his purchasing experience. These timelines change substantially and sometimes follow different patterns if evaluated in an in-store shopping experience compared to an online one. During a purchase the percentage of time most used is given by the evaluation. The over-choice to which the user has access today leads to an expansion of the evaluation phase which becomes even more extensive in the case of an online purchase since the purchasing possibilities are almost infinite. If on the one hand the offer in quantitative terms takes on considerable proportions, on the other hand the insecurity on the part of the buyer grows, who is lost in a sea of products from which he cannot choose. The user finds himself losing orientation among all the products offered due to the lack of a figure who in some way guides the user to make an informed purchase. This figure in a physical store is identified by specialized personnel who are able to understand the customer's needs and help him in the product evaluation process. In online purchases, the figure of the purchase assistant disappears and is compensated for by the opinions of those who have already bought that product. Therefore, the reviews of other users become fundamental, the evaluations already expressed by users on a particular product and who have themselves received a quotation in terms of reliability from the online buying community. Reviews fill those experiences that cannot be made online due to the limitations of the medium. Therefore, we will not have the ability to physically evaluate a product in its entirety but only visually based on the suggestions that the seller has decided to publish online. For some users, the physical store becomes a showcase, for others the place of purchase but in both cases, the modern buyer uses the online service to achieve his goal. In recent years, two new ways of shopping have emerged. The first way (showrooming) the buyer culminates the sales cycle in the store, in the second way the buyer culminates the sales cycle in the physical store (webrooming) [2]. What these new purchasing trends have in common is the technological tool which, through the connections it generates, leads to a new awareness of the buyer and makes him increasingly independent in his buying experience. In the case of showroomers, buyers make use of the help of the physical store to overcome all those obstacles related to electronic commerce, such as, for example, the test of a certain item of clothing but, their purchase, is concluded through the internet, evaluating the convenience between the various online stores. On the contrary, webroomers first visit virtual stores to find out about the products, compare their prices, analyze their advantages and look for the best offers on the market and then go to the physical store and finish the purchase there. Webrooming is preferred over showrooming as the physical store still offers advantages over the online store such as the absence of shipping costs, zero delivery times, the ability to see, touch and physically try a product before purchasing, finally, greater ease of return to the store if necessary. Both methods now take advantage of the internet means of communication and new technologies to bring the customer to purchase, obtaining data that the data centers will process to make their shopping experience more and more personalized. The technological tool must certainly be integrated into the customer's purchase path and new experiences must be devised that make the buyer interactive in the purchasing experience. Brands, for their part, could exploit this link between user and technology that has become increasingly welded through the internet, to attract customers to physical stores, making them more attractive, not only as places of purchase but also as places of experience. linked to the product you want to promote.

3 Retail Design in the Epidemiological Shopping Experience

The health emergency we are experiencing in this period leads us to reflect on what are the repercussions in the field of physical retail we are used to. The health regulations on Coronavirus have changed the physical dimension of the trade, affecting the physical store but also on goods and services for the consumer, favoring the development of e-commerce and digital purchasing but which brings with it advantages and disadvantages both for the market that for the user. Digital shopping has numerous advantages for the consumer, such as the immediacy of the purchase and the convenience of being able to place an order from the comfort of home and the possibility of being able to customize the product to be purchased, strengthening the link between brand and buyer. On the contrary, the long waiting times for product delivery and the overload of information on the products on sale, leads to a sense of loss in consumers who find themselves lost in an infinite world of offers. The convergence of the flow of consumers in the world of digital commerce leads to an ever lower turnout in physical stores and leads to an inevitable failure. The consequences are the shopping centers that see their influx of consumers reduced, emptying themselves and losing their attraction for users. The small shops and neighborhood shops derive a relative benefit from it, which become essential for the consumer who cannot leave home due to compliance with the limitations according to health regulations. Retail design must intervene to ensure that physical stores can "reinvent themselves" in a hybrid guise between analog and digital, thus redesigning physical spaces with the help of technology. This new design approach would thus allow brands to attract consumers to physical stores to involve them and make them the protagonists of an interactive shopping experience that pushes them to return to the physical store even in full compliance with social distances and health regulations. In light of the new purchasing methods used by users and the continuous search for emotions and experiences by consumers, the need arises on the part of companies to provide satisfaction to these latent consumer needs. One of the possible strategies to attract the customer to the store is to offer experiences that cannot be replicated with online shopping by leveraging environmental stresses to arouse the user's psychological involvement and facilitate their propensity to purchase. The store evolves from a commercial space to a "place of experience" that involves the user in a moment of emotional and multisensory stimulation. The design of the store environment acquires enormous importance as it appears to be the point of communication link between the brand and the consumer in order to facilitate the emotional involvement of the user and therefore to entice him to purchase. The digital transformation we are facing represents an opportunity for the evolution of distribution to enhance customer-dedicated services. Technological innovation enters retail design as an attraction for the customer but also as a benefit for the company in improving interactions with buyers, forging loyal ties with the company that offers new services in step with the wishes and needs of the user.

4 The Virtual Dimension of Shopping

The buyer is no longer enough with the product but wants to recognize himself in a brand identity that leads him to an elevation of his status. The physical store retains its hegemony for the purchase of products but is increasingly influenced by the online world and above all by the smartphone. The retailer can take advantage of this means of communication and display to bring the user to a new experiential level within the store. The integration of the smartphone with the purchase possibilities is now in common use, the smartphone and new technologies in general have become the tools to attract the customer into the store not only to make a purchase but also to be able to have an experience which further strengthens its link with the brand. We are historically facing a time of alteration and change in social habits and the last few months have led to a change in the approach to products from the point of view of touch. The total or partial impossibility of knowing a product through the sense of touch leads us to reflect on how we can increase the perception of the product through the other senses. Virtual reality (VR) and augmented reality (AR) can be exploited to the advantage of retailers to make the product visible, more interactive and to provide the customer with more reference information. Large companies have started experimenting with virtual reality as a means of displaying the products offered, making the user increasingly safer not only from the point of view of purchase but also from the health point of view as there is no need to interact with them by touch. The developments from a technological point of view in the retail field are infinite and expand the possibilities of the stores, making them places for the spectacularization of the experience and not just for buying. The customer is attracted to the store to interact with the brand through technological expedients that aim at the best promotion of the product and the highest customer satisfaction. Often the customer already knows the product he is interested in and what drives him to enter the store is his desire to immerse himself in a new dimension, albeit virtual, that makes him the protagonist of a moment of interaction with the product on sale, enriching his experience of entertainment and additional information. The holographic projection is useful for displaying the product in a promotional store that does not involve the use of service personnel. The store becomes "aseptic" and devoid of tactile interaction between user and product. Contact with the products inside the store is reduced to a minimum and the customer will have the opportunity to experience the holographic tool by making a three-dimensional scan of himself inside the store as an interactive experience. Augmented reality becomes a resource to save time, inserting an object in a context by simply framing it with the smartphone camera, the choice of a dress could be done by flipping through a digital mirror that shows us the different options directly on you thanks to scans three-dimensional or based on previously purchased products. The *internet of things* (IoT) [3] represents a resource for companies that can make commonly used tools such as shelves or mirrors normally used in traditional stores more efficient in communicating useful information to the user in choosing a product or in suggesting a particular promotional offer. The major online commerce companies are experimenting with stores equipped with digital tools that limit the intervention of service personnel to almost zero and where the product is brought home without all the operations that usually require a purchase such as paying at the checkout as everything comes uploaded to the customer's online account which is recognized at the store entrance. We are heading towards a new era of shopping that envisions technology as a cornerstone in the customer's shopping experiences of the future.

5 Conclusions

The future of retail will lead to an increase in the relationship with the digital world and new technologies that will increasingly integrate into the world of retail trade. The current and increasingly emerging health need with respect to an aseptic nature of the places of sale will lead to new reflections on spaces, distances and interactivity between things and people. The way in which stores are used is changing. The possibility of having an in-store experience could attract more than the product itself, buyers will be increasingly smart and demanding in search of an ever new and interactive shopping experience that makes them protagonists of a memorable moment linked to their favorite brand. Retail design will have to adapt to continuous changes not only in fashions and technologies, but also to future and unpredictable states of emergency. Probably the only physical element inside a store will be the buyer himself and the store will become more a place of experience than of purchase. In this digital metamorphosis of retail, the user will benefit from an increasingly "tailored" and personalized shopping experience that will lead the buyer to have an increasingly symbiotic relationship with the brand. Digital connections will be more and more dense and integrated with the omnichannel service offered by the brand for an ever stronger customer loyalty through a constantly evolving digital experience. The health emergency has changed the habits of buyers by drastically reducing the time spent in the stores. Retail design can exploit this digital acceleration as an opportunity to design new retail spaces to be more functional, rational and meet the needs of the consumers of the future.

Acknowledgments. This contribution is the result of a discussion on the current health emergency with respect to the retail design sector. The topic was addressed during the lockdown in the Interior Retail Design online course at the Architecture Department of the University of Chieti-Pescara in Italy and conducted by Stefania Camplone, with the collaboration of Emidio Antonio Villani. The final writing of the paragraphs is attributed to: Emidio Antonio Villani for "paragraphs 3-4-5" and Stefania Camplone for "Abstract, paragraphs 1–2".

References

1. Statista. https://www.statista.com
2. Techopedia. https://www.techopedia.com
3. Ashton, K.: That 'internet of things' thing. RFID J. **22**(7), 97–114 (2009)

Combined Method for Accessibility Evaluation in Tele-Rehabilitation Platforms for Low Vision Users

Patricia Acosta-Vargas[1(✉)], Jorge-Luis Pérez-Medina[1], Gloria Acosta-Vargas[2], Belén Salvador-Acosta[1], Wilmer Esparza[1], Karina Jimenes-Vargas[1], and Mario Gonzalez[1]

[1] Intelligent and Interactive Systems Lab (SI2 Lab), Universidad de Las Américas, Vía a Nayón, Quito, Ecuador
{patricia.acosta, jorge.perez.medina, maria.salvador.acosta, wilmer.esparza, karina.jimenes, mario.gonzalez.rodriguez}@udla.edu.ec

[2] Facultad de Medicina, Pontificia Universidad Católica del Ecuador, Quito, Ecuador
gfacosta@puce.edu.ec

Abstract. According to the World Health Organization, about 15% of persons who have a disability. Existing web applications are not easily accessible, particularly for people with disabilities. In this research, we present a combined method that consists of applying an automatic inspection with a WAVE tool and manual method. In this study, we employed an examination of the accessibility assessment of ePHoRt, a web platform for hip arthroplasty patient rehabilitation, from the physical therapist's perspective. In this research, we included the application of 12 accessibility barriers for people with a low vision based on the Web Content Accessibility Guidelines (WCAG) 2.1 and 2.2. The results obtained suggest that digital ramps and WCAG have more accessible and inclusive web platforms.

Keywords: Accessibility · Combined method · Evaluation · Low vision · Tele-rehabilitation · Web content accessibility guidelines · WCAG 2.1

1 Introduction

Today, the use of technology and intelligent electronic solutions for health have become standard tools. However, when applications are not easily accessible, they can create barriers that exclude people from using them. According to World Health Organization statistics, 15% of the world's population has a disability of some kind [1].

Designing accessible applications is a challenge for experts around web accessibility, so they recommend applying accessibility standards so that a more significant number of people can access the applications.

For these reasons, our research presents a combined method [2] that incorporates: 1) an automatic assessment with the WAVE [3] and 2) a manual examination considering the guidelines for users with low vision.

D. Russo et al. (Eds.): IHSI 2021, AISC 1322, pp. 632–638, 2021.
https://doi.org/10.1007/978-3-030-68017-6_93

The accessibility assessment was applied to the ePHoRt platform [4], a web-based platform for improving patient rehabilitation after hip arthroplasty. To make the platform inclusive and accessible to all users. In this research, we used the Web Content Accessibility Guidelines (WCAG) 2.1 [5] and 2.2 [6] that consist of 1) perceivable, 2) operable, 3) understandable, and 4) robust. To complement the analysis, we also consider accessibility barriers based on accessibility principles for users with low vision.

An accessibility expert in eight steps applied the evaluation process: 1) Select the ePHoRt screens to evaluate. 2) Explore and interact with each screen. 3) Select the type of users. 4) Define barriers for users. 5) Evaluate each screen with WAVE. 6) Manually assess each screen with the barriers defined according to WCAG 2.1and 2.2. 7) Recording and examine the outcomes. 8) Suggest recommendations to improve the platform.

This research can motivate software developers to apply WCAG 2.1 and 2.2 during the software development cycle. This combined approach can be replicated on other platforms if the guidelines for different disabilities are considered. As future work, we suggest: Evaluate and compare the performance of ePHoRt with other platforms with similar characteristics. Finally, we recommend including in the ePHoRt platform assistive technologies that allow the adaptation of external devices and the achievement of universal access.

This document is constructed as follows: in Sect. 2, we present a literature review of associated studies, in Sect. 3, we describe the method and the case study, in Sect. 4, we offer the outcomes and discussion of them, and finally, in Sect. 5 we include the conclusions, recommendations, and future research.

2 Literature Review

Today, e-health has generated considerable attention in web platforms that facilitate the fast recovery of the patient. Rybarczyk et al. [7] suggest that ePHoRt can support tele-rehabilitation treatments for people who have undergone hip surgery.

A recent systematic study confirms that accessibility has been the subject of a few studies [8]. The literature review details that preceding researches by Acosta-Vargas et al. [9] recommend the use of an inspection method using an automated tool to assess the accessibility of the ePHoRt platform. It also suggests the application of WCAG in educational resources and videos used on the platform.

Furthermore, accessibility is necessary for web applications so that more people can easily navigate the platform. Previous studies by the authors Acosta-Vargas [10–12] argue that it is essential to apply the WCAG-EM 1.0 [13], which considers fundamental aspects for WCAG 2.0 [14].

Preceding studies [15] argue that it is not enough to review accessibility barriers with automatic tools when users suffer age-related changes such as low vision problems that make it difficult for users to access the resources offered by the web easily.

Therefore, in this research, we defined 12 accessibility barriers for low vision users based on WCAG 2.1 [5] and 2.2 [6]. A barrier refers to the impediment that the user experiences when navigating an application. The barriers were determined by the evaluators based on the four accessibility principles 1) perceivable, 2) operable, 3) understandable, and 4) robust.

3 Case Study and Method

We applied the case study in ePHoRt [7], a design based on a web platform. The evaluation was carried out on the physiotherapist perspective; we evaluated seven screens selected at random.

This research uses an adaptation of WCAG-EM 1.0 [13] and WCAG 2.1 and 2.2. Our study applies the evaluation to the tele-rehabilitation platform with a combined method: 1) automatic inspection with the WAVE tool (uses the Google Chrome plug-in); 2) the manual review of the 12 accessibility barriers summarized in Table 1.

Table 1. Accessibility guidelines for low vision users

ID	Barrier	WCAG 2.1-2.2	Success criteria	Level
B01	Accessible keyboard	Operable	2.1.1	A
B02	Flash and luminance settings	Operable	2.3.1	A
B03	Easy to read font	Perceivable	1.1.1	A
B04	Text alternatives	Perceivable	1.1.1	A
B05	Sensory characteristics	Perceivable	1.3.1	A
B06	Adjusting the display settings	Perceivable	1.3.4	AA
B07	Color configuration	Perceivable	1.4.1	A
B08	Well-spaced elements	Perceivable	1.4.12	A
B09	Images as sharp as possible	Perceivable	1.4.5	AA
B10	Visual presentation	Perceivable	1.4.8	AAA
B11	Consistent navigation	Robust	4.1.3	AA
B12	Help	Understandable	3.3.5	AAA

This combined method consists of eight steps depicted in Fig. 1.

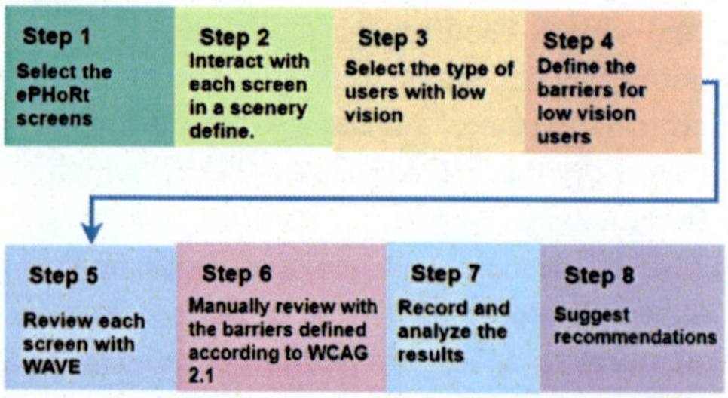

Fig. 1. A combined method for measuring accessibility.

Step 1: The seven screens were randomly chosen, from the physiotherapist's perspective, to evaluate them with the combined method and are: 1) Exercise list, 2) Patient list, 3) Physiotherapist menu, 4) Suspended therapeutic programs, 5) Therapeutic program form, 6) Recommendations list, 7) Messages.
Step 2: The experts define the tasks to be performed on each panel related to the physiotherapist to apply the combined method.
Step 3: The type of user for the application of accessibility barriers; in our case, we describe for users with low vision that have a degree of partial vision to read, learn and achieve information that limits the capabilities in everyday activities, but can perform them with the help of lenses to recognize objects.
Step 4: The 12 barriers for low vision users are described in Table 1, including WCAG 2.1 and 2.2.
Step 5: The accessibility experts evaluate each screen with the WAVE tool detailed in Table 2.
Step 6: The accessibility specialists manually analyze each screen with the barriers for low vision users specified in step 4.
Step 7: The data from the WAVE inspection and manual review are documented in a spreadsheet to analyze the outcomes.
Step 8: The scores of the combined method are analyzed, and solutions are sought to reduce the accessibility barriers found.

4 Outcomes and Discussion

In Table 2, we present the results of the evaluation of the seven ePHoRt screens. We found that: 1) All screens show a frequent problem related to "Missing form label" that must be rectified to eliminate accessibility barriers. 2) Alerts, structural elements, features, and ARIA may or may not be fixed by the programmers; they do not influence accessibility to a high degree. 3) "Contrast Errors" are directly related to users with low vision. Even though they are registered in zero, there are other characteristics that the WAVE tool does not detect and must be manually reviewed by accessibility experts.

Table 2. Errors found with WAVE tool

ID	Screen	Errors	Alerts	Structural elements	Contrast errors	Features	ARIA
S01	Exercise list	1	2	30	0	11	11
S02	Patient list	1	2	27	0	4	11
S03	Physiotherapist menu	1	1	10	0	8	14
S04	Suspended therapeutic programs	1	1	9	0	4	7
S05	Therapeutic program form	1	2	27	0	4	11
S06	Recommendations list	1	2	33	0	11	11
S07	Messages	1	1	8	0	1	12

Figure 2 presents the outcomes of the manual assessment with the 12 accessibility barriers defined based on WCAG 2.1 and 2.2. The value of one (1) is recorded if it exceeds the barrier, and of zero (0) if it does not exceed the barrier. The barriers that are absent in all the screens evaluated correspond to "Flash and luminance settings", "Sensory characteristics", "Color Configuration", and "Help. Data and analysis for replication of the experiment are available in the Mendeley repository[1].

ID	Screen	B01	B02	B03	B04	B05	B06	B07	B08	B09	B10	B11	B12
S01	EXERCISE LIST	1	0	1	1	0	1	0	1	1	1	1	0
S02	PATIENT LIST	1	0	1	1	0	1	0	1	1	1	1	0
S03	PHYSIOTHERAPIST MENU	1	0	1	1	0	1	0	1	1	1	1	0
S04	SUSPENDED THERAPEUTIC PROGRAMS	1	0	1	1	0	1	0	1	1	1	1	0
S05	THERAPEUTIC PROGRAM FORM	1	0	1	1	0	1	0	1	1	1	1	0
S06	RECOMMENDATIONS LIST	1	0	1	1	0	1	0	1	1	1	1	0
S07	MESSAGES	1	0	1	1	0	1	0	1	1	1	1	0

Fig. 2. Data of the manual evaluation

Figure 3a shows that the principle that is most met is perceivable with 75%, supported by the operable and robust with 12.5% each principle, the understandable principle, in this case, is zero per percentage. In Fig. 3b, it shows that the success criterion that is most met is 1.1.1, which corresponds to 25% related to "Easy to read Font". The level that offers the highest compliance is "A" with 50%, followed by "AA" with 37.5% and "AAA" with 12.5%.

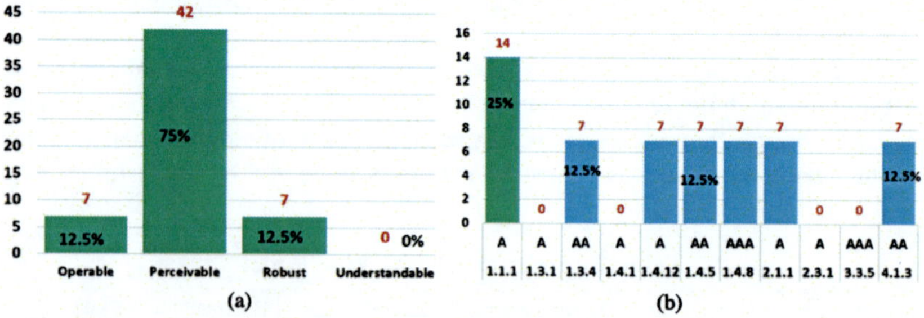

Fig. 3. WCAG principles 2.1 and 2.2 (a). Success criteria and levels (b).

To make the platform more accessible, we suggest incorporating support products or digital ramps on all screens that allow access to more people with low vision. The digital ramps can be supported by software and hardware. For example, for vision deficits, even blindness, outstanding audio players can be used as stand-alone devices that allow reading books in MP3, Daisy, and NISO format.

[1] https://data.mendeley.com/drafts/6bffhgjrz3/1.

Another alternative is to include electronic magnifying glasses such as telescope magnifiers. These devices allow you to enlarge any object placed under your lens; they are portable devices, becoming pocket-sized. With the controls, a combination of zoom, contrast, and helpline can be obtained, so that texts can be read with some comfort.

5 Conclusions and Future Works

This combined method can be used to evaluate any web platform, considering the accessibility barriers affording to the kind of disability of the people. The perceivable principle is related to visual disability; according to the analysis, it is fulfilled by 75%. The level "AA" is fulfilled in 37.5%, which implies that the telerehabilitation platform allows an acceptable level of accessibility for people with low vision. With the application of the manual method that includes the 12 accessibility barriers for low vision users, we found new issues that were almost invisible to the automatic WAVE tool.

As future work, we suggest evaluating the accessibility of other telerehabilitation platforms, including ePHoRt, making public reports with best practices to improve these types of platforms. Finally, we recommend including in the ePHoRt platform assistive technologies that allow the adaptation of external devices and achieve universal access. We suggest evaluating the platform with end-users to assess the performance of people who use ePHoRt. In this way, the effectiveness, efficiency, and satisfaction of the incorporated technology can be evaluated.

Acknowledgments. The Universidad de Las Américas has financed this work. Quito - Ecuador, grant numbers: SIS.JPM.18.01, and, by the NATIONAL RESEARCH FINANCING program (Senescyt-INÉDITA), under the Convention No.: 20190020CI and project code: PIC-18-INE-UDLA-001.

References

1. World Health Organization (WHO): World report on disability (2019). https://www.who.int/disabilities/world_report/2011/report/en/. Accessed 20 Sept 2020
2. Salvador-Ullauri, L., Acosta-Vargas, P., Gonzalez, M., Luján-Mora, S.: Combined method for evaluating accessibility in serious games. Appl. Sci. **10**, 6324 (2020). https://doi.org/10.3390/app10186324
3. WAVE Web Accessibility Tool. https://wave.webaim.org/. Accessed 20 Sept 2020
4. Acosta-vargas, P., Rybarczyk, Y., Pérez, J., et al.: Towards web accessibility in telerehabilitation platforms. In: ETCM, pp 1–6. IEEE (2018)
5. World Wide Web Consortium: Web Content Accessibility Guidelines (WCAG) 2.1 (2018). https://www.w3.org/TR/WCAG21/. Accessed 20 Sept 2020
6. World Wide Web Consortium (W3C): Web Content Accessibility Guidelines (WCAG) 2.2 (2020). https://www.w3.org/TR/WCAG22/. Accessed 20 Sept 2020

7. Rybarczyk, Y., Deters, J.K., Gonzalvo, A.A., et al.: ePHoRt project: a web-based platform for home motor rehabilitation. In: Advances in Intelligent Systems and Computing, pp. 609–618 (2017)
8. Pérez-Medina, J.L., Acosta-Vargas, P., Rybarczyk, Y.: A systematic review of usability and accessibility in tele-rehabilitation systems. In: Rybarczyk, Y. (ed.) Assistive and Rehabilitation Engineering. IntechOpen (2019)
9. Acosta-Vargas, P., Jadán-Guerrero, J., Guevara, C., et al.: Technical contributions to the quality of telerehabilitation platforms: case study—ePHoRt project. In: Assistive & Rehabilitation Engineering, pp. 1–22. IntechOpen (2019)
10. Acosta-Vargas, P., Luján-Mora, S., Salvador-Ullauri, L.: Quality evaluation of government websites. In: 2017 4th International Conference on eDemocracy and eGovernment, ICEDEG 2017, pp. 8–14. Institute of Electrical and Electronics Engineers Inc. (2017)
11. Acosta-Vargas, P., Hidalgo, P., Acosta-Vargas, G., et al.: Designing an accessible website for palliative care services. CCIS, vol. 1193, pp. 371–383 (2020)
12. Acosta-Vargas, P., Hidalgo, P., Acosta-Vargas, G., et al.: Challenges and improvements in website accessibility for health services. AISC, vol. 1131, pp. 875–881 (2020)
13. World Wide Web Consortium (W3C): Website Accessibility Conformance Evaluation Methodology (WCAG-EM) 1.0 (2014). https://www.w3.org/TR/WCAG-EM/
14. World Wide Web Consortium (W3C): Web Content Accessibility Guidelines (WCAG) 2.0. In: 30-04-2018 (2008). https://www.w3.org/TR/WCAG20/. Accessed 19 Sept 2020
15. Acosta-Vargas, P., Salvador-Ullauri, L.A., Luján-Mora, S.: A heuristic method to evaluate web accessibility for users with low vision. IEEE Access **7**, 125634–125648 (2019). https://doi.org/10.1109/ACCESS.2019.2939068

Bio-plastic: The Challenge of a Perpetual Material

Stefania Camplone(✉)

University of Chieti-Pescara, Viale Pindaro 42, 65127 Pescara, Italy
stefania.camplone@unich.it

Abstract. Man has always believed that the natural resources available were unlimited. In particular, in the last century, the extraordinary technological development and the resulting economic development have helped to spread the perception that the availability of products and materials was almost infinite. These were also the years in which the use of plastic proliferated as a "material of the possible" derived, moreover, from a raw material such as oil, available and transportable all over the planet. Today, unfortunately, the unbridled use of plastics leads to negative and out of control environmental consequences. The resulting negative scenario, however gradually accelerating, immediately "obliges" everyone to identify new paths of sustainability, also through a change in the relationship with natural resources and with the environmental heritage. The technological challenge we face, in addition to requiring more efficient transformation processes in terms of consumption, also aims to generate new materials whose mechanical features deteriorate minimally and which can therefore be reused many times. This contribution focuses on the opportunities offered by research on innovative materials. In particular, it deals with bio-plastics, obtained from waste and organic waste derived from other productions: for these, research aims to obtain an ever greater reliability, both with respect to physical and mechanical properties and to biodegradability characteristics.

Keywords: Bio-plastics · Circular economy · Physical and mechanical properties · Biodegradability

1 Introduction

The extraordinary technological development of the last century, together with the resulting economic development, has contributed to spreading the perception that the availability of products and materials was almost unlimited. This concerned in particular the production of plastics, considered as "materials of the possible" due to the infinite methods of use associated with them, all derived from a raw material such as oil, available and easily transportable all over the planet.

Today, unfortunately, we are all aware of the negative and out of control environmental consequences due to the unbridled use of plastics. Add to this is the unsustainability of the current exploitation of the planet's resources also on a quantitative level: various raw materials are running out, thus determining not only an environmental problem but also supply difficulties, with strong repercussions on the

D. Russo et al. (Eds.): IHSI 2021, AISC 1322, pp. 639–644, 2021.
https://doi.org/10.1007/978-3-030-68017-6_94

entire current economic and industrial system. This negative scenario, moreover in progressive and accelerating degradation, immediately "obliges" everyone to identify new paths of sustainability, also through a change in the relationship with natural resources and with the environmental heritage. If until some time ago even companies, deluded by a perception of inexhaustible resources, were oriented towards a "linear" production process (production-use-disposal), today they are forced to face the challenge (and opportunities) of a "circular" economy. It is an economic and production model that aims to reintroduce the maximum possible amount of resources already used into the production cycle, including those useful for production such as air, water and energy.

In addition, the inevitable rethinking of lifestyles, as well as the methods of production of consumer objects, inevitably also leads to a reconsideration of both the current design processes and the materials used. Thus, if on the one hand the effort is to limit the environmental damage deriving from the use of "conventional" non-biodegradable polymers and derived from non-renewable resources (such as polypropylene, polyethylene, polyvinyl chloride, polyethylene terephthalate and polystyrene), through the application of Life Cycle Design criteria, on the other hand new materials are identified that have a lower impact on the environment.

In particular, in recent years the plastics supply chain has also opened its doors to materials of plant and animal origin. Among these are polymers with "bio" characteristics, which can be classified into three main subgroups: non-biodegradable polymers, derived from plant biomass and therefore biological (bio-based: bio-PE, bio-PP, bio-PET), and compostable or biodegradable polymers, which in turn are derived from both renewable biomass (PLA, PHA, PHB) and non-renewable fossil raw materials (PBAT, PCL, PBS). The technological challenge that arises in this case is that the new materials have more performing mechanical characteristics that deteriorate minimally "in operation" and that for this reason they can better withstand cyclic and thermal stresses on the products during their use, to allow applications also on products "not only disposable" but where high performance is required.

2 From Recyclable Plastics to Biodegradable Plastics

The development of new plastic materials with "bio" characteristics has evolved starting precisely from thermoplastic synthetic polymers as they are considered recyclable. The main characteristic of thermoplastics, in fact, is above all that of being fusible, to be modeled and return to being rigid when cooled, a cyclic process that can occur numerous times without the material degrading significantly.

2.1 Recycling as a Starting Point for Environmental Awareness

Recycling represents the recovery of a material present in a product that has already been used. It can be performed in various ways: mechanical homogeneous primary recycling, mechanical heterogeneous secondary recycling, tertiary or chemical recycling, quaternary or energy recycling. In the first case, the homogeneous and uncontaminated thermoplastic material derived from the production scraps is recovered and can be processed as a pure material while maintaining the same characteristics. In the second case, materials with lower characteristics are considered because they are regenerated from discarded products that may be homogeneous or heterogeneous. Furthermore, the presence of different polymeric materials can greatly increase the difficulty of the recycling operation. In the third case, a subdivision of the plastic waste is required in order to be reconverted into the starting monomer. The fourth case, finally, is applied in cases of difficulty in separating the different mixed plastics and using them through energy recovery. The recycling of plastics can cause some problems due both to their degradation during the processing phase (with the relative loss of physical and mechanical characteristics), and to the incompatibility between the different polymers for shape or for different fusion temperatures.

Nevertheless, the possibility of being recycled has allowed the countless products made of thermoplastic materials to be recovered after their disposal, in order to subsequently be transformed into recycled raw material, reusable in a new life cycle. If apparently this may appear to be a completely inadequate process for dealing with environmental problems due to the immeasurable production of plastic products, recycling however requires careful and capillary separate collection, which is the first step towards building a sustainable economy, and, even before that, a widespread sensitivity oriented towards sustainability.

2.2 Biopolymers: Biomass, Biodegradability and Compostability

Recent research on ecological plastics is oriented towards new "bio" material scenarios, deepening the biodegradability and compostable characteristics of thermoplastic materials derived from renewable resources such as vegetable oils or starches, also commonly called biopolymers. The term "biopolymer" is often used as a synonym for biodegradable plastic, even if the term includes materials with different origins, features and behaviors.

However, the meaning of some recurring terms needs to be clarified.

A **"bio-based" material** is entirely or partially obtained from plant biomass, therefore it is of biological origin and does not include components of fossil origin (coal or oil).

The term **"biodegradable"** instead means a material, both of natural and synthetic origin, which, through microorganisms (bacteria or fungi), can be degraded into water, into natural gases (such as carbon dioxide and methane), or in biomass. In particular, the ASTM (American Society for Testing and Materials) defines a biodegradable material when it is "capable of completely degrading into carbon dioxide, methane, water and inorganic compounds, or in biomass for which the dominant mechanism is the 'enzymatic action of microorganisms, an action that can be monitored by standard tests over a certain period of time, which reflect actually available environmental conditions" [1]. In practice, biodegradable polymers should easily decompose, in a certain time (90% in less than 6 months), by enzymatic action, into simple molecules that are normally found in the environment.

On the other hand, a more rigorous definition is linked to the concept of **compostability**, since a material to be compostable, in addition to having to be biodegradable, must also be:

- disintegrable, i.e. fragmentable and no longer visible in the final compost;
- not have negative effects on the composting process and on the final compost (for example by reducing its agronomic value or causing toxicological effects on the plant growth process).

Therefore, a biopolymer is not necessarily also biodegradable.

In particular, the Italian Association of the sector, Assobioplastics, simply defines with the term "**vegetable plastics**" all those thermoplastic materials deriving (partially or entirely) from biomass but which are not also biodegradable and compostable [2].

2.3 The Bio-plastics

According to the European Bioplastics Association, by **biopolymer** or **bioplastic** we mean a type of plastic material which, derived from renewable sources or of fossil origin, can be both biodegradable and bio-based, or possess both characteristics [3]. A biopolymer can in fact derive:

- (partially or entirely) from biomass and not be biodegradable (for example: bio-PE, bio-PP, bio-PET)
- entirely from non-renewable raw materials and be biodegradable (for example: PBAT, PCL, PBS)
- (partially or wholly) from biomass and be biodegradable (for example: PLA, PHA, PHB, starch-based plastics).

With respect to the sources, however, it is possible to classify biodegradable polymers in different families, the main of which are:

Starch and PLA (Polylactic Acid) Based Biopolymers: The first generation of biopolymers used corn as a renewable natural source, from which starch was obtained, which can be used directly, to obtain a thermoplastic polymer, or fermented to obtain lactic acid first and then polylactic acid (PLA). These two types of biopolymers found their first applications in the packaging sector, in disposable products (plates, glasses, cutlery) and in the mulching sector, which still remain the most common uses today. Over the years these biopolymers, especially those based on PLA, have also been used in other sectors, such as cosmetics, textiles and paper converting. In addition to corn, the main renewable sources currently used to make biopolymers are potatoes, sugar beets, sugar cane and agricultural waste products.

Biopolymers from Bacterial Fermentation: A family of biodegradable biopolymers that recently appeared on the market is that of plastics obtained from bacterial fermentation. Thanks to bioengineering, it has been discovered that a particular category of bacteria is able to feed on sugars and lipids, coming from renewable natural resources, and transform them, through fermentation, into molecules and substances useful for obtaining plastic materials. The resulting biopolymers, indicated with the initials PHA, PHB, PHBH, are processable like traditional thermoplastics and offer properties similar to polyesters, such as PET, and polypropylene.

Synthetic Polymers from Renewable Resources: Several chemical companies are now able to synthesize molecules identical to those obtained from oil processing, starting from renewable resources. Examples are propanediol, butanediol, adipic acid, succinic acid and ethanol, from which different types of polyesters are obtained (for example polybutenterephthalate PBT) and polyethylene. However, only some of these 'synthetic' naturally derived polymers are biodegradable.

3 Towards Techno-Biopolymers

One of the major limitations associated with some uses of biodegradable polymers, such as PLA, derive from their brittle and rigid mechanical behavior. In some cases, a possible solution to the problem is the addition of plasticizers that allow you to change the mechanical properties.

On the other hand, as mentioned above, not all polymers derived from renewable resources (bio-based) are biodegradable: there is in fact a category of polymers partially or totally derived from natural resources that are made by combining a variable percentage of a naturally derived component with classic fossil-derived components. These polymers, unlike the previous ones, are characterized by high mechanical performances, which are comparable to those of traditional polymers; for this reason they are often defined as "technical biopolymers" and their eco-sustainability is based not only on the fact that they are naturally derived, but also that their use allows a significant reduction of greenhouse gases and non-renewable energy used for their production (energy efficiency, low environmental impact). Among these, for example, polyamides derived totally or partially from castor vegetable oil (PA11), they are very similar in characteristics to other polyamides, presenting high resistance to chemical agents, in particular to hydrocarbons, dimensional stability, relative low density, good processability (molding, extrusion, roto-molding, blow-molding) and withdrawal from the mold. In addition to the electrical, electronic and transport sectors, bio-polyamides are used in the medical, sports and leisure sectors.

In particular, in the medical sector this material is used for chemical stability, low permeability to gases such as oxygen and carbon dioxide and certification for food and medical contact. In the sport and leisure sector, on the other hand, lightness, resistance to abrasion, the possibility of having different degrees of hardness, good impact behavior are the key characteristics that allow the production of components with a high level of performance such as shoe soles, components for tennis rackets or mechanical parts for skis and boots.

Finally, in the field of fabrics, a new generation of products that derive from renewable or natural resources deserve particular attention: fibers derived from corn, milk or bamboo, soybeans and even chitin obtained from crab. These fabrics are united by high comfort and moisturizing and soothing properties. Among these types of materials, the fibers, yarns and polyester fabrics containing activated carbon obtained from the pulverization of the coconut shells are particularly interesting.

4 Conclusions

The limited availability of non-renewable natural resources, together with the environmental emergencies that contemporary society must face quickly and with decisive and resolute ways, raises a decisive question about the excessive use of plastic materials in the production of industrial products. While it is possible to act both on user behavior (waste reduction, separate collection) and on design processes (starting from Life Cycle Design criteria), a decidedly significant role is played by research related to more eco-friendly materials.

With respect to this, it is important to act on the conscience of all the actors of the economic and productive systems, starting with the awareness of the problems at stake and the correct interpretation of the possible solutions. In this it is useful first of all to be aware of the meaning of some terms, sometimes abused, but often not fully understood. Biopolymers, biomass, biodegradability, compostability: these are some of the terms and concepts on which the circular economy models are based and therefore on which it is appropriate to clarify, in order to be able to correctly address the environmental challenges we face.

References

1. ASTM International (American Society for Testing and Materials). http://www.astm.org
2. Assobioplastiche. http://www.assobioplastiche.org
3. European Bioplastics Association. http://www.european-bioplastics.org

Design and Construction of a Prototype for Measuring the Thermal Conductivity of Insulating Materials, Plastics and Metals According to ASTM C177 with Staged Heating

Alex Meza[1], Grimaldo Quispe[2], and Carlos Raymundo[3](✉)

[1] Facultad de Ingenieria, Universidad Continental, Huancayo, Peru
60659958@continental.edu.pe
[2] Facultad de Ingenieria, Universidad Nacional Autonoma Altoandina de Tarma, Tarma, Peru
gquispe@unaat.edu.pe
[3] Direccion de Investigacion, Universidad Peruana de Ciencias Aplicadas, Lima 15023, Peru
carlos.raymundo@upc.edu.pe

Abstract. The design of a better device for the measurement of thermal conductivity is described, which consists of systems such as: cooling system, heating system, cooling system, SCADA system, which for the measurement we will mix the advantages of stationary and transients, speeding up the measurement of this parameter achieved with stepped heating, this device specializes in the measurement of solid insulating materials, widely used in construction as cost reduction or alternative to the use of home heaters, samples can achieve weighing up to 12 kg, the ASTM C177-13 standard and those related to it were taken as a fundamental basis for the design of the device and the specimens to be measured, concluding that this design can perform measurements quickly, efficiently controlling the losses of the radiation and convection heat fluxes.

Keywords: Thermal conductivity · Insulators · Thermal properties · Conduction · Convection · Radiation

1 Introduction

The value of the thermal conductivity (k) of a material is a thermal parameter that is used to design or manufacture materials for a better and low consumption of power or electric energy flow [1]. This parameter is dependent on several values, essentially on the physical properties of the material, such as density (ρ), its crystalline structure, etc. [2]. Today there are many methods to calculate the conductivity and thermal diffusivity of materials. That is why, so the appropriate method must be selected and designed for that, it is necessary to take into account some criteria. 1. If you don't want reliable data with a high mar-gin of error, then elaborate methods are unnecessary. 2. Stationary methods last about 6 to 24 h, but inflict a large number of samples. 3. If you do not prepare samples or specimens with a symmetric or definite geometry, methods that use

D. Russo et al. (Eds.): IHSI 2021, AISC 1322, pp. 645–650, 2021.
https://doi.org/10.1007/978-3-030-68017-6_95

point and constant sources should be preferred [3]. Both national and international standards that require energy efficiency, in many of them it is necessary to know the thermal properties of some of them, that is why it is necessary to have technological devices that allow us to calculate these properties not only for the educational field if not for any project belonging to the industrial sector, construction sector among others [4, 5]. One of the materials that has an influence on the reduction of thermal conductivity It is the crushed glass added to the mortar achieving a reduction of 18% compared to a common reference mortar for testing [6], which indicates that this factor is widely used in the aforementioned sectors and the control of this factor must be taken as a reliable and precise value for use at the engineering level, there are studies on this calculated value, using ASTM standards (C177-13) and ISO standards (8302), which must comply with the values according to the regulations in force in each country, and a few are directly related to heat absorption, such is the case of massive concrete with polypropylene and steel fibers [7, 8]. As mentioned, the magnitude of this parameter is dependent on the physical properties, temperature and pressure are influential in the calculation process for this parameter, therefore, it is also important to synthesize a mathematical model that complies with the approach to the value. real and this is accurate and reliable, having a minimum margin of error [9].

2 Methodology

The design of the thermal conductivity measuring device is based on the recommendations and indications of the German association of engineers VDI 2221 and VDI 2225 (Varein Deutscher Ingenieure) called, systematic approach to the design of technical systems and products. The German standard VDI 2221.2225 consists mainly of the elaboration of the list of requirements, morphological matrix to determine a more efficient solution concept and an economical technical validation quantified in terms of ease of use, among others.

2.1 Sizing

According to the ASTM C177 - 13 standard, it illustrates the main components of the idealized system: which are two isothermal sets with a cold surface and a protection plate, it is possible that some devices have more than one guard, the protected hot plate is composed of a section Thermally isolated measurement of a concentric primary shield by a defined gap, the sample is sandwiched between these three units as shown in the figure, in the double-sided measurement mode, the sample is composed of two pieces and therefore these pieces have to be very identical.

Sample Width. The width of the sample influences how uniform the unidirectional and transversal heat flow is, therefore, we will take as a reference the ISO 8302 standard, which is specific for the measurement of thermal conductivity and general use, a dimensioning of 300 to 500 mm will be taken, For this project, samples of 30 x 30 cm will be made so that it is much easier for us to control certain parameters such as area, and the calculation of heat flow (Fig. 1).

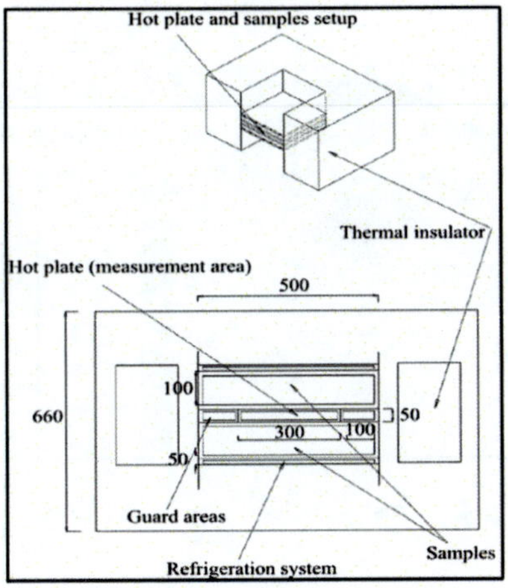

Fig. 1. Sample dimensions

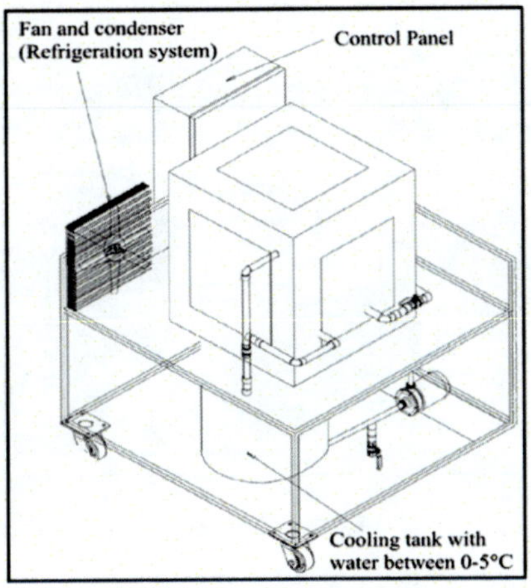

Fig. 2. General assembly

The proposed design will consist of a cooling system driven by a pump and a data acquisition and one driven by push buttons located on a control board that will contain our PLC, among others (Fig. 2).

2.2 Stages of the System

The elaboration of the measurement system by a heating system, a cooling system and a SCADA system (Control and data acquisition software). We depend on the Fourier equation in steady state, where it says that the thermal conductivity is related to the area, thickness, length and thermal flux, where the latter can be quantified with an instrument called a power supply or power source, the range operation will depend on the measurement temperature, in this study values of (10A and 10 V) will be taken, where the heat acts on the resistance in the 30 cm × 30 cm sample, both the thermocouples and the frequency inverter are connected to the PLC TWDLCAE40DRF, the variable speed drive controls the flow of water that circulates for the cooling of cold plate insulating materials according to ASTM C117-13, the voltage difference of the thermocouples are proportional to the temperatures, which thanks to the ethernet communication are obtained with a computer (Fig. 3).

3 Results

3.1 Verification of Resistance of the Structure

With the help of SolidWorks software and its Simulation section, we will perform the stress calculation by analysis using the finite element method and a fine mesh to obtain better results then we have the forces acting on the device support (Fig. 4).

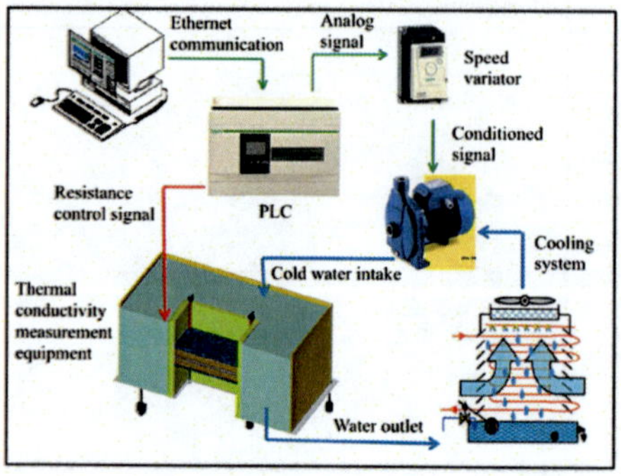

Fig. 3. General system

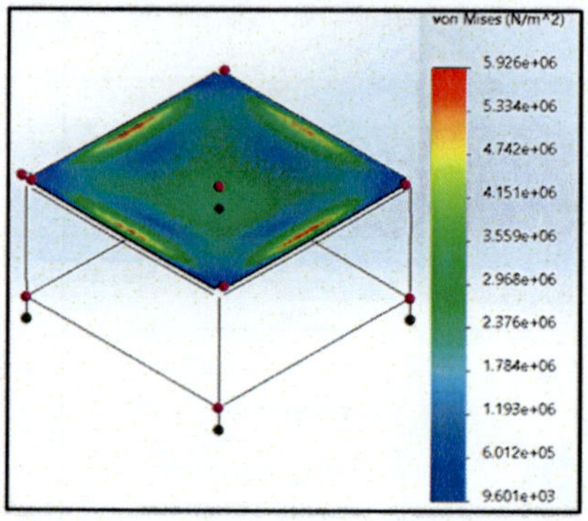

Fig. 4. Stress simulation by Von Mises

We observe that the Von Mises analysis offered by the SolidWorks software we can note that the maximum stress occurs in the rectangular profiles, consequent to the fact that they are responsible for supporting the weight of the measurement system, which has a maximum value of 5,926 MPa, as long as it does not exceed The value of the yield strength of A36 steel which is 210 MPa, we can affirm that the designed structural system will efficiently support the weight of the measurement system.

3.2 Verification of the Deformation of the Structure

When introducing the samples that may have a maximum weight value of 30 to 35 kg, we must avoid any deformations, since the stipulated standard assumes a unidirectional heat flow that states that there are no deviations in the samples or in the structure, since if it existed, the mathematical model proposed will not be effective so the heat flow varies and it would not be possible to calculate quantitatively consequent to this deviation, therefore we analyze the value of the deformation of our structure that supports approximately 700 N of force. Below is the deformation analysis performed by the SolidWorks software Von Mises static analysis tool (Fig. 5).

3.3 Analysis of Air Flow by Convection

For our analysis by convection as mentioned, at an initial moment the circulating heat flow will be approximately 6 Watts, 3 V input with an amperage of approximately 2 A, we performed the simulation with SolidWorks software with the Flow Simulation section to corroborate the temperature of the fluid and the solid and detail how the behavior is when a heat flow is active (Fig. 6).

3.4 Analysis of Heat Flow by Conduction

We can see that the flow of heat or power is 6 W at first, which is why the sample theoretically reaches 16.7 °C in a certain time as we can see in Fig. 7.

3.5 Adjustments to the Final Model

According to the analysis made with the SolidWorks software and its Flow Simulation extension, we note that there is a strong heat flow by convection that causes the air to heat up, as noted in Fig. 8.

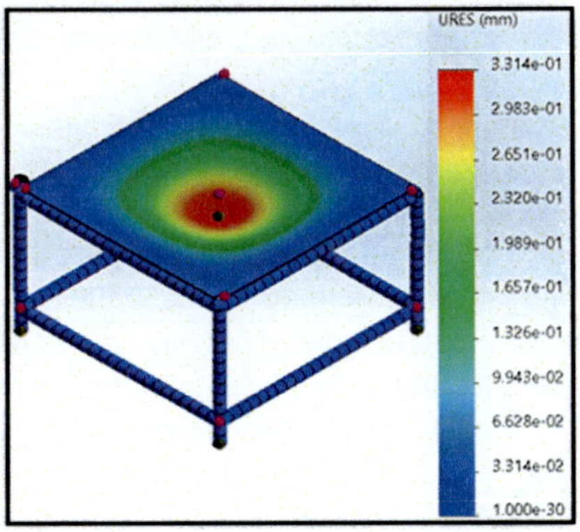

Fig. 5. Structure deformation simulation

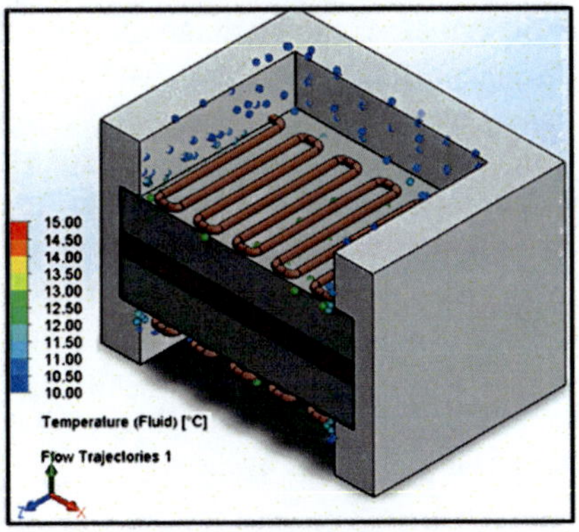

Fig. 6. Airflow behavior

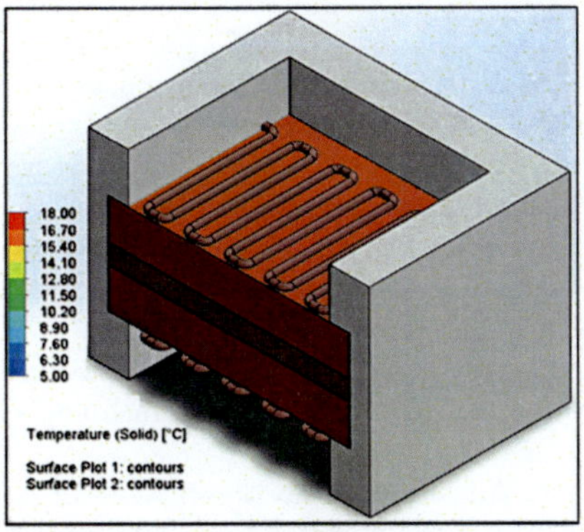

Fig. 7. Heat flow by conduction

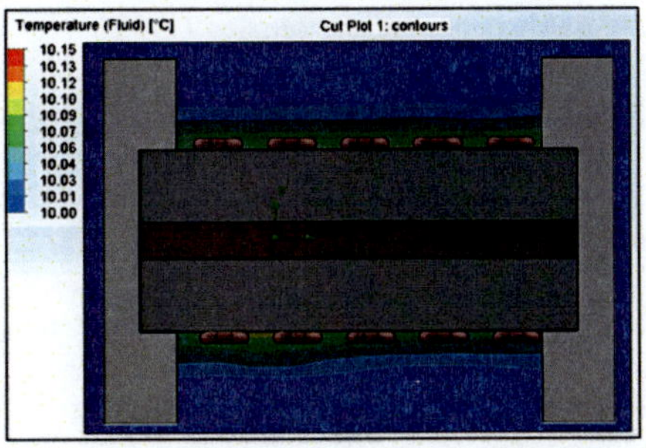

Fig. 8. Cutplot in the middle of the solid

4 Conclusions

The sizing of the equipment is highly optimal, complying with the requirements of the list of requirements to measure the thermal properties of building materials with specimens of 30 cm × 30 cm side (range by ASTM C117-13 standard), guaranteeing good measurement of this type of materials.

The device is designed for the achievement of measuring samples up to 12 kg and the budget is set for around 5740.00 Nuevos soles, which includes the manufacture in case outside and the purchase of materials, accessories excluding the cost of design engineering.

The chosen thermal conductivity parameter measurement technique was performed on the basis that it is a primary measurement method, thanks to its high precision and the probability of measuring other materials.

A finite element analysis was developed to achieve results of the real behavior, as fundamental to those pieces of complex geometry and those that are in contact with the heat flow in order to determine the critical regions of the structural, measurement and space equipment. of samples.

The general design consists of a heating system, a SCADA system, a mechanical system, a cooling system, the mechanical system consists of the structure and the installation of the welding of the coils together with the respective installations of the bomb. The heating system consists of a resistance by wires connected to a 10A and 10 V power source (DC sources). The SCADA system consists of a PLC connected with thermocouples and a frequency inverter that controls the flow of cooling water and the cooling system consists of copper tubes welded to aluminum plates through which water circulates passing through a cooling system and again circulating from the tank thus fulfilling a closed cycle.

References

1. Mills, A.F.: Heat Transfer. Prentice Hall, Los Angeles (1998). ISBN-13: 978-0139476242
2. Welty, J.R., Wicks Ch.E., Wilson, R.E.: Fundamentos de transferencia de momento, calor y masa. México. Limusa, México (1999). 968-18-1306-5
3. Hechavarría, R., López, G., Pazmiño, F., Ureña, M., Hidalgo, A.: Cálculo numérico de las propiedades térmicas efectivas de un commpuesto por elementos finitos. Ingenius, 18, Ambato (2017). 1390-650X
4. Wieser, M., Onnis, S., Meli, G.: Conductividad térmica de la tierra alivianada con fibras naturales en paneles de quincha. ProTerra, La Antigua Guatemala (2018)
5. Torres, R.P.R., Cetina, R.E.C., Lira-Cortés. L.: Equipo para la medición de la conductividad térmica de materiales homogéneos y heterogéneos. ResearchGate, Simposio de metrología, Yucatán, vol. 1 (2014). 978-607-96162-9-8
6. Flores-Alés, V., Jiménez-Bayarri, V., Pérez-Fargallo, A.: Influencia de la incoporación de vidrio triturado en las propiedades y el comportamiento a alta temperatura de morteros de cemento, vol. 57. Cerámica y Vidrio, España (2018). 0366-3175
7. Serrano Somolinos, R., Escamilla, A.C., Prieto Barrio, M.I.: Study of thermal conductivity in mass concrete with polypropylene and steel fibers, vol. 93. DYNA, Ingeniería y industria, Madrid (2017). 978-84-16397-56-3
8. C1043: Diseño de un dispositivo de medición de la conductividad térmica de materiales de edificación según la norma ASTM. http://tesis.pucp.edu.pe/repositorio/handle/20.500.12404/7065. [En línea] 02 de Julio de 2016. [Citado el: 05 de Mayo de 2020] http://hdl.handle.net/20.500.12404/7065
9. Lira-Cortés, L., González Rodríguez, O.J., Méndez-Lango, E.: Sistema de Medición de la Conductividad Térmica de Materiales Sólidos. Simposio de Metrología, Querétaro (2008)

Development of a Methodology for the Learning-Teaching Process Through Virtual and Augmented Reality

Stalyn Ávila-Herrera(✉) and Cesar Guevara

Universidad Tecnológica Indoamérica, Bolivar 2035 y Guayaquil, Ambato, Ecuador
{stalynavila, cesarguevara}@uti.edu.ec

Abstract. This paper presents the development of a learning-teaching methodology, based on the application of virtual and augmented reality. Its aim is to develop a teaching system based on immersive technologies containing 3D virtual environments. Lack of specialized workspace can be partially conciliated with the use of simulators for different areas of knowledge. Technical contributions of researchers in engineering areas are considered. The quality of the methodological design is verified through virtual laboratory sessions in simulated environments. Benefits of having these spaces include greater ease of access to laboratories when student populations are numerous, relieve in conducting controlled experiments without fear of experiencing equipment failures. For the execution of this work, bibliographic search on immersive methodologies applied to learning has been done. Works that contribute to virtual education, enhanced by the application of augmented reality, were selected. The second step has been the construction of a proposal of the teaching methodology for virtualized learning.

Keywords: Augmented reality · Laboratory · Teaching methodology · Virtual reality

1 Introduction

University education, specifically training in the areas of engineering, has the responsibility to train professionals who solve real problems at the industrial level. Therefore, the execution of experimental practice sessions or work that emulate industrial conditions are necessary in vocational training. Having physical laboratories that resemble an industrial environment is difficult to achieve, mainly because this infrastructure involves a high investment. Due to the limited number of workstations, it is necessary that the practices be executed in a group, and with limited times. These problems can be conciliated with the use of specialized simulators for each of the areas of knowledge.

In the field of engineering, computational tools have been used to understand the dynamic behavior of industrial processes. Different software currently allow the simulation of industrial processes, among them are: GASP IV, for sequence and event simulation, CONTROLP for simulation of processes in simple cascade control and re-

D. Russo et al. (Eds.): IHSI 2021, AISC 1322, pp. 651–656, 2021.
https://doi.org/10.1007/978-3-030-68017-6_96

built, SiPAQ-9 used for industrial process, SIMIT developed by SIEMENS, in addition to the mostly used in the academy, such as SIMULINK and LabVIEW. However, these tools do not have a realistic 3D environment that supports the teaching-learning process in an attractive and intuitive way for students [1].

The application of virtual reality (VR) and augmented reality (AR) in different educational projects seeks to achieve synergy between theory and practice, where the teaching of teachers is facilitated, as well as the learning of students through the virtualization of specialized laboratories. Relevant educational applications include *augmented reality in experimental teaching applied at universities*. [2] explored the relationship between learner perception and presence in an AR-facilitated learning environment. A game-type AR learning process was developed, and an experiment was implemented employing a one-group pretest–posttest method. *Teaching systems for industrial automation*. [3] describes a solution built on an open source platform, which underscores the low cost of this system. A combination of learning information and real-world images allows real-time process information, as in a real environment. *Manufacturing systems training laboratory*. [4] presented the combination the elements of the CPS technology concept with modern IT features, emphasizing on simplicity of solution and hardware ease. *Anatomy teaching* [5] presents a work related to the design and implementation of a collaborative augmented system, that introduces a multi-users experience in a face-to-face configuration, used by students in anatomy course. *History teaching* [6] presents the feedback of an experimental protocol using AR as an additional support for a History lesson. *Learning practices in physics*. [7] presents the results of interviews with secondary school teachers about their experience with AR. To garner meaningful information, a prototype application was developed and presented to the teachers. These developments have enabled validating the significant support of the use of VR and RA-based tools in the teaching-learning process, and the support it can provide for applications in industrial environments.

Global development in the industrial field has led to the concept of Industry 4.0 (consists of the introduction of digital technologies into the industrial processes of factories) and thus, to the evolution of other aspects of their environment such as the educational field. Emerging trends in education in relation to Industry 4.0 requirements have been geared towards the use of VR and RA as part of modular learning. In this context, this project proposes development of a specific methodology that orients the learning teaching process through virtual and augmented reality in proposed simulated spaces.

This paper consists of the following sections: literature review, dealing with theory of knowledge and immersive education; the applied methodology, were the model that was considered for the proposed methodology is described; results, and conclusions and future work.

2 Epistemology of the Immersion of Knowledge

From a learning point of view, behaviorism is a change in behavior and the mind is like a black box [8]. Cognitivism poses that learning includes symbolic mental constructs in the mind of the apprentice. The learning process is how these symbolic representations are recorded in memory. Constructivism, for its part, points out that learning is an active process in which apprentices build new ideas or concepts based on their current or past knowledge.

Advances of the sciences are significant in the changes of teaching, learning, and capturing the world in different ways. However, the roles of new technologies in education are of vital importance. Learning may be established as the process of connecting nodes or sources of information; in addition, not only do humans learn as knowledge can reside outside the human being. For this, it is necessary to nurture and maintain connections to facilitate continuous learning. Cognition and emotions should be integrated into decisions to organize knowledge [9]. Thought and emotions influence each other. A learning theory that considers only one dimension excludes a large part of how learning happens. Connectivism is defined as a theory of learning for the digital age. Therefore, one can understand the emergence of this new trend in a social context, characterized by the creation of economic value through networks of human intelligence to create knowledge. Current knowledge is based on an economization of content to give greater emphasis to pre-professional practice. Consequently, the revolution of technology transforms traditional pedagogy around work leading to learning processes. Connectivism supports the approach of learning through the product of sciences and technologies. It attributes open, creative, and inclusive learning oriented to incorporate society using networks that, in turn, are built collaboratively, in educational practice within society.

2.1 Immersive Education

The use of ICT in higher education streamlines the teaching-learning process in face-to-face or non-face-to-face education; technological supports facilitate the development of student skills. Immersive virtualized environments are real or imaginary three-dimensional spaces generated by computer with which the user can interact. They give them the feeling of being inside. These spaces must be suitable so that the user of this type of intervention feels comfortable. In this way, they generate learning that encourages more academic knowledge, as well as social knowledge that helps in the professional life and in the environment in which it will be developed.

University settings are the most benefited from having virtualized environments. This integrates students from different geographical areas who may be limited in conditions to face-to-face education. The learning obtained goes beyond a simple reading of books, by proposing real exercises that challenge the cognition of the student. Augmented reality presents spaces of communication and work, to integrate knowledge and to carry out these principles it is necessary to put in place spaces such as Virtualmind or TeKnoArtia.

2.2 Immersive Methodologies

A methodology should be a set of techniques by which the information process is facilitated, mainly promoting virtual learning. Which methodologies can be used in this type of virtual education? Despite technological advances as a society, computer gaps remain that prevent the subject from understanding these processes. Therefore, the use of methodological strategies for this type of learning has been considered important [10].

One of the innovative methodologies for three-dimensional teaching has been Virtual School Suitcase, which has been referred to in some countries as a didactic-media learning medium; this tool has a pedagogical vision established through communication and interaction. Even as a technological component in the physical separation, it remains as one of the most widely used means today to shorten communication boundaries [11]. Virtual education establishes that content is distributed in electronic formats, this facilitates the creation of a community of active students in training and teachers who share their educational experience in a motivating and effective way, reducing barriers to accessibility of communication [12].

3 Methodology

For the execution of this work, a bibliographic search was developed. Firstly, on the immersive methodologies applied to learning. From this search the works that contribute to the proposed theme were selected, especially because of their commitment to virtual education, improved by the application of AR. The second step has been the construction of a proposal – still subject to revisions – of the teaching methodology for virtualized learning, depending on the roles of the educational actors. The outline of this design is based on [13], as shown in Fig. 1. This model summarizes the fundamentals of instructional design, exploring concepts and procedures for analyzing, designing, developing, and evaluating instruction.

Fig. 1. The systematic design of instruction.

Learning is regarded as a social communicative act - which now has technological support - in which a subject can achieve the development of his intelligence and thoughts through the intervention of another, more competent subject. In this scenario, learning must ensure a change in the learning experience and thinking, based on learning tools and strategies within the framework of structured work that stimulates cognitive, affective, and attitudinal processes. Consequently, in teaching it will be necessary to reflect on the cognitive process because it is the core of all learning. Developing these cognitive processes in students is the fundamental task of the teacher.

4 Results

A virtual laboratory practice guide, in the form of a checkpoint scale, was used to validate the first trials of the methodology. A total of five sessions were conducted, two for robotic arms simulations and three for industrial processes. With the help of a group of volunteer students and teachers, the steps in the guide were reviewed as a practice session took place. This activity allowed to observe a) if the proposed steps were clear, logical, and aided in attaining the session's objectives and b) any possible adjustments that could be made to improve both the methodology and guide.

Hopefully, after testing the final proposed virtual environments, as well as the completion of a final version of a laboratory practice guide, this methodology could be socialized with a wider audience. Thus, it is expected that it becomes a benchmark for the proper application of VR and RA in simulated 3D learning environments. Trials have showed promising results. However, training for both teachers and students before they can use the virtual environments has been necessary. As it has been recorded, not all the participants possess the same abilities to perform appropriately in this type of environments. As this proposal develops further, more attention should also be given to the assessment of the methodology as a whole.

5 Conclusions and Future Work

This work has explored the development of a methodology for the learning-teaching process through virtual and augmented reality. From a pedagogical point of view, this research benefits universities dedicated to the implementation of VR and AR as part of their teaching practices. In the case of Ecuadorian universities, this is a rather unexplored field, which opens opportunities for plenty of research on technology-based educational interventions. Nonetheless, implementation of these initiatives to a greater scale needs the commitment of administrators of higher education and governments. Cooperation among institutions is also to be considered so that efforts are conducive to better implemented developments. Research should now concentrate on how immersive technologies could be matched with sound pedagogical principles. In this way, the methodology mentioned herein, and other models, could serve as efficient methodologies for work in educational settings.

Acknowledgment. The authors are thankful for the support of the Ecuadorian Corporation for the Development of Research and Scholarship (CEDIA) Ecuador to the research project CEPRA-XIV-2020-08, RVA.

References

1. Monroy Reyes, A., Vergara Villegas, O.O., Miranda Bojórquez, E., Cruz Sánchez, V.G., Nandayapa, M.: A mobile augmented reality system to support machinery operations in scholar environments. Comput. Appl. Eng. Educ. **24**(6), 967–981 (2016)
2. Chen, Y.H., Wang, C.H.: Learner presence, perception, and learning achievements in augmented–reality–mediated learning environments. Interact. Learn. Environ. **26**(5), 695–708 (2018)
3. Martin, J., Bohuslava, J.: Augmented reality as an instrument for teaching industrial automation. In: 2018 Cybernetics & Informatics (K&I), pp. 1–5. IEEE (2018)
4. Martin, J., Bohuslava, J., Igor, H.: Augmented reality in education 4.0. In: 2018 IEEE 13th International Scientific and Technical Conference on Computer Sciences and Information Technologies (CSIT), vol. 1, pp. 231–236. IEEE, September 2018
5. Touel, S., Mekkadem, M., Kenoui, M., Benbelkacem, S.: Collocated learning experience within collaborative augmented environment (anatomy course). In: 2017 5th International Conference on Electrical Engineering-Boumerdes (ICEE-B), pp. 1–5. IEEE, October 2017
6. Schiavi, B., Gechter, F., Gechter, C., Rizzo, A.: Teach me a story: an augmented reality application for teaching history in middle school. In: 2018 IEEE Conference on Virtual Reality and 3D User Interfaces (VR), pp. 679–680. IEEE, March 2018
7. Pittman, C., LaViola, J.J.: Determining design requirements for AR physics education applications. In: 2019 IEEE Conference on Virtual Reality and 3D User Interfaces (VR), pp. 1126–1127. IEEE, March 2019
8. Voogt, J., Knezek, G.: Rethinking learning in a digital age: outcomes from EDUsummIT 2017. Technol. Knowl. Learn. **23**(3), 369–375 (2018)
9. Goldie, J.G.S.: Connectivism: a knowledge learning theory for the digital age? Med. Teach. **38**(10), 1064–1069 (2016)
10. Mourão, A.B., Netto, J.F.M.: Inclusive Model for the development and evaluation of accessible learning objects for graduation in computing: a case study. In 2018 IEEE Frontiers in Education Conference (FIE), pp. 1–8. IEEE, October 2018
11. Barvinski, C., Ferreira, G., Machado, L., Longhi, M., Behar, P.: Construction of a socio-affective profile model of students in a virtual learning environment. In: Smart Education and e-Learning 2019, pp. 159–168. Springer, Singapore (2019)
12. Kiat, L.B., Ali, M.B., Abd Halim, N.D., Ibrahim, H.B.: Augmented reality, virtual learning environment and mobile learning in education: a comparison. In: 2016 IEEE Conference on e-Learning, e-Management and e-Services (IC3e), pp. 23–28. IEEE, October 2016
13. Dick, W.: A model for the systematic design of instruction. Instructional Design: International Perspectives I: Volume I: Theory, Research, and Models: volume Ii: Solving Instructional Design Problems, p. 361 (2013)

Electric Circuit Simulator Applying Augmented Reality and Gamification

Vinicio Burgos(✉), Cesar Guevara, and Lorena Espinosa

Maestría en Educación Mención en Pedagogía en Entornos Digitales, Universidad Tecnológica Indoamérica, Bolivar 2035 y Guayaquil, Ambato, Ecuador
eburgos3@indoamerica.edu.ec, {cesarguevara, lorenaespinosa}@uti.edu.ec

Abstract. In Ecuador, the statistics presented by Educational Evaluation National Institution EENI register that 22.8% of the students from the Andean region and 18.3% from the coast, had insufficient grades in areas that include the study of science (Physics, Chemistry, and Biology). This study proposes the development of a physics mobile application that applies the methodology of gamification and increased reality to improve creativity, academic performance, the interaction between students and teachers (collaborative learning) and that is motivating and interesting for reinforce students' knowledge during use. For the development of the application, the theme will focus on the study of direct current electrical circuits. The proposed application will be made using the development environment called Scratch, which will allow us to create applications in a simpler and faster way with quite efficient results. On the other hand, the gamification methodology is applied which allows the student to use the application, learn, and reinforce their knowledge continuously. The functionalities that will make up the application will consist of the ease and versatility of building various associations of electrical resistances (series, parallel and mixed), circuits (batteries, variable and fixed resistors, motors, etc.) and the use of other components that allow students to set challenges and problems who will be able to solve them in a defined time. This application greatly encourages collaborative work and allows constant information by applying in-creased reality to electronic components. The application creates a safe and dynamic environment for learning since multiple tests can be carried out, fostering creativity, and developing reinforcement learning. The tests carried out have been satisfactory with students, carrying out a control group (traditional teaching) and a test group (with the use of the proposed application). The expected results include an increase in the motivation towards the study of the subject, an increase in teamwork, and an improvement in the academic performance of the students by 20%, through evaluation using questionnaires.

Keywords: Augmented reality · Gamification · Electric circuit · Simulator

D. Russo et al. (Eds.): IHSI 2021, AISC 1322, pp. 657–662, 2021.
https://doi.org/10.1007/978-3-030-68017-6_97

1 Introduction

The teaching of Physics in the context of science and technology constitutes one of the greatest problems and challenges, which may be due to the limited knowledge of teachers in training, in this sense different methodological alternatives should be proposed, giving it a sociocultural and research character. Teaching and learning science means connecting theory with practice, knowing the connections between phenomena and processes, as well as their historical con-text [1].

The International Association for the Evaluation of Educational Achievement (IAEEA) is an entity made up of 70 members, between research institutions and educational evaluation organizations, which conducts studies and reports aimed at understanding and improving education around the world. In this way, it develops the assessments TIMSS (Trends in International Mathematics and Science Study), which are international in nature and have been carried out every four years since 1995; being a valuable tool to monitor international trends in mathematics and science achievement, as they evaluate students' knowledge in all are-as, including algebra, physics and chemistry.

Based on the results of 2015, carried out on a sample of 7,764 students from 358 educational centers in Spain and 276,641 international students from 10,265 educational centers, through multiple-choice and open-response tests, they concluded that the percentage of students who have a high level of learning in science is 5.2% and a very low level of 5.2%, with the best scores the students from Singapore (36.7%). Also that the percentage of students who say "I don't like studying science" is 12% in Spain and in the average of the countries that belong to the OECD (Organization for Economic Cooperation and Development), as well as 13% In the total of countries that belong to the European Union, Educational Evaluation National Institution.

Cantador [2] says that the insertion of technology in the educational field produces an increase in motivation, a better collaboration between students and the management of emotions as facilitating elements of the teaching and learning process, for which a transition from ICT (Information and Communication Technologies) to LKT (Learning and Knowledge Technologies).

According to UNESCO (2013), both students and teachers progressively use mobile technologies in various contexts and for a wide range of teaching and learning purposes, currently, the presence of cell phones in the classroom has produced divisions among the teaching community, since there are those who pro-mote its prohibition and others visualize it as a pedagogical tool. The results show that 43.9% of students and 19.6% of teachers use it as elements to down-load applications related to games and recreation, which could and should be used to improve student learning in science [3].

The Educational Evaluation National Institution EENI in Ecuador, is a public entity that periodically collects information regarding the academic performance of students, in different areas, including Physics, Chemistry and Biology. In 2018, the results analyzed indicated that 22.8% of the students from the sierra and 18.3% from the coast, obtained insufficient grades in those areas, while 47.2% (sierra) and 51% (coast) of the students achieved an elementary level; In the Physics subject, it is due in large part to the difficulty that students have in representing physical objects and situations that allow them to understand various phenomena [4].

As a possible solution to this problem [5], propose to improve the interaction between the object and the subject, through the graphic and real representation of physical phenomena, using animation, that is, augmented reality. Authors such as Cabero and Barroso [6] point out that augmented reality (AR) promotes the abstraction and understanding of physical concepts that may be complex, by dividing them into simpler stages, thus achieving their under-standing from multiple perspectives.

The present work develops a mobile application in the subject of physics, focused on the subject of electrical circuits, for high school students, through the use of augmented reality and the application of the gamification methodology. This work is presented as a contribution to improve the learning and motivation of students in this experimental science.

2 Methods and Materials

The effectiveness of a learning methodology lies in several factors such as the characteristics of the student, the teacher and the subject to be taught, as well as the objectives set.

Gamification

The term gamification is originally used by the British software designer and programmer Nick Pelling in 2002, from the business field and it was until 2010 when it began to be used in the educational field. It is designed to allow the trans-formation of products, social or educational aspects into games with clearly established and delimited objectives [7].

Gamification refers to the use of game mechanics in non-playful environments and applications, as a fundamental tool to improve motivation, concentration and the development of values [8].

Augmented Reality

As far as augmented reality is concerned, in 1962 the cinematographer Morton Heiling developed the Sensorama motorcycle simulator, which recreated a ride through the streets of New York, in this regard, the first term coined was virtual reality, which arises in 1965 when Ivan Sutherland publishes the scientific article entitled "A Head-mounted Three-Dimensional Display" and describes the screen as a window through which it is possible to identify a virtual world, allowing it to be seen and perceived as the real world [9].

The beginning of augmented reality (AR), according to Sabarís and Brossy [9] dates back to 1992, where David Mizell and Tom Caudell published their work called "Augmentative reality: an application of heads up display technology to manual manufacturing processes" in response to a variation of virtual reality technology and as an enrichment of the real world, being applied for the first time in the manufacturing industry.

3 Prototype Proposal

The learning circuits prototype was developed for the learning of electrical circuits in series, parallel and mixed with high school students and consists of an application that can be used from any device with data, fixed or mobile.

The topic of electrical circuits was selected because a large number of students have difficulties in understanding how electric current circulates through bulbs, motors or any other resistive element, for this reason, it is considered important and relevant to motivate fourth year high school students around achieving a better understanding of the subject, in a playful and entertaining way.

To develop the prototype, augmented reality was used, that is, beside of simulating an electrical circuit with its corresponding connotations of voltage, intensity and resistance, sound or visual signals are supplied, as indicators that the connection to be made is not secure, it is not correct or that may involve some type of risk, as it happens in a short circuit, high voltage or an overheating of any electrical element.

There are countless limitations for young people to carry out this kind of experimentation in real laboratories, among them are the fact that several educational institutions do not have the economic resources to implement them, and there are also those who show fear that, for a bad connection, a thermal effect is generated that can burn them, or an electric shock that faces some type of risk.

When analyzing the range of technological tools for the development of the application, the programmatic environment of "Scratch" was chosen, as it allows to carry out block programming, in a synthetic, versatile, easily understood and adaptable way for all users, this from a playful perspective, with a gamified teaching-learning system.

Figure 1 shows part of the Scratch programming environment, which makes it easier for an early age student, becomes familiar with block programming and to follow simple instructions to modify it, incorporating game parameters to their liking.

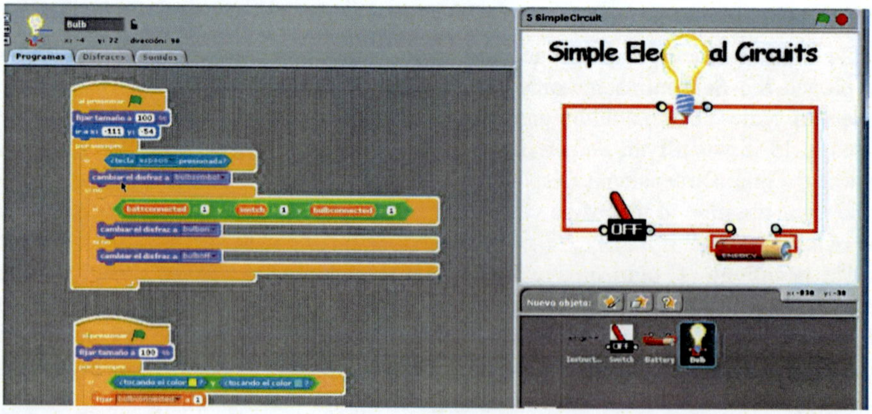

Fig. 1. Scratch block programmatic environment, with application to the diagramming of electrical circuits

The actions undertaken around developing the circuit prototype included defining the type and quantity of electrical devices that could be included, such as bulbs, batteries, resistors, switches, etc. Later establish if the connection that is required to represent is series, parallel or a combination of both, which is known as mixed.

The next step consisted of drawing the objects to represent and how they will interact, in addition to including the necessary cables and switches, after which the blocks, sentences and appropriate coordinates were chosen for the circuit to work under the Scratch programming logic and with the fundamentals of physics regarding the circulation of electrical charge (current) along conductors and other electrical elements.

In the programming dynamics, it is essential to declare the variables and their conditions within the chosen scenario, which is done considering the number of switches present, assigning them a name or symbol and setting their position at 45 o 90°, depending on whether they are open or closed respectively; It should be noted that the switches will allow or prevent the normal circulation of electric charge through one or more parts of the circuit, as the case warrants.

On the other hand, it is necessary to program the operation for the rest of the electrical elements such as lights, LEDs and motors, declaring the conditions in which they should be turned on or off (ON/OFF) in a coordinated and synchro-nous way with the positions open or closed of the switches that are part of the circuit. It is also essential to foresee all the possible combinations that can be made, in such a way as not to incur errors that violate fundamental physical principles, such as the conservation of charge or energy.

In the same sense, messages with their respective associated sound signal were included, which identify them as informative, control, motivational, instructional or safety and that synergistically correspond to alerts on, off, danger, poorly con-nected, you did excellent, you are on the right track and others like those.

Among the functionalities that the application presents, it is included that the student becomes familiar with the two possibilities of presentation of electrical circuits, one provided with the symbology of the constituent devices such as sources, lights, motors, resistors, switches, breakers, conductors and the other in pictorial form with real images incorporated of those elements, all of this in order to effectively and pertinently strengthen the learning of the proposed topic. Logically, the formation of series, parallel or mixed arrangements with these elements is allowed.

The results that are intended to be achieved with the mentioned application include an improvement in the learning of circuits and the capacity for logical-deductive reasoning, in a playful and entertaining way, in addition to awakening motivation and creativity in the student with the proposed topic, which without doubt will undoubtedly lead to a significant advance in a holistic way, in the physical understanding of the phenomenon studied and therefore in the academic performance of the subject, which has been established at 20%, measured through evaluation with various questionnaires.

References

1. El tratamiento a la relación ciencia, tecnología sociedad y medio. https://www.eumed.net/rev/atlante/2018/09/ciencia-tecnologia-sociedad.html. Accessed 29 Sept 2020
2. Gamificación en las aulas. https://www.researchgate.net/profile/Ruth_Contreras_Espinosa/publication/319629646_Gamificacion_en_aulas_universitarias/links/59c8b4cc458515548f3be1d7/Gamificacion-en-aulas-universitarias.pdf#page=67. Accessed 29 Sept 2020
3. Silva Calpa, A.C., Martínez Delgado, D.G.: Influencia del Smartphone en los procesos de aprendizaje y enseñanza. Suma Negocios **8**(17), 11–18 (2017)
4. Jara-Reinoso, A.: Realidad Aumentada aplicada a la enseñanza de la Física de Primero de Bachillerato, February 2020
5. El aprendizaje de la física y las matemáticas en contexto | Flores-García | Cultura Científica y Tecnológica. http://148.210.132.19/ojs/index.php/culcyt/article/view/415. Accessed 29 Sept 2020
6. Cabero Almenara, J., Barroso Osuna, J.: Los escenarios tecnológicos en Realidad Aumentada (RA): posibilidades educativas. Aula Abierta **47**(3), 327 (2018)
7. Education in the Knowledge Society
8. Gamificación: alcances y perspectivas en la ciudad de La Plata. Universidad Nacional de La Plata (2016)
9. Martín-Sabarís, R.M., Brossy-Scaringi, G.: La realidad aumentada aplicada al aprendizaje en personas con Síndrome de Down: un estudio exploratorio. Rev. Latina Comun. Soc. **72**, 737–750 (2017)

Digital Transformation of Education: Technology Strengthens Creative Methodological Productions in Master's Programs in Education

Noemí Suárez Monzón[1(✉)], Janio Jadán-Guerrero[2], Maritza Librada Cáceres Mesa[3], and María Verónica Andrade Andrade[4]

[1] Centro de Investigación de Ciencias Humanas y de la Educación (CICHE), Universidad Tecnológica Indoamérica, Ambato, Ecuador
noemisuarez@uti.edu.ec

[2] Centro de Investigación en Mecatrónica y Sistemas Interactivos (MIST), Universidad Tecnológica Indoamérica, Quito, Ecuador
janiojadan@uti.edu.ec

[3] Departamento de evaluación y planificación curricular del área académica de Ciencias de la Educación, Universidad Autónoma del Estado de Hidalgo, Pachuca, Mexico
maritza_caceres3337@uaeh.edu.mx

[4] Carrera de Educación Básica, Universidad Tecnológica Indoamérica, Ambato, Ecuador
mandrade29@indoamerica.edu.ec

Abstract. Teachers who enter postgraduate training programs in the field of education get prepared for the use of technology to create better conditions for learning. The objective of this work is to identify the potentialities of research and pedagogical training in the use of technologies as a form of professionalization of teachers in Postgraduate Programs in Education. A bibliometric analysis of the degree works reports on the high involvement of postgraduate teaching staff in creative methodological productions, using different technologies to bring children, adolescents and young people closer to learning, but also on the differences in their application in the face of multiple learning needs and knowledge areas. In future degrees, it is necessary to move towards the training of teachers in the applications of artificial intelligence in the educational field.

Keyword: Digital transformation · Higher education · Technology · Teacher training · Creative production

1 Introduction

The digital transformation of education has created better conditions for learning and thus high student motivation [1]. A previous research has confirmed its impact, when used correctly, on academic performance, on commitment, on the active role and on the attitude of students towards the teaching-learning process [2, 3]. However, not all

D. Russo et al. (Eds.): IHSI 2021, AISC 1322, pp. 663–668, 2021.
https://doi.org/10.1007/978-3-030-68017-6_98

teachers have the skills to handle technology. The training has not specified the methodological creative production because the teachers do not have sufficient preparation for research in the field of educational technologies, their pedagogical use and less for the attention to diversity [4, 5].

Traditional teaching methodologies are still valid, even though the importance of technology has been demonstrated to achieve quality education today [6]. It has also been identified that its use does not respond to active teaching methodologies [7].

Despite these limitations, the contribution of technology to education is undeniable. Although it is necessary to advance in the consolidation of competences and skills for use among teachers and students, technology has transformed the teaching-learning scenarios. Thanks to its expansion, new study programs have emerged in innovation-research and educational technologies that contribute to teacher improvement. They allow the transformation and updating of the teaching-learning process from the design of methodological productions on the adequate use of resources, tools and technological applications [8]. These training offers, which are growing in Latin America, mainly in postgraduate education, contribute to the maintenance and strengthening of professional skills.

In this regard, the present study identifies the potentialities of investigative and pedagogical training in the use of technologies as a form of professionalization of teachers in a Postgraduate Program in Education. Specifically, a bibliometric analysis of the scientific production is carried out on the incorporation of technologies in this field, based on the review of the degree works of the Master in Education, Innovation and Educational Leadership Mention of the Technological University Indoamerica in Ecuador.

2 Theoretical Foundations

A series of profound changes are experienced in the educational field as a consequence of technological development and the incorporation of its means, resources and tools in the teaching-learning process. This educational dynamic responds more to the demands of the students and there is a change in the traditional pedagogical paradigm [9], students assume an active role in the proper use of technology that benefits their learning by promoting autonomous work, exploration, visualization of real and multisensory situations, skills development and experimentation during learning [10].

The widespread use of technology in the field of education has created better conditions for interactive processes and diversified attention to students [11]. The crucial nature of the exchanges between students and teachers has been notably clarified, through the analysis of previous research that underlines the influence of motivational and cognitive tutorial processes in the development of a learning model different from the traditional one [12]. On the other hand, other approaches point to the need to generate specific interventions, to create a positive learning space among students from vulnerable sectors of society and the expansion of opportunities for reinforcement in learning [13, 14].

These technologies contributions to the learning process do not result from spontaneity or from the simple incorporation of means and resources to the educational dynamics. The use of technology must be intentional, planned and evaluated, so research is required in this regard to understand the realities of students and their effect on the training process [15]. Based on this, it is possible to design innovative teaching methodologies or strategies that respond to real problems in a given setting. The process of building knowledge at different levels of teaching requires teacher training in technology use. When the teacher investigates and gets involved in this topic, he is able to know the possibilities and limitations of the options that are presented for their integration in the teaching-learning process [16].

Given the technological development, the training offer has also been expanded, mainly through postgraduate programs, which are intended to contribute to education development from options for the development of teaching skills [17]. These training offers that combine innovation, research and technologies have increased worldwide to respond to the problems and tensions of the educational context from the didactic use of technology in creative methodological proposals.

3 Methodology

At present, a bibliometric analysis is carried out on the product of the investigative activity of theses of Master in Education, Innovation and Educational Leadership Mention of the Indoamérica Technological University in Ecuador. This postgraduate program focuses on research skills development and the pedagogy of integration of technologies in learning is treated through the Design Thinking methodology. The interest is to contribute to training action, during three semesters, of a professional who leads innovative educational processes and projects.

To meet the study objective, after action research, a sample of 48 degree papers about use of technology in education from the last three years was obtained to evaluate the results. In these studies, the last three years are quantitatively analyzed in terms of the problem or learning need that they solve, the area of knowledge, the educational level for which the methodological product is proposed and the technological resources used. Qualitatively, the particularities of creative methodological productions were evaluated.

4 Analysis and Discussion of the Results

Scientific production on the use of technology by students of Master's program under study has increased as it is consolidated. If in 2018 only nine methodological productions were contributed, these contributions increased progressively, adding 48 until September 2020. This trend responds, on one hand, to the use of the Design Thinking methodology by the teachers of the program. On the other hand, the high interest of postgraduate students to contribute with innovations on the use of technology to respond to learning needs in contextual educational settings. The methodological products have been proposed for all levels of education, which shows the value of the training offer being open to all teachers in the educational system so that research and pedagogical training in the use of technologies has a positive impact. Although there is diversity in the recipients, the proposals are mainly focused on Basic Education (Fig. 1).

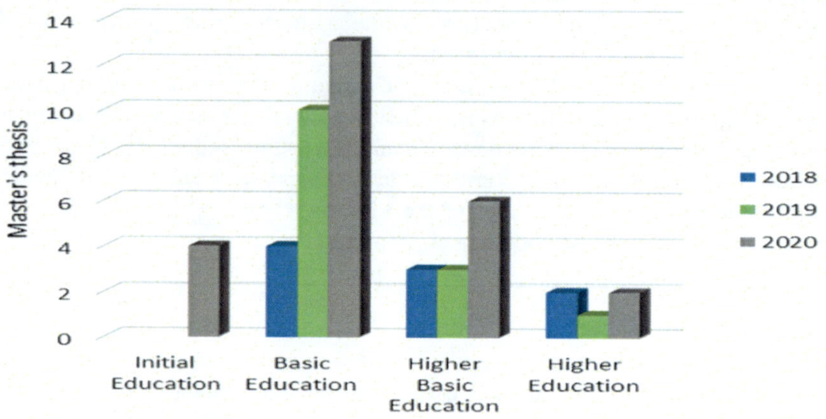

Fig. 1. Scientific production on the use of technology in the Master

Social Sciences is the area of knowledge that benefits from methodological products (Fig. 2). This may be due to the high number of teachers who access the program from this area of knowledge. Despite this, these means, tools and resources can benefit all areas of knowledge and this seems to be fostered little by little through the research and pedagogical training offered by the postgraduate program.

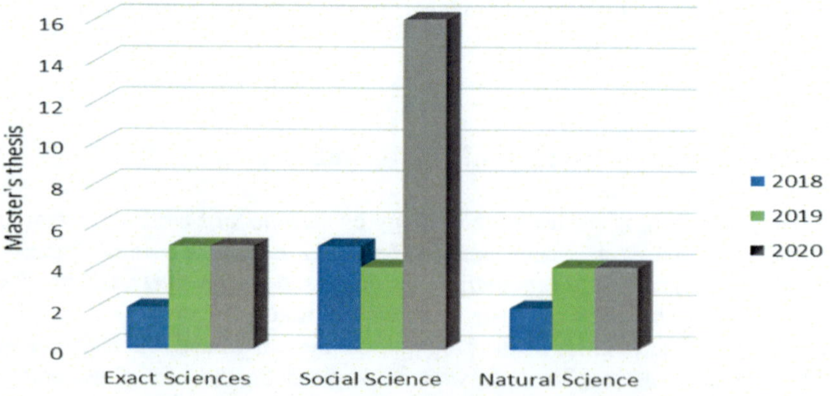

Fig. 2. Areas of knowledge of methodological productions

The methodological productions are specific in their objectives, activities dynamics, resources to be used and forms of evaluation. In each one of them, technology is used in an innovative way that includes from a better management of Platforms to the use of interactive resources, multimedia, videos, audios, content management images and for evaluation. This shows that they make optimal use of active teaching methodologies to respond to identified learning needs. Other proposals focus on promoting the motivation and attitude of students before learning or developing specific skills.

5 Conclusions

After the results presentation, it is concluded that there are differences in the use of technologies in terms of knowledge areas, teaching levels and educational needs. Thus, it is evident that it is possible to articulate educational research with teaching didactics from technology use to achieve a creative product in graduate student that brings these resources, tools, media and applications closer to students. The training favors the design of specialized intervention proposals for the different areas where teachers work.

Finally, it is argued that training interventions carried out on research in educational technologies and the teaching of its use promoted creative methodological productions that respond to the needs of specific educational contexts. It is evident that training offer has a contextualized approach that seeks to improve processes in times when the advancement and use of technology are undeniable.

In the future, the expansion of these training offers and the study of their impact should continue in order to respond to the demands of students, teachers, educational settings and society in general. As technology is constantly advancing, it must also move towards training in artificial intelligence applications in education.

Acknowledgments. The authors thank Universidad Tecnológica Indoamérica for funding this study through the project "Transformation of postgraduate teacher training" and to students of Master in Education program, mention Innovation and Leadership.

References

1. Nickerson, R.S., Zodhiates, P.P.: Technology in Education: Looking Toward 2020. Routledge, New York (2013)
2. Alammary, J.: Educational Technology: a way to enhance student achievement at the University of Bahrain. Soc. Behav. Sci. **55**, 248–257 (2012)
3. Higgins, K., Huscroft-D'Angelo, J., Crawford, L.: Effects of technology in mathematics on achievement, motivation, and attitude: a meta-analysis. J. Educ. Comput. Res. **57**(2), 283–319 (2019)
4. Delgado, A.J., Wardlow, L., McKnight, K., O'Malley, K.: Educational technology: a review of the integration, resources, and effectiveness of technology in K-12 classrooms. J. Inf. Technol. Educ.: Res. **14**, 397–416 (2015)
5. Vickers, R., Field, J., Melakoski, C.: Media culture 2020: collaborative teaching and blended learning using social media and cloud-based technologies. Contemp. Educ. Technol. **6**(1), 62–73 (2015)

6. Buchanan, T., Sainter, P., Saunders, G.: Factors affecting faculty use of learning technologies: implications for models of technology adoption. J Comput. High Educ. **25**, 1–1 (2013)
7. Spiteril, M., Chang, C.N.: Rundgren literature review on the factors affecting primary teachers' use of digital technology. Technol. Knowl. Learn. **25**, 115–128 (2020)
8. Admiraal, W., Vugt, F., Kranenburg, F., Koster, B., Smit, B., Weijers, S., Lockhorst, D.: Preparing pre-service teachers to integrate technology into K–12 instruction: evaluation of a technology-infused approach. Technol. Pedagog. Educ. **26**(1), 105–120 (2017)
9. Law, K., Geng, S., Li, T.: Student enrollment, motivation and learning performance in a blended learning environment: the mediating effects of social, teaching, and cognitive presence. Comput. Educ. **136**, 1–2 (2019)
10. Dunna, T., Kennedy, M.: Technology enhanced learning in higher education; motivations, engagement and academic achievement. Comput. Educ. **137**, 104–113 (2019)
11. García-Tudela, P.A., Prendes, M.P., Solano, I.M.: Smart learning environments and ergonomics: an approach to the state of the question. J. New Approaches Educ. Res. **9**(2), 245–258 (2020)
12. Azer, S.: Interactions between students and tutor in problem-based learning: the significance of deep learning. Kaohsiung J. Med. Sci. **25**(5), 240–249 (2009)
13. Kapasia, N., Paul, P., Roy, A., Saha, J., Zaveri, A., Mallick, R.: Chouhan, P: Impact of lockdown on learning status of undergraduate and postgraduate students during COVID-19 pandemic in West Bengal. India. Child. Youth Serv. Rev. **116**, 105–194 (2020)
14. Zhang, Y., Chu, B., Shu, Z.: A preliminary study on the relationship between iterative learning control and reinforcement learning. This project is sponsored by the China Scholarship Council (CSC). IFAC-PapersOnLine **52**(29), 314–319 (2019)
15. Twining, P.: Planning to use ICT in schools? Education **29**(1), 9–17 (2001)
16. DiPietro, K.: The effects of a constructivist intervention on pre-service teachers. Educ. Technol. Soc. **7**(1), 63–77 (2004)
17. Markauskaite, L., Goodwin, N., Reid, D., Reimann, P.: Modelling and evaluating ICT courses for pre-service teachers: what works and how it works? In: Mittermeir, R.T. (ed.) Informatics Education – The Bridge between Using and Understanding Computers. Springer, Berlin (2006)

Method for Assessing Accessibility in Videoconference Systems

Patricia Acosta-Vargas[1(✉)], Javier Guaña-Moya[2], Gloria Acosta-Vargas[3], William Villegas-Ch[1], and Luis Salvador-Ullauri[4]

[1] Facultad de Ingeniería y Ciencias Aplicadas, Universidad de Las Américas, Quito, Ecuador
{patricia.acosta,william.villegas}@udla.edu.ec
[2] Facultad de Ingeniería, Pontificia Universidad Católica del Ecuador, Quito, Ecuador
eguana953@puce.edu.ec
[3] Facultad de Medicina, Pontificia Universidad Católica del Ecuador, Quito, Ecuador
gfacosta@puce.edu.ec
[4] Universidad de Alicante, San Vicente del Raspeig, Alicante, España
lasul@alu.ua.es

Abstract. The World Health Organization reveals that over one billion persons worldwide living with some disability. On the other hand, videoconferencing applications have become massively popular in our daily lives, from the domestic to the professional, even for the management of public services. Therefore, the use of video conferencing systems by any person, especially the elderly, with a disability or little training, is essential. Furthermore, not all videoconferencing applications comply with accessibility standards. In this study, the six most used videoconferencing systems were selected to evaluate compliance with accessibility guidelines, related to the principles of the Web Content Accessibility Guidelines (WCAG) 2.1 and 2.2. This study has limitations, depending on the experience of the evaluators in accessibility. This study can serve as a guide for expert and non-expert users in the selection of accessible applications.

Keyword: Accessibility · Assessment · Method · Videoconference · WCAG 2.1

1 Introduction

Nowadays, the massive consumption of videoconferencing systems primarily affects the population with sensory disabilities. Following the World Health Organization (WHO) [1] indicates that more than 15% of the world's population suffers from a disability.

The situation caused by the COVID-19 pandemic has accelerated the paradigm shift in the way people interact, in all activities, and at all professional and personal

D. Russo et al. (Eds.): IHSI 2021, AISC 1322, pp. 669–675, 2021.
https://doi.org/10.1007/978-3-030-68017-6_99

levels. Many countries that are yet to have structures designed to deal with this crisis have addressed it as a matter of urgency.

Many applications facilitate teleworking in everyday activities, but not all are accessible, which implies that many users cannot easily access them. This research evaluates the accessibility of five applications most used by users.

The researchers used a manual inspection because no tool evaluates the parameters of accessibility in video conferencing systems. The evaluation applies the Web Content Accessibility Guidelines (WCAG) 2.1 [2] and 2.2 [3]. The WCAG involves four principles that are: perceptible, operable, understandable, and robust. Besides, it comprises three accessibility levels. In most cases, up to the "AA" level of compliance is accepted.

The remainder of the document is constructed as follows: Sect. 2 shows the literature review and background, Sect. 3 describes the methodology and case study, and Sect. 4 presents the outcomes and debate, and finally, Sect. 5 offers conclusions and proposals for future work.

2 Literature Review

Due to the mandatory quarantine decreed in most countries of the world, a high impact on professionals and on the communities where we live has been generated, so people have had to adapt to technology in an accelerated manner, breaking the barriers of time and space.

The WHO disability reports argue that 15% of users suffer from some disability [1]. Therefore, an accessible video conferencing system will allow access to all users, regardless of their disability, to easily navigate the application, facilitating access to many people [4].

Braun [5] argued on 1) the description of audio for people with visual impairment that points to mental shaping, and 2) recommend researching the various forms of communication focused on the methods of understanding and creation.

Acosta-Vargas et al. [6] suggest applying WCAG 2.0 for the educational subject, according to the Web Accessibility Initiative (WAI). The authors analyzed the accessibility of resources in the teaching-learning processes for older adults. They suggested applying methods to help generate accessible resources, including videos and comprehensively.

Funes et al. [7] recommend applying WCAG 2.0 for the construct of video playback and focusing research on the accessibility of videos for people with disabilities.

Salvador-Ullauri et al. [8] aim to use the WCAG 2.1, suggest combining a method with the automatic tool to evaluate the photosensitivity in patients with epilepsy and the manual evaluation in the assessment of serious games.

Previous studies by the authors Acosta-Vargas [9, 10] argue that the application of WCAG-EM 1.0 [11] is fundamental, considering fundamental aspects for the development of software based on WCAG 2.0 [12].

Preceding studies [13] argue that it is not enough to review accessibility barriers with automatic tools when users suffer age-related changes such as low vision problems that make it difficult to access the services offered by computer applications easily.

Therefore, in this research, we defined 21 accessibility barriers for users based on WCAG 2.1[2] and 2.2 [3]. A barrier refers to the impediment that the user experiences when navigating in the videoconference systems. The barriers were determined by the evaluators based on the four accessibility principles 1) Perceivable, the information and application components are presented to users in a way they can easily understand. 2) Operable, the user interface elements and interactions are manageable. 3) Understandable, the information and operations performed by users are easy to understand. 4) Robust, the content is strong and can be interpreted by any user, including assistive technologies.

3 Case Study and Method

As a case study, it was applied to six videoconferencing systems that were randomly selected, among the most used in 2020, according to [14, 15].

Currently, there are no studies related to the evidence of automatic review tools for the evaluation of accessibility in videoconferencing systems. Therefore, a manual evaluation method is proposed considering WCAG 2.1 and 2.2; the process consists of eight phases summarized in Fig. 1.

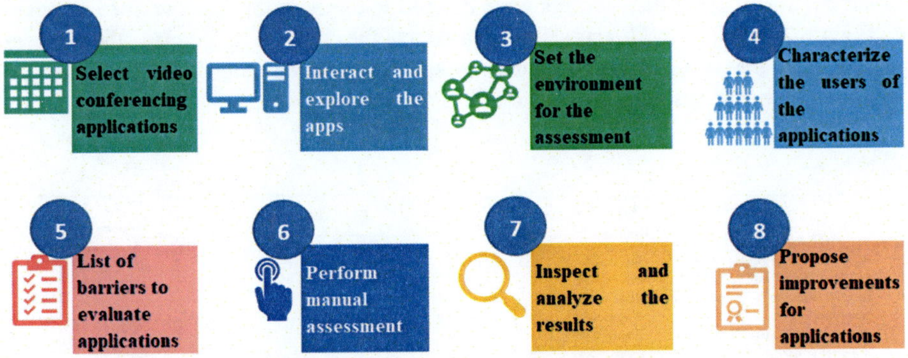

Fig. 1. Method for assessing accessibility in videoconference systems.

Phase 1: The videoconferencing systems that were randomly selected among the most used in 2020 [14, 15], are detailed in Table 1, including the ID, the name, the icon of the tool, the license type, and the URL.

Phase 2: The videoconferencing systems detailed in phase 1 were installed. The experts generated interaction with the videoconferencing systems to identify possible barriers for users using the application.

Phase 3: Define the activities and scenarios that users must perform while running the video conference. The activities were 1) enter the videoconferencing system; 2) navigate the menu; 3) share the whiteboard during a 15-min videoconferencing session.

Table 1. Selected videoconferencing systems to evaluate

ID	Tools	Icon	License	URL
VCA01	Google Meet		Free	https://apps.google.com/meet/
VCA02	GoToMeeting		Payment	https://www.gotomeeting.com/es-es
VCA03	Jitsi		Free	https://jitsi.org/jitsi-meet/
VCA04	Teams		Free	https://products.office.com/
VCA05	Webex		Payment	https://www.webex.com/es/index.html
VCA06	Zoom		Payment	https://zoom.us/

Phase 4: The evaluators selected users who experience interference with daily activities such as reading and driving, including people over 50. Five users were selected who are sufficient to apply in the evaluation [16] with an average age of 55 years.

Phase 5: The accessibility barriers were defined to apply to the users defined in phase 4. Figure 2 (a) details 21 accessibility barriers, the WCAG principles, the success criteria, and the level of conformance applied in the evaluation of accessibility to videoconferencing systems.

ID	Barrier	WCAG 2.1 y 2.2	Success criteria	Level
B01	Content when passing the mouse	Perceivable	1.4.13	AA
B02	Easy to read font	Perceivable	1.1.1	A
B03	Text alternatives	Perceivable	1.1.1	A
B04	Subtitled	Perceivable	1.2.4	AA
B05	Automatic transcriptions	Perceivable	1.2.5	AA
B06	Sign Language	Perceivable	1.2.6	AAA
B07	Sensory characteristics	Perceivable	1.3.1	A
B08	Adjust screen settings	Perceivable	1.3.4	AA
B09	Interface rearrangement	Perceivable	1.3.5	AA
B10	Use of color	Perceivable	1.4.1	A
B11	Well-spaced elements	Perceivable	1.4.12	A
B12	Good audio techniques	Perceivable	1.4.2	A
B13	Improved contrast	Perceivable	1.4.6	AA
B14	Images as sharp as possible	Perceivable	1.4.5	AA
B15	Visual presentation	Perceivable	1.4.8	AAA
B16	Accessible keyboard	Operable	2.1.1	A
B17	Pause, stop, hide	Perceivable	2.2.2	A
B18	Language	Understandable	3.1.1	A
B19	Consistent navigation	Understandable	3.2.3	AA
B20	Help	Understandable	3.3.5	AAA
B21	Screen reader support	Robust	4.1.2	A

(a)

ID	Teams	Zoom	Webex	Jitsi	GoTo Meeting	Google Meet
B01	✓	✓	✓	✗	✗	✓
B02	✓	✓	✓	✓	✓	✓
B03	✓	✓	✗	✗	✗	✓
B04	✓	✓	✗	✗	✗	✓
B05	✓	✓	✗	✗	✗	✗
B06	✗	✓	✗	✗	✗	✗
B07	✓	✓	✗	✗	✗	✗
B08	✓	✓	✓	✓	✗	✓
B09	✓	✓	✓	✗	✗	✓
B10	✓	✓	✗	✓	✓	✓
B11	✓	✓	✗	✓	✗	✓
B12	✓	✓	✗	✗	✗	✗
B13	✓	✓	✗	✗	✗	✗
B14	✗	✗	✗	✗	✗	✗
B15	✓	✓	✗	✓	✗	✓
B16	✓	✓	✓	✓	✓	✓
B17	✓	✓	✓	✓	✗	✓
B18	✓	✓	✓	✓	✓	✓
B19	✓	✓	✓	✓	✗	✓
B20	✓	✓	✓	✓	✓	✓
B21	✓	✓	✓	✗	✗	✓

(b)

Fig. 2. Accessibility barrier according to WCAG 2.1 and 2.2 (a) Evaluation of accessibility in videoconferencing applications (b)

Phase 6: Accessibility experts manually reviewed each videoconferencing tool for compliance with the accessibility guidelines defined in Fig. 2(b). If the videoconferencing systems satisfy the guideline, a check was recorded; otherwise, an X was recorded. The disagreements of the experts were solved with the intervention of a third expert to reach a consensus.

Phase 7: The evaluators recorded the evaluation data of the six videoconferencing systems. Figure 2 reveals the results of the assessment; the dataset is available in the Mendeley repository[1]. The results are discussed in more detail in the results and discussion section.

Phase 8: The authors propose the following suggestions:1) Automatic transcriptions, the system should include an automatic program or bot to perform the transcription of video or audio to text without person intervention. 2) Sign language is a natural language of expression and gesture-spatial configuration and visual perception, through which users with hearing disabilities can establish a communication channel with their social environment. 3) To make the system more accessible, it should include an automated mechanism or bot that provides for live sign language or in the edition of the conference. 4) Include an option so that the user who has photosensitivity problems can control the application's configuration.

4 Results and Discussion

Figure 3 (a) Zoom, Teams, and Google Meet complies with the highest WCAG 2.1 and 2.2.

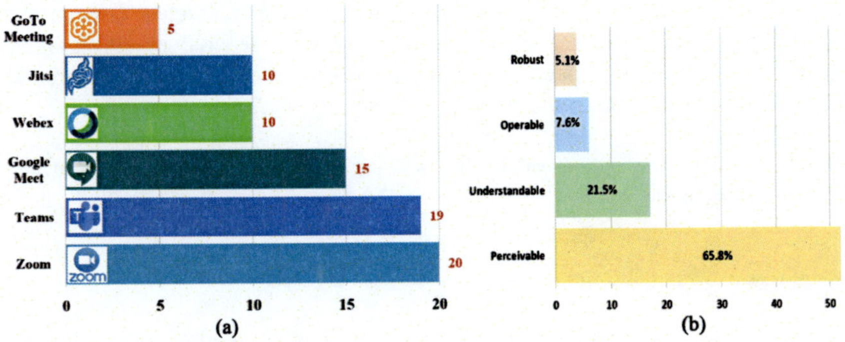

Fig. 3. Videoconferencing systems with a total number of barriers (a). Accessibility principles found in videoconferencing systems (b).

[1] https://data.mendeley.com/datasets/jjngcn3j62/1.

Figure 3 b) shows the accessibility principles found in the evaluated videoconferencing systems. The highest percentage corresponds to the perceptible one with 65.8%, followed by the comprehensible one with 21.5%, then the operable one with 7.6%, and the robust one with 5.1%.

The results of the accessibility evaluation of video conferencing systems indicate that the videoconferencing systems are not accessible because they do not connect the minimum accessibility necessities based in the WCAG 2.1 and 2.2. At the same time, it is suggested to include options that allow: 1) managing the contrast and color in the images, 2) operate in situations with low light and flash activation, 3) guiding the user with an audio assistant, 4) providing an alternative text, 5) ensuring the use of the accessible keyboard. This investigation can serve as a recommendation for less experienced accessibility users to identify the guidelines to choose a platform that meets the most accessibility guidelines.

5 Conclusions and Future Works

The applied method has its limitations; it requires a lot of time to evaluate an application. The level that is most met is A with 54.4%, followed by the AA with 31.6% and AAA with 13.9%. The barriers that are least met are related to the success criterion 1.4.5 and corresponds to *images as sharp as possible*, followed by 1.2.6 referring to *sign language*, 1.2.4 related to *subtitling*. Then the 1.2.5 related to *automatic transcriptions*. In terms of accessibility, the best ones are undoubtedly Zoom, Teams, and Google Meets. They can generate subtitles, but while Zoom allows them to be created in Spanish, Teams and Google Meets develops them in English, which dilutes their potential. We hope that the next versions of video conferencing systems will solve the problems encountered that disrupt the accessibility of millions of new users. This investigation can serve as a character reference for future researches associated with combined and heuristic approaches of accessibility for videoconferencing systems to achieve more inclusive software. The manual method can be reproduced for other kinds of disabilities, contemplating the various accessibility barriers.

Acknowledgments. The authors would like to thank the Pontificia Universidad Católica del Ecuador for funding this work.

References

1. World Health Organization (WHO): 10 facts on disability (2017). https://www.who.int/features/factfiles/disability/en/. Accessed 22 Sep 2020
2. World Wide Web Consortium (W3C): Web Content Accessibility Guidelines (WCAG) 2.1 (2017). https://www.w3.org/TR/WCAG21/. Accessed 20 Sep 2020
3. World Wide Web Consortium (W3C): Web Content Accessibility Guidelines (WCAG) 2.2 (2020). https://www.w3.org/TR/WCAG22/. Accessed 20 Sep 2020

4. Acosta-Vargas, P., Luján-Moras, S., Salvador-Ullauri, L.: Web accessibility policies of higher education institutions. In: ITHET2017, Ohrid, 10th, 11th and 12th of July, 16th International Conference on Information Technology Based Higher Education and Training, Ohrid, Macedonia, pp. 1–7 (2017)
5. Braun, S.: Audio Description from a discourse perspective : a socially relevant framework for research and training (2006)
6. Acosta-Vargas, P., Esparza, W., Rybarczyk, Y., et al.: Educational resources accessible on the tele-rehabilitation platform. In: AHFE International Conference on Human Factors System Interaction 2018, vol. 781, pp. 210–220 (2019)
7. Funes, M.M., Sp, S.C., Sp, S.C., et al.: Gesture4All : a framework for 3D gestural interaction to improve accessibilit of web videos. In: Proceedings of the 33rd Annual ACM Symposium on Applied Computing, pp. 2151–2158. ACM (2018)
8. Salvador-Ullauri, L., Acosta-Vargas, P., Gonzalez, M., Luján-Mora, S.: Combined method for evaluating accessibility in serious games. Appl. Sci. **10**, 6324 (2020). https://doi.org/10.3390/app10186324
9. Acosta-Vargas, P., Luján-Mora, S., Salvador-Ullauri, L.: Quality evaluation of government websites. In: 2017 4th International Conference on eDemocracy and eGovernment, ICEDEG 2017, pp. 8–14. Institute of Electrical and Electronics Engineers Inc. (2017)
10. Acosta-Vargas, P., Hidalgo, P., Acosta-Vargas, G., et al.: Challenges and improvements in website accessibility for health services. AISC **1131**, 875–881 (2020)
11. World Wide Web Consortium (W3C): Website Accessibility Conformance Evaluation Methodology (WCAG-EM) 1.0 (2014). https://www.w3.org/TR/WCAG-EM/
12. World Wide Web Consortium (W3C): Web Content Accessibility Guidelines (WCAG) 2.0 (2008). https://www.w3.org/TR/WCAG20/. Accessed 26 Sep 2020
13. Acosta-Vargas, P., Salvador-Ullauri, L., Luján-Mora, S.: A heuristic method to evaluate web accessibility for users with low vision. IEEE Access **7**, 125634–125648 (2019). https://doi.org/10.1109/ACCESS.2019.2939068
14. HubSpot: Los mejores 6 programas para videoconferencias en 2020 (2020). https://blog.hubspot.es/sales/programas-videoconferencias
15. Business Insider: 9 aplicaciones con las que puedes organizar videoconferencias y reuniones de trabajo online (2020). https://www.businessinsider.es/9-aplicaciones-organizar-videoconferencias-reuniones-online-597077
16. Nielsen, J., Rolf, M.: Heuristic evaluation of user interfaces. In: Proceedings of the SIGCHI Conference on Human Factors in Computing Systems, pp. 249–256 (1990)

Co-creation of Pediatric Physical Therapy Environments: Humanistic Co-design Process

Hadeel Alomrani[1(✉)], Rana Aljabr[1], Rneem Almansoury[1], and Abduallah Alsinan[1,2]

[1] Human-Computer Interaction (HCI) Design Lab, Riyadh, Saudi Arabia
Hadeel.S.Alomrani@gmail.com, aljabr.rana@gmail.com, raneemalmansouri@gmail.com, A.m.alsinan7@gmail.com
[2] King Abdullah University for Science and Technology (KAUST), Thawal, Saudi Arabia

Abstract. In this paper, we describe the co-design process for creating interactive pediatric physical therapy (PT) spaces by applying gamification techniques, with a focus on helping children with gross motor delay which impact their mobility, independence, quality of movement, and balance coordination. The co-design of activities for age-appropriate gross motor skills was conducted through iterative cycles of co-creation with physical therapists and occupational therapists. Consequently, we propose exergaming interactive user interfaces, designed in a way to support improving a child's gross motor performance. The design involves personalized illustrative graphic characters drawn on the room's surfaces. The efficacy of such immersive space will be examined in experimental user acceptance studies and the design implications will be discussed.

Keywords: Design methodology · Human-systems integration · Assistive technology · Gait training

1 Introduction

According to the University of Delaware Physical Therapy department [1], which is ranked as number one in the Nation by US News & World Report, that one of the most essential key points that form the success of a physical therapy session, is the satisfaction of both the patient and the therapist. Physical therapists are usually satisfied by what they do, as it is what they signed up for and aim to do by helping the others; but how can therapists delight the patient, especially if the patient is a child? Many children need physical therapy for various reasons such as, but not limited to, improving movement, managing pain, trying to prevent serious injuries, and most importantly recovering from injuries by doing daily/life activities, or self-monitor and self-manage symptoms and pain. However, most of the children in physical therapy need a long recovery time in physical therapy centers which results in drawbacks like boredom, pressure, and lack of commitment. When it comes to physical therapists when it is a high season, they will be fully occupied and overwhelmed to have more patients, this is why the limited manpower is one of the downsides healthcare systems are currently facing. Instead of only relying on the physical therapists we can further utilize the user

D. Russo et al. (Eds.): IHSI 2021, AISC 1322, pp. 676–682, 2021.
https://doi.org/10.1007/978-3-030-68017-6_100

interface as well. Using user interfaces will revolutionize physical therapy sessions by its higher flexibility, cheaper available setup, and most importantly lower need for manpower with the use of smaller spaces. Approximately, 4.1% of the Saudi population suffers from a physical disability, and around 30 thousand individuals aged from 1 to 9 years, have some form of difficulty [2]. The prevalence of developmental delays in Saudi Arabia among children with 36, 48, and 60 months of age is 16.4% with 1.5% of them dealing with gross motor delay [3]. Gross motor delay is a state in which a child cannot perform tasks involving large muscle groups, such as, crawling, walking, kicking, and jumping. These skills are crucial as they provide the child with independence, mobility, quality of movement, and balance coordination [4, 5]. Thereby, treatments aiding developmental delays are critical as they form the foundation of the individual's future health and social life. Gross motor delay can be treated through physical therapy sessions that include various activities, with gait training being the target of this study, which involves exercises to learn and improve walking. Despite the presence of many physical therapy centers, technologies are limited as they are costly and hard to install. This is one of the many challenges that the Saudi healthcare system is facing in general. Other challenges included limited capacity and lack of research in the field. With pediatric physical therapy sessions, the biggest obstacle is keeping the child engaged. Accordingly, physical therapy interventions play a significant role as an assistive technology during the treatment journey [6–10].

Interactive surfaces can be achieved using conductive inks, acting as sensors. All previously used interactive technology, does not support a large area. However, with the usage of spraying as conductive inks, large areas can be covered, with the flexibility of designing individual patterns for different applications. as well as the ability to spray on a variety of surfaces and irregular geometries [11].

2 Related Work - Designing Pediatric PT and OT Spaces

This section supplies background on interventions used in physical therapy rooms, in developmental delay cases specifically. Alongside with their effectiveness and affordability. Following that we will provide background related to the technology used in this study, user-interface using conductive inks.

2.1 Interventions in Physical Therapy Rooms

Different interventions have been done to improve physical therapy treatment. Body weight-supported treadmill training (BWSTT) is an example for a technology used in gait training. BWSTT is used to carry the body's weight, while providing assistance in the child's posture and stepping. This technology has the advantage of allowing patients to start training early during treatment [6, 10]. Despite all the advantages, this technology is not affordable, and hard to install.

Other lower cost technologies have been used and tested out in physical therapy rooms. Such as, virtual-reality, video gaming and Nintendo Wii™. They all are recommended to be used in the treatment process of developmental delay. Having a variety of complementary interventions to traditional physical therapy exercises, would help significantly in developing skills aimed for during the treatment plan [12].

2.2 Interactive Surfaces Using Conductive Inks

Human-computer interaction (HCI) is becoming an essential part of our daily lives. Many approaches have been used to fulfill the purpose of HCI. Interactive surfaces using conductive inks is one way, which is what we are focusing on. This approach was recently introduced by MIT CSAIL. Where they explored usage of fabrication techniques to make large-scale user interfaces. Single-Layer and Multi-Layer Stencils have been demonstrated. And application examples have been given, such as, Smart home applications, Interactive furniture and Smart cities [11].

3 Co-design of Pediatric PT Spaces

In this section, we describe the Design Innovation (DI) modules that were conducted in this study, which is essentially a part of the humanistic co-design initiative of assistive tech [13]. The modules were presented by SUTD-MIT International Design Center (IDC), these modules fall under the 4 main phases called the 4D process: Discover, Define, Develop, and Deliver, carrying the following mindsets: Joy, Empathy, Mindfulness, and Non-attachment. In the Discover phase, exploring the field of physical therapy, researching about the previous assistive tech, and interviewing the experts were done. Moving to Define, scenarios and sketches were created, in addition to determining the functional requirements. Reaching to develop, iterative cycles of developing designs and receiving feedback occurred. Lastly, we seek for further developments in order to move to the Deliver phase.

3.1 Personas and Scenarios

During the Discover process, Personas and Scenarios were made. Which are used during user-centered design. Scenarios are considered important both for designing assistive technologies and for usability testing with people with disabilities [14]. The goal of the scenario-based personas was to allow the team to sense the user's emotions, interaction, and reactions. Further, it was used to imagine and visualise situations where interactive surfaces could be implemented, which helped shed focus on applying the technology where it is most needed.

3.2 Interviews with PTs

physical therapists (PTs) were met to explore the difficulties they encounter while working with patients in the physical therapy setting to gain a full understanding of their needs in order to design the game. The second iteration of interviews were conducted in aim to receive feedback in regards to the game design. Moreover, the first PT works with Paeds and has experience with developing assistive technologies, PT 2 works with disabled adults, and PT 3 is a pediatric-cases expert and a founder of a physical therapy center that supports child-related cases.

3.2.1 Exploratory Phase

In the ideation phase, PT 1 was met for exploratory purposes, seeking to understand the environment of the therapy rooms. Particularly, issues that hinder the flow of the therapeutic sessions were investigated. PT 1's answers mainly highlighted a need and potential of applying interactive technologies to enhance the commitment of kids. Gait training exercise was recommended to consider.

The previous interview alongside past investigations from literature mapped the design of interactive footprints, in which sprayable user interfaces will be shaped as colorful footprints in the floor, and then, it will be associated with sounds and lights inside the physical therapy room. Two different horizontal sketches were considered as shown in Fig. 1.

Fig. 1. Two different design themes.

3.2.2 Verification Phase

General feedback regarding the design elements was provided by PT 2 and 3 as illustrated in Table 1.

3.3 Collaborative Sketching of Prototypes: C-Sketching

C-Sketching is a rapid method to brainstorm, generate, and build upon ideas. The team iterated through three sketching sessions, all of which supporting the purpose of this project. Samples of the sketches that have been done are shown in Fig. 2. Some of these were considered and developed.

3.4 Real? Win? Worth?

Real? Win? Worth? Is a risk management strategy that assesses the marketability of the project. This method allowed the team to recognize and manage threats to the project's

Table 1. This table illustrates physical therapists' feedback related to the design

Element	PT 2 suggestions	PT 3 suggestions
Design theme	Designing something fun like animals is recommended	Any design would be good as far as it's colorful & attractive for kids
Dimensions & measurements	Suggested to develop 3 sizes for the different age groups	a. suggested to customize the size of the steps according to the kid's age b. suggested to focus on one size, which is the kids between 1.5 years and 3 years
Associate sounds	Include kind and calm sounds	Enable the PT to choose the sounds according to each kid's preferences
Associate lights	Avoid using lights since it might distract the kid	Avoid using lights since it might distract the kid

Fig. 2. Horizontal sketches done during c-sketching sessions.

success, which further resulted into considerations of new iterations of the brainstorming phase. Comparisons between different technologies lead up to choose Sprayable User-Interface. as well as questioning various conditions and users, we chose to focus on gait training in paeds.

4 Design Signature

The design signature of the work described in this Fig. 3. This illustrates the stages that this study went through.

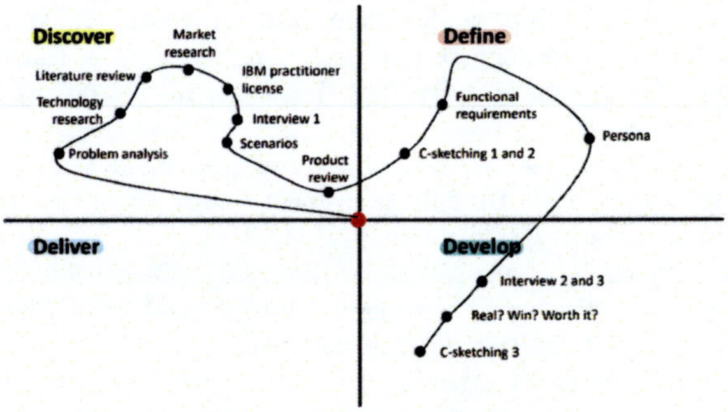

Fig. 3. Design signature.

5 Conclusions

Initial findings, that were extracted from the literature and the series of applied design innovation modules, showed a need of developing an attractive and interactive tool inside PT rooms, providing a smooth user experience as well as serving therapeutic purposes. It has been found that many difficulties occur with pediatric cases including lower body motor delay ones. To illustrate, these were reported in terms of kids' response and commitment with the therapeutic plan, which will also consume the PT efforts and times. Sprayable user interface has the potential to be applied for the gait training exercises since its flexible features. Final design is presented in Fig. 1B.

Future works will include developing the system, then the design implementation in a physical therapy room. Moreover, a number of target users will experiment it in order to evaluate and record their experience and reaction toward it.

Acknowledgments. We thank the Humanistic Co-Design Initiative and the Human-Computer Interaction (HCI) Lab for supporting this work. We also thank the Saudi Authority for Intellectual Property (SAIP) and the Saudi Health Council's National Lab for Emerging Health Technologies for hosting Cocreate's workshop and mentoring. This work is part of the authors' project that is carried out under the CoCreate Fellowship for Humanistic Co-Design of Access Technologies.

References

1. DEPARTMENT OF PHYSICAL THERAPY (n.d.). https://sites.udel.edu/pt/. Accessed 05 Nov 2020
2. Disability Survey (2020). https://www.stats.gov.sa/sites/default/files/disability_survey_2017_ar.pdf. Accessed 05 Nov 2020
3. Shatla, M., Goweda, R.: Prevalence and factors associated with developmental delays among preschool children in Saudi Arabia. J. High Inst. Public Health 10–17 (2020). https://doi.org/10.21608/jhiph.2020.79318

4. Lucas, B.R., Elliott, E.J., Coggan, S., Pinto, R.Z., Jirikowic, T., Westcott McCoy, S., Latimer, J.: Interventions to improve gross motor performance in children with neurodevelopmental disorders: a meta-analysis. BMC Pediatrics **16**, 193 (2016). https://doi.org/10.1186/s12887-016-0731-6
5. Matheis, M., Estabillo, J.A.: Assessment of fine and gross motor skills in children. Handb. Childhood Psychopathol. Dev. Disabil. Assessment Autism Child Psychopathol. Ser. 467–4842018). https://doi.org/10.1007/978-3-319-93542-3_25
6. Damiano, D.L., Dejong, S.L.: A systematic review of the effectiveness of treadmill training and body weight support in pediatric rehabilitation. J. Neurol. Phys. Therapy **33**(1), 27–44 (2009). https://doi.org/10.1097/npt.0b013e31819800e2
7. Poli, P., Morone, G., Rosati, G., Masiero, S.: Robotic technologies and rehabilitation: new tools for stroke patients' therapy. Biomed. Res. Int. **2013**, 1–8 (2013). https://doi.org/10.1155/2013/153872
8. Schmitt, Y.S., Hoffman, H.G., Blough, D.K., Patterson, D.R., Jensen, M.P., Soltani, M., Sharar, S.R.: A randomized, controlled trial of immersive virtual reality analgesia, during physical therapy for pediatric burns. Burns **37**(1), 61–68 (2011). https://doi.org/10.1016/j.burns.2010.07.007
9. Reinkensmeyer, D., Lum, P., Winters, J.: Emerging technologies for improving access to movement therapy following neurologic injury. In: Emerging and Accessible Telecommunications, Information and Healthcare Technologies. RESNA Press (2002)
10. Lowe, L., Mcmillan, A.G., Yates, C.: Body weight support treadmill training for children with developmental delay who are ambulatory. Pediatric Phys. Therapy **27**(4), 386–394 (2015). https://doi.org/10.1097/pep.0000000000000172
11. Wessely, M., Sethapakdi, T., Castillo, C., Snowden, J.C., Hanton, O., Qamar, I.P., Fraser, M., Roudaut, A., Mueller, S.: Sprayable user interfaces: prototyping large-scale interactive surfaces with sensors and displays. In: Proceedings of the 2020 CHI Conference on Human Factors in Computing Systems (2020). https://doi.org/10.1145/3313831.3376249
12. Salem, Y., Gropack, S.J., Coffin, D., Godwin, E.M.: Effectiveness of a low-cost virtual reality system for children with developmental delay: a preliminary randomised single-blind controlled trial. Physiotherapy **98**(3), 189–195 (2012). https://doi.org/10.1016/j.physio.2012.06.003
13. Humanistic Co-design Initiative. (n.d.). https://www.hc-initiative.org/. Accessed 06 Nov 2020
14. Camburn, B.A., et al.: Design innovation: a study of integrated practice. In: International Design Engineering Technical Conferences and Computers and Information in Engineering Conference (IDETC), ASME 2017, pp. V007T06A031–V007T06A031. American Society

Applications and Future Trends

The Openness of Open Innovation in the Product Development Process

Afnan Zafar[1,2](✉)

[1] University of Gävle, Kungsbäcksvägen 47, 80176 Gävle, Sweden
[2] LAB University of Applied Sciences, Mukkulankatu 19, 15210 Lahti, Finland
afnan.zafar@lab.fi

Abstract. Open innovation has strong roots in both academia and industry and resulted in the adoption of this practice across all departments of companies. This paper discusses different criteria needed to adopt open innovation in the product development process. The research explores open innovation in product development based on company needs, strategic planning and managerial preferences. The conceptual framework of this paper was developed from open innovation and management science-related literature. A total of 60 intensive open innovation dependent companies were surveyed in 20 countries concerning the product development process. Statistical techniques were used to analyse the data. The collected data helped to identify six crucial preferences of companies to adopt open innovation. The results provided validation of the conceptual framework and ranked the priorities based on managers' opinions.

Keywords: Open innovation · Companies · Product development

1 Introduction

Development of new products as well as the capture of existing and new markets are typical topics discussed in recent literature and industry [1]. It is also true that new product development and innovations complement each other when it comes to holding on to market shares [2]. As markets becoming more competitive, it is getting harder for innovative companies to regularly produce winning innovations in their closed R&D units. Companies are opening up to collaboration with external partners to find innovation and enhance their product development process [3]. The pursuit of best practice and time-saving methods leads to the openness of companies and forces them to source their product development innovation from an open innovation environment.

Henry Chesbrough [4] defined open innovation as "the use of purposive inflows and outflows of knowledge to accelerate internal innovation, and expand the markets for external use of innovation, respectively. [This paradigm] assumes that companies can and should use external ideas as well as internal ideas, and internal and external paths to market, as they look to advance their technology" [4]. The open innovation environment plays a vital role in product development for companies which are unable to come up with innovative ideas internally.

Past research on the relationship between open innovation and product development has concentrated on two broad issues: success factors of products developed in an

D. Russo et al. (Eds.): IHSI 2021, AISC 1322, pp. 685–690, 2021.
https://doi.org/10.1007/978-3-030-68017-6_101

open innovation environment [2] and speed of generation of new products [2, 3]. There are very few studies available regarding the reasons and needs of open innovation for companies while in product development [5]. Consequently, the picture is incomplete regarding the way companies decide to involve open innovation in product development. Against this background, the purpose of the research is to answer the following **research question (RQ)**: "What are the reasons/needs which companies have that result in their use of open innovation in the product development process?".

More specifically speaking, this paper has two objectives:

- To explore the reasons/needs for open innovations in product development
- To explore the level of importance of the identified reasons/needs for open innovations in product development

This paper responds to the call of exploring reasons from the perspective of actual managers who are developing new products in open innovation environments. The research draws inspiration from Grönlund et al. [2] and Huizingh [5] who have stressed the role of open innovation in product development. The findings of this research are expected to assist professionals in product development units of companies in evaluating and relating their reasons and needs to adopt open innovation.

The paper has four parts. First, it describes related background research and then research methods are presented. Next, the findings and analysis are summarised. The article ends with a discussion and conclusion which contain managerial implications and scope for future research.

2 Research Background

The concept of open innovation was coined by Chesbrough [6] when companies were struggling to come up with inhouse innovations independently. The main areas in companies which were struggling for innovations were product development units [2]. Growing competition in markets and closed R&D departments in the companies were not favourable to the generation of innovative product developments. At this point of saturation, companies started to look for external innovation resources which can be other companies and academic organisations [7]. These developments lead to a functional open innovation environment where internal and external R&D sources started to interact. These interactions resulted in many new, innovative products and pushed the companies to officially adopt the open innovation concept in product development for the good of all stakeholders [2, 8].

Open innovation and product development literature provides an understanding of two groups of companies, first, those who are looking to innovate and do not have the required capabilities, these are referred to as parent companies [9]. The second group is comprised of those innovation providers who have the required capabilities and are ready to sell or collaborate; they are called contract research organisations (CRO) [3]. These innovation providers, in recent decades, become billion-dollar companies. The value of these companies indicates the dependence new product development has on open innovation settings [3]. Research also indicated that the process of open innovation collaboration is getting more complex as the market and competition expand [10].

Nevertheless, today, companies are more dependent on external innovation providers for the development of new products than ever [11]. This trend helped some companies compete in their respective markets, but in many cases, the R&D capabilities of companies were compromised [11, 12]. Companies are looking into stronger regulations for such open innovation product development collaborations. The starting point to make such regulations is to study and analyse the reasons and needs for acquiring open innovation settings for product development [2, 13]. Previous studies have mentioned different needs and reasons up to some extent, but it lacks clarity due to their varying contexts [5, 14]. Some notable reasons to go into such open innovation settings for product development include the parent company's weak R&D, lack of focus and competitiveness in the market, a process of transformation or inadequate strategic direction of the company [3]. Literature shows that all of these reasons have been inconsistently documented as having an influence on open innovation partnerships for product development [2, 5]. There is a lack of clarity in grouping and ranking these reasons and needs based on their importance.

3 Methods

The dataset for this project consisted of professionals with substantial experience in open innovation and product development. The respondents in the research project were from 20 different countries around the world. A total of 60 companies were involved in this research project and the research was conducted throughout 2016–2018. The unit of analysis was each respondent. The sampling frame consisted of 260 randomly selected individuals with a final realised sample of 112 usable questionnaires, which shows a 43% response rate. The initial survey was pretested with a convenience sample of 10 professionals using the participant pretesting method explained by Cooper and Schindler [15]. The research was conducted incentive free, and data were handled according to the General Data Protection Regulation (GDPR). The survey had four parts and a total of 31 questions [3]. Data used for this paper is extracted from question number 17 of the survey, which was stated as follow:

Q17: My company outsource innovations in the product development process because of the following reasons: (list of options provided and explained in Table 1).

Six options and an open text area were provided in the survey form for the answers. The question and optional answers were derived from the literature related to the Chesbrough's open innovation concept and other product development studies. These options include the company's R&D long-term planning is weak, the R&D department is frail and faces difficulties to compete in the market and the company is focusing on other departments (i.e. sales & marketing, manufacturing & production) and is less concentrated on R&D (Table 1). Another three options stated that the company knows the present situation very well and is implementing new plans to tackle the issues related to outsourcing innovations, well aware of the current situation but do not have any strategy to tackle the problem and the decision was made by higher management (Table 1). The respondents were allowed to select more than one options and write open-ended texts. The six options were divided into six groups, and each group was named alphabetically (group A, group B and so on). Table 1 shows the group name, description of each group and the corresponding number of responses.

Table 1. Description of six groups with the corresponding number of respondents based on needs/reasons for open innovation in the product development process

Group name	Group description	Number of responses
Group A	Company's R&D long-term planning is weak	21
Group B	At present, the R&D department is very weak and faces difficulties to compete in the market	20
Group C	Company is focusing on other departments (i.e. sales & marketing, manufacturing & production) and is less focused on R&D which can be readily outsourced	39
Group D	They know the present situation very well and is implementing new plans to tackle the issues related to outsourcing innovations	36
Group E	They know the present situation but do not have any strategy to tackle the issue	11
Group F	A decision by higher management (CEO/GM/MD/Owners/Directors)	53

4 Analysis and Findings

The data were collected from 10 industrial segments with the help of a self-designed research instrument. The industrial elements include sectors such as the pharmaceutical industry (42%), the ICT industry (20%), life sciences (6%), engineering services (11%), financial services (6%), robotics (3%), logistics (3%), telecommunication (2%), the chemical industry (5%) and the automotive sector (2%). The final realised sample consisted of 180 responses to this specific question as each of the 112 participants was allowed to select more than one response. Figure 1 illustrates the distribution of the 180 responses, excel tools were used to develop the pie chart showing the six groups of reasons.

Figure 1 shows the percentage of responses in each group; Group A has 12%, Group B 11% and Group C 22%. Similarly, Group D, E, F have 20%, 6% and 29% responses respectively. The percentage of each group shows the relevance of the reason or need to acquire an open innovation environment in product development. The highest number of responses were for Group F, which means 29% of respondents accepted the influence of higher management (CEO/GM/MD/Owners/Directors) in the decision-making process involving open innovation settings.

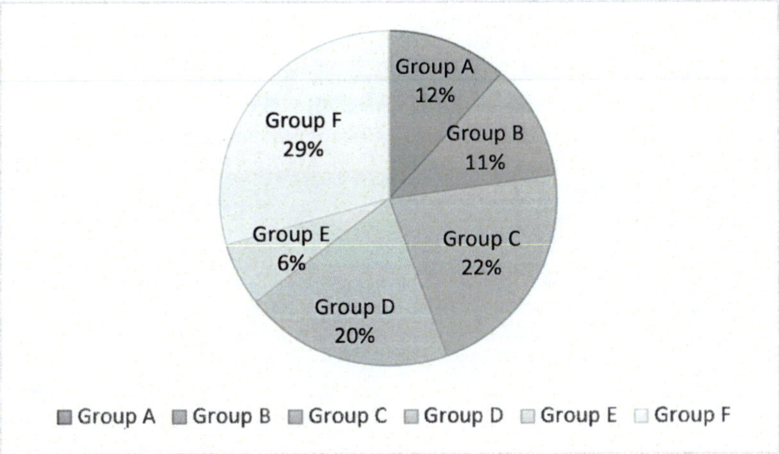

Fig. 1. Data distribution of six groups of respondents based on reasons/needs for adopting open innovation in the product development process

5 Discussion and Conclusion

This paper focused on the potential reasons and needs that motivate the decision to collaborate in an open innovation environment. The main contribution of the paper is to group and rank the reasons and needs to collaborate in an open innovation setting in product development. The relevance of the mentioned reasons complements the work of Grönlund et al. [2] and Huizingh [5] and additionally ranks them in order of importance as the contribution of this study. It was believed that a company's weak R&D planning (Group A) and difficulties in competing in the market (Group B) and lack of strategy to tackle issues (Group E) are the most important reasons to enter into open innovation settings [2, 5]. This study complements the three reasons (Group A, B, E) but expands the literature, indicating that these are not the most critical reasons involved in such contracts. The findings of this study show that the lack of focus towards a company's R&D (Group C) and awareness of the present situation and intentionally entering into well planned open innovation settings (Group D) are important reasons. Another contribution of this research is the validation of Principle Agent Theory, showing that decisions of open innovation in product development are influenced by the company's higher management [3] (Group F).

These findings help to validate the theoretical framework and offer valuable rankings of the reasons and needs involved in open innovation settings in product development. The influence of higher management in decision making can be the turning point of such collaborations. The company should make decisions based on expert opinion or panels of experts to get reliable and comprehenssive results.

References

1. West, J., Salter, A., Vanhaverbeke, W., Chesbrough, H.: Open innovation: the next decade. Res. Pol. (2014). https://doi.org/10.1016/j.respol.2014.03.001
2. Grönlund, J., Sjödin, D.R., Frishammar, J.: Open innovation and the stage-gate process: a revised model for new product development. Calif. Manage. Rev. (2010). https://doi.org/10.1525/cmr.2010.52.3.106
3. Zafar, A.: The Outsourcing Innovation Paradox : A Company's Growth Option or a Risk to R&D Capabilities. Univ. Vaasa Ser., no. 418, p. 128 (2019)
4. Chesbrough, H.: Open innovation: a new paradigm for understanding industrial innovation. In: Open Innovation: Researching a New Paradigm (2006)
5. Huizingh, E.K.R.E.: Open innovation: state of the art and future perspectives. Technovation (2011). https://doi.org/10.1016/j.technovation.2010.10.002
6. Chesbrough, H.W.: The new imperative for creating and profiting from technology (2003)
7. Enkel, E., Gassmann, O., Chesbrough, H.: Open R&D and open innovation: exploring the phenomenon. R D Manage. (2009). https://doi.org/10.1111/j.1467-9310.2009.00570.x
8. New Frontiers in Open Innovation (2014)
9. van de Vrande, V., de Jong, J.P.J., Vanhaverbeke, W., de Rochemont, M.: Open innovation in SMEs: trends, motives and management challenges. Technovation (2009). https://doi.org/10.1016/j.technovation.2008.10.001
10. Chesbrough, H.: The future of open innovation. Res. Technol. Manage. (2017). https://doi.org/10.1080/08956308.2017.1255054
11. Gassmann, O., Enkel, E.: Towards a theory of open Innovation: three core process archetypes. In: RADMA (2007)
12. West, J., Bogers, M.: Leveraging external sources of innovation: a review of research on open innovation. J. Product Innov. Manage. (2014). https://doi.org/10.1111/jpim.12125
13. Bogers, M., et al.: The open innovation research landscape: established perspectives and emerging themes across different levels of analysis. Ind. Innov. (2017). https://doi.org/10.1080/13662716.2016.1240068
14. Lee, S., Park, G., Yoon, B., Park, J.: Open innovation in SMEs-an intermediated network model. Res. Pol. (2010). https://doi.org/10.1016/j.respol.2009.12.009
15. Cooper, D.R., Schindler, P.S.: Cooper Schindler 2006. Business (2003)

Mouse Tracking IAT in Customer Research: An Investigation of Users' Implicit Attitudes Towards Social Networks

Merylin Monaro(✉), Paolo Negri, Francesca Zecchinato, Luciano Gamberini, and Giuseppe Sartori

Department of General Psychology, University of Padova, Via Venezia 8, 35131 Padova, Italy
{merylin.monaro, luciano.gamberini, giuseppe.sartori}@unipd.it, paolonegri0@gmail.com, francesca.zecchinato@studenti.unipd.it

Abstract. The Implicit Association Test (IAT) has been widely used over the past as an implicit measure of cognition. In a recent attempt to better understand the mechanisms underlying the IAT, Yu et al. (2012) modified the classical IAT paradigm introducing a novel classification method based on mouse dynamics analyses (Mouse Tracking IAT; MT-IAT). The present study sought to empirically evaluate the feasibility of applying the MT-IAT to the consumer research field. Specifically, the analysis of mouse movements was applied to explore users' implicit attitudes towards two popular social networks: Facebook and Twitter. Forty participants performed a MT-IAT task, where they were asked to classify Facebook/Twitter and positive/negative images. Results replicated the IAT effect, demonstrating that the mouse response time was significantly shorter in the compatible block as compared to the incompatible block. These findings successfully extended the implementation of the MT-IAT to a novel field of consumer research.

Keywords: Human-mouse interaction · Mouse tacking · Mouse Tracking IAT (MT-IAT) · Implicit Association Test (IAT) · Consumer research

1 Introduction

Classical works in social cognition extensively demonstrated that attitudes and preferences can be activated implicitly, through automatic and unconscious processes [1]. These findings, together with the fact that self-reported measures (e.g., questionnaires) suffer from several weaknesses (e.g., being vulnerable to social desirability biases), led researchers to introduce methods to measure attitudes and preferences at implicit levels, such as the well-known Implicit Association Test (IAT) [2].

The IAT was originally designed to detect the strength of automatic associations between concepts measuring the latency of responses in a classification task [2]. During the task, participants are instructed to classify visual stimuli (e.g., words or images) belonging to four different categories (e.g., flowers/insects and negative/positive images or words). The assumption underlying the IAT is that the subject is faster in the

D. Russo et al. (Eds.): IHSI 2021, AISC 1322, pp. 691–696, 2021.
https://doi.org/10.1007/978-3-030-68017-6_102

categorization when stimuli belonging to mentally associated categories share the same response key (flower/positive or insect/negative; compatible block), as compared to stimuli belonging to non-associated categories (insect/positive or flower/negative; incompatible block). The difference between the response latency in the compatible and the incompatible blocks is defined as the "IAT effect" and indicates the direction of the subject's implicit attitude or preference.

The IAT has been widely used over the past as an implicit measure, mainly to investigate attitudes towards social groups [2]. Moreover, studies in consumer research showed that people develop implicit attitudes and preferences also towards brands and products [3, 4], making the IAT an useful tool in the marketing field [5]. Specifically, Maison and colleagues tested the classical IAT in three different contexts to measure attitudinal preferences of groups of consumers: the first and the second study revealed differences in implicit attitudes between users of two yogurt brands and two fast foods; implicit preferences were consistent with the explicit ones. In the third study, the IAT effect of consumers of two soft drinks predicted brand preference, product usage, and brand recognition in a blind taste test [5].

In a recent attempt to better understand the mechanisms underlying the IAT, Yu and colleagues modified the classical IAT paradigm introducing a novel classification method based on human-computer interaction analyses: instead of using the traditional keyboard keys, participants were instructed to choose their answer via a computer mouse [6]. This new method, known as Mouse Tracking IAT (MT-IAT), replicated the classical IAT effects, while the kinematic analysis of the mouse movements allowed to shed some light on the real-time cognitive processes underlying implicit preferences. Moreover, a first attempt to apply mouse movements analysis to the marketing field was made by Monaro et al. [7].

The present study sought to empirically evaluate the feasibility of applying the MT-IAT to the consumer research field. Specifically, for the first time in this paper, the analysis of the human-computer interaction, through the study of mouse movements, was applied to investigate users' implicit attitudes towards two popular social networks, namely Facebook and Twitter. We hypothesized that the analysis of response times and mouse movement trajectories would be indicative of participants' implicit preferences towards either Facebook or Twitter. Moreover, we expected these implicit preferences to be coherent with the preferences explicitly expressed in self-reported measures, as we have no reason to believe that these are biased.

2 Method

Forty volunteers were recruited for this study (20 males and 20 females; average age = 24.05, *SD* = 3.11; range 19–32). All participants had normal or corrected-to-normal vision and gave their informed consent before the experiment. The experimental procedure was designed in accordance with the Declaration of Helsinki.

Firstly, a preliminary six-item questionnaire was administered. It was aimed at measuring both the frequency of Facebook and Twitter usage and the explicit preference for the former or the latter: (1) I have a Facebook account ("Yes" or "No"); (2) I have a Twitter account ("Yes" or "No"); (3) I utilize: (five-point Likert scale where

1 = "Much more frequently Facebook than Twitter" and 5 = "Much more frequently Twitter than Facebook"); (4) I like Facebook (1 = "Not at all" and 5 = "Very much"); (5) I like Twitter (1 = "Not at all" and 5 = "Very much"); (6) I prefer (1 = "Definitely Facebook than Twitter" and 5 = "Definitely Twitter than Facebook").

Next, to measure their implicit preference for either Facebook or Twitter, participants were asked to perform a MT-IAT. As in Yu and colleagues [6], the MT-IAT included 7 blocks, with 20 or 40 trials in each block. Blocks 4 and 7 represented the critical ones. We created two versions of the MT-IAT so that, for half of the participants block 4 was the compatible block in case they preferred Facebook to Twitter (Facebook + positive images/Twitter + negative images) and block 7 was the incompatible one (Twitter + positive images/Facebook + negative images), while for the other half of the participants the opposite was true. A total of 40 stimulus images were included. Ten images were related to the Facebook logo and ten to the Twitter logo. Additionally, we used ten images having a positive valence (i.e., depicting positive objects), and ten images having a negative valence (i.e., depicting negative objects) from the International Affective Picture System (IAPS) [8]. The stimuli are available upon reasonable request to the authors. A "start" button appeared at the bottom-center of the screen (see Fig. 1). After participants clicked on it, an image was shown in the middle of the screen: participants had to classify it moving the mouse towards one of the two response labels located in the top-left and right corners. As in [6], the interval between trials was 750 ms. To capture the cognitive process underlying the task, participants were solicited to start moving the mouse just after the stimulus appeared, even if they were not sure about the answer.

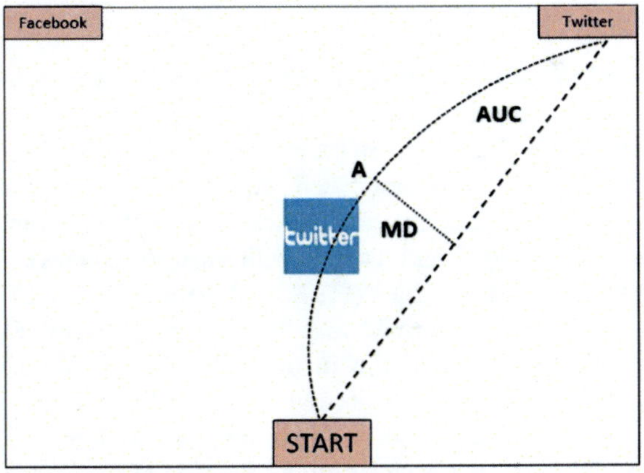

Fig. 1. The figure shows the main spatial measures captured by the software during the MT-IAT task and the experiment interface as it appeared to participants.

The MouseTracker (MT) software [9] was used to present the stimuli and record temporal and spatial features of mouse movements. As for temporal features, the Initiation Time (IT), the time to reach the point of Maximum Deviation (MD-time), and the overall response time (RT) were recorded. As for spatial features, the software registered the Maximum Deviation (MD), the Area Under the Curve (AUC), and the number of changes in direction on the two axes (y-flip and x-flip). For a more thorough description of the features see the original paper [9]. Given that the trajectories had different lengths, liner interpolation was used to normalize them into 101 time steps before conducting the statistical analyses [9]. The analyses focused only on trajectories and times in critical blocks (4 and 7). Therefore, for each subject, we had 80 trials. As in Yu et al. [6], we discarded trials with incorrect response or with initial time exceeding 500 ms, as well as those with an overall response time exceeding 4000 ms. Finally, for each feature, we computed the average value of the 80 trials.

3 Results

3.1 Questionnaire

All participants were familiar with both Facebook and Twitter. Responses to the questions 1 and 2 showed that 19 participants out of 40 had both a Facebook and a Twitter account, while the remaining 21 had only a Facebook account. Thirty-nine participants declared using Facebook much more frequently than Twitter, one participant stated using them in equal measure (midpoint of the scale to question 3). With regard to questions 4 and 5, the overall average rating for Facebook was 3.13 (SD = 0.88), while for Twitter was 2.45 (SD = 1.13). The answers to question 6 revealed that 6 participants preferred Twitter, 30 preferred Facebook, while 4 did not express a preference (midpoint of the scale).

3.2 MT-IAT

First, compatible and incompatible blocks were defined for each participant according to the preference explicitly expressed in the preliminary questionnaire for either Facebook or Twitter (e.g., if the participant preferred Facebook, the compatible block was that in which Facebook was associated with positive images and Twitter with negative images, while the incompatible block was that in which Facebook was associated with negative images and Twitter with positive images).

Next, for all the collected variables (IT, RT, MD-time, MD, AUC, x-flip, y-flip) a paired samples t-test (t) was performed to compare mean values related to the compatible block with those related to the incompatible block. Note that a Wilcoxon signed-rank test (W) was run when the normality assumption was not met according to the Shapiro-Wilk test. To resolve the multiple testing problem, the Bonferroni correction was applied, dividing the p-value by the number of tested features (n = 7) and setting the significance level at 0.007. Results are reported in Table 1.

Comparison Between Implicit and Explicit Measures. Implicit preferences towards one of the two social networks was quantified by the D-index [10], calculated by subtracting the mean MD-time in the compatible block from the mean MD-time in the incompatible block and then dividing this difference by the pooled standard deviation of the two blocks. Participants with an implicit preference towards the social network that they explicitly reported to prefer were expected to have positive D-index values.

Results showed that just 4 out 36 participants (11%) obtained a negative D-index, meaning that, in line with our hypotheses, most of our participants held coherent implicit and explicit attitudes (i.e., positive D-index).

Table 1. Results from the paired samples *t*-test.

MT measure	*M (SD)* compatible	*M (SD)* incompatible	$t_{(df)}$/W	*p*	Effect size
IT	234.36 (104.11)	254.92 (114.17)	$t_{(35)} = -2.772$	0.09	$d = -0.462$
RT	1030.99 (174.91)	1075.49 (188.69)	$W = 153.00$	0.004	$r_B = -0.541$
MD-time	577.60 (88.71)	619.34 (114.35)	$W = 100.00$	<0.001	$r_B = -0.700$
MD	0.297 (0.214)	0.302 (0.197)	$t_{(35)} = -0.317$	0.753	$d = -0.053$
AUC	0.551 (0.455)	0.542 (0.439)	$t_{(35)} = 0.228$	0.821	$d = 0.038$
x-flip	6.543 (1.711)	6.585 (1.592)	$t_{(35)} = -0.268$	0.790	$d = -0.045$
y-flip	5.338 (1.439)	5.378 (1.341)	$t_{(35)} = -0.286$	0.777	$d = -0.048$

Note. Table reports the results of the *t*-test or Wilcoxon signed-rank test comparing the mean values related to the compatible block with those related to the incompatible block. Effect sizes are also reported (Cohen's *d* for *t* and rank-biserial correlation – r_B, for *W*)

4 Discussion and Conclusion

The present study tested the feasibility of applying the kinematic analyses of mouse movements to a MT-IAT conceived to investigate users' implicit preferences towards the social networks Facebook and Twitter.

Although no significant differences between compatible and incompatible blocks emerged regarding the spatial features of the mouse trajectories, the analysis of the temporal parameters (RT and MD-time) yielded statistically significant results. Indeed, participants' mouse movements were overall faster in compatible blocks (i.e., when the preferred social network was associated with positive images and the non-preferred one with negative images) as compared to incompatible blocks. These findings suggest that in compatible blocks the decision process is perceived as easier, implicitly reflecting the preference of the participant. Furthermore, participants' implicit attitudes, quantitatively measured via the mouse tracking analyses, were demonstrated to correlate with their explicit (i.e., self-reported) preferences.

In summary, our results replicated previous findings [5, 6] and successfully extended the implementation of the MT-IAT to a novel field of consumer research, demonstrating the effectiveness of this tool to assess implicit preferences towards social networks. We argue that this study represents an important contribution to the field of human-computer interaction (HCI), introducing new possible applications of the IAT to the marketing field. Indeed, given that this method does not require explicit responses from users, the kinematic analysis of data collected through a MT-IAT has the potential to represent a brief, informative and easy-to-administer procedure to measure consumers' implicit attitudes and preferences as they interact with a wide range of services, technologies and products. For instance, it could be extremely beneficial for Social Networking Services (SNSs) developers and other digital platforms (e.g., Netflix, Spotify, Amazon), especially considering the recent growth of the offer in this field and the consequent competition [11]. Furthermore, the mouse tracking adaptation of the IAT, compared to its traditional version, allows a better understanding of the real-time cognitive processes underlying consumers' decision-making processes.

References

1. Bargh, J.A.: The automaticity of everyday life. In: Wyer Jr., R.S. (ed.) The Automaticity of Everyday Life: Advances in Social Cognition, t. 10, s. 1–61 (1997)
2. Greenwald, A.G., McGhee, D.E., Schwartz, J.L.K.: Measuring individual differences in implicit cognition: the implicit association test. J. Pers. Soc. Psychol. **74**, 1464–1480 (1998). https://doi.org/10.1037/0022-3514.74.6.1464
3. Branthwaite, A., Lunn, T.: Projective techniques in social and market research. In: Walker, R. (ed.) Applied Qualitative Research, pp. 101–129. Gower Publishing Company, Aldershot (1985)
4. Hussey, M., Duncombe, N.: Projecting the right image: using projective techniques to measure brand image. Qual. Mark. Res. Int. J. **2**, 22–30 (1999). https://doi.org/10.1108/13522759910251918
5. Maison, D., Greenwald, A.G., Bruin, R.H.: Predictive validity of the implicit association test in studies of brands, consumer attitudes, and behavior. J. Consum. Psychol. **14**, 405–415 (2004). https://doi.org/10.1207/s15327663jcp1404_9
6. Yu, Z., Wang, F., Wang, D., Bastin, M.: Beyond reaction times: incorporating mouse-tracking measures into the implicit association test to examine its underlying process. Soc. Cogn. **30**, 289–306 (2012). https://doi.org/10.1521/soco.2012.30.3.289
7. Monaro, M., Cannonito, E., Gamberini, L., Sartori, G.: Spotting faked 5 stars ratings in e-commerce using mouse dynamics. Comput. Hum. Behav. **109**, 106348 (2020). https://doi.org/10.1016/j.chb.2020.106348
8. Lang, P.J., Bradley, M.M., Cuthbert, B.N.: International affective picture system (IAPS): technical manual and affective ratings. NIMH Cent. Study Emot. Atten. **1**, 39–58 (1997)
9. Freeman, J.B., Ambady, N.: Mousetracker: software for studying real-time mental processing using a computer mouse-tracking method. Behav. Res. Methods. **42**, 226–241 (2010)
10. Greenwald, A.G., Nosek, B.A., Banaji, M.R.: Understanding and using the implicit association test: I. an improved scoring algorithm. J. Pers. Soc. Psychol. **85**, 197–216 (2003). https://doi.org/https://doi.org/10.1037/0022-3514.85.2.197
11. Kleineberg, K.-K., Boguñá, M.: Competition between global and local online social networks. Sci. Rep. **6**, 25116 (2016)

Uncanny Valley in 3D Emoji: Investigating User Perceptions of Realistic Representations

Juhee Chung(✉) and Soojin Jun

Graduate School of Communication and Arts, Yonsei University, 50 Yonsei-ro, Seodaemun-gu, Seoul, Republic of Korea
{chuhii815, soojinjun}@yonsei.ac.kr

Abstract. In this study, we examined the effects of user perception of 3D emojis' facial expressions in regards to levels of likeability and human likeness. To create the scale for measuring realism levels, six 3D emoji face samples were created depending on levels of realism. Based on the results (N = 151), we found that likeability and human likeness were significantly reduced in inverse proportion according to the levels of realism for the 3D emoji. These findings revealed the implications of 3D emoji or avatar design in message communication.

Keywords: 3D emoji · Uncanny valley · Face perception · Human likeness · Likeability · Realism consistency

1 Introduction

Emojis have become one of the most popular means of SNS communication because of the advantage of conveying emotions in an engaging way. While the existing emojis, mostly 2D representations, have been used by finding an appropriate match in the emoji list to express one's feelings, 3D emojis, such as Apple's Memoji and Samsung's AR Emoji [1], enable users to send personalized emojis to others that reflect their own faces. With the emergence of 3D emojis, the emoji's role is expanding, going from being an alternative to conveying emotions indirectly in text messages to using avatars that project a user's identity in a direct and personalized way [2]. In the field of human-computer interaction (HCI), several studies have investigated human reactions to realistic replicas (e.g., robots and avatars) [3], and they have found that unpleasant sensations occur when facial images are very realistic [4]. In particular, when some features of the replicas are recognized as human and others as non-human (e.g., realistic eyes and computer-rendered skin), they are perceived to be the most unpleasant [5]. This inconsistency in the realistic representation of a human-like replica causes disconcerting feelings in viewers, which is called an uncanny valley [6]. In addition, the aforementioned previous studies have shown that the proporitons of the face and the level of realism can affect perception and influence recognition. This is particularly true in regards to the eyes and mouth as humans usually focus on the eyes and mouth when communicating with others [7]. As shown by Mori's (1970) uncanny valley, the degree of realism of an emoji can influence users' perceptions. However, despite the widespread use of emojis in message communication, little research has been conducted on

D. Russo et al. (Eds.): IHSI 2021, AISC 1322, pp. 697–702, 2021.
https://doi.org/10.1007/978-3-030-68017-6_103

users' perceptions of them, particularly the extent to which they affect reality. Therefore, specific verification is needed. The purpose of this study was to examine the effects of 3D emojis on users' perceptions based on phases of realistic facial appearance. In particular, the study aimed to determine whether there is a difference in likeability and human likeness depending on different levels of realism. We aimed to answer two research questions:

2 Method

Our experiment aimed to find differences in perception according to the level of realism of a 3D emoji (see Fig. 1).

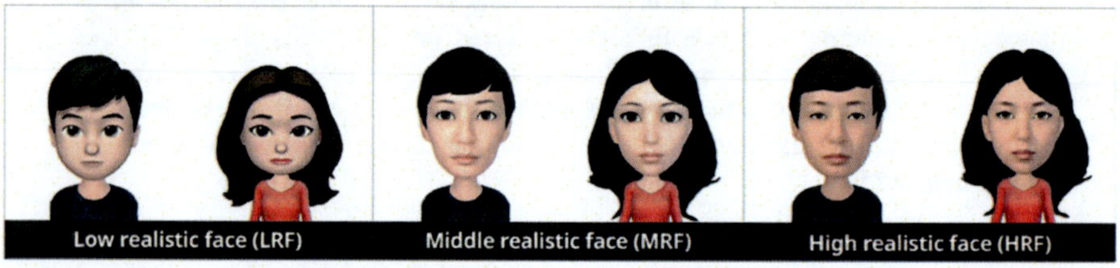

Fig. 1. For evaluation, each female/male 3D emoji was selected according to its level of realism: *(left) low realistic face, (middle) middle realistic face, and (right) high realistic face.*

2.1 3D Emoji Conditions

The three 3D emoji conditions tested in this study were as follows:

a. Low Realistic Face (cartoon-like)
b. Middle Realistic Face (cartoon-like + human-like)
c. High Realistic Face (similar to a real face)

The sample images chosen were Apple's Memoji and Samsung AR emoji, both of which are widely used in the market [8]. All the 3D emojis were designed in male and female versions.

2.2 Procedure

Overall, the survey procedure consisted of three parts: (1) Participants were given a mock-chat conversation in which one of the three sets (LRF, MRF, or HRF) was used, having been randomly assigned (see Fig. 1). (2) After viewing each 3D emoji sample in a dialog, all the participants answered a short questionnaire. The questionnaire consisted of eight items under two categories: human likeness and likeability (see Table 1). Each question was evaluated on a seven-point scale [9, 10]. (3) After the survey, an in-depth, semi-structured interview was conducted for deeper insights. The interview questions covered overall 3D emoji experiences, preferences, and feelings for 3D emojis.

Table 1. Questionnaire series.

Category	Questions
Human-likeness	Machine-like/human-like Unconscious/conscious Artificial/lifelike Fake/natural
Likeability	Nice/awful Friendly/unfriendly Kind/unkind Pleasant/unpleasant

2.3 Participants

We recruited participants through social media and online survey communities. In all, 151 participants completed in the online survey (65 males, 86 females; M = 27.56 years, SD = 8.66). After the survey, we interviewed 15 of the 151 participants who reported using 3D emojis more than three times a week.

3 Results

Identifying non-normality in the questionnaire data, the non-parametric Kruskal-Wallis test and Mann-Whitney test were conducted to determine whether there were differences in human likeness and character likeability based on the types of 3D emoji. In addition, follow-up interview data was subjected to thematic analysis and open coding [11].

3.1 Human-Likeness

As a result, human likeness had a significant effect based on the realism level ($H(2) = 33.93$, $P < 0.001$). Pairwise comparisons using the Mann-Whitney test further identified that the LRF emoji ranked significantly higher than both the MRF and HRF (LRF-MRF, $P < 0.001$; LRF-HRF, $P < 0.001$). No significant differences were identified in comparing the constructs between the MRF and HRF (MRF-HRF, $P > 0.05$). This result can be interpreted as a significant decrease in the level of human likeness recognized by the user according to realism levels (see Fig. 2).

The interview results showed a trend similar to that of the quantitative data analysis. In all, 10 out of the 15 participants reported that higher levels of realism created more artificial, rather than human-like, feelings [6]. This result supported the hypothesis of the uncanny valley. In particular, participants replied that the reason they found the Low Realistic Face (LRF) more relatable was because of the emoji's big eyes and exaggerated facial emotions. This was also found in some of the interview responses: *"Rather, the Low Realistic Face (LRF) looks more like a person… It feels like emoji has feelings…" (p4)* In contrast, two of the rest of participants reported that the more realistic the 3D emoji was, the more humanlike it felt. Participants were observed to

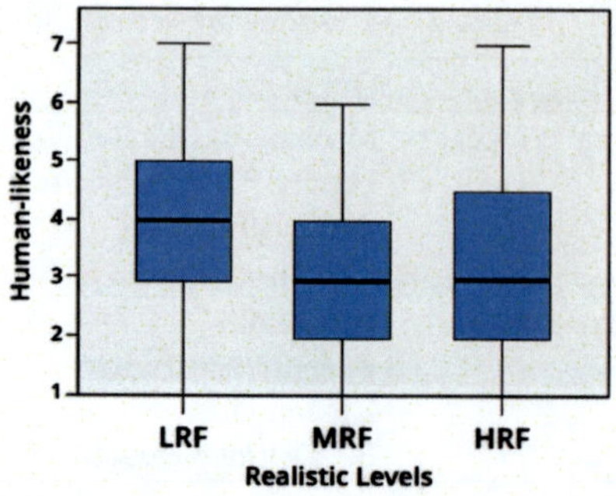

Fig. 2. Human-likeness interaction according to level of realism.

rate the 3D emoji as being similar to people but negative. Overall, the Low Realistic Face (LRF) emoji had a variety of expressions, making it highly regarded as a human-like emoji. However, the High Realistic Face (HRF) emoji was recognized as artificial. This verified that 3D emojis are more likely to be perceived as human-like based on diversity and naturalness of facial expressions rather than a high level of realistic representation.

3.2 Likeability

The Kruskal-Wallis results showed a significant difference in likeability responses on the basis of the 3D emoji's level of realism ($H(2) = 138.69$, $P < 0.001$). Moreover, post-hoc Mann-Whitney U tests identified the LRF preference (LRF-MRF, $P < 0.001$; LRF-HRF, $P < 0.001$) as compared to the MRF and HRF groups. As shown in Fig. 3, preference gradually declined according to the level of realism. This can be interpreted as supporting the argument of Bartneck (2007) who proposed an uncanny cliff rather than an uncanny valley [12].

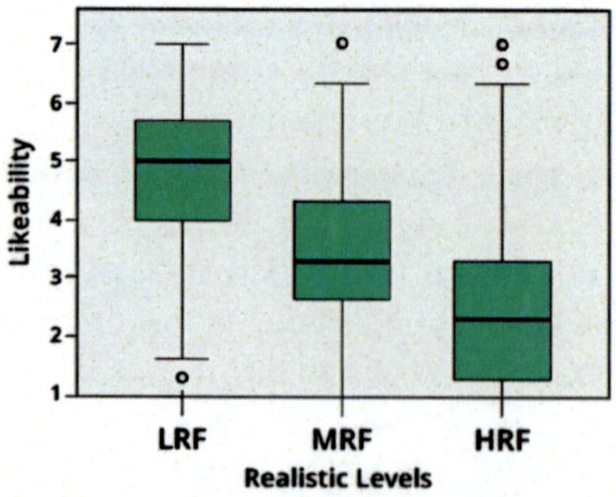

Fig. 3. Likeability interaction according to level of realism.

According to the interviews, we found that the Low Realistic Face (LRF) was the most appealing because of its ability to express emotions, which is the main purpose of an emoji. In all, 8 of the 15 participants reported that the 3D emojis reflecting reality were MRF and HRF, but their preferred image was LRF. This is because the components of the LRF are harmoniously organized without any heterogeneity (i.e., round face and computer-rendered skin texture, big eyes): *"I liked the Low Realistic Face (LRF) character because it seemed to focus on the round shape and cute details" (p10).* In contrast, 4 of the 15 participants mentioned that they felt unnerved because the High Realistic Face (HRF) looked human-like while its body looked like an animated graphic. Unintended discomfort was caused because the reality of the 3D emoji was not consistent (i.e., realistic face and computer-rendered body), which further supported previous research [1]. In particular, the participants reported that the High Realistic Face's (HRF's) eyes were unpleasant, and that they felt as though the eyes were staring at them. Participant 13 mentioned, *" felt HRF was weird and unpleasant because the character's eyes are too realistic, and I feel like they're staring at me" (p13).* This finding also reflected the users' perceptions of communicating through the shape of the eyes and mouth [7], and the human trait of expressing emotions through eyes. In summary, we found that realism inconsistency increased the participants' discomfiture and decreased the degree of an emoji's likeability. These results suggested that proportions and consistency of realism were key variables to evaluate affinity with a 3D emoji.

4 Design Implications

Based on these findings, we have two suggestions that could be applied to the 3D emoji design: First, the human likeness of a 3D emoji should reflect reality, but a consistent expression of reality should be applied [5]. If the degree of reality of the object is not designed consistently, it will lead to an uncanny valley, so all the expressions should be designed by accurately understanding the characteristics of the object.

Second, the use of archetypes and unrealistic proportions (i.e., big eyes and foreheads, small noses and chins) can instill a character with a familiar image [13]. In particular, it is important to consider that the eyes and the mouth of a character can be an effective means of expressing the feelings of an emoji.

5 Conclusions and Future Works

In this study, we examined the effects 3D emojis' facial expressions on user perceptions of likeability and human likeness and found that users' preferences differed according to the levels of realism of the emoji. We also observed that unconditional realism in 3D emoji design did not reflect the degree of human-likeness. Therefore, it is necessary to elaborate the design of a 3D emoji in consideration of an uncanny valley. Despite the implications, in this study, we tested the default values of 3D emojis and did not apply avatars that reflected the faces of the real participants in the experiment. In the future, we intend to conduct a quantitative study on whether users reflect the self-image when their real face is reflected in a 3D emoji.

Acknowledgments. This work was supported by the Ministry of Education of the Republic of Korea and the National Research Foundation of Korea (NRF-2019S1A5A2A01045980).

References

1. How to create, customize, and use Memoji in Apple's iOS 13. https://www.digitaltrends.com/mobile/how-to-make-a-memoji/
2. Apple Memoji vs Samsung AR Emoji: How do they work and which is best? https://www.gearbrain.com/apple-memoji-samsung-ar-emoji-2579721007.html
3. Seo, Y., Kim, M., Jung, Y., Lee, D.: Avatar face recognition and self-presence. Comput. Hum. Behav. **69**, 120–127 (2017)
4. Seyama, J.I., Nagayama, R.S.: The uncanny valley: effect of realism on the impression of artificial human faces. Presence: Teleoperators Virtual Environ. **16**(4), 337–351 (2007)
5. MacDorman, K.F., Chattopadhyay, D.: Reducing consistency in human realism increases the uncanny valley effect; increasing category uncertainty does not. Cognition **146**, 190–205 (2016)
6. Mori, M., MacDorman, K.F., Kageki, N.: The uncanny valley [from the field]. IEEE Robot. Autom. Mag. **19**(2), 98–100 (2012)
7. Bateson, M., Nettle, D., Roberts, G.: Cues of being watched enhance cooperation in a real-world setting. Biol. Lett. **2**(3), 412–414 (2006)
8. Apple Memoji Vs. Samsung's AR Emoji: Comparison Review. https://www.ibtimes.com/apple-memoji-vs-samsungs-ar-emoji-comparison-review-2688394
9. Monahan, J.L.: I don't know it but I like you: the influence of nonconscious affect on person perception. Hum. Commun. Res. **24**(4), 480–500 (1998)
10. Powers, A., Kiesler, S.: The advisor robot: tracing people's mental model from a robot's physical attributes. In: Proceedings of the 1st ACM SIGCHI/SIGART Conference on Human-Robot Interaction, pp. 218–225 (2006)
11. Charmaz, K.: Constructing Grounded Theory: A Practical Guide Through Qualitative Analysis. Sage (2006)
12. Bartneck, C., Kanda, T., Ishiguro, H., Hagita, N.: Is the uncanny valley an uncanny cliff? In: RO-MAN 2007-The 16th IEEE International Symposium on Robot and Human Interactive Communication, pp. 368–373. IEEE (2007)
13. Etcoff, N.: Survival of the prettiest: the science of beauty. Anchor (2011)

Blended Design Strategies to Plan Again the New Territorial Networking and Its Interconnections

Irene Fiesoli(✉)

University of Florence, Via della Mattonaia 14, 50121 Florence, Italy
irene.fiesoli@unifi.it

Abstract. Today, sustainable innovation is designed to develop from the lineal model to the circular one. Attempts are made to offer – starting from an upgrade of knowledge and educational paths – a new innovative society (known as Society 5.0 or Super Smart Society) where companies and territorial stakeholders are a part of complex relation systems leading to the creation of new sustainable and intertwined supply chains (territorial networking), also with the support of properly designed tools such as the service platforms. In this scenario, design develops new planning visions based on relational paradigms and represents the bridge between technological innovation and social contexts since it is able to plan technology by harmonizing cultural, social, economical and political elements.

Keywords: Strategic design · Networking territorial · Innovation technologies · IoT · Platform · Circular economy · Knowledge · Super smart society 5.0

1 Introduction

Today, sustainable innovation is designed to regenerate, from the linear model to the circular one, and, considering all the production and design stages, aims to seize every opportunity trying to prevent negative environmental externalities and to realize new social and territorial value. In fact, the linear model typical of our current economy – production, consumption and disposal – has reached its limit since the beginning of energy and natural resources depletion is more and more visible. For this reason, it is necessary to focus on the transition from this linear model to a circular one (circular economy) which, considering all the stages – from the planning, to the production and the consumption, up to the end of products' life –, knows how to seize every opportunity to limit material and energy input and to minimize losses and waste, paying attention to the prevention of negative environmental externalities and especially to the creation of a new social and territorial value [1].

Starting from this concept, sustainable innovation offers a new strategic model of society which links the environmental aspects with the economical and social ones, creating deep changes especially in the relationship among producers, consumers and territorial stakeholders. Companies will be part of a complex relation system and will

D. Russo et al. (Eds.): IHSI 2021, AISC 1322, pp. 703–708, 2021.
https://doi.org/10.1007/978-3-030-68017-6_104

develop, more and more interrelated, new design visions based on a new relational paradigm allowing the creation of innovative, sustainable and intertwined supply chains.

Moreover, including the Industry 4.0 scenario which is already foreseen in the new Society 5.0 (or Super Smart Society), the relationships among the active subjects in the today's productive landscape and the birth of new forms of interrelation among them are essential in a model of circular economy which adds different methods of production-use and alternative ways of social interaction among users, allowing to find new unsatisfied and challenging demands for both technological innovations and our society.

The "philosophical" thought of Society 5.0 – promoted by Keidanren, the powerful Japan Business Federation – develops the German concept of Industry 4.0, not stopping at the complete digitalization of manufacturing processes but prefiguring a model of society which represents the last stage of civilization. After the hunter-gatherer society followed by the agrarian, industrial and information societies, it also suggests the arrival of a Super Smart Society characterized by self-driving vehicles, smart cities, digitalized tourism, Finthech, smart houses, cyber security, smart agriculture, 5G, data mining, open data and digital transformation of healthcare [2] (Fig. 1).

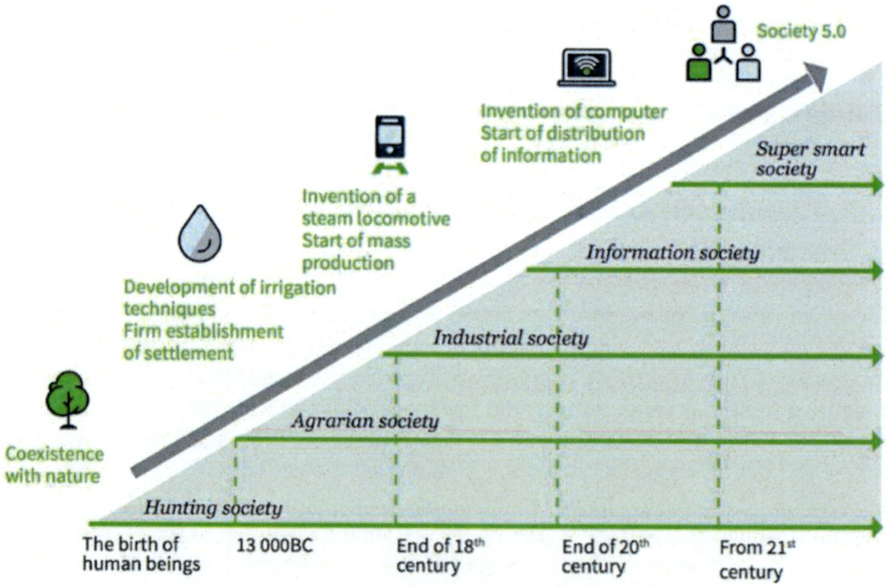

Fig. 1. Dimension Data – The birth of human beings. [3]

So, the challenge is posed by the fact that there will be a change not only in the manufacturing technologies, but also in the cultural and business contexts. The territory is therefore designed as an area including social contexts and even the attitude of human settlements in terms of, for example, civic education, observance of the laws, hospitality: all these qualitative elements will substantiate the future identity of territories.

Particular attention needs to be paid at the role of design as a discipline traditionally able to manage – and plan – the complexity which is currently developing, as expressed by Manzini, in a project scenario full of new forms of creative realization which uses different methods, such as Open Design and Co-Design [4], and for which, as Verganti explains, the innovation is not only the one induced by technological progress or push marketing, but also the one resulting from a third strategy, that is a radical change in perspective introducing a new way of competing: the Design-Driven innovation, headed by design, which produces new meanings [5]. These meanings will make it possible to understand the role of future designers, if they will always be knowledge catalysts or if they will turn into other figures, and which methods and tools can or will be used [6].

In this context, the role of the designers has changed: they design not only products but also complex systems, services, interactive products, models of integrated communication and relationships. This modality will take place both within the business management of the production cycles and outside the companies, facilitating the smart production by dividing it among all the present stakeholders in the territorial manufacturing system, acting on the territorial networking and on the birth of future supply chains.

Considering digital and other kind of tools aimed at the creation of territorial networks and collaborative synergies, the service platforms are certainly standing out as the most efficient way to achieve these objectives with significant results in terms of: activation of new connections among groups of stakeholders; revitalization of existing (although very weak) synergies and collaborations; development of innovative work/training practices; creation of an active and collaborative territorial network.

Analyzing the main Research and Development Projects at a national (Italian) and European level, it is evident that there is a tendency to plan smart platforms which try their best to contribute to achieve goals shared not only by the scientific community. But, especially because of this wide distribution and application, it is strange that there is not a systematic use – both locally and globally – of these strategic systems at a public, training and productive level. They would help to project effectively our society towards the globalization and cross-contamination era, taking advantage of trans-disciplinary approaches which would spread out from the management of the supply chains to the administration of the public systems, up to the daily and training activities.

These assumptions are the basis of two research projects, used as best practice. The first one is the PhD Research Project in Architecture – curriculum Design (University of Florence), "The (Sixth) Sense of Design: the Ability of Mediating and Linking the Territories with the Local Systems and the Knowledge with the New Forms of Innovation" [7], by the PhD student Irene Fiesoli, with the supervisor Giuseppe Lotti and the co-supervisors Debora Giorgi, Gianni Sinni and Mauro Lombardi; the second one is the Strategic Research and Development Project "MixedRinteriors - AR/VR, Enabling Technology for Industry 4.0 in RVs, Yatch, Furniture and Furnishing Sectors", funded by the Call for Tenders in the region of Tuscany POR FESR 2014–2020, with a partnership of: Richard Ginori, lead partner of the project for the complements of house furnishing; Savio Firmino and Marioni, for the house furnishing sector; SEA-Camper for the sector of campers and caravans; 7Stars for the nautical sector; the University of Florence – Department of Architecture DIDA, in particular the Sustainable Design Laboratory (scientific director, Giuseppe Lotti); University of Siena – Department of Social, Political and Cognitive Sciences DISPOC; Interior and Design District (Region of Tuscany); ITIA-CNR and Mediacross as technological partners specialized in the development and the application of innovative technologies, such as VR and AR, in different contexts.

2 Discussion

The general objective of both projects is to contribute to increasing the connections on the territory among the actors of the current production system, developing a tool to exchange services in a territorial, productive and innovative network projected into the future. Hence, the idea of a service platform whose main objective is to widen the possibility of matching supply with demand for services and performances, creating a system where users can insert general or specific supply and demand, and send them directly to the system and not to a single user. In this way, the platform effectively becomes smart and interactive, producing itself – thanks to an algorithm – notifications which link the users among them. The platform acts as an independent and impartial intermediary between who offers the service and who searches for it, representing a tool with obvious benefits because it helps not only to regenerate and recover territorial competitiveness, but also to reactivate cooperation processes on the territory which are the basis for an effective social and economical development. So, the platform is intended to be a strategic tool to encourage future synergies and the creation of both physical and virtual meeting points at a local and global level, in order to make different realities communicate with each other and to facilitate the mutual exchange of knowledge and experiences.

Through these possible interactions, the world of Research and University tries to join the ecosystem of territorial centers and companies thanks to the creation of a digital tool which represents an opportunity to move towards different physical and virtual places, trying to reconnect meanings and actions which often follow one another in a discontinuous and fragmented way. This is an attempt to link, within the same system, the contexts of formal and informal learning with the framework of project activities and entrepreneurial business, seeking to truly understand how each one of the different types of actors, in this complex scenario, can bring added value in an experience of shared development (Fig. 2).

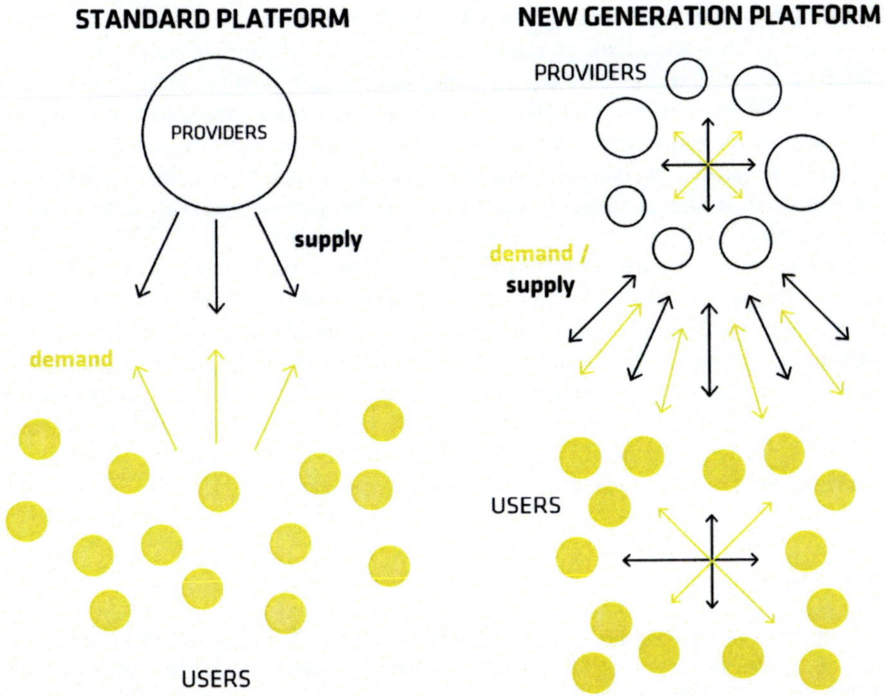

Fig. 2. Illustrative chart with the interactions of standard platforms (on the left-hand side) and platforms developed from the projects analyzed (on the right-hand side). It is important to highlight the different relations and exchanges between supply and demand of services.

3 Conclusion

To sum up, it is fair to say that technology is designed also as a social fact and its origin implies a complex harmonization process among cultural, social, economical and political elements, despite the differences that arise by comparing them and that make it so difficult to carry on this kind of action. In order to spread the application of the new technological breakthroughs in various social contexts, it will be necessary to expand their use inside the educational contexts where they turn out to be a strategic tool for the creation of an emerging class with greater cultural ability and able to avoid negative repercussions in the future [8].

Therefore, knowledge is the basis to start the regeneration of the society. However, in order to facilitate and design a permanent industrial rebirth based on cultural and also technical-scientific backgrounds, these assumptions will acquire meaning in the future only if they fit into an organic framework [9], promoting a rearrangement of the available resources towards trajectories of industrial leadership with virtuous relationships among universities, companies and territories.

In the last decades, society has moved away from the ensured progress to head towards an extraordinary uncertainty [10]. In fact, especially in the light of social distancing experienced because of the COVID-19 pandemic, the risk of a collapse of values and of the current social structure is tangible, and for this reason, today it is important to perceive the progress as a challenge and a possible achievement, as a product of our choices and our will, to develop renewal skills proper to humanity, It is necessary to reconsider the idea of progress through a new culture and a new anthropology, since humanity is asked to get out of the age of war and unconditional exploitation in nature to create a collaborative game paradigm, finally with positive-sum outcomes.

Obviously, it is not enough to regenerate a culture of tolerance, it is essential to go further, trying to conceive and live the global community in a positive way. A 'culture of encounter' has to grow, characterized by comprehension, connection, hybridization and constructive dialogue. Without these features, there is no culture, no social life, no spirituality or planning. Consequently, it is crucial to develop and redesign the sense of belonging shared by all the populations and territories, as an ethical task fundamental to build, following Edgar Morin's phrase, a "Earth civilization" which promotes an anthropological development with a sustainable coexistence [11].

References

1. European Commission: an EU action plan for the Circular Economy (2015). https://eur-lex.europa.eu/legal-content/EN/TXT/?qid=1453384154337&uri=CELEX:52015DC0614
2. Carrer, S.: Il Giappone alza la posta: rotta verso la Società 5.0. Il Sole 24 Ore, Rapporto Impresa 4.0, sezione Rapporti24 (2017). https://www.ilsole24ore.com/art/il-giappone-alza-posta-rotta-la-societa-50-AEdJyJZC?refresh_ce=1
3. Teruzzi, E.: inno3 (2018). https://inno3.it/2018/03/06/bordignon-societa-5-0-un-maturity-model-per-dimension-data/
4. Manzini, E.: Design, When Everybody Designs: An Introduction to Design for Social Innovation. The MIT Press, Massachusetts (2015)
5. Verganti, R.: Design-Driven innovation. Cambiare le regole della competizione innovando radicalmente il significato dei prodotti e dei servizi. Harvard Business Press, Rizzoli Etas, Milan (2009)
6. Germak, C.: Uomo al centro del progetto. Design per un nuovo umanesimo. Umberto Allemandi, Turin (2008)
7. Fiesoli, I.: The (Sixth) Sense of Design: The Ability of Mediating and Linking the Territories with the Local Systems and the Knowledge with the New Forms of Innovation. Flore (2020). https://flore.unifi.it/handle/2158/1191414#.X3sjMmgzZPY
8. Ceruti, M.: Il tempo della complessità. Raffaello Cortina Editore, Milan (2018)
9. Bellandi, M., Caloffi, A.: I nuovi distretti industriali. Rapporto di Artimino sullo sviluppo locale 2012–2013. Il Mulino, Bologna (2010)
10. Santachiara, D.: Dowload design. Manutenzione straordinaria della cultura materiale. 24 ORE Cultura srl, Milan (2016)
11. Ceruti, M., Morin, E.: La nostra Europa. Raffaello Cortina Editore, Milan (2013)

Smart Retail and Phygital Customer Journey in the Kids and Toys Stores

Benedetta Terenzi[1(✉)] and Arianna Vignati[2]

[1] Department of Civil and Environmental Engineering, University of Perugia, 06125 Perugia, Italy
benedetta.terenzi@unipg.it

[2] Department of Design, Politecnico Di Milano, 20133 Milan, Italy
arianna.vignati@polimi.it

Abstract. The paper present the results of the research conducted in 2019 by Kids&Toys Observatory of Politecnico di Milano, which aims to understand the impacts of digital transformation on parents and children's behaviours and habits, with a particular focus on the processes of choosing and buying toys. Thanks to qualitative and quantitative analysis, the study focus on digital transformation's, social effects and user's emerging needs to analyze how much is changing in the dialogue between companies in toys sector and their customers (adults and children). The results open valuable discussions on the role of design in terms of tools, methods and approach in supporting companies in the Kids&Toys sector to develop hybrid scenario including new in-store experiences (ISE) and on-line experience both in line with traditional needs and the emerging behaviors of children, parents and care-givers.

Keywords: Human-centred design · Product service system design · Customer journey · Phygital experience · Smart retail

1 Introduction

The increased deployment of new technologies such as smart mobile devices and social networks and the growing importance of in-store technology (IST) solutions create new opportunities and challenges for retailers [1]. We are witnessing an emerging retail scenario based on the smart usage of technologies in which customers are omnichannel both for attitude and behaviour. Since as more and more people have easy interaction with multiple touchpoints throughout their shopping journey, in the same way, retailers must consider all channels holistically, to provide them with a unique, hybrid, inclusive shopping experience [2]. Smart Retailing (SR) [3] is defined as technology-supported retailing practices aimed at meeting the needs of new smart shopper [4, 5].

According to this change of purchasing habits, the research has defined "Kids&Toys Experience Journey" the new hybrid experience that involves parents and children and which sees them at the centre of a complex network of relationships.

There is a multiform and adaptive path, physical and virtual, made of information and data exchange that is established between children and parents and the whole system of actors and stakeholders around the toy world (social media and

D. Russo et al. (Eds.): IHSI 2021, AISC 1322, pp. 709–714, 2021.
https://doi.org/10.1007/978-3-030-68017-6_105

communication channels, influencers, producers, brick-and-mortar stores, e-commerce platforms, dedicated communities, etc.). The research has translated this experience into 7 macro phases, based on which the collected data have been analysed and interpreted.

The research adopted the following integrated methods of analysis:

- a direct observation through international shadowing activities of 49 retail shops and 19 bookshops;
- a collection of case studies for 17 e-commerce platforms specialized in toys and children's products or with a dedicated section;
- collection of 157 surveys questionnaire shared with Italian parents to investigate their behaviours and needs to purchase products for children.

The research, highlighting the relevance of the product-service system design (PSSD), has analyses how much space, services and communication today influence the experiences of children and adults in physical and virtual retails to understand what kind of customer experience is now offered physically and digitally for the specific target of children, and how much the offer responds to the real and/or potential demand. We must also consider that the pervasive use of devices has an important impact in the daily kid's experience, refered to playfulness in relating to products and services [6, 7]. In this sense, the design process has to consider the playful approach in defining experience dedicated to them. This means that to make every activity more engaging and pervasive, is important to involve all senses in the process. From this assumption, with the diffusion of digital innovation [8], comes the need to implement the new digital experiences with a physical and material part, or vice versa. From this the term phygital was born: it concerns the concept of fusion of digital experiences with physical one taking the best aspects from each one, to create a new and augmented user experience [9].

For companies rethink and redesign all touchpoints experience means building a conversation with customers but to be able to do so they will need to understand what need is driving the interaction and where and when the interaction takes place in fact in the case of kids-oriented sectors these means consider two different types of consumers: kids and adults.

2 Customer Experience and Digital Transformation.

In Observatory, research activity has been conducted to understand the impacts of digital transformation on the behaviours and habits of parents and children with a particular focus on the processes of choice, purchase and relationship with the world of toy stores. The research aims to build a framework with guidelines useful for designers who want to innovate store and distribution channels of companies in sectors with children as a target group.

The research has put children and parents at the centre of a new experience: the kids&toys experience journey. There is a path, a relationship experience that is established between children, parents and the complete system of actors and stakeholders around the toy world (social media and communication channels, influencers, producers, distributors, dedicated communities, etc.). This experience has 7 macro steps based on which the research data was analyzed and interpreted.

The 3 key research focuses have been the following:

1. Spaces - different types of sales spaces have been analyzed to understand the offer experiences and the involvement of children and their families in different specific activities (eg learning, relaxing, playing, etc.).
2. Services - the different types of services offered to children and their families have been analyzed through e-commerce platforms (dedicated to the kids' world or with dedicated sections).
3. Behaviours – the research has evaluated the impact of the omnichannel scenario in the purchase behaviours of target by administering a survey to 157 parents.

3 Findings: Customer Journey in the Kids&toys Sector

The store is becoming a place of valuable relationship with customers, by expanding its scope of action, integrating with other touchpoints in which digital and physical entities can be linked, employing appropriate information and communication technologies, to enable a whole new class of applications and services [10]. New technologies facilitate digital interactions in the physical and digital store the use of different devices allows the omnichannel fusion of the shopping experience.

On the one hand, online is easy to track every data, every click, but also the physical behaviour of customers in the store is measurable. The store's infrastructure becomes a valuable source of information and data, complementing those collected digitally. Retail, therefore, is not only a place to welcome its customers but becomes an effective tool with which to evaluate their choices and reactions to a given product, for example. The store is now able to collect a huge amount of data on the behaviour of its customers through the introduction of tools such as digital walls and multimedia totems, Near Fiel Communication technology, beacon marketing, Virtual Reality and Augmented Reality, wearable technology, virtual dressing rooms, Mobile POS. In this scenario the IoT's play a strategic role [11].

On the other hand, however, the traditional brick-and-mortar stores must not overlook the importance of physical experience and the effectiveness of place attachment to offer a complete experiential approach focusing on cognitive processes that can involve the customer in all spheres, especially considering that the primary target is children and therefore have the need to express themselves freely.

Intending to evaluate the actual and potential changes favoured by digital transformation in the purchasing process of children's products, the research analyzed the phygital approach during the 7 steps of the kids and parents' customer journey.

Need: What Children Want. The research involved 109 children in a creative game session. The results show how progressive increase in age corresponds to a progressive increase in complexity of the toy: in the 3–5 year old group, moving toys are preferred (37%); in the 6–8 years age group we prefer toys that move (33%), with App (26%), sound (21%), to assemble (20%); in the 9–12 age group we prefer toys that move (28%), voice control (23%), sound (19%), with App (17%).

Trend: Advice and Inspiration. A panel of parents (124) was involved to share the experiences and then co-designing new solutions for the toy product system. The data show that 93% of them buy toys for their children and in the search for trends and inspirations 43% get advice from friends and relatives, 34% seek advice on forums and social networks network and 22% would like to be advised by a smart shopping assistant.

Idea: Looking for Product Information. After deciding to buy a specific product, parents go looking for tools and ways to go into more information. The questionnaires show that 66% look for information and reviews on forums/social networks, 19% ask for information in the store and 14% would rely on the smart shopping assistant that collects and selects all the online information for them. The results clearly demonstrate the important role of social channels, as well as specialized forums and communities, as well as systems integrated with e-commerce platforms that allow you to leave reviews.

Choice: to Decide Where to Buy. In the complex panorama in which the adult finds himself moving during his journey of experience, he must decide the purchase channel: physical or online? The interviews show that 60% buy online or with the smartphone, 27% buy in the store, 8% buy online and buy in the store and 5% see in the store and complete the online purchase. If we look at the analysis of the international stores and take into consideration the methods of purchase, we find that in 59% it is possible to buy only in-store. In the stores where there is the possibility to make purchases also online, for 39% it is possible to buy online from the website and receive at home, for 29% it is possible to buy online and collect in the store, 22% allows you to buy in store and receive at home. Crossing the data of the demand with the offer it is found that from the analysis of the stores 44% do not foresee activities, while 56% offers various entertainment possibilities such as 28% product testing, 32% gaming, 22% workshops and laboratories, 26% animated reading, relaxation and recreation 4%.

Purchase: Online/Offline Stores. Most of the physical stores analyzed to offer a varied range of products, not specifically just toys: 96% have a prevalence of toys, 81% also books and publishing, 37% including clothing, 25% offer hi-tech, 19% personalized merchandising, 12% foods and 10% also video games. Even for the e-commerce platforms observed, 82% offers different categories of products (furniture and accessories, clothing, accessories, electronics, books, etc.) while only 18% specialize in the toy. In addition to the traditional method of display on shelves, the child-friendly store offers games in a context of real use and takes advantage the support of e-commerce in store to allow customers to access an even wider selection of products and options.

Problems: Customer Care. The experience of relationship and interaction does not end with the purchase. Customer care is increasingly passing through some after-sales services. The e-commerce analysis shows that 71% of them have return and refund management services and 62% have a customer newsletter. In the activity carried out shows parents' preference towards solve problems through access to forums / social networks and call centres (46%), while 30% return to the store, only 24% would turn to the smart shopping assistant to get advice and suggestions.

Sharing Experience. Parents wish to share their experiences and the observation shows that 57% prefers word of mouth by voice, 22% share it on social networks and 21% automatically share their experience online in real-time. If we consider that only 69% of the stores have dedicated social channels and only 26% have active customer loyalty policies, it is interesting to consider this need as a new space of planning for businesses.

4 Guidelines for a New Phygital Customer Experience

The data collected from the research clearly show how the on-line to off-line (O2O) experience is the plot and the warp on which to design the customer journey. The results highlight a strong re-evaluation of the Human to Human (H2H) experience in the shop, at the time of purchase and customer care. On the contrary, in the researching of the idea and product information phases, as well as in the post-purchase phase and sharing and feedback phase, the support of technology allows responding more precisely and immediately to emerging needs.

The research has supported the identification of five guidelines for designers who need to overcome the challenges in define new customer experience able to put children and parents at the centre and at the same time catch the opportunities of the digital technologies:

- Physical-digital in a unique experience: the distinction between the concept of physical (brick-and-mortar stores) and digital (e-commerce) experience must be overcome since in the store of the future the digital and physical entities will be merged and perfectly linked to each other in one unique experience lived by the families.
- Technology with empathy: the upcoming increase in the use of technology in the purchasing process, make the store shift towards optimization and efficiency, and that opens up to new forms of retailtainment and in-store gamification.
- Not only products: the store atmosphere, the store design, and pre and post-sales support totally impact the discovering, testing and purchasing experience. The product is no longer the real focus that moves people to visit a store, but it is a set of experiences and emotions that consumers expect to live.
- Store as space with product-service system design: the store of the future will be a place of belonging, with its own identity, where kids and caregivers feel 'at home"'. The store will be a hybrid and augmented system based on human comprehension, data collection for a bespoke service, place of smart retailing strategies.
- Interaction and collaboration: the digital era brings new forms of human-human, human-space, human-product interaction. Within this scenario, collaboration, co-design and co-creation will be the basis of any creative process towards the integration of personalized services able to enhance the phygital purchasing experience with specific contents and ad hoc services by involving different stakeholders with consolidated and emerging skills.

5 Conclusions

The integration of on-line/off-line experience may represent the next big leap ahead in the ICT sector. The possibility of seamlessly merging the real and the virtual world, through the massive deployment of embedded devices, opens up new exciting directions for both research and business. In this article, we provided an overview of the key issues related to the development of retail technologies and services. Several research challenges have been identified, which are expected to become major research trends in the next years. The most relevant application fields have been presented, and many use cases identified. We do hope that this research will be useful for researchers and practitioners in a different field, helping them to understand the huge potential of phygital products and services and what are the major issues to be tackled, devising innovative technical solutions able to turn in innovative and inclusive Product-Service System design for kids and care-givers.

References

1. Piotrowicz, W., Cuthbertson, R.: Exploring Omnichannel Retailing: Common Expectations and Diverse Reality. Springer, Cham (2019)
2. Savastano, M., Bellini, F., D'Ascenzo, F., De Marco, M.: Technology adoption for the integration of online-offline purchasing: Omnichannel strategies in the retail environment. Int. J. Retail Distrib. Manag. **47**(5), 474–492 (2019)
3. Pantano, E., Timmermans, H.: What is smart for retailing? Procedia Environ. Sci. **22**, 101–107 (2014)
4. Atkins, K., Kim, Y.K.: Smart shopping: Conceptualization and measurement. Int. J. Retail Distrib. Manag. **40**(5), 360–375 (2012)
5. Voropanova, E.: Conceptualizing smart shopping with a smartphone: Implications of the use of mobile devices for shopping productivity and value. Int. Rev. Retail Distrib. Consumer Res. **25**(5), 529–550 (2015)
6. Prensky, M.: Digital natives, digital immigrant. In: On the Orizon, vol. 9, no. 5, pp. 1–6. MCB University Press (2001)
7. Marsh, J.: Online and offline play. In: Burn, A., Richards, C. (eds.) Children's Games in the New Media Age, pp. 109–312. Ashgate, Cambridge (2014)
8. Cianfanelli, E., Crescenzi, P., Goretti, G., Terenzi, B.: Playful learning for kids with special educational needs. In: Bagnara, S., Tartaglia, R., Albolino, S., Alexander, T., Fujita, Y. (eds.) IEA 2018, AISC 826, pp. 732–742. Springer, Cham (2019)
9. Terenzi, B., Vignati, A.: Digital transformation in Product Service System for kids. Design tools for emerging needs. In: Ahram, T. et al. (Eds.), IHSI 2020, AISC 1131, pp. 228–234, Springer, Cham (2020)
10. Miorandi, D., Sicari, S., De Pellegrini, F., Chlamtac, I.: Internet of things: Vision, applications and research challenges. Ad Hoc Netw. **10**, 1497–1516 (2012)
11. Grewal, D., Noble, S.M., Roggeveen, A.L.: The future of in-store technology. J. Acad. Mark. Sci. **48**, 96–113 (2020)

Co-design of Gesture-Based Arabic Sign Language (ArSL) Recognition

Abeer Al-Nafjan[1,2(✉)], Layan Al-Abdullatef[1], Mayar Al-Ghamdi[1], Nada Al-Khalaf[1], and Wejdan Al-Zahrani[1]

[1] Computer Science Department, College of Computer and Information Sciences, Al-Imam Muhammad Ibn Saud Islamic University, Riyadh, Saudi Arabia
annafjan@imamu.edu.sa, {lsalabdullatef,mnsalghamdi, namalkhalaf,weaalzahrani}@sm.imamu.edu.sa
[2] Human-Computer Interaction (HCI) Design Lab, Riyadh, Saudi Arabia

Abstract. Deaf and hard of hearing (DHH) individuals face barriers to communicating with others using sign language. Accordingly, the need for sign language translators is growing significantly. Technology plays an important role in our lives because it allows communication to be more convenient and effective, especially for DHH individuals, who desire privacy and independence. Many rely on technology to communicate. In this paper, we propose SignSpeak, an Arabic sign language (ArSL) recognition system using a leap motion controller. SignSpeak provides a framework for two-way communication by translating ArSL into spoken language and transcribing spoken language into text.

Keywords: Arabic sign language (ArSL) · Deaf and hard of hearing (DHH) · Assistive technology · Sign recognition · Translation

1 Introduction

Four hundred and sixty-six million people worldwide are deaf or hard of hearing (DHH), according to the World Health Organization [1]. Furthermore, the results of the 2017 disability survey conducted by the General Authority for Statistics in Saudi Arabia indicate that 1.4% of the Saudi Arabian population (i.e. approximately 456,576 people) is DHH [2, 3]. In recent years, many researchers and developers have presented sign language recognition and translation systems [4–6].

Our primary aim is to develop a novel Arabic sign language (ArSL) system that provides two-way communication between DHH individuals and hearing individuals. Specifically, the project aims to build a system that can correctly identify ArSL signs that correspond to hand gestures. In this paper, we propose SignSpeak, an ArSL recognition system. So far, SignSpeak can recognize the 28 letters of the ArSL alphabet using a leap motion controller (LMC). And transcript the spoken language into text. This paper contributes a framework for two-way communication to provide DHH individuals with an alternative communication method. Moreover, a prototype of the system was developed and later evaluated by ArSL interpreters to test its feasibility.

D. Russo et al. (Eds.): IHSI 2021, AISC 1322, pp. 715–720, 2021.
https://doi.org/10.1007/978-3-030-68017-6_106

2 Related Work

Sign language is a visual language used by DHH people based on hand gestures. Many approaches have been proposed in the literature to develop a sign language recognition system to detect these gestures, such as a vision-based approach that uses a camera [7], a sensor-based approach that uses sensors [8], and a hybrid approach that utilizes cameras with sensors [9].

LMCs and Microsoft Kinect are examples of motion sensors used in sign language recognition. They provide accurate results and are natural to use [10]. Thus, by using the existing tools and technology, we can focus on improving sign language recognition. A comparison between Kinect and an LMC was conducted by [10], and multiple aspects were considered, such as accuracy. The LMC suffers from a tracking problem with the fingers when they are too close to each other. Nevertheless, both LMCs and Kinect are considered to be effective low-cost visual sensors.

Concerning ArSL recognition systems, [6] presented a system for static hand gestures and discussed the limitations regarding the LMC, which included a limited detection range and the inability to detect the hand in the presence of obstacles. Moreover, the system proposed in [5] tested several classification algorithms, for both static and dynamic gestures. Lastly, [4] proposed a framework using dual LMCs for dynamic gesture recognition. The two LMCs were used to address hand occlusion (Table 1).

Table 1. Recognition systems using leap motion controller

Ref	Motion	Sign language	Type	Signs	# of hands
[4]	Two LMCs	Arabic	Dynamic	100 words	Two
[5]	LMC	Arabic	Static and dynamic	38 static and 20 dynamic signs Alphabet and words	Two
[6]	LMC	Arabic	Static	28 letters and Numbers 0–9	One
[9]	LMC and Microsoft Kinect	American	Static	26 letters	Two
[11]	LMC	Arabic	Static	28 letters	One

*All mentioned systems provided one-way communication and were PC based.

3 System Description

SignSpeak aims to serve DHH and hearing individuals. Therefore, the system is further divided into two subsystems intended for translation and transcription, as shown in Fig. 1. In translation, the LMC captures the sign performed by the DHH user. Then, gesture features are extracted and sent to the classifier. Based on the features, the sign is mapped to a class. In this version of SignSpeak, the classes are the 28 Arabic letters. Finally, the result is sent to Google's Text-To-Speech (TTS) API to output a spoken language for the hearing user. In transcription, Google's Speech-To-Text (STT) API transcribes the hearing user's input speech. The text equivalent of the speech is presented to the DHH user.

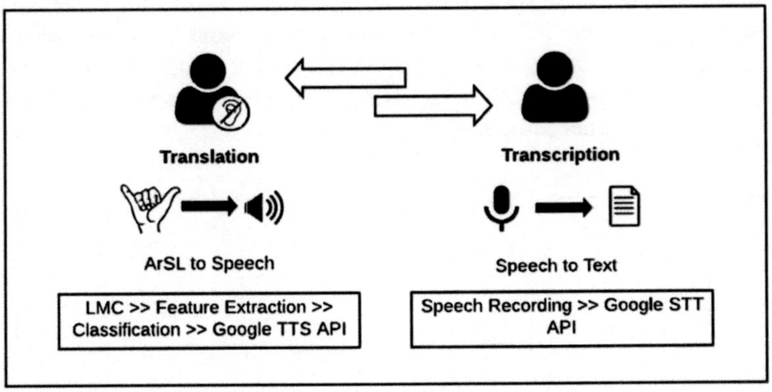

Fig. 1. SignSpeak system framework

4 Hardware and Dataset

Leap Motion Controller: The LMC is a specialized sensor by Ultraleap that combines two near-infrared cameras and three LEDs to prevent overlap. The LMC can detect hand movement as well as other hand elements, including fingers, joints, and bones, within an interaction zone of 60 cm extending from the device in a 120 × 150° field of view. Furthermore, the detection range is at arm's length, which is up to 80 cm.

Dataset: In our prototype, SignSpeak's database consists of a flat-file. The file stores the 28 letters of the Arabic alphabet, and the sign language features for each letter. The data were collected from two signers who repeated each letter 10 times. Thus, the file includes 560 samples. The selected features were hand direction, palm position, finger direction, fingertip position, bone direction, the base, and the end of the bone. All the features are given as vectors (i.e. with respect to x, y, and z).

4.1 Prototype

Using a co-design approach at the early stage helped us to understand the needs, gain more insights, realize challenges and opportunities, and explore possibilities of technology to serve sign language. Moreover, having interpreters and sign language experts engaged in the design development process helped us to propose a design project and implement a rough prototype that could enable DHH individuals to communicate with hearing individuals.

The application's interface was developed with the intended users in mind. In brief, we developed a GUI that presents two options: Sign and Speak. The Sign option as shown in Fig. 2, allows the user to record hand movements and re-input a hand movement in case of a mistake. Then, the system checks whether an ArSL sign can be detected from the movement. If no sign is detected, the user is asked to re-input the hand movement. Finally, the system translates the detected sign into spoken language. The Speak option as shown in Fig. 2, allows the user to record the input speech and re-input the speech in case of a mistake. If the speech cannot be detected, then the user is asked to re-input the speech. Otherwise, if a comprehensible speech is detected, the system transcribes the speech into text.

Fig. 2. SignSpeak system GUI

Nevertheless, recognizing ArSL in itself is a challenge because ArSL employs numerous gestures to convey words. The physical distinctions differentiating one gesture from another are subtle. These gestures involve different parts of the body, making ArSL challenging for the system to interpret.

The prototype was tested by our co-designer to find out whether it has the potential to solve the challenges the design process seeks to address.

5 Classification

SignSpeak uses two machine learning classification models. Both classifiers are supervised learners, namely, a support vector machine (SVM) and a K-nearest neighbor (KNN) with the Scikit-learn library [12]. The purpose of using two classifiers is to compare and analyze the results. The data were acquired from a CSV file and split into training and testing sets using cross-validation: 80% for the training set and 20% for the testing set. Grid search was used to tune the hyperparameters. Furthermore, the SVM and KNN were trained and tested on the same sets.

6 Discussion

The system produces an output in an average of 4 s. Co-designers' feedback was positive and promising. However, the system's limitations are mainly related to LMC's performance. Problems include a limited range of motion and lag between movements. Common issues of LMCs in similar studies include limited range of motion, a small number of features, and in some cases the inability to detect both hands [6]. However, to reduce these issues, a few studies used both the LMC and the Kinect [9, 10]. Others, such as [4], used two LMCs. To improve hand gesture recognition technology in VR and motion tracking sensors, future efforts should consider a wider tracking range and easily deployable systems.

7 Conclusion

SignSpeak provides a reasonable starting point for building a motion-based ArSL translation tool. The need for two-way communication is critical for DHH individuals. Moreover, end-user involvement is crucial for designing systems that are user-centered. The co-design approach enabled us to make a creative contribution to solve the problem and resolve a particular challenge.

Accessible possibilities for future work include the following: The addition of new static or dynamic gestures will allow DHH individuals to communicate more naturally. Furthermore, it will enhance SignSpeak's effectiveness for both ArSL learning purposes and translation. Facial expressions are essential in ArSL. Therefore, future iterations of SignSpeak will consider facial expressions using sensors or cameras. Moreover, the lack of an ArSL dataset is considered one of the technological challenges for developing effective technological solutions to create collective impact. An ArSL dataset should be built to solve the lack of uniformity problem in ArSL to serve real-world use cases and account for real-world constraints. Additionally, a large dataset with diverse signers is essential for training software to perform well for diverse users. Moreover, public availability of the dataset is important for spurring developments.

Acknowledgments. We thank the Humanistic Co-Design Initiative and the Human-Computer Interaction (HCI) Lab for supporting this work. We also thank the Saudi Authority for Intellectual Property (SAIP) and the Saudi Health Council's National Lab for Emerging Health Technologies for hosting and mentoring. This work is part of the authors' project that is carried out under the CoCreate Fellowship for Humanistic Co-Design of Access Technologies.

References

1. Deafness and hearing loss (2019). https://www.who.int/news-room/fact-sheets/detail/deafness-and-hearing-loss
2. General Authority for Statistics.: Disability Survey 2017 (2020). https://www.stats.gov.sa/en/5669

3. General Authority for Statistics.: Population Estimates (2020). https://www.stats.gov.sa/en/43
4. Deriche, M., Aliyu, S., Mohandes, M.: An intelligent arabic sign language recognition system using a pair of LMCs with GMM based classification. IEEE Sens. J. **19**, 18 (2019)
5. Hisham, B., Hamouda, A.: Arabic static and dynamic gestures recognition using leap motion. J. Comput. Sci. 337–354 (2017)
6. Khelil, B., Amiri, H.: Hand gesture recognition using leap motion controller for recognition of Arabic sign language. In: 3rd International Conference on Automation, Control Engineering and Computer Science (ACECS), pp. 233–238 (2016)
7. ElBadawy, M., Elons, A.S., Shedeed, H.A., Tolba, M.F.: Arabic sign language recognition with 3D convolutional neural networks. In: Eighth International Conference on Intelligent Computing and Information Systems (ICICIS) (2017)
8. Lee, B.G., Lee, S.M.: Smart wearable hand device for sign language interpretation system with sensors fusion. IEEE Sens. J. 1224–1232 (2018)
9. Marin, G., Dominio, F., Zanuttigh, P.: Hand gesture recognition with leap motion and Kinect devices. In: IEEE International Conference on Image Processing, Institute of Electrical and Electronics Engineers Inc., pp. 1565–1569 (2014)
10. Guzsvinecz, T., Szucs, V., Sik-Lanyi, C.: Suitability of the Kinect Sensor and Leap Motion Controller—A Literature Review (2019)
11. Mohandes, M.A., Aliyu, S.O., Deriche, M.: Arabic sign language recognition using the leap motion controller. In: 2014 IEEE 23rd International Symposium on Industrial Electronics (ISIE), pp. 960–965 (2014)
12. Pedregosa, F., Varoquaux, G., Gramfort, A., Michel, V., Thirion, B., Grisel, O., Blondel, M., Prettenhofer, P., Weiss, R., Dubourg, V., Vanderplas, J., Passos, A., Cournapeau, D., Brucher, M., Perrot, M., Duchesnay, É.: Scikit-learn: machine learning in python. J. Mach. Learn. Res. 2825–2830 (2011)

Developing an Application for Walking in Nature for Post COVID-19

Yuko Hiramatsu[1(✉)], Atsushi Ito[1], Jinshan Luo[2], Madoka Hasegawa[2], and Akira Sasaki[3]

[1] Chuo University, 742-1 Higashinakano, Hachioji-shi, Tokyo 192-0393, Japan
{susana_y,atc}@tamacc.chuo-u.ac.jp
[2] Utsunomiya University, 7-1-2 Yoto, Utsunomiya-shi, Tochigi 321-8505, Japan
mc196875@cc.utsunomiya-u.ac.jp,
madoka@is.utsunomiya-u.ac.jp
[3] GClue Inc., 134-3 Ikkimachi Turuga, Aizu-wakamatsu-shi, Fukushima 965-0006, Japan
akira@gclue.jp

Abstract. We developed an application for Nikko National Park, which offers natural scenery with marshlands, streams, and hot springs. People living in cities tend to visit rural areas after COVID-19. However, they know little about local plants, insects and other creatures of the forest. Some internet map has little information about such rural areas, too. In addition, tourists sometimes cannot use the Internet in the deep forest. The signal is weak. Therefore, we developed an application which they can use the application without internet access. We set solar Bluetooth Low Energy (BLE) beacons on the boards among the national park's woody roads. It sends them messages and is needless to make a group with a tour guide using our application. This application brings people to live environmentally friendly according to be interested in nature on and after COVID-19.

Keywords: BLE beacon · Post COVID tourism · National park · Application

1 Introduction

The report by UNWTO told that 67 million fewer international tourists up to March translate into US$80 billion in losses, and if tourism brings us together, then travel restrictions keep us apart (Aug. 2020). On the other hand, it told that for both developing and developed nations, all governments and international organizations have a stake in supporting tourism [1]. Tourism is forced to change. Some countries denied permission to enter foreign countries. However, people have chances to visit domestic areas alone or with a small group. We suggest our new local application for post COVID-19 tourism.

We have researched and developed applications in Nikko, the traditional place in Japan, from 2014. Now we developed an application for Nikko National Park, which offers natural scenery with marshlands, streams, and hot springs. People living in cities tend to visit rural areas under COVID-19. However, they know little about local plants,

D. Russo et al. (Eds.): IHSI 2021, AISC 1322, pp. 721–727, 2021.
https://doi.org/10.1007/978-3-030-68017-6_107

insects, and other characteristic creatures in the area. Some internet map has little information about such rural areas, too. In addition, tourists sometimes cannot use the Internet in the deep forest. Trees sometimes block radio waves. Therefore, we developed an application which they can use without internet access. We set solar Bluetooth Low Energy (BLE) beacons on the boards among the national park's woody roads, with the cooperation of the Ministry of the Environment in Japan. When tourists walk near the board, they get messages from the beacon. According to our experiments, the wave sends massages 5–10 m in the humid areas. Contents of our application are creatures of the forest, mountains, maps etc. Tourists can check bus schedules near their location, too. It is needless to make a group with a tour guide for visiting there.

This application brings people to live environmentally friendly according to be interested in nature. Engineers have created ubiquitous systems. Now, we developed a system for local nature. We suggest a new application fitting to the place based on several articles of "Global Guidelines to Restart Tourism" by UNWTO [2] for post COVID-19.

2 Related Works

2.1 BLE Beacon

BLE Beacon is one of the tools for near field communication. It is mainly used inside. For example, the "Indoor Autonomous Positioning System on Smartphones using Compass Beacon Device [3]" described demonstration by using an autonomous positing system, "Indoor Positioning Method Based on Bluetooth Low Energy Beacon" [4], is one of the papers using BLE beacon in Japan. Casa Batlló in Spain [5] has a virtual and augmented reality application using a beacon system by the University Research Institute Robotics Information Communication Technologies.

We use BLE Beacons for outdoor. We have developed an application for Nikko main street, which brought tourists to the world heritage site in Nikko and also provided messages from shop owners on the street [6]. It was useful for foreign tourists and young tourists who knew little about the traditional place.

2.2 Agritourism and IT

Agritourism is drawing attention in eco-tourism. When people left cities and went to farms to visit their relatives for a short period at the late 1800s, agritourism was started [7]. We can find papers and some articles about agritourism in the world. For example, Samoan business development has put the island nation's farming communities on the world map as producers of high-value organic products (The 2nd Pacific Agribusiness Forum in 2016 [8]). Havlicek et al. told "Agritourism facilities in the Czech Republic are several thousand, but most of them have non-quality web sites or none at all. It is proposed that the new portal solution, according to the principles of Web 2.0, will provide greater information sharing among the partners in the field of agritourism [9]". Information & Communication Technology (ICT) brought Agritourism to the next stage. Mpiti & Harpe told about Agritourism in Africa in "ICT factors affecting agri-tourism growth in rural communities of Lesotho" [10].

Agritourism is not a new style, but it will be one new lifestyle after COVID-19 using ICT. ICT connects places and makes our living place freely. We focus on people living in the city who would like to visit rural areas.

2.3 Our Previous Works in Nikko

We developed an application using BLE Beacons for Nikko main street in 2015. We set them about 1.5 km on the route from the railway station to the main shrine in the world heritage site. Beacons sent location-specific information within a range of 100 m. They sent advertising messages every 500 ms. BLE requires only 1/10 to 1/100 the power of classic Bluetooth. In this case, we used a beacon with a coin battery whose capacity is 1,000 mAh. The beacon works a year or more without a battery change. This system has been operating year-round.

3 Application for National Park in Oku-Nikko for Post COVID-19 Tourism

3.1 Requirements for the Application

3.1.1 Requirements for National Park in Nikko

After the use case on the main street in Nikko, we planned to develop this system for National Park. We had to meet requirements for setting beacons at Nikko National Park, which offers natural scenery with marshlands, streams, and hot springs. In the forest that is registered under the Ramsar Convention [11].

- Poor signals
- Coldness in winter (snow)
- The sulfur of hot springs

In addition, our solar panels for beacons don't get much sun in summer because of thick woods.

3.1.2 Requirements for Post COVID-19 Tourism

"Global Guidelines to Restart Tourism" by UNWTO on 28 May 2020 [2] tells us the requirements to restart tourism. In the guideline, it is told that develop segmented and sustainable products focused on nature, rural areas, and culture: eco-tourism, small group or individual sports, history, birds-watching tourism, traditional routes.

The articles as follows in the guideline especially conform our application.

- Implement health and safety protocols & promote digital communication
- Create personalized and small group tours and packages.

3.2 Our BLE Beacon Application

The application we have developed confirmed the requirements both for solutions of National Park and for the guidelines by UNWTO. We created solar beacons in the waterproof case and set them on the signboards by the Ministry of the Environment, which were cooperated with us. Figure 1 is a picture of our beacon on the board. Figure 2 is a block diagram to explain a mobile phone application's behavior for our sightseeing support system. Figure 3 is a map to show the location of beacons in the national park. Thin dotted lines are wooden roads that tourists can walk on.

A beacon has a solar battery since it is difficult to check a beacon's lifetime in the wide mountain area. In the forest and wetland, the BLE beacon signal becomes very weak so that the reach of the beacon signal is about 10 m that is very shorter than in the city area, such as the load side between the station and shrines.

Fig. 1. One of the BLE beacon we set in the National park in Nikko (upper right)

Main contents of the application are a map, bus schedules, picture books in the national park. Those contents work without an internet connection. Also, tourists can use SNS, weather information using the Internet. (The wave is not so stable, but tourists can access the Internet at some grassland). The system is suitable for personal use or a small group to guide in the Nikko National Park. Tourists can know about information about their location, plants, animals, and mountains around them in nature without some tour guide. Also, they can see pictures of different seasons at the same place. Figure 4 shows screenshots of the application (picture book of birds in the national park).

3.3 Results of the Experiment

We set beacons and performed experiments for several times. The radio wave reached about 10 m in the moist forest air. It reached less than ordinary streets in Nikko town areas, in which the wave reached 80–100 m in our experiments in 2015–2018 [12]. However, it was enough to get information for tourists who were walking in the forest. There are wooden roads in the national park. As the area is a national park's wetland, it was not easy to walk freely anywhere. Tourists have to walk several limited routes, such as the route of the wooden paths. Therefore, it was easy to determine the beacon

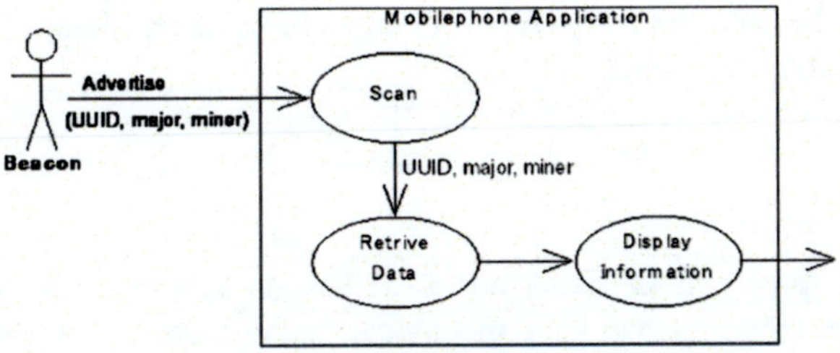

Fig. 2. Our Beacon system

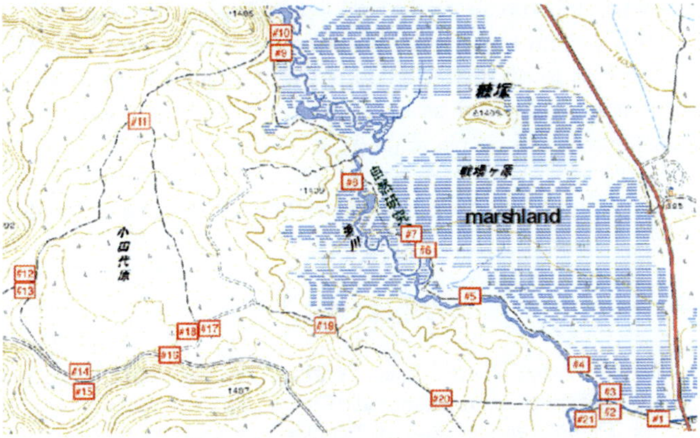

Fig. 3. Beacon points in Nikko National Park

Fig. 4. Screenshots of the application

place on the hiking routes. We set 21 Beacons and checked them in 2019 and 2020. The beacon at the bus stop is still working normally. However, we found dew condensation in most of the beacon cases. After setting beacons for half a year, several batteries gave out beneath the leafy trees. We have to consider about the battery and more impactive waterproof cases.

4 Conclusion

We are facing the problems of fitting our BLE beacon system in the harsh natural environments now. However, we have methods to solve them. We are testing to use larger solar panels and will change cases to prove airtight more. In addition, we will change beacon cases for more impactive waterproof. Such a system is useful for Post COVID-19 in the world. Small group walks in nature. They have little information now. We will make their travels fertile and safe using our system.

In addition, this system will be useful during times of disaster, letting people know the nearest safe place's location via the map. Communication traffic often occurs after the disaster. Our application can be used without requiring Internet access. We will continue our research in Nikko National park.

Acknowledgment. This research is supported by JSPS KAKENHI Grant Number JP17H02249.

References

1. UNWTO: COVID-19 and Transforming Tourism, 28 Aug 2020. https://www.unwto.org/news/un-policy-brief-on-tourism-and-covid-19
2. UNWTO: Global Guidelines to Restart Tourism, 28 May 2020. https://webunwto.s3.eu-west-1.amazonaws.com/s3fs-public/2020-05/UNWTO-Global-Guidelines-to-Restart-Tourism.pdf
3. Inoue, Y., Ikeda, T., et al.: Indoor autonomous positioning system on smartphones using ComPass beacon device. Information Processing Society of Japan Transactions Ubiquitous Computer System Report 2008 (18(2008-UBI-017)), pp. 195–202, 27 February 2008
4. Furudate, T., Horikawa, M., et al.: Indoor positioning method based on Bluetooth low energy Beacon. Forum Inf. Technol. **14**(4), 311–312 (2015)
5. Casa Batlló HP. https://www.casabatllo.es/ca/novetats/visita-con-realidad-aumentada-y-virtual/
6. Hiramatsu, Y., Sato, F., Ito, A., Hatano, H., Sato, M., Watanabe, Y., Sasaki, A.: A study of designing service model for sightseeing using BLE beacons -to provide tourism information of traditional cultural sites-. J. Adv. Intell. Syst. **9**(3&4) (2016). https://www.iariajournals.org/intelligent_systems/
7. Hatch, D.: Agri-tourism: a new agricultural business enterprise. LSD AgCentre Research & Extension (2006). https://www.caes.uga.edu/topics/sustainag/documents/LSUAgritourism_Pamphlet1.pd. Accessed 21 May 2015
8. The Technical Centre for Agricultural and Rural Cooperation (CTA): Outcomes of the 2nd Pacific Agribusiness Forum in 2016. https://www.cta.int/en/digitalisation/article/outcomes-of-the-2nd-pacific-agribusiness-forum-sid0b833fb64-3e5c-4035-8779-44024c633e45
9. Havlicek, Z., Lohr, V., Benda, P.: ICT and agritourism in Czech Republic. Agribus. Commerce **3**(3–4) (2009). https://www.researchgate.net/publication/46535726_ICT_and_agritourism_in_Czech_Republic

10. Mpiti, K., Harpe, A.: ICT factors affecting agritourism growth in rural communities of Lesotho. Afr. J. Hosp. Tour. Leisure **4**(2), 2. https://www.ajhtl.com/uploads/7/1/6/3/7163688/article60vol.4_2_july-nov2015.pdf
11. OkuNikko. https://nikko-travel.jp/english/attract/oku_nikko.html
12. Hiramatsu, Y., Ito, A., Sasaki, A., Ueda, K., et al.: A survey to create attractive contents for tourism -to comprehend other cultures-. In: Computer Aided Systems Theory – EUROCAST 2019, pp. 407–415, April 2020. https://doi.org/10.1007/978-3-030-45096-0_50

A Study on the Attitude and Practice of Chinese Teachers and Students Towards VR Teaching for Foreign Language Learning

Weihua Du(✉) and Zhongli Hu

German Department, Guangdong University of Foreign Studies,
Baiyun dadao bei Nr.2, Guangzhou 510420, China
duwh@gdufs.edu.cn, edithhuzhongli@qq.com

Abstract. In the field of foreign language education in China, virtual teaching is still in its infancy. The unique modal reconstruction function of virtual reality technology can provide new ideas for foreign language teaching. Virtual reality technology can create scene experience for learners, adjust modal information, and give them a new identity. VR Teaching for Foreign Language Learning is helpful for learners to participate in multi-sensory collaborative learning, deepen the effect of perception and experience, and improve the learning effect and language application ability. Based on the practice of foreign language teaching in China, this paper analyzes the efforts of Chinese foreign language education in the field of VR, discusses the difficulties in the process of reconstructing teaching mode assisted by virtual reality technology, and puts forward feasible suggestions for realizing foreign language learning objectives through VR teaching.

Keywords: Attitude and practice · Chinese teachers and students · VR teaching · Foreign language learning

1 Introduction

For foreign language learners, a real and natural target language environment can provide comprehensible input for language learning, but foreign language learning in a non-native environment usually lacks the necessary context to serve as a construction of meaning. Virtual reality (VR) provides a new perspective on the environment and process for effective teaching and learning, and its modal re-construction feature provides a simulated environment with interactive functions for language learning, so that language competence is generated in it.

The first year of VR in China was 2015, and since then, with the rapid iteration of VR products in various fields, the product and market have gradually matured. Relying on information technology, it creates information-based teaching environment, promotes the reform of teaching philosophy, teaching mode and teaching content, promotes the depth and wide application of information technology in daily teaching, and meets the demand for cultivating high-quality talents in the information age.

D. Russo et al. (Eds.): IHSI 2021, AISC 1322, pp. 728–732, 2021.
https://doi.org/10.1007/978-3-030-68017-6_108

Virtual technology has a special value in addressing the contextualization of language learning, greatly extending the space between learning and hands-on training, when students learn English or other languages in simple sentence drills, they have no prediction of when they will use the conversation. Learning language in a context leads them to predict the environment in which it will be used.

2 Existing Applications

In the conceptualization and design of teaching and learning situations, VR technology has assisted in reconstructing learning modalities that, in addition to traditional sensory experiences, it also emphasizes a shared sensory environment such as equilibrium and somatosensory, mobilizes more exciting areas of the cerebral cortex, creates a virtual space from both physical and psychological aspects, and brings learners into an immersive space to observe and experience, and learns to embody values in this space.

Professor Chris Dede's team at Harvard's Graduate School of Education has been a pioneer in the development of virtual resources. Funded by the State Department of Education, they have developed a course that uses virtual environments to teach ecosystem knowledge. Stanford University and the State University of New Jersey have also constructed a virtual reality lab and a MIX 3D printing lab, respectively, for the development of new forms of educational resources. For foreign language teaching, there are the EU VILL@GE (Virtual Language Learning through Edutainment Activities) project (Ma, Chongyu, Chen, 2013), the Russian Language Learning through Virtual Contexts (RTL) project and the Russian Language Learning through Edutainment Activities (RTL) project. VR has also been used in speech-language therapy for patients with communication disorders [1].

VR's modal re-construction function can be applied in the field of language and culture teaching for:

(1) Scenario experience. VR is particularly suitable for scenarios that are not normally available, such as a lecture where the audience and jury are present. The uses, such as business English, exhibition translation, etc. are all language communication in a specific environment.
(2) Cognitive experience. For example, it can be used to adjust the modal information and disturbance information in the lecture to increase the cognitive load of disturbance, so that the learners can anticipate the scene and improve the ability to cope with the situation.
(3) Identity construction. In the interpretation of foreign languages, VR can also create mirroring relationships. For example, based on the scenes of classical literary works, learners are given new identities in virtual simulations and are cognized in a similar way to the protagonist.

Depending on the form of user participation, VR applications currently include desktop, distributed, immersive, augmented, and hybrid. Among them, immersive VR is more suitable for foreign language teaching, placing the user's hearing and vision into the scene to provide a fully immersive experience. The language used for communication is not static, but a dynamic cognitive process that develops with changes in the communication context. The characteristics of the communication task, the purpose of communication, and the characteristics of the parties are all factors that determine the effectiveness of language communication. The learner's motivation for using a foreign language in the actual interaction is clear, and the language output is an act with real communication intention. Immersive VR recreates the scenes required for non-verbal communication, enabling learners to develop the linguistic responses they would expect in the real world.

3 VR Practice in Chinese Language Teaching

VR re-construction function simulates real-world for foreign language learners, such as competition, work and social environments, especially the presentation and interaction of non-verbal information, and provides new immersive spaces for the appreciation of literary works. Restructuring the physical space of learning improves the perception of teachers and students, it provides learners with multimodal real-world contexts, and promotes deep interaction between teachers and students with the learning environment and learning content.

China's Ministry of Education plans to identify about 1,000 exemplary virtual simulation experimental teaching projects by the end of 2020. The proposed teaching objectives are: to improve students' practical ability and innovation spirit as the core, relying on modern information technology, the information content of experimental teaching for relevant professional categories as the direction, based on complete experimental teaching projects, to build exemplary virtual simulation experimental teaching projects, and promote universities to actively explore a new mode of personalized, intelligent and ubiquitous experimental teaching combining online and offline teaching.

Guangdong University of Foreign Studies is a key university of Guangdong Province with distinctive international characteristics, and is an important base for the cultivation of international talents and the study of foreign language and culture, foreign economy and trade, and international strategy in China. It has 19,749 full-time undergraduates, 3,549 doctoral and postgraduate students, and more than 23,000 adult undergraduates, specialist students, advanced training students and international students. The university has one national experimental teaching demonstration center, 10 provincial experimental teaching demonstration centers and laboratories with a total area of more than 52,000 m^2. In 2019, the university has built a virtual reality application platform, but the utilization rate is not high. In 2020, the university has identified five virtual teaching projects through bidding, so as to promote virtual foreign language teaching (Table 1).

Table 1. Experimental teaching projects.

No	Name of faculty	Name of experimental teaching projects
1	Faculty of English language and culture	English drama scenario interactive interpretation virtual simulation experimental teaching project
2	School of innovation and entrepreneurship education	Virtual simulation experimental teaching project on entrepreneurship financing
3	Business school	Virtual simulation experimental teaching project of cross-border E-commerce startups operation
4	Faculty of western language and culture	Spanish language communicative and interpreter virtual simulation training experimental teaching project
5	Faculty of Chinese language and culture	Virtual simulation teaching project of creative writing under cultural space visualization

Source: https://jwc.gdufs.edu.cn/info/1039/6862.htm

The teachers of Guangdong University of Foreign Studies (GDUFS) have shown different attitudes towards the construction of the courses. Most of the teachers do not have more knowledge about virtual reality technology, and they do not have a correct concept of the funds and resources needed for virtual reality courses. Some students themselves have already experienced virtual teaching in other social institutions, and they have some of their own judgments about virtual foreign language teaching. Most of the students think that virtual teaching is something fun and can be done if playing games and having adventurous and exciting virtual experiences. If it is used for foreign language teaching, there is some expectation about its application, but they are not optimistic about it. Some students think that virtual foreign language teaching cannot replace the current mode of foreign language teaching, and that the main way for students to learn is still the traditional way of teaching and learning.

4 Discussion

With the continuous development of science and technology, VR immersion technology has an important role in the future development, especially in the field of education it is a kind of advanced teaching methods change, virtual learning scenarios can bring learners a good learning experience. At the same time, VR products are a powerful resource library, can provide rich learning resources for teaching and learning, so as to maximize the advantages of good teaching and resources. VR technology can play a role in facilitating teaching and learning in the midst of a neo-coronavirus epidemic. However, there are some points to note about VR immersion in foreign language teaching.

(1) Economic problems: The more functional VR products are, the higher the price will be. The cost of introducing VR products to universities will be too high, and although VR products can bring a good learning experience to learners, the high cost of the equipment will bring pressure on schools and learners, which may

prevent the teaching from achieving real results. This is a financial strain on schools with small budgets.

(2) The role of the teacher: Traditionally, the teacher is the learner's guide as well as the practitioner and transmitter of teaching and learning information. By making the learner the subject of knowledge acquisition, the teacher is relegated to the role of facilitator of technology or facilitator of student connections.

(3) Student experience: For a short period of time, students maintain a certain level of interest, but if they use VR technology for a long time, they will be under too much pressure, and their eyes will become tired, so they cannot concentrate for a long time. At the same time, most of the current VR products have a very low resolution, it is difficult to immerse learners, and secondly, the dizziness and fatigue caused by the delay in the picture lead to brain overload.

References

1. Li, Y.: An analysis of the modal reconstruction in virtualized foreign language teaching and learning. TEFLE **191**, 24–30 (2020)
2. National Center for Biotechnology Information Ministry of Education of the People's Republic of China. https://www.moe.gov.cn/fbh/live/2020/51987/sfcl/202005/t20200514_454117.html
3. Norris, S.: Analyzing Multimodal Interaction: A Methodological Framework. Routledge, London (2004)
4. Gros, B.: Digital games in education: the design of games-based learning environments. J. Res. Technol. Educ. **40**(1), 23–38 (2007)

Lean Manufacturing for Production Management and Strategic Planning to Increase Productivity in SMEs Engaged in Manufacturing Chemicals

Hernan Zegarra-Mendez[1], Víctor Nuñez[1], and Carlos Raymundo[2(✉)]

[1] Ingenieria Industrial, Universidad Peruana de Ciencias Aplicadas, Lima 15023, Peru
{u201420302,pcinvnun}@upc.edu.pe

[2] Direccion de Investigacion, Universidad Peruana de Ciencias Aplicadas, Lima 15023, Peru
carlos.raymundo@upc.edu.pe

Abstract. This research proposes productivity scenarios in a small- and medium-sized enterprise (SME) dedicated to manufacturing coatings (e.g., lacquers, resins, emulsions, and varnishes). The problem identified in these SMEs is that they experience high downtime because of low productivity compared to companies in the same industry, which is important because it increases cost and inefficiencies. Thus, this study aims to increase the productivity of coating manufacturing SMEs. We studied the labour productivity variable, and the techniques found in the literature to address this issue are as follows: lean manufacturing and strategic planning tools such as value-stream mapping, Single-Minute Exchange of Dies, and standardized work. Based on these tools, we aimed to eliminate non-value-added (NVA) activities and to improve business productivity.

Keywords: Lean manufacturing · Productivity · SMEs · Chemicals

1 Introduction

The chemical industry of coatings and paints has undergone continuous technological changes in the last 10 years. Thus, the industry was subdivided to two levels: large manufacturers, competing in the international market, and small manufacturers, dedicated to specialized applications. The last National Census of Manufacturing Establishments showed that in Peru there are 172 companies engaged in manufacturing paints, varnishes, and lacquers. In the last 15 years the paint, varnish, and lacquer manufacturing sector has had a favourable production growth, with an average annual rate of 15.8% from 2004 to 2008.

The purpose of this study is to increase the productivity of coating manufacturing SMEs using lean manufacturing tools to improve processes and activities. First, we identified the main causes for low productivity by quantifying its impact. Then, we analysed ways to address these causes and proposed the use of standardized work,

D. Russo et al. (Eds.): IHSI 2021, AISC 1322, pp. 733–739, 2021.
https://doi.org/10.1007/978-3-030-68017-6_109

value-stream mapping (VSM), and Single-Minute Exchange of Dies (SMED). Using the first tool, we identified activities that do not add value to the process to reduce or eliminate them. Then, by using SMED, we aimed to reduce changeover or reactor cleaning times.

2 State of the Art

2.1 Production Management Model in the Sector SMEs

Studies agree that SMEs need to continuously enhance their performance to survive in the current competitive business environment [1, 2]. They conclude that the operational performance of an organization lies in its core business compliance with industry standards in terms of quality, costs, and delivery (QCD) [3, 4].

2.2 Lean Manufacturing in the Sector MGP

Research has shown that lean manufacturing tools can reduce costs by up to 15% by increasing productivity, particularly labour productivity [5, 6]. In this regard, other studies observe the significance of waste elimination from manufacturing processes in reducing the amount of work in process (WIP) [6, 7]. The lean manufacturing principle focuses on cost reduction by identifying and eliminating non-value added activities.

2.3 Strategic Planning in the Sector

The relation between lean manufacturing culture and the management of key performance indicators (KPIs) provides an in-depth understanding of essential factors for successful implementation [8, 9]. Thus, research has identified approaches through which lean instructors and professionals can avoid dilemmas and problems derived from its implementation [3, 10] to change the general mindset of students or employees and ultimately facilitate successful process improvement and replication on a larger scale [3, 11]. The results show that dynamic strategic planning has a negative but not significant association with financial performance.

3 Contribution

3.1 Proposed Method

It focuses on the active cooperation between operators and the production manager because they coordinate and execute the production planning activities, supported by KPIs management as a strategic control element (Fig. 1).

The model is complemented by lean manufacturing and strategic planning tools to reduce or eliminate waste that negatively influences the company's work process. This is mainly based on standardized work and resource optimization.

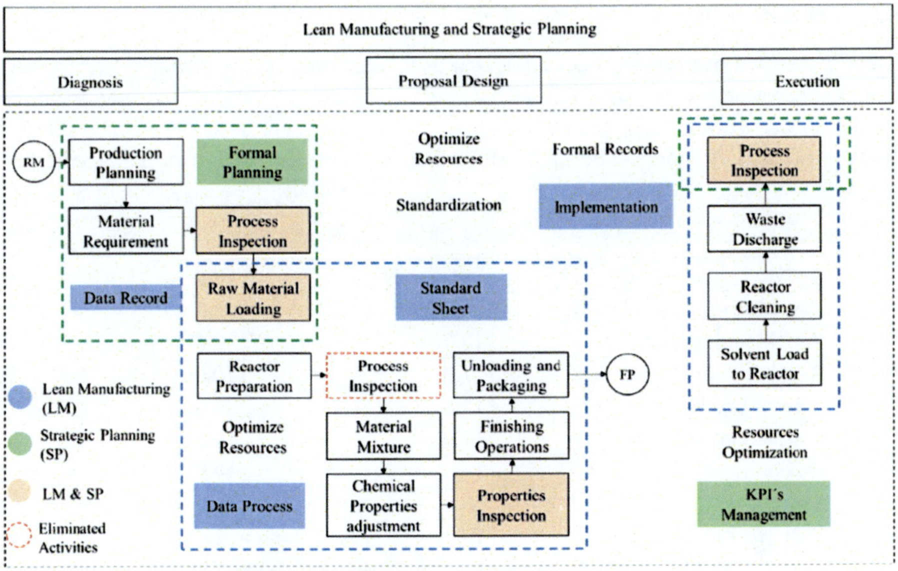

Fig. 1. Lean manufacturing and strategic planning

3.2 Method Components

Standardized work directly addresses the specific cause of non-standardized activities to identify and eliminate non-value-added activities and generate a standard procedure to reduce downtime. It also classifies activities by their type and by time spent on their execution, which is a practice that is not currently used by SMEs engaged in manufacturing coatings.

Table 1 identifies and classifies activities by their type and time needed, as well as the proposed improvements. This shows the standardized activities after the implementation of the improvement proposals and allows for a defined work method to eliminate the waste that causes the downtime identified in the analytical flowchart.

Table 1. Format for activity record

Activity	Initial value	Goal
Operation	24.0	22.0
Transport	11.0	7.0
Wait	5.0	4.0
Inspection	15.0	9.0
Total	55.0	42.0
Data process	Format helps to determine the number of activities and m-h to be reduced	
Decision	To be implemented in every product family	

3.3 Process View

Table 2 shows a queueing process diagram as a basic production management model. The process starts with the required resource planning, followed by raw material transfer to the working area of reactor No. 4. Then, pallets are used to load and unload the raw material cylinders. The measuring of reactor cleaning activities is also shown.

Table 2. Format for activity record

Activity	Initial value	Goal
Raw Material Weight	4.0	3.0
Raw Material Load	18.0	9.0
Raw Material Unload	21.0	10.0
Solvent Load	12.0	6.0
Solvent Unload	17.0	9.0
Others	16.0	12.0
Total	55.0	42.0
Data process	Format helps to determine the number of activities and m-h to be reduced	
Decision	To be implemented in every product family	

3.4 Indicators

The following productivity indicators were used:

Productivity (Hours). Determines direct labor productivity by relating productive man-hours to the total time available for production.

$$\text{DL Productivity} = \frac{\text{Productive time}}{\text{Available Time}} \times 100\% \quad (1)$$

Transport Time. Measures productivity by relating the manufactured product (kg) to the productive hours used during the process.

$$\text{Inspection times} = \frac{\text{Initial hours} - \text{Final Hours}}{\text{Initial hours}} \times 100\% \quad (2)$$

Inspection Time. Determines labor productivity by relating the inspection hours before and after implementing any improvement.

$$\text{Transport time} = \frac{\text{Initial hours} - \text{Final Hours}}{\text{Initial hours}} \times 100\% \quad (3)$$

Loading and Unloading Time. Determines labor productivity by relating the inspection hours before and after implementing any improvement.

$$\text{Loading and unloading time} = \frac{\text{Initial hours} - \text{Final Hours}}{\text{Initial hours}} \times 100\% \qquad (4)$$

4 Validation

4.1 Scenario Results

We developed the proposed model based on the queuing process because of the amount of activities and resources needed for its execution. The intention is to reduce non-value-added inspection activities, which are maintained in high-risk activities such as temperature or chemical density. Figures 2 and 3 show the cylinders that will be replaced by higher capacity cubitainers to reduce the number of man-hours employed during changeover and material movement operations.

Fig. 2. Area of work

Fig. 3. Containers and cylinders that will be replaced by cubitainers

The enterprise has a productivity of 72% in the last 6 months, while the standard established by the National Society of Industries is 80%.

4.2 Validation Development

The process proposed in Arena software was simulated to dimension the production queue and set-up activities.

Then, we placed the number of operators and their proportionality in time to estimate the number of man-hours generated in the production run, shown in Fig. 4. The result indicates a system time of 19.4 h, compared to 9.6 man-hours.

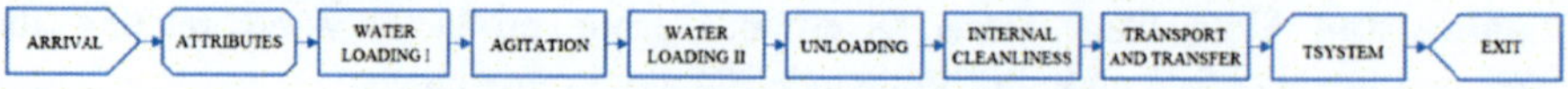

Fig. 4. Simulation of the proposed process

The results show the system time and the percentage used by the operator; whose product indicates the number of man-hours. Table 3 compares the initial situation with the simulation. We found a 45% reduction in downtime and an increase in the productivity level, taking as reference the production process of the queueing product and the cleaning of reactor No. 4.

Table 3. Initial vs. final values

Category	Initial Value	Final Value	Variation
Productivity	72.0%	80.0%	+8.0%
Transport time (m-h)	23.1	2.5	−48.0%
Inspection time (m-h)	4.4	3.7	−39.4%
Material's load time (m-h)	6.0	3.7	−37.8%
Material's unload time (m-h)	5.3	3.2	−36.3%

5 Conclusions

Lean manufacturing, implemented through the standardized work, SMED, and VSM tools, increased labour productivity by at least 8% in SMEs in this sector.

Standardized work reduced inspection times by at least 44% by decreasing the number of NVA inspections.

Standardized work and SMED reduced transfer times by at least 30%. These tools also reduced loading and unloading times by at least 36% because of the change in the system and structure, using containers to replace cylinders.

SMED complements the standardized work tool creating synergies that reduce or eliminate activities without interrupting process operation.

References

1. Hrubý, V., Koráb, V.: Performance of small and medium-sized enterprises in the chemical industry in the Czech Republic – an exploratory analysis. Trends Econ. Manag. **12**(31), 9–18 (2018)
2. Abolhassani, A., Layfield, K., Gopalakrishnan, B.: Lean and US manufacturing industry: popularity of practices and implementation barriers. Int. J. Prod. Perform. Manag. **65**(7), 875–897 (2016)
3. Burenina, I.: Integral assessment of the development of Russia's chemical industry. Treatise Electr. Magn. **2**(3) (2018)

4. Gambhir, D., Sharma, S.: Productivity in Indian manufacturing: evidence from the textile industry. J. Econ. Adm. Sci. **31**(2), 71–85 (2015)
5. Singh, S., Dixit, S., Sahai, S., Sao, A., Alonia, Y., Subramanya, R.: Key benefits of adopting lean manufacturing principles in Indian construction industry. In: MATEC Web of Conferences, vol. 147 (2018)
6. Dora, M., Kumar, M., Van Goubergen, D., Molnar, A., Gellynck, X.: Operational performance and critical success factors of lean manufacturing in European food processing SMEs. Trends Food Sci. Tech. **31**, 156–164 (2013)
7. Hama Kareem, J., Mohamad Al Askari, P., Muhammad, F.: Critical issues in lean manufacturing programs: a case study in Kurdish iron & steel factories. Cogent Eng. **4**(1), 1–15 (2017)
8. Vinayagasundaram, R., Kannan, V.: Implementation of lean manufacturing practices and its impact on productivity in Coimbatore foundries. J. Entrep. Manag. **1**, 4 (2016)
9. Mirdad, M., Eseonu, C.: A strategy of resilient systems thinking to sustain a lean organizational culture: promoting conceptual change in lean class. In: IIE Annual Conference Proceedings, pp. 1472–1478 (2017)
10. Antosz, K., Stadnicka, D.: Lean philosophy implementation in SMEs – study results. Proc. Eng. **182**, 25–32 (2017)
11. Alhaqbani, A., Reed, D., Savage, B., Ries, J.: The impact of middle management commitment on improvement initiatives in public organisations. Bus. Process Manag. J. **5** (22), 924–938 (2016)

Implementation of Lean Manufacturing Tools to Improve the Flow of Productivity in a Craft Shoe Workshop

Nicol Manrique[1](✉), Stephanie Nuñez[1], Grimaldo Quispe[2], and Carlos Raymundo[3]

[1] Facultad de Ingenieria, Universidad Continental, Huancayo, Peru
{75163923, 72410034}@continental.edu.pe
[2] Facultad de Ingenieria, Universidad Nacional Autonoma Altoandina de Tarma, Tarma, Peru
gquispe@unaat.edu.pe
[3] Direccion de Investigacion, Universidad Peruana de Ciencias Aplicadas, Lima 15023, Peru
carlos.raymundo@upc.edu.pe

Abstract. This paper focus on improving productivity in a shoe production workshop in Bonssa using the tools of Lean Manufacturing. The factors that affect the productivity of the company have been evaluated, such as recent delays in order delivery, disturbances and unrest in the workplace and difficult working hours. The production processes and their implications were recognized. Balance of lines, 5'S and design were applied for a better distribution in the factory, having as results, the reduction of delivery time from 12.6 days to 5.9 days, The reduction of the added value from 27.516 to 23.14 min, 30% of Saving time searching for models, reducing machine operating time by 70%, improving communication with employees, a pleasant and safe work environment.

Keywords: Lean manufacturing · Productivity · Craft shoe · Productivity · Improve

1 Introduction

In recent years, there has been a brutal change due to the emergence of China in the market with an import of 53% of the total, so much so that in 2018 there was a 29% drop in the production of Peruvian footwear, according to the National Society of Industries (SNI).

According to Gestión, Peru is the fourth largest producer of footwear in South America; However, most are micro and small companies that are characterized by low levels of productivity and limited financing, in which it is necessary for footwear companies to adopt different tools and techniques to improve their productivity, quality standards and satisfaction the client's [1, 2].

Currently, there are various tools that serve to make companies more efficient, one of them is the "Lean Manufacturing" philosophy, which means Lean Manufacturing or

D. Russo et al. (Eds.): IHSI 2021, AISC 1322, pp. 740–746, 2021.
https://doi.org/10.1007/978-3-030-68017-6_110

Lean Manufacturing, which seeks to eliminate waste and activities that use unnecessary resources without creating value, thus achieving benefits in productivity, competitiveness and profitability of the company [3–5]. According to Manzano, the word lean refers precisely to that company or process free of inefficiencies or waste and that works with the minimum resources that the company has [6].

Due to the aforementioned, the purpose of this research is to apply VSM as part of the Lean Manufacturing tools, according to the problems identified in the Bonssa footwear company, to improve its productivity and make it much more competitive compared to its competitors. local and international [7, 8].

2 Methods

2.1 Current Situation

Currently, the company has four types of products listed below: Shoes for daily use in ladies, Shoes for common men, School shoes, Daily shoe for children (Figs. 1 and 2).

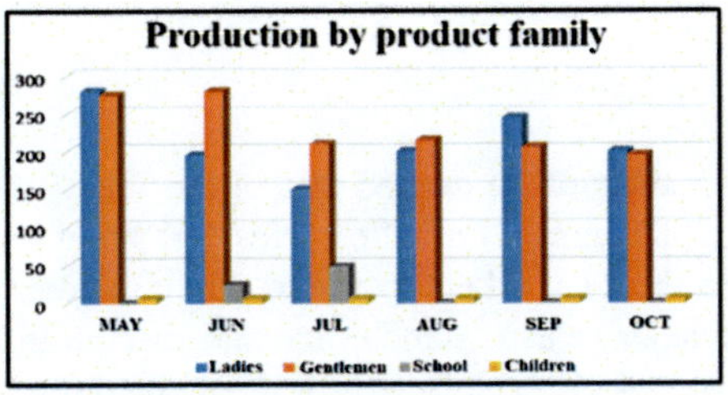

Fig. 1. Pairs produced by product lines.

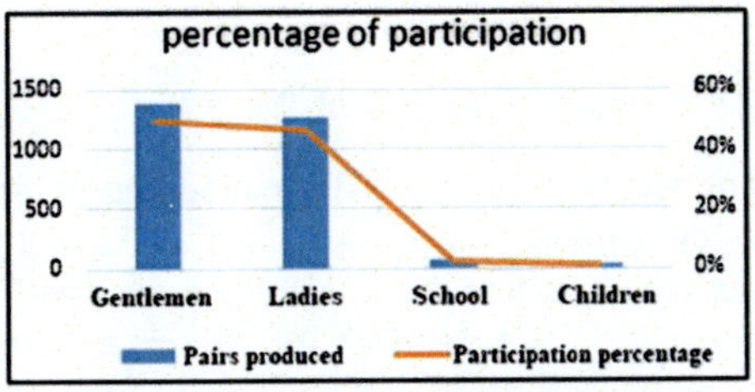

Fig. 2. Most produced product line.

With the Pareto analysis, we conclude that the product family of men's shoes has the highest demand for which it will be the one that we will analyze in the VSM.

2.2 Study of Times

The distribution of time is carried out cumulatively, that is, from the moment of its beginning to the end of each activity. The employee's work rate is then evaluated at any time to obtain the basic work time. Applies the evaluation criteria in Appendix 4, which is added to the base time applications and is the standard time of service, as defined in the standard time calculation, as the sum of the standard time of each activity. As a result, the standard time for each job and production process (Table 1).

Table 1. Calculation of samples

Company Bonssa			***Area*** Production	
Method TEST			***Process*** Dress shoes for men	
Produced by Nicol Manrique			***Products*** 1 pair of dress shoes	
ITEM	**ACTIVITY**	**Σx**	**Σx²**	**n**
1	Cut	54.16	2933.3	6
2	Roughing and Marking	10.19	103.84	5
3	Union	66.95	4482.3	4
4	Plastered	20.95	438.90	4
5	Armed	35	1225	2
6	Sole	27.45	753.50	2
7	Finish	23.74	563.59	10

Where:

$$n = \left(\frac{40\sqrt{n^2 \sum x^2 - \left(\sum x\right)^2}}{\sum x}\right)^2 \tag{1}$$

2.3 Implementation of the 5 S

For the application of the methodology, an audit will be carried out in the areas of assembly and flooring because they have a greater impact on deficiencies compared to the others; where the questions were asked to the 11 workers from the different areas about the 5 segments of the 5S: classify, organize, clean, standardize and discipline. Then the responses will be scored based on the criteria established for the audit shown in Table 2.

Table 2. Criteria for the audit

Level 1	Initial	It runs sporadically, there are no procedures
Level 2	Managed	The 5 s standard is evolving
Level 3	Definite	Standards are implemented and action plans are executed
Level 4	Predictable	It is controlled and measured
Level 5	Optimized	Continuous improvement is sought, corrective actions are proposed

Likewise, a summary of the evaluation was made of the assembly and flooring areas of the Bonssa company, where an average of 2.05 was obtained; that is, 41% of the range of the goal set. This shows that it is necessary to make a good plan for the implementation of the 5S¨s (Table 3).

Table 3. Evaluation results

N°	Criterion	Achievement	Goal
1	Classify	2.17	5.00
2	Organize	1.89	5.00
3	Clean	2.00	5.00
4	Standardize	2.40	5.00
5	Discipline	1.80	5.00

The objective of the Bonssa company is to reach the goal with a rating of 5, which is continuous improvement and establish corrective measures for an optimal production process.

3 Results

3.1 Preparation of VSM Future

To create a future VSM, you need to identify existing VSM enhancements and make suggestions that help reduce and/or eliminate waste, and most importantly: continuous flow in all processes reduces load.

Takt time calculation: Calculation of the available work time between the client's requirements.

$$Takt\ time = \frac{Time\ available\ in\ shift}{Customer\ requirements\ per\ day} \tag{2}$$

Available time = 9 h × 60 min = 540 min − 10 min = 530 min.

Customer requirement × day = 504 pairs/22 days = 22.90 pairs/day.

The resulting Takt time is 23.14 min/pair, which means that each time a pair of shoes must be produced to satisfy customer demand.

The future VSM is shown in Fig. 3 (Fig. 4).

3.2 5 S Tool Application

After an internal audit, regarding compliance with the provisions above, a noticeable and favorable improvement was obtained in the application of the 5S methodology. Ascending to 80% the improvement that was obtained through this Lean methodology with respect to the initial results (Table 4).

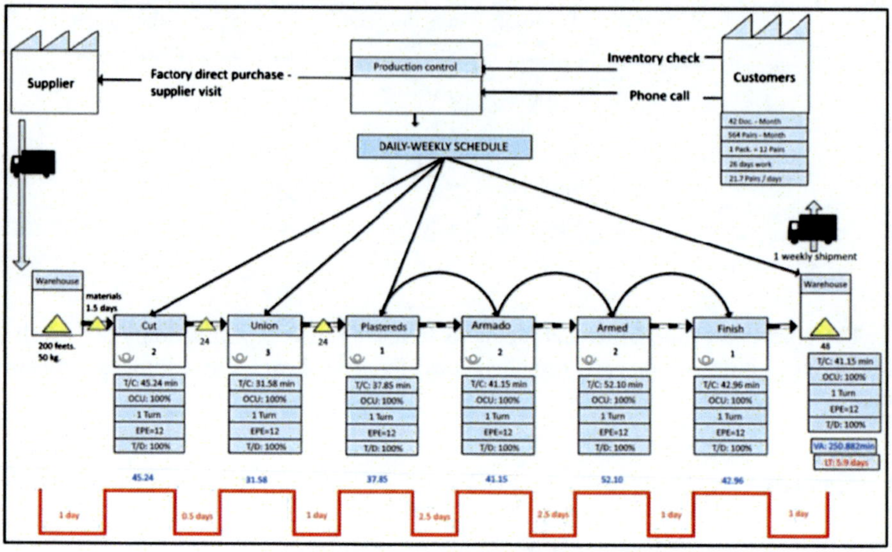

Fig. 3. Future VSM

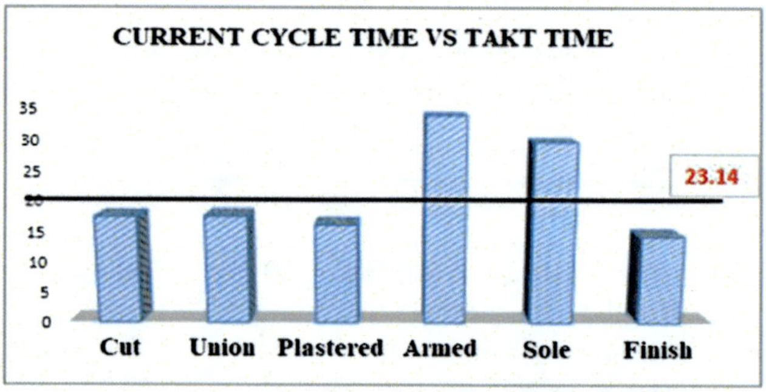

Fig. 4. Cycle times vs Takt time

Table 4. Results after implementation

Sole and assembly area				
Pillars	Qualification before	Qualification after	Improvement	%Improvement
Classify	2.17	4.40	2.23	49.32
Organize	1.89	4.50	2.61	42.00
Clean	2.00	3.70	1.70	54.05
Standardize	2.40	3.60	1.20	66.67
Discipline	1.80	3.90	2.10	46.15

3.3 Discussion of Results

ORTIZ Triana and CAICEDO- Rollón - The best thing about this program is the footwear - Francisco de Paula Santander, Faculty of Engineering of San José de Kucheta-Colombia. 15p.p.

The spirit of this research has developed an optimal production program in a small footwear company in San José de Cuquette, Colombia, and has been based on the theory of the limitations of the production system for mathematical modeling. restrictions according to the purpose of the machinery used.

Production planning is an immediate response to the production of goods or services. There are different production methods. However, many researchers admit that they are using mathematical programming methods to solve the problem optimally. Today's investigations are an important decision-making tool, so it's important to take action.

Colombia's production structure is characterized by the presence of smaller productive devices, where the size of the company affects the ability to innovate, compete, export and finance. The results of the analysis correspond to the design of the models for the semester of 2011.

4 Conclusions

With the help of the company, the study of espresso for the production of men's footwear and the products for which the rubber factory was valued was identified. Likewise, people with disabilities, in this process, comply with the original work method, determine the type of activities that constitute 53% of the added value of the added value; On the other hand, the initial time will change when the standard time is 27,516 min/pairs and 19 pairs per day. At the same time, the company did not properly distribute raw materials, as well as a lack of cleaning materials.

It was clear that improving productivity could improve working methods and reduce time, improve process and achieve results; 71% of the total events took place, the new schedule was established at 23.14. Minutes/pairs of shoes, 24 pairs per day. All the above factors affected productivity growth by 23.37% at the footwear firm Bonssa.

A good resource to improve our productivity has been the introduction of 5S methodologies aimed at agitation and displacement; It should also be noted that this is a simple way to apply zero costs that are attractive to any organization. This was very useful for the company, mainly because it was more productive, thanks to its 80% focus on continuous improvement in consolidation.

References

1. Aplicación del método 5S. Capítulo 8 Universidad de Sevilla. Mejora de la distribución de cultivos de hortalizas y cereales. Universidad de Sevilla. http://bibing.us.es/proyectos/abreproy/5055/fichero/8.+APLICACI%D3N+DEL+M%C9TODO+DE+LAS+5S+AL+TALLER%252F8+Aplicación%D3N+DEL+M%C9ALL+DE+LAS+5S+AL+WORKSHOP.pdf

2. Ghodrati, A., Zulkifli, N.: The impact of 5S implementation on industrial organizations performance. Int. Bus. Manage. Invent. J. **2**(3), 43–49 (2013)
3. Czifra, G: Implementation process of 5S for a company in real life - problems, solutions, successes. Res. Pap. Fac. Mat. Sci. Technol. Slovak Univ. Techno. **25**(41), 79–86 (2017)
4. García, R.: Estudio del Trabajo: Ingeniería de métodos y medición del trabajo, 459 pp. 2nd edn. McGraw-Hill, México (1998). ISBN: 9788479782306
5. Purohit, S.R., Shantha, V.: Implementation of 5S methodology in a manufacturing industry. Int. J. Sci. Eng. Res. **6**(8), 225–231 (2015)
6. Manzano Ramnrez, M., Gisbert Soler, V.: Lean Manufacturing: Introducción a 5S. Tecnología 3C: glosarios innovadores para pequeñas y medianas empresas **5**(4), 16–26 (2016). http://dx.doi.org/10.17993/3ctecno.2016.v5n4e20.16-26
7. Nafizah, U.Y., Tiara, A.R.: Value stream mapping in Indonesia's shoe industry: an alternative tool to map innovation ecosystem. Int. J. Technol. Learn. Innov. Dev. **10**(2), 113–130 (2018)
8. Paredes-Rodríguez, A.M.: Aplicación de la herramienta Value Stream Mapping a una empresa embaladora de productos de vidrio. Entramado **13**(1), 262–277 (2017)

Gestalt Prototyping Framework - Evaluation Tool

Daniel Ripalda[1](✉), César Guevara[1], and Alejandra Garrido[2]

[1] Mechatronics and Interactive Systems Research Center, Universidad Tecnológica Indoamérica, Machala & Sabanilla, Quito, Ecuador
{danielripalda, cesarguevara}@uti.edu.ec

[2] LIFIA, Facultad de Informática, Universidad Nacional de La Plata and CONICET 50 & 120, 1900 La Plata, Argentina
garrido@lifia.info.unlp.edu.ar

Abstract. The Gestalt Prototyping Framework linked the Nielsen Usability Heuristics to the principles of human perception and presents some parameters that can be used in developing high and low-fidelity prototypes for mobile application interfaces. The link between the fundamentals of usability and Gestalt principles focuses on the graphical components of the interfaces and the functions that they fulfill in the development of different functional actions. Previous articles have presented promising experimental results in reducing trial-error regressions in the interface design process and improving interface redesign processes; There have been positive results in parameters such as learnability, ease of use, perception of simplicity, and user preference. Within this same line of research, the present work describes the development process of an application that summarizes the fundamental aspects of the Gestalt prototyping framework in an evaluative model, which helps application designers, software engineers, and usability experts, to assess the prototypes of the interfaces and the incidence of the different graphic components in usability interactions. The tool is based on the most widely used usability parameters and is structured based on the responses that the development team fills in an automated LIKERT assessment. The results are processed based on an algorithm that simulates and predicts the results that could be obtained in high-fidelity prototype tests. This application issues specific recommendations on the visual components of the interfaces, to obtain better results in the production of high and low-fidelity prototypes.

Keywords: Gestalt · User interfaces · Usability · Mobile · Agile software development

1 Introduction

The Gestalt Prototyping Framework links the Nielsen Usability Heuristics with the principles of human perception and presents a series of parameters that can be used by designers, software developers, and usability experts, during the development stage of high and low fidelity prototypes in the production of mobile applications. A link was found between the statements of the heuristics, the graphical components of the

D. Russo et al. (Eds.): IHSI 2021, AISC 1322, pp. 747–752, 2021.
https://doi.org/10.1007/978-3-030-68017-6_111

interfaces, the principles of human perception of Gestalt, and some interactions of the user experience.

The first stages of this research describe the Framework from a constructive model that guides the Development Team in the production of mockups, providing specific guidelines for the use of iconography, menu configuration, button interaction and content organization, factors that affect the learnability, memorability, efficiency, and usefulness of an interface; the relationship between visual components and interactions facilitate that a heuristic can be understood from specific visual representations, reducing the developer's margin of interpretation in the early stages of a project, without interfering with the methodology used in the project. In the experimental phase of the constructive model, he obtained promising results in the redesign of an interface prototype, increasing the perception of speed and use by 33.1% and 97% in preference of use, having measured experience, learnability, and ease.

This paper describes in Sect. 2 some works related to usability measurement tools, Sect. 3 describes the concepts that were used to structure the application, Sect. 4 describes the development process and general operation, and Sect. 5 describes the results of tests performed during application production.

2 Related Work

With the rapid development of the mobile application market, there has been a notable interest in developing tools to measure different criteria in the Interfaces of mobile devices, one of the most interesting systems is described in Automated model-based Android GUI Testing, using multi-level GUI comparison criteria [1], this tool allows checking the bugs of an application by configuring test cases based on the actions that occur in an interface and could be used to test high-quality models.

Another tool is integrated automated test case generation for safety-critical software [2], which is a patented tool that among its various functions can generate GUI tests and verify the usability of an industrial application. This program is based on a device that obtains the requirements of the users, generates an intermediate model, and prepares test cases for an application, This work has made it possible to obtain important data on the hardware functions of the devices and the instructions of the program that controls it. Other programs [3] use this methodology to detect errors in the requirements, using an analysis module that compares other requirements to detect conflicts and repetitions.

One of the most recent revisions regarding this type of application was developed by Azham Hussain [4], In his work, ten different applications are reviewed such as Perfect Mobile, Device Anyware, Monkey Runner, Robotium, among others. According to this author, one of the tools with the best performance is Micro Focus Silk Mobile, a Micro Focus product; This multi-platform application performs tests based on the recognition of images of the interface and captures information about the performance of the device, validating the quality of the construction of the software, it can save, process and simulate user interactions with generating detailed reports that are later analyzed.

Another quite interesting tool is HUI Analyzer, which was reviewed by Simon Baker [5], The tool is capable of interpreting data related to Learnability, Understandability, Efficiency, Memorability, and Satisfaction; and it can find different defects in the usability of an application using mathematical models; The technical documentation available explains that the quantitative evaluation scheme could minimize the effectiveness of the results since, for example, the contrast factor of the interface is difficult to evaluate because at the same time that it can increase performance, it can compromise aesthetics.

In general, the tools to measure heuristics still have limitations, in the article Finding Usability Problems Through Heuristic Evaluation [6], Nielsen points out that some of the guidelines are abstract and leave room for interpretation by usability experts using their criteria and personal experience. Some studies show that automated tools can be 50% effective in detecting usability problems due to false positives [7]. However, in some experimental tests [8] in applications such as pCloudy, Test Object, Device Anywhere, and Perfect Mobile, they obtained an 85% effectiveness in identifying different types of failures, such as application crashes, failures in the loading of interfaces, or incomplete information in the display; although these applications are quite effective in detecting functional problems, we cannot say that they detect errors or make recommendations in aspects directly related to usability.

3 Concepts Applicable to the Evaluation Tool

The Gestalt Prototyping Framework [9], It has an evaluative model that is based on the idea that the graphic components of the interfaces affect specific usability interactions so that the heuristics can be perceived; therefore, if the perceptual fundamentals are used in these components, they can be associated with the heuristics, providing information during the low and high fidelity prototype development phase.

The usability heuristics [10] are seen as guides to develop the functional interactions of the applications, while the Gestalt [11] provides guidelines to develop the visual aspect. Using conditional logic to structured tools [12], these concepts can be used to evaluate prototypes and detect errors in components, predict errors in usability tests, and measurement of aesthetics. Figure 1 explains the concept of the evaluative model of the Framework applied to a software tool.

According to Baker's analysis, this tool is a model-based Guideline Checker [13], focused on evaluating the interfaces for specific usability parameters; during the product development phase of a mobile application, the usability experts can use it to interpret qualitative the data in a for a static analysis [14], that detects problems relating the visual aspect of the GUI with the functional properties of the application.

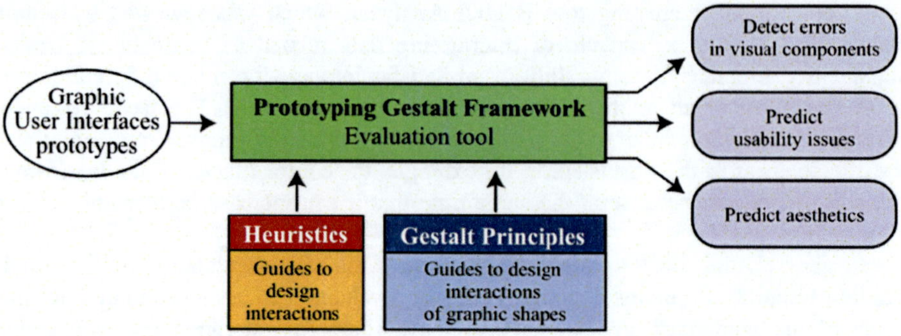

Fig. 1. Concepts of the Gestalt Prototyping Framework applied to an evaluative tool

4 Application Development

A team of students and teachers from the Indoamérica University developed the application taking into account the established parameters, it is a mobile application for the Android platform, two types of tests can be executed, one focused on specific components of the interfaces, and the other that covers a complete test. The application asks a series of questions related to the graphic components of the mobile application interfaces to be evaluated, the usability expert answers the questions through a Likert scale, the answer to each question assigns a score to the heuristic and the specific usability interaction; for example, in the iconography category, the question "Icons can be easily recognized?" assigns a point to the heuristic " Match between system and the real world", at the same time assigns a point to the interaction "Learnability" and another point to "Efficiency". The perception of aesthetics is calculated based on the statement that there is a correlation between usability, the perception of aesthetics [15], and credibility [16]; therefore, the aesthetic assessment is given based on the general score assigned to each component. The results are quantified according to Table 1.

Table 1. Relation of interface components, heuristics, and usability interactions

Questions related to	The score is assigned to the heuristic	The score is assignedto the usability interactions
Iconography Colour Buttons	Match between system and the real world	Learnability Efficiency
Menus Submenus Pop up Slide buttons	Recognition rather than recall	Learnability Efficiency Memorability
Backgrounds Dividers Graphic style	Aesthetic and minimalistic design	Memorability Easy

Once the user completes the test, the system performs a calculation to quantify the results by heuristics, by usability interaction, and by the perception of aesthetics; To present the final values, an algorithm is used to adjusts the results to the parameters of real usability tests carried out by human users.

5 Results

The name that was selected for the Project and the application is GEPROF-EMAT, (Gestalt Prototyping Framework - Evaluative Mode Automated Tool), the beta version has been developed with the SCRUM methodology, using Unity technology, for the Android platform and run locally. In the preliminary tests, it was possible to see that the calculation carried out by the application is consistent with the applied concepts and the guidelines of the Framework. In the evaluation of prototypes, it was possible to appreciate positive evaluations around some parameters that have been measured in similar applications, according to the parameters evaluated by Baker, the application seem to meet expectations such as: Not being intrusive to the user, providing sufficient information to perform static analysis and maintaining an independent evaluation process. Figure 2 shows the main application interfaces.

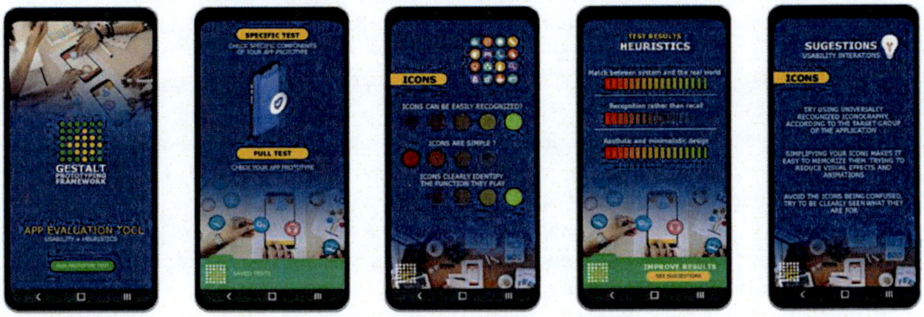

Fig. 2. Application interface

6 Conclusions and Future Work

The application GEPROF-EMAT is in the testing phase, has been complying with the different interactions typical of the development methodology that has been applied, the first results are encouraging and there is a projection that the margin of error for tests with real users is currently 35%, this value will decrease as continue using the tool and comparing the results to make adjustments to the prediction algorithm. It is important to point out that this tool even in its final version will not replace usability tests or the current measurement instruments, because just provides referential information for the construction of prototypes and establishes approximate values of the real tests. The quantitative assessment of the tool prompts the discussion of the development team of a mobile application, the results require human intervention and interpretation, as a determining factor in identifying and correcting errors. The developer's vision, intuition, background, and experience are required to positively validate the interfaces and find functional solutions that consider abstract aspects such as aesthetics. We expect that in the near future, the effectiveness of the application is expected to reach 85%, equaling automatic error detection tools; for this, it is necessary to make adjustments in the error prediction algorithm based on the values obtained in usability metrics measured with real users.

References

1. Baek, Y.M., Bae, D.H.: Automated model-based Android GUI testing using multi-level GUI comparison criteria. In: Proceedings of the 31st IEEE/ACM International Conference on Automated Software Engineering, pp. 238–249 (2016)
2. Li, M.: Integrated automated test case generation for safety-critical software, U.S. Patent No 10,108,536 (2018)
3. Manolios, P.: U.S. Patent No. 9,639,450, Washington DC: Patent and Trademark Office (2017)
4. Saad, N.H.: The perceived usability of automated testing tools for mobile applications. 12 edn. J. Eng. Sci. Technol. (JESTEC) 89–97 (2017)
5. Baker, S., Au, F., Dobbie, G., Warren, I.: Automated usability testing using HUI analyzer. In: 19th Australian Conference on Software Engineering (aswec 2008) (2008)
6. Nielsen, J.: Finding usability problems through heuristic, Monterrey, California USA: In: Proceedings of the SIGCHI Conference (2002)
7. Dix, A., Finlay, J., Abowd, G.D., Beale, R.: Human-Computer Interaction. Prentice-Hall, Harlow (2004)
8. Vilkomir, S., Marszalkowski, K., Perry, C., Mahendrakar, S.: Effectiveness of multi-device testing mobile applications. In: ACM International Conference on Mobile Software Engineering and Systems, pp. 44–47 (2015)
9. Ripalda, D., Garrido, A., Guevara, C.: Gestalt framework in the design of interfaces for mobile devices: Theoretical approach. In: 13th Iberian Conference on Information Systems and Technologies (CISTI), Cáceres, Spain (2018)
10. Nielsen, J., Molich, R.: Heuristic evaluation of user interfaces. In: Proceedings of the SIGCHI Conference on Human Factors in Computing Systems. ACM (1990)
11. Hartmann, G.W.: Gestalt Psychology: A Survey of Facts and Principles (1935)
12. Vessey, I., Weber, R.: Structured tools and conditional logic: an empirical investigation. Commun. ACM **29**, 48–57 (1986)
13. Beine, M.: A model-based reference workflow for the development of safety-critical software (2010)
14. Engler, D., Musuvathi, M.: Static analysis versus software model checking for bug finding. In: International Workshop on Verification, Model Checking, and Abstract Interpretation, pp. 191–210 (2004)
15. Frison, A.K., Wintersberger, P., Riener, A., Schartmüller, C., Boyle, L.N., Miller, E., Weigl, K.: UX we trust: investigation of aesthetics and usability of driver-vehicle interfaces and their impact on the perception of automated driving. In: Proceedings of the 2019 CHI Conference on Human Factors in Computing Systems (2019)
16. Oyibo, K., Adaji, I., Vassileva, J.: Mobile web design: the effect of education on the influence of classical and expressive aesthetics on perceived credibility. In: International Conference on Human-Computer Interaction (2019)

Social Media Happiness Expression Through the Virtual Reality: Cultural Differences on Instagram

Young Ae Kim[1(✉)] and Qiuwen Li[2]

[1] School of Design, South Dakota State University, 117 Grove Hall, Brookings, SD 57007, USA
YoungAe.Kim@sdstate.edu

[2] Department of Art and Art History, Santa Clara University, 500 El Camino Real, Santa Clara, CA 95053, USA
Qli2@scu.edu

Abstract. The social media happiness expression through the virtual reality analyzes Instagram photographic data to identify factors predictive of happiness from the Western and the Eastern cultures. The emotional expressions (photographs), hashtags (words), demographics (gender, age, and locations), and pose (head tilt, body turn, posture) were used as factors to detect an individual's various happiness expression. This research observed the patterns of the photographic dataset to discover the universal happiness expression today through social media, in particular Instagram. We integrated the computational methods from image processing and machine learning to obtain practical psychological indicators from photographic data. We observed one's perception of happiness was significantly influenced by the communication methods both photographic and hashtags data as well as their own culture through social media. This research centralized disparate archives of happiness expressions with photographic data from Instagram using a theoretical, artistic, and quantitative methods. The photographic data and written data are in the process of compiling into divergent visualization as a Virtual Reality, which will provide users the opportunity to identify the thought-provoking patterns and navigate the photographic data and written data at the extensive perspective from an individual perspective. This research addresses the representation of one's social media image with the pressure on the external viewpoint and creates an interactive platform of the direct comparison in diverse social and cultural contexts of happiness expression on social media.

Keywords: Happiness · Instagram · Emotional contagion · Cultures · Virtual reality · Visualization · Envy · Social comparison · Self-concept · Self-image · Body image · Vanity validation

1 Introduction

Social media has become an important daily communication and social interaction tool. Instagram users spend 28 min every day on the platform in 2020 [1]. Social media allows users easy access to observe personal lifestyles and information shared via

D. Russo et al. (Eds.): IHSI 2021, AISC 1322, pp. 753–759, 2021.
https://doi.org/10.1007/978-3-030-68017-6_112

social media. In the previous research, we found that communication types influence on sensing the emotions of others and shapes one's perceptions of being happy. We were able to observe that the value of being happy in young adults has been shifted from invisible value (e.g. relationship, achievement, etc.) to visible value (e.g. attractive physical appearance) [2].

The social media happiness expression through the virtual reality project is to identify significant patterns of individuals' emotional expression, value of life, presentation, and communication from the Western and the Eastern cultures. The research provides the opportunity to be conscious of one-sided and overstressed information through social media when exploring Virtual Reality photographic data and written data points. However, the usage of app with photo filters were not concerns in this study as it does not distort the foundational information of photographic data.

Social Comparison Theory. The social comparison theory detects individuals' evaluation of themselves and comparison themselves with others to dimmish uncertainty and insecurity and understand how to define themselves [3]. Contrasting and comparing one's life with others are a part of the identity formation for adolescents and young adults [4]. People seek out others, who are closer to their abilities and beliefs for comparison because correct comparisons are challenging when others are too divergent from themselves. Thus, people take actions to decrease divergences in attitudes by changing others to bring closer to them or by changing themselves to bring closer to others.

Today, social media platforms allow people to make direct comparisons by increasing proximity, by rejecting concealment, and by eliminating encapsulation. It inspires a tendency to be competitive and jealous, which cause a powerful feeling of envy. Envy is intertwined into the experience of using social media. It can be encouraged by the picturesque vacations and social success of others as well as by shallow and materialistic. Social media envy occurs as people compare themselves to others, who have social success. However, it provokes negative psychological and behavioral impact.

Upward and Downward Social Comparisons. Downward social comparison is defensive tendency self-evaluation and can promote self-respect. People look to others' worse off than themselves in order to feel better about personal situation and their self. However, upward comparison can lower self-respect. People make upward comparisons in order to improve their self and their personal reality. People want to believe that they belong to the part of the elite or superior in order to emphasize similarities between themselves and the comparison group [5].

Social media plays a significant role in social comparisons. People tend to involve in upward social comparisons with others (social ideal) in social media, which fuels in negative feelings about themselves. Social comparisons have become a substantial instrument to discover the appearance standards of social expectations among peers and for evaluating the self. Thus, self-perceived likenesses with others (social ideal) on social media can impact self-esteem. Social comparison with peers on social media develops a feeling of self-pity or satisfaction.

Different Types of Happiness. Happiness has a complex concept. Happiness classified in to three types: 1) pleasant sensation (represents momentary feeling of joy and pleasure); 2) feeling (represents judgements about feelings); and 3) attitude (represents a higher meaning of life, flourishing and fulfilling one's potential) [6]. The first type of happiness represents short-term positive emotion with physical symptoms and behavioral disposition. It is immediate feelings of experience and easy to measure and compare. The second type of happiness is thoughtful and needs an evaluation that goes beyond the momentary feelings of happiness. It is related in the consciousness of an individual comparing their situation with others and feeling about the past. The third type of happiness represents the self-realization and accomplishments with a higher meaning. It provides deeper values of happiness [7].

There is a trade-off between different types of happiness. It is difficult to get all three types of happiness all the time. If one gives an attention on temporary pleasures, it is going to be difficulty to be successful on long-term objectives in life. In other words, one will neglect to enjoy the simple things in life if they only motivate themselves working towards long-term objectives in life.

Cultural Variation in the Self-concept. Along with confirmations of cross-cultural resemblance in people's behavior, there is some dominant variances. In particular, it is true for the differences between the Western and the Eastern culture [8]. The Western culture gives the value of standing out and being noticed as an individual rather than group consensus. For instance, American politicians praise success to believing their own instincts, self-confidence, and ability to make decisions. Moreover, the Western culture promotes on high arousal emotion (e.g. joy and anger) expression [9].

The Eastern culture emphasizes on interconnectedness with group members. Asians pay attention to the external influences or contexts of individuals with a concept of human as social beings. The group focus is more important than the individual focus in the context of interdependent Eastern culture. In addition, it is advised to keep one's emotions to oneself in Asian culture, in particular, when the negative emotions and social separation emotions are involved [10]. For example, Japanese politicians trace their political success to the interpersonal relationship and harmony first, because they believe that a fraction can become whole when it fits to the proper place in a social unit. Accordingly, overt behavior is less appreciated and less function in the concept of interdependent culture.

Desire for Sharing Information on Social Media. Social media is the most universal cultural phenomena. People share to inform others about what they care and focus or to influence new ideas and concepts or stimulates actions from others. One of these factors plays a role in social media every day. They are spontaneous and can be strengthened by other desires and urges, which are the core of sharing.

In fact, topics about feeling good about something receives drastically more attention than negative topics, such as death or disasters. There is a validation-seeking component behind sharing. Often, excitement of sharing personal experience with photographs is the most popular format in social media, because people enjoy talking about themselves and making themselves seem positive in front of large audiences. However, the pressure to get the positive reaction from social media negatively influences to one's positive experience due to the rise of self-presentational concern [11].

2 Method

The research observes individuals' perception of happiness through photographic data and written data (hashtags) from Instagram. Users will be exposed the Virtual Reality platform that has the 1200 photographic data with four associated hashtags, such as #beautiful, #relationship, #achievement, and #travel. In addition, they have control of detail demographic information and emotional expressions that are associated with photographic data. Users get the opportunity to discover the thought-provoking pattern of photographic data and written data, navigate and explore the photographic data and written data at the extensive perspective from an individual perspective.

2.1 Data Collection

We focused on four cities with diverse climate environments, including Seoul, South Korea, Beijing, China, New York, New York, and Los Angeles, California. The photographic data was collected from fifty percent of data points from both the Western and Eastern culture for the reliability and validity of the data. The dataset of photographic images was coded based on locations, associated hashtags, features, pose, and emotions, which required several steps. In particular, majority of Instagram post has a long list of hashtags as Instagram accepts thirty hashtags each post. The four hashtags (#beautiful, #relationship, #achievement, and #travel) often listed on the same post. Consequently, a closer examination was required to identify association of image and hashtags as well as type of images whether it was taken by others or themselves.

2.2 Computer Analysis

A total of 1200 photographic data was coded by emotions, features, pose, people, and hashtags. They are the main user interface elements in VR to explore the data with any combination of coded condition (see Fig. 1). To provide precise and intuitive control of compound data point, each coded category distributed by detail sections. For example, emotions have five sections (e.g. happy, sad, angry, fear, and surprise), hashtags have four sections (e.g. #beautiful, #relationship, #achievement, and #travel), features have

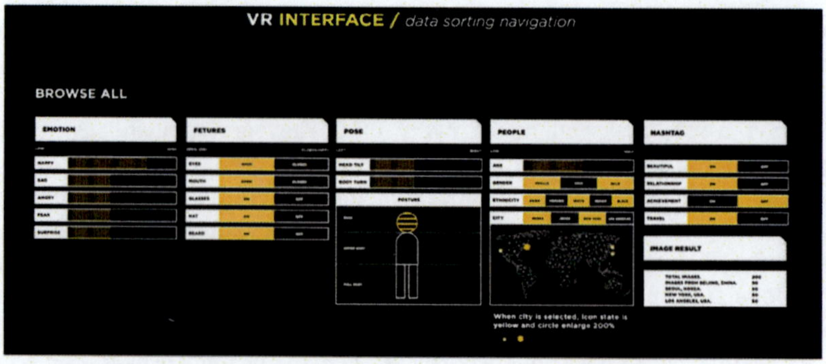

Fig. 1. Virtual reality user controller interface.

five sections (e.g. mouth (open/closed), glasses (on/off), eye (open/closed), hat (on/off), and beard (on/off)), pose has three sections (e.g. head tilt, body turn, and body posture), and people have four sections (e.g. age, gender, ethnicity, and city).

3 Outputs

3.1 The 3D Visualization Environment: Virtual Reality (VR)

Virtual Reality technology provides the first-person's perspective with authentic engaging learning experience. VR requires gazing at particular user interface (UI) elements when users invoke an action, which allows them to access compound data at digestible self-guided exploration with intuitive and precise full control of the data points. Virtual Reality experience constructs a multi-dimensional space with the unforgettable data visualization experience to discover the difference and similarity of how people render their happiness from different cultures through Instagram.

Users can control the self-guided user interface (UI) elements, which allows precision control of complex photographic data and written data. Users have several data exploration options to increase the focus on particular data exploration conditions (see Fig. 2) as well as full control of data exploration (see Fig. 1). A user can control details and conditions of photographic data characteristic to identify the pattern of data in terms of differences and similarities.

Certain colors have an influence on performance and moods. It can enhance the effectiveness of the creations. Color yellow and black were carefully chosen to increase the actions in VR environment as the gazing is the key interaction in VR. Color yellow is utilized to represent positive function in UI elements, such as on, yes, and open. Yellow is the most attention-grabbing color, which get noticed from a small amount use. It increases metabolism and willingness to take risks. Color black used to represent negative function in UI elements and written elements, such as off, no, and closed. Black is the absorption of all color and absence of light. It represents the unknown and unwilling to show the real feeling.

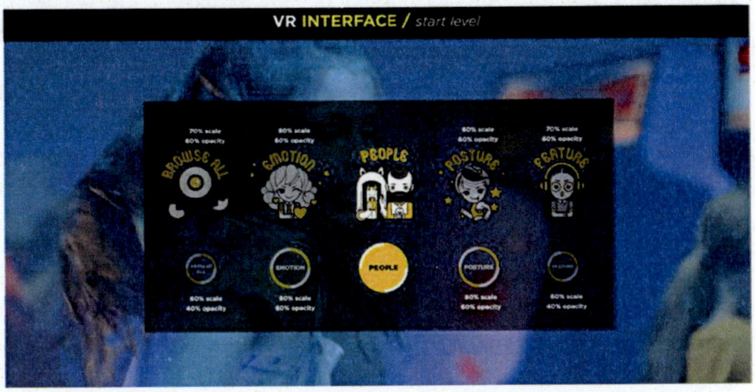

Fig. 2. The virtual reality environment user interface

The user interface control element displays to users to invoke an action and control the data points at each coded category (see Fig. 3, left). Users can control each category and characteristics of data points based on provided detail sections. The observed photographic data and detail summary displays instantaneously as the control of data point changes to help users to understand the exploration results. Once the desired data sorting has made from control panel, users see the data output (see Fig. 3, right).

Fig. 3. VR UI element of each coded category (left) and UI element of emotion category (right).

4 Limitations and Future Direction

We identified the preferred communication types and the cue of sensing the feeling of others in the previous research. The rapid changes in daily communication method through social media manipulated individual's perceptions and value of happiness. It has triggered shallow and materialistic (short-term positive emotion) as significant value of happiness as one of our research found that the physical appearance was the most critical factor of being happy for young adults in Instagram [2]. Today, the self-realization and accomplishments with a higher meaning (long-term positive emotion) became less important value of happiness to young adults.

This research is to reach general public (non-specialist audience). It creates the opportunity to users to observe divergent perspectives from others and identify the provoking patterns of photographic data and written data from Instagram. It can deliver something intangible and tangible individual perspective from the observed photographic data and written data. Exploring a compound sample from the Western and Eastern culture helps to see the data pattern objectively to identify how social media defines values of happiness and meaning of life rather than subjectively perceiving data based on personal agenda.

All photographic data have restrictions and are standardized in terms of appearance, format (612 $\times$ 612 pixels), and resolution of posted images on Instagram. Photographs are most likely edited by photo editing apps. The UI element control panel benefits for researchers and users to inspect extensive scopes of data patterns.

More research is required to develop a scale for measuring detail level of happiness value based on hashtags that is robust across diverse domains (e.g. #beautiful, #relationship, #achievement, and #travel). There might be some biases due to the nature of self-reporting as a measure of emotions. This research only focused on the emotions that are expressed on Instagram.

References

1. Instagram Stats That Matter to Marketers in 2020. https://blog.hootsuite.com/instagram-statistics/
2. Li, Q., Kim, Y.A.: Happiness on Instagram–content analysis and engagement based on attention theory. In: International Conference on Applied Human Factors and Ergonomics, pp. 13–20. Springer, Cham, July 2019
3. Festinger, L.: A theory of social comparison processes. Hum. Relat. **7**(2), 117–140 (1954)
4. Social Media Comparison & Teen Mental Health. https://www.newportacademy.com/resources/empowering-teens/theory-of-social-comparison/
5. Gibbons, F.X.: Social comparison and depression: company's effect on misery. J. Pers. Soc. Psychol. **51**(1), 140 (1986)
6. Goldman, A.H.: Happiness is an emotion. J. Ethics **21**(1), 1–16 (2017)
7. Goldman, A.H.: Life's Values: Pleasure, Happiness, Well-Being, and Meaning. Oxford University Press, Oxford (2019)
8. Mesquita, B., Karasawa, M.: Different emotional lives. Cogn. Emot. **16**(1), 127–141 (2002)
9. Lim, N.: Cultural differences in emotion: differences in emotional arousal level between the East and the West. Integr. Med. Res. **5**(2), 105–109 (2016)
10. Qu, L., Zhu, L.: Influence of culture diversities in emotional control and emotional reaction. J. Jiyuan Vocat. Tech. Coll. (2019)
11. Diehl, K., Zauberman, G., Barasch, A.: How taking photos increases enjoyment of experiences. J. Pers. Soc. Psychol. **111**(2), 119 (2016)

Flexible Manufacturing Systems: A Methods Engineering and Operations Management Approach

María-Cristina Herrera-García(✉) and Claudia-Yohana Arias-Portela

Faculty of Engineering, Panamericana University, Augusto Rodín 498. Col. Benito Juárez, 03920 México City, Mexico
macristinahega@outlook.com, cyarias@up.edu.mx

Abstract. This paper covers a significant extent of the literature published in the field of operations and production research related with Flexible Manufacturing Systems (FMS). There is currently limited knowledge published for the implementation of FMS for mixed-models assembly; which can present a relevant and critical analysis of the effects such a system has in terms of productivity, supply chain, and operation process, as well as the effects on human factors and methods engineering within the operation. Findings include advantages for the FMS effectiveness, management process, productivity, process analysis and technological advancement, while contrasting them against disadvantages on efficiency, learning curve, Kaizen principles and investment risks. Culminating on the importance of integrating intercommunicated systems throughout the company beyond the manufacturing system in order to control and profit from such implementation. The paper illustrates recommended measures to utilize the advantages mentioned while reducing the possible negative aspects. Finally, proposing some outlooks on future research.

Keywords: Flexible Manufacturing Systems · Methods engineering · Productivity · Efficiency · Economy of scale · Profitability · Automated guided vehicles · Kaizen

1 Introduction

Mixed-models assembly lines (MMAL) have been developed as a result of the third industrial revolution; when computer technology started mixing with manufacturing technology. As a result, these systems produce several models on the same line in an intermixed sequence without a turnover. Moreover, what further devolved the need for mixed models manufacturing systems was and still is, the personalized product trend. In fact, MMAL were developed to enable the industry to offer a diversified product portfolio while ensuring an efficient and effective production [1].

Nevertheless, a new era of manufacturing layout technology is upon our times, with the development of FMS. These systems integrate computer-controlled material handling devices and machine tools that simultaneously can produce different parts of several types of products [2]. The mix between workers and machinery not only leads to product portfolio variation, but also, a decrease in set up times and costs. On the

D. Russo et al. (Eds.): IHSI 2021, AISC 1322, pp. 760–765, 2021.
https://doi.org/10.1007/978-3-030-68017-6_113

other hand, in a conventional mixed-model assembly line, when trying to modify or change a station, the whole production must stop, which means loss of time and money in order to modify the production. In an FMS, the production does not need to stop to change a station, meaning that when demand fluctuates, the layout flexibility saves time and, in consequence, money to the manufacturer; increasing then the effectiveness of resources [2].

This paper contains a structure of four sections. First section contains a literary review to evaluate the current state of the art in the field of FMS in the context of operations management, Methods Engineering (ME) and implementation of flexible layouts in a manufacturing system (MS). The second section presents a new critical analysis for FMS where the high compatibleness for individualized products has advantages for the productivity and effectiveness indexes, as well as for operations management while mentioning the disadvantages this system has in terms of efficiency, learning curve and logistics. Followed by a managerial insights section of the improvements that could be made within the automotive manufacturing systems, and the investment risks and technological advantages such layouts have. In the last section, an outlook on future research for new concepts that are worth researching into for future technologies is presented.

Finally, the relevance of the present paper lies in the importance of having a deeper analysis from the ME perspective, as it is a subspecialty of Industrial Engineering that combines manufacturing aspects with management aspects, resulting in several subjects that determine the manufacturing productiveness of a given company; which is quite relevant when a new MS is proposed. Specifically, the present paper presets such analysis for the novel FMS layout proposed in Hottenrott, Grunow (2019) [2].

2 Literature Review

Discussions regarding FMS have dominated research in recent years, however, this specific matrix layout interconnected by automated guided vehicles (AGVs) has not been address, or even analyzed as often given its recent proposal. Which is why the literature review was carried out among international referred journals in the interval of 2000–2019 using a combination of the keywords FMS, optimization, manufacturing systems, AGVs, flexibility, process layout and productivity, in the databases of EBSCO and Springer. Determined the inclusion criteria of the articles as: year, keywords and FMS related subjects, a total of 10 articles were chosen and 6 articles were selected after the review.

For AGVs be implemented in FMS, as presented in Hottenrott, Grunow (2019) [2], the pathways and fleet size must be determined. Moreover, buffers are needed in order to attenuate the blocking and starvation of stations apart from reestablishing the outgoing sequence of a flexible segment. Because the AGVs movements are not continuous, buffers are also required within the flexible segments and at the entry and exit points [2]. It is important to note that the AGVs and the stations must be in line with the industry 4.0, because real-time updates are needed in order to optimize the movements of the AGVs.

Fleet size and decomposition of AGVs is discussed in more detail by Hall, Sriskandarajah, Ganesharajah (2001) [3], in this paper, not only tandem configurations were analyzed but also, they proposed both analytical and computational methods to compare state cycle times of the single and decomposed AGV-served system.

In order to tackle effectively the machine layout problem in FMS, the definitions of single-row layout problem (SRLP) and multi-row layout problem (MRLP) are stated. The solution presented by Tubaileh (2017) [4], iterates between two steps in order to find the optimal layout design. The method presented effectively achieved to minimize transportation costs between machines, treating the problem as an unconstrained optimization problem. This method is quite relevant for FMS because it may be applied to single-row, double-row, and multi-row layouts.

A different approach is presented in Zhao, Tseng (2007) [5]. Where they focus on the layout problem in the operational phase instead of in the design phase; a concept which counteracts the importance given to the consideration of the layout problems in the early stages of the design phase stated in Tubaileh, Siam (2017) [4]. This allows considering the problem over the entire operational period of the facility. Managerial options are presented through numerical tests, where the effectiveness of these increases profit and service quality.

3 Critical Analysis Using Methods Engineering

Flexible layout linked by AGV's for the assembly of heterogeneous vehicles is the industry for which the solution of the layout design is proposed by Grunow. The growing tendency of personalized products is forcing industries to adapt the way products are made, which implies that manufacturing processes must adapt to recent tendencies and innovate.

3.1 Productivity, Effectiveness and Efficiency

The productivity rate of an industry has a high impact on its ability to satisfy recent demand, but also enables the company to grow and increase its rentability; which is why analyzing FMS in terms of productivity is imperative.

Productivity is an economic index that links production with supplies used in order to obtain said production. Productivity may be mathematically calculated as the total production divided by the sum of utilized supplies [6]. As a clear advantage of implementing FMS, when having to modify a station, there is no need to stop the whole production as the AGV may direct the vehicle to other stations able to perform the task at hand while that station is being modified. Also, as the vehicles are produced by demand, the use of unnecessary supplies is lower; which means that a bigger production will be divided by a smaller amount of supplies utilized. Therefore, the productivity index for the manufacturing plant will increase substantially [2].

Because the primary objective of the model is to minimize the number of opened stations, there will be stations, and therefore, space that is not going to be used frequently. Even though that could be seen as a reduction of resources, it could also negatively impact the overall efficiency index.

Efficiency and effectiveness are considered as the main productivity indicators [6]. These indexes are analyzed together with the productivity index in order to grasp a complete picture of the manufacturing performance. Efficiency focuses on resources used; the efficiency index will not be as high as expected due to the amount of stations available but a minimum amount of them being used for production. On the other hand, the effectiveness index will be close, if not, a hundred percent as the expected production will be obtained given the nature of the FMS.

3.2 Supply Chain System

Another aspect from which we can assess an FMS is the nature of the supply system. The automotive industry then thrives from implementing a pull system, which is the supply chain system that matches the flexible layout matrix as is presented by Grunow [2]; it is worth noting that the optimization model presented also contains a constraint that guarantees demand fulfillment. By changing from a push to a pull system, manufacturing processes start when a client's order is received. This system reduces significantly the risk of the bullwhip effect and stock shortage, apart from eliminating the strong dependency on demand forecasts.

Nowadays, it is imperative to adapt the supply chain nature in order to satisfy the tendency for personalized vehicles. What is more, the fast-developing vehicle technology will also force manufacturing plants to constantly change their processes, materials, and tasks; which implies that flexible layouts are in line with the future outlook for the industry.

3.3 Kaizen Principles

Starting with sustainability principle [7], for FMS in the matrix proposed by Grunow, assembling materials will not be wasted. By personalizing each vehicle at the assembling stage, the manufacturing processes change and not all vehicles will have the same equipment. Meaning that wasting materials to personalize vehicles after assembly will no longer be necessary. Moreover, the flexibility of each station to perform several tasks and the ability to modify any station with AGVs also implies high sustainability for the future while still being able to comply with the everyday demands.

Process standardization is another Kaizen principle that can be difficult when talking about flexible layouts with the matrix proposed; however, there should be an effort to maintain standardization in every station despite having the ability to perform more than one task. Moreover, standardization ensures the vehicle's quality, which then translates to the value of the company.

A flexible layout matrix will be able to handle workload efficiently given its ability to use several stations for different tasks; when needing more time to perform a task or the production time is slowing down, new stations may be modified in order to perform the tasks that are taking longer. The constraint algorithm presented considers those eventualities when the second objective function strives to minimize the flow intensity. As well as having a constraint that ensures flow balance and another one that ensures the satisfaction of precedence relations.

3.4 Operation Analysis and Learning Curve

From a ME perspective, the constant goal is to increase productivity index by time unit, reduce unit costs and maintain or improve quality. In order to achieve that, there should be a constant operational analysis [6], where mainly an operation should be eliminated or combined before there is any attempt at improving it.

The fundamental reason is to save resources and optimize time. As the matrix and the novel use of AGVs enable the manufacturing process to be flexible with the operations that are carried out in each vehicle, the constant analysis becomes much less needed because, for each product, only the necessary tasks will be carried out. This analysis will be required when technology evolves, however, as flexibility in the model allows adaptability, the complexity of the process exponentially reduces.

Lastly, the learning curve, is a subject meaningful enough to be mentioned. Although it is mainly applied to human resources, it relates the production time per unit and the number of units produced consecutively [6]. What affects the learning curve is the process design, materials, product design, working methods and workers' abilities. The learning curve sheds some light on production times and could be helpful for the overall analysis of the MS.

Moreover, having a new matrix with AGV's being able to reroute, workers would have to adapt to the new configurations of the system and learn how to work around it. This becomes quite important when addressing manufacturing safety conditions and efficiency of the human workforce that is still an important factor in any manufacturing system; which is why considering the results of the learning curve when analyzing a new MS becomes a task worth doing.

4 Managerial Insights

Constant innovation for MSs is endlessly being demanded by new-coming trends as customers become more informed. Being able to satisfy these demands forces operations management to keep a close eye on their productivity indexes. Having a good productivity index is not only going to maintain the business profitable but also helps maintain quality. Therefore, it is pivotal to understand how applying FMS matrix connected by AGV's will impact the productivity indexes of a company; additionally, when implemented, constant reports must be made to ensure the highest quality.

Implementing FMS as a matrix and having "*t*" interconnected stations with AGV's would imply a substantial investment; not only regarding equipment but also with space required for optimal performance. Furthermore, this model calls for the use of buffers that allow for resequencing, which are quite expensive. This amount of investment urges the company's chiefs to assess how gainful would the investment be compared to the innovation's risk. Strategically speaking, having an FMS for the automotive industry is worth the investment given the fact that Audi, a leading automotive company, has already implemented such MS to produce one of their models. Grunow [2] quantifies the industry's FMS benefits.

The information technology (IT) department is also a concern, as it would have to be updated when adapting to an FMS. As the whole system needs to be interconnected and real-time updates are required in order to optimize AGV's routs [8] and station tasks, a good part of the investment would have to be used for technologically updated equipment. However, this updated system goes in line with the most recent industrial revolution, so investing in updating the IT department is an intelligent move whether the company is adapting to a flexible layout or not. Having a MS technologically developed translates to a strategic move that will, in the long and short term, benefit the company's profits.

Finally, a challenge produced by changing the supply chain to a pull system is stock management; the logistics team within the company will have to optimize lead time, as the production will only start when the customer has personalized their product.

5 Outlook on Future Research

FMS have been already addressed in literature even though they are considered the most advanced technological systems in manufacturing. However, as flexible layouts using a matrix connected by AGV's are in fact a novel idea, there still needs to be quite a deeper research as most of the papers regarding flexible layouts address specifically the automobile industry, whereas there are several industries that could benefit from the implementation of flexible layouts because of the heterogeneity of their products' assembly.

Furthermore, the solution for a method of determining the optimal number of stations depending on the number of tasks performed needs to be developed. As well as the amount of space needed in order to have an appropriate functioning system. This method should be applicable to any kind of industry and adapt itself to their limitations by entering specific constraints. This not only would lower the investment risk of adapting to FMS, but also would reduce the time of layout design.

References

1. Boysen, N., et al.: Sequencing mixed-model assembly lines: survey, classification and model critique. Eur. J. Oper. Res. **192**(2), 349–373 (2009)
2. Hottenrott, A., Grunow, M.: Flexible layouts for the mixed-model assembly of heterogeneous vehicles. OR Spectr. **41**(4), 943–979 (2019)
3. Hall, N., et al.: Operational decisions in AGV-served flowshop loops: fleet sizing and decomposition. Ann. Oper. Res. **107**(1/4), 189–209 (2001)
4. Tubaileh, A., Siam, J.: Single and multi-row layout design for flexible manufacturing systems. Int. J. Comput. Integr. Manuf. **30**(12), 1316–1330 (2017)
5. Zhao, T., Tseng, C.: Flexible facility interior layout: a real options approach. J. Oper. Res. Soc. **58**(6), 729–739 (2007)
6. Niebel, B., et al.: Ingeniería industrial. McGraw-Hill Interamericana Editores, México (2014)
7. Godínez, A., Hernández, G.: Poder Kaizen. Ignius Media, México (2018)
8. Raj, T., Gothwal, S.: Optimisation of AGVs path layout in flexible manufacturing system using 0–1 linear integer programming. Int. J. Process Manag. Benchmarking **8**(2), 182 (2018)

Lean Manufacturing Production Management Model Under a Change Management Approach to Enhance Production Efficiency of Textile and Clothing SMEs

Reddy Heredia-Mercado[1], Shaly Flores-Piñas[1], Pedro Chavez[1], and Carlos Raymundo[2(✉)]

[1] Ingenieria Industrial, Universidad Peruana de Ciencias Aplicadas, Lima 15023, Peru
{u201412603,u201315729,pedro.chavez}@upc.edu.pe

[2] Direccion de Investigacion, Universidad Peruana de Ciencias Aplicadas, Lima 15023, Peru
carlos.raymundo@upc.edu.pe

Abstract. The combination of Kaizen technologies and Total Productive Maintenance (TPM) systems is proposed with the introduction of elements of change management, which will enable to improve productive efficiency. Kaizen technologies enable waste reduction throughout the manufacturing process, and this is translated into a well-organized and standardized environment. Meanwhile, the TPM processes will allow to reduce losses related to machine and equipment breakdowns as well as quality and costs in industrial production processes. The company object of this investigation is a small- and medium-sized enterprise engaged in the manufacture of blouses that has production issues, which causes a high level of non-fulfilment of orders. As a result from applying the selected tools, failure to deliver the orders decreased from 12% to 5%; hence, production efficiency increased from 88% to 95%.

Keywords: Lean manufacturing · Textile · Smes · Change management

1 Introduction

Globalization has offered a window of opportunities to the textile sector companies in countries under development where working conditions feature low labour costs and poor environmental policies [1]. At domestic levels, it represents 8.9% of the local production, which means 1.3% of the country's gross domestic product, representing one of the primary sources of domestic economic growth [2]. In the last few years, domestic production and exportation has decreased significantly in contrast to the increase in labour costs, primarily attributable to the highly competitive Asian industry [3]. In addition, the poor production efficiency of micro-enterprises has been identified as the major problem, with it being the chief reason for the non-fulfillment of orders.

Today, production management is focused on cycle time management as a productivity measure to ensure this ratio can be streamlined with a view to reduce delivery

D. Russo et al. (Eds.): IHSI 2021, AISC 1322, pp. 766–772, 2021.
https://doi.org/10.1007/978-3-030-68017-6_114

times and meet clients' needs [4]. Certain studies reveal the importance of indicators such as efficiency and productivity in manufacturing processes to minimize reprocessing and production costs that are the most significant in the sector.

This research intends to prove that the proposed model features better results in terms of cycle time, waste levels, and delivery of in comparison to other existing models. In view of the above, the research proposes a Lean Manufacturing Production Management under a change management approach, which allows to offer the company the capacity to adapt to the transformation through learning.

2 State of the Art

2.1 Lean Manufacturing

To create processes based on Lean techniques, first, it is necessary to recognize the Total Productive Maintenance (TPM) system and Kaizen technologies (focused improvements) as the pillars of the methodology because these offer a new and technological approach to understand behaviours and reduce delivery times. There are studies on the implementation of Kaizen events with respect to the production management model through the Value Stream Mapping (VSM) method. This process identifies the inefficiencies and wastes that appeared in the company's processes and proposes solutions to eliminate them effectively and permanently [5–7].

2.2 Lean Manufacturing and Change Management in the Textile Industry

Lean Manufacturing and change management in textile manufacturing management are focused on waste reduction and productivity enhancement through several tools; namely, VSM, line balancing, and work standardization [8, 9]. The synergies obtained through tools such as the 5S's system, line balancing, standardized work, and VSM lead to an improvement in productivity, both as regards labour intensity and flexibility of the teams. Studies show that the successful implementation of the tools in small- and medium-sized enterprises (SMEs) improves the performance of the workforce, machines, and methods in its manufacturing processes [10, 11].

3 Contribution

3.1 Proposed Model

The model proposed in Fig. 1 provides new components, which were included in the basis of the research performed and the needs identified.

3.2 Indicators

This research will use indicators that measure the actual performance of the project. The indicators and metrics to be evaluated are:

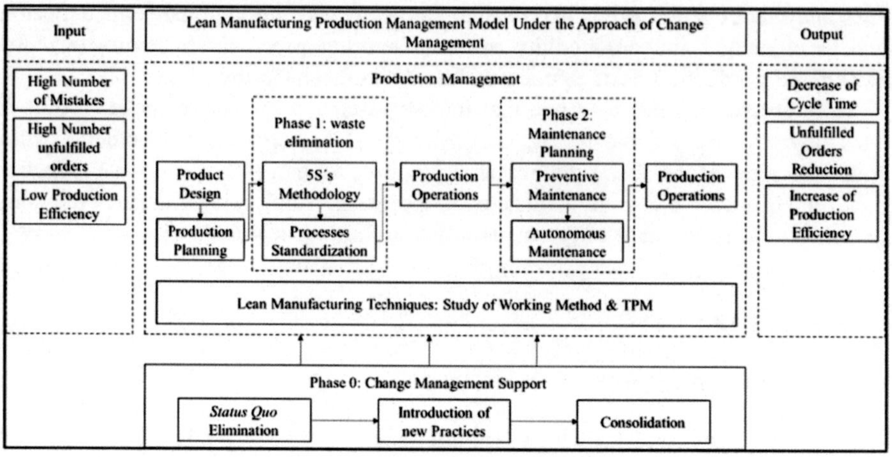

Fig. 1. Proposed research model

Overall Equipment Efficiency (OEE): This is related to the performance of the critical equipment assessed during the diagnosis. This indicator measures production enhancement through the following three factors: Availability, Performance, and Quality.

$$\text{OEE} = \text{Availability} \times \text{Performance} \times \text{Quality} \tag{1}$$

$$\text{OEE} = \frac{\text{Production time}}{\text{Available time}} \times \frac{\text{Actual output}}{\text{Output capacity}} \times \frac{\text{Units without defects}}{\text{Actual output}} \tag{2}$$

Cycle Time (TCT): This indicator refers to the time it takes to produce a garment throughout all production stages.

$$\text{TCT} = \text{Drawing TC} + \text{Cutting TC} + \text{Sewing TC} + \text{Finishing TC} \tag{3}$$

Waste Level in the Cutting and Sewing Area: This indicator refers to the percentage of waste produced in the cutting and sewing area.

$$\%\text{Waste} = \%\text{Design and cutting} + \%\text{Joining parts} + \%\text{Sewing} + \%\text{Finishing} \tag{4}$$

Level of Unfulfilled Orders: This indicator measures the efficiency level in the production area through the percentage of orders not fulfilled.

$$\%\text{Unfulfilled orders} = \frac{\text{Number of orders delivered late}}{\text{Total delivered orders}} \tag{5}$$

4 Validation

4.1 Case Study

The research was developed in a SME dedicated to the production and sale of blouses for women made of challis, gauze, and rayon. This organization is located in Lima, Perú and has nine employees.

The validation was performed considering a single work shift, with six days per week, the operators and the company manager.

4.2 Initial Diagnosis

The current situation of the enterprise was measured based on the selected indicators, and goals were set as indicated in Table 1.

Table 1. Initial diagnosis

Indicator	Initial	Goal
Cycle time at cutting and sewing stage	Cutting: 370.13 s Sewing: 548.19 s	Cutting: 299 s Sewing: 438.55 s
Waste level at the cutting area	26.9 kg/month	22.87 kg/month
OEE	67%	77%
Unfulfilled orders	12%	7%
Efficiency	85%	94%

4.3 Validation Development

Phase 0; Change Management: The enterprise's staff and the research team held several meetings according to a program on specific topics. Measures were perfromed to promote change management and reduce resistance to it: convey a sense of urgency, build a strong coalition, create a vision for change, and spread such vision.

Phase 1; Waste Elimination: First, the staff was trained to use the Lean tools. Component 1: Training and Education: The entire staff was trained on the theory of the Lean techniques and tools regarding the basic function of TPM, pillars, machine cleaning and handling in hard-to-reach areas, management of minor failures, and Autonomous Maintenance.

Component 2: Application of the 5S's System. First, unnecessary elements were sorted and classified to organize the workplace through an organization method and red tags were placed for elimination and relocation purposes. Second, the necessary elements at the workplace were identified and then their location was arranged according to quantity. Furthermore, areas were defined to separate corridors and locate machines, among others. Third, the equipment, machines, and workplace were cleaned on the basis of an operating manual and a schedule. Fourth, the process was standardized through specific guidelines for organizing and cleaning the workplace and responsibilities were delegated for proper compliance. Finally, the process was sustained through frequent audits to check its implementation.

Component 3: Processes Standardization. First, the cutting and sewing processes were identified as critical, as they are above the take time identified in the VSM. The current method was registered using a Diagram of Activity Process so it could be further analyzed through interrogatory methods aimed at flagging improvement opportunities and activities that do not add value. Then, an enhanced method was designed, the passive activities were eliminated, other activities were grouped to execute them simultaneously, and, lastly, these were reassigned to other processes. The method was validated, and the processes required for its execution were designed. The method was first implemented in a controlled environment, and once its functioning was validated, it was extended to the entire process. Finally, periodic revisions were arranged to validate its correct follow-up.

Phase 2; Maintenance Planning: Component 1: Preventive maintenance. To start Phase 2, we conducted a test on the workload per hour of each machine in the plant, with the purpose of comparing it to the maintenance technical specifications described in the user manual of each machine. Thus, it was possible to determine the time intervals between the relevant inspections of each machine. On that basis, the schedules for the inspection are prepared. In turn, the long-term inspection schedules are also outlined.

Component 2: Autonomous Maintenance. To develop the Autonomous Maintenance, substantial documents were produced, which allowed the drafting of a description of all the activities that should be conducted. This was implemented through charts of One Point Lessons that must be developed by each operator in the plant.

Once the implementation stage was over, the information about the last month of application was collected and the results exhibited in Table 2.

5 Results

The average of all validations for Indicator 1 was 342.21 s in the cycle time at the cutting stage, so in comparison with the final result of Scenario 0, which was 297.47 s, it was inferred that there has been an improvement due to its application in the said scenario.

The standard deviation for the cycle time at the sewing stage was 96.11 s, which means that the error margin of the scenarios was valid in ±96.11 s. The comparisons were made between the initial and final scenarios and they were within the range.

The goal of the market as to the efficiency of the machines was approximately 80%. Thus, in comparison with the market standard, 82% of efficiency was obtained, which was within the accepted standards.

Table 2. Comparative results chart

	I1: CT of Cutting Stage (s)	I2: CT of Sewing Stage (s)	I3: Waste level in Cutting area (kg/month)	I4: OEE	I5: Level of Unfulfilled orders
S0	297.47	353.49	19.5	82%	4.70%
S1	315.20	386.47	18.3	82%	6.30%
S2	355.80	344.49	20.8	82%	7.00%
S3	400.43	550.32	17.2	82%	7.60%

6 Conclusions

The average of all validations for Indicator 1 was 342.21 s in the cycle time at the cutting stage; hence, when compared with the final outcome of Scenario 0, which was 297.47 s, it was concluded that there has been an improvement as a result of its application in the said scenario.

The standard deviation for the cycle time at the sewing stage was 96.11 s, meaning that the error margin of the scenarios was valid in ±96.11 s, i.e., the initial and final scenarios were compared and were found to be within the range.

The total waste level decreased in August by approximately 21%. This amount was significant considering the amount of fabric used per month.

The OEE increased by 15% and the level of unfulfilled orders decreased to 7%. However, the integration of components in our case study enabled a 45% reduction during the first month of the implementation; afterwards, a 30% reduction took place in August.

References

1. Slović, D., Tomašević, I., Radocić, M.: Improving productivity in the apparel industry through gain sharing and continuous process improvement: the case of a Serbian manufacturer. Fibres Text. Eastern Eur. **24**(2), 15–22 (2016)
2. Al-Hassan, K., Chan, J., Metcalfe, A.: The role of total productive maintenance in business excellence. Total Qual. Manag. **11**, 596–601 (2000)
3. Braglia, M., Frosolini, M., Gallo, M.: SMED enhanced with 5-Whys Analysis to improve set-upreduction programs: the SWAN approach. Int. J. Adv. Manuf. Technol. **90**, 1845–1855 (2017)
4. Guariente, P., Antoniolli, I., Ferreira, L., et al.: Implementing autonomous maintenance in an automotive components manufacturer. Procedia Manuf. **13**, 1128–1134 (2017)
5. Gupta, S., Jain, S.: An application of 5S's concept to organize the workplace at a scientific instruments manufacturing company. Int. J. Lean Six Sigma **6**(1), 73–88 (2015)
6. Moscoso, C., Fernandez, A., Viacava, G., Raymundo, C.: Integral model of maintenance management based on TPM and RCM principles to increase machine availability in a manufacturing company (2020). https://doi.org/10.1007/978-3-030-25629-6_137
7. Abhishek, J., Rajbir, S., Harwinder, S.: OEE enhancement in SMEs through mobile maintenance: a TPM concept. Int. J. Qual. Reliab. Manag. **32**(5), 503–516 (2015)

8. Kalluri, V., Kodali, R.: Reliability and validity of new product development practices in Indian manufacturing industries. J. Adv. Manag. Res. **1**(11), 82–101 (2016)
9. Ranjan, R., Mishra, A.: Evaluation and optimization of overall equipment effectiveness on a pasting machine in a battery manufacturing industry. Int. J. Perform. Eng. **12**(6), 503–512 (2016)
10. Arévalo, M., Montenegro, J., Viacava, G., Raymundo, C., Dominguez, F.: Proposal for process StandarDization for continuous improvement in a peruvian textile sector company (2020). https://doi.org/10.1007/978-3-030-27928-8_136
11. Ahmad, N., Hossen, J., Ali, S.: Improvement of overall equipment efficiency of ring frame through total productive maintenance: a textile case. Int. J. Adv. Manuf. Technol. **94**, 239–256 (2018)

Supporting Collective Intelligence of Human-Machine Teams in Decision-Making Scenarios

Alexander Smirnov and Andrew Ponomarev(✉)

St. Petersburg Federal Research Center of the Russian Academy of Sciences (SPC RAS), 14th Line, 39, 199178 St. Petersburg, Russia
{smir,ponomarev}@iias.spb.su

Abstract. Efficient collaboration of humans and machines has the great potential for improving many knowledge-intensive processes in variety of applications. Therefore, developing means supporting such collaboration and making it efficient is an important area of research. The paper presents a part of research aimed on the development of a collective intelligence environment that would support joint work of humans and machines on decision support problems, allowing participants to self-organize (define and adapt the plan of actions). In particular, it describes an approach to solving semantic interoperability issues in supporting human-machine collective intelligence for decision-making scenarios. The proposed approach is based on using multi-aspect ontologies and ontology-based smart spaces.

Keywords: Collective intelligence · Human-machine systems · Ontologies · Multi-aspect ontologies · Decision support · Decision-making

1 Introduction

Many of the complex problems arising in modern dynamic business and societal systems require both creative human intelligence as well as reasoning and data-processing capabilities of artificial intelligence techniques. The development of global communication networks has resulted in the emergence of new crowd-based solutions, consolidating human resources and providing them on the on-demand basis. However, current developments in the area of supporting work of loosely coupled collectives on complex problems (e.g., [1–3]) typically lack:

- collaboration between team members (e.g., in crowd computing work is typically distributed into a set of independent, isolated fragments),
- the ability of participants to define the work plan (rather then follow a workflow defined by the problem setter),
- the integration of AI agents that can enrich the collective endeavor with new information/knowledge.

At the same time, each of these features has significant potential in improving the quality of collective work and the quality of the solutions obtained as a result of it. For example, the support for collaboration and self-organization (to some degree) can obviate the need for a problem setter to decompose the task into a set of subtask (which

D. Russo et al. (Eds.): IHSI 2021, AISC 1322, pp. 773–778, 2021.
https://doi.org/10.1007/978-3-030-68017-6_115

is one of the burdens of using crowd-based approaches to information processing). Seamless integration of AI agents to the collective work will allow one to take the best both from human creativity, problem-solving and social skills as well as from data-processing capabilities of the AI.

Rather generic type of problems that usually require both human and computer effort is decision-making (especially, decision-making in complex problems and complex environments). It frequently demands multiple expert evaluations of different aspects of the problem (therefore, human involvement), relies on the analysis of some factual data and leverages a range of domain models (therefore, computer/AI involvement). Besides, in complex decision-making it is hard (or even impossible) to build a detailed plan due to inherent uncertainty of the process (later stages may be significantly altered by some findings of the earlier stages). In this sense, complex decision-making makes a very good exemplar problem that can benefit from the new kind of collaborative environments.

The goal of our research is to develop a collective intelligence environment that would support joint work of humans and machines on decision support problems, allowing participants to self-organize (define and adapt the plan of actions).

An important problem that has to be solved in order to build such environment is providing semantic interoperability among heterogeneous peers, which is necessary to form a shared representation of both problem-specific information and process-specific information (planned activities, responsibilities distribution, etc.). The paper describes the proposed ontology-based collaboration approach serving as a cornerstone of the human-machine collective intelligence.

2 Ontology-Based Human-Machine Collaboration

Any collaboration is based on two foundations: semantic and syntactic interoperability (agreements, allowing peers to understand each other's messages) on the one hand, and communication protocols and mechanisms (allowing peers to exchange messages) on the other hand. The protocols and mechanisms, in their turn, can be basic and problem-specific. Basic protocols allow delivering messages from one peer to another. Problem-specific mechanisms are built upon the basic ones and support complex kinds of collaborative activities (e.g., voting, negotiation, etc.). Here, we describe the approaches to achieve semantic interoperability and basic communication protocols employed in the proposed environment to support human-machine collaboration.

2.1 Semantic and Syntactic Interoperability

Ontologies defined in [4] as an "*explicit specification of a conceptualization*" have proved to be one of the most convenient and effective mechanisms for both syntactic (as there are widely used representations for them, described in W3C standards [5, 6]) and semantic interoperability. They have two advantages justifying their choice as an interoperability enabler for the human-machine collective intelligence:

- they provide means to specify concrete semantics associated with some term and reason about properties of objects using inference engines based on description logics;
- they can be effectively processed by both humans and machines. In this sense, they can provide a kind of *lingua franca*, acceptable by both human and software participants of the collaborative decision-making process.

The second thesis requires a comment. Processing ontologically-represented knowledge in a native form (a set of axioms expressed in one of OWL syntaxes, for example) requires some qualification in formal logics and may be too complicated and not convenient for non-expert users. Therefore, the environment implements several techniques to make the use of ontologies friendlier for non-expert participants:

- Implicit ontological representation of the structure of problem information. Human participants see the description of the decision-making problem as a set of constraints, alternatives, criteria, alternative evaluations, arguments (in other words, in terms that are widely used in decision-making literature). They see organization aspects (plans, assignments) as typical widgets in GUI. However, all these information is encoded with the help of relevant ontologies and is available to software agents, who can also analyze the problem structure and organization issues of the team.
- Natural language processing. Recent advances in this area make it possible to infer the role of some information pieces, its connection to the goal of decision-making and/or to some line of argumentation and so on. Based on natural language processing techniques it is possible to make a structured summary of a textual message (resembling to some ontological statement) and verify it during the interaction with the author.
- GUI-based nudging participants to encode problem structure in an ontology-compatible way. Digital nudging [7] alongside with the natural language processing helps to disambiguate terms and relations in the problem representation.

There are two principal kinds of information in collective decision-making activity: problem information (what is known about the problem at the specified point in time) and process information (describing the structure of the team and plan of the activities). It is also convenient to describe problem on two levels: abstract level (with general terms, relevant to any decision-making process) and concrete level (defined by the domain area where the decision-making is performed). Correspondingly, the environment deals with three ontologies:

- Decision-making ontology that defines main concepts used during decision-making (criterion, alternative, evaluation etc.) and their interactions. The ontology is provided by the environment and is based on the analysis of existing decision-making methodologies (e.g., [8]) to support the majority of them.
- Collaboration and coordination ontology that defines concepts employed in describing distributed collaborative activities (responsibility, role, dependency, etc.). This ontology is also provided by the environment.
- Problem domain ontology that defines concepts and relations of the problem domain where decision is being made. In contrast to the previous two ontologies, this ontology is not provided by the environment; the team may reuse any existing ontology of the problem domain introducing necessary concepts.

These ontologies capture orthogonal views on the decision-making process. To tie these views into a complex, consistent structure, we propose to use the apparatus of multi-aspect ontologies [9]. It allows connecting multiple conceptualizations (possibly, built using different formalisms); so that changes performed in one view (aspect) are automatically reflected in other views (aspects). For example, if the environment is used in smart tourism applications and the decision to be made is choosing a tourist route, then *route* (defined in the problem domain ontology) is equivalent to the *alternative* (defined in the decision-making ontology). The presence of these mappings allows the software participants of the decision support process to treat routes as alternatives (e.g., in visualizing them or building Pareto optimal set [10]). It also allows problem-specific services to contribute to the decision-making (e.g., estimate typical traffic of the route, its length). Obviously, these mappings between different aspects, joining aspect ontologies into a multi-aspect ontology are problem-specific and building these mappings is a part of problem definition (Fig. 1).

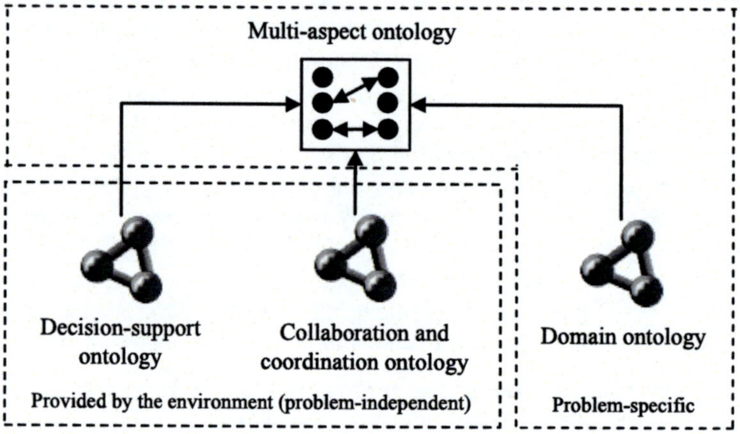

Fig. 1. Relationship between ontology aspects.

2.2 Communication Protocols and Mechanisms

The proposed environment is based on ontology-based smart space technology [11, 12], meaning that every participant of a heterogeneous team working on a decision support problem has access to an ontological representation of the current state of the problem and modify it (computer agents can apply transformations of the ontological description according to their purpose and functions). Besides, participants can subscribe to certain structures in the smart space, expressed in terms of ontology subgraphs (via SPARQL). It means that a participant can register a subscription to a certain construction in the smart space describing some state of the problem solving process, and when this construction appears, the participant will receive notification and will be able to offer some services.

The explicit modification of problem-relevant information makes it possible to formally identify many specific situations in collective work and propose effective problem-specific protocols and mechanisms of resolving them. This is done as a part of soft guidance of collective work, offered by the environment. For example, typical activity during decision-making is evaluating some alternative with respect to certain criterion. Evaluations made by different team members are connected to the alternative instance in the smart space. However, one can easily define a formal pattern of contradiction (several evaluations are connected to the same alternative). Once a contradiction is detected according to this pattern, the environment recommends one of the protocols for resolving a contradiction (e.g., Delphi method).

In an upper level, the environment supports two processes taking place when team works on a problem: *solution preparation* and *decision support (re)organization*. Solution preparation is a main productive process, during which the problem is enriched with new information and artifacts created by team members. General scheme of the solution preparation process is driven by decision support methodologies (e.g., DECIDE [8]). These methodologies provide the general scheme of the solution preparation, main stages that have to be performed, while execution of each stage is done via activities relevant to the problem at hand. Decision support (re)organization process represents all the activities aimed on planning and organization of teamwork (e.g., deciding whether additional resources are required, assigning team member responsibilities, setting task deadlines and identifying new tasks to be solved in order to reach the goals of the whole process). To support (re)organization process the environment proposes process assistance mechanisms proposing efficient modes and scenarios of the collaboration depending on the status of the problem solving process.

A typical scenario of using the environment is the following. End-user (decision-maker) describes the problem in a natural language. The problem description is processed with natural language processing modules to extract domain area, decision type, and possible competencies required during decision support. These pieces of information are verified by the decision-maker and encoded with a help of ontology. A decision support team for the problem starts to form. The coordinator of the team decides on what kind of activities should be performed and what roles are required to perform them. Team members work on the description, enriching it, and possibly making changes to the plan. The solution elaboration process continues until the decision problem is detailed enough. After that, decision-maker makes an informed decision based on the results obtained by the team.

3 Conclusion

The paper describes an approach to solving semantic interoperability issues in supporting human-machine collective intelligence for decision-making scenarios. The proposed approach is based on using the apparatus of multi-aspect ontologies (connecting decision-making, collaboration and domain aspects in a task-specific manner) and ontology-based smart spaces (supporting team's shared knowledge and indirect message passing).

Technologically the environment is as a web application where each problem has its own workspace, supporting both informal text-based communication, but also have integrated 'nudging' modules allowing to tie the information into decision-making ontology formalized view, making it accessible for software services.

The proposed environment can be used for decision-making support in a variety of domains characterized by high levels of uncertainty and dynamics (emergency and natural disaster response, government and business scenarios).

Acknowledgments. The research is funded by Russian Science Foundation, project 19-11-00126.

References

1. Kulkarni, A., Can, M., Hartmann, B.: Collaboratively crowdsourcing workflows with turkomatic. In: Proceedings of the ACM 2012 Conference on Computer Supported Cooperative Work - CSCW 2012, p. 1003. ACM Press, New York (2012). https://doi.org/10.1145/2145204.2145354
2. Retelny, D., Bernstein, M.S., Valentine, M.A.: No workflow can ever be enough: how crowdsourcing workflows constrain complex work. Proc. ACM Hum.-Comput. Interact. **1**, Article 89 (2017). https://doi.org/10.1145/3134724
3. Karacapilidis, N., Tampakas, V.: On the exploitation of collaborative argumentation structures for inducing reasoning behavior. In: Proceedings of the 18th International Conference on WWW/Internet 2019, pp. 78–84. IADIS Press (2019). https://doi.org/10.33965/icwi2019_201913L010
4. Gruber, T.R.: Toward principles for the design of ontologies used for knowledge sharing? Int. J. Hum. Comput. Stud. **43**, 907–928 (1995). https://doi.org/10.1006/ijhc.1995.1081
5. OWL 2 Web Ontology Language Document Overview. https://www.w3.org/TR/owl2-overview/
6. RDF 1.1 Concepts and Abstract Syntax. https://www.w3.org/TR/rdf11-concepts/
7. Schneider, C., Weinmann, M., vom Brocke, J.: Digital nudging. Commun. ACM **61**, 67–73 (2018). https://doi.org/10.1145/3213765
8. Guo, K.L.: DECIDE: a decision-making model for more effective decision making by health care managers. Health Care Manager **27**, 118–127 (2008). https://doi.org/10.1097/01.HCM.0000285046.27290.90
9. Smirnov, A., Shilov, N., Parfenov, V.: Building a multi-aspect ontology for semantic interoperability in PLM. In: Product Lifecycle Management in the Digital Twin Era. IFIP Advances in Information and Communication Technology, pp. 107–115. Springer (2019). https://doi.org/10.1007/978-3-030-42250-9_10
10. Luc, D.T.: Pareto optimality. In: Pareto Optimality, Game Theory and Equilibria. Springer Optimization and Its Applications, vol. 17. pp. 481–515. Springer, New York (2008). https://doi.org/10.1007/978-0-387-77247-9_18
11. Korzun, D.: On the smart spaces approach to semantic-driven design of service-oriented information systems (2016). https://doi.org/10.1007/978-3-319-40180-5_13
12. Roffia, L., Azzoni, P., Aguzzi, C., Viola, F., Antoniazzi, F., Salmon Cinotti, T.: Dynamic linked data: a SPARQL event processing architecture. Future Internet **10**, 36 (2018). https://doi.org/10.3390/fi10040036

Calculation of the Level of Service in the Gondola in Supermarkets

Juan Ñaupari[1], Anghela Urbina[1], Grimaldo Quispe[2], and Carlos Raymundo[3(✉)]

[1] Facultad de Ingenieria, Universidad Continental, Huancayo, Peru
{73175558, 72948776}@continental.edu.pe
[2] Facultad de Ingenieria, Universidad Nacional Autonoma Altoandina de Tarma, Tarma, Peru
gquispe@unaat.edu.pe
[3] Direccion de Investigacion, Universidad Peruana de Ciencias Aplicadas, Lima 15023, Peru
carlos.raymundo@upc.edu.pe

Abstract. "On-Shelf Availability" (OSA) or "Out of Stock" (OOS) in a store is a key factor for customer satisfaction. Shelf availability is a critical issue for both manufacturers and retailers, fixing it improves consumer value, fosters consumer brand loyalty and shopper loyalty to the store, increases sales, and most importantly, increases the profitability of the category. Likewise, the logistics rationalization in the supply chain is a key factor for the profitability of the business. The present research establishes an OSA methodology that through the reception and replacement processes with Category Management (CATMAN) will increase profits and the level of customer satisfaction, also achieving a better inventory. From January to October 2019, the percentage of OSA was raised from 78.3% to 92.4%. This methodology can be applied to different supermarkets, adapting CATMAN to avoid OOS.

Keywords: Service level · Category management · Availability · Satisfaction

1 Introduction

During the last 10 years, the retail business has grown substantially in Peru, becoming a very important source of employment and a fundamental sales channel for national and foreign companies. In this work we highlight how a good reception and replenishment help to improve the inventory by applying Category Management (CATMAN), and how this increases the OSA, which was measured in percentage to observe the behavior of the change with respect to the applied improvements. The OSA percentage was also compared with the Fill Rate percentage, which is the delivery evaluation for suppliers, understanding that there will be no replacement if we do not have merchandise.

Retail is about offering the right product in the right place at the right time at the right price [1]. However, in-store operations pose complicated issues related to back-room control, staff optimization, product assortment, and consumer response [2]. Due to their complicated nature, store run failures such as misreporting sales and incorrect

D. Russo et al. (Eds.): IHSI 2021, AISC 1322, pp. 779–785, 2021.
https://doi.org/10.1007/978-3-030-68017-6_116

product placement are common even in financially successful retailers [3]. Poor store execution puts supply chain efforts and marketing strategy at risk by compromising product availability. As a result, operations researchers have been troubleshooting execution issues to increase service levels and minimize lost sales.

Product availability has long been considered a critical issue in retail due to its impact on future demand and it is not surprising that the number of out of stock (OOS), that is, unavailable items, has become a integrated measure of performance for retail workforce, processes and technology. Retail OOS can be classified into two types: Store OOS, where the item is not available; or shelf-OOS, where the item is in the store but customers cannot find it [4].

2 State of the Art

Since one of the fundamental parts of our research focuses on the fact that we will apply a Reengineering in the reception and replacement process to increase the stock in the Tottus supermarket shelves, it is necessary to set a conceptual axis, therefore, we have that reengineering is the fundamental overhaul and radical redesign of business processes in order to bring about dramatic improvements in performance and results [5].

On the other hand, ensuring "On-Shelf Availability" (OSA) or avoiding "Out of Stock" events (OOS) in a store is a key factor for customer satisfaction. Likewise, logistics rationalization throughout the supply chain is a key factor for business profitability [6].

Supermarkets such as Tottus, Metro, Plaza Vea, etc. they have annual losses that add up to about $ 40 million due to lack of replacement on time (on time) of merchandise in the gondolas, which is clearly unrealized sales. The causes that originate it are diverse, but the truth is that about 60% have to do with products with high turnover in the categories of basic foods and beverages [7]. At the national level, it was recorded in 2014 that 38.24% of the products not replaced in the gondola are due to supplier responsibilities while 61.76% are due to chain responsibilities causing losses in sales of $ 359 million.

Due to its complicated nature, store run failures such as erroneous sales reporting and incorrect product placement are common even in financially successful retailers. Poor store execution jeopardizes supply chain efforts and marketing strategy by compromising product availability [8]. Below are the causes that make the product not be displayed in the gondola, related to the responsibility of the supermarket, the most representative reason being the high turnover of the product versus its designated space in the gondola (Fig. 1).

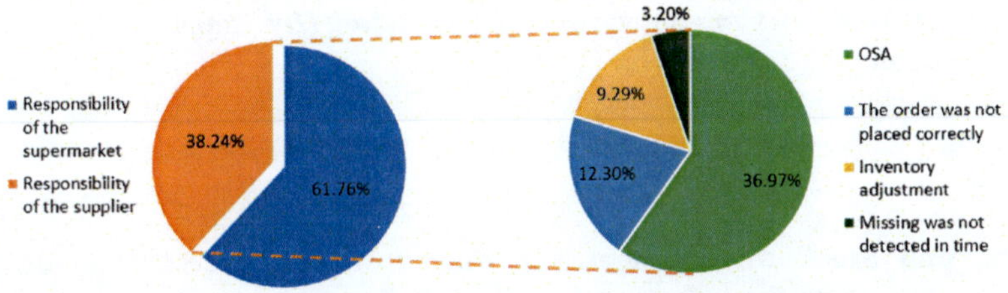

Fig. 1. Causes for which the product is not replaced in the Gondola [9]

3 Method

A lot of data was processed in the Excel tables for which it is necessary to clarify the following: It must be remembered that the Service Level in Shelf (OSA) measures the presence of the product in the shelf. That is, if the CATMAN (that is, the plan of the gondola where it indicates how many products should go) says that there must be 20 jars of milk and during the measurement process only 10 are found, it means that the product only has 50% presence; that is, OSA of 50%.

The FILL RATE must also be taken into account. This indicator helps us measure the supplier's order delivery efficiency. For example, if we ask ALICORP to bring us 100 bags of noodles and deliver the 100 bags, it means that its compliance is 100%. On the other hand, if you give us 20 noodles only, it means that your compliance is 20%

Both the OSA and the Fill Rate must be complemented in order to evaluate if the OSA is low due to a shortage. In other words, if there should be 10 jars of milk in the gondola and the supplier only sent 2 of an order of 10, it is logical that their presence in the gondola will be deficient because they did not comply with the delivery of products. An acceptable fill rate for the business is 85% and an optimal OSA 90%.

Based on these two indicators, considering that the replacement process is already implemented and approved, we propose the following actions (Table 1):

Table 1. Legend

Excellent		FR >= 90% and OSA >= 90%
Good		FR >= 90% and 90% >= OSA >= 85%
Internal action		FR > 85% and OSA < 85%
External action		FR < 90% and OSA >= 85%
Joint action		FR < 85% and OSA < 85%

We will be able to increase the OSA with:

CATMAN: Exhibition guidelines, Indicator dashboard: to monitor compliance with CATMAN.

Planning: Bankruptcy and overstock management: To send the store what it needs based on demand and avoid bankruptcies, which impact the OSA.

Replacement Models: Constant flow of merchandise (By implementing routines), Layout, Structure.

3.1 Applications

Planning and Stock Management: For this, it will be necessary to define an inventory policy, with the following definition:

- Products that constitute 90% of Share of the sub-Class, 20/80, MMPP that have a shelf coverage of less than 5 days, must have a warehouse stock, considering a maximum range of coverage in store of 7 days – 10 days.
- The products that make up 80/20 of the assortment must not have stock in the warehouse

Objective:

- Initially reduce inventory days J01 and J02 by 20%.
- Establish the ASR setting for the maximum supply quantity, OUTL.
- Achieve greater efficiency in the replacement of the 90% share of the sub-class and 20/80 products.
- Stock in the warehouse must be carried out at the level of single-product pallet and casepack as established in the previous point

Change of Gondola Capacities and Identification of Palletizable SKUs on the Sales Floor

- Products that must be palletized on the sales floor.
- 953 products can have warehouse storage (90% sub-class share), of which 12% must be "pallet or half euro pallets mono product".
- 576 products to which the maximum shelf coverage must be extended to an estimated range of 2–5 days (90% sub-class share, 20/80 in units, MMPP). Example: Vegetable Oils / Soy
- For the vegetable oils subclass, the days of coverage in the gondola have been reduced by 31%, enhancing those that make up 90% of the share and reducing the days for the rest of SKUs.

Implementation of the Product Life Cycle Policy: It is suggested to formalize the following conditionals in an internal policy (Table 2).

Table 2. Internal policy conditionals

Status change	Politics	Considerations
Inactive	Active Product with Fill Rate 0 during the last 4 weeks and that has had at least 4 OC generated	
Discontinued	In the case of seasonal SKUs: Inactive products with 30 days of inactivity, without OC in process and OH 0 in CD Rest of SKU: Product with more than 30 days of inactivity	It does not apply to Check out products that are purchased by Break Set, nor to imported products without receipt on the CD Discontinued products are not scheduled and are settled at 50% off the list price
Purged	Discontinued SKU or Class 100 SKU with OH equal to zero at chain level	

*Note: OH refers to On Hand = store stock

Optimize Replenishment: Management of daily break walk: This walk consists of identifying daily which are the product breaks in the gondola. The guideline is proposed to respect the CATMAN and not cover the gaps with other products to sincere the breaks, with it the breakage factor.

Objective:

- Reduce the breaks of active SKUs with OH, strengthening the adjustment processes to the one available in stores.
- Reduce active SKU breaks without OH, daily work of the EO team with planning.
- Reduce the number of Inactive and discontinued SKUs with the "Product Life Cycle Policy" in coordination with the commercial area.

Organizational Structure of Warehouses and Replacement:

- In the structure of the platform personnel (where the warehouse is included), there are three positions: boss, assistant and assistant.
- The actual total endowment varies even within the stores that belong to the same format and cluster, since the amount of FTE's necessary for the operation is not standardized.
- The tasks to be carried out by job are not properly defined.

4 Results and Validation

A growth was observed at the OSA level, which applying the aforementioned in October 2019 reached its highest peak as can be seen in Fig. 2.

On the other hand, the % of Fill Rate did not have a very significant rise, but we are still improving acting jointly, since this problem is clearly the supplier.

The evolution of the %Without Presence is another indicator that shows the results of our application, this percentage, if possible, has to tend to 0 (Fig. 3).

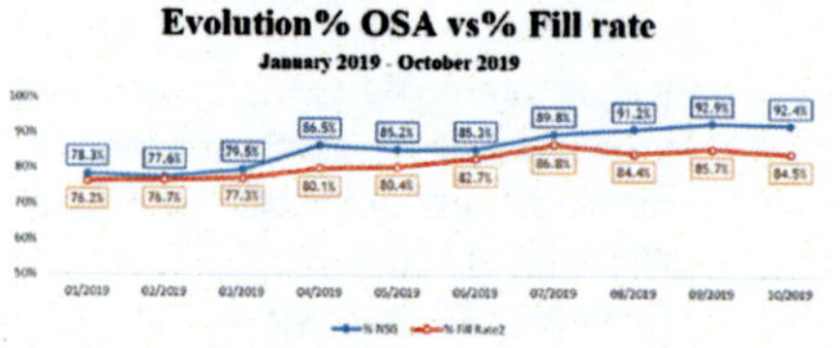

Fig. 2. OSA vs Fill Rate

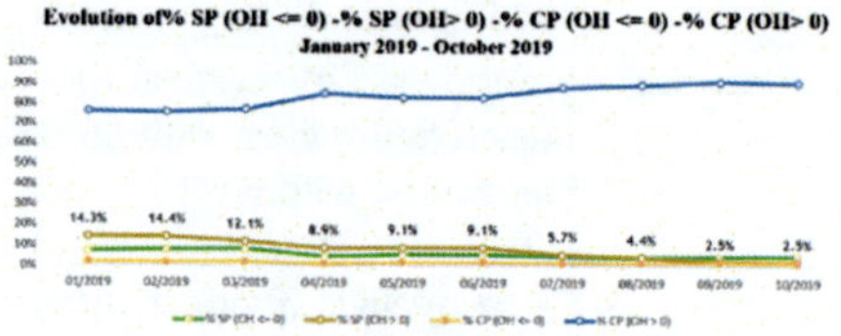

Fig. 3. Evolution SP - CP

We can also observe a ranking of stores with the highest OSA and be able to compare how our progress is going at the store chain level (Fig. 4).

To complement the previous graph (Fig. 4), we can also present the graph of the% SP that have a stock greater than 0 (Fig. 5).

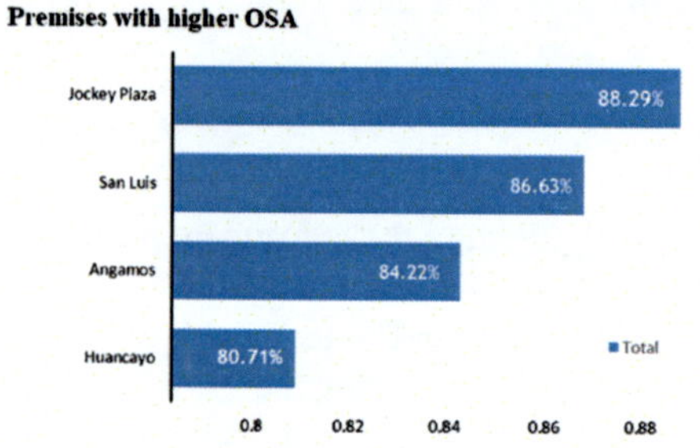

Fig. 4. OSA vs Fill Rate

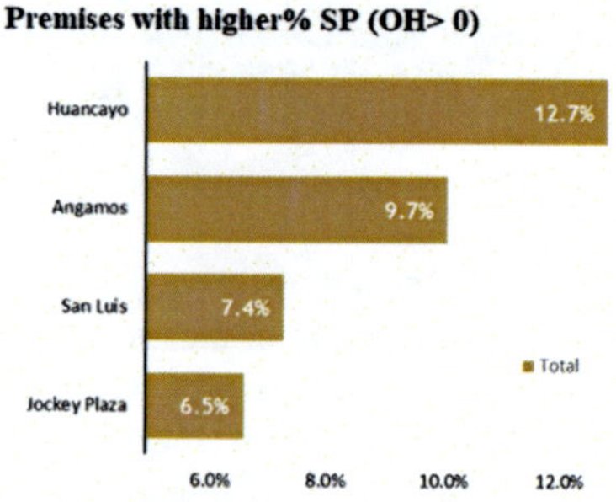

Fig. 5. Evolution SP - CP

5 Conclusions

The Level of Service in the Gondola is one of the most important indicators for Tottus, since it measures the availability of the product in the gondola.

The fact of having a shortage gondola represents a lost sale for the company and a reduction in customer satisfaction. This has a direct impact on market share.

The increase in OSA has a positive impact on sales, since it represents an opportunity to offer the product when the customer needs it; as well as the opportunity to increase participation.

From an operational point of view: To date, 2 solutions have been implemented:

- Change of the replacement flow: It ranges from the receipt of the product to the replacement in the gondola, itself. Before, first the products had to be stored, and then they were searched in the clutter of the warehouse, according to the need for replacement. Now, before the arrival of the truck, those products that must be replaced directly in the gondola are identified, without the need to store them (direct replacement from the arrival of the products).

- Alignment of the Organizational Structure: For the warehouse, platform and sales floor replacement areas. This led to the creation of the list of Activities per position and the replacement routines.

The implementation of the aforementioned solutions allowed to increase the OSA of the group of 4 stores in 12.9% (Month of start of implementations: March) Base month (March): 79.5%, Current month (October): 92.5%, OSA Target: 95%.

Having an OSA measurement routine requires stores to promote replenishment and to comply with operational guidelines, since the indicator represents the store's "health" in terms of replenishment.

Finally, it is expected that by implementing all the solutions the objective of 95%, which is the Voice of the Customer, will be achieved. In this case, the Central Operations manager.

References

1. Fisher, M., Raman, A.: The New Science of Retailing: How Analytics are Transforming the Supply Chain and Improving Performance, vol. 5, no. 2, pp. 1–30. Harvard Business Press (2010)
2. Raman, A., Dehoratius, Z.: Execution: the missing link in retail operations. Calif. Manag. Rev. **43**(3), 136–152 (2001)
3. Guerrero-Martínez, D.G.: Factores clave de éxito en el negocio del retail. Ingeniería Industrial, no. 030, pp. 189–205 (2012)
4. Hammer, M., Stanton, S.: La Revolución de la reingeniería: un manual de trabajo, Madrid, Ediciones Díaz de Santos (1997). ISBN 84-7978-309-5
5. García-Arca, J., Prado-Prado, J.C., Garrido, A.T.G.-P.: On-shelf availability and logistics rationalization. A participative methodology for supply chain improvement. J. Retail. Consum. Serv. **52**, 101889 (2020)
6. Moussaoui, I., et al.: Drivers of retail on-shelf availability: systematic review, critical assessment, and reflections on the road ahead. Int. J. Phys. Distrib. Logist. Manag. **46**(5), 516–535 (2016)
7. Chuang, H.H.-C., Oliva, R., Liu, S.: On-shelf availability, retail performance, and external audits: a field experiment. Prod. Oper. Manag. **25**(5), 935–951 (2016)
8. Aastrup, J., Kotzab, H.: Forty years of out-of-stock research–and shelves are still empty. Int. Rev. Retail Distrib. Consum. Res. **20**(1), 147–164 (2010)
9. Mary, G.: Exhibicion de productos en tiendas de autoservicios ¿Como mejorarla? 1 Julio 2006, https://www.gestiopolis.com/exhibicion-de-productos-tiendas-de-autoservicio-como-mejorarla

Using Drones for Tourism: Exploring Exciting Places in Ecuador

Santiago Uribe-Montesdeoca[1(✉)], Hugo Arias-Flores[2], Carlos Ramos-Galarza[3], and Janio Jadán-Guerrero[4]

[1] Carrera de Diseño Digital y Multimedia, Universidad Tecnológica Indoamérica, Quito, Ecuador
santiagouribe@uti.edu.ec

[2] Carrera de Administración de Empresas, Centro de Investigación en Mecatrónica y Sistemas Interactivos (MIST), Universidad Tecnológica Indoamérica, Quito, Ecuador
hugoarias@uti.edu.ec

[3] Facultad de Psicología, Pontificia Universidad Católica del Ecuador, Quito, Ecuador
caramos@puce.edu.ec

[4] Carrera de Ingeniería en Ciencias de la Computación, Centro de Investigación en Mecatrónica y Sistemas Interactivos (MIST), Universidad Tecnológica Indoamérica, Quito, Ecuador
janiojadan@uti.edu.ec

Abstract. Drones (unmanned aerial vehicles) are one of the newest trends for presenting tourism attractions using aerial photographs and videos. Drones can take high quality pictures and videos that can intensify the emotional response of a viewer by displaying a very special view and unusual perspective of various tourism destinations. Generating a film aimed at promoting tourism entails five stages: development, pre-production, production, post-production, and distribution. This paper explains how these five stages were used in the making of aerial drone films used to promote tourism in Ecuador and explores the potential use of drones in tourism in general. The results of the project were honored in two world events related to audiovisual production with drones. The film "Mindo, a Hidden Paradise" was awarded first prize in the Drone Focus Film Festival held in North Dakota, United States in 2018, and the film "Galapagos, the Enchanted Islands" won first prize in the same event in 2019 and second place in the Kharkiv International Festival of Arts and Tourism in 2019. These unique films brought worldwide attention to some of Ecuador's most beautiful and unique tourist destinations.

Keywords: Drone · Unmanned aerial vehicles · Tourism · Ecuador · Audiovisual production · Drone focus film festival

D. Russo et al. (Eds.): IHSI 2021, AISC 1322, pp. 786–791, 2021.
https://doi.org/10.1007/978-3-030-68017-6_117

1 Introduction

The audiovisual industry is constantly evolving as techniques and equipment are being replaced by more efficient tools and processes. Current filmmakers require timely and adequate information regarding the tools at their disposal; this is how, in the audiovisual field, new technologies have been developed for capturing images or videos, such as unmanned aerial vehicles (UAVs), better known as drones. The incursion of these devices has positively influenced competitiveness and positioned agencies and production companies in many markets to add value and creativity to their products [1].

Within this context, this paper highlights the use of drones in tourism by describing the work carried out in the five phases of production (development, pre-production, production, post-production, and distribution) for specific tourism destinations in Ecuador [2].

The idea was born from a tourism campaign known as "All You Need is Ecuador," which was carried out by the Ecuadorian government in 2014. This tourist promotion showed wonderful places, such as the mountain region, the coast, the Amazon, and the insular area. Through camera shots taken with drones, the majesty of the volcanoes, the stunning sunsets, and the paradisiacal landscapes of the Galapagos Islands were appreciated worldwide [3]. The success of this campaign motivated the authors to create audiovisual productions as part of the degree work for the Digital and Multimedia Design career at Universidad Tecnológica Indoamérica of Ecuador. Three promotional films were produced—"The Quilotoa Lagoon," "The Caldera and Crater Lake Quicocha," and "Mindo, a Hidden Paradise," the last of which won first prize in the landscape category at the Drone Focus Film Festival held in North Dakota, United States in 2018. The following year, "Galapagos, the Enchanted Islands," in which scenes of landscapes considered unique to the Ecuadorian archipelago were shown, won first place in that festival and was awarded second place at the Kharkiv International Festival of Arts and Tourism in Ukraine. These achievements allowed people around the world to see the beauty of Ecuador.

2 Background

The development of the audiovisual industry is constantly evolving; the initial techniques and equipment are being replaced by more efficient tools and processes. Current generations of filmmakers require timely and adequate information regarding the tools at their disposal, including new technologies in the audiovisual field for capturing images or HD videos with UAV or UAS, better known as drones [4]. These technological developments are becoming common today, and they can often be seen across the skielines of our cities. Richard Miller (2015, p. 9) presents an extensive list in terms of civil applications, such as cartography, logistics, public and private security, aerial photographs, and videos used in the press, advertising, art, and entertainment [5].

The growing interest of users in drone technology have developed new fields of application. Today, drones are serving in many areas, and the continuous advances in the technologies of these machines will facilitate even more applications in the future. One of the fields in which this technology is being developed is the tourism sector, where there are already new prototypes that provide tourism experiences based on the combination of drone technology and virtual reality. This tool is getting closer to being available to any tourist destination worldwide [6].

In recent months, due to the coronavirus (COVID-19) pandemic, the use of drones in virtual tourism has opened up new possibilities for creating or extending virtual experiences that tourists may accept and use as alternatives for real visitation. Several tourism associations are supplying avid travelers with real-time video images of different places in the world that they can enjoy from home [7]. Consecuently, virtual tourism is a new paradigm that aims to reduce the limitations created by COVID-19 by recreating real tourism venues as 3D models using UAVs at the target locations to be distributed live to people in their homes [8].

3 Method

In this paper, we describe the methodology developed in the audiovisual production for tourism using drones. First, we describe the natural places in Ecuador selected to capture on video; second, we describe the drones used in the production, and finally, we address the five-stages used to produce the films.

3.1 Selection of Tourism Places

Ecuador is located in South America and has many paradisiacal places and a variety of microclimates due to the Andes mountain range that crosses its entire territory. The glacial volcanoes, tropical forests, wonderful beaches, and diverse and exuberant animals of the Amazonia and Galapagos Islands were considered in this work. Four iconic locations were selected for the promotional videos: 1) Mindo, also known as the Mindo Valley, a mountainous watershed in the western slopes of the Andes, 2) Quilotoa, a water-filled crater lake and the most western volcano in Cotopaxi Province, 3) Quicocha, a three-kilometer wide caldera and crater lake at the foot of Cotacachi Volcano in Imbabura Province, and 4) the Galapagos Islands located in the Pacific Ocean, an archipelago of volcanic islands known for their large number of endemic species.

3.2 Type of Drones

The drones used in the project were all a special kind of UAV featuring a high speed, high definition camera and intelligent sensors with flexible operation. Based on the recommendations and the literature, three drones were selected: 1) Phantom 4 Pro; 2) Inspire 2; and 3) Mavic pro 2.

3.3 Procedure

This project was carried out during an approximately six-month period between 2018 and 2019 as part of the degree work toward a career in Digital and Multimedia Design at Universidad Tecnológica Indoamérica. In the project, the five stages of audiovisual production were considered: development, pre-production, production, post-production and distribution [2].

The development stage included the creation, writing, organizing, and planning of the film project. In this sense, it was important to consider the budget for buying the drones and the cost of mobilization to the selected places. It was important to select the correct type of drone considering the desired goal. For example, a racing drone was necessary to record area shots at high speeds. On the other hand, to record landscapes or open spaces, it was necessary that the drone have an image stabilizer and good camera resolution.

In the pre-production stage, the filmmakers learned how to fly a drone according to the instruction manual. However, it was necessary to practice and practice to gain the necessary skills. Like any sport, it was necessary to train with the drone every day so as not to cause an accident or crash the drone. Initially, the filmmakers tested the drone using the master control for short distances and heights so that they maintained visual contact. Before piloting a drone, operators must know the restrictions in the locality where they will be filming; sometimes, a license is required, but in some locations, drones are allowed to fly freely. Also, it was necessary to be careful about the safety of all the people around because the drone could become a weapon if it was not used properly.

In the production stage, also known as principal photography, all the actual shooting and recording takes place. To record with the drone, it was necessary to know how to perform the various movements to allow different shots, movements, angles, and planes. The movements used to record several videos were, for example, panning, this movement was done on the axis of the camera which lets go from left to right.

In post-production, it was usual to take subjective camera shots; this type of movement is a first-person view as if the drone were a character, such as a bird or an airplane. Various perspectives must be applied at the time of recording, and one must be ready to capture images of ever-changing landscapes.

In the distribution stage several tourist videos were entered in the contests at two drone festivals, the "Drone Focus Film Festival" in Unites States and the Festival of Arts and Tourism in Ukraine.

4 Results

The three promotional films ("Mindo, a Hidden Paradise," "The Quilotoa Lagoon," and "The Caldera and Crater Lake Quicocha") were awarded the first prize in the and scape category at the Drone Focus Film Festival held in North Dakota, United States in 2018 [9] (Fig. 1).

Fig. 1. Caldera and Crater Lake Quicocha

Fig. 2. Sunset in the Galapagos Islands

The video "Galapagos, the Enchanted Islands" (Fig. 2) won first prize in the landscape category at the Drone Focus Film Festival held in North Dakota, United States in 2019 and won second place at the Kharkiv International Festival of Arts and Tourism in Ukraine [10].

5 Conclusions

Audiovisual production has benefited greatly from the use of drones, since nowadays including different perspectives draws the attention of the viewer. It is therefore necessary to study the audiovisual language and its syntactic aspects to be able to apply them in the best way possible.

This project aimed to demonstrate the benefits and advantages of the uses and applications of UAVs as tools for the realization of commercial audiovisual productions. The use of different syntactic aspects of audiovisual language provides a guideline for the advantages that these can bring to a production.

This experience and the work carried out brought worldwide attention to Ecuador and allowed many people to see its beauty and unique landscapes.

Acknowledgments. The authors thank Universidad Tecnológica Indoamérica for funding this project and to the people that contributed to facilitate the records.

References

1. Uaslogic: Using Drones for Tourism, Explore the Many Exciting Uses of Drones and UAS Vehicles (2016). https://www.uaslogic.com/drones-for-tourism.html
2. Buehler, Ch.: What are the phases of film production, IPR College of creative arts (2020). https://www.ipr.edu/blogs/digital-video-and-media-production/what-are-the-phases-of-film-production
3. Segura, S.: The spread of the campaign "All you need is ecuador" and its impact on tourism incoming, Revista Empresarial (2016)
4. Messina, A., et al.: The future of media production through multi-drones 'EYES' (2018)
5. Miller, R.: Les Drones, La Nuvelle Révolution Technologique, Paris, Francia: Centre Jean Gol (2015)
6. Brian, H.: Drone tourism: a study of the current and potential use of drones in hospitality and tourism. Blue Mountains International Hotel Management School, Sydney (2016)
7. Mirk, D., Hlavacs, H.: Using Drones for Virtual Tourism, Intelligent Technologies for Interactive Entertainment. Springer International Publishing (2014)
8. Mirk, D., Hlavacs, H.: Using Drones for Virtual Tourism, Intelligent Technologies for Interactive Entertainment, Social Informatics and Telecommunications Engineering, vol. 136. Springer (2014)
9. El Universo: Joven ecuatoriano gana premio en Festival de Drones en EE.UU. con producción sobre Mindo (2018). https://www.eluniverso.com/noticias/2018/06/06/nota/6796430/joven-ecuatoriano-gana-premio-festival-drones-eeuu-presentar
10. El Comercio, El ecuatoriano Santiago Uribe ganó un concurso de drones en Estados Unidos (2019). https://www.elcomercio.com/tendencias/ecuatoriano-santiago-uribe-gano-concurso.html

Visual Perception Based on Gestalt Theory

Zhiyuan Ye, Chenqi Xue(✉), and Yun Lin

School of Mechanical Engineering, Southeast University, Nanjing, China
{220180377, ipd_xcq, 230159629}@seu.edu.cn

Abstract. Gestalt psychology puts forward the laws of similarity, proximity and common destiny. These laws all point out that visual stimuli work together to form a similar pattern organization, which results in the visual elements conforming to the same rules to be regarded as a group by the observer. However, the attributes of visual elements are now more diverse, including the dynamic changes of visual elements. In this paper, based on Bertin's five visual variables, we will explore the impact of the coordinated changes of these visual variables on human visual perception. The experiment produced the ranking of visual variables on the strength of grouping ability. In practical application, the relative grouping strength of these visual elements has high application value for interface design and data visualization research.

Keywords: Animation · Visual perception · Gestalt theory · Cognition

1 Introduction

Animation is increasingly used in the presentation of visual content. The state of this animation allows users to track and observe the changes of the screen without having to reinterpret the visual content or interface every time [1]. However, it is difficult to design animation and make it transmit smooth and consistent with the change of user cognition [2]. For example, it involves the speed of animation, the form of motion, the tracking of visual objects and so on.

In this paper, we investigate the effects of these dynamic and static visual elements on users' cognition through a large-scale graphic perception experiment. This experiment involves 50 subjects performing a group task of graphic perception. Our experiment was designed to compare five static visual elements (position, size, opacity, color, shape) with five dynamic visual elements (position, motion, opacity, size, color, shape). In each trial, four visual objects are grouped according to two attributes: two pairs are grouped according to one element, and two pairs are grouped according to another element. This experiment can not only study the individual grouping strength of each visual element, but also rank these elements according to their relative grouping strength.

D. Russo et al. (Eds.): IHSI 2021, AISC 1322, pp. 792–797, 2021.
https://doi.org/10.1007/978-3-030-68017-6_118

2 Research Background

Here we introduce Gestalt theory. Visual cognition is an important perceptual component of human cognitive system used to express and understand the environment. As the most important sense organ, human visual system has evolved for millions of years, allowing human to recognize and track objects visually [3]. In Gestalt theory, the law of proximity refers to the perception that close objects are divided into a group. The law of similarity means that objects with similar visual appearance are considered to be combined. The law of common destiny refers to the perceptual combination of objects with the same motion behavior [4].

3 Materials

We focus on the static and dynamic versions offive visual elements proposed by Bertin [5], which meet our criteria as follows:

3.1 Static Visual Elements and Dynamic Visual Elements

Static visual elements do not change over time, so they do not follow the law of common destiny. Considering these factors, we can study the relationship between the common destiny law and other Gestalt laws through static visual elements.

Static position (SP): close visual elements are considered to be grouped, a phenomenon known as proximity. Geometric location is generally considered to be the most accurate attribute of visual elements. Static color (SC): visual elements of the same color are considered to be combined according to similarity. Static shape (SF): visual elements of the same shape are grouped together according to similarity. Static size (SS): visual elements of the same size are considered to be the same group based on similarity. Static transparency (ST): groups elements with the same transparency according to the same similar row law.

The characteristic of dynamic visual elements is that they change over time. This means that they may show the law of common destiny. By studying the dynamic visual elements, we can answer whether the law of common destiny also applies to other forms of dynamic change, not just simple positional movement.

Dynamic position (DP): a typical example of the law of common destiny in which objects moving at the same speed and direction are considered to belong to the same group. Dynamic color (DC): changes the color of the visual object in the same way. Dynamic shape (DF): changes the shape of the visual object in the same way. Dynamic dimensions (DS): linear changes in the size of the visual object in the same way. Dynamic transparency (DT): Previous studies have suggested that visual objects become deeper or more transparent in the same way are considered to belong to the same group, but these experiments did not consider the relationship between DT and other visual elements [6].

4 Research Objectives

4.1 Experimental Principle

For the integrity of the experiment, we also tested the static visual elements. Through the competition between dynamic visual elements and static visual elements, dynamic visual elements and dynamic visual elements, static visual elements and static visual elements compete with each other, we can quantitatively measure the ability of different visual elements to perceive grouping. The relative grouping strength of each visual element can be directly measured and sorted by task comparison of non conflicting visual element pairs.

4.2 Task Summary

Table 1 summarizes all possible pairwise comparisons of the 10 visual variables we selected. In all cells, we discard all self comparisons (diagonals) and eliminate duplicates.

Table 1. Comparison tasks generated from the ten visual elements

	DP	DS	DF	DC	DT	SP	SS	SF	SC	ST
DP	—	DS- DP	DF- DP	DC- DP	DT- DP	—	SS- DP	SF-SP	SC- DP	ST- DP
DS	DP-DS	—	DF- DS	DC- DS	DT- DS	SP- DS	—	SF- DS	SC- DS	ST- DS
DF	DP-DF	DS- DF	—	DC- DF	DT- DF	SP- DF	SS- DF	—	SC- DF	ST- DF
DC	DP-DC	DS- DC	DF- DC	—	DT- DC	SP- DC	SS- DC	SF- DC	—	ST- DC
DT	DP-DT	DS- DT	DF- DT	DC- DT	—	SP- DT	SS- DT	SF- DT	SC- DT	—
SP	—	DS- SP	DF- SP	DC- SP	DT- SP	—	SS- SP	SF-SP	SC- SP	ST- SP
SS	DP-SS	—	DF- SS	DC- SS	DT- SS	SP- SS	—	SF- SS	SC- SS	ST- SS
SF	DP-SF	DS- SF	—	DC- SF	DT- SF	SP-SF	SS- SF	—	SC- SF	ST- SF
SC	DP-SC	DS- SC	DF- SC	—	DT- SC	SP- SC	SS- SC	SF- SC	—	ST- SC
ST	DP-ST	DS- ST	DF- ST	DC- ST	—	SP- ST	SS- ST	SF- ST	SC- ST	—

5 Methods

5.1 Task Generation

The default diameter of the visual object we generated is 75px, the default transparency is 0.5 (transparency range from 0 to 1) black circle, static color selected red (R), green (G), blue (B). For static shapes, circles, squares and equilateral triangles with an outer diameter of 80px are selected. For the selection of color and shape style, we consider that the experimental results will have guiding significance for the visualization of scatter diagram, so we choose the representative color and shape in the scatter diagram. The consistency of the outer diameter ensures that the visual object looks the same size in size. In order to maintain the consistency between groups, we adopted the following strategies:

There are only two kinds of visual interface observed by the subjects: one is static visual interface, and the other is cyclic animation lasting for 2S. In the process of dynamic change, continuous cycle animation can ensure that the visual object is never static and avoid cognitive bias. For the dynamic changes of DP, DS and DL, we use linear interpolation to achieve smooth transition. When visual objects are given visual elements of DP, the two groups of visual objects will continue to move 400px in different directions. When the visual object is given DS, the diameter of one group will increase from 75px to 100px, and the diameter of the other group will decrease from 75p to 50px. When the visual objects are given DL, the transparency of one group will increase linearly from the initial value of 0.5 to 0.8, and the transparency of the other group will linearly decrease from the initial value of 0.3 to 0.2. When the visual object is given DC, the color of the two groups of visual objects will be converted from neutral gray to red and green through the interposition animation. When the visual object is given DF, the shapes of the two groups of visual objects will be smoothly transformed into different shapes through the shape gap animation (Fig. 1).

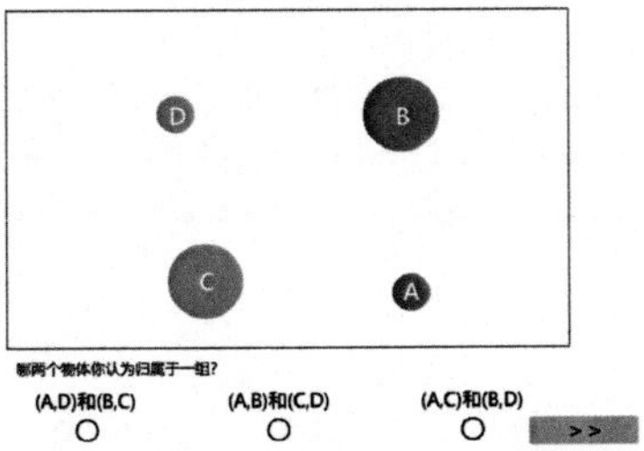

Fig. 1. We studied the experimental interface during the ds-dl task. Here, A-B and C-D are DL similar, while A-D and B-C are DS similar, forming two parallel possible groups.

5.2 Procedure

Participants were first asked to give informed consent. We first maximized the window of the experimental interface to avoid scrolling. Participants were told that their task was to select the visual object that best fitted their intuition from three radio buttons and grouped them in pairs. Participants were asked to select a group before moving on to the next trial. They can change their minds until they press the "next" button, with a short break in between. Participants were told "don't spend too much time thinking because there are no right or wrong answers" and they are encouraged to stick to their first impressions. After learning the details of the task, participants were asked to participate in a pre experiment to practice. Then the participants were asked to do the experiment. Participants could see a static visual interface, or a two second animation, which would automatically play and loop infinitely until they chose the answer. Participants cannot control the animation and have no feedback on the submitted answers. After completing all trials, participants had the opportunity to comment.

5.3 Participants and Apparatus and Experimental Design

We recruited 50 participants (including 26 women) in schools and in the community. They were between 20 and 30 years old and did not know the purpose of the experiment before the experiment. At the end of the experiment, each person was awarded $2. All the materials used in the experiment are created by using Adobe animate, and the whole experiment is completed by using this software. Our experimental environment is a quiet laboratory, using 1920 * 1200px display equipment.

Our independent variables were visual variables DP, DS, DC, DF, DT, SP, SS, SC, SF and ST in pairs to form the above 40 tasks. Each type of task has three repeating groups.

	50	Crowdsourced participants
X	40	Tasks (pair of DP, DS, DF, DC, DT, SP, SS, SF, SC, ST)
X	3	Repetitions
6000	Trials (excluding practice trials)	

The experiment is divided into three bolcks, each bolck has 40 trials. Each trial appears randomly in the experimental sequence. After the experiment is completed, the software background will automatically record the participants' options.

6 Results

Figure 2 shows a detailed summary of our experimental results. The left most column shows the overall grouping strength of each visual element. The higher the value, the stronger the grouping ability. It can be seen from the figure that DP, DS, DC, DT, ST, SC (CIs > 0.5) are stronger than DF, SP, SS, SF (CIs < 0.5), but there is no strict order between these 10 visual elements. Reviewing all the visual elements, we can see that the grouping ability of DF is significantly lower than that of other visual elements (CIs < 0.5), other dynamic visual elements still have strong grouping ability.

In static visual elements, SP, SF and SS are significantly lower than other visual elements in grouping ability, but ST and SC show amazing grouping ability, and SC's grouping ability surpasses all dynamic visual elements. Reading the table column by column, it is found that SP and DF have the weakest grouping ability compared with other elements. The grouping ability of DF is slightly stronger than that of SP.

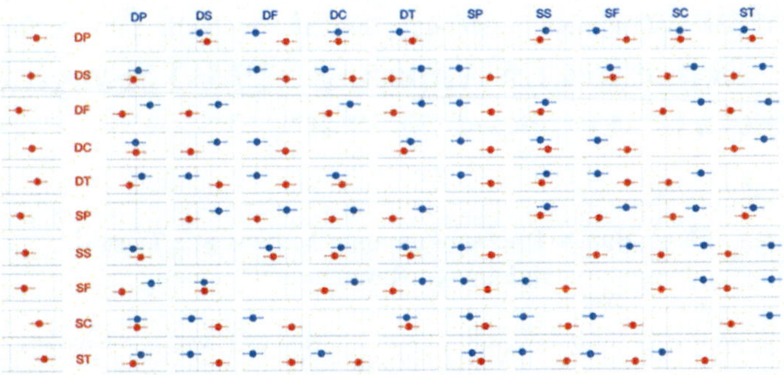

Fig. 2. The average grouping intensity of all visual variables in the experiment was summarized. The left column shows the average grouping strength of each variable.

7 Discussion

Reviewing the feedback of participants at the end of the experiment, we found that the difficulty of grouping was different among the participants. Some people have certain groups that are easy to complete, while others find it difficult. Participants are hesitant about what visual elements will produce the best combination. Most of the participants reported that color and dynamic position changes significantly attracted their attention, and they thought that exercise was the highest standard for group division. However, the dynamic shape change is not noticeable. The possible situation is that the dynamic shape change is different from the size, position and transparency, and the nonlinear change is not easy to cause the perception of the participants.

Acknowledgments. The authors would like to gratefully acknowledge the reviewers' comments. This work was supported jointly by National Natural Science Foundation of China (No. 71871056, 71471037).

References

1. Baecker, R., Small, I.: Animation at the interface. In: Laurel, B. (ed.) The Art of Human-Computer Interface Design, pp. 251–267. Addison-Wesley (1990)
2. Dragicevic, P., Bezerianos, A., Javed, W., Elmqvist, N., Fekete, J.-D.: Temporal distortion for animated transitions. In: Proceedings of the ACM Conference on Human Factors in Computing Systems, pp. 2009–2018. ACM, New York (2011). https://doi.org/10.1145/1978942.1979233
3. Jerison, H.J.: Evolution of the Brain and Intelligence. Academic Press (1973)
4. Maddox, W.T.: Perceptual and Decisional Separability, chap. Multi-dimensional Models of Perception and Cognition. Lawrence Erlbaum, Hillsdale (1992)
5. Bertin, J.: Sémiologie graphique. Mouton/Gauthier-Villars, Paris, France (1967)
6. Sekuler, A.B., Bennett, P.J.: Generalized common fate: grouping by common luminance changes. Psychol. Sci. **12**(6), 437–444 (2001). https://doi.org/10.1111/1467-9280.00382

Maintenance Management Model for Cost Reduction by Applying TPM Through a DMAIC Approach in SMEs in Service Sector

Hugo Bazan-Torres[1], Fernando Maradiegue-Tuesta[1], and Carlos Raymundo[2](✉)

[1] Ingenieria Industrial, Universidad Peruana de Ciencias Aplicadas, Lima 15023, Peru
{u201316487, fmaradie}@upc.edu.pe
[2] Direccion de Investigacion, Universidad Peruana de Ciencias Aplicadas, Lima 15023, Peru
carlos.raymundo@upc.edu.pe

Abstract. The studied organization is dealing with a challenge faced by every new company in the market: reducing waste as part of its maintenance processes to expedite its own processes and to optimize these for the client. To solve this problem, an improvement is proposed based on lean manufacturing (LM) tools that include standardized work, 5S's and total productive maintenance (TPM) using the DMAIC approach. This article aims to implement LM tools to reduce the waste in the mini loader maintenance process. The improvement proposal has a direct impact on the key process indicators, obtaining favorable results in late work, rework by guarantee, and service levels of 8.8%, 4.5%, and 7%, respectively. Finally, the analysis performed show favorable obtained by reducing waste through LM with the DMAIC approach.

Keywords: DMAIC · Lean manufacturing · Total productive maintenance

1 Introduction

Outsourcing of services involves 92% of industries' operations worldwide, and Latin American countries are no strangers to this phenomenon. At the top of the list, Brazil outsources 33.5% of services, followed by Colombia with 30.8%, Argentina with 19%, and Peru with 5.1% [1].

Such outsourcing of services is carried out with SME's; the mining and retail sectors have the highest demand for outsourcing services with 50% and 30% respectively. In addition, as to why companies outsource, the majority, 54.9%, aims to optimize their productivity; 21.4% opt for it to automate services to streamline processes; 8.1% do so to reduce costs; and 4% see it as an opportunity to update [2]. Thus, it is of great relevance for suppliers—specifically, for them to take appropriate steps toward developing good quality services in companies that outsource their maintenance processes. The problem results from the fact that the majority of these suppliers are micro, small, and medium-sized companies that are newly formed or with little time in

D. Russo et al. (Eds.): IHSI 2021, AISC 1322, pp. 798–804, 2021.
https://doi.org/10.1007/978-3-030-68017-6_119

the market, and therefore, they do not have any management systems applied to their processes, which affects their capacity for advice, guidance, and operational assistance in the different operational areas of the company [3].

2 State of the Art

2.1 TPM for Maintenance in the Service Sector

An analysis was carried out on literature related to the tools associated with the research project. Studies have focused on companies that lack adequate maintenance management. These articles not only implement the TPM based on each of the Lean Six Sigma pillars but also include other complementing tools such as 5S's, VSM, Kaizen, and SMED. The results of these studies are optimal because previous actions are implemented when applying full maintenance [4–6].

2.2 DMAIC in the Service Sector

Results of improving the process performance are validated not only with other researches performed but also with multicriterial simulation techniques that included cost savings. Proposals were suggested and resulted in maintaining quality through measurable indicators [7–9]. After several weeks of implementing the DMAIC methodology through each of its steps, the following validation results were obtained: data analysis, documentation system improvement, modification of some components, improvements in some of the machine repair methods, and reduction of rejections in the manufacturing unit, obtaining 1.88% for the first article and an improvement of 12.01% for the second [10, 11].

2.3 Maintenance Management of SMEs in the Service Sector

Researches state that there is a need to minimize operational costs and logistics processes by applying lean concepts, modifying the maintenance model and validating it through simulation. This resulted in an increase of the service level by up to 10% and the maintenance efficiency increased by 15%. In addition, the importance of optimizing resources in small businesses is outlined, where good quality, product delivery, and timely service is guaranteed to customers [12].

3 Contribution

The contribution of this project is that it attempts to combine TPM, 5S's and standardization of work in an innovative way under the DMAIC framework where it's five steps aim to reduce the problems caused by waste in the mini-loader maintenance process.

3.1 Proposed Model

The model focuses on improving processes that consume too much time and costs for the organization (Fig. 1).

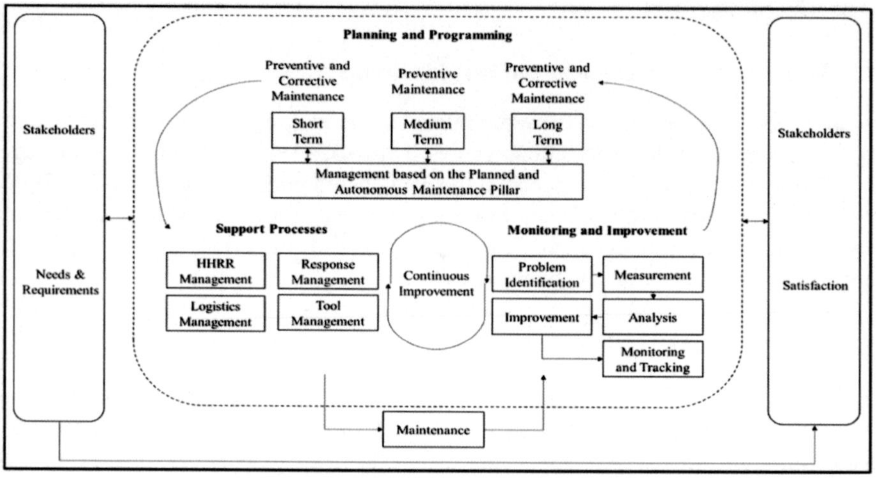

Fig. 1. Proposed Model design

3.2 Model Components

Artefact 1: VSM aims to outline the complete flow of information in detail to assess the conflicting parts of the maintenance process.

Artefact 2: The purpose of the validation sheet of standardized work is to validate each of the activities to be followed for equipment maintenance and to carry out appropriate monitoring.

Artefact 3: The red card aims to identify and organize objects in the work area that are not in their correct place; this activity is developed to create a culture of tidiness and cleanliness in the area where maintenance is performed.

Artefact 4: The tracking sheet of unnecessary objects is intended to track the objects found in the work area that cause time waste. The reason why the objects were in that place is assessed to prevent it from happening again.

Artefact 5: The sequence of the maintenance planned for mini loaders aims to perform each of the necessary steps for total equipment maintenance and to prevent it from being damaged at the time of being sent for hire.

4 Validation

The project will be validated based on the following three indicators that are appropriate to verify the improvements made and, therefore, also to verify the increase in company productivity.

Late Work. The indicator of the work carried out and not delivered on time. The purpose of this indicator is to quantify the problem of lateness and minimize it. The formula is as follows.

$$\frac{\text{Out of time work orders}}{\text{Received orders}} \tag{1}$$

Rework by Guarantee. The indicator of rework carried out because of a problem in the service performed and the company having responsibility. The purpose of this indicator is to quantify the problems during the term of guarantees to minimize them. The formula is as follows.

$$\frac{\text{Rework orders by guarantee}}{\text{Received orders}} \tag{2}$$

Service Level. The indicator of services performed correctly and with customer satisfaction. The purpose of this indicator is to quantify the service level to maximize it. The formula is as follows.

$$\frac{\text{Orders managed correctly}}{\text{Received orders}} \tag{3}$$

4.1 Scenario Description

The company in which the project is carried out is a SME with less than 5 years in the market, which seeks to provide good quality services to its customers. However, it does not have proper organization, both in terms of staff and operations. Their maintenance planning for their equipment and tools is poor.

4.2 Description of Initial Values

The time taken (min) for activities carried out by the technician when performing maintenance tasks as part of the service provided by the company were detailed. Table 1 shows the initial values of the diagnosis.

Table 1. Initial diagnosis activity time

Activities	Summary current calculation by technician	Current simulation
Reception and diagnosis	20	239.14
Report and budget	30	
Maintenance	105	
Test and delivery	15	

4.3 Procedure from de the Proposed Model

After the initial diagnosis, with the help of the VSM and the problem tree, the correction of the activities that lead to long operations was started to solve the main root causes.

After the activities and processes to be improved were identified, work standardization was carried out with the new and modified sheets that speed up the operation time. It is at this point that artifact 2 was used, validating that the technician carries out each of the maintenance activities with the support of the form sheets.

The 5S's manual, in which meetings were included for trainings with follow-ups, was implemented. The first steps of this methodology are carried out for the proper organization of materials and tools.

For control and monitoring, the 5S's culture is being followed, and an action plan is developed in case something is wrong. Finally, with the help of the planned maintenance manual, each of the steps evaluated for the hiring of equipment is performed, which help to mitigate future problems with customers.

For the validation method of the project, a simulation was used in which the optimum "n" size of 385 was obtained, with a confidence level of 95%, a Z confidence level value of 1.96, and a probability of 50% that the sample is (p) and is not (q) similar to the results. For this project, 500 data were obtained, with 115 additional data than the minimum required to be considered reliable.

Then, the simulation was run in the Arena program, and the results obtained from the 15 iterations were validated according to the following statistical theoretical formula, which finds the optimal "n" of replicas. Therefore, in the current situation n = 260 replicas were needed and for the improvement situation n = 409 replicas were required.

For the NPV analysis, an initial return on investment was obtained in the first four years of implementation. This is because of the subtraction of income from the expenses associated with the implementation which resulted in PEN 1,400/year.

Thus, to calculate the current net value, the division between the initial cost of implementation and the company's weighted average cost of capital (WACC) was performed to evaluate the NPV through an indefinite time (∞) and where a resulting in PEN 10,769.00 as the project's profit exhibited in Fig. 2.

Finally, for the initial result vs. the proposed one, supported by simulation results, it was found that the time taken was substantially reduced, and there was an opportunity to take advantage of the time lost in new business methods 25% of customer service time was reduced as shown in Table 2.

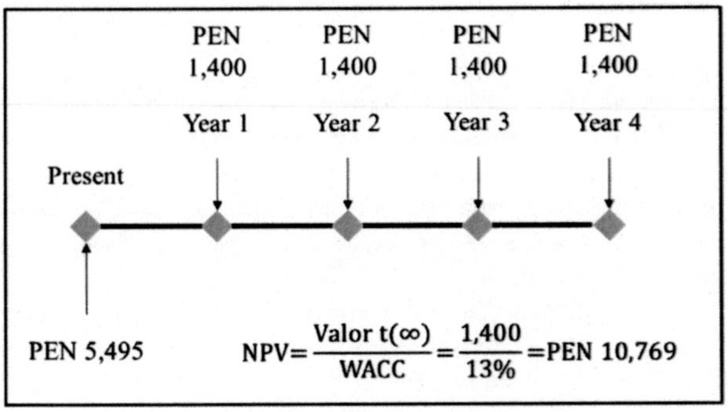

Fig. 2. NPV value from the project

Table 2. Summary of results after implementation

Activities	Summary				
	Current Calculation by Technician	Current Simulation	Improvement Calculation by Technician	Current Simulation	Variation % Current vs. Improvement
T system	170	239.14	136	196.52	25%
Reception and diagnosis	20		15		33%
Report and budget	30		18		67%
Maintenance	105		90		17%
Test and delivery	15		13		15%

5 Conclusions

With the implementation of lean and quality tools, it is possible to demonstrate operational improvements and make the following more efficient:

Quality of maintenance services.
Increased efficiency of hired equipment and the hours lost decreases from 47 to 30.2 in the last 3 months.
The cost-benefit ratio of the project indicates that it is profitable, with PEN 1.54 for each PEN 1.00 invested.
The post implementation indicators after the improvement show that late work KPI es equal to 4.7%, rework by guarantee KPI is 3.8% and service level raised to 96%.

References

1. Gestión: Outsourcing: ¿Qué países tercerizan más sus operaciones?. 17 October 2017. https://gestion.pe/tendencias/management-empleo/outsourcing-paises-tercerizan-operaciones-220898-noticia/. Accessed 21 Oct 2019
2. Gestión: Outsourcing: 86% de empresas en Perú tercerizan servicios, 27 Abril 2018. https://gestion.pe/economia/outsourcing-86-empresas-peru-tercerizan-servicios-232422. Último ac Last access ceso 16 Enero 2019
3. Instituto Nacional de Estadistica e Informática: Clasificación Industrial Internacional Uniforme, Junio 2009. https://webinei.inei.gob.pe/anda_inei/index.php/catalog/154/download/1056. Accessed 14 Enero 2019
4. Moscoso, C., Fernandez, A., Viacava, G., Raymundo, C.: Integral model of maintenance management based on TPM and RCM principles to increase machine availability in a manufacturing company (2020). https://doi.org/10.1007/978-3-030-25629-6_137
5. Ames, V., Vásquez, W., et al.: Maintenance management model based on Lean Manufacturing to increase the productivity of a company in the Plastic sector. In: Proceedings of the LACCEI International Multi-conference for Engineering, Education and Technology (2019)
6. Sharma, R., Sachdeva, A., Gupta, A.: Commonality amongst various lean manufacturing techniques: an investigation in the Indian automobile industry. IUP J. Oper. Manag. **16**(2), 21–35 (2017)
7. Gupta, A., Sharma, P., Malik, S., Agarwal, N., Jha, P.: Productivity improvement in the chassis preparation stage of the amplifier production process: a DMAIC six sigma methodology. Int. J. Reliab. Qual. Saf. Eng. **23**, 1640012 (2016)
8. Simanová, Ľ., Sujová, A., Gejdoš, P.: Improving the performance and quality of processes by applying and implementing six sigma methodology in furniture manufacturing process. Wood Ind./Drvna Industrija **70**(2), 193–202 (2019)
9. Garza, R., González, C., Rodríguez, E., Hernández, C.: Aplicación de la metodología DMAIC de Seis Sigma con simulación discreta y técnicas multicriterio. Revista de Metodos Cuantitativos para la Economia y la Empresa **22**(17), 19–35 (2016)
10. Singh, J., Singh, H., Pandher, R.P.S.: Role of DMAIC approach in manufacturing unit: a case study. IUP J. Oper. Manag. **16**(4), 52–67 (2017)
11. Phruksaphanrat, B.: Six sigma DMAIC for machine efficiency improvement in a carpet factory. Songklanakarin J. Sci. Technol. **41**(4), 887–898 (2019)
12. Vasanthakumar, C., Vinodh, S., Vishal, A.W.: Application of analytical network process for analysis of product design characteristics of lean remanufacturing system: a case study. Clean Technol. Environ. Policy **19**(4), 971–990 (2016). https://doi.org/10.1007/s10098-016-1293-x

Design of Travel Auxiliary Products and APP for People with Mobility Impairments

Wen Shao[1], Jiong Fu[1](✉), and Yingjue Dai[2]

[1] School of Design, Shanghai Jiao Tong University, Shanghai, China
redfox78@sjtu.edu.cn

[2] School of Art Design and Media, East China University of Science and Technology, Shanghai, China

Abstract. People with mobility impairments have become a serious problem in their travel due to their deficiencies and the influence of external environmental factors. Considering to meet the physical and psychological needs of this group of people traveling, this study proposes a feasible travel-assisted navigation scheme in an interactive way to help this group of people get a convenient and fast travel experience. The program includes AR glasses and APP to provide corresponding services. By using AR glasses, navigation reminders are obtained from visual and tactile aspects. While APP provides navigation services for this group of people, it also provides them with a platform for social interaction. This research is oriented to the needs of people with mobility impairments and provides insightful explorations for their travel plans. It not only assists in the design practice of navigation aids for the disabled but also provides solutions for the disabled to socialize.

Keywords: Mobility impairments · Interaction design · Wearable auxiliary products · Accessible navigation · Travel auxiliary design

1 Introduction

In China, according to the statistics of the second national sample survey of persons with disabilities, the proportion of people with disabilities in the total population of the country is estimated to be 6.34%, of which 24.12 million people are physically disabled, accounting for 29.07%. However, they are actually in urban public places. We rarely see people with lower limb disabilities traveling autonomously. In recent years, the Chinese government has gradually increased its investment in accessible facilities. However, the current research on the travel of people with disabilities is only limited to sociology and other theories. In this field, there is a lack of actual attention to the travel needs of people with mobility impairments (MIPs), and they still face the influence of various factors such as the environment and themselves that make them unable to travel normally [1]. As a result, we thought of tapping the travel needs of the disabled and conducted design evaluations to find solutions.

D. Russo et al. (Eds.): IHSI 2021, AISC 1322, pp. 805–810, 2021.
https://doi.org/10.1007/978-3-030-68017-6_120

2 Basic Understanding of MIPs

2.1 Travel Barriers of MIPs

As a part of the traffic vulnerable groups in cities, the disabled are not always considered in mainstream traffic planning to achieve the level of planning required for these people to achieve a seamless journey [2]. Due to personal reasons or transportation system reasons, their travel demand is often suppressed, including combining multiple trips into one trip, abandoning the trip, or being late or even missing the activity time despite the trip behavior. When public transportation services are not available, they usually rely on family members or other shuttle services [3]. Besides, MIPs cannot always use the shortest route, which means that itineraries will be longer, more complicated, and expose them to more risks [4]. MIPs must be more cautious than able-bodied people because they must process more information during the journey [5]. While MIPs pay attention to the road conditions, their attention will divert from the wayfinding information, which will increase the difficulty of travel [6].

2.2 Psychological Needs of MIPs

Some studies have found that changes in self-cognition and social evaluation caused by physical disabilities have led to psychological barriers to travel activities for people with disabilities, which are also further hindering the travel choices of them. At the same time, people with disabilities still need to travel. Cai Yan used Maslow's hierarchy of needs theory to explain the needs of persons with disabilities [7]. Satisfying these psychological needs can enable people with disabilities to get a good experience during travel, and help them overcome psychological barriers or even eliminate some of them, thereby increasing the possibility of future travel.

Through literature research, we found that the situation of people with lower limbs is different, their income, social class, physical condition, etc. are very diverse, but they all face very similar problems-"cannot find the entrance to accessible facilities".

3 Pre-design Analysis

3.1 User Research

Based on the research on the travel and psychological barriers of MIPs, we interviewed 5 MIPs, all of whom have varying degrees of mobility impairment, and recorded their views on their conditions and travel. Summarizing and analyzing the content of the interviews, it is found that the interviewees have a desire to travel, but due to physical, psychological, and environmental obstacles, they have become reluctant to leave the house. Based on the concept of "Accessible Journey Chain" proposed by Park and Chowdhury [8], the existing technologies are investigated and analyzed, corresponding plans are diverted, and the feasibility and influence of these plans are evaluated, and the results can meet the travel needs of MIPs. The initial plan is to design navigation that can help MIPs complete point-to-point path planning.

3.2 Empathy Design

After determining the preliminary design direction, the Baidu map was used as a test prototype, and the empathy experiment was carried out in Binjiang Tianjie, Longhu Hangzhou by setting corresponding tasks. Through this experiment, it is found that the traditional navigation APP cannot meet the basic needs of MIPs.

4 Prototype Iteration

4.1 Prototype Test 1: Accessible Map

According to the analysis of the results of the empathy experiment, we used a paper prototype to mark out all the accessible facilities in the original map highlight and generated a route to simulate the way of traveling as a accessible map. Through experiments, it is found that through the "point-to-point" approach of accessible maps to guide the travel route of the disabled, the travel time of the disabled can be effectively shortened, and the travel process becomes safer and more reliable. At the same time, we have also found that when people with disabilities use their mobile phones to navigate, they will see beyond the map in front of them and ignore the outside world.

4.2 Prototype Test 2: Accessible Map + Wearable Device

As a result, we conducted a second prototype test to introduce wearable devices into the guidance of the accessible map. The use of the mobile phone part of the accessible map only exists in the setting stage before travel, so that the disabled can free their hands. Use other senses to perceive the route to achieve the "goal of safe travel".

Table 1. Advantages and disadvantages of different sensory combinations

Sensory combination	Advantage	Disadvantage
Vision + Hearing	Can provide users with more informative content	The information provided by hearing will interfere with visual information and easily divert users' attention from the road
Vision + Touch	The user's perception of tactile mapping is more direct, and the prompt effect of tactile vibration will not make the user fall into the predicament of multitasking	If the instruction of the tactile prompt is not accurate enough, the product fails to attract the user's attention, it may be ignored by the user
Hearing + Touch	Can free the user's vision	The processing of auditory information is not intuitive and may cause users to fall into short-term thinking, so they can only complete some simple tasks

There are many combinations of ways people perceive routes. Therefore, we tested different sensory combinations and summarized the advantages and disadvantages of different combinations in Table 1. In the end, we found that the "vision + touch" approach is the best choice. For users, vision is the most familiar way to obtain information, and touch can be used as a light interaction method to assist users in guiding their travel.

5 Design Development

Based on the analysis before the design, we proposed an intelligent design scheme to help this group of people gets a convenient and fast travel experience. The final product design includes two parts: AR glasses and APP to provide corresponding services.

5.1 Design of AR Glasses

This product uses prism technology. On both sides of the glasses are vibration sensing devices, combined with AR navigation to provide corresponding reminder functions. The user connects the glasses to the mobile phone via Bluetooth before use. During use, the prompt information will be reflected and presented on the lens to form an AR prompt image to guide the user to the next step. When the image is presented, it will be accompanied by a certain amount of vibration to match the transmission of visual information (Fig. 1).

Fig. 1. AR glasses usage scenarios

5.2 Design of APP

The APP is divided into 3 major modules: navigation module, community module, and homepage module.

Navigation Module. In the navigation module, we recommend accessible walking routes and accessible site routes for users based on the "point-to-point" route planning characteristics of barrier-free maps (for example recommended routes for accessible facilities from communities to subway stations). To motivate users to travel and further improve the definition of accessible routes, we have designed an obstacle error reporting function. When the user leaves the navigation route, the navigation will actively ask whether it is an obstacle. If you encounter obstacles, we encourage users to take photos of the roadblocks and upload them to our platform. Doing so can remind more of the road conditions for disabled people appearing nearby and help the platform plan accessible routes (Fig. 2).

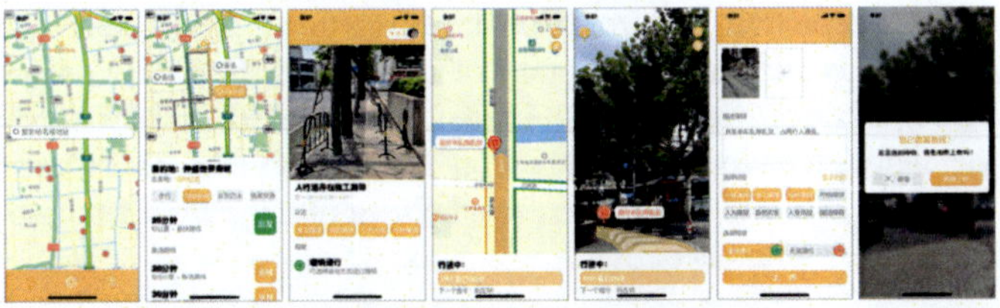

Fig. 2. Navigation interactive interface

Community Module. The community module is regarded by us as an exchange platform for people with disabilities to share travel routes, evaluate and publish accessible facilities, and share life. Users can post the problems encountered in their lives or their travel thoughts in the community through videos, pictures, texts, etc., so that more interested people can see them, promote communication between users, establish social interaction, and better social integration (Fig. 3).

Fig. 3. Community interactive interface

Homepage Module. The homepage module mainly displays the number of trips, travel routes, and the number of upload obstacles for users in a visual manner, providing users with more sense of accomplishment (Fig. 4).

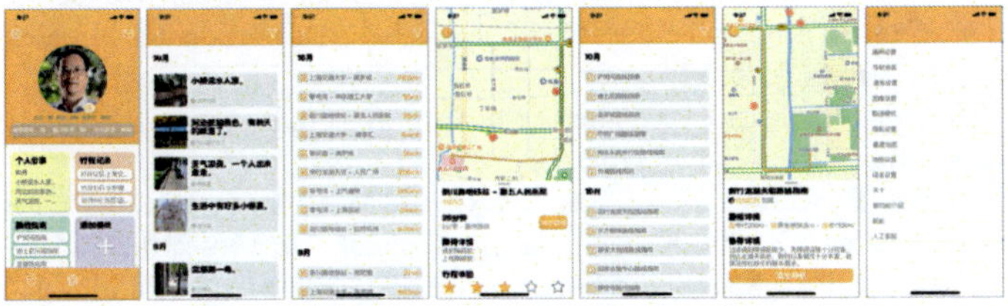

Fig. 4. Homepage interactive interface

6 Design Evaluation

The evaluation includes user testing and expert evaluation. First, set up the task flow according to the functional characteristics of the product, record the completion time, and ask 5 groups of test subjects to use the ten principles of interaction design to evaluate product interaction and functional satisfaction. The results show that the product can effectively improve the quality of travel, and the user's desire to travel is greatly enhanced. Besides, we have also invited senior interaction designers who have worked in the field of navigation interaction for many years as experts to conduct heuristic evaluations of our products. Experts believe that the products can help people with disabilities travel without barriers.

7 Limitations and Future Work

Based on the travel needs of MIPs, this study designed a navigation tool for such users to help them travel. The research has achieved certain results in the interactive design of intelligent hardware, but there are still many insufficient details. Existing products are still not perfect for the service model design of the APP part, and more consideration is needed for the handling of obstacles and the promotion of the plan, such as whether they can cooperate with the current mature navigation apps and use their accumulated data for many years. Define a route suitable for the disabled, or whether to design some functions to enable other healthy residents in the city to participate in the update of app data and the maintenance of accessible facilities, to improve the awareness of healthy residents to the protection of accessible facilities and the disabled Attention to travel needs.

All in all, these studies focus on the travel needs of MIPs and provide a smart product that can help them travel. In the future, it is hoped that the concept of accessible travel will serve a larger user group and help everyone prompt travel experience.

References

1. Prescott, M., et al.: Factors that affect the ability of people with disabilities to walk or wheel to destinations in their community: a scoping review. Transp. Rev. **40**, 646–669 (2020)
2. Maynard, A.: Can measuring the benefits of accessible transport enable a seamless journey? J. Transp. Land Use **2**(2) (2009)
3. Deka, D.: The role of household members in transporting adults with disabilities in the United States. Transp. Res. Part a-Policy Pract. **69**, 45–57 (2014)
4. Kasemsuppakorn, P., Karimi, H.A., Ding, D., Ojeda, M.A.: Understanding route choices for wheelchair navigation. Disabil. Rehabil.-Assist. Technol. **10**, 198–210 (2015)
5. Pecchini, D., Giuliani, F.: Street-crossing behavior of people with disabilities. J. Transp. Eng. **141** (2015)
6. Charette, C., Routhier, F., McFayden, B.: Visuo-locomotor coordination for direction changes in a manual wheelchair as compared to biped locomotion in healthy subjects. Neurosci. Lett. **588**, 83–87 (2015)
7. Cai, Y.: Review of travel psychology of the disabled and suggestions for urban traffic planning. In: 2016 China Urban Planning Annual Meeting, vol. 12 (2016). (in Chinese)
8. Park, J., Chowdhury, S.: Investigating the barriers in a typical journey by public transport users with disabilities. J. Transp. Health **10**, 361–368 (2018)

Design of VR Games Safety Criteria Based on the Method of Ergonomics and Behavior Simulation in Family

Wei Lu[1(✉)], Xiacheng Song[2], Hanjie Cao[3], and Binhong Zhai[4]

[1] Guangdong Polytechnic College, Guangzhou, China
lw960426@163.com
[2] The Guangzhou Academy of Fine Arts, Guangzhou, China
sinksongleo@163.com
[3] Hunan University of Science and Technology, Xiangtan, China
1094783721@qq.com
[4] Southeast University, Nanjing, China
864244934@qq.com

Abstract. At present, virtual-reality industry is still in the early exploration period. Safety problems exists in various VR devices provided by mainstream manufactures due to the serious safety designing drawbacks. Therefore, some new safety laws are needed to reduce safety accidents. This research based on the safety behavioral science, systematically studied the safety issues in human-machine interface, auditory alarm, and vision fields by behavior simulation and other methods, providing safety designing criterions for VR helmet game in family environment, as well as prototype support and reference for products upgrade to relevant manufactures.

Keywords: Virtual-reality · Safety designing criterions · Behavior simulation · VR games

1 Introduction

Born in the middle of the 20th century, ergonomics is currently used in scientific research and design in many industries. It mainly studies the systematic relationship between operators, interfaces and machines [1], and explores appropriate methodology to improve the work efficiency of mankind. After years of development, ergonomics has been divided into several subjects, one of them is safety ergonomics. Safety ergonomics studies safety factors between operators and machines during operation, laws, and essences, in order to prevent and eliminate any risks that are caused by improper human behavior during operation. Safety ergonomics focuses on safety-related factors in human activities and work and prevention of various hazards [2]. According to the Safety Ergonomics by Wang Baoguo: In the process of product design and development, safety-related specifications should be considered in the first place [3]. The development process of product design should incorporate safety specification design.

D. Russo et al. (Eds.): IHSI 2021, AISC 1322, pp. 811–822, 2021.
https://doi.org/10.1007/978-3-030-68017-6_121

According to the safety psychology in the field of industrial psychology, design suggestions can be proposed from the perspective of users to study human psychological laws when accidents occur in human activities, and provide scientific basis to prevent such accidents [9]. This subject originally studied the situation of mankind in operation, including analysis of human factors in accidents, research on characteristics of perpetrators in industrial accidents, and psychological countermeasures to prevent accidents [4]. These theories help us propose design suggestions or solutions at the users' psychological level. In addition to eliminating the potential safety hazards directly brought about by the products, this subject also needs to consider the interaction behavior of users with the external environment during the use of the products. By analysis on users' psychological activities and observation of their behavior, products' safety can be better designed. The safety design of HTC Vive got improvement in the product release in 2019. Originally, when users exceeded a certain limit, a virtual prompt grid would appear on the game screen. After the upgrade, gamers will be gradually transit to the perspective of the "real environment" after they walk out of the game safety boundary. This experience improvement is designed based on the safety psychology.

Compared with traditional game devices, common VR devices come with more controllers and the displays must be worn on heads. Immersion, Interaction, and Imagination are the three basic characteristics [5] of VR systems. These VR devices use electronic computer technology to generate game screens and project them onto corresponding head-mounted displays. Sound, smell, and vibration mobilize people's five senses to produce the feeling of "virtual reality". In terms of behavior, the biggest difference from traditional games is that gamers have changed in space. The fixed activities of gamers will provide guidance to game development and design. On the contrary, the game interaction of game manufacturers will also affect the gamer's actual behavior. For games applicable to indoor environment or narrow space, hardware manufacturers and game manufacturers have proposed their own solutions. Roughly, VR games are divided into static operation and dynamic operation. Static operation sacrifices certain game experience in exchange for a relatively safe game process: During play, gamers control movement of game characters with handles. This type of operation applies to narrow environments or game devices with weak performance. In dynamic operation, gamers can adjust the relative position of the game characters by changing their own positions. This type of operation more complies with the requirements of immersive experience, but it often reduces the game experience by repeated safety prompts or unexpected accidents. Under such circumstance, third-party manufacturers have designed a "treadmill" game. Gamers only need to walk and run on the "treadmill" to control the game characters. Gamers' bodies will be fixed on the "treadmill" to avoid relative displacement. This type of product is often expensive and covers a large area, so it is generally used in game centers.

As far as the current VR game market is concerned, various forms of game interaction emerge, immersive experience rises continuously, and market level lowers. For example, household games require a cheaper and safer VR game solution to adapt to, but main manufacturers apply their own "flagship" equipment directly to the home environment, resulting in difficulty for users to install and operate the professional equipment and making the narrow home space crowded. The issues with VR game experience in home environment remain existent.

2 Principle of Safety Design

The key of human-machine interface (HMI) design is the coordinated design between a display and related devices. The design is supposed to simplify operation on the HMI, reduce learning time, and improve operation efficiency. To design an HMI, you need to understand the three ways [1, 5] for mankind to receive physical information: touch, hearing, and vision. The physiological mechanism of mankind makes a person unconsciously alert upon hearing a special sound. Therefore, the auditory pathway will be more efficient and intuitive than the visual pathway and tactile pathway in an emergency. For the tactile pathway, mankind skin or any other organ is stimulated to receive prompt information. Nowadays, most VR systems are equipped with handle controllers to realize information transmission via the tactile pathway.

2.1 Principle of HMI Safety Information Presentation by Head-Mounted Display

A VR head-mounted display is used to present the specified information [6] a gamer wants to see, and it is an important medium for the gamers to use and operate a game. Therefore, the content presented by this head-mounted display with safety reminders must be efficiently and quickly identified, and accurately read without leading misjudge in order to minimize misleading information during transmission. The principles of safety information presentation by the head-mounted display are as follows [7]:

(1) Display and transmit information in simple and concise way to minimize misleading information during transmission.
(2) Use unified data units, rule descriptions, and terms to shorten users' information reading time.
(3) Display information in a way that adapts to gamers' operation habit to improve information readability.
(4) Give safety reminders in a way that makes information easily readable according to the play habits of the gamers so that the gamers have sufficient time to respond.

When designing a safety reminder message [8], specify a goal: the graphic label and color must be readable at first glance for gamers, including the new gamers. Therefore, this article proposes four design principles [9] of display content:

Accuracy. "Quantify" the degree of danger visually, rather than simple warnings. For example, warnings fall into high, medium, and low levels. Add relevant data for users to judge during warning.

Simplicity. Deliver as simple information as possible, draw simple labels, and simplify parameters. Warning colors can be used to replace a part of text or labels to reduce the reading time.

Consistency. Set warning information and color signs to be consistent regardless of the game being played to avoid misjudgment.

Arrangement. The most commonly used and most important display area should be located within 3° of the center of field of vision. Low-level warning messages can be placed on the periphery of the screen to avoid affecting game experience.

2.2 Principle of Warning Sound Selection and Design

This article studies VR head-mounted displays for home gaming, so they are generally used in a closed indoor space such as living room or bedroom, which is private and quiet. A warning is supposed to avoid collisions with surrounding obstacles or walls during the play. Therefore, the warning will be alerted when the gamer approaches the obstacles or walls. Considering the home environment, abide by the following principles [4, 10, 11] when selecting or designing the warning:

(1) Select high-frequency and low-loudness sound. High frequency tends to improve the penetrating power of sound, making it different from the sound made during gaming. In home environment, environmental noise is low. So, do not increase the loudness of the sound.
(2) Use fast and non-continuous sound. You are not advised to use sound similar to horns and siren to alert the gamer. This will make the gamer panic when he/she encounters an emergency.
(3) Keep the design of the screen and sound consistent. Giving sound warning and screen warning at the same time will enable the gamer to immediately understand the warnings in a short time, reducing the understanding cost of the gamer in first prompt.
(4) The sound warning and screen warning can take effect at any time after device start, even if no game is being played. Upon startup, the head-mounted display runs this set of safety system preferentially, including prompts.

2.3 Study on Eyes' Vision Scope and Field of Vision of Head Mounted Display

Eyes' Vision Scope and Design Suggestion
The vision scope usually refers to the scope of mankind vision, which is related to the physiological factors of the body, and it is limited [10]. The measurement parameters of the vision scope indicate the vertical angle and horizontal angle generally. For healthy adults, the monochrome horizontal vision scope is about 200°, and the monochrome vertical vision is about 130°. Mankind's eyes are more sensitive to the vision scope where the angle of the visual center is less than 20°. Therefore, based on the research and analysis on the vision scope of standard eyes, the design scope and specifications for the interface layout of the head-mounted display are roughly determined [3].

Field of Vision and Design Suggestion
Static field of vision indicates the visual spatial information captured by your eyes when you look at a front target with the eyes and keep your head and body remain still.

Field of vision is also referred to static field of vision. Accordingly, dynamic field of vision is the visual spatial information [1, 3] captured by your eyes when you rotate the eyes and keep your body and head still.

The information captured by human eyes is 3D information [12], which can be divided into horizontal information and vertical information. Based on the information, two related concepts: horizontal field of vision and vertical field of vision emerge. The maximum range of angle that can be captured by human eyes is 94°–104°. The field of vision for a single eye is −62°–94° or −94°–62°. The binocular visual area is 60°, respectively, on each side of the standard line of sight, totaling about 120°. The identifiable angle of English/Chinese is 10°–20° on both sides of the standard line, and the identifiable angle of a single character and letter is 5°–30° on both sides of the standard line. Beyond this range, the human eyes will gradually fail to recognize the characters and words. Within the range of 30°–60° on both sides of the standard line, the human eyes can recognize color of images. Where, the identifiable range is the range that is about 10° away from the standard line on both sides.

According to the analysis and research on the horizontal field of vision and the purpose of VR head-mounted display interface design, the horizontal field of vision is divided into the superior viewing zone, optimal viewing zone, transient viewing zone, effective viewing zone, and identifiable viewing zone, as shown in Table 1.

Table 1. Division of horizontal field of vision.

Horizontal field of vision angle	Field of vision
5	Superior viewing zone
10	Optimal viewing zone
20	Transient viewing zone
30	Effective viewing zone
60	Identifiable viewing zone

Set a standard line of sight to 0° within the vertical line of sight. The upper half of the standard line of sight is 50°, and the lower half of the standard line of sight is 70°. In the upper half of the line of sight, the maximum eyeball rotation angle is 25°; in the lower half of the line of sight, the maximum eyeball rotation angle is 30°; a total of about 55°. The color identifiable limits are 30° in the upper half of the standard line of sight and 40° in the lower half of the standard line of sight, a total about 70°. If a person maintains a relaxed standing or sitting posture, the person's field of vision will be lower than the line of sight 0°. For example, when the person maintains a relaxed sitting posture, the person's field of vision is about 15° below the standard field of vision. When the person maintains a relaxed standing posture, the person's field of vision will be about 10° below the standard field of vision. In a completely relaxed state, the person's field of vision is 38° below the standard line when he/she is seated, and the person's field of vision is 30° below the standard line when he/she is standing.

According to the analysis and research on the vertical field of vision and the VR head-mounted display worn by gamers (primarily in standing posture, and secondarily in sitting posture), the vertical field of vision is divided into the following zones, as shown in Table 2.

Table 2. Division of vertical field of vision.

Horizontal field of vision angle	Field of vision
5	Superior viewing zone
15	Optimal viewing zone
30	Transient viewing zone
40	Effective viewing zone
50	Identifiable viewing zone

For the visual display, in addition to the horizontal field of vision and vertical field of vision, the gamer's response speed and time should also be considered. Whether the safety reminders or related operations can be seen or operated by the gamer depends entirely on whether the gamer sees this information. The speed and accuracy of people's response to the displayed information in different fields of vision are not the same, the response speed of eyes can be as fast as 280 ms, which corresponds to the area that is formed by 8° away from the center line in the upper and lower direction, 45° to the right, and 10° to the left of the visual center line. In this area, the response speed of human eyes is much faster than that in other areas. Therefore, important information must be displayed in this area.

Eyes' Latency and Design Suggestion
The field of vision that is formed by 15° based on the basic field of vision in the up and down directions is the least-error area of human eyes. Setting information in this area will reduce the gamer's misjudgment rate. Otherwise, the gamer's misjudgment rate will increase.

Principle of Game Space Design
During play with a VR head-mounted display, the home environment is the gamer's action space. An action space is a space environment designed to ensure stable work and convenient use during work. Therefore, the design of the home environment space for VR gaming should meet the following requirements [3]:

(1) Meet the requirements of anthropometrics. This requirement is supposed to ensure that the design of the environmental space complies with the nearest-operation principle. For example, when the gamer rotates in situ with his/her arms fully open, he/she should not touch the surrounding obstacles and walls. In other words, the space provided must be larger than 8 m^2. If the game requires the gamer to move within certain space, then the required space will be larger.
(2) Provide guidance to non-gamers. In gaming, the gamer will make a series of actions according to the game instructions. Unlike traditional games, VR games require the gamer to perform large range of actions. At this time, the non-gamers

must be reminded by corresponding instructions to avoid physical collisions with the gamer and unnecessary accidents.

Because VR games are immersive games, the gamer's eyes can only see the game screen, and he/she cannot judge everything in the real world. External reminders such as carpets, warning stickers, and signal lights can be used to inform the non-gamers. It is important to note that the HTC Vive VR system is equipped with an infrared laser lighthouse, which is used to locate the gamer's location. This article tries to use this infrared laser lighthouse as a start point of design to locate the gamer's position and remind the non-gamers to keep a safe distance from the gamer.

(3) Use special materials as floor. In addition to the electronic solution that uses infrared laser lighthouse for positioning, using physical means such as laying game carpets can also limit the gamer's activity area. With a game carpet, the gamer can know the safe area of the game through the feel of feet and avoid going out of the safe area, even if he/she cannot see the external environment; the non-gamers can avoid going into the safe area to have physical collision with the gamer.

Compared with traditional electronic solutions, this solution is simpler and easier to implement, but it does not apply to all game environments and all gamers. In Sect. 4, testees will give their evaluation about the physical solution and electronic solution. According to the experimental results of the testees and experts' opinions, an ultimate solution will be determined.

(4) Make proper layout of equipment. Based on the HTC Vive Pro released in 2019, VR head-mounted displays should be wirelessly connected. This prevents the probability of gamer falling down by entangled wires. Game equipment should be placed outside the gamer's safe area to avoid damage to the game equipment by the gamer's actions.

3 Theoretical Study on Behavioral Simulation

Behavioral simulation means the procedure of building an experimental environment that meets the requirements or using a computer for experiments to simulate the possible conditions [13] of the final experiment. During the experiment, observe the testees or view the running results on the computer, and then analyze and measure the overall experiment based on the observation results and running results to get reliable experimental data, and use this data to prove the conclusion. Though the behavioral simulation technology can be proved theoretically, the behavioral simulation method must be used in real environment. Otherwise, simulation test may not be able to operate [4] due to safety, economic, ethical, technical and other reasons.

As the main method used in this study, behavioral simulation has the following five characteristics [14]: (1) Users can perform simulation tests on a computer in a short time to predict experimental results and design effects. (2) Behavioral simulation requires more accurate computer software or test site in order to be realized.

(3) Behavioral simulation must not be interrupted externally upon start. (4) Behavioral simulation must be performed repeatedly in order to close to the final experimental results. (5) For the same test item, evaluation results will vary with tests or computer programs, so simulation can be used as a reference only, and should not be used as a unique standard to determine behavior [15].

The behavioral simulation method used in this article is a physical simulation, in which a simulated test environment is set to observe gamers in gaming. Experiment.

3.1 Experiment Design

In this study, build a more realistic living room environment and put a game carpet into the living room to carry out a behavioral simulation test. The test devices are two head-mounted displays: HTC Focus Plus and Oculus Quest, and the test games are Quest-Arizona Sunshine (shooting), Quest-epic roller coasters (roller coasters), and Quest-Ninja legends (fighting). The games are supported by the head mounted displays.

During gaming, the gamer will be recorded by two cameras in different angles to record his/her behavior, which is used as experiment material. The two head-mounted displays will be connected to a PC so that the game screen is projected onto the PC screen. The game screen will be used as experiment material and saved and studied.

3.2 Experiment Recording

By multiple games, the game behavior diagram as shown in below Tables. The gaming procedure is divided into three phases: pre-gaming, gaming, and post-gaming, and gamer behavior is recorded in each phase (Tables 3, 4, 5, 6 and 7).

Table 3. Game Behavior - Steps

Phase	Description
Pre-gaming	1. Clear surrounding articles
	2. Wear the head-mounted device
	3. Use the handle to outline the game scope
	4. Select a game and start the game
Gaming	1. Select a game mode
	2. Start the game
	3. Adjust game operations
Post-gaming	1. Exit the game
	2. Power off and take out the head-mounted displays
	3. Put surrounding articles back to original positions

Table 4. Game Behavior - Behavior Instructions or Reminders

Phase	Description
Pre-gaming	1. Adjust the light brightness so that the controller is visible to the gamer
	2. Use the handle to draw the game space in the safety area
	3. If the game space is at least 1 m × 1 m, the game mode will be changed to the in situ mode
	4. Confirm the size of the game safety wall
	5. Open the control page
Gaming	1. Select a game mode
	2. Start the game
	3. Adjust game operations
Post-gaming	1. Exit the game
	2. Power off and take out the head-mounted displays
	3. Put surrounding articles back to original positions

Table 5. Game Behavior–Possible Operations of Gamer

Phase	Description
Pre-gaming	1. Configure the physical wall as the safety wall of the game
	2. Draw the game borderlines close to the furniture
	3. Draw a small scope
Gaming	1. Do not notice that a gesture causes a misoperation
	2. Trigger a safety alert when the head-mounted display or handle closes to the safety wall
	3. The head-mounted display surpasses the safety wall from time to time
Post-gaming	1. Have no idea of how to exit the game
	2. Exit the game forcibly
	3. Power off the device

Table 6. Game Behavior - Gamer's Psychology

Phase	Description
Pre-gaming	1. Start the game as soon as possible
	2. Light needs to be set. So troublesome
	3. Draw the safety area as large as possible to guarantee game experience
Gaming	1. I am fed up with the anti-virus prompts
	2. Why does the screen change back to the homepage suddenly?
Post-gaming	1. It is tedious to exit the game
	2. It is troublesome to confirm the game exit operation when I want to exit the game

Table 7. Game Behavior - Discovering Issues

Phase	Description
Pre-gaming	1. Before the gamer is approaching a safety wall, the gamer touches the wall with hands
	2. The gamer cannot act freely
Gaming	1. The game experience cannot be guaranteed when the gamer makes an incorrect gesture
	2. The virtual space is too narrow, and the gamer touches the virtual safety frequently wall during gaming
Post-gaming	1. The game exit procedure is tedious

3.3 Experiment Results

Analysis on VR Game Player's Behavior

Experience the three basic gestures in VR games: Sitting, Standing and Jumping. The safer gestures are standing and sitting, because the player may easily get involved or lose balance when there is any jumping actions in games. The observer found in several game experience halls here are some games requires the companion of the staff, for keeping player safe.

Besides the potential danger caused by the collision between the players' own behavior and the surroundings, another one is the interference from others when playing the game. The immersion in VR games indicate the player separated from the actual environment. In the gaming, the passer-by likely get injured if they failed to get a clear reminder, about the safe distance from the player. The accidents such as physical collisions may happened when some onlooker make fun of the player. Thus, protecting the player and reminding the passer-by in design are necessary accounts for safety.

Psychoanalysis on VR Game Player

Although the head-mounted display system has been designed with a security grid with security prompts, some players choose to turn off security-related settings or even turn off the security system for pursuing a more realistic gaming experience The discussion with several players indicate that the security prompts ruined their immersion and fluent experience of games. That is why they shield or disenable the security-related settings.

Commonly, the VR game player seek for the immersive gaming experience, the last thing they want to do is being interfered by the outside. The research found that some game anchor would turn off the security grid during live games, and some players reported the emergence of the security grid, which easily obstructed the game screen and affected the immersion. The security grid design is required to be optimized for the better user experience.

Possible Improvements
Based on the analysis, it can be inferred that:

(1) Before gaming, provide gamers with more accurate behavior prompts, simplify the operation process, and ensure a safe distance between the virtual safety wall and actual wall. Upgrade the light tracing technology so that the head-mounted display can work normally without being affected by the light strength.
(2) Ensure that prompts will not affect the game experience, and take effective measures to prevent gamers from hitting the wall, for example, designing protective devices such as protective fences.
(3) Give clear and concise prompts about exiting a game and powering off a display, and simplify the operation process.

4 Conclusion

Based on ergonomics, this thesis explores the safety design principles of VR games, including HMI safety design, alerting sound design, and vision scope and field of vision design of the head mounted display. By analyzing the gaming experience using the HTC Focus Plus and Oculus Quest head-mounted displays, summarize the shortcomings and improvement directions of the integrated head-mounted display products, and provide design prototype support for game companies to upgrade their products. Meanwhile, expect to provide references for industrial designers and user experience designers in related industries.

References

1. Ding, Y.: Ergonomics. Beijing Institute of Technology Press (2017)
2. Zhao, J.: Safety Ergonomics. Xidian University Press (2019)
3. Wang, B.: Safety Ergonomics. China Machine Press (2016)
4. Li, J.: Safety Psychology. China University of Mining and Technology Press (2012)
5. Zhang, F., Dai, G., Peng, X.: A survey on human-computer interaction in virtual reality. Sci. China (Inf. Sci.) **46**(12), 1711–1736 (2016)
6. Wang, S.: Research on Virtual Product Design Based on Ergonomics. Xi'an Polytechnic University (2016)
7. Zhao, F.: Research on Industrial Design of Head Mounted Displays (HMDs). Southeast University (2016)
8. Sagot, J.-C., Valérie, G., Samuel, G.: Ergonomics in product design: safety factor. Saf. Sci. **41**(2), 137–154 (2003)
9. Editorial Committee Green Cross Safety Infrastructure New Knowledge Series. Application Knowledge of Safety Psychology. China Labor and Social Security Press (2014)
10. Zhao, X.: The application of ergonomics in interactive media. Interface design. Art Sci. Technol. **28**(04), 59 (2015)
11. Wang, S.: Research on Three-Dimensional Display Based on Human Visual Specialties. University of Science and Technology of China (2016)

12. Hillmann, C., Babader: Comparing the gear VR, oculus go, and oculus quest. 2018: In this chapter, we look at the key differences between the Gear VR, the Oculus Go, and the Oculus Quest (based on prerelease information and subject to change when it ships). Even though each one of these headsets runs on an Android operating system, hardware differences play an important role when targeting the individual headset and its use. We will compare hardware specifications and different input events, and learn how to assign them in Unreal (2018)
13. Yu, A.: Design and Research on Immersive Virtual Reality Game in Application of Mental Relaxation. Harbin Institute of Technology (2016)
14. Li, J.: Safety Behavioristics. China Machine Press (2009)
15. Li, M.: Research on Safety Behavior Methods and Models. Documentation and Information Center, Chinese Academy of Chengdu, (Issue 05, 2016) (2016)

Ergonomic Method for Redesigning Workstations to Reduce Musculoskeletal Disorders Among Workers in Textile SMEs

Brenda Miranda-Rodriguez[1], Luis Saenz-Julcamoro[1], Edgardo Carvallo-Munar[1], Claudia Leon-Chavarri[1], and Carlos Raymundo[2(✉)]

[1] Ingenieria Industrial, Universidad Peruana de Ciencias Aplicadas, Lima 15023, Peru
{u201510525, u201510661, pcinecar, claudia.leon}@upc.edu.pe

[2] Direccion de Investigacion, Universidad Peruana de Ciencias Aplicadas, Lima 15023, Peru
carlos.raymundo@upc.edu.pe

Abstract. Several studies indicate the importance of proper design and of the correct use of ergonomic methodologies to assess and validate them. However, these studies focus on the validation of the methodology and not the process, ranging from the development of proposals to the validation of the process. This study deals with a workstation redesign model covering varied aspects of the process, from design proposals to design validation, by using observational ergonomic techniques such as REBA, RULA, and NIOSH and engineering tools such as a material handling system, methods study, systematic questioning technique, and motion economy. A 44.42% decline in MSDs was achieved in spreading and overlocking jobs and in the roll supply process, as well as a decrease in absenteeism by 849 h, along with savings of PEN 5096 per year.

Keywords: Musculoskeletal diseases · Ergonomic · Textile · SMEs

1 Introduction

In Peru, the garment industry is one of the main non extractive productive activities, with a contribution of 1.3% to the GDP and 8.9% to manufacturing production; it generates employment that accounts for 8.9% of the economically active population [1].

In 2017, 15.8% of the accidents reported were musculoskeletal disorders (MSD) involving general morbidity [2]; workers are vulnerable to developing such disorders because of repetitive and sedentary activities, poor posture, and postural loads, which are problems existing in an inadequate work environment [3]. Redesigning workstations helps to reduce risks and MSDs, as well as comparisons using observational methods to certify improvements [4]. In some garment factories, the costs associated with MSDs increased to 40% of the total cost.

This study proposes a method to reduce MSDs and absenteeism by using ergonomic techniques; further, it proposes assessing the situation in different scenarios,

D. Russo et al. (Eds.): IHSI 2021, AISC 1322, pp. 823–829, 2021.
https://doi.org/10.1007/978-3-030-68017-6_122

implementing the proposals developed with engineering tools, and finally validating the improvements achieved.

2 State of the Art

2.1 Station Design Methods in the Textile Sector

Several initiatives have been implemented by the textile industry to redesign workstations based on ergonomic criteria, with the aim of improving working environment conditions [5]. Such designs seek to reduce the risks of MSDs by eliminating factors such as machine tilt, optimal distance, height, and load on the job tasks [6].

Studies demonstrate that a good ergonomic design is essential to improve working conditions and that it contributes to posture correction and prevention of MSDs, resulting in a significant boost in productivity. A correct layout also helps to eliminate unnecessary movements and postures at workstations [7].

2.2 Ergonomic Techniques

The results of ergonomic assessments give rise to a number of ideas for improving workstations to reduce MSD risk factors, such as the implementation of furniture designs, ergonomic interventions, and the distribution of postural loads [8, 9].

Observational techniques such as RULA, REBA, and NIOSH are useful to assess the current situation of a given position or task to identify inefficient designs and reduce MSDs [6, 10]. The results of ergonomic assessments of both upper limbs and full body with RULA and REBA tools indicate higher incidences of musculoskeletal symptoms in knees, feet, thighs, and legs [9, 10].

2.3 Engineering Tools

The methods study is a key tool in the redesigning of job positions with a sequence of activities because it provides guidelines for starting, modifying, or eliminating processes; its proper use results in a more efficient flow of activities and a new working method [11].

In addition, there are tools that are used to delimit workstations based on the subject's own specifications called "motion economy." They are intended to identify suitable areas in the workstation to avoid improper postures or overexertion [12, 13]. Anthropometry allows to define the principle of redesign where the station adjusts to the subject and not the other way around, resulting in the perfect adjustment for each job position [14].

3 Contribution

3.1 Basis

The redesign of job positions considers aspects such as layout, change of machinery, and work method [6]. For a design to be ergonomically appropriate, it is necessary to identify factors that are directly related to the human factor through the use of ergonomic techniques [9].

3.2 Proposed Method

Phase 1 involves the ergonomic assessment of a specific task using a technique that matches the characteristics of the task. Phase 2 comprises the design of the method, location, machines or furniture based on engineering tools and establishing visual controls and procedures. Finally, phase 3 seeks to implement the proposals defined in phase 2.

Figure 1 shows the process as a cycle of continuous improvement; after its implementation, there is a reassessment of ergonomics to demonstrate that the proposals deliver on reducing ergonomic risks, obtaining a lower score than the current situation.

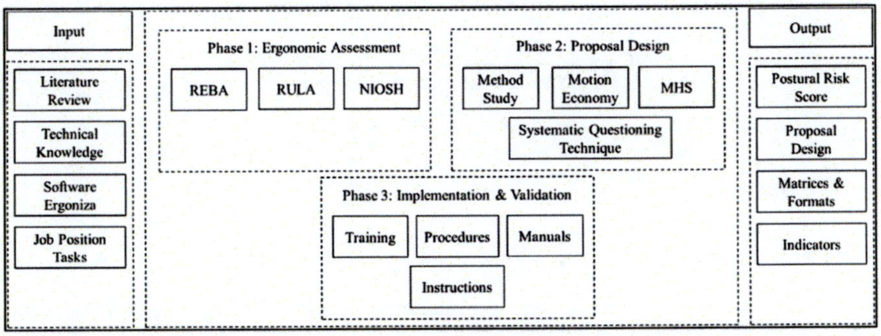

Fig. 1. Proposed method

3.3 Method Components

The development of the described phases of the proposed method results in the following components.

Results Comparison Table. It is generated in phase 1, comparing the results of the initial assessment and the proposal for improvement.

Training Record. Executed during the implementation and validation phase. Personnel attendance to training sessions is verified.

Daily Attendance. The attendance of collaborators was controlled to identify any improvements in absenteeism with respect to the initial situation.

3.4 Indicators

Components serve as an information storage source for the development of indicators as shown below:

MSD Reduction. It shows any reductions in risks after comparing the results of the ergonomic assessments located in the comparison table.

The initial and final scores are the results that generate the assessment of the initial and proposed situation, respectively. The reduction percentage is the comparison of both results.

$$RE = \frac{\text{Current Score} - \text{Final Score}}{\text{Current Score}} \times 100\% \tag{1}$$

Reduction of Absenteeism in the Workplace. This is the reduction percentage with respect to employee absenteeism.

$$A = \frac{\text{Final Period of Absenteeism}}{\text{Initial Period of Absenteeism}} \times 100\% \tag{2}$$

Compliance. It demonstrates the percentage of attendance to training sessions. The indicator is used to control the percentage of collaborators who attend training sessions.

$$C = \frac{\text{Attendances}}{\text{Training Sessions Held}} \times 100\% \tag{3}$$

4 Validation

4.1 Case Study

The workstation design method was implemented in a Peruvian garment SME with an average income of PEN 500,000.0/year, serving customers in Germany, England, and the United States. It has a total of 20 employees. It has areas for the storage of raw materials and finished products, cutting–spreading, sewing and overlocking, and finishing. Production is focused on the manufacture of polo shirts, trousers, and shirts for men and women.

4.2 Validation Development

In phase 1, the initial diagnosis of workstations was carried out, identifying ergonomic problems in the areas due to various root causes classified as postural load, material handling, and poor posture. Data were analyzed using ERGONIZA software, evaluating the postural risks listed in Table 1. Fabric spreading, cutting, and positioning tasks were assessed with REBA, while RULA was used for overlocking and NIOSH for roll supply.

Table 1. Initial Diagnosis

Task	Current score	Observation technique
Spreading	9.00	REBA
Cutting	9.00	REBA
Overlocking	5.00	RULA
Roll Supply	1.36	NIOSH
Fabric Positioning	12.00	REBA

For phase 2, tools such as methods study and systematic questioning technique were used to improve ergonomically improper processes to find the procedures with the least postural risk. The principles of motion economy were applied, delimiting the work areas to identify the workspace and material distribution. Finally, the criteria of the material handling system were used, establishing methods for transporting, holding, and distributing heavy objects. Table 2 summarizes the proposals developed.

Table 2. Design proposal

Task	Proposal	Support
Spreading	Method of spreading	Simultaneous spreading between two persons
Cutting	Use of end-cutting machine	Replaces manual cutting by and operator
Overlocking	Ergonomic chair method or overlocking	Ergonomic chair under specific conditions. Method or overlocking using motion economy
Roll Supply	Method of supply	Roll supply with two persons applying MHS
Fabric Positioning	Method of positioning	Fabric placed over the table

For phase 3, the results of the first ergonomic assessment were compared with the results obtained after the implementation of proposals and assessed with the same observational techniques. The results are shown in Table 3.

Table 3. Comparison of postural risks before and after the implementation of proposals

Task	Current score	Improvement score	Improvement percentage
Spreading	9.00	4.00	55.50%
Cutting	9.00	3.00	66.67%
Overlocking	5.00	3.00	40.00%
Roll Supply	1.36	1.19	12.50%
Fabric Positioning	12.00	3.00	75.00%

Finally, the percentage of reduction of the causes found in the initial diagnosis was calculated, with a total decrease of 44.42% as shown in Table 4.

Table 4. MSD Reduction

Task	Reduction (%)
Spreading	14.96%
Cutting	5.13%
Overlocking	20.00%
Roll Supply	1.44%
Fabric Positioning	2.89%
Total	44.42%

5 Conclusions

A redesign of the workstation using ergonomic techniques and engineering tools to develop proposals reduces MSDs by 44.42% and absence hours from 1912 to 1062, with a total of 849 h per year.

The percentages of ergonomic risk were reduced, the highest reduction being in the new method of fabric positioning, which is considered the most risky activity solved, showing a total reduction of 75%.

The continuous improvement cycle ensures the reduction of MSDs continuously because it must show that the developed proposals work to be able to proceed to the next process.

Engineering tools such as methods study supplemented by ergonomic assessments focus on a complete study of workstations; therefore, a workstation design can be made of methods, machine, or furniture changes.

The proposed method for redesigning workstations may be replicated in other companies with similar specifications or in those that can incorporate a system of continuous improvement.

References

1. de Producción, M.: Industria Textil y Confecciones (2017). https://ogeiee.produce.gob.pe/images/oee/docTrab_Textil.pdf. Accessed 05 Oct 2019
2. Seguro Social de Salud
3. De la Cruz, N., Viza, G.: Factores de riesgos ergonómicos que inciden en la salud de los trabajadores del área de producción de la Empresa Andes Yarn S.A.C., Arequipa – 2016. Universidad Nacional San Agustín (2017)
4. Shah, Z., Amjad, A., Ashraf, M., Mushtaq, F., Sheikh, I.: Ergonomic Risk Factors for Workers in Garments Manufacturing - A Case Study from Pakistan
5. Colim, A., Carneiro, P., Costa, N., Arezes, P., Sousa, N.: Ergonomic assessment and workstation design in a furniture manufacturing industry - a case study. In: Studies in Systems, Decision and Control, vol. 202, pp. 409–417 (2019)

6. Tondre, S., Deshmukh, T.: Guidelines to sewing machine workstation design for improving working posture of sewing operator. Int. J. Ind. Ergon. **71**, 37–46 (2019)
7. Van Niekerk, S., Louw, Q., Hillier, S.: The effectiveness of a chair intervention in the workplace to reduce musculoskeletal symptoms. A systematic review. **6**(1), 145 (2012)
8. Sakth, T., Reyapaul, R., Mathiyazhagan, K.: Evaluation of ergonomic working conditions among standing sewing machine operators in Sri Lanka. Int. J. Ind. Ergon. **70**, 70–83 (2019)
9. Van, L., Chaiear, N., Sumananont, C., Kannarath, C.: Prevalence of musculoskeletal symptoms among garment workers in Kandal province. Cambodia. J. Occup. Health **58**(1), 107–117 (2016)
10. Sanjog, J., Baruah, R., Patel, T., Karmakar, S.: Redesign of work-accessories towards minimizing awkward posture and reduction of work cycle elements in an Indian shop-floor workstation. In: Advances in Intelligent Systems and Computing, vol. 485, pp. 219–232 (2016)
11. Shin, W., Park, M.: Ergonomic interventions for prevention of work-related musculoskeletal disorders in a small manufacturing assembly line. Int. J. Occup. Saf. Ergon. **25**(1), 110–122 (2019)
12. Mahdavi, N., Motamedzade, M., Jamshidi, A., Darvishi, E., Moghimbeygi, A., Moghadam, R.: Upper trapezius fatigue in carpet weaving: The impact of a repetitive task cycle. Int. J. Occup. Saf. Ergon. **24**(1), 41–51 (2018)
13. Castellucci, H.: Anthropometric characteristics of Chilean workers for ergonomic and design purposes. Ergonomics **62**(3), 459–474 (2019)
14. Veisi, H., Choobineh, A., Ghaem, H., Shafiee, Z.: The effect of hand tools' handle shape on upper extremity comfort and postural discomfort among hand-woven shoemaking workers. Int. J. Ind. Ergon. **74** (2019)

Check for updates

Design of a Semi-automatic Machine for Processing Ecological and Antibacterial Bricks to Save Concrete Based on Polyethylene and Copper Terephthalate Residues in Huancayo

Antony Ibarra[1], Grimaldo Quispe[2], and Carlos Raymundo[3(✉)]

[1] Facultad de Ingenieria, Universidad Continental, Huancayo, Peru
70236442@continental.edu.pe

[2] Facultad de Ingenieria, Universidad Nacional Autonoma Altoandina de Tarma, Tarma, Peru
gquispe@unaat.edu.pe

[3] Direccion de Investigacion, Universidad Peruana de Ciencias Aplicadas, Lima 15023, Peru
carlos.raymundo@upc.edu.pe

Abstract. The brick-making activity that is commonly carried out manually and that requires a large amount of physical effort and that has problems within the design of brick-making such as the model and drying, which is why the development of Ecological bricks which is today a new improvement to reuse plastics. The problem of manufacturing time for ecological bricks, as well as their quality, can be overcome with the elaboration of a machine and the design of the machine to process bricks based on PET-COPPER is proposed. To be applied by construction companies or artisan companies, to improve the production of a new model of mold that can withstand the compaction and heating process. The development of the machine will help to generate bricks based on PET to minimize contamination and copper, which as a material is an anti-pathogenic agent to prevent virus infections.

Keywords: Design · Copper anti pathogen · Block machine · PET

1 Introduction

Pollution from the use of plastic is in excess due to its consumption, therefore, plastic will increase, reducing the amount of fish in the ocean by 2050. These residues harm more than 600 marine species. The vast majority of fish will be affected by ingestion, become stuck and in danger of extinction, which reflects 15% of these species [1] It could be said that "In Peru, on average around 29 kilos a year are used by each citizen of these types of PET plastic. Today they generate 890 tons of plastics per day within Lima and the Callao area, which is representing 45% of this PET-type waste at the national level [2] On the other hand, the brick-making activity that people commonly think of is carried out manually and that requires a large amount of physical effort and

D. Russo et al. (Eds.): IHSI 2021, AISC 1322, pp. 830–835, 2021.
https://doi.org/10.1007/978-3-030-68017-6_123

that has problems within brick-making such as the model and drying, which is why, Nace the elaboration of ecological bricks that is today a new improvement to reuse plastics. The problem of ecological brick manufacturing time, as well as its quality can be overcome with the elaboration of a machine and the design of the machine to process bricks based on PET is proposed. To be applied by construction companies or artisan companies, to improve the production of a new model [3].

In a new plastic economy, plastic never becomes waste or pollution. Actions are required to achieve this vision and create a circular economy for plastic that innovates to ensure that the plastics we need are reusable, recyclable or compostable [4].

The design of the semi-automatic machine for processing bricks based on polyethylene and copper terephthalate will help pollution and avoid virus contagion, making the production of ecological bricks profitable, help to practice more on recycling and be responsible for the pollution generated by society, the environmental impact and the dangers they may represent.

2 Theoretical Framework

The raw material supply system that the operator is responsible for making the mixture in ideal proportions that can be developed by a mixing machine to then pass to an inlet or hopper by manual means.

In the compacting system, the operator ensures that the raw material reaches the conduit through the hopper then takes care of the manipulation of the machine to effect a force by means of a lever system.

The product extraction system, here the operator is in charge of removing the finished brick manually using gloves to avoid contact with the skin.

VDI 2221 Standard. The method used is the design structure according to the VDI 2221 standard that proposes a methodology. Four stages based on seven steps. definition and clarification of the task, functional structure, solution principles, modular structure, preliminary design, final design and detailed design [5].

Task:

- Stage 1: Clarify and define the objective DETAIL EACH STAGE IN DETAIL
- Stage 2: Determine functions and their structure
- Stage 3: Look for principles of the Sun.
- Stage 4: Divide feasible modules
- Stage 5: Develop the key modules
- Stage 6: design
- Stage 7: provide production and operation instructions

Finite Element Analysis. The finite element analysis gives us an optimal technique to analyze engineering designs, as a first point it creates a geometric model, for which the general model is divided into small elements that are connected by nodes. Meshing is the division of the tiny parts, thus the element analysis programs consider the model as a network of interconnected elements [6].

SOLIDWORKS Simulation. Currently, for a static, frequency and thermal study simulation we have SOLIDWORKS simulation software. In the field of engineering we can appreciate the problems that can be solved with scientific methods that present benefits, but certain limitations. In this sense, the simulation with SOLIDWORKS aims to serve as an introduction to the reader in computer-assisted simulation in mechanical and structural design engineering, starting from the essential fundamentals of the finite element method.

Design Analysis System. Design analysis system that offers simulation solutions for linear and nonlinear static, frequency, buckling, thermal, fatigue, pressure vessel, drop, linear and nonlinear dynamics, and optimization analyzes. It also allows you to solve problems intuitively while you design, as it presents accurate and fast problem solving programs (solvers). This saves time and effort in the search for optimal design results; in this way, the product's time to market is reduced [7].

3 State of the Art

(See Table 1).

Table 1. Contributions and your article results

Author - Design	Contribution
KRUY SHOTEA (2010): Development of crusher and mixer that make bricks. It develops finite element analysis on critical parts such as its crusher shaft, backing plate tool crusher, Plate tool crusher, Round tool crusher [8]	Through the simulation of the Abaque CAE software, which compound comes out successfully since I consider the ergonomic parameters, safety factors of the parts and the calculations for the design
KOLAWOLE (2013): In his scientific article that designs a manual clay brick mold, he mentions that his machine can produce bricks from the mild steel that he uses as a material, therefore, they are suitable for the use that tries to improve the production of the clay type of 215 × 102.5 × 65 mm [9]	The mold is four compartment with 216 × 103 × 66 mm Bearing resistance to buckling failures
DIEGO MISE (2014): Designs his machine to create building blocks mentions that in terms of complex elements he performs a simulation of his hopper with a steel material [10]	Hopper with an ASTM A-36 steel material in which results are obtained from the thickness of said material.
RICARDO GARCÍA LEÓN (2015): In the structural analysis of a pressing machine to produce bricks, the article read mentioned that by means of the SOLIDWORKS software they carried out the analysis in parts of their pressing machine [11]	Analysis of the VON MISES tension, which is evidence that there will be no failure since the elastic limit of the material in this case is AISI 1020 is 205 Mpa

4 Contribution

(See Fig. 1).

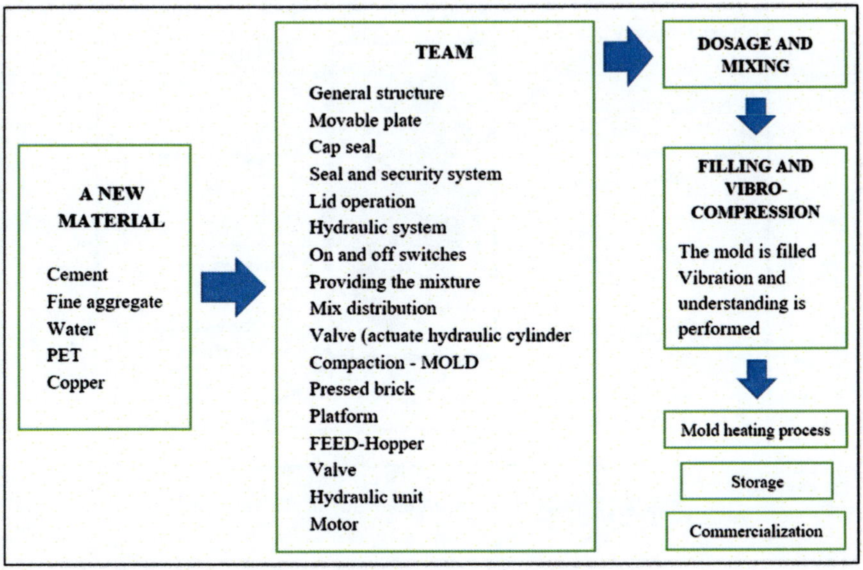

Fig. 1. Design of the investigation

5 Validation

5.1 Validation Design

It was designed in SOLIDWORKS SOFTWARE from data obtained by previous design calculations, the general model and each of its components (Table 2).

Table 2. Pieces to design

Piece	Description	Quantity
Structure	ASTM A36 Steel	1
Mold	DIN Steel (For hot work) 1.2343 (X38CrMo V5-3)	1
Spring base	Vibration base	1
Piston press type pressure base	Compaction	1
Mixer	PET-Copper-Cement-Aggregate Mix	1

5.2 Test Instances

Details of the main elements of the machine are show in Fig. 2.

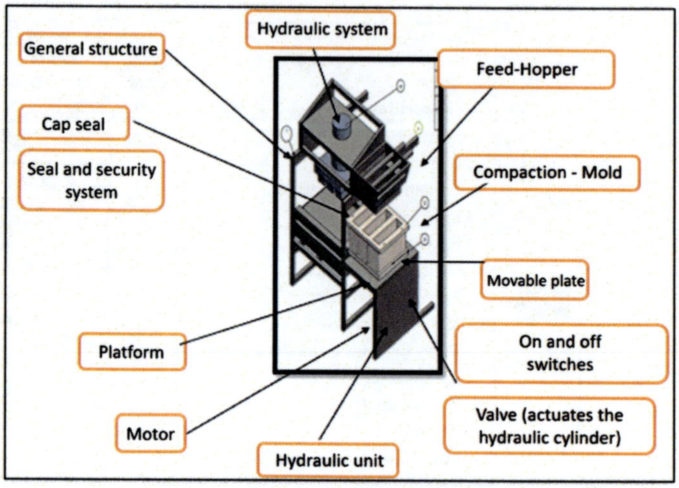

Fig. 2. Details of the main elements of the machine

5.3 Results-Comparison with Other Contributions

The mold can produce 3 COPPER-PET bricks, in its process the mixture is added in its housing, which when compacting continues the heating for its final process.

As an optimal design criterion, the piece must not exceed the elastic limit of the construction material, which in this case is 2120 N/mm^2 (MPa).

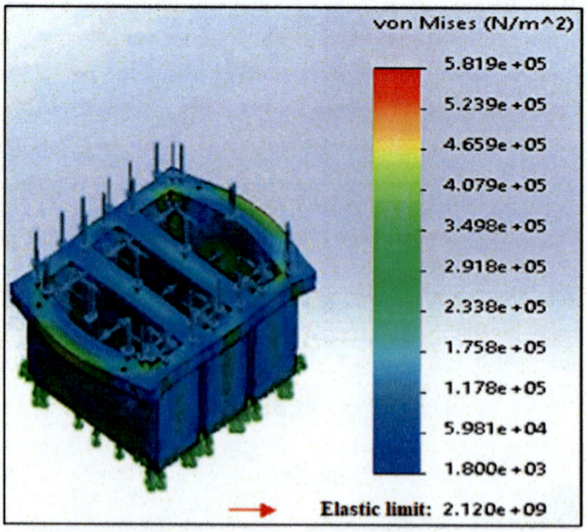

Fig. 3. Mold simulation static analysis - stresses

The simulation approves the correct type of material to use for a more effective compaction, it also develops a safety factor of 2.3 and the processes that we improve to save the cost of the machine (Fig. 3).

6 Conclusions

The design of the machine is S /. 10,000 compared to the value granted by national industries or models is S /. 40,000, thus generating a saving of 25%.

It is inexpensive and minimizes plastic contamination, helps combat new threats from viruses such as COVID.

The design has improvements in terms of mechanical characteristics in its design and adaptation of materials, a piston-type compacting system was adapted to help avoid air bubbles.

Mechanisms were implemented to accelerate the production of Copper-PET bricks with the induction process to accelerate drying in a shorter time.

When carrying out the simulation, we solved the compacting process of the new COPPER-PET mixture, a safety factor of 2.3 was also obtained.

References

1. Macarthur, E.: Ellen MacArthur Foundation. Obtenido de EL PERÚ PRIMERO, 21 October 2016. www.newplasticseconomy.org
2. UNEP: 24 de JUNIO de 2018. http://www.minam.gob.pe. Obtenido de Campaña Menos PlásticoMásVida. http://www.minam.gob.pe/menos-plastico-mas-vida/campana-menosplasticomasvida/#_ftnref1
3. Amariz, A.D.: Diseño y Fabricación de Ladrillo Reutilizando Materiales a Base de PET. Inge Cu Educosta, pp. 76–80 (2014)
4. Bullon, A.Y.: ladrillos ecológicos con material PET. lima: conocimiento (2017)
5. Sianipar, C.: Metodologías del diseño, 6 April 2016. https://estudiantesavp.ucontinental.edu.pe/pluginfile.php/1272663/mod_resource/content/1/METODOLOGIA%20DE%20DISE%C3%91O%20PARA%20MAQUINAS.pdf
6. Mirlisenna, G.: Contenido de simulación computacional en américa. Método de elementos finitos, 5 March 2018. https://www.esss.co/es/blog/metodo-de-los-elementos-finitos-que-es/
7. Quintela, A.: Tipos de estudio que ofrece SOLIDWORKS simulation, 11 April 2018. https://easyworks.es/tipos-de-estudios-que-ofrece-solidworks-simulation/
8. Sothea, K.: Development of a crush and mix machine for composite brick fabrication. In: AMPT, pp. 1437–1442 (2010)
9. Kolawole, S.K., Odusote, J.K.: Research article design, fabrication and performance evaluation of a manual clay brick moulding machine. J. Eng. Sci. Technol. Rev. **6**(1), 17–20 (2013)
10. Mise, D.: Diseño y construcción de una máquina semiautomática para la fabricación. Maskana **5**, 13–24 (2014)
11. Cusihuallpa: Diseño de una vibro-compactadora para la producción de prefabricados de concreto. Cusco: cusco (2015)

Gaze Based Interaction for Object Classification in Reconnaissance Missions Using Highly Automated Platforms

Joscha Wasser[1(✉)], Marten Bloch[1], Konrad Bielecki[1], Daria Vorst[1], Daniel Lopez[1], Marcel Baltzer[1], and Frank Flemisch[1,2]

[1] Fraunhofer FKIE, Fraunhoferstraße 20, 53343 Wachtberg, Germany
{joscha.wasser,marten.bloch,konrad.bielecki, daria.vorst,daniel.lopez,marcel.baltzer, frank.flemisch}@fkie.fraunhofer.de

[2] RWTH Aachen University, Bergdriesch 27, 52062 Aachen, Germany
f.flemisch@iaw.rwth-aachen.de

Abstract. A gaze based interface was used as the basis for a interaction system supporting a remote operator of a ground reconnaissance robot when classifying objects previously identified by a software. Using the gaze point as a cursor, a pre-selection of the object in the sightline is made, allowing the user to effect the symbol via a haptic interface. An extension of the system enables the user to manually re-adjust the pre-selection if highlights are overlapping. The results from a first exploration were encouraging; the system was easily understood and used but does require further testing with scenarios that are more complex.

Keywords: Gaze based interaction · Human-systems · Integration · Remote operation · Human systems exploration

1 Introduction

In the military domain, there is a growing trend towards more capable digital systems creating an increasing amount of data and information. To reap the benefits of this trend, the integration into existing processes or the development of new processes and systems needs to be carefully managed. Within these increasingly complex automated systems, a balance needs to be found between the capabilities and requirements of both human and software to create a holistic and beneficial system [7]. Especially in the context of cooperative automation and vehicle guidance, a well-designed interplay between the system and the human operator is crucial to the success and safety of a mission. This project therefore explored a gaze based human machine interaction method to be used in military reconnaissance missions in the near future. These are likely to involve remote operators controlling a platoon of highly automated vehicles traveling through unknown terrain supported by a system capable of identifying relevant objects. The proposed system then requires the human operator to make the final classification of the identified object, as this is a critical decision potentially marking an object for combative engagement. In order to make the classification as intuitive and efficient as possible for the operator, this project explored a gaze-based interaction with

D. Russo et al. (Eds.): IHSI 2021, AISC 1322, pp. 836–842, 2021.
https://doi.org/10.1007/978-3-030-68017-6_124

a software providing the operator with a video feed of a scene in which an identified object is highlighted.

An extension of a point-and-click methodology was devised, allowing the operator to use his gaze to direct a cursor to the object in question, which is subsequently marked as pre-selected. If the operator intends to classify this object, it can be done immediately by pressing a colour-coded button. If a re-adjustment is needed, the operator can navigate the scene with an arrow pad, which allows him to switch to the next closest identified object. The reconnaissance and classification of objects in the operational area are key [2]; knowing where other friendly forces are operating is one of the most fundamental aspects of situational awareness in military operations, e.g. to prevent fratricide and to coordinate manoeuvres.

Eye Tracking. The technology that makes gaze based interaction possible, eye tracking, has recently experienced big advancements and the introduction into consumer grade electronics as a component of head mounted displays (HMDs) [4, 5]. The availability of inexpensive systems to track the eye movement of a user and the growing number of possibilities to integrate them into other systems, has allowed a number of new interaction concepts to emerge. Typically, the eye movement of a user is tracked with a system integrated into a HMD, AR glasses or in this project, a system that is mounted in front of the user underneath a screen array. This system projects a pattern of near infrared light onto the eyes of the user, which is then captured together with the eyes by a high-resolution camera. These images are processed into a gaze point that can be displayed in the view field (digital or as an overlay in a real scene) [15].

Cooperation and Interaction. A number of different interaction methods for efficient human-machine cooperation, based on eye tracking were already explored, e.g. using the tracking data to move a cursor across a screen to select an item by dwelling on it for a predefined time [1, 11]. These methods however, are often slow and imprecise, which leads to false positives and thus, are often combined with a secondary interaction, either on screen in the form of a virtual button or with another haptic element such as a button [10, 14]. This two-step interaction results in a higher successful task completion rate, which for this application is key, but does also slow down the process. Therefore, one of the interaction methods for this project is an extension of the "MAGIC" method [16], where the user initially directs the cursor towards the object of interest using his gaze and then is able to do a fine adjustment of the cursor with an additional hardware interface.

2 Methodology

The project followed the bHSI approach; brainstorm, literature research, modelling, system exploration, implementation and test & evaluation [8]. The explorative aspect provided the freedom required within the concept and design stage to discover novel concepts. By describing the meta-system, the stakeholders, design and use space, a sphere within which the new concept can be developed is created. In the iterative process, the researchers start with the "bigger picture" and add detail with each iteration and trial [6–8].

3 System Concept

To process objects highlighted in a field of view, to analyse and classify them, a system needs to combine an intuitive and focused display of information and a naturalistic interaction [9]. Content in the view field of the user should therefore be easily understood and provide only the information required to gain situational awareness without losing or obscuring any additional information. A naturalistic approach means that previous training and procedures, as well as normal human behaviour, should also be considered. Livingstone et al. [9] highlight that in typical military operations, the user's hands are often occupied with existing equipment and processes; therefore, this project explores eye tracking as a means of interacting with the system. The operator is monitoring the camera view of a UGV and within which identified objects are marked as "Unknown". The operator can then move the gaze tracker, which functions akin to a cursor, onto a highlight, pre-selecting it for an interaction. To avoid the gaze tracker moving off the target, the tracker can be "locked" into position with a button. The object can then be classified using colour-coded buttons on a controller; Yellow – Unknown, Red – Hostile, Green – Neutral, Blue – Friendly (Fig. 1). A reclassification is done by first selecting "Unknown" before classifying it as a different group. The aim is to avoid introducing a delete button as this could be misinterpreted as deleting the highlight and erasing the information from the system. If there are overlapping highlights in the view field, such as a group of vehicles, a precise pre-selection using only gaze may not be possible. Thus, a fine adjustment method can be triggered by locking the view cursor on one object and then using the arrow keypad on the controller to jump to the nearest object to classify it.

Fig. 1. Point of view in the simulation with the gaze cursor focused on an object

Symbol Design. The symbol concept (Fig. 2) proposes that highlights are displayed in two stages, depending on the distance of the vehicle to the identified object, to avoid cluttering the view. Identified objects are marked with the NATO Symbol for

"Unknown" [12] with an additional line below, lifting the symbol above the object, to avoid occluding it. If the operator moves closer to the object, within 100 m of it, a bracket appears around the identified object, providing a more detailed marker. Initially it will be the standard size but will grow when approached, always framing the object. When the operator moves his view tracker across the view field and onto one of the highlighted objects, the software will pre-select this object.

This is indicated by increasing the line thickness of the bracket as well as decreasing the opacity of the unselected markers. When pre-selected, the operator can then classify the object, which changes the highlight according to the selected classification. Should the operator move away from the object again, the frame decreases again based on the object size and, beyond 100 m, only the symbol and line remain. If there are a number of identically classified objects, which are close together, the software will group them under one frame and ultimately under one symbol with a small marker indicating the number of objects grouped together.

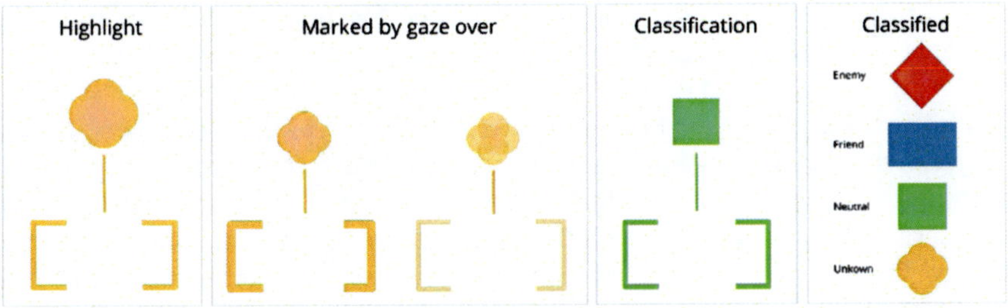

Fig. 2. Initial symbol design scheme

Simulation. The system consists of three components: the user, the hardware setup and the simulation software (Fig. 3). The hardware consists of a 3-monitor system with a field of view of 120°, a Tobii Eye Tracker 4C [15] and a gamepad. The eye tracker, mounted beneath the central display, enables the system to monitor the user's gaze as well as its head position; the latter however is not required for this setup. The software simulation, i.e. the vehicle and the map, were modelled in Unity3D from Unity Technologies. The UGV is based on the HUSKY from Clearpath Robotics [3] and is able to freely move across the rural environment, which is loosely based on the robotic proving ground at Fraunhofer FKIE. The eye tracker provides Unity with gaze data, which is then converted to a cursor displayed on the screen, giving the user the position of his gaze as a feedback. The single eye tracker limits the interaction to the central screen; an extension is prepared for the two outer screens. Properties such as the characteristics of the eye tracker cursor (i.e. size, shape, and colour) and the properties of the AR highlights (i.e. activation radius and size of the selectable area), are controlled and recorded through a separate UI, to facilitate the exploratory approach. The success rate and the time elapsed are also recorded within the system.

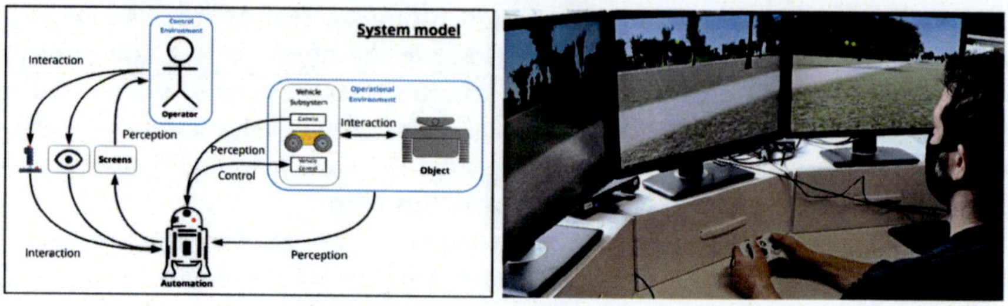

Fig. 3. System model (left) & participant in the simulator setup (right)

4 Evaluation

The preliminary study was conducted with six participants, starting with an interview before informing them that they will be filmed to document their comments during the trial. A familiarisation with the system followed in a practice stage whilst receiving instructions on how to categorise a marked object. Once confident with the system, participants were asked to complete the first scenario with one object. The UGV, on which the camera is located through which the participants are viewing the scene, was manually controlled using the theatre method [13] by one of the researchers, to create the impression of a highly automated vehicle. A second scenario followed containing four objects. The simulation was then paused to discuss the symbol and interaction design, such as the distance at which the markings become visible, when the frame is added and the size of the area in which the software registers the view cursor. Afterwards, another interview was conducted to gather information about the general impression and experiences with the system as well as potential improvements.

Results. Due to the qualitative nature of the explorative study and the small number of the participants, the interpretation of the results focussed on identifying the key issues with, as well as the potential of the proposed system. Only one eye-tracker was available which limited the interaction to the central monitor and even though the "automation" tried to keep the highlights on the central screen, this was not always possible. Consequently, all participants reported that this impeded their ability to interact with the highlights. Some also disliked the extra step to reclassify objects via the "Unknown" symbol and suggested that under stressful conditions this might lead to errors. Positive feedback was given by all regarding the initial object marker describing it as helpful and unobtrusive. Another positively evaluated system property was the, apparently oftentimes surprising, high accuracy of the eye tracking. Consequently, some users relied solely on it to mark and classify objects, without using any fine adjustment. They generally acknowledged however that the fine adjustment would be useful in situations that are more complex. The colour coding of the buttons was found to be helpful, even though most looked down to identify the correct button, something they felt they would do less with time. The interpretation of the frame was discussed most, as some falsely concluded that an interaction with an object is only possible once the frame appears and or thought of it as a marker for the borders within which they had to place the cursor. The distance at which the frame appears (100 m) was often stated be to too short, especially when there was a line of sight to the object, but no frame present yet. The frame's transparency and the change when hovering over it as well as the line thickness were criticised by some who stated that these properties make it too intrusive and clutters the view, with one participant suggesting the frames could disappear completely once classified.

5 Conclusion

The overall positive feedback indicates that gaze based classification is a method worth further investigation, as it is easily used and understood. The highlights were quickly spotted and the symbol changes were considered helpful, as were the fine-adjustment options. These however, require further testing as most participants did not use them in the scenarios, as they were able to complete them by using their gaze input alone. Based on the feedback, a more detailed classification option needs to be included into the system, which will add another layer of complexity.

Outlook. A second iteration is currently being developed. As the simple scenario and the high usability of the eye tracking limited the feedback regarding the manual interaction methods, this will be amended by presenting a more complex scenario in order to collect valid usability data.

References

1. Bednarik, R., Gowases, T., Tukiainen, M.: Gaze interaction enhances problem solving: effects of dwell-time based, gaze-augmented, and mouse interaction on problem-solving strategies and user experience. J. Eye Mov. Res. **3**, 1–10 (2009)
2. Bolstad, C.A., Endsley, M.R.: Tools for supporting team SA and collaboration in army operations. Technical report, SA Technologies (2002)
3. Clearpath (2020). Husky: https://clearpathrobotics.com/husky-unmanned-ground-vehicle-robot/. Accessed 01 Aug 2020
4. Duchowski, A.T.: Gaze-based interaction: a 30 year retrospective. Comput. Graph. **73**, 59–69 (2018). https://doi.org/10.1016/j.cag.2018.04.002. ISSN 0097-8493
5. Duchowski, A.T., Medlin, E., Cournia, N., Gramopadhye, A.K., Melloy, B., Nair, S.: 3D eye movement analysis for VR visual inspection training. In: ETRA 2002: Proceedings of the 2002 Symposium on Eye Tracking Research & Applications, New Orleans, LA, pp. 103–110, 155. ACM (2002)
6. Flemisch, F., Baltzer, M.C.A., Sadeghian, S., Meyer, R., Hernández, D.L., Baier, R.: Making HSI more intelligent: human systems exploration versus experiment for the integration of humans and artificial cognitive systems. In: Advances in Intelligent Systems and Computing, pp. 563–569. Springer International Publishing (2019)
7. Flemisch, F., Semling, C., Heesen, M., Meier, S., Baltzer, M., Krasni, A.: Towards a balanced Human Systems Integration beyond time and space: exploroscopes for a structured exploration of human-machine design spaces. In: NATO STO Meetings Proceedings of the STO Human Factors and Medicine Panel Symposium: Beyond Time and Space (2013)

8. Kelsch, J., Heesen, M., Löper, C., Flemisch, F.: Balancierte Gestaltung kooperativer multimodaler Bedienkonzepte für Fahrerassistenz und Automation: H-Mode beim Annähern, Notbremsen, Ausweichen. 8. Berliner Werkstatt MMS: 7.-9, Berlin, Deutschland, October 2009
9. Livingstone, M.A., Ai, Z., Karsch, K., Gibson, G.O.: User interface design for military AR applications. Virtual Reality **15**, 175–184 (2011). https://doi.org/10.1007/s10055-010-0179-1
10. Lutteroth, C., et al.: Gaze vs. mouse: a fast and accurate gaze-only click alternative (2015)
11. Miniotas, D., et al.: Eye gaze interaction with expanding targets (2004)
12. NATO Joint Military Symbology APP-6(C) 2011. https://www.awl.edu.pl/images/en/APP_6_C.pdf. Accessed 15 Sept 2020
13. Schieben, A., Heesen, M., Schindler, J., Kelsch, J., Flemisch, F.: The theater-system technique: agile designing and testing of system behavior and interaction, applied to highly automated vehicles, pp. 43–46 (2009). https://doi.org/10.1145/1620509.1620517
14. Stellmach, S., Dachselt, R.: Look & touch: gaze-supported target acquisition (2012)
15. Tobii (2020). This is eye tracking. https://www.tobii.com/group/about/this-is-eye-tracking/
16. Zhai, S.: What's in the eyes for attentive input Commun. ACM **46**(34), 39 (2003)

Comparative Analysis of Emotional Design Based on Web of Science and CNKI

Yongkang Chen, Yuhao Jiang, Xingting Wu(✉), and Renke He

School of Design, Hunan University, ChangSha, China
344600768@qq.com

Abstract. Through the analysis and comparison of the literature of WOS and CNKI data-bases, based on objective data and intuitive graph analysis, the current hot issues and development trends of emotional design research in China and the world are analyzed and summarized. Comprehensive use of Cite Space and VOSviewer, with scientific bibliometric methods, draw a knowledge map from the co-citation status of documents and keyword clustering. The results show that there are many research topics in emotional design, mainly in the fields of Kansei engineering, emotional computing, human-computer interaction, and experience de-sign. Traditional culture and rural feelings are the unique research direction of Chinese emotional design research; the cooperative relationship between research scholars and research institutions is relatively small, and it is the current research status that needs to be strengthened.

Keywords: Emotional design · Bibliometrics · Vosviewer

1 Introduction

Although China is a rising country in the research of emotional design, its attention has gradually increased in academic circles. In terms of statistical analysis of WOS literature, China ranks first in the world in terms of the amount of articles published in the research field of emotional design. Therefore, it is of great significance to clarify the current research status, knowledge evolution and frontier trends of Chinese emotional design, and to compare and analyze it with worldwide research. In order to obtain a more comprehensive exploration of the current research status, research hotspots and development trends of motional design, the WOS database and CNKI are used as data sources in the research, and VOSviewer, CiteSpace and other software are used to visualize the knowledge structure of existing documents, and then Provide reference and overall overview for scholars in this field [1].

2 Research Design

The research literature in this article comes from CNKI (China Knowledge Network) and WoS (Web of Science) core collection databases. To ensure the quality of the literature, CNKI data sources are limited to Peking University core, CSSCI, and CSCD source journals. Subject search is carried out with "emotional design" as the keyword.

D. Russo et al. (Eds.): IHSI 2021, AISC 1322, pp. 843–848, 2021.
https://doi.org/10.1007/978-3-030-68017-6_125

The search time is not limited, and a total of 275 articles are obtained; in the core of WOS Search in the collection database, the search strategy is set to TS = ("emotional design") or TS = ("affective design"), in order to collect all relevant articles, the search time span is set to all years; to avoid the loss of interdisciplinary literature, The source of the literature was not condensed; the retrieved literature was exported as a plain text file in the format of "full records and cited references", and interference articles and duplicate data were eliminated. Finally, a total of 775 pieces of literature were obtained.

This paper adopts scientific bibliometric methods and comprehensively uses software such as VOSviewer and CiteSpace to obtain more comprehensive data indicators. The academic map (or scientific knowledge map) is supported by the theories and methods of scientometrics and bibliometrics, through the visualization technology of scientific and technological texts to show the emerging research direction of knowledge distribution, structure, and relevance of research topics. Pritchard proposed in 1969. VOSviewer was developed in 2009 by van Eck and Waltman of the Leiden University Science and Technology Research Center in the Netherlands. It has a powerful user graphical interface and mapping visualization functions [2]. CiteSpace was developed by Professor Chaomei Chen from the School of Information Science and Technology, Drexel University, and has been widely used in literature citation network analysis in recent years [3].

3 Measurement Results and Analysis of Emotional Design Literature

3.1 The Basic Characteristics of Emotional Design Research

Figure 1 is obtained by field extraction of the data after WOS and CNKI are cleaned and deduplicated. It can be seen from the situation of WOS publications that since 1996–2020, the number of annual articles published has increased from 2 in 1996 to 66 in 2019, which peaked in 2017, with a total of 78 articles published annually, and the overall document output has been Upward trend. The research on emotional design in China started relatively late, and the first article was published in 2003. Compared with the number of publications of WOS, the overall fluctuations in the output of Chinese literature are positively correlated with WOS. From one content in 2003 to a peak of 34 in 2019, there is an overall upward trend. It can be seen from the output of CNKI and WOS that the research theme of emotional design is constantly developing, and has been the focus of attention of scholars for the past five years.

3.2 Representative Author

The statistics of high-yield authors in the CNKI and WOS literature (take five) are formed in Table 1. Through the comparative analysis of the co-occurrence knowledge map of high-yield authors, we can find that there are not many high-yield authors and the cooperative research between scholars is not close, and Sporadic cooperation between institutions is the current status of emotional design research.

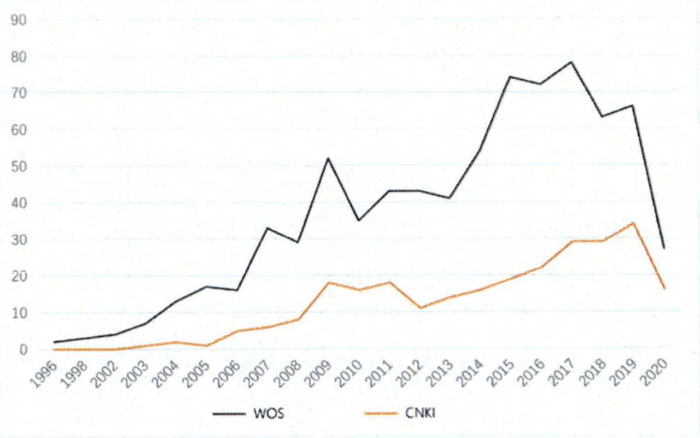

Fig. 1. Distribution of emotional design literature annual distribution

Table 1. CNKI and prolific authors in the WOS literature.

CNKI	The frequency	norm. Citations	WOS	The frequency	norm. Citations
Yi zhang	4	3.2521	kwong, c. k	8	4.5258
Xinxin sun	3	0.0437	yang, chih-chieh	7	3.2084
Bochu xu	3	0.0437	zhou, feng	7	1.4102
Lin wang	3	2.1128	chu, chih-hsing	6	0.3683
Jingyan tan	3	0.2096	guo, fu	6	3.1572

3.3 Emotional Design Research Keyword Time Zone View

Through the literature analysis of CNKI, the time zone view of Chinese emotional design research is obtained (Fig. 2). It can be found that after 2014, the research content has gradually shifted to the frontiers of artificial intelligence, human-computer interaction, experience economy, three-level interaction, and emotional models, and the virtual and sustainable research direction has shifted. Especially the research on traditional culture and rural feelings is the biggest difference between Chinese emotional design research and international research objects.

Through the literature analysis of the WOS core database, the knowledge map is obtained (Fig. 3), N = 61, E = 88, Density = 0.0481. The time period is from 1996 to 2020. Through the time zone chart, it can be found that the research frontiers of WOS literature are emotion recognition, loyalty, facial recognition, emotion recognition, human-computer interaction, eye tracking, gestures and other multi-channel interaction methods and emotional design Research.

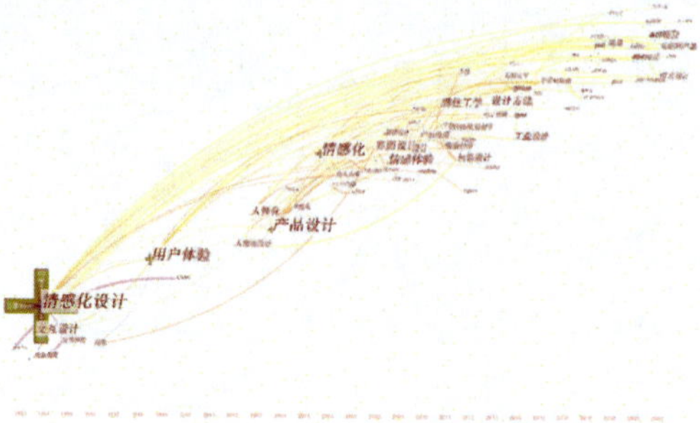

Fig. 2. CNKI time zone

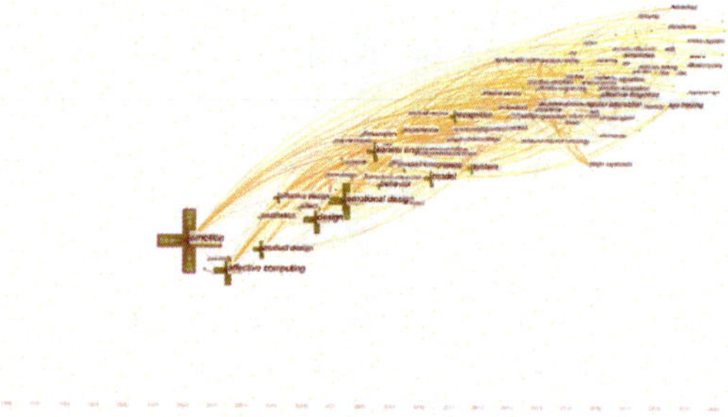

Fig. 3. WOS time zone

3.4 Research Hotspots of Emotional Design

Due to the different co-occurrence intensity of themes, clustering phenomenon will be formed and specific theme clustering will be formed [4]. From the analysis results, the hot research topics of CNKI emotional design can be divided into 5 categories, cluster #1 – inclusive design, mainly including design concept, barrier-free design, the elderly, emotion, function, and other keywords. Cluster #2 – Kansei Engineering, mainly including Kansei Engineering, Kansei intention, product form, user demand, emotional expression, etc. As a representative method in emotional design, Kansei Engineering can guide and evaluate the design process by acquiring users' perceptual images of product attributes [5]. Cluster #3 – Interaction design, mainly including user experience, interaction design, interface design, human-computer interaction, unconscious behavior, etc. Interaction design is the design field that defines and designs the behavior of an artificial system. It defines the content and structure of the communication between two or more interacting individuals and makes them cooperate with each other [6]. Cluster #4 – Service and Experience Design mainly includes service design, experience design, features, emotional needs, costume design, scene, etc. Cluster #5 – Space Design, mainly studying the role of emotional design in public space (Fig. 4).

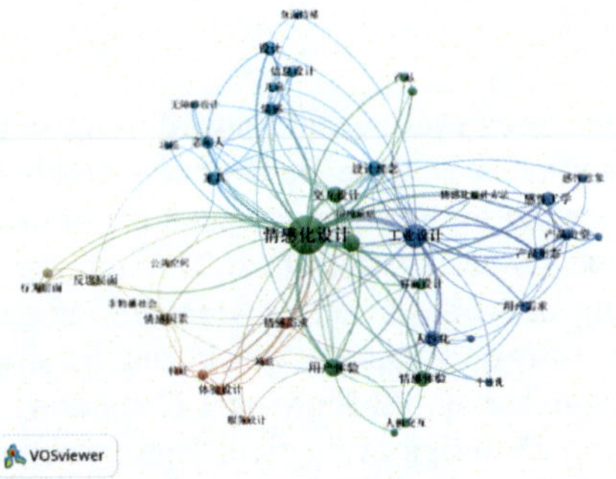

Fig. 4. CNKI keyword co-occurrence clustering network

The same method was used to analyze the WOS core collection database literature, and five clusters were also obtained (Fig. 5). Cluster #1 – Kansei Engineering, which mainly includes Kansei Engineering, product design, Form design, user-centered design, neural network, and decision-making, etc. Cluster #2 – Affective computing mainly includes affective computing, emotion recognition, emotional expression, Data Mining, machine learning, Physiological signals, etc. "Affective computing" is defined as "the calculation of factors related to emotion, triggered by emotion or capable of affecting emotion [9]", and its main researches are acquisition of emotional signals, recognition of the emotional state, understanding of emotional information and expression of emotion [10]. Cluster #3 - Affective and experience design, mainly including experience design, Interaction design, emotional Response, affective haptics, and Depression. Cluster #4 – Man-machine multi-channel interaction, mainly including human Robot interaction, recognition, facial expression, motion, eye Tracking and agents, etc. Cluster #5 – Design evaluation consists of 18 members, including model, Usability, satisfaction, aesthetics, quality and consequence, etc.

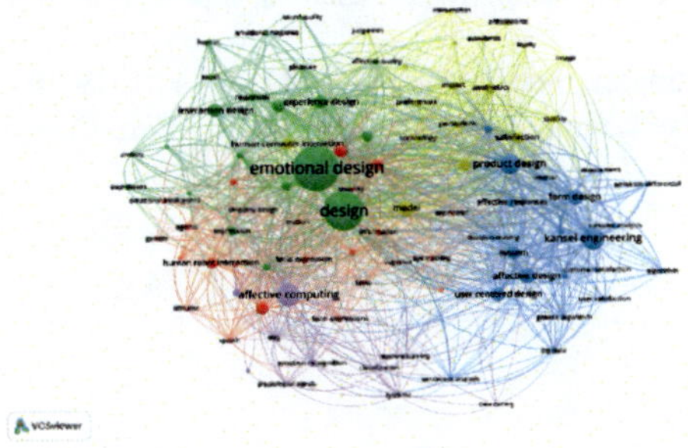

Fig. 5. WOS keyword co-occurrence clustering network

4 Conclusion

Emotional design research shows an overall upward trend in the output of time series papers. In the development process, there are not many highly productive scholars, and the research cooperation is mostly based on cooperation within the institution, which is scattered and needs to be further strengthened. Through the synthesis of CNKI and WOS keyword clustering network, it can be seen that emotional design is comprehensively diversified in research content, which can be divided into 5 categories, namely #1 Perceptual Engineering, #2 Emotional Computing, #3 Experience Design, #4 Interaction Design And #5 design evaluation. These clusters together constitute the research hotspots and themes of emotional design, including not only the shape and color characteristics of tangible products but also the intangible interaction, experience, and service process. Through the keyword time zone diagram, it can be concluded that the future research trend is toward the virtualization of research objects, the comprehensive research perspective, and the way of multi-channel natural interaction.

References

1. Luo, S., Pan, Y.: Research progress of perceptual image theory, technology and application in product design. J. Mech. Eng. **03**, 8–13 (2007)
2. van Eck, N.J., Waltman, L.: Software survey: VOSviewer, a computer program for bibliometric mapping. Scientometrics **84**(2), 523–538 (2010)
3. Chen, C.: CiteSpace II: detecting and visualizing emerging trends and transient patterns in scientific literature. J. Am. Soc. Inf. Sci. Technol. **57**(3), 359–377 (2006)
4. Li, H., et al.: Evolutionary features of academic articles co-keyword network and keywords co-occurrence network: based on two-mode affiliation network. Phys. A: Stat. Mech. Appl. **450**, 657–669 (2016)
5. Jiao, J., Zhang, Y., Helander, M.: A Kansei mining system for affective design. Expert Syst. Appl. **30**(4), 658–673 (2006)
6. Picard, R.W.: Affective computing. MIT Technical report, 1995(# 321) (2009). Accessed 29 Aug 2009
7. Russell, J.A.: A circumplex model of affect. p. 1161–1178 (1980)
8. Picard, R.W.: Affective computing: challenges. Int. J. Hum.-Comput. Stud. **59**(1–2), 55–64 (2003)

Design of a Ventilation System for the Uchucchacua-Buenaventura Oyón-Lima Mine

Luis Espinoza[1(✉)], Grimaldo Quispe[2], and Carlos Raymundo[3]

[1] Facultad de Ingenieria, Universidad Continental, Huancayo, Peru
45222415@continental.edu.pe
[2] Facultad de Ingenieria, Universidad Nacional Autonoma Altoandina de Tarma, Tarma, Peru
gquispe@unaat.edu.pe
[3] Direccion de Investigacion, Universidad Peruana de Ciencias Aplicadas, Lima 15023, Peru
carlos.raymundo@upc.edu.pe

Abstract. One of the very important activities in underground mining is ventilation, because air has to be supplied in most of the cases that is fresh and clean for the tasks in which they are in operation, taking into account the conditions that the product already has of the previous activities, such as: implementation of alternative methods such as the use of fans, extractors or work for ventilation purposes in order to achieve an adequate, healthy and comfortable environment for all employees who are in the mine; is that constant improvements are made to the "ventilation system." That is why this article aims to simulate a new ventilation system that enhances the current ventilation system of the "Uchucchacua" mine of the Buenaventura company located in the province of Oyón-Lima, taking into account, that sufficient air flow must be guaranteed for persons and equipment authorized to work on a watch.

Keywords: Ventilation system · Air flow · Ventilation system design

1 Introduction

The fundamental objective of an adequate ventilation system is to be able to achieve a work environment for the mine employees, as well as for the equipment; The latter is not only affected in its mechanical efficiency by the insufficient air flow, but also, causing the equipment to reach high temperatures that impair its operation and, on the other hand, when the air is insufficient, the combustion of the diesel fuel is not performed completely [1].

The problems that the new ventilation system proposal will solve must improve the supply of air by mechanically assisted ventilation to the mining operations in operation, such as the following: supply of oxygen to adequate standards, extract the toxic gases produced by drilling and blasting, attenuate the fine suspended dust particles found in mining operations, reduce the temperature for a suitable work environment, the latter

D. Russo et al. (Eds.): IHSI 2021, AISC 1322, pp. 849–855, 2021.
https://doi.org/10.1007/978-3-030-68017-6_126

for optimal equipment performance and finally to determine the optimal flow rates in the ventilation of underground mining operations [2].

This research work will be carried out, because there are cases of intoxication of the collaborators in the underground mining works and the decrease in the mechanical performance of the equipment due to overheating due to high temperatures, also due to the loss of pressure due to abandoned tasks that do not were suitably sealed and in order to achieve the reduction of costs by mechanically assisted ventilation [3].

It is convenient to carry out research work to be able to develop a proposal for a ventilation system that improves the environment of the work fronts, and thus achieve the reduction of accidents due to intoxication caused by the gases of the blasting and also the reduction of eventualities due to mechanical failure equipment due to temperature rise or insufficient air flow [4].

2 Theoretical Framework

2.1 Mining Ventilation

Mining ventilation is mentioned, "to the action that is carried out, to cool the environment with air, achieving a clean, safe place without the presence of toxic gases, all to achieve the optimal performance of workers and mechanical equipment in their activities. The ventilation system must meet and maintain the standards of permissible limits on exposure to chemical agents" [5].

Considering what is mentioned in the regulation of safety and occupational health in mining (DS N ° 023–2017-EM), the following is taken into account: "In all underground work, a circulation of clean and fresh air in quantity and quality will be maintained. Sufficient according to the number of workers, with the total of equipment with internal combustion engines, as well as for the dilution of gases that allow to have a minimum of nineteen point five percent (19.5%) in the work environment oxygen [5].

In Article 247 (Supreme Decree No. 023–2017-EM) "it also mentions. In the workplaces of mines located up to one thousand five hundred (1,500) m above sea level, the minimum amount of air required per man shall be three cubic meters per minute. At other altitudes the amount of air will be according to the following scale [5].

From 1,500 to 3,000 masl it will increase by 40% which will be equal to 4 m^3/min.

From 3,000 to 4,000 masl it will increase by 70% which will be equal to 5 m^3/min.

Above 4,000 masl it will increase by 100% which will be equal to 6 m^3/min.

For no reason "the speed of the air should be below 20 or higher than 250 in mining operations, which includes preparation and development. When bulk blasting agents such as ANFO are used, the air must be less than 25 as dictated by the Mining Occupational Health and Safety Regulations" [5].

2.2 Underground Ventilation

Process by which air is injected into the mining workings in order to refresh the interior of the mine with adequate air to achieve a workable and very safe atmosphere, ventilation will consist of a system that allows air to reach properly to all tasks. That is why it is required that the mine have tasks exclusively to carry out this action, that later magas or pipes will be used to get to each place that is in operation or in worst cases use small diameter pipes that are used for the pierce [6, 7].

3 Materials and Methods

3.1 Study Site

The present work was carried out in the Uchucchacua mine, it is located in the province of Oyón, department of Lima, at an altitude between 4,000 and 5,000 m above sea level. Its geographical coordinates are the following: South latitude 10° 26′34″ and west longitude 76° 50 '56″. It currently has a treatment capacity of 2,200 TCS / day. This production is supported by the Carmen and Socorro mines, and the Casualidad, Lucrecia and Huantajalla mines are prospective areas.

3.2 Experimental Design

This research work "Feasibility study to develop a proposal to improve the ventilation system of the Uchucchacua Buenaventura Oyón-Lima 2020 mine", is a descriptive research since it will collect information on the current ventilation system and based on the data collected there will be a new ventilation system.

The population was taken from Level 4150 of the Uchucchacua Buenaventura Oyon-Lima mine, for which 9 monitoring points were chosen for each work front, for a total of 54 monitoring points in a total of 4 levels and 2 sublevels. Which does 6 jobs.

The monitoring points that will be taken for the sample will be simple random in each work front, then for each work front a total of 8 monitoring points will be taken (simple random).

The technique used to collect the data was used electronic instruments, establishing 46 monitoring points at level 4450 of the Uchucchacua Buenaventura Oyon – Lima mine, the data processing was carried out as follows.

Measurements were made at 86 monitoring sites at level 4150 of the Uchucchacua Buenaventura Oyon – Lima mine. For the measurement of gases, the instrument "Detector Multigas Altair 5x" was used.

3.3 Analysis of Data

The data obtained in the monitoring of the mine work indicates the following. Current flow, actual requirement and how much is required for said ventilation system (Table 1).

Table 1. Flow balance of the Uchucchacua mine

Air intake			Air vent		
Work	Q m^3//min	Q CFM	Work	Q m^3//min	Q CFM
Pithead 4450	4985.24	176052.1	Fireplace 4452	4862.03	171701
			Variation	123.21	4351.12
			Request	5380.20	190000
				394.96	13947.88

4 Results

For the evaluation of the ventilation system, a balance was made with air intakes and outlets that involves all the tasks, this research will focus on these levels with the support of the Uchucchacua Mine workers, for which the measurements were carried out.

The average temperatures for the 48 points sampled at 4 different times as shown in the previous tables, indicate that for the day watch the temperature is higher, this is typical of the time (Fig. 1).

According to the regulation of safety and occupational health in mining (D.S. 024 - 2017 - EM), it is established that the maximum permissible limits for oxygen are: Minimum 19.5% and maximum 22.5%.

This would indicate that we are in the standards for all 4 averages taken at different times (Fig. 2).

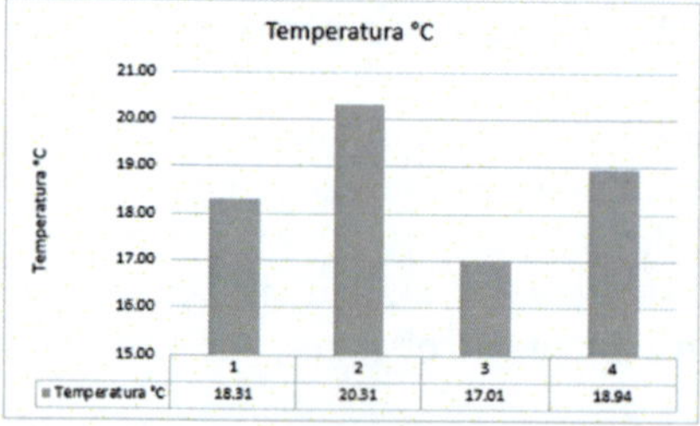

Fig. 1. The average temperature taken at 4 different times for both watches.

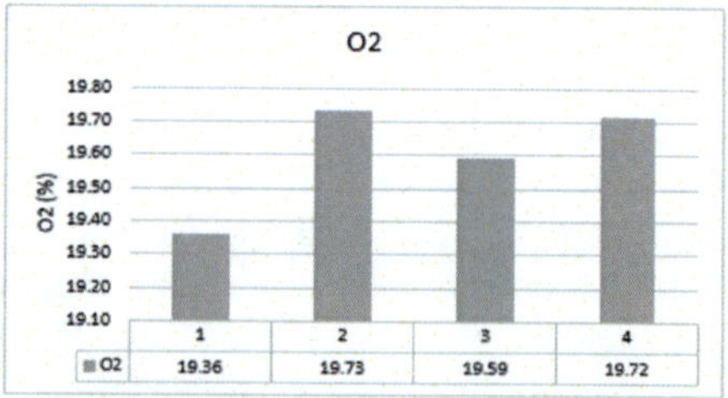

Fig. 2. Bar graph of the average Oxygen for the two watches

"According to the regulations on occupational health and safety in mining" (MINEM 2017), it is established that the maximum permissible limits for carbon dioxide are:

A maximum of 5000 PPM or 0.5%. For this case, it is shown that the CO2 content is less than the allowable limit (Fig. 3).

"According to the regulations on occupational health and safety in mining" (MINEM 2017), it is established that the maximum permissible limits for carbon monoxide are: Minimum 0.005% and maximum 0.0025% (Fig. 4).

In the same way, in the case of carbon monoxide it is among the permissible ranges.

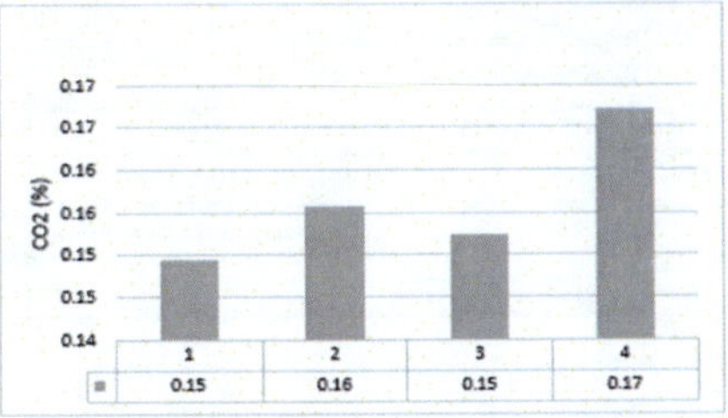

Fig. 3. Bar graph of CO2% in the mine

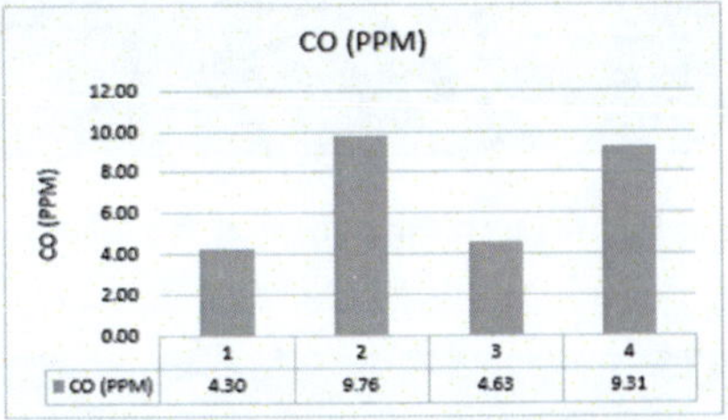

Fig. 4. Bar graph of CO content by PPM

"According to the regulations on occupational health and safety in mining" (MINEM 2017), it is established that the maximum permissible limits for nitrogen dioxide are: Minimum 0.0005% and Maximum 5 PPM.

In this case, nitrogen dioxide has already exceeded 5 PPM in almost all areas, which is why a redesign of the ventilation system is required (Fig. 5).

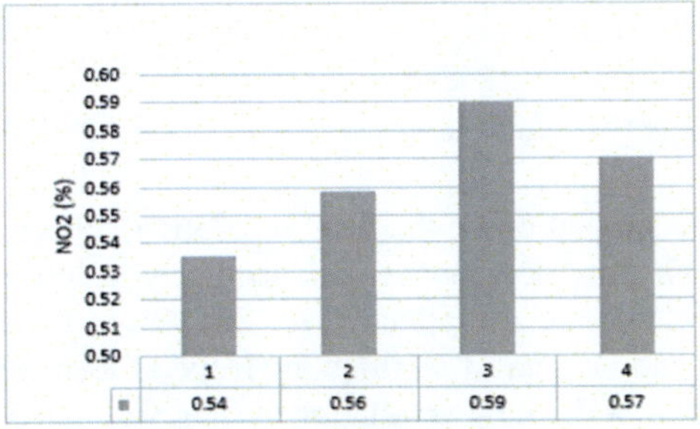

Fig. 5. Bar graph of NO2 (%)

After monitoring the "ventilation system" at the levels of the "Uchucchacua" mine, the air requirements for said levels were calculated and it is proposed to embed a fan with a capacity of 15,000 cfm in order to improve the ventilation system levels and improve productivity by minimizing the presence of toxic gases on the work fronts.

Supported by the VENTSIM ™ software, we will select which type of fans are optimal, in order to meet the air demand. The fan selection took into account the following: the current fan capacity and areas where there is no adequate air flow (Fig. 6, Fig. 7, Fig. 8).

Fig. 6. Increase in flow at the mine

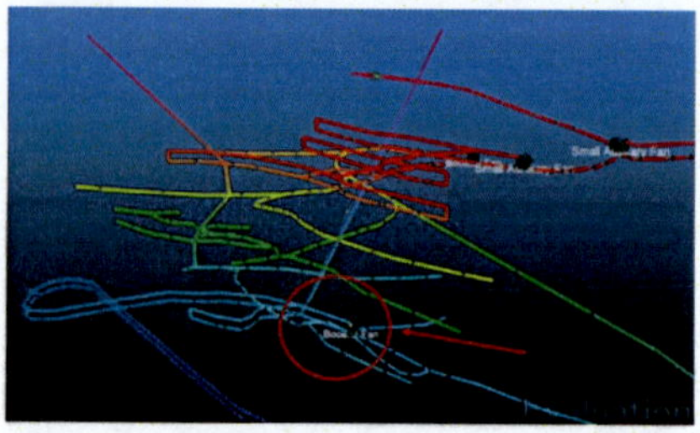

Fig. 7. Change of ventilation sleeve at level 4352

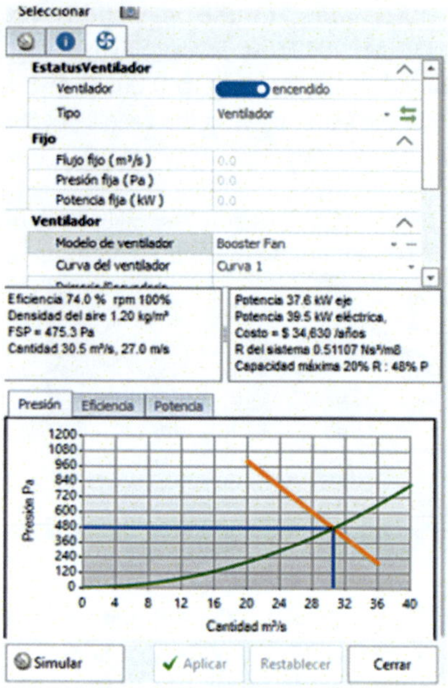

Fig. 8. Fan specifications

5 Conclusions

When carrying out the simulation of the new ventilation system in the VentSim Lite, the points where the air flow does not reach adequately at lower levels such as level 4352 of the Uchucchacua mine are reflected, for this reason the air requirement in the mouth was increased and change of ventilation sleeve at level 4352.

When the simulation of the new ventilation system was carried out in the VentSim Lite of the Uchucchacua mine, the feasibility of its realizations was confirmed due to the high levels of nitrogen dioxide that exceeded the permissible limits, this product of a bad ventilation system and insufficient flow.

The flow required for the Uchucchacua mine showed us that 15,000 cfm was lacking to meet the minimum allowable standards for the number of authorized people, quantity of authorized diesel equipment, consumption of wood, loss due to friction and the use of explosives.

Once the analysis had been carried out, the new ventilation system in the VentSim Lite software began to improve and increase the flow rate at the entrance to the manhole to 190,000 cfm, at a capacity of 78% of the fan, which allowed us to satisfy the requirement of all levels of the "Uchucchacua Buenaventura Oyón - Lima" mine.

References

1. Baltzar, R.: Influencia De Los Ventiladores En El Sistema Ventilación De La Mina Kazán De La Compañía Minera Paraíso (2016)
2. Condori Condori, V.H.: Optimización del sistema de ventilación con el software VENTSIM VISUAL en la U.E.A. Heraldos Negros de la compañía minera San Valentín S. A. – Huancavelica. Universidad Nacional Del Altiplano (2016)
3. Córdoba Quiceno, C.C., Molina Escobar, J.M.: Characterization of ventilation systems in underground mines. Boletín de Ciencias de La Tierra **29**, 73–86 (2011)
4. Delgado Ramos, G.C.: Ecología Política de la Minería en América Latina Aspectos socioeconómicos, legales y ambientales de la mega minería (2010)
5. MINEM. Reglamento de Seguridad y Salud Ocupacional en Minería D.S. No 023–2017-EM. EL Peruano, p. 22 (2017). http://www.minem.gob.pe/_legislacionM.php?idSector=1&idLegislacion=10221
6. Ibanez Zamudio, V.: Diseño del Sistema de Ventilación en el NV4050 Veta Don Ernesto Unidad Minera el Porvenir-MILPO (Universidad Continental, vol. 53) (2019). https://doi.org/10.1017/CBO9781107415324.004
7. Kerguelen Bendeck, J.L., González Martínez, V.A., Jiménez Builes, J.A.: Techniques for determining parameters in the preparation of a fan circuit in underground coal mining using programming structured. Boletín de Ciencias de La Tierra **33**, 155–162 (2013)

Designed for Designer: An Online Co-design Workshop

Di Zhu, Anni Li, Nan Wang, Jiazhen Wu, and Wei Liu(✉)

Beijing Normal University, Xinjiekouwai Street 19, Haidian District, Beijing 100875, China
{di.zhu,wei.liu}@bnu.edu.cn, {anni.li,nanwang, wujiazhen}@mail.bnu.edu.cn

Abstract. One fundamental challenge for designing the human-machine interface of the future is to understand users' needs. This online form provides the possibility to expand the scope of recruitment. The setting of workshops can maximize the enthusiasm of designers, deeply explore their computer usage behavior and habits, and verify previous user research results. This paper reports on a study iterating the use of an online co-design workshop that recruited designers (n = 30) for computer and laptop design. The online workshop has three phases: exploring, brainstorming, and designing with useful toolkits. Finally, the research team collected 88 interaction design solutions to improve the designer's user experience when using the computer. Future work will look in-depth, reorganize, and redesign these solutions and test them in a real working context.

Keywords: Co-design · Online collaboration · Designer computers

1 Introduction

Co-design is one of the user-centered design methods. It invites end-users to attend research and design sessions, helps designers or researchers to understand contexts and experience of them [1]. Co-design often open to external users to stay open to innovative ideas, and it also provides a possibility to feed forward to other design activities [2]. Besides, co-design will activate users' tacit to shape the design and come up with the most useful solutions [3]. These solutions and findings are more persuasive to stakeholders, and the team will also get a better understand of their stakeholders [4]. When it comes to designers, they have abilities to analyze their needs and contexts, and they can visualize their ideas by drawing, building quick prototypes. At the same time, they require higher quality design than other user groups [5]. In other words, this form of collaboration allows end-users to the participant and build a common language across the boundaries between manufacturers and users.

The internet creates possibilities of innovative online collaborations, and the design team can design the user interface remotely by using Figma. It can allow users in a different location to join in the same virtual conference room. However, it needs support from tools or platforms that can provide collaboration service, with audio, video, whiteboards [6]. The team should maintain trust and share information

D. Russo et al. (Eds.): IHSI 2021, AISC 1322, pp. 856–861, 2021.
https://doi.org/10.1007/978-3-030-68017-6_127

effectively [7], it has more challenges when applying it online, even though participants are designers.

This research wants to enable designers to contribute and involve further in the design research process by launch an online design workshop so that the team can find it easier to recruit qualified designers to participate in design sessions. The workshop will provide useful toolkits such as the persona user journey map to help them understand the primary findings of user research so that they can follow the instruction to think deeply. Compared with traditional methods, these tools encourage and support participants to express themselves, including preference, feelings, and motivations [8]. Finally, they will propose several design solutions to the team.

2 Method

The online co-design workshop process research design is under the collaboration between JD (JD.com) and the User eXperience Program of Beijing Normal University. This project will study the designers' needs for designer's computers. The research process follows a user-centered design process and focuses on design activities [9]. And the objective of the design project is to develop an online co-design workshop process and toolkits to involve designers to design sessions. These design activities are based on previous user research results. The team provides a scope that stakeholders want to focus on but not limit their creativity. The project provides an empirical context for the research because it applied a user-centered design process for designer computer design research addressing online co-design problems while creating business value for the stakeholders [10]. Stakeholders will benefit from manufacturing, user data, brand competency, and get first-hand data from users [11]. The project also offers a useful context for studying how combining a user-centered design process, and toolkits can contribute to addressing online co-design challenges.

At first, the team review works of literature in which users participate in a design session, and select the most user-friendly and stable online meeting software. Based on research findings, the team builds a workshop process and several useful toolkits [12]. There are three toolkits in Shimo spreadsheet, user journey, How-To brainstorming, C-box. Then the research team launches two pilot workshops to test the process and toolkits. The whole process is recorded. After iterating these tools and arrangement, the team conducts four formal tests. The team used WeChat public account that recruited eighty-six designers and selected thirty designers to participate in the in-depth discussion and brainstorming on designer computers. There are 16 males and 14 females, including 13 designers and 17 design students. Their age is between 22 to 28. And the research team collect behavior data and their feedback on the workshop. After analyzing these data, the workshop process was iterated.

3 Findings

This section reports the empirical findings on how designers experienced the online co-design workshop. The results are structured according to the user-centered design process, exploring, brainstorming, and designing.

3.1 Online Co-design Workshop Process

As Fig. 1 shows, there are three phases of the online co-design workshop: exploring, brainstorming, designing. At the exploring stage, designers understand user research results, and explore their needs, design challenges in a specific context. Finally, they will select a few most essential design problems. These problems will be addressed with inspiration toolkits at the brainstorming phase. In the last step, the group will visualize their design solutions with drawings or collages.

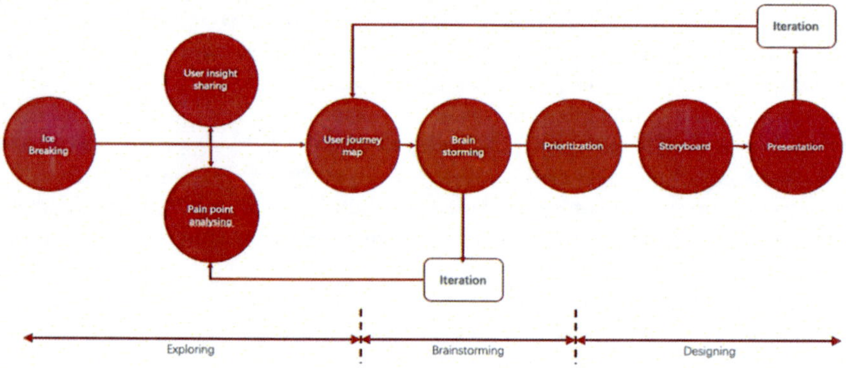

Fig. 1. Online co-design workshop process.

3.2 Exploring

Before the workshop, the research team adapts user research results to a persona and scenario cards that include users' needs, transcription, and a picture to show the context. At first, the facilitator will present these findings to designers. After that, there will be a journey map in Shimo, an online collaborative tool to show every detail in a particular scenario, including user behavior, need, and design objectives. The research team divided user behavior when they are using a computer at night into five stages, before working, working for a long time, getting dark, getting late at night, before sleeping. Each step has detail behavior such as designers may need to switch the computer to night mode, transfer to work in bed. Designers are encouraged to add new behavior and needs in the journey map, and the facilitator will help them to express their feedback on user research findings. It will be a group recorder to take down notes that every team member could see it on their screen. Designers provide brand new perspectives for the whole team. After the designers expand and improve the content, the facilitator should review the original design objectives carefully. It will affect the next brainstorming phase. The design objectives should be clear and specific.

3.3 Brainstorming

Based on selected design objectives, the research team copy all goals to the next form and use the How-to method to support idea generation. The objectives are How-To questions that will guide designers to think aloud, analyze what problems they are facing and try to brainstorm more solutions. As Table 1 shows, The facilitator instructs the team to select issues to cover different stages of working at night. When the team screen How-To questions, they want to focus on in the next step.

Table 1. How-to brainstorming toolkit example.

How to determine that the brightness?	How to let the computer automatically evaluate your eye health?	How to enter the silent mode naturally?	How to enter the silent mode naturally How to ensure the user's status?	How to give an early warning of immobility?
Judging the user's adaptation to light by eye movement data, such as pupil size	Detection of pupil state by biometric technology	Add a key for mute, and the state after pressing the key can be set	Feedback misoperation information in time through system notification	Connect the smart bracelet and judge with health data
Generate health analysis report	The condition of detecting the constantly enlarged text or image	Remind users of the possible impact of the volume of current software/hardware operations	Use the rest time to let users play small games and judge the user's state according to the game results	Linkage with smart homes and feedback through other devices
…	…	…	…	…

The facilitator will send the technical toolkit to the group chat. This set of cards has 12 pages in total, each of which shows the overview and application of the most advanced technology, providing creative inspiration for designers. The same number of rounds is brainstorming with selected objectives until every designer answers all purposes. Designers will face a new goal, and they can refer to the former ideas on each round. After 5 rounds of brainstorming, the team produced a total of 88 ideas. These ideas are mainly focusing on how to solve the problem of designer's eye health and how to carry out efficient design work in unconventional office places such as a bed.

3.4 Designing

There are a lot of ideas, so designers cluster the schemes which will be sorted by innovativeness since the research team wants to understand their criteria of innovation. Since these features are independent, the facilitator will guide them to link features to the design solutions. Finally, each group has a series design solution that can match to all design objectives in working at night context. One of the groups selected anthropomorphic animals linking all design schemes. The computer obtains the user's environmental information through geographic location information, photosensitive sensors, etc., and recommends users to choose appropriate brightness with the animation of a small animal avatar. After group discussion, the facilitator leads them reorganizing and using storyboards to improve and visualize the ideas (as Fig. 2 shows).

Fig. 2. Storyboard: design of interaction mode between open permission and selection mode

The key features and interactions should be present clearly by hand drawing or design software. A member draws the user's interaction interface when opening the camera and positioning authority, and the interaction mode of switching and selecting mode, as shown in Fig. 2. Through the division of labor and cooperation, the members complete the design work together and finally elaborated on the use process of the product in the form of oral presentation.

4 Conclusion

This paper focuses on combining the user-centered design process with co-design activities. Leading designers, the target users to participate in the design session according to user research findings. The research team designs an online co-design workshop process and toolkits to inspire participants to understand motivations and needs behind their behavior. The paper consequently shows how these approaches can be combined into practice through a designers' computer project collaboration with JD. While there are many methods and toolkits supporting practitioners design by themselves, this is not the case for tools for integrating users online. The process depicted aims at filling this gap. Mainly, one of the few tools that support practitioners in online co-design allows users to understand design background and existing user research findings fully. In the future, the research team will analyze user feedback and recorded videos to iterate the online co-design workshop process. Based on design schemes from the workshop, the team will redesign the schemes and build a high-fidelity prototype to test its' usability performance.

References

1. Marc, S., de Jenny, B., Lilliane, K., Froukje, S.V.: Co-design: practices, challenges and lessons learned. In: Proceedings of the 10th International Conference on Human Computer Interaction with Mobile Devices and Services (MobileHCI 2008), pp. 561–562 (2008)
2. Gabriela, A., Luigina, C., Simona, S., Deborah, R., Petrelli, D.: Co-design goes large. Interactions **26**(5), 58–63 (2019)
3. Katja, B., Andrea, B.C., Tuuli, M., Francesca, R.: Designed for co-designers. In: Proceedings of the Tenth Anniversary Conference on Participatory Design 2008 (PDC 2008), pp. 299–300 (2008)
4. Toni, R., Jesper, S.: Participatory Design: An Introduction. Handbook of Participatory Design (1st edn.) p. 18, Routledge (2013)
5. Kristin, C, Thecla, S.: Designing interaction for designers: defamiliarization in user's creative decision-making. In: Proceedings of the 9th ACM Conference on Creativity & Cognition (C&C 2013), pp. 300–303 (2013)
6. Ricarose, R., Yasmin, K., Deborah, F.: From tools to communities: designs to support online creative collaboration in scratch. In: Proceedings of the 11th International Conference on Interaction Design and Children (IDC 2012), pp. 220–223 (2012)
7. Luigina, C., Gabriela, A., Laura, M., Nick, D., Mark T.M., Dick, D., van Fiona, M.: Articulating co-design in museums: reflections on two participatory processes. In: Proceedings of the 19th ACM Conference on Computer-Supported Cooperative Work & Social Computing (CSCW 2016), pp. 13–25 (2016)
8. Juha, K., Anna, S.: Co-designing (with) organizations: human-centeredness, participation and embodiment in organizational development. In: Proceedings of the 2011 Conference on Designing Pleasurable Products and Interfaces (DPPI 2011), pp. 1–8 (2011)
9. Yu-Chang, H., Kerry, R., Lisa, D.: Empowering educators with Google's android app inventor: An online workshop in mobile app design. Br. J. Educ. Technol. **43**(1), E1–E5 (2012)
10. Baldassarre, B., Calabretta, G., Bocken, N.M.P., Jaskiewicz, T.: Bridging sustainable business model innovation and user-driven innovation: A process for sustainable value proposition design. J. Clean. Prod. **147**, 175–186 (2017)
11. Lowes, L., Robling, M.R., Bennert, K., Crawley, C., Hambly, H., Hawthorne, K., et al.: Involving lay and professional stakeholders in the development of a research intervention for the depicted study. Health Expect. **14**(3), 250–260 (2011)
12. Christos, F., Vassilis, K., Nikolaos, A.: Design of collaboration-support tools for group problem solving. In: Proceedings PC HCI, pp. 263–268 (2001)

Design of Equipment for a Forage and Agricultural Waste Processor to Improve Livestock Feed

Abelardo Limaymanta[1(✉)], Carlos Cantalicio[1], Grimaldo Quispe[2], and Carlos Raymundo[3]

[1] Facultad de Ingenieria, Universidad Continental, Huancayo, Peru
{73181430,47824246}@continental.edu.pe
[2] Facultad de Ingenieria, Universidad Nacional Autonoma Altoandina de Tarma, Tarma, Peru
gquispe@unaat.edu.pe
[3] Direccion de Investigacion, Universidad Peruana de Ciencias Aplicadas, Lima 15023, Peru
carlos.raymundo@upc.edu.pe

Abstract. The present study focus on designing better system for forage and agricultural waste processing machine to improve the quality of the final product, in addition to reducing the processing time of the balanced feed. As specific objectives are the following: Achieve the design of a forage chopping system that helps to optimize resources. Use current solutions to improve the processing time of forage and agricultural residues. Chop time improvement compared to traditional. The following results were obtained: The tests carried out on the machine were of quality, cut, length of forage and cut agricultural waste, in addition to the efficiency of the machine, achieving a production of 50 kg/10 min With these results It is concluded that the use of this machine reduces the processing time in addition to a reduction of losses of the forage and of the agricultural waste when being chopped.

Keywords: Cattle raising · Cutter · Forage · Machine · Improvement

1 Introduction

In Peru, agricultural activity, the producers of this activity constitute one of the pillars that with their knowledge, attitudes and practices on agricultural and livestock management contributes to the development of this economic activity, whose contribution to national production reaches 6, 0%, where the livestock activity has a greater influence in said sector with 73.1%, of which one of the most important pillars of the country's production. The agricultural producers, as natural persons, declared that they owned 4 million 993 thousand 685 head of cattle. This population of cattle is concentrated in the Sierra of the country, with 3 million 713 thousand 417 heads, representing 74.4% of the total [1, 2].

D. Russo et al. (Eds.): IHSI 2021, AISC 1322, pp. 862–868, 2021.
https://doi.org/10.1007/978-3-030-68017-6_128

But for the most part, the use of resources for feeding cattle are not efficient for a good use of the feed, it is there where a need arises to establish new alternatives for feeding cattle, as a way to also face the months of drought [3–5].

There is a great variety of forage chopping machines on the market, with which the aforementioned resources can be chopped, but they have a high economic cost in the local market [3–8]. The problem is that small producers do not have adequate knowledge of how to take advantage of resources efficiently for a correct use of food [2] that is why, when the cultivation process ends, few farmers use it as food for livestock having to cut them into dimensions of 5 to 15 cm [8]. This is chopped with a machete or sickle, so that the cattle can digest them and another part is burned or discarded [8, 9] (approx. 10% of the cultivated product), there is the need for how to process and use the maximum agricultural waste this happens to a lack of knowledge of the potential of the mentioned element if this element were used correctly they would obtain an extra economic income since the final product obtained after being chopped has many applications, among which is included the feeding of livestock cattle especially for fattening and milk production [6, 7]. With the advancement of technology, a number of machines have been created, but they are machines that are used to chop pastures, and in order to carry out this process, they require a tractor, this means that small farmers cannot acquire them. Due to the need to improve production and avoid the waste of corn stalks, it has been proposed to implement an efficient automatic machine capable of competing with existing machines on the market.

2 State of the Art

2.1 State of the Art

Adjustable Helix Angle Blade. The study was carried out by simulating the blades to be used, obtaining a maximum resistance to cutting alfalfa at a moisture content of 20% in the range of 7.9 to 16.5 mm N − 1 of the length of the fixed blade. If the force applied to the stationary knife unit is 16.5 mm N − 1, since the length of the fixed knife is 400 mm, the force against the blade is 6,600 N. According to Newton's third law, this force is equal to the amount of force applied to the edge of the cutting blades [10].

Forage Chopper Cylinder. The cutterhead is configured in such a way that when the movable helical blades are actuated, it generates the interaction of the fixed blade, ensuring that it reaches a scissor cut. On the other hand, mobile and fixed blades require greater rigidity in the cutting components, this reflects the fact that there is less damage to the cutting head [2].

Stationary Chopper Blower. This machine is characterized to chop dry material obtained from the harvest. These are called blowing machines since they have blades in the blade holder, which causes the chopped material to be transferred by the turning force towards the tube where it will be transported to later be taken to the place where the animals are fed [11].

2.2 Operation of the Machine

The purpose of the forage and agricultural waste chopper machine is the equitable chopping of the elements to be processed, this consists of 4 essential steps for proper operation, all part of the engine ignition that will be the one that will generate movement to the entire system, then of that a person will be in charge of the machine with which the forage chopping and agricultural waste or waste will be carried out, the staff will take charge of the continuous supply of the food to carry out the respective chopping, later the chopped food will be verified in case does not have uniformity will have to return to the mincer and later to storage.

3 Methodology

3.1 Calculations

Gear Ratio

$$i = \frac{n1}{n2} \tag{1}$$

$$i = \frac{d1}{d2} \tag{2}$$

Band Selection

$$Lt = 2E + \frac{\pi}{2}(D + d) + \frac{(D - d)^2}{4E} \tag{3}$$

Where:

Lt: belt length
D: diameter of the largest pulley
d: diameter of the smaller pulley
E: distance between centers.

Contact Arc

$$\beta = Arc(^\circ) = 180 - 57\frac{(D - d)}{E} \tag{4}$$

Belt Speed (Vt)

$$V_t = \frac{\pi \cdot d \cdot N}{60 \cdot 1000} \tag{5}$$

Axis Calculation

$$\sigma_a = K_f \frac{M_a c}{I} \qquad \sigma_m = K_f \frac{M_m c}{I} \tag{6}$$

$$\tau_a = K_{fs} \frac{T_a c}{J} \qquad \tau_m = K_{fs} \frac{T_m c}{J} \tag{7}$$

ED-ASME Elliptical

$$\frac{1}{n} = \frac{16}{\pi d^3}\left[4\left(\frac{K_f M_a}{S_e}\right)^2 + 3\left(\frac{K_{fs} T_a}{S_e}\right)^2 + 4\left(\frac{K_f M_m}{S_y}\right)^2 + 3\left(\frac{K_{fs} T_m}{S_y}\right)^2\right]^{1/2} \tag{8}$$

$$d = \left\{\frac{16n}{\pi}\left[4\left(\frac{K_f M_a}{S_e}\right)^2 + 3\left(\frac{K_{fs} T_a}{S_e}\right)^2 + 4\left(\frac{K_f M_m}{S_y}\right)^2 + 3\left(\frac{K_{fs} T_m}{S_y}\right)^2\right]^{1/2}\right\}^{1/3} \tag{9}$$

Bearing Selection

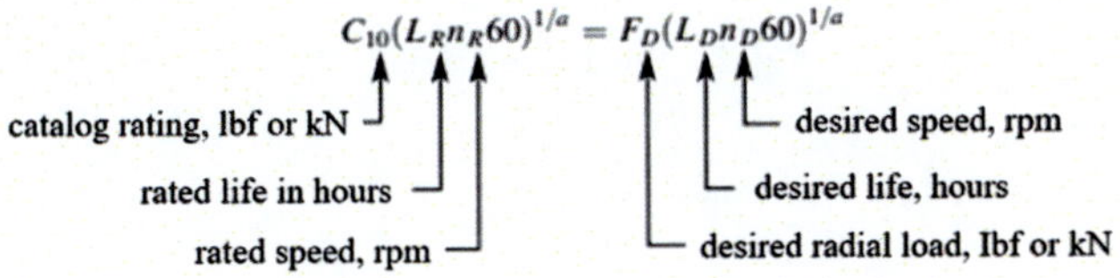

$$C_{10} = F_D\left(\frac{L_D n_D 60}{L_R n_R 60}\right)^{1/a} \tag{10}$$

Bolt Selection

(See Fig. 1).

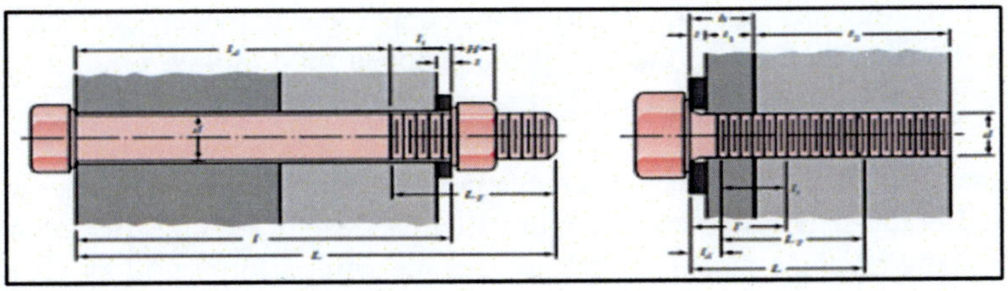

Fig. 1. Bolt selection [12]

Grip is thickness I,
Hasp length: L > I + H.

Weld Calculation
(See Fig. 2).

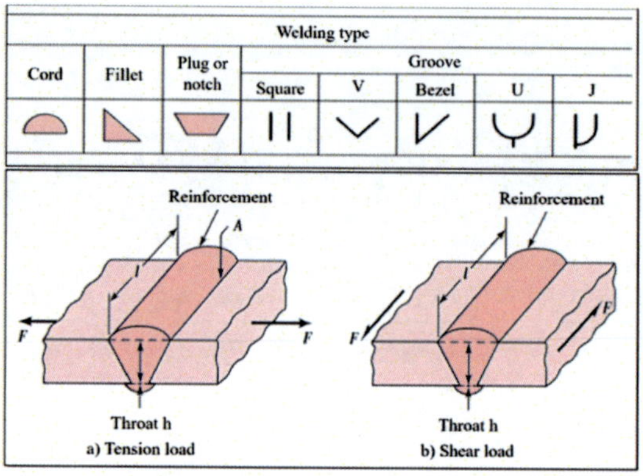

Fig. 2. Weld calculation [12]

Von Mises Effort for Welding

$$\sigma' = \left(\sigma^2 + 3\tau^2\right)^{1/2} = \frac{F}{hl}\left[\left(cos^2\theta + sen\,\theta\,\cos\,\theta\right)^2 + 3\left(sen^2\theta + sen\,\theta\,\cos\,\theta\right)^2\right]^{1/2} \quad (11)$$

4 Results

After having made the design and calculations of the machine, the construction of the machine was carried out, which could make the following measurements such as the chopping capacity, the experiment was carried out in a time of 10 min which could chop 60 kg During this time, the quality of the chopping is homogeneous in addition to reducing losses at the time of the exit of the chopped food in 98% of all the material used 2% is the particles that remain impregnated in the blades and in the food outlet duct.

On the other hand, this machine is much more efficient compared to other machines mentioned in the state of the art which in its thesis Design and construction of a corn cane chopping machine mentions that its machine will perform a 50 kg chopping in 10 min. which our machine turns out to be more efficient at the time of the chopping, on the other hand one of the most important aspects and that generally that people usually ask is the cost of this, the price of the machine approaches 2500 new soles, which in the national market has a higher cost that consists of 3500 new soles as a minimum cost in addition to being much more complicated at the time of being able to replace the blades since this has a high cost since they are removed with all the support in which they are held And many times you do not find the parts on the market, unlike the new design it makes it more accessible in addition to the fact that the blades are independent since they are removable and it is much easier. 1 of acquiring them, in addition to their economic impact, another aspect is the easy maintenance since all the parts are removable, easy to transport and that it is ideal for small farmers in the District 3 de Diciembre and the entire Mantaro valley.

Today many of the projects directed to the field of agriculture and therefore to the field of livestock are those that facilitate the reduction of time in certain activities and in livestock help feed the livestock, today the machines for the forage chopping focus only on that process, the chopping while our proposal represents the operation as a chopping machine, grinding the residues of different types of plantations in the main in the corn cane either dry or wet and agricultural waste improving the yield in regarding its forms of use (Fig. 3).

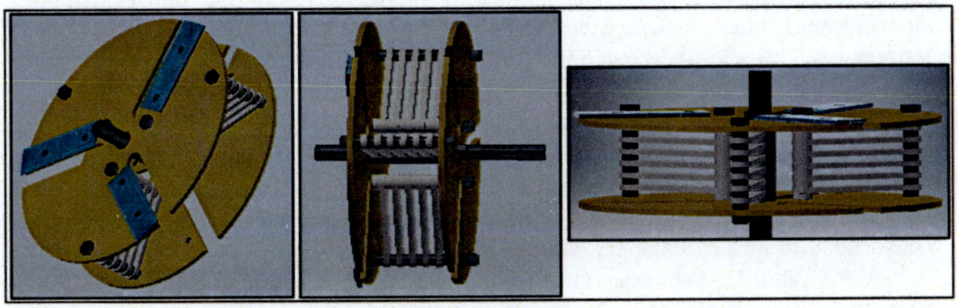

Fig. 3. Blade design

With a change of the inner covers for the grinding and with the coupling of the fixed hammers during the chopping, it works as a fan that expels the forage and when the hammers are released, they pulverize the matter or forage introduced inside the machine. Thanks to a grinding filtering system with perforated plates that help to pulverize agricultural waste according to the calibrated holes.

5 Conclusion

The main objective of this project has been to design and manufacture a machine capable of chopping forage and agricultural waste, optimizing the resources mentioned, for this the type of element to be used must be taken into consideration, otherwise it will not have a correct performance At the time of chopping, when the tests were carried out in the Cantalicio Meza family home in the December 3 district, the tests were carried out with husk where it produced 60 kg in 10 min, being the most resistant material to the chopping in addition to much more is produced with elements that have a much lower resistance to pitting, showing that the machine fulfilled the main function. Finally, the machine worked normally without any defect for 8 h, which is why there was no need to make adjustments to the machine.

References

1. IV CENAGRO: EN EL PERÚ IV Censo Nacional Agropecuario 2012. Instituto Nacional de Estadistica e Informatica, p. 388 (2012)
2. Flores Tamayo, L.P.: Diseño De Un Cilindro Picador De Forraje. Estudio de Factibilidad Para La Creación de Una Operadora de Ecoturismo En La Ciudad de Otavalo, vol. 244 (2008). http://bibdigital.epn.edu.ec/handle/15000/8658pdf
3. Delgado, C.A., Trigueros, V.A., Tang, P.J., Angelats, M.R., Gavidia, C.C.: Efecto De Un Modificador Orgánico En La Ganancia De Peso En Ganado Cebú En El Trópico Peruano. Revista de Investigaciones Veterinarias Del Perú **23**(2), 153–159 (2012). https://doi.org/10.15381/rivep.v23i2.895
4. Hernández, P.A.V., Rodríguez, A.M., Valencia, Y., Brito, E.: Validación del modelo de cálculo de la potencia consumida por las picadoras de forraje del tipo de tambor con alimentación manual. Revista Ciencias Técnicas Agropecuarias **21**(2), 5–10 (2012)
5. Barragán, A., Ingeniero Mecánico, Á.: 8o Congreso Iberoamericano De Ingenieria Mecanica Confiabilidad Operacional Para La Ingeniería Del Mantenimiento (2007)
6. Guzman, F.D., Caiza, G.: CHIMBORAZO Espoch, vol. 242 (2011). http://dspace.espoch.edu.ec/bitstream/123456789/1484/1/85T00206.pdf
7. Acad, U., Ciencias, M.D.E., Ingenier, D.E.L.A., Ingenier, A.C.D.E., Investigaci, N.P.D.E., Una, D.E., Autom, Q., Autores, I., Paneluisa, A., Le, A., Tutor, B.X., Quevedo, M.: Universidad técnica de cotopaxi (2016)
8. Adapa, P., Tabil, L., Schoenau, G.: Grinding performance and physical properties of non-treated and steam exploded barley, canola, oat and wheat straw. Biomass Bioenergy **35**(1), 549–561 (2011). https://doi.org/10.1016/j.biombioe.2010.10.004
9. Ortiz, M.E.J.: Universidad Nacional De Cajamarca. Universidad Nacional de Cajamarca, pp. 1–55 (2018). http://repositorio.unc.edu.pe/bitstream/handle/UNC/2987/TesiscompletaRonaldRomero.pdf?sequence=1&isAllowed=y
10. Jamshidpouya, M., Najafi, G., Tavakoli Hashjin, T.: Design, fabrication and evaluation of electric forage chopper with adjustable helix angle. J. Agric. Sci. Technol. **20**(5), 923–938 (2018)
11. Nogales, J., Andrade, J.: Diseño y construcción de una maquina picadora de hierba y caña para alimentar el ganado vacuno en el sector el porvenir perteneciente a la parroquia Gualea. Tesis (ingeniero mecánico). Universidad Politécnica Salesiana, Quito (2017). Disponible en https://dspace.ups.edu.ec/bitstream/123456789/14432/4/UPS-KT01409.pdf
12. Budynas, R., Nisbett, K.: Shigley Mechanical Engineering Design, 8th edn. The McGraw-Hill Companies Inc, Mexico (2008)

CAD/CAE Tools and Additive Manufacturing to Reduce the Impacts of Critical Equipment Shutdown on Production Planning

Byron Remache-Vinueza, Kévin Dávila-Cárdenas, and Mireya Zapata(✉)

Research Center of Mechatronics and Interactive Systems - MIST, Facultad de Ingeniería y Tecnologías de la Información y la Comunicación, Universidad Tecnológica Indoamérica, Machala y Sabanilla, Quito, Ecuador
{byronremache,mireyazapata}@uti.edu.ec, kdavila2@indoamerica.edu.ec

Abstract. When a breakdown affects critical equipment, spare parts must be supplied in the shortest possible time. If local distributors are not available, a common practice is to replace the mechanical element with a similar replacement from a different manufacturer. However, this can lead to systematic failures. If the spare part fails periodically, it can be supplied so that it is in the warehouse for replacement. As a consequence, inventory costs rise. In this project, Computer Aided Design (CAD), Computer Aided Engineering (CAE) and Additive Manufacturing were combined to solve this issue. The results show that the manufactured spare part tolerates the stresses in working conditions and the failure due to wear is cushioned. The saving in the cost of the spare part is approximately 96% and the reduction in the supply time reaches 98%. The replacement time is reduced by 46% due to the decrease in the number of parts.

Keywords: Additive manufacturing · Computer Aided Design (CAD) · Computer Aided Engineering (CAE) · Corrective maintenance

1 Introduction

Corrective maintenance that is not managed correctly can affect a production process as it involves unplanned equipment stoppages [1], which can in turn lead to a shortage in the supply chain [2]. To solve an unexpected failure, it is necessary to have tools, materials, and spare parts in case a change is required. If the element is not available, the time between detecting the failure and the solution increases, and thus also economic losses [3]. However, keeping a large number of spare parts in inventory generates additional costs [4]. Additionally, when there are no local distributors, the supply of spare parts results in high costs due to the implicit logistics [5]. A common practice is to replace the mechanical element with some local solution; either by a similar part from a different manufacturer [3] or the manufacture of the part; in both cases, the parts

D. Russo et al. (Eds.): IHSI 2021, AISC 1322, pp. 869–875, 2021.
https://doi.org/10.1007/978-3-030-68017-6_129

may not meet the mechanical or geometry specifications [6]. The consequences of the use of these spare parts are, for example, impact on product quality, delayed production delivery, long times due to corrective maintenance, occupational and environmental accidents.

In this context, computer-aided design (CAD), computer-aided engineering (CAE), and additive manufacturing tools, in combination with a systematic design process can offer an efficient solution. The 3D design allows the visualization of the mechanical elements to identify opportunities for improvement [7], while with the simulation of efforts, fast iterations can be generated until reaching the optimal design [8]. On the other hand, additive manufacturing is a flexible manufacturing process for rapid prototyping that allows a design to be tested without incurring high manufacturing costs for plastic injection molds [9]. These three tools are combined to find the solution to the problem of replacing mechanical elements of machinery whose acquisition is not profitable. Other authors have explored the CAD/CAE tools and additive manufacturing tools for the mechanical elements design that have requirements such as resistance to friction wear and the application of stress, and that in turn must be functional elements within a mechanical assembly. For example, in [10] the authors implement digital manufacturing tools in the design of extraction and drilling dies for body parts. The modeling in CAD software is coupled with a system to generate manufacturing diagrams; as a result, the waiting times of these parts for assembly are reduced by 8%. In [11], CAD/CAM tools were used in the turbine blade design and manufacturing process. The authors modeled and simulated the manufacturing process of the blades; As a result, in a 30-min simulation, all cutting patterns can be precisely defined. In this research, the design and engineering of a mechanical element is carried out using CAD/CAE tools as well as its additive manufacturing to solve the requirement of a spare part that generates import costs, and that affects the productivity of equipment in a production line. A study of the materials and methodologies used to achieve the objectives of the research is carried out. Subsequently, the results are presented and analyzed, to finally develop conclusions and recommendations.

2 Materials and Methods

The project was developed using CAD/CAE software and additive manufacturing equipment that allows multi-material printing with soluble supports and offers high reliability and precision for printing mechanical elements.

The methodology consists of identifying the design requirements, the redesign of the mechanical part and functional tests (see Fig. 1).

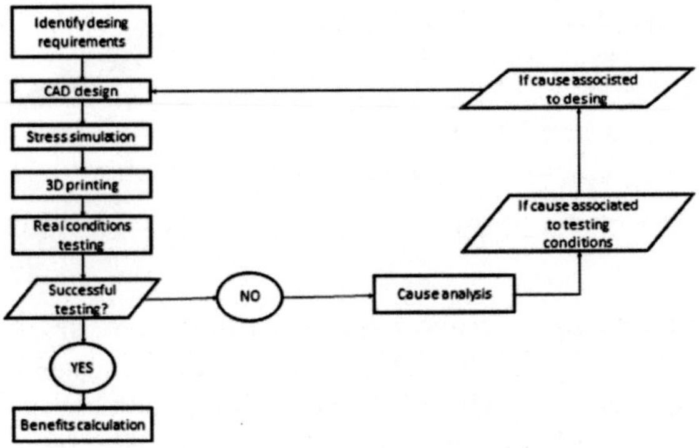

Fig. 1. Methodology flow chart

2.1 Design Requirements

The mechanical element called pusher arm is an element of the product dragging system of a wrapping machine. Its function is to hold the pusher arm on the drag chain. For the operation of the machine, around 35 of these parts are required, with the characteristics presented in Table 1.

Table 1. Characteristics of the arm

Original part	
Weight (gr)	4.5
Length (mm)	105.53
Width (mm)	15.27
Thickness (mm)	5
Material	Stainless steel 304

Under normal working conditions, this element deforms in such a way that it hits secondary parts, which generates wear and unbalance in the main drag chain, causing the product to lose its ideal position in the packaging.

2.2 3D Modelling (CAD)

The mechanical element was designed taking into account the type of printing material and nozzle, with the dimensions increased 0.5 mm due to the contraction of the material when it is drying. The two elements necessary for the separation of the mechanical part from the chain, are incorporated directly into the redesigned arm to reduce the number of elements and optimize the assembly process (See Fig. 2).

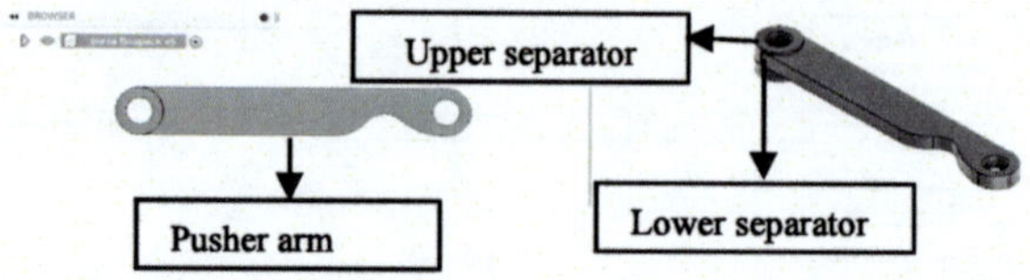

Fig. 2. Pusher arm 3D model with separators included

An important factor that is taken into account in the redesign of the mechanical element is the material. For 3D printing there is a variety of materials such as wood filaments, metallic filaments, polymers of different types and high-performance materials [12]. In the local market the most common materials for 3D printing are nylon, PLA and carbon fiber. Nylon was selected for this project due to its low cost, adequate mechanical resistance, and low coefficient of friction [13].

2.3 Stress Simulation (CAE)

Before manufacturing the element, stress simulations were performed using CAE software. For this, it was determined that the pusher arm is subjected to compression stress, whose magnitude was obtained from the torque generated by the servomotor that drives the chain. Through an iterative process, simulations were carried out until the optimal combination of material and geometry was achieved. Figure 3 shows that the safety factor obtained by replacing the mechanical element with one made of nylon is 6.11, which, although it is lower than the safety factor obtained with stainless steel (15), shows that the element will resist to tests under operating conditions.

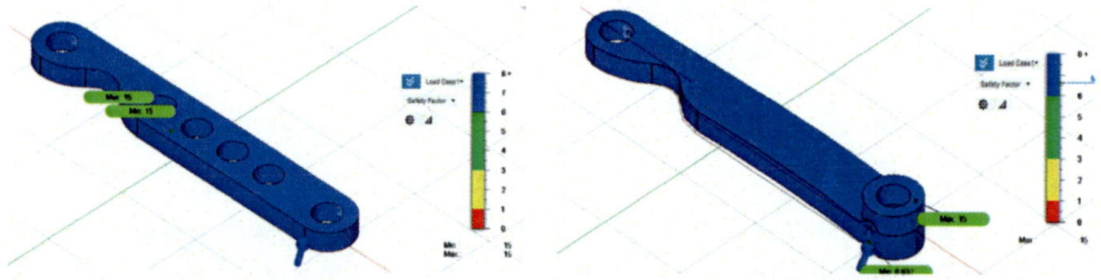

Fig. 3. Verification of the safety factor of the element designed in nylon (right) and original element in stainless steel (left).

3 Results

3.1 3D Printing

The nylon printed arm has a weight of 0.5 gr and can be seen in Fig. 4. The arm was tested under normal working conditions to compare its performance with that of the original spare part, with a packing speed of 250 products/minute. The coupling between the redesigned arm, and the other components of the equipment was verified by means of a caliper. The correct functioning of the element was confirmed, obtaining a reduction in wear thanks to the fewer number of parts and the optimization of the friction factor (see Fig. 5).

Fig. 4. Redesigned pusher arm manufactured using additive manufacturing or 3D printing.

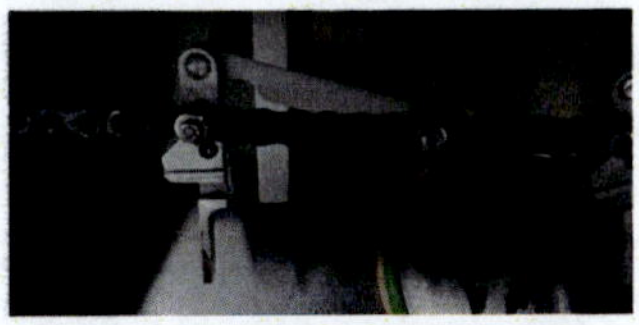

Fig. 5. Brazo rediseñado instalado en el equipo

3.2 Benefits

According to the cost calculation, a saving in the replacement of the part of 96% is achieved. On the other hand, the logistics process for the acquisition of the spare part shows a reduction in time of 98%, which is mainly due to the elimination of import processes and customs procedures. Furthermore, by reducing the number of parts, the replacement of the spare part is facilitated. Therefore, the machine downtime is reduced by 46%. Table 2 shows a summary of the benefits.

Table 2. Economic and administrative benefits

Benefit	Original	New design	Saving
Cost (USD)	197	7,4	96%
Purchase order time (horas)	720	24	
Manufacturing time (horas)	0	6	
International transport time (horas)	1440	0	
Customs time (horas)	504	0	
National transport time (horas)	8	1,5	
Delivery total time (hours)	2672	30	98.9%
Production process stoppage time (min)	15	8	46.0%

4 Conclusions

In this project, Computer Aided Design (CAD), Computer Aided Engineering (CAE) and Additive Manufacturing, pillars of Industry 4.0, were combined to solve the problem of replacing a critical machinery part whose acquisition generates high logistics costs. The 3D design allowed visualizing the mechanical elements in such a way that opportunities were identified to improve the geometry, the manufacturing process, and its subsequent assembly. Stress simulation was used as a strategy to reduce the risk of failure during the testing sessions and to generate various design options as an iterative process until the optimal combination of material and geometry was reached. Additive manufacturing, being a flexible manufacturing process, allowed rapid prototyping to test the design without incurring high costs of designing and building plastic injection molds. The results show that the manufactured spare part tolerates the stresses in working conditions and the failure due to wear is cushioned. The saving in the cost of the spare part is approximately 96%, and the reduction in the supply time reaches 98%. The replacement time is reduced by 46% due to the decrease in the number of parts. In future work, it is expected to have a history of failures to determine the frequency of failures in order to optimize the preventive maintenance plan. Furthermore, once the design has been validated under working conditions during a specific period of operation, the option of migrating to a plastic injection manufacturing process can be analyzed and the cost-benefit ratio determined. Finally, life cycle analysis and design strategies for manufacturing and assembly could be added as strategies to generate options with less environmental impact.

References

1. Flores, M., Medina, D., Vargas, D., Remache-Vinueza, B.: Asignación de modelos de mantenimiento basada en la criticidad y disponibilidad del equipo. CienciAmérica **9**(4), 27–34 (2020)
2. Nikoueghbal, A., Valibeigi, H.: Technology transfer in developing countries, challenges and strategies: case study of Iran's auto industry. Iran. Econ. Rev. **10**, 57–78 (2005)
3. Garrido, S.G. (ed.): Mantenimiento correctivo Organización y gestión, p. 28. Renovetec, Madrid (2009)
4. Botero, G.C.: Manual de Mantenimineto. In: R. R. B. SENA Digeneral, Bogota (1991)
5. Paz, R.C.: Loguistica empresarial. In: Gomez, D.G. (ed.) Marde Del Plata: facultad de ciencas economicas y sociales (2013)
6. Laberenz, A., Sain, G.: Central repair workshops in developing countries. Development Digest (1971)
7. Alvarado, B.J.: Estudio para el modelado e impresion 3D de autoparte. UIDE, Guayaquil (2018)
8. Lia, L., Zhengd, Y., Yange, M., Lengf, J., Chengd, Z., Xieg, Y., Ma, Y.: A survey of feature modelling methods: historical evolution and new development. Robot. Comput. Integr. Manuf. **61**, 101851 (2020)
9. Vallejo, F.M.: estrudio sobre la apliacacion de las tecnologuias de fabricacion aditiva al sector aeronautico y espcecioal. sevilla (2016)

10. Armillotta, A., Moroni, G., Rasella, M.: Computer-aided assembly planning for the diemaking industry. Robot. Comput.-Integr. Manuf. **22**, 409–419 (2006)
11. Olama, I., Besant, C.B., Ristic, M.: A CAD/CAM system for die design and manufacture. Int. J. Adv. Manuf. Technol. **3**, 21–29 (1988)
12. Lua, Y., Xua, X., Wangb, L.: Smart manufacturing process and system automation – a critical review of the standards and envisioned scenarios (2020)
13. Mena, J.C., Ricardo, G.E., Torres, E.D.: Effect of the filling percentage on tensile strength in 3D desktop printing for different printing patterns, using a randomized design of experiments. Enfoque UTE revista **10**, 13–27 (2019)

Comparison of Accessibility Tools for the Evaluation of E-Commerce Websites

Freddy Paz[1(✉)], Freddy A. Paz[2], Arturo Moquillaza[1], Luis Collantes[2], Juan Jesús Arenas[1,3], and Daniela Villanueva[1]

[1] Pontificia Universidad Católica del Perú, San Miguel Lima 32, Peru
{fpaz, amoquillaza}@pucp.pe,
{jjarenas, dvillanuevab}@pucp.edu.pe
[2] Universidad Nacional Pedro Ruiz Gallo, Lambayeque, Peru
freddypazsifuentes@yahoo.es, lcollantes@unprg.edu.pe
[3] Pontificia Universidad Javeriana, Bogotá, Colombia
jjesusarenas@javeriana.edu.co

Abstract. Accessibility is an aspect that should be considered when building software products. All people have the right to access the interactive systems that exist in the market. Even in certain countries, there are laws and government regulations that guarantee access to new products and services to citizens with different abilities and capabilities. This fact has generated a current growing concern on the part of companies to provide accessible applications, especially in the web domain where competition is high. Likewise, the importance of accessibility has led to establishing standards and guidelines to ensure minimum considerations in the development of software products. The W3C provides a list of tools that automate the accessibility evaluation process according to international standards. This study aims to compare the results offered by five tools when used to inspect the accessibility of e-commerce websites. The results show that differences are depending on the tool that is employed.

Keywords: Accessibility · Automated tools · Evaluation · Comparative study · E-commerce domain

1 Introduction

Accessibility is an important software quality attribute that must be considered during the development process. This consideration of assembly is currently relevant because it allows the construction of accessible software products to anyone regardless of their abilities or capabilities [1]. Contemplating that around one billion people or 15% of the world's population experience some form of disability [2], software development teams must have special consideration on this quality attribute. The E-Commerce domain is a particular scenario because there are several available options on the web, and the market is highly competitive [3]. If the website does not offer accessible mechanisms, then there is a high probability of losing a representative number of customers. Besides, web accessibility is considered a human right in some countries. There are government regulations; for example, in the United States, the Section 508 of the Rehabilitation Act

D. Russo et al. (Eds.): IHSI 2021, AISC 1322, pp. 876–881, 2021.
https://doi.org/10.1007/978-3-030-68017-6_130

of 1973 establishes that electronic and information technology from all federal agencies must be accessible to people with disabilities [4]. In the same way, in Latin America, for example, in Peru, there is a Ministerial Resolution No. 126-2009-PCM [5], and two laws, the law No. 29973 [6] and the law No. 28530 [7], that favor the participation of people with disabilities on the Internet, contributing to the reduction of the digital gap and opening possibilities for inclusion, providing people the opportunity of interacting with equal opportunities.

Given the relevance of accessibility, there are several regulations that force developers and designers to consider this quality attribute during the software construction process. The infringement of these regulations can meet with financial penalties [8]. In this study, we examine the accessibility degree of some recognized E-commerce websites. The purpose of this study is to determine whether the main companies are aware of the importance of providing accessible websites that can subsequently be used with assistive technology such as screen readers, alternative keyboards, scanning software, and other devices. For this evaluation, we have used software tools that automate the accessibility inspection process and verify if the web content meets specific guidelines. The evidence shows that the results can vary depending on the software tool that is used, even though these tools are focused on inspecting the same standard.

2 Accessibility Evaluation Process

In a previous study [9], we conducted a systematic literature review (SLR) according to the protocol proposed by Kitchenham and Charters [10] to identify the most reported methods that are used by the specialists in Human-Computer Interaction (HCI) to perform accessibility evaluations. According to the findings, we could establish that there are three approaches:

- **Automated methods**: There are software tools that have been designed to determine if a website meets accessibility guidelines. These tools only require the URL of the website to execute an in-depth analysis of the HTML code in search of infringements to the guidelines. Most of these tools are programmed to conduct the evaluation according to the standard *Web Content Accessibility Guidelines* (WCAG) proposed by the Web Accessibility Initiative (WAI) [11]. Versions 2.0 and 2.1 are currently the most used. However, the W3C WAI is working in a new draft that will be next version 2.2 to be published in 2021.
- **Inspection methods**: This approach involves having a set of specialists highly experienced in accessibility evaluations who inspect each of the graphical user interfaces in search of infringements to a specific standard. The quality of the results will depend on the expertise of the evaluators and the granularity level of the inspection. In this type of assessment, the code HTML must be verified manually, and the specialists can also make use of assistive technology to corroborate their findings.

- **Tests with users**: Possibly, the most effective method given that it allows obtaining real and direct information from users. However, this can represent, at the same time, the most expensive approach since it is mandatory to recruit users of different profiles. In general, these tests are usually focused on determining if the software is suitable and accessible to a single type of user profile. According to our literature review, most of these studies are centered on tests to guarantee the visual and auditory accessibility. The evaluation must be led by specialists in HCI that can collect, interpret, and discuss the opinions, reactions, comments, and responses of the users.

Like usability tests, there is no one accessibility evaluation method that is better than the other [12]. The approaches complement each other for a better understanding and analysis of the results. In the case of the web accessibility evaluation tools, these have the capability to process multiple web pages and provide in a short time a detailed report about the guidelines that are infringed. Most of these software tools have been designed to perform the review according to the standard WCAG 2.0, which is widely accepted internationally. Some countries have even adopted this standard to institute it as a technical norm, specifically for the design of websites that offer e-government services [13].

The WCAG standard establishes three levels of compliance: A, AA, and AAA. The level AAA is the highest level and represents that the website fully complies with all the aspects indicated in the standard, for which it is determined that it is unquestionably accessible. However, an accepted, appropriate, and recommended level of compliance is level AA. Regarding level A, although it ensures a minimum level of accessibility in the websites, the quality attribute is not widely covered in many situations. Likewise, to establish that a website meets a conformance level, the previous level must be entirely satisfied. In the same way, the WACG standard does not recommend level AAA as a general policy because it is not possible to meet all the success criteria for certain types of content [14].

The inspection presented in this study was conducted following the Web Accessibility Conformance Evaluation Methodology (WCAG-EM) [15] which is an approach to determine the degree in which a website meets the WCAG standard. This methodology defines five phases to execute the evaluation:

- **Define the scope of the evaluation**: In this phase, the evaluators must define the purpose of the evaluation, the sections of the websites that will be assessed, and the WCAG conformance level towards which the inspection is addressing. For this case study, the websites of the main international E-commerce stores have been selected. The purpose of the evaluation was to determine if the main companies in this field concerns about accessibility. Given the representative number of users with different abilities and skills, this research is focused on identifying the awareness degree of international enterprises and how accessible are their websites. Regarding the conformance level, the recommendation of WAI was followed. Therefore, the websites have been evaluated up to the AA conformance level.

- **Explore the website**: In this phase, the evaluators must identify and select the relevant web pages to examine. Besides, the inspectors can explore the application to determine the critical functionality and web content. In this study, a preliminary examination of the pages was performed to recognize the purchase process on the E-commerce websites to be evaluated.
- **Select a representative sample**: The evaluators must determine the pages that will be part of the inspection as well as the automated accessibility evaluation software tools. For this study, the scope was the main page of the E-Commerce websites, and the selected tools were:
  - aChecker (www.achecker.ca)
  - WAVE (wave.webaim.org)
  - TAW (www.tawdis.net)
  - SortSite (www.powermapper.com/products/sortsite/)
  - AInspector Sidebar WCAG (addons.mozilla.org/en-US/firefox/addon/ainspector-sidebar/).

All these tools are approved by W3C WAI [16] and provide support to evaluate according to the standard WCAG 2.0.

- **Evaluate the selected sample**: In this phase, the evaluators must carry out the evaluation according to the specifications established in the previous phases. The websites of the following companies were selected for this study:
  - Amazon, Inc. (www.amazon.com)
  - Walmart (www.walmart.com)
  - Best Buy Co., Inc. (www.bestbuy.com)
  - eBay (www.ebay.com).

- **Report the evaluation findings**: In this phase, once the accessibility assessment has been executed the evaluators must report the findings. The specialists can establish the aspects of improvement as well as the number of errors and infringements to the international standard WCAG. The findings must be discussed and analyzed with the aim of turning the software product into an accessible application for people regardless of their skills or abilities.

3 Results of the Accessibility Evaluation

The results of the accessibility evaluation are described in Table 1. The evaluation was conducted on September 26, 2020. The findings evidence that depending on the tool, a dissimilar result can be achieved. This fact can be explained because there is a different implementation of the tools to verify compliance with the guidelines. Some tools provide their results in a generic way, while others are specific and can detail several errors for the same infringed guideline. Likewise, some tools are not capable of reviewing all the guidelines established in the standard WCAG 2.0 and request a manual review that can complement the identified results. On the other hand, with respect to the level of accessibility of the websites, it is possible to determine that although the results are not optimal, they are not discouraging either. The gap to reach the recommended level AA is not high. In fact, few errors are identified at the AA conformance level, and the reported problems with respect to level A are minor. Finally, considering the results of a previous research [1], it is possible to demonstrate that the level of accessibility in websites with an international approach is better than those websites that have a more limited audience such as at the local companies in a country.

Table 1. Number of reported accessibility problems by the multiple tools cataloged by website.

Website	aChecker		WAVE		TAW		SortSite		AInspector	
	A	AA	A	AA	A	AA	A	AA	A	AA
www.amazon.com	10	17	3	3	68	0	12	4	58	12
www.walmart.com	11	1	2	0	2	0	4	1	35	12
www.bestbuy.com	11	29	0	2	17	0	5	1	38	12
www.ebay.com	10	58	5	4	436	0	18	3	42	12

4 Conclusions and Future Works

Accessibility is an important quality attribute that must be considered in the software development process to ensure that people, regardless of their abilities or skills, can use the final product. Given the importance of designing accessible websites, some standards have been elaborated. WCAG is one of the most recognized proposals in the field. Likewise, WCAG-EM is a methodology that allows specialists the execution of accessibility evaluations to determine if a website meets minimum guidelines. In this study, it was possible to determine that although the main international E-Commerce websites meet a high degree of accessibility, there are aspects to improve. The number of errors reported can vary depending on the evaluation tool that is used. It is important to highlight that some tools provide a detailed report of errors, while others basically indicate the guidelines that are infringed. Some guidelines could not be reviewed automatically, and the participation of specialists could be required in these cases. As future work, it would be interesting to complement the results of this evaluation with assessments conducted in other software domains and with ad-hoc examinations.

Acknowledgments. This research is highly supported by the *Department of Engineering* of the *Pontifical Catholic University of Peru* (PUCP) in Peru and the *HCI, Design, User Experience, Accessibility and Innovation Technology Research Group* (HCI-DUXAIT). HCI-DUXAIT is a research group of PUCP.

References

1. Paz, F., Paz, F.A., Villanueva, D., Moquillaza, A., Arenas, J.: Accessibility evaluation of Peruvian E-commerce websites. Int. J. Adv. Trends Comput. Sci. Eng. **9**, 558–561 (2020). https://doi.org/10.30534/ijatcse/2020/76912020
2. World Bank: Disability inclusion (2020). https://www.worldbank.org/en/topic/disability. Accessed 26 Sep 2020
3. Paz, F., Paz, F.A., Pow-Sang, J.A.: Experimental case study of new usability heuristics. In: Proceedings of the 17th International Conference of Human-Computer Interaction (HCI International 2015), Los Angeles, United States, pp. 212–223 (2015). https://doi.org/10.1007/978-3-319-20886-2_21
4. Gibson, E.: Web accessibility and section 508 compliance. J. Comput. Sci. Coll. **17**, 257–259 (2002)
5. Presidencia del Consejo de Ministros del Perú: Resolución Ministerial No. 126-2009-PCM (2009). https://www.gob.pe/institucion/pcm/normas-legales/292536-126-2009-pcm. Accessed 26 Sep 2020
6. Government of Peru: Law No. 29973 – general Law of the people with disabilities (2012)
7. Government of Peru: Law No. 28530 – law for the promotion of Internet access for people with disabilities and the adaptation of physical space in public Internet stands (2005)
8. Kuzma, J., Price, C.: Analysis of UK parliament web sites for disability accessibility. In: Proceedings of the 9th European Conference on e-Government (ECEG 2009), London, United Kingdom, pp. 416–421 (2009)
9. Nuñez, A., Moquillaza, A., Paz, F.: Web accessibility evaluation methods: a systematic review. In: Proceedings of the 21st International Conference on Human-Computer Interaction (HCI International 2019), Orlando, United States, pp. 226–237 (2019). https://doi.org/10.1007/978-3-030-23535-2_17
10. Kitchenham, B., Charters, S.: Guidelines for performing systematic literature reviews in Software Engineering. Technical report EBSE 2007-001, Keele University and Durham University (2007)
11. W3C Web Accessibility Initiative (WAI): Web content accessibility guidelines (WCAG) overview (2020). https://www.w3.org/WAI/standards-guidelines/wcag/. Accessed 26 Sep 2020
12. Paz, F., Paz, F.A., Villanueva, D., Pow-Sang, J.A.: Heuristic evaluation as a complement to usability testing: a case study in web domain. In: Proceedings of the 12th International Conference on Information Technology New Generations (ITNG 2015), Las Vegas, Nevada, United States, pp. 546–551 (2015). https://doi.org/10.1109/ITNG.2015.92
13. Sanchez-Gordon, S., Lujan-Mora, S., Sanchez-Gordon, M.: E-government accessibility in Ecuador: a preliminary evaluation. In: Proceedings of the 7th International Conference on eDemocracy and eGovernment (ICEDEG 2020), Buenos Aires, Argentina, pp. 50–57 (2020). https://doi.org/10.1109/ICEDEG48599.2020.9096766
14. Acosta-Vargas, P., Luján-Mora, S., Salvador-Ullauri, L.: Quality evaluation of government websites. In: Proceedings of the 4th International Conference on eDemocracy and eGovernment (ICEDEG 2017), Quito, Ecuador, pp. 8–14 (2017). https://doi.org/10.1109/ICEDEG.2017.7962507
15. W3C Web Accessibility Initiative (WAI): Website accessibility conformance evaluation methodology (WCAG-EM) 1.0 (2014). https://www.w3.org/TR/WCAG-EM/. Accessed 26 Sep 2020
16. W3C Web Accessibility Initiative (WAI): Web accessibility evaluation tools list (2016). https://www.w3.org/WAI/ER/tools/. Accessed 26 Sep 2020

Research on Inheritance and Innovation of Bamboo Weaving Technology in Zhejiang Province Based on Cultural Gene Extraction and Transformation

Yixiang Wu[1(✉)] and Xinhui Kang[2]

[1] Yiwu Industrial and Commercial College, Yiwu, Zhejiang, China
1951601836@qq.com
[2] School of Art and Design, Nanchang University, No. 999 Xuefu Avenue, Nanchang 330031, Jiangxi, China
nbukxh@163.com

Abstract. Zhejiang Province, China, has abundant bamboo resources and profound bamboo weaving culture deposits. In recent years, due to changes in lifestyles, the living utensils made by the traditional bamboo weaving technology have been gradually replaced, and the bamboo weaving technology has turned to the exquisite arts and crafts route, becoming a niche of art appreciation. The bamboo weaving cultural and creative design mostly focuses on the inheritance mode and development strategy, and design practices are rarely guided from the perspective of cultural genes. Therefore, by taking the development of bamboo-woven tourist souvenirs as the study case, Kano Model as the theoretical basis, and tourists of tourist attractions as the research object, the positive and negative questionnaire was used to investigate the satisfaction of tourists' demands, in order to obtain the genes of the traditional bamboo weaving culture for the key demands of tourists. The traditional Kano model ignores the uncertainty of the tester's thinking. In order to avoid this situation, the key genes that attract tourists were selected based on the fuzzy Kano model, and the corresponding development strategy for bamboo-woven tourist souvenirs was proposed to provide reference for the development and design of bamboo-woven tourist souvenirs, which is of high academic value to the integration of culture and tourism.

Keywords: KANO model · Cultural gene · Bamboo weaving · Tourist souvenirs

1 Introduction

Zhejiang Province, China, has abundant bamboo resources and profound bamboo weaving culture deposits. Bamboo weaving is a handicraft that is processed and woven with bamboo as the raw material. The local rainfall in Zhejiang is sufficient and the green bamboo grows into forests. The working people create practical and ornamental folk arts and crafts with local materials. However, due to changes in lifestyle in recent years, the traditional bamboo weaving technology has turned to the exquisite arts and

D. Russo et al. (Eds.): IHSI 2021, AISC 1322, pp. 882–888, 2021.
https://doi.org/10.1007/978-3-030-68017-6_131

crafts route, becoming a niche of art appreciation. In the context of China's intangible cultural heritage inheritance and innovation, starting from the activation creation practice of Zhejiang bamboo weaving, an intangible cultural heritage, the traditional bamboo weaving was combined with the fashion culture as well as the modern fashion trend, so as to develop the bamboo-woven products that conform to modern life aesthetics, show the new thinking of Zhejiang's intangible cultural heritage inheritance and development, and make a new attempt to the modern transformation of traditional handicrafts. Highlighting the symbiosis of Chinese traditional handicraft and modern design is an important idea to realize the modern transformation and activation of traditional handicrafts. This helps promote the inheritance and innovation of Zhejiang traditional bamboo weaving technology in modern life, and make the bamboo weaving technology return to modern life.

In this study, the development of bamboo weaving in Zhejiang Province, the technique and application of bamboo weaving were studied to explore the design of bamboo-woven tourist souvenirs. In the development and design of bamboo-woven tourist souvenirs, the consumer demand is often unpredictable. How to accurately grasp the various needs of consumers is the key to the development and design of tourist souvenirs. In 1984, Noriaki KANO, a professor at Tokyo University of Science, put forward a study on service quality and user satisfaction in Japan, and obtained the KANO user demand model [1]. At present, most foreign researchers employ the Kano model to study emotions. For example, Ishardita Pambudi Tama et al. [2] designed ceramic souvenirs using the KANO model, Kuo-Liang Huang et al. [3] analyzed the correlation between function and reading satisfaction using the KANO model, and Shyh-Huei HWANG et al. [4] studied the relationship between user satisfaction and cultural products using the KANO model. By taking the development of bamboo-woven tourist souvenirs as the study case, Kano Model as the theoretical basis, and tourists of tourist attractions as the research object, the cultural genes of bamboo-woven tourist souvenirs that attract tourists were selected based on the fuzzy Kano model, and the corresponding development strategy for bamboo-woven tourist souvenirs was proposed to carry out the design practice of bamboo-woven tourist souvenirs.

2 Method

2.1 Kano User Demand Model

Through the analysis of user demands by Kano model, user demands that affect satisfaction can be divided into five attributes (see Table 1): Attractive Quality, One-Dimensional Quality, Indifferent Quality, Reverse Quality and Must-be Quality.

2.2 Determination of KANO Demand Category

According to the KANO method, the positive and negative questions were set for each demand, and then the KANO demand category was determined according to the KANO Evaluation Table (Table 3). However, the traditional Kano model only allows customers to choose the most satisfactory answer, which may lead to inaccurate results,

Table 1. Five user demands affecting satisfaction

kano attributes	Relationship between demand and satisfaction
Must-be Quality (M)	The prerequisite demand attribute for a product. With this attribute, user satisfaction will not be significantly improved; Without the basic attributes of the product, user satisfaction will be reduced
One-Dimensional Quality (O)	The functional attribute that users desire. There is a one-dimensional linear relationship between the realization degree of user demands and user satisfaction
Attractive Quality (A)	The demand beyond most users' expectations. In the case of little difference in other properties, the attractive quality is more appealing to consumers, and can increase the product competitive advantage. It is often the part that enterprises are trying to develop, as well as the innovation point of products
Indifferent Quality (I)	The demand attribute that users do not care about. Satisfaction of needs has no effect on satisfaction
Reverse Quality (R)	The demand attribute that causes user dissatisfaction. Satisfaction of needs results in decreased satisfaction

while the fuzzy Kano model can overcome the ambiguity of consumer demand for products. The fuzzy Kano, adopted for investigation in this paper, allows consumers to answer the fuzzy satisfaction value in two options, and express their satisfaction with a certain decimal point satisfaction value between [0, 1]. The sum of the two values is 1 (Table 2).

Questionnaire:

Table 2. Fuzzy Kano questionnaire

	Like	Merited	Neutral	Tolerable	Dislike
Realizable	0.4	0.6			
Irrealizable				0.8	0.2

Finally, the results were statistically analyzed, and the KANO attribute of each evaluation index question item in each survey sample was determined against the KANO Model Evaluation Table (Table 3). The frequency of various KANO attributes (M, O, A, I, R, Q) in each evaluation index question item was also statistically analyzed.

Table 3. KANO model demand classification evaluation table

Positive questions	Negative questions				
	Like	Merited	Neutral	Tolerable	Dislike
Like	Q	A	A	A	O
Merited	R	I	I	I	M
Neutral	R	I	I	I	M
Tolerable	R	I	I	I	M
Dislike	R	R	R	R	Q

3 Case Study

3.1 Sorting of Bamboo Weaving Cultural Archetype

A comprehensive and in-depth investigation of the bamboo weaving in Zhejiang Province was carried out using the literature review method and interview method. The relevant real pictures were collected via literature review, online materials and field trips to Zhejiang Bamboo Weaving Cultural Center, Bamboo Weaving Factory, Family Workshop, etc. Interviews with locals and intangible cultural heritage inheritors in Zhejiang Province helped understand the cultural background, craftsmanship and local folk customs of bamboo weaving in Zhejiang Province. Moreover, the bamboo weaving cultural elements were divided in detail to facilitate the subsequent extraction of cultural genes.

3.2 Extraction of Bamboo Weaving Cultural Genes

Zhejiang Province has rich bamboo weaving culture. The bamboo weaving cultural gene of Zhejiang Province was extracted from the dominant and recessive aspects by analyzing and comparing the previous data. In terms of shape, by deconstructing the shape of bamboo-woven products such as bamboo steamers and bamboo baskets, the shape characteristics of bamboo-woven products were obtained; in terms of color, software tools were used to extract color values; in terms of materials, bamboo-woven utensils were classified in materials, such as bamboo + porcelain, bamboo + wood, bamboo + cloth. Recessive genes were mainly extracted from bamboo weaving techniques and other aspects. The bamboo weaving techniques were listed and sorted according to their development history, including various weaving methods derived from the original cross weaving and twill weaving, to complete the extraction of cultural genes.

3.3 Establishment of Satisfaction Evaluation Indexes for Tourist Souvenirs

For the establishment of various satisfaction evaluation indexes for tourist souvenirs, the evaluation indexes affecting tourist satisfaction were proposed through literature review, questionnaire survey, interviews and other methods.

3.4 Questionnaire for KANO Demand Model

Through interviews and exchanges with tourists, the tourist demands were matched with the category of Kano attributes. There were a total of 50 survey subjects, of which 25 were males and 25 were females. Among them, 20 were tourists related to the product design, students from Zhejiang Yiwu Industrial and Commercial College, and 40 were temporary tourists in tourist attractions.

4 Data Analysis

After the questionnaire survey of KANO demand category, the data of the questionnaire were analyzed and sorted, and the KANO attributes were classified. First, the survey results were statistically analyzed, and the KANO attribute of each evaluation index question item in each survey sample was determined against the KANO Model Evaluation Table (Table 3); The frequency of various KANO attributes (M, O, A, I, R, Q) in each evaluation index question item was also statistically analyzed, and the maximum was taken as the final demand attribute of the evaluation index. The results are shown in Table 4.

Table 4. Fuzzy Kano questionnaire

NO	Evaluation indexes	KANO attribute statistics						Demand attribute
		M	A	I	O	R	Q	Category
1	Bamboo + other materials	20	51	7	15	5	2	A
2	Environmental protection	11	16	10	61			O
3	Prominent cultural genes	10	60	16	13	1		A
4	Portable	11	28	21	40			O
5	Experience design	22	23	45	10			I
6	Bamboo weaving and function integration	23	20	43	12	2		I
7	Attractive shape	10	19	51	16	4		I
8	Life demands based	15	65	6	14			A
9	Simple shape	17	54	16	10	3		A

5 Development Strategy of Bamboo-Woven Tourist Souvenirs Based on Kano User Demand Model

According to the relative customer satisfaction coefficient list and the principle of demand screening, the Must-be Quality (M) and Indifferent Quality (I) were removed, and the Attractive Quality (A) and One-Dimensional Quality (O) were retained, to summarize the development strategy of bamboo-woven tourist souvenirs.

5.1 Bamboo + Other Materials

The design was carried out by combining the traditional bamboo weaving technology with modern life demands. Based on the traditional bamboo weaving technology, an attempt was made by combing bamboo with ceramics, woven fabrics, metal and other materials to realize the form shaping between bamboo weaving and other materials, and achieve a coordinated effect.

5.2 Integration of Bamboo Weaving and Function

According to the actual functional needs of bamboo-woven works, the design of the shape is mainly stable, and the image of bamboo weaving tends to the aesthetic needs. Based on the classic bamboo weaving techniques such as insertion and pattern weaving as well as the random weaving, the work can be completed more perfectly without the restriction of bamboo materials, thereby enhancing the weaving texture of the work.

5.3 Combination with Modern Life and Aesthetics

The bamboo weaving technology is complicated and has higher requirements. Therefore, it is necessary to design bamboo-woven tourist souvenirs that integrate modern life and aesthetics based on the modern design, design expertise and modern lifestyles.

5.4 Prominent Cultural Genes

The traditional bamboo weaving technology needs to be inherited and developed urgently. To enhance the competitiveness of products, cultural elements should be extracted and transformed. The added value of bamboo-woven products lies in the cultural genes. Designers should emphasize the function of bamboo culture identity and inject cultural genes into modern designs using the creative design. The research results show that the cultural property has higher demand attributes of Attractive Quality (A). According to the existing homogeneity of tourist handicrafts design, the cultural property should be stressed in the design of tourist souvenirs.

5.5 Design Practice Based on KANO User Demand Model

Starting from the Attractive Quality (A) and One-Dimensional Quality (O) of user demands, the bamboo-woven tourist souvenir with Chinese cultural deposits was designed. As shown in Fig. 1, the material is mainly bamboo, which restores the bamboo texture and is also energy-efficient and environmentally friendly. Because of higher requirements of traditional handicrafts, the modern technology can be employed for processing commemorative and practical tourist souvenirs. The focus was stayed on meeting tourists' needs for midnight snacks, snacks, fruits and drinks. By taking the Fujian earthen building of the Hakkas, a beautiful earth high-rise building, as the prototype, the strong cultural connotation of the earthen building was incorporated for appearance design. It also has the retractable function, which is convenient for transportation, space-saving and portable in traveling.

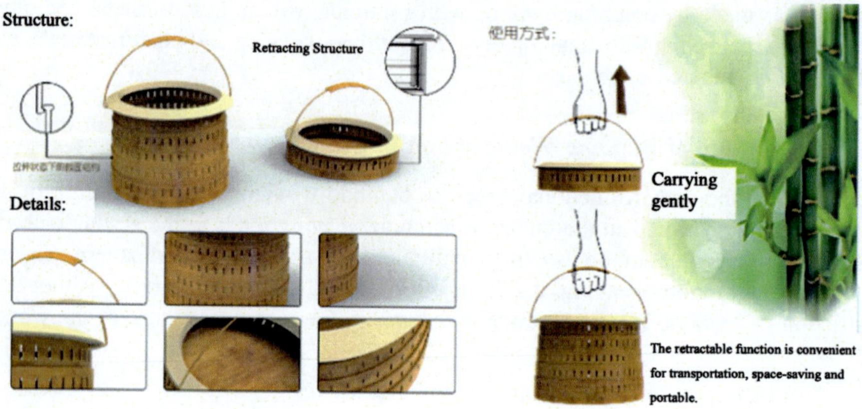

Fig. 1. Design of Bamboo-woven tourist souvenir based on KANO user demand model.

References

1. Kano, N., Seraku, N., Takahashi, F., Tsuji, S.: Attractive quality and must-be quality. J. Jpn. Soc. Qual. Control **14**(2), 147–156 (1984)
2. Tama, I.P., Azlia, W., Hardiningtyas, D.: Development of customer oriented product design using Kansei engineering and Kano model: case study of ceramic souvenir. Procedia Manuf. **4**, 328–335 (2015)
3. Huang, K.-L., Chen, K.-H., Ho, C.-H.: Promoting in-depth reading experience and acceptance: design and assessment of tablet reading interfaces. Behav. Inf. Technol. **33**(6), 606–618 (2014)
4. Hwang, S.-H., Tsai, I.-C., Mitsuhashi, T., Miyazaki, K.: The application of Kano model on exploring the attractive attributes of community culture products. Bull. Jpn. Soc. Sci. Des. **61** (1), 27–36 (2014)

Automation Design for the Dispersion of Pigments and Homogenization of Water-Based Paint

Estefano Arias-Ponce, Blanca Topón-Visarrea, and Mireya Zapata(✉)

Research Center of Mechatronics and Interactive Systems - MIST, Facultad de Ingeniería y Tecnologías de la Información y la Comunicación, Universidad Tecnológica Indoamérica, Machala y Sabanilla, Quito, Ecuador
ariasivan@indoamerica.edu.ec, {blancatopon, mireyazapata}@uti.edu.ec

Abstract. A worldwide production process is of utmost importance, since it increasingly encourages the industry to automate its processes to improve quality and productivity, reduce costs, increase operational safety, etc. In this frame, productivity is one of the most important factors to be competitive, which is related to working methods, efficient processes, continuous improvement, and the use or implementation of automation in production processes. In Ecuador, the polymer industries dedicated to the manufacture of paint lack automated processes, which causes delays in the delivery of production batches and a lack of standardization in cycle times. In particular, we focus on the Wesco S.A company dedicated to the manufacture and marketing of paints where the waterborne paint is one of the most elaborate products in its portfolio. Despite being a product with an established standard formulation, it lacks automated controls in one of its main processes as the dispersion and mix process. It is currently managed through the activity and clearly operational experience of its operators, which causes delays in the delivery of production batches and waste of time, since each manufactured batch has an estimated time of 1 h and 30 min for the dispersion and mix process. In this paper, we propose implementing an automated design for controlling both dispersion and mix process, based on automated tools and industrial control through PLCs, in order to standardize cycle times, improve manufacturing efficiency, and thus productivity.

Keywords: Automation · Standard time · Paint · Process improvement

1 Introduction

At a global level, the construction and manufacturing industry uses paint as an input, which must fulfill with an effective formulation and mixture, to obtain the maximum performance, and quality of raw materials. According to [1] paint is a dispersion of pigments in latex and solvent, which forms a continuous and adherent barrier when drying. Currently in Ecuador, paintings are regulated under the NTE INEN 1544 "Architectural Paintings. Water-based emulsion paint (latex)", this regulation aims to define the requirements that paints use to coat exterior and interior surfaces of masonry,

D. Russo et al. (Eds.): IHSI 2021, AISC 1322, pp. 889–894, 2021.
https://doi.org/10.1007/978-3-030-68017-6_132

concrete, plasters, stucco, asbestos-cement and others must meet, for protection and decorative purpose [2].

In the manufacture and elaboration of architectural paint, the main processes are the dispersion of pigments and the homogenization of the mixture, these processes must control the absence of lumps as well as its homogeneity, many times caused by the humidity and compaction to which raw materials are subjected [3, 4] and if they are also done manually, they cause serious delays, generating high production times. In this paper it is proposed to reduce these times by automating both processes through the use of programmable automatons in order to reduce operator intervention, achieving with this the increment of efficiency in the process and substantially reduce production times.

2 Materials y Methods

In paint-making process, the dispersion is intended to improve the brightness, texture, opacity of white pigments, tinting ability of colorants, flotation of the particle, rheological properties, adhesion, etc. To achieve an adequate dispersion, the selection of the disperser axis is important to obtain a high performance of the final product [3].

On the other hand, homogenization encompasses the integration of the formulation components, phase in which the elements are added to a master homogenization tank, on the one hand, the dispersed raw materials are pumped to the tank, then the rest of the formulation as resins or additives are also added. To carry out this process, a "U" anchor stirrer is used which incorporates inclined radial blades to enhance the movement of the product. Mixing should be done at low speeds between 400 and 550 rpm and for a time not less than 15 min [4].

In Fig. 1 the steps to be followed to meet the objectives proposed in this research are presented. It begins with an analysis of the dispersion and homogenization process, taking a time cycle of each activity carried out to compare it with the established standard times, thus identifying the most critical stages. Once identified, the structure of the automatic process is carried out using programming software and control elements [5]. Once the automatic process was implemented, the cycle times were checked again to establish whether there was an improvement in efficiency and savings in production times.

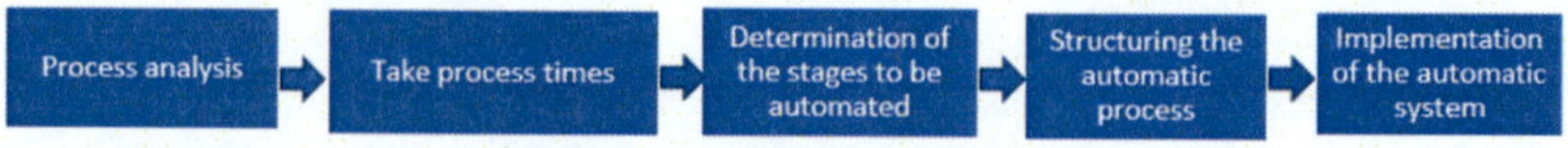

Fig. 1. Operative model

At the end of the paint production process, it is necessary to carry out a quality control of it, this means a control of viscosity, gloss, tonality, drying time and coverage according to the parameters shown in Table 1. Once the production batch is approved, it goes to the stage of packaging, packing and storage of the finished product for its subsequent commercialization.

Table 1. Paint quality control parameters. Taken from [2]

Requirements	Unit	Water-based paint minimum values to comply	Testing method
Brightness	% (85° angle)	<10	NTE INEN 1003
Viscosity 25 °C	U. Krebs	85	NTE INEN 1013
Fineness of dispersion	Hegman	3	NTE INEN 1007
Drying time	h (touch)	1	NTE INEN 1011
Tonality	% (with pattern)	80	-

2.1 Automatic Process Design

Dispersion: Dispersion process involves seven stages: reception of raw materials, filling of disperser tanks, start-up of disperser propellers, level regulation of disperser shafts, addition of pigments and extenders, level regulation of shafts dispersers, pigment dispersion and extender.

Homogenization: According to [6], this process includes four stages: pumping of dispersed product, homogenizer shaft start-up, addition of resins and additives, final mixing. In Fig. 3 you can see the structure of dispersion and homogenization system. Figure 2 shows the flow diagram of the dispersion and homogenization process.

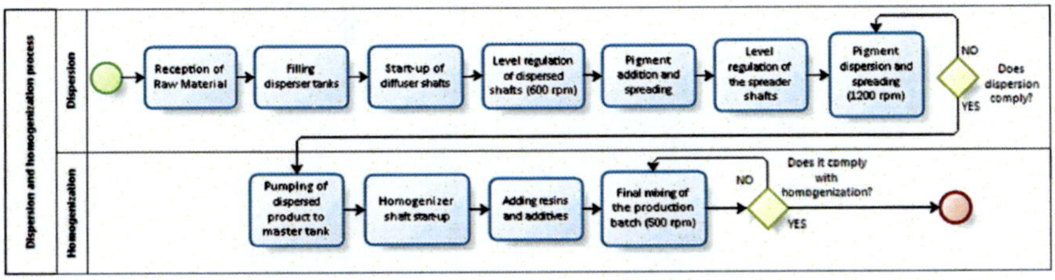

Fig. 2. Dispersion and homogenization process

3 Results

The automatic system design will significantly improve production times, achieving standardization of the dispersion stage as well as the final homogenization stage. For this design, the variables used in the automatic system are identified in Table 2.

Table 2. Programming variables

Inputs	Function	Outputs	Function
I0.0	Start	Q0.0	Water solenoid valve
I0.1	Emergency stop	Q0.1	Pumping solenoid valve
I0.2	Start dispersion	Q0.2	Motor propellers going up
I0.3	Start homogenization	Q0.3	600 rpm disperser motors
I0.4	Water level sensor	Q0.4	1200 rpm disperser motors
I0.5	Charge level sensor	Q0.5	Homogenization motor
I0.6	Dispersion level sensor	Q0.6	Propellers going down
I0.7	Product level sensor		
I1.0	Propeller initial position level sensor		
I1.1	Homogenization tank level sensor		-

In Fig. 3 the ladder and GRAFCET scheme used to program the programmable control system according to the previously established parameters is presented. Figure 4 presents the system simulation and the way in which the tanks, valves and dispersers are structured.

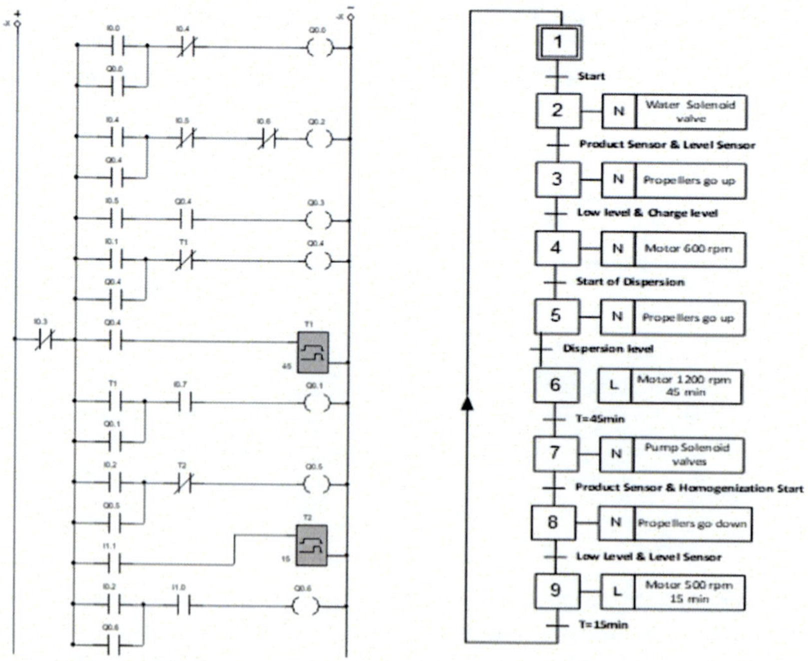

a) b)

Fig. 3. Program for the homogenization and dispersion system in: a) ladder language, b) GRAFCET language

The dispersion and homogenization system starts when the start button I0.0 is activated. The dispersion tanks will be filled by activating the Q0.0 solenoid valve until reaching the level sensor I0.4, with the activation of this solenoid valve the product level sensor I0.7 will be activated, reaching the specified level for mixing, this causes that the solenoid valve activates the dispersing shafts Q0.2 which will automatically rise to the level indicated by the sensor I0.5, this will cause the propeller initial position level sensor to activate I1.0. The Q0.3 disperser motors will automatically start at 600 rpm.

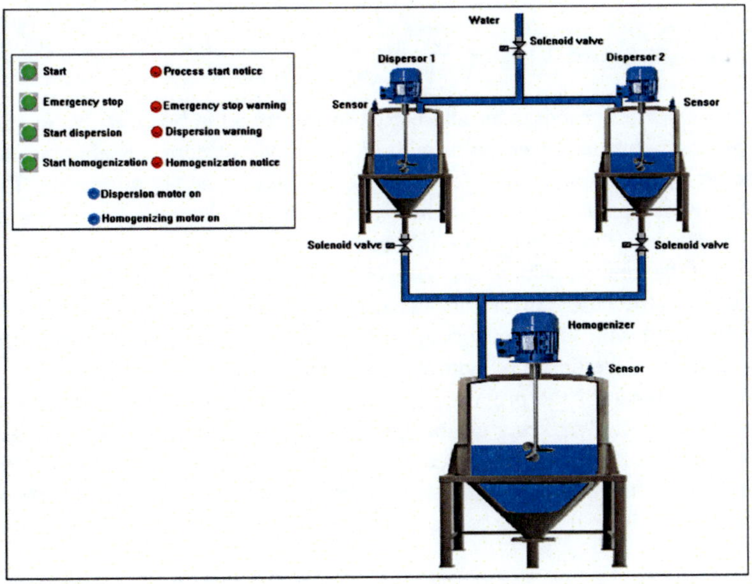

Fig. 4. Dispersion and homogenization system simulation

At this point the operators can begin to add pigments and fillers. The spreading stage will start with the activation of the I0.1 button. The Q0.4 disperser shaft motors will be activated at 1200 rpm and the propellers will rise again to the I0.6 dispersion level, the motors will remain on for 45 min. At the end of this time, the Q0.1 solenoid valve will be activated starting the pumping of the dispersed product towards the master tank. Here the operator will place the final additives and resins until reaching the sensor I1.1 where the level will be 5000 ltrs. The homogenization stage will start when the I0.2 button is activated. The Q0.5 homogenization motor will start running at 500 rpm. The spreading propellers will return to their position through Q0.6. The homogenizing motor will stay on for 15 min, then it will turn off automatically. Using the emergency button I0.3, the machine can be stopped at any time.

Once the process is finished, the standard times have been taken and compared with the historical data determined by the company, which are shown in Table 3, which illustrates that the standard time established by the company is adjusted to the times taken in the automatic process, helping to improve the production system.

Table 3. Standard time in the dispersion and homogenization process.

Production times	Dispersion minutes	Homogenization minutes	Total minutes
Standard time	140′	70′	210′
Current time	168′	105′	273′
Automated process time	143′	72′	215′

4 Conclusions and Future Work

The present research proposes an alternative to improve the homogenization and dispersion process of water-based paint with an automatic system, maintaining the standard time set by the company, but reducing the workforce used considerably. The tasks performed by the operators were repetitive and tedious, which will now be replaced by automated elements that impulse the process response and start-up.

Important factors on which the sample quality depends in the homogenization process, such as the height of the disperser, speed in rpm, amount of mixture entered, will be controlled with level sensors, valves, motors and an adequate speed configuration by means of the designed automatic system.

Finally, the result of the proposed process can be improved by following a better production planning. By having established times and completed processes, it will be possible to carry out a greater number of production batches per day, in this way we will be improving the company productivity and a better control will be carried out in production.

References

1. Geant, C., Vasquez, S.: Modification proposals for the elaboration of architectural emulsified paint based on the study of intermediate systems obtained during its manufacture. Engineering UC, pp. 49–58 (2020). https://www.redalyc.org/articulo.oa?id=707/70723245007. Accessed 18 Sep 2020
2. Ecuadorian Standardization Service: NTE INEN 1544. https://www.normalizacion.gob.ec
3. Giudice, C., Pereyra, A.: Paint and coatings technology: components, formulation, manufacturing and quality. Universidad Tecnologica Nacional, pp. VIII–XI (2009)
4. Inoxpa: Agitators for standard tanks. https://www.inoxpa.com
5. Gonzalez, G., Vidal, L.: Automation system of an industrial plant for the preparation and bottling of some liquid products. In: Proceedings of the XV Seminario Anual de Automática, Electrónica Industrial e Instrumentación, vol. 8–11 (2008). https://doi.org/10.13140/RG.2.1.2879.4322
6. Toapanta, E.: Optimización del Proceso de Homogeneización en la planta de masillas de la empresa Pinturas Condor S.A. Bachelor thesis, Universidad Politécnica de Chimborazo, Riobamba, Ecuador (2009). https://dspace.espoch.edu.ec/handle/123456789/246. Accessed 20 Sep 2020

Analyzing Selfie Opportunities and Trends in a Chinese Context

Ruilin Wang, Dawei Wang, Ziwei Zhang, Di Zhu, and Wei Liu(✉)

Faculty of Psychology, Beijing Normal University,
Beijing 100875, People's Republic of China
{202028061037,202028106033,
ziwei.zhang}@mail.bnu.edu.cn,
{Di.Zhu,Wei.Liu}@bnu.edu.cn

Abstract. While technology has improved to a great extent in recent years, users have more channels to express themselves. Selfie, as the latest fad, has rapidly risen into public view and formed a related culture. This study focuses on trends and opportunities arising from selfie behavior in a Chinese context that promotes the User eXperience. It aims to understand using the scenario of taking, editing, retrieving, and sharing selfies. By reviewing scientific literature, industry reports, and popular selfie applications, the research team analyzes and presents six trends of the selfie: 1) selfie booth with entertainment features, 2) double-scene selfie, 3) groupie, 4) selfie and recognition of emotion, 5) selfie with covering the face, 6) interactive guide of selfie expression and action. The outcomes reveal potential user needs in the current background of trends in the selfie industry, as well as assist researchers and designers in getting inspirations for providing a better selfie experience.

Keywords: Selfie scenario design · Selfie industry trend · User needs

1 Introduction

The selfie is a form of self-portrait mainly created by smartphones or webcams and shared on social networks. It is one of the most direct ways for someone to express themselves to others as well as to record their daily life. Nowadays, taking selfies has become a pervasive activity in people's social life [1].

With the advent of the 5G era, selfies have brought more possibilities. People can take selfies in more scenes in various ways. Taking selfies provides personalized expression opportunities to users, and user needs are becoming more and more diverse. Meanwhile, it is hard to meet up with friends, due to COVID-19, which also puts forward new demands for the development of the selfie industry. As the current research on China's selfie market mainly focuses on the technical level, less attention is paid to the needs of users. At the same time, research has shown that there are cultural differences in the way of personal presentation of selfies. Chinese selfie users not only show themselves in selfies but also show their "ideal self". There are also many differences in self-portrait situations and self-portrait methods [10]. Therefore, the

D. Russo et al. (Eds.): IHSI 2021, AISC 1322, pp. 895–900, 2021.
https://doi.org/10.1007/978-3-030-68017-6_133

research team mainly use China as the background to conduct research around users and situations and analyze the new opportunities and trends of selfies.

This search presents six trends of selfies. First of all, people are becoming more and more immersed in the virtual world and lack of real offline communication. So, the research team presented a selfie booth with entertainment features. In the currently existing research, the selfie booth can be combined with community opinion surveys and cafes, respectively, or these booths could be entertainment facilities in public places. Secondly, E-commerce live streaming is playing an increasingly important role in China's consumer market [2], but a single selfie picture can hardly meet the needs of live broadcasts. At the same time, Huawei introduced the double-scene shooting feature. Then the research further explored the double-scene selfie based on the specific needs and usage scenarios of users in live broadcasts. Thirdly, taking selfies with friends is also an essential activity in our social life. Even in the same selfie, different people will have different needs. Therefore, the research team presented a new trend for groupies. According to the self-portrait trend data report released by Meitu, among the selfie group in China, young users under the age of 25 take up half of the total. At the same time, the results of the Study on the Main Social Stressors and Mental Health of Different Occupational Groups in the Period of Social Transition by the Institute of Psychology of the Chinese Academy of Sciences show that people in their 20s to 30s have the highest mental stress. The group that is most keen on selfies and the group with the most pressure have overlapped in age, so the research team further explored the issue of selfie and recognition of emotion. Next, there is a specific group of users who usually like to cover their faces while taking selfies. What they show is not their face, but their style and taste. This self-portrait method is especially popular among Gen Z. Therefore, it has become the latest trend for selfies. Finally, with the popularity of selfies, people want to post excellent selfies on social platforms, but some people encounter problems with not posing or having a stiff expression during selfies. Therefore, how to improve the stiffness of people's facial expressions and actions when taking selfies through the interaction between people and mobile phones has also become an urgent problem to be solved [13].

2 Method

To obtain a meticulous and comprehensive report, the research team took 2 weeks to read a great number of papers which studied selfie-and related technology published in academic journals within 10 years and more 170 various reports focusing on Chinese users. Besides, the research team also made mainstream selfie applications' competitive analysis through statistics web and app store from aspects of core competitive, market share, user distribution, and function iteration. Based on these data and analyses, the research summarized 11 selfie industry trends, after several discussions the following 6 most important points are picked up for further study.

3 Trends

3.1 Selfie Booth with Entertainment Features

As online social networks becoming more popular, a significant number of people have used selfies to express their opinions, thoughts, and emotions on social networks. After all, social networks cannot satisfy people's desire to communicate with others in real life, which seems to make social communication more efficient and straightforward. How can selfies enhance social interaction in real life? According to some researches, physical selfie booths located in public areas can help people increase real communication and interaction with others, and create a more enjoyable selfie atmosphere. For example, Andre O. Bueno et al. set up some selfie booths in a café [3]. When people drink coffee in the cafe, they can take selfies with other people here. The selfie booth extends their stay in the cafe and improves the fun of the cafe. What's more, Nemanja Memarovie [4] and others have deployed some Moment Machines on the streets that allow residents living in local communities to take and share photos of themselves at that moment and view others' pictures on the machine's screen. The Moment Machine has dramatically increased the interaction and communication among the residents of the local community. Meanwhile, it brings the residents more pleasure with a selfie. Selfie booths can not only enable people to express themselves through selfies but also enhance real communication and interaction among users, which satisfies the need for interaction with others in real life.

3.2 Double-Scene Selfie

In China, online shopping is becoming increasingly developed. The mature logistics system and e-commerce platform make shopping more convenient and time-saving, so that online shopping has become the shopping choice of most people. With the popularization of the 4G network, live video broadcast has become a common way of information transmission. Therefore, e-commerce stores have also started to introduce and advertise their products through live video streaming. However, when displaying the details of the product, the mobile lens will focus on the product so that people can only see the product on the screen without the anchor.

What's more, on the one hand, it is very inconvenient for anchors to contact with the audience; on the other hand, it will affect the audience's experience because the audience wants to watch detailly about the products and presentation of anchors. Based on this situation, the emergence of the double scene selfie solves the above problems well. In e-commerce live streaming, as it comes to displaying product details, the front-facing camera will focus on the product and divide it into half of the screen. In contrast, the other half of the screen can show the anchor with the size, shape, and filter can be adjusted separately according to needs. For example, Huawei has launched a series of mobile phones with the dual-view selfie function, which allows photographers in different time and space to take a dual-view selfie with the same frame. With the advent of the 5G era, e-commerce live streaming will become progressively popular. The application of double Scene selfie in e-commerce live streaming can show not only the details of goods but also allows anchors present in the screen, which shortens the distance between anchors and fans and enhances consumers' loyalty.

3.3 Group Selfie

According to studies, the users of beautifying selfie applications have already found the common that enhance photos is necessary to share selfies. But because of the heavy similar and stereotyped repetition within decorating filters, there are more and more people such as males, elders, and citizens who are looking for their style. So, it is easy to observe that different types of users have a significant gap in the beautify parameters. For example, a large part of male users prefers to contain a natural face without too much enhance filter. Besides, some senior users of these applications grew bored by the sameness of the filters, begin to pursuit individual selfies. In group selfie situations, there remain two central paradox. The one is that people accept beautify their faces or not; the other one is broad parameters' needs gap, on account of personal differences. The industry's moving directions are the followings: firstly, the algorithm should provide individually targeted beautify parameters to different users to meet their every demand. Second, this function is better also to support short videos.

3.4 Selfie and Recognition of Emotion

The pace of life of young people today is accelerating, and the pressure of social competition is also increasing. As a result, young people face growing pressure, anxiety, and negative emotions. The research of the Institute of Psychology of the Chinese Academy of Sciences has shown that people in their 20s and 30s have the highest mental stress, and depressed patients in China are getting younger and younger. The finding coincided with the age of the selfie group. Thus, the daily act of taking selfies to relieve young people's anxiety and negative emotions is also an opportunity for attention in the future. At present, the recognition rate of facial emotion recognition technology in the ImagaNet data set has been improved to 95.06% through a deep learning algorithm. Moreover, the recognition technology of facial feature capture does not need to be actively expressed by the recognized object. The use of this technology in selfies may help ease the negative emotions of young people [5].

3.5 Selfie with Covering the Face

Sharing selfies with hidden faces in various social platforms such as the Red Book and WeChat are very common in recent years. People often directly use physical objects to block their faces or conceal their faces by using edit applications process photos. Sometimes they may just do not show faces when taking pictures. One of the most crucial reasons for this phenomenon is users are having more privacy concerns in their selfie behaviors [6]. People are afraid of leaking privacy through these selfies to the people who are not acquaintances. So, they choose to hide the faces in the selfies to protect their data, this boundary privacy can effectively enhance their sense of safety [7]. Social media platform users also block faces in other situations, including a paradox that they expect to express themselves. Still, at the same time, they are a shame to do that and when they want to show their fashions and styles. According to the report, the central part of these social media users is girls who were born after the 2000s. There are several potential needs should be met to improve the User eXperience. At first, currently, selfie applications can process photos, but most of them do not have unique and convenient editing photo functions for hiding faces. Besides, even if they use objects to block faces, camera applications are also better to can recognize face and beauty photos.

3.6 Interactive Guide of Selfie Expression and Action

Nowadays, selfies have become popular among different types of people. Selfies are no longer exclusive to one group but increasingly popular in a different kind of person. People all want to get the right pose or expression, but not all of users are good at taking selfies. Most of the time, users can't help but smirk and have trouble posing. To solve the problem, the phone can guide users to adjust their movements, expressions, or head position by visual and voice to take good pictures. For instance, there is a body posture recommendation system. The system can recognize the target person in the shooting and extract key points of the human body based on the described person. The system can obtain one or more matching attitude templates from the posture library based on critical aspects of the human body and guide the user to adjust the body based on the initial posture to get the appropriate position.

Another example is the Baidu Magic, which can take selfies with just a blink. The shot aims at your face after turning on the camera. The magic shot will start to detect your eyes. You can take a selfie by closing and opening your eyes, and then posing in an attractive expression and pose. The whole process doesn't need users to use their hands to operate [1, 8, 9]. Under the help of an interactive guide of selfie expression and action, people with low self-esteem and willingness to take selfies but who are not good at posing take selfies with personal characteristics, helping them take selfies more naturally.

4 Conclusion

The purpose of this study is to explore the future development trend and potential opportunities of selfie in the Chinese context, which can be summarized into three aspects: particular using scenarios, user's emotions and selfie experience optimization. The results show that the diversity of selfie users and using scenarios will lead to more personalized and detailed needs for selfies in the future.

Firstly, from the perspective of scenarios, the selfie booth can enhance the emotional interaction among users in real life and reduce user's loneliness. Group selfies focus on how people set different parameters for group photos, and the double-scene selfie caters to the current needs of e-commerce live broadcasting in China. What's more, for selfies that focus on user's emotions, face-covering selfies and emotion recognition pay attention to user's privacy issues and negative emotions respectively. Last but not least, interactive guidance can be applied in various scenarios to help users take more natural and satisfying selfies.

All in all, in the future selfie market, the new social trend will constantly raise new claims on the selfie industry, and more attention should be paid to satisfy different users' needs and optimize User eXperience.

References

1. Fang, N., Xie, H., Igarashi, T.: Selfie guidance system in good head postures. In: Joint Proceedings of the ACM IUI 2018 Workshops. SymCollab, pp. 1–8 (2018)
2. Lin, J., Lu, Z.: The rise and proliferation of live-streaming in China: insights and lessons. In: Stephanidis, C. (ed.) HCI 2017. CCIS, vol. 714, pp. 632–637. Springer, Cham (2017)
3. Callum, P., Joel, F., Soojeong, Y., Marius, H.: "Let's take a selfie": design considerations for public selfie booths. In: 31st Australian Conference on Human-Computer-Interaction (OZCHI 2019), Fremantle, WA, Australia, pp. 560–562. ACM, New York (2019)
4. Nemanja, M., Ava, F., Schieck, H.S.: Capture the moment: "in the wild" longitudinal case study of situated snapshots captured through an Urban screen in a community setting. In: Proceedings of the 18th ACM Conference on Computer Supported Cooperative Work & Social Computing, CSCW 2015, pp. 242–253, Vancouver (2015)
5. Wioleta, S.: Emotion recognition using physiological signals. In: Proceedings of the Mulitimedia, Interaction, Design and Innnovation, MIDI 2015, pp. 1–8 (2015)
6. Amandeep, D., Torbjorn, T., Stale, P., Cecilie, S.A.: Do online privacy concerns predict selfie behavior among adolescents, young adults and adults? Front. Psychol **8**, 815 (2017)
7. Petronio, S., Altman, I.: Boundaries of privacy: dialectics of disclosure, 8–11, 30–31 (2002)
8. Yeh, M.-C., Lin, H.-W.: Virtual portraitist: aesthetic evaluation of selfies based on angle. In: Proceedings of the 22nd ACM International Conference on Multimedia (MM 2014), pp. 221–224. ACM, New York (2014)
9. Zhu, J.-Y., Agarwala, A., Efros, A.A., Shechtman, E., Wang, J.: Mirror mirror: crowdsourcing better portraits. ACM Trans. Graph. **33**, 6 (2014). (SIGGRAPH Asia 2014)
10. Ma, J., Yang, Y.: What can we know from selfies - an exploratory study on selfie and the implication for marketers. In: 2016 Global Marketing Conference at Hong Kong Proceedings. Global Alliance of Marketing & Management Associations, pp. 597–601 (2016)

Analysis of the Activation Volume and the Pressure Resistance for Handling Kinetic of Processing Food

William Rolando Miranda Zamora[1(✉)], Nelly Leyva[2], Manuel Jesus Sanchez Chero[1], Marcos Timaná-Alvarez[1], Leandro Alonso Vallejos More[1], Priscila Estelita Luján Vera[1], and Marlon Martín Mogollón Taboada[1]

[1] Universidad Nacional de Frontera, Sullana, Peru
wimiza23@hotmail.com, manuelsanchezchero@gmail.com, priscilavera930@gmail.com, {mtimana, lvallejosm, mmogollon}@unf.edu.pe
[2] Universidad Nacional de Piura, Piura, Peru
nleyvap@gmail.com

Abstract. The mathematics used in handling logarithmic kinetic data often involves unnecessary steps. Determination of activation volume (V) and pressure resistance (z) often becomes laborious. In this paper we illustrate the simplest way to determine activation volume and pressure resistance needed for evaluating high pressure processing de foods. Activation volume and pressure resistance are determined in one step by linear fit with least squares on computers. Four new equations were proposed for the prediction of the activation volume and pressure resistance in a single step in the case of evaluation of processes of food with high pressure.

Keywords: Simplified procedure · Pressure treatment · Activation volume · Pressure resistance

1 Introduction

The pascalization process (as a tribute to Blaise Pascal) [1], bridgmanization, (as a tribute to Percy Williams Bridgman Nobel Prize winner in physics in 1946 for his work in high pressure physics) [2], high pressure (HP) [3], ultra high pressure (UHP) [4] or high hydrostatic pressure (HHP) [5–9] is a method of preserving food. High hydrostatic pressure processing has been considered an innovative technology in food processing due to its minimal processing of the technical characteristics of food products [10]. The high hydrostatic pressure process, as a non-thermal process, can be used to inactivate microorganisms while minimizing chemical reactions (color, flavor and nutrition), thus preserving the freshness of food [11, 12]. High hydrostatic pressure enables safe, high-quality food to be processed. Determination of parameters of the microbial, enzymatic inactivation rate or retention such as D and z values is given to characterize microbial reduction or nutrient retention under hydrostatic pressure.

D. Russo et al. (Eds.): IHSI 2021, AISC 1322, pp. 901–906, 2021.
https://doi.org/10.1007/978-3-030-68017-6_134

2 Methods

The simplification methodology has been achieved after the analysis of several examples collected from the scientific literature [13–20]. The simplification allows calculations and data handling in an agile and simple way using an Excel spreadsheet to determine the z value and the activation volume.

3 Case Study and Results

Exponential regression analysis belongs to non-linear regression, Excel has this tool incorporated [21] that can be used to adjust the data of reaction rate constants, k versus pressure, and then find z and V. Excel has many built-in functions and tools that we can use [22, 23] to find the pressure resistance and the activation volume. Up to four ways of determining the V and z value are shown in this work, which are summarized in Fig. 1, Fig. 2, Fig. 3 and Fig. 4. There are many ways to find the activation volume, V and the z value from the data of the reaction rate constants, if these were the years in which Bert Holmes Hite used high hydrostatic pressure to process milk [12], I would use semi-log paper, which has one axis on a logarithmic scale and the other axis on a linear scale, to find the slope between the activation constants, k, and the pressure at a constant temperature, T, as seen in Fig. 1. Its use in scientific literature is common.

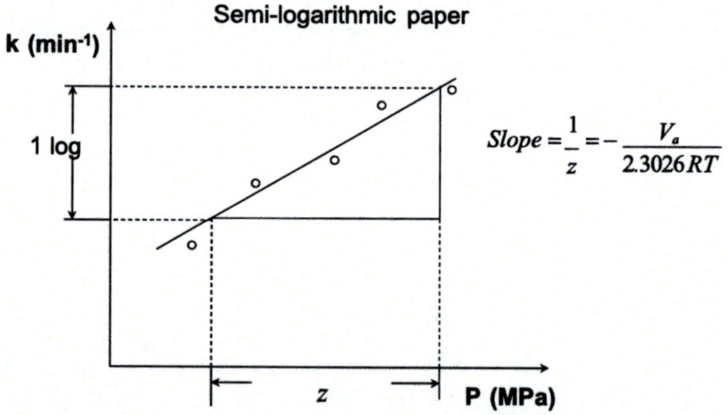

Fig. 1. Determination of the z value and the activation volume. Graph between the activation constant, k and the pressure.

The second way to find the activation volume and the z value using semi-log paper is by plotting 1/k versus pressure, at a constant temperature, T, as shown in Fig. 2. A third way to find the activation volume and the z value using semi-log paper is by plotting the D value versus the pressure, at a constant temperature, T, as shown in Fig. 3. Its use in scientific literature is common.

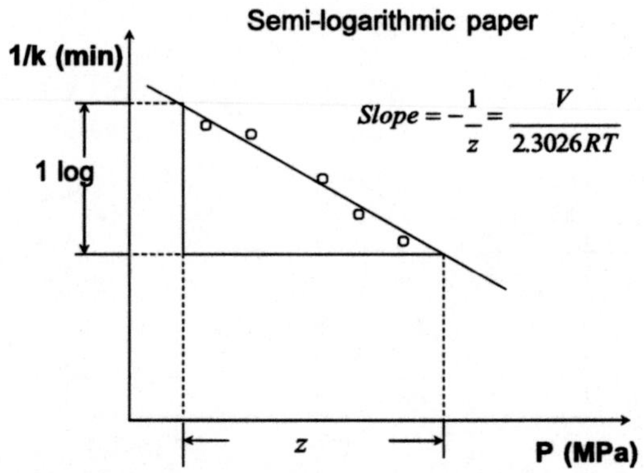

Fig. 2. Determination of the z value and the activation volume. Graph between the inverse of the activation constant, 1/k and the pressure. Its use is considered.

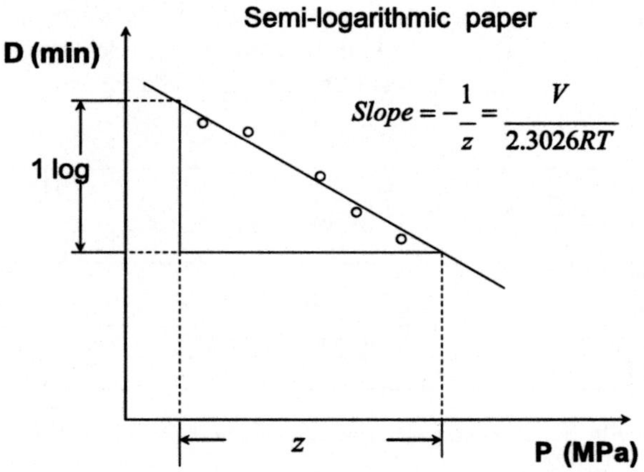

Fig. 3. Determination of the z value and the activation volume. Graph between D value and pressure.

A fourth way to find the activation volume and the z value using semi-log paper is by plotting the 1/D value versus the pressure, at a constant temperature, T, as shown in Fig. 4.

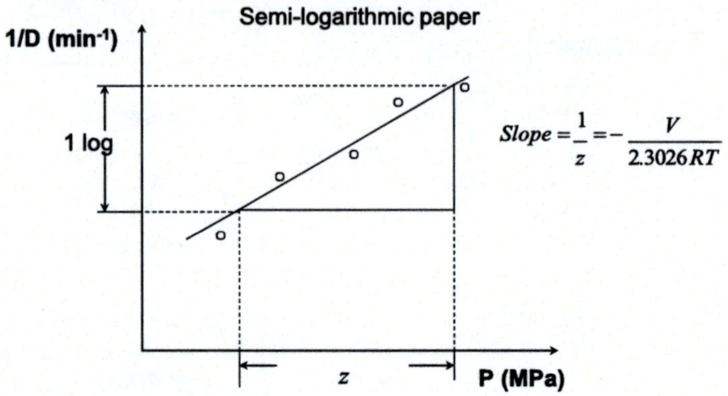

Fig. 4. Graphic determination of the z value and the activation volume. Graph between 1/D value and pressure. Its use is considered.

If you had an Excel spreadsheet, the pressure and reaction rate data ordered in the way shown in Fig. 5 and Fig. 6. To find the slope in an Excel spreadsheet and then proceed to find the value z and V according to Fig. 5 and 6.

	D	E	F
7	**Pressure, P (MPa)**	**Reaction rate, k (min^{-1})**	
8	.	.	
9	.	.	
10	.	.	
11	Slope	=SLOPE (LOG(E8:E10);D8:D10)	
12			

Fig. 5. Pressure and reaction rate data in an Excel sheet.

	G	H	I
7	**Pressure, P (MPa)**	**D value (min^{-1})**	
8	.	.	
9	.	.	
10	.	.	
11	Slope	=SLOPE(LOG(H8:H10);G8:G10)	
12			

Fig. 6. Pressure and D value data in an Excel sheet.

From the review of the scientific literature, for example, [16] works with the inactivation of a group of microorganisms in foods, calculating the slope of the log D curves versus pressure and ln k versus pressure using two equations, when it can be simplified, and from one of the two curves both V and z can be obtained.

4 Conclusions

An analysis of the activation volume and the pressure resistance (z) for handling kinetic data has been presented. A procedure with four new equations are presented and allow prediction (activation volume and pressure resistance) in one step.

References

1. Largeteau, A., Prakasam, M.: Trends in high pressure developments for new perspectives. Solid State Sci. **80**, 141–146 (2018). https://doi.org/10.1016/j.solidstatesciences.2018.04.012
2. Heremans, K., Smeller, L.: Protein structure and dynamics at high pressure. Biochimica et Biophysica Acta (BBA) - Protein Struct. Mol. Enzymol. **1386**(2), 353–370 (1998). https://doi.org/10.1016/s0167-4838(98)00102-2
3. Yu, Y., Zheng, Z., Wang, C., Hu, L., Ramaswamy, H.S., Zhu, S.: Melting endothermic technique for establishing different phase diagram pathways during high pressure treatment of liquid foods. Innov. Food Sci. Emerg. Technol. **62**, 102361 (2020). https://doi.org/10.1016/j.ifset.2020.102361
4. Gallego, R., Bueno, M., Chourio, A.M., Ibáñez, E., Saldaña, M.D.A., Herrero, M.: Use of high and ultra-high pressure based-processes for the effective recovery of bioactive compounds from Nannochloropsis oceanica microalgae. J. Supercrit. Fluids **167**, 105039 (2020)
5. Cava, R., García, J., Ladero, L.: Effect of high hydrostatic pressure processing and storage temperature on food safety, microbial counts, colour and oxidative changes of a traditional dry-cured sausage. LWT **128**, 109462 (2020). https://doi.org/10.1016/j.lwt.2020.109462
6. Huang, H.-W., Hsu, C.-P., Wang, C.-Y.: Healthy expectations of high hydrostatic pressure treatment in food processing industry. J. Food Drug Anal. **28**, 1–13 (2020). https://doi.org/10.1016/j.jfda.2019.10.002
7. Marciniak, A., Suwal, S., Naderi, N., Pouliot, Y., Doyen, A.: Enhancing enzymatic hydrolysis of food proteins and-production of bioactive peptides using high hydrostatic pressure technology. Trends Food Sci. Technol. **80**, 187–198 (2018). https://doi.org/10.1016/j.tifs.2018.08.013
8. Pérez-López, E., Mateos-Aparicio, I., Rupérez, P.: High hydrostatic pressure aided by food-grade enzymes as a novel approach for Okara valorization. Innov. Food Sci. Emerg. Technol. **42**, 197–203 (2017). https://doi.org/10.1016/j.ifset.2017.07.012
9. Lee, N.-Y., Koo, J.-G.: Effects of high hydrostatic pressure on quality changes of blends with low-protein wheat and oat flour and derivative foods. Food Chem. **271**, 685–690 (2019). https://doi.org/10.1016/j.foodchem.2018.07.171
10. Orel, R., Tabilo-Munizaga, G., Cepero, Y., Reyes-Parra, J.E., Badillo, A., Pérez-Won, M.: Effects of high hydrostatic pressure processing and sodium reduction on physicochemical properties, sensory quality, and microbiological shelf life of ready-to-eat chicken breasts. LWT **127**, 109352 (2020). https://doi.org/10.1016/j.lwt.2020.109352
11. Tucker, G.S., Featherstone, S., Miranda-Zamora, W.R.: Fundamentos del procesamiento térmico de alimentos. AMV ediciones (2020)
12. Miranda-Zamora, W.R., Heldman, D.R.: Diseño de procesos térmicos y alta presión de alimentos. AMV ediciones (2018)

13. Nakaura, Y., Morimatsu, K., Inaoka, T., Yamamoto, K.: Listeria Monocytogenes cells injured by high hydrostatic pressure and their recovery in nutrient-rich or -free medium during cold storage. High Press. Res. **39**, 324–333 (2019). https://doi.org/10.1080/08957959.2019.1584895
14. Buckow, R., Weiss, U., Knorr, D.: Inactivation kinetics of apple polyphenol oxidase in different pressure–temperature domains. Innov. Food Sci. Emerg. Technol. **10**(4), 441–448 (2009). https://doi.org/10.1016/j.ifset.2009.05.005
15. Raj, A.S., Chakraborty, S., Rao, P.S.: Thermal assisted high-pressure processing of Indian gooseberry (Embilica officinalis L.) juice – impact on colour and nutritional attributes. LWT **99**, 119–127 (2019). https://doi.org/10.1016/j.lwt.2018.09.051
16. Verbeyst, L., Crombruggen, K.V., Van der Plancken, I., Hendrickx, M., Van Loey, A.: Anthocyanin degradation kinetics during thermal and high pressure treatments of raspberries. J. Food Eng. **105**(3), 513–521 (2011). https://doi.org/10.1016/j.jfoodeng.2011.03.015
17. Terefe, N.S., Yang, Y.H., Knoerzer, K., Buckow, R., Versteeg, C.: High pressure and thermal inactivation kinetics of polyphenol oxidase and peroxidase in strawberry puree. Innov. Food Sci. Emerg. Technol. **11**(1), 52–60 (2010). https://doi.org/10.1016/j.ifset.2009.08.009
18. Terefe, N.S., Delon, A., Buckow, R., Versteeg, C.: Blueberry polyphenol oxidase: characterization and the kinetics of thermal and high pressure activation and inactivation. Food Chem. **188**, 193–200 (2015). https://doi.org/10.1016/j.foodchem.2015.04.040
19. Ravichandran, C., Purohit, S.R., Rao, P.S.: High pressure induced water absorption and gelatinization kinetics of paddy. Innov. Food Sci. Emerg. Technol. **47**, 146–152 (2018). https://doi.org/10.1016/j.ifset.2018.02.012
20. Shkolnikov, H., Belochvostov, V., Okun, Z., Shpigelman, A.: The effect of pressure on the kinetics of polyphenolics degradation – implications to hyperbaric storage using Epigallo-catechin-gallate as a model. Innov. Food Sci. Emerg. Technol. **59**, 102273 (2019). https://doi.org/10.1016/j.ifset.2019.102273
21. Delgado-Aguilar, M., Valverde-Som, L., Cuadros, L.: Solver, an Excel application to solve the difficulty in applying different univariate linear regression methods. Chemom. Intell. Lab. Syst. **178**, 39–46 (2018). https://doi.org/10.1016/j.chemolab.2018.04.018
22. Grech, V.: WASP (write a scientific paper) using Excel – 13: correlation and regression. Early Hum. Dev. **122**, 60–63 (2018). https://doi.org/10.1016/j.earlhumdev.2018.04.009
23. Romeo, G.: Excel VBA, solver, and other advanced worksheet tools. Elements of Numerical Mathematical Economics with Excel, pp. 3–44 (2020). https://doi.org/10.1016/b978-0-12-817648-1.00001-3

Correction to: Analysis and Modeling of Factor Determinants for Ukraine Hotels and Tourist Sphere

Nazariy Popadynets, Iryna Hryhoruk, Mariana Popyk, Olha Bilanyuk, Oleksandr Halachenko, Inna Irtyshcheva, Natalia Batkovets, Nataliia Lysiak, Yevheniya Boiko, Nataliya Hryshyna, Mariana Bil, and Tetiana Nezveshchuk-Kohut

Correction to:
Chapter "Analysis and Modeling of Factor Determinants for Ukraine Hotels and Tourist Sphere"
in: D. Russo et al. (Eds.): *Intelligent Human Systems Integration 2021*, AISC 1322,
https://doi.org/10.1007/978-3-030-68017-6_76

In the original version of the book, the following belated correction has been incorporated: Table 2 of Chapter 76 has been amended. The book and the chapter have been updated with the change.

The updated version of this chapter can be found at
https://doi.org/10.1007/978-3-030-68017-6_76

D. Russo et al. (Eds.): IHSI 2021, AISC 1322, p. C1, 2021.
https://doi.org/10.1007/978-3-030-68017-6_135

Author Index

D. Russo et al. (Eds.): IHSI 2021, AISC 1322, pp. 907–911, 2021.
https://doi.org/10.1007/978-3-030-68017-6